Africa from MIS 6-2

Vertebrate Paleobiology and Paleoanthropology Series

Edited by

Eric Delson
Vertebrate Paleontology, American Museum of Natural History
New York, NY 10024, USA
delson@amnh.org

Eric J. Sargis
Anthropology, Yale University
New Haven, CT 06520, USA
eric.sargis@yale.edu

Focal topics for volumes in the series will include systematic paleontology of all vertebrates (from agnathans to humans), phylogeny reconstruction, functional morphology, Paleolithic archaeology, taphonomy, geochronology, historical biogeography, and biostratigraphy. Other fields (e.g., paleoclimatology, paleoecology, ancient DNA, total organismal community structure) may be considered if the volume theme emphasizes paleobiology (or archaeology). Fields such as modeling of physical processes, genetic methodology, nonvertebrates or neontology are out of our scope.

Volumes in the series may either be monographic treatments (including unpublished but fully revised dissertations) or edited collections, especially those focusing on problem-oriented issues, with multidisciplinary coverage where possible.

More information about this series at http://www.springer.com/series/6978

Africa from MIS 6-2

Population Dynamics and Paleoenvironments

Edited by

Sacha C. Jones

McDonald Institute for Archaeological Research, University of Cambridge, Cambridge, England, UK

Brian A. Stewart

Museum of Anthropological Archaeology, University of Michigan, Ann Arbor, MI, USA

Springer

Editors
Sacha C. Jones
McDonald Institute for Archaeological Research
University of Cambridge
Cambridge, England
UK

Brian A. Stewart
Museum of Anthropological Archaeology
University of Michigan
Ann Arbor, MI
USA

ISSN 1877-9077 ISSN 1877-9085 (electronic)
Vertebrate Paleobiology and Paleoanthropology Series
ISBN 978-94-017-7519-9 ISBN 978-94-017-7520-5 (eBook)
DOI 10.1007/978-94-017-7520-5

Library of Congress Control Number: 2015957232

Cover illustration: Cover image created by Katherine Clahassey.

Printed on acid-free paper

This Springer imprint is published by SpringerNature
The registered company is Springer Science+Business Media B.V. Dordrecht

Preface

Africa, from Marine Isotope Stages (MIS) 6 to 2, approximately 190–12 ka (thousand years ago), witnessed our species' biological development and behavioral florescence. To date, archaeological, paleontological, and genetic research on this period has been dominated by efforts to classify "modernity" and chart the geographic spread of *Homo sapiens* out of Africa. While such themes remain important, reconstruction and explanation of the demographic processes that occurred within Africa from MIS 6-2 are issues that are consistently under-addressed. This is a major drawback given that population dynamics within Africa played a central role in our species' genetic and cultural evolutionary trajectories. Genetic analyses suggest that African populations experienced multiple bottlenecks over the last 200 ka, interspersed with population expansions (releases). The former almost certainly engendered population isolations, extinctions, genetic and cultural drift, and biological and behavioral adaptation; while the latter saw humans radiate and colonize, with some groups eventually dispersing from, and re-entering, the continent. Analogously, recent theoretical advances in cultural evolution suggest that demography was perhaps the single most important factor underlying modern human innovation.

Yet until these insights can be contextualized, dated, and elaborated using the increasingly refined archaeological and paleoenvironmental records for late Middle and Late Pleistocene Africa, they remain mere glimpses of the complex paleodemographic processes that made us what we are today. To begin redressing this problem, we held a conference at the McDonald Institute for Archaeological Research, University of Cambridge in July 2010, entitled *Africa from Stages 6 to 2: population dynamics and paleoenvironments*. This conference was a first attempt to explore the potential of using diverse African datasets—archaeological, paleoenvironmental, paleontological, and genetic—to reconstruct spatiotemporal population histories during MIS 6-2. The principal aim was to investigate continent-wide population dynamics during MIS 6-2 by fostering interdisciplinary discourse between experts working in ecologically comparable zones across Africa. Twenty-three leading researchers in African archaeology, paleoenvironments as well as genetics and paleontology were thus invited to present their work at the conference. The main issues that were raised included: (1) the impact of this glacial-interglacial-glacial cycle on human group sizes, movements, and distributions throughout the continent; (2) the macroevolutionary and microevolutionary processes underpinning our species' anatomical and behavioral evolution; (3) an initial assessment of the state of the relevant data for addressing these issues and; (4) setting an agenda whereby Africa can benefit from, and eventually contribute to, the increasingly sophisticated theoretical and methodological paleodemographic frameworks developed on other continents.

The scope of the conference was pan-continental. This was not because we expected to pin down population dynamics on this grand scale, but rather because we hoped to address three further objectives: (1) to encourage a more balanced geographical coverage than is typical in MSA/early LSA research (wherein eastern and southern Africa are focal points); (2) to facilitate trans-continental comparisons in order to evaluate the current status of the data; and (3) to foster discourse among and between researchers working in different regions and biomes of Africa. The conference was thus organized according to the biome in which the research was taking place, rather than according to a specific region or time period. This organization—wherein research results from the Kalahari, for example, were presented alongside those from the Sahara, or those from highland Lesotho discussed in relation to highland Ethiopia—proved fruitful. This book, therefore, is similarly organized according to the broad ecological zone into which the different authors' research falls. We have subdivided the papers into: (1) coasts; (2) deserts; and (3) grasslands, woodlands and rainforests. In addition to papers focusing on regional African archaeological and palaeoenvironmental records, we also include two broader scale papers by researchers with expertise in physical anthropology and genetics. This multidisciplinary breadth is essential for addressing questions regarding palaeodemography. In the final chapter of the book, Peter Mitchell raises a number of important theoretical issues and addresses how we can begin to model past population dynamics in Africa from MIS 6-2.

Scholarship on Africa has previously lagged behind that of other continents, particularly Europe, in generating models of prehistoric population dynamics that can be tested against high resolution archaeological and paleoenvironmental records. There are many reasons why this is so, some of which have clearly influenced the composition of this book. First, archaeological and paleoenvironmental research coverage on this enormous continent is extremely patchy, dictated by a host of factors including the variable preservation and exposure of ancient remains, differing research traditions and recurrent political instability. The prehistoric records of some areas (e.g., East, North, and southern Africa) are more thoroughly researched, and thus more highly resolved, than others (e.g., central and West Africa), making our chapter balance inevitably, and regrettably, skewed toward the former. A related reason for Africa's dearth of paleodemographic research compared to other continents is the relative paucity of African researchers working on MIS 6-2. Instead, most research teams and funds come from international institutions with geographically circumscribed research foci instead of trans-regional comparative aims or inter-project collaborations. Particularly rare in research on MIS 6-2 are black African scientists, who have generally tended to concentrate more on earlier (Plio-Pleistocene) or later (Holocene) periods of their continent's past. The result is another unavoidable imbalance in this book. By taking a pan-continental, trans-regional approach to the population dynamics of prehistoric Africans, however, we hope this book will inspire researchers in Africa to investigate MIS 6-2 not only in the continent's best researched regions and archaeologically more remote corners, but also, crucially, in the areas connecting these two extremes. For paleodemographic research in Africa, it is still early days. This book, while perhaps raising more questions than providing answers, takes a critical first step toward elucidating the demographic processes that underpinned our species' development during this formative evolutionary phase.

The *Africa from Stages 6 to 2: population dynamics and paleoenvironments* conference was generously funded by the D.M. McDonald Grants and Awards Fund at the McDonald Institute for Archaeological Research at the University of Cambridge. We are sincerely grateful to the McDonald Institute for funding the conference. This three-day event was also sponsored by the Centre of African Studies at the University of Cambridge. We thank the speakers who presented their research at the conference, most of whom have contributed to this volume, as well as the conference's attendees, who contributed to stimulating, important discussions on the topics raised in this book. We are grateful to Katherine Clahassey of the University of Michigan's Museum of Anthropological Archaeology for skillfully creating the cover illustration. Finally, we dearly thank all those who have supported the creation of this book and who have been instrumental in its completion. First, we are grateful to Springer's Series Editors, Eric Delson and Eric Sargis, for their commitment, help and support throughout all stages of this volume's development. We thank all fifty-two peer-reviewers for their comments on the book's chapters. Last but not least, we are indebted to all the chapter authors who have contributed so generously to this volume. We are most grateful for their time and patience and we sincerely hope that they will enjoy being part of this book.

Sacha C. Jones
Brian A. Stewart

Contents

Contributors

Note * indicates preferred address for correspondence

Lucy F. Allott Institute of Archaeology, University College London, London, UK

Annita Antoniadou School of Geography, Archaeology and Palaeoecology, Queen's University, Northern Ireland, UK

Graeme Barker McDonald Institute for Archaeological Research, University of Cambridge, Cambridge, UK

Huw Barton School of Archaeology and Ancient History, University of Leicester, Leicester, UK

Paul Breeze Department of Geography, King's College London, London, UK

George A. Brook Department of Geography, University of Georgia, Athens, GA, USA

Sallie L. Burrough School of Geography and the Environment, Oxford University, Oxford, UK

Alec C. Campbell Crocodile Pools, Gaborone, Botswana

Emanuele Cancellieri *Dipartimento di Scienze dell'Antichità, Sapienza Università di Roma, Rome, Italy; Istituto Italiano di Paleontologia Umana, Rome, Italy

Andrew S. Carr Department of Geography, University of Leicester, Leicester, UK

Brian M. Chase Centre National de la Recherche Scientifique, UMR 5554, Institut des Sciences de l'Evolution de Montpellier, Université de Montpellier, Montpellier Cedex 5, France

Els Cornelissen Heritage Studies, Royal Museum for Central Africa, Tervuren, Belgium

Mauro Cremaschi Dipartimento di Scienze della Terra "A. Desio", Università degli Studi di Milano, Milan, Italy

Genevieve Dewar Department of Anthropology, University of Toronto, Scarborough, Toronto, ON, Canada

Savino di Lernia *Dipartimento di Scienze dell'Antichità, Sapienza Università di Roma, Rome, Italy; School of Geography, Archaeology and Environmental Studies, University of the Witwatersrand, Johannesburg, South Africa

Nick Drake Department of Geography, King's College London, London, UK

J. Tyler Faith School of Social Science, Archaeology Program, University of Queensland, Brisbane, Australia

Lucy Farr McDonald Institute for Archaeological Research, University of Cambridge, Cambridge, UK

Robert A. Foley Department of Archaeology and Anthropology, Leverhulme Centre for Human Evolutionary Studies, University of Cambridge, Cambridge, UK

Frederick E. Grine Departments of Anthropology and Anatomical Sciences, Stony Brook University, Stony Brook, NY, USA

Chris Hunt School of Natural Sciences and Psychology, Liverpool John Moores University, Liverpool, UK

Robyn Inglis Department of Archaeology, University of York, York, UK

Andrew H. Ivester Department of Geosciences, University of West Georgia, Carrollton, GA, USA

Sacha C. Jones McDonald Institute for Archaeological Research, University of Cambridge, Cambridge, UK

Alex Mackay School of Earth and Environmental Sciences, Centre for Archaeological Science, University of Wollongong, Wollongong, NSW, Australia

Marta Mirazón Lahr Department of Archaeology and Anthropology, Leverhulme Centre for Human Evolutionary Studies, University of Cambridge, Cambridge, UK

Peter Mitchell *School of Archaeology, University of Oxford & St Hugh's College, Oxford, UK; GAES, University of the Witwatersrand, Johannesburg, South Africa

Mike W. Morley Centre for Archaeological Science, University of Wollongong, Wollongong, Australia

Michael L. Murphy Kalamazoo Valley Community College, Kalamazoo, MI, USA

Adrian G. Parker Human Origins and Palaeoenvironments Research Group, Faculty of Humanities and Social Sciences, Department of Social Sciences, Oxford Brookes University, Headington, Oxford, UK

Philip Van Peer Prehistoric Archaeology Unit, Institute of Geo-Sciences, Katholieke Universiteit Leuven, Louvain, Belgium

Daniel J. Peppe Terrestrial Paleoclimatology Research Group, Department of Geology, Baylor University, Waco, TX, USA

Luísa Pereira *IPATIMUP (Instituto de Patologia e Imunologia Molecular da Universidade do Porto), Porto, Portugal; Faculdade de Medicina da Universidade do Porto, Porto, Portugal

Tim Reynolds Department of History, Classics and Archaeology, Birkbeck, University of London, London, UK

Martin B. Richards Department of Biological Sciences, School of Applied Sciences, University of Huddersfield, Huddersfield, UK

Teresa Rito IPATIMUP (Instituto de Patologia e Imunologia Molecular da Universidade do Porto), Porto, Portugal; *Life and Health Sciences Research Institute (ICVS), School of Health Sciences, University of Minho, Braga, Portugal; ICVS/3B's—PT Government Associate Laboratory, Braga/Guimarães, Portugal

Lawrence H. Robbins Department of Anthropology, Michigan State University, East Lansing, MI, USA

Judith Sealy Department of Archaeology, University of Cape Town, Rondebosch, South Africa

Pedro Soares IPATIMUP (Instituto de Patologia e Imunologia Molecular da Universidade do Porto), Porto, Portugal; *CBMA (Centre of Molecular and Environmental Biology), Department of Biology, University of Minho, Braga, Portugal

Brian A. Stewart Museum of Anthropological Archaeology, University of Michigan, Ann Arbor, MI, USA

Nicholas Taylor Department of Archaeology, Classics and Egyptology, University of Liverpool, Liverpool, UK

Christian A. Tryon Department of Anthropology, Harvard University, Peabody Museum of Archaeology and Ethnology, Cambridge, MA, USA

Kevin White Department of Geography, University of Reading, Reading, UK

Andrea Zerboni Dipartimento di Scienze della Terra "A. Desio", Università degli Studi di Milano, Milan, Italy

Chapter 1
Africa from MIS 6-2: The Florescence of Modern Humans

Brian A. Stewart and Sacha C. Jones

Abstract Africa from Marine Isotope Stages (MIS) 6-2 saw the crystallization of long-term evolutionary processes that culminated in our species' anatomical form, behavioral florescence, and global dispersion. Over this ~200 kyr period, Africa experienced environmental changes on a variety of spatiotemporal scales, from the long-term disappearance of whole deserts and forests to much higher frequency, localized shifts. The archaeological, fossil, and genetic records increasingly suggest that environmental variability profoundly affected early human population sizes, densities, interconnectedness, and distribution across the African landscape – that is, population dynamics. At the same time, recent advances in anthropological theory predict that such paleodemographic changes were central to structuring the very records we are attempting to comprehend. The book introduced by this chapter represents a first concerted effort to assess modern human population dynamics throughout Africa, whether these changed with environmental fluctuations, and how they contributed to our species' evolutionary trajectory.

Keywords Cultural transmission • Human evolution • Middle and Late Pleistocene • Middle and Later Stone Age • Paleodemography • Paleoenvironment • Population dynamics

B.A. Stewart (✉)
Museum of Anthropological Archaeology, University of Michigan, Ruthven Museums Building, 1109 Geddes Avenue, Ann Arbor, MI, USA
e-mail: bastew@umich.edu

S.C. Jones
McDonald Institute for Archaeological Research, University of Cambridge, Downing Street, Cambridge, CB2 3ER, UK
e-mail: scj23@cam.ac.uk

Introduction

The last three decades represent a watershed in our understanding of modern human origins. In the mid-1980s, evolutionary genetics established that the most ancient human lineages are African (Cann 1988; Vigilant et al. 1991). Since then, steady streams of genetic, paleontological and archaeological insights have converged into a torrent of evidence that Africa is our species' evolutionary home, both biological and behavioral. When these changes occurred, however, remains less well understood, and much less so how and why. Where within Africa modern humans and our suite of behaviors developed is also problematic. One thing seems clear: the changes that shaped our species and its behavioral repertoire were gradual, rooted deeper in the Pleistocene than previously imagined. Anatomically, the accrual of modern traits and disappearance of those deemed archaic now appears to have been a phased process lasting some 400 kyr (Grine 2016; Lahr and Foley 2016), a biological blurring underscored by genetic evidence for later admixture with archaic forms outside (Green et al. 2010; Reich et al. 2010) as well as inside (Hammer et al. 2011; Mendez et al. 2013) the continent (but see Eriksson and Manica 2012). Similarly, the pedigrees of many complex behaviors – including hafting (~500 ka; Wilkins et al. 2012; Barham 2013), projectile technologies (~279 ka; Sahle et al. 2013), and pigment use (~270 ka; Barham 1998, 2002a) – seem to attenuate with nearly every year that passes.

Yet it is not until 200 ka – roughly the start of Marine Isotope Stage (MIS) 6 – that the first fully derived modern human anatomical features appear (McDougall et al. 2005). We must wait another 100 kyr, until late MIS 5, for the first compelling archaeological expressions of behaviors characteristic of living people, including external symbolic storage (e.g., Bouzouggar et al. 2007; Henshilwood et al. 2009). Thus Africa from MIS 6-2 seems to have witnessed the crystallization of complex, longer term evolutionary processes set in motion at or beyond the advent of Middle Stone Age

Sacha C. Jones and Brian A. Stewart (eds.), *Africa from MIS 6-2: Population Dynamics and Paleoenvironments*,
Vertebrate Paleobiology and Paleoanthropology, DOI 10.1007/978-94-017-7520-5_1

(MSA) technologies ~300 ka (McBrearty and Brooks 2000; McBrearty 2007) or possibly earlier (cf. Wilkins et al. 2012). How and why this happened at this particular time are challenging questions indeed, involving constellations of interconnected variables that are difficult to parse. With recently accelerated research on this topic, however, important variables are being successfully isolated. Of particular significance are population dynamics and their relationship to paleoenvironments, and how both changed through time and space.

Population Dynamics and Paleoenvironments

The end of the Pleistocene epoch witnessed increasingly volatile climates with major implications for changes to African environments and their resource structures. Globally, MIS 6-2 saw a particularly warm interglacial period (MIS 5, but particularly 5e) bookended by some of the Pleistocene's most severe glacials (MIS 6, 4 and 2), with an intervening stage (MIS 3) so erratic that researchers disagree over which category it should fall into (Fig. 1.1). Humans in Africa were faced with immense changes in humidity, temperature, windspeeds, vegetation, animal biomass and landscape geomorphology that occurred at a variety of spatiotemporal scales, from the long-term disappearance of whole deserts to much higher frequency, localized shifts that were perceptible to people over the course of their own lifetimes. The human responses to these pressures were undoubtedly as diverse and complex as the changes themselves, engendering a wide range of biological and cultural selective outcomes. As with all humans, but particularly with hunter-gatherers due to their subsistence practices, local and regional shifts in resource base have demographic consequences and affect, for example, population sizes, densities, interconnectedness and distribution across the landscape – that is, population dynamics. Changes in human population dynamics – often but of course not exclusively initiated by environmental change – can be seen as part of a suite of adaptive tools that were mediated by, and in turn influenced, human biological and cultural evolution and history.

The evidence for widespread and rapid paleoenvironmental changes through late Middle and Late Pleistocene Africa that has accumulated over the past three decades has rendered major population flux over this period probable (e.g., Blome et al. 2012). Crucially, this has recently begun to be confirmed by numerous evolutionary genetic studies of modern African populations throughout the continent (Tishkoff et al. 2009; Henn et al. 2011; Soares et al. 2016). Such data are particularly significant in light of recent insights from cultural evolutionary theory that suggest effective population size influences rates at which innovations are generated and transmitted (e.g., Powell et al. 2009; Collard et al. 2013). Thus paleodemography may have a direct bearing on how we interpret the behavioral evolution of our species, particularly the wax and wane of complex expressions seen most famously in the Still Bay and Howiesons Poort of southern Africa (Henshilwood and Dubreuil 2011). Such expressions seem to surface only intermittently throughout MIS 6-2, before culminating explosively in the relative climatic stability of the Holocene. The implications of this argument, though enticing, are extremely contentious not least because inferring information about metapopulation size from archaeological data is often problematic (see Chamberlain 2006). Beyond this specific issue, however, the weight of genetic, paleoenvironmental, fossil, and archaeological evidence is making it increasingly clear that the stability of African populations from MIS 6-2 vacillated. Any understanding of modern human origins and early history is therefore contingent upon the exploration of African population histories.

While timely, the topic of Africa's demographic history is far from new. From its outset until relatively recently, African prehistory was cast within a culture historic framework in which archaeological changes were explained with reference to unfounded assumptions about successive waves of incoming populations (Robertshaw 1990; Mackay 2016). Beginning around the mid 20th century, attempts were made to more explicitly relate hypothetical population dynamics to climatic changes (Clark 1959, 1960; Lee 1963), the latter as postulated in the subsequently disproven pluvial-interpluvial schema (e.g., Brooks 1914; Wayland 1934; Leakey 1951). With characteristic incisiveness, John Desmond Clark (1959, 1960) provided the most penetrating treatment. Seeking to explain gross patterns of spatiotemporal variability in the African paleoanthropological record, Clark invoked climate shifts and their impact on human population dynamics against a background of longer term, macroevolutionary trends in human cognition and technical skill. He hypothesized that important changes were ushered in with the MSA, when behavioral advancements allowed regionally distinct technologies to develop during wet 'pluvial' periods as populations became isolated and adapted their toolkits to varying ecologies. Punctuating these phases of prolonged stasis and diversification were shorter and drier 'interpluvials', which provoked rapid cultural change ("speed-up") stemming from enhanced innovation and population dynamism. Clark (1960: 310) wrote that during such times,

> "…decreasing rainfall may be expected to have opened corridors and sometimes highways from north to south and from east to west, which seem to have invited, even though they also sometimes sidetracked, movement by groups forced to migrate by the deterioration of their traditional environment. Sometimes biological proof, in the shape of human fossils, confirms this. At other times we have to rely on different evidence, for example

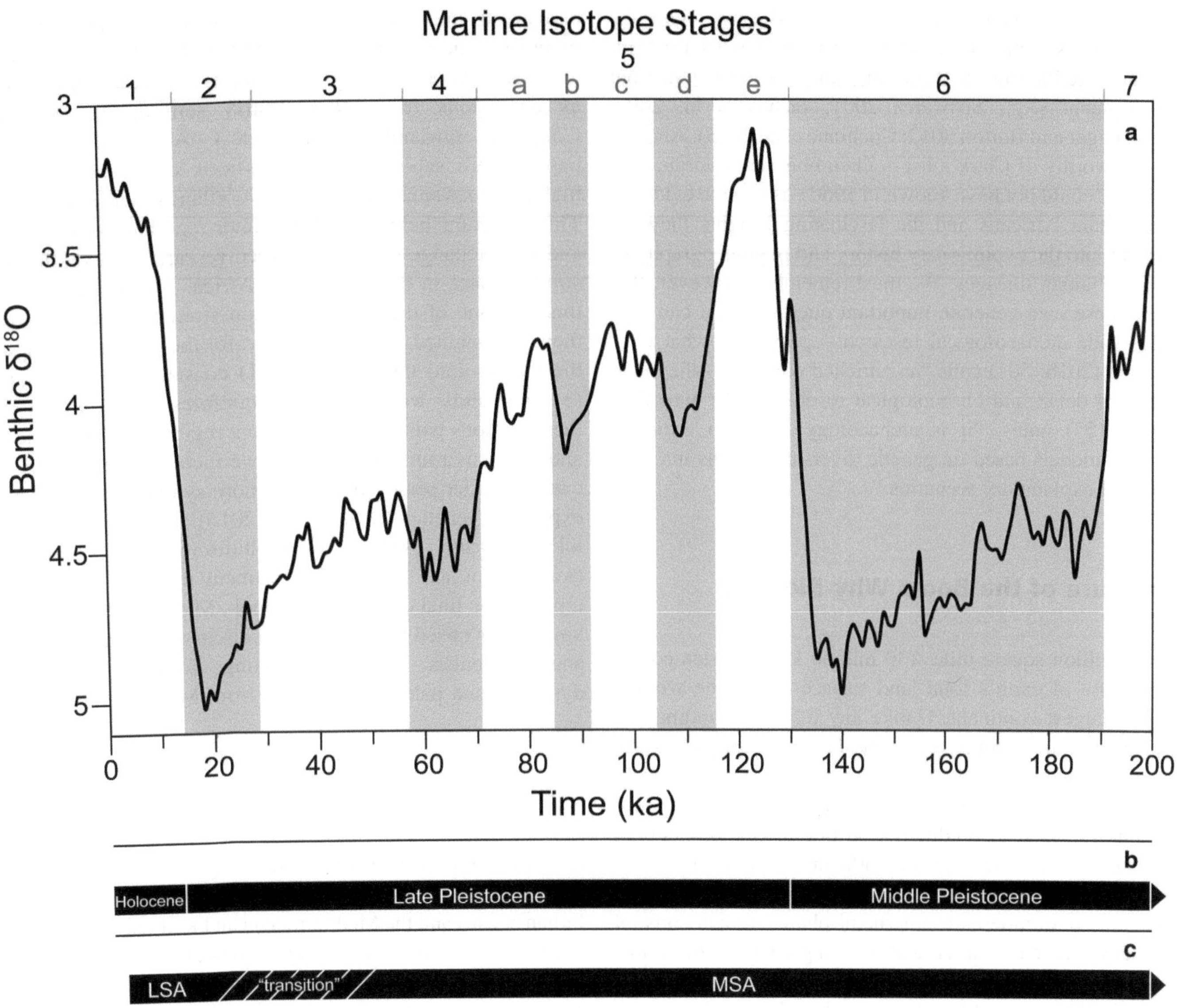

Fig. 1.1 Climatic, geological, and cultural frameworks used in this chapter and book. **a** Marine Isotope Stages (MIS) 6-2 (191–14 ka) differentiated according to Lisiecki and Raymo's (2005) LR04 stack, an average record of global benthic $\delta^{18}O$ measuring global ice volume and deep ocean temperature; MIS 6: 191–130 ka; MIS 5: 130–71 ka; MIS 4: 71–57 ka; MIS 3: 57–29 ka; MIS 2: 29–14 ka. **b** Geological ages (Middle and Late Pleistocene) and epoch (Holocene) correlating to MIS 6-2. **c** Major divisions of the African Stone Age archaeological record correlating to MIS 6-2; *MSA* Middle Stone Age; *LSA* Later Stone Age

> the patterns of spread of new items of cultural equipment. Contact between bands naturally encourages the spread of cultural traits, and new inventions or techniques may be surmised to have been speedily transmitted at times of ecological change and population movement."

Clark's observations – though predicated on a flawed climatic model, lacking a chronological framework and inherently speculative – were nonetheless prescient. Interpreting MSA diversification, the pace of behavioral evolution, the role of climatic fluctuations on population dynamics and cultural transmission, the opening and closing of biogeographical corridors and barriers, the meaning of similar yet widely dispersed artifact forms and human fossils, remain as resonant today as they were a half-century ago. What is fundamentally different about today are the data. In 1960, Clark listed six datasets "essential to any study of the life of early man, if the results are to have lasting value" (Clark 1960: 307). These were: (1) a chronological framework; improved knowledge of (2) artifact typologies and (3) the spatial distributions of stone tool industries; (4) improved paleoenvironmental data and its incorporation into paleoanthropology through interdisciplinary collaboration; (5) careful site selection and excavation; and (6) the appropriate use of ethnography. In the succeeding half-century, massive strides have been made toward each, with the ongoing construction of a chronometric framework for Middle and Late Pleistocene Africa arguably the most significant development (e.g., Schwarcz 2001; Jacobs et al. 2008; Bouzouggar and Barton 2012; Cochrane et al. 2013). Another dataset worthy of Clark's list – albeit one the importance of which he could not have known in 1960 – is genetic evidence from extant Africans and the fascinating insights they are offering into the evolutionary history and phylogeography of modern human lineages. We must remember, however, that while these data generate important questions that can usefully guide archaeological research – providing what Cornelissen (2016: 301) calls "an enriched context" – they lack historical detail and chronological resolution. As Van Peer (2016: 157) notes, "it is archaeology's role to critically develop models based on genetic diversity patterns into historic and explanatory scenarios."

Structure of the Book: Why Biomes?

At 12 million square miles (30 million km^2), Africa covers some 20% of earth's total land surface and is the world's second largest continent. Unlike any other, its landmass is distributed fairly evenly across the Equator, reaching ~35° latitude both north and south. The upshot is a 'sandwich' configuration, with mirrored rainforest to woodland to savanna to desert to Mediterranean littoral environments as one moves from the Equator north or south (Barham and Mitchell 2008). This broad ecological symmetry presents unique opportunities for interhemisphere comparisons of paleoenvironmental shifts, and resulting adaptive behaviors of early modern human groups operating in different yet ecologically comparable areas of the continent. This is particularly interesting with respect to mounting evidence for the asynchronous nature of environmental change in North *versus* sub-Saharan Africa, as it implies that at certain times what are today similar environments either side of the Equator presented humans with very different conditions. Because of its size and colonial history, however, the study of Africa's past has traditionally been pursued in different areas with relatively minimal inter-regional communication. While this problem is more acute for Africa's recent archaeology, deeper periods including MIS 6-2 are also affected, most notably by the conventional division between North and sub-Saharan Africa but also by that between largely francophone West/Central and anglophone northeast, East and southern Africa.

For all these reasons, we have organized the book, like the 2010 Cambridge conference from which it derives, according to biome rather than by region. Here we use biome in its original sense as referring to large-scale ecosystem units with similar biotic structure and macroclimate (e.g., deserts, rainforests etc.), rather than as a proxy term for subordinate, geographically specific units such as ecoregions (e.g., Nama-Karoo semi-desert, Guineo-Congolian lowland rainforest etc.). At the conference, papers were subdivided into sessions on coasts, deserts, highlands, savannas and mixed woodlands, and rainforests. This structure helped facilitate inter-regional comparisons and the exchange of ideas between researchers working on similar issues in distant parts of Africa. Attendees felt that this was one of the meeting's main strengths and we have therefore retained it here. However, for the book we collapse the biomes into three sections: (1) coasts; (2) deserts; and (3) grasslands, woodlands, and rainforests. In addition to these sections with chapters detailing regional archaeological and paleoenvironmental records, we include a section dedicated to wider perspectives with more synthetic chapters by experts in genetics (Soares et al. 2016) and human paleontology (Grine 2016). This multidisciplinary breadth is essential for any successful treatment of such a complex topic. In the final chapter of the book, Mitchell (2016) traces key themes raised throughout, and discusses future prospects and challenges for reconstructing African population dynamics and paleoenvironments from MIS 6-2.

Coasts

Extending for ~30,000 km excluding Madagascar (Orme 2005: 9), Africa's coastline is surrounded by the Atlantic and Indian oceans and the Mediterranean and Red seas. The lower reaches of some of Africa's major rivers dissect this coastline (e.g., the Congo and the Zambezi), with some creating complex deltaic environments (e.g., the Nile and the Niger). A diversity of coastal ecologies exists along Africa's littoral, dictated by varying latitudes, regional geographies and climatic regimes, with vegetation varying from one extreme (e.g., desert) to another (e.g., tropical rainforest). For example, Mediterranean vegetation biomes are found at the extreme north and south of the continent, with mangrove habitats along West and East Africa. Coastal zones commonly consist of complex ecotonal and mosaic habitats, in places alternating between desert, subdesert steppe, savanna grasslands and woodland, and lowland tropical forest. Fluctuations in Pleistocene sea level resulted in variable exposure of Africa's continental shelf; for example, sea-level lowstands would have exposed little land in northeast Libya (Jones et al. 2016) yet extensive areas in southern Africa (van Andel 1989; Fischer 2010; Dewar and Stewart 2016), particularly in the area of the Agulhas Bank (Carr et al. 2016). Relative to other large continents, however, Africa's continental shelf is narrow, averaging only 25 km in width (Orme 2005: 9).

Uniting the chapters on coastal records from MIS 6-2 is the relationship between coastal habitats and human occupation with respect to three issues: behavioral complexity, coasts as refugia, and coasts as dispersal corridors. The archaeological records of coastal sites in the Maghreb of northwest Africa and South Africa are well known for exhibiting at least relatively brief periods of seemingly enhanced behavioral complexity (e.g., shell beads, osseous technologies, and engraved ochre, bone and ostrich eggshell). This was particularly the case during late MIS 5 and MIS 4 in Aterian, Still Bay and Howiesons Poort contexts (e.g., Henshilwood et al. 2002; d'Errico et al. 2005, 2008, 2009; Bouzouggar et al. 2007; Texier et al. 2010), but also in earlier MSA contexts in mid-MIS 5 ($\sim$100 ka) at Blombos Cave (Henshilwood et al. 2009, 2011) and mid-MIS 6 ($\sim$164 ka) at Pinnacle Point 13B (Marean et al. 2007; Brown et al. 2009), both coastal caves along South Africa's southern Cape. The spatiotemporal distribution of these so-called cultural indicators of behavioral complexity seems to suggest they are more frequently associated with coastal or near-coastal settings, and are more common from late MIS 5. Robbins et al. (2016), for example, note that comparable innovations have yet to be found in similarly early MSA contexts in the Kalahari. Findings from the southern Cape coast have been the basis of arguments that link shellfish exploitation with behavioral complexity and the evolution of modern human cognition, where the southern extremity of Africa is central to these evolutionary developments. Parkington (2001, 2003, 2006, 2010) has suggested that regular shellfish consumption contributed to encephalization, through the benefits of a diet rich in omega-3 long-chain polyunsaturated fatty acids. Encephalization resulting from this dietary shift perhaps enabled various innovations that we currently associate with 'behavioral modernity'. In contrast, Marean suggests that humans were able to exploit this new dietary resource as a consequence of modern human cognition. Given the earliest evidence for shellfish exploitation c.164 ka at Pinnacle Point 13B, and for other advanced behaviors of a comparable age (e.g., ochre use, heat treatment of rock for tool manufacture), he argues that modern human cognition was coincidental with the origin of *Homo sapiens*, rather than there being a decoupling of anatomical and cognitive modernity (Marean et al. 2007; Marean 2010, 2011, 2014). Both Parkington and Marean see the south coast of South Africa as playing a central role in the origins of our species (Parkington 2010; see also Compton 2011), with the region's plant diversity, resource-rich coastline and moderate climate all providing "the ideal conditions for a refuge for the bottlenecked modern human lineage during the long cold MIS6" (Marean 2011: 434).

Both this MIS 6 southern Cape refugium scenario (Wurz 2013; Lombard et al. 2013) and the importance of South Africa in species origins models (Lahr and Foley 2016) have nevertheless been questioned. Two issues are of particular importance. The first of these touches on a greatly debated subject: are we looking at behavioral (or cognitive) complexity in the right way, or are we constrained by our own assumptions? Perhaps diversity in MSA lithic reduction strategies (Jones et al. 2016; Mackay 2016; Van Peer 2016) is a viable measure of behavioral complexity, reflecting modern cognitive capacities for finding adaptive solutions to ecological, social, and demographic changes. Perhaps, stone artifacts – the most abundant material remains available to us – are somewhat overlooked in favor of more obvious yet less frequent signs of behavioral complexity (e.g., pierced shells, engraved ochre etc.). An absence of evidence is not necessarily evidence for the absence of cognitively complex populations, particularly when our understanding is hampered by a severe dearth of certain datasets, such as the intangible (e.g., body decoration) and unpreserved (e.g., wood and hide). It is important to continue challenging how we define and think about behavioral and cognitive complexity (e.g., Shea 2011; Wadley 2013), and to acknowledge and assess the biases in our records.

This brings us to the second question: could these patterns simply reflect a geographic sampling bias, created by an intense research focus on certain regions – the Maghreb and South African coasts – driven in part by coastal geomorphologies where mountainous areas rich in caves and rockshelters are situated close to the coast? Like the chapters in this volume, most studies of ancient African coastal occupation and human exploitation of marine resources focus almost exclusively on the far north and south, which together constitute a tiny proportion of Africa's vast coastline (e.g., Volman 1978; Walter et al. 2000; Erlandson 2001; Parkington 2003, 2010; Klein et al. 2004; Marean et al. 2007; Avery 2008; Dewar 2008; Steele and Klein 2008; Ramos et al. 2008; Jerardino and Marean 2010, 2011, 2014; Sealy and Galimberti 2011; Steele and Álvarez-Fernández 2011; Dewar and Stewart 2012; Langejans et al. 2012; Dusseldorp and Langejans 2013; Will 2013; Kyriacou et al. 2015). As fascinating as these records are, this geographic bias inevitably contracts our view of African coastal adaptations in prehistory. An intriguing prospect would be to discern how the MIS 6-2 records of the coastal habitats discussed in this volume contrast with those of Africa's non-Mediterranean shorelines (e.g., coastal rainforests, mangroves, deserts, steppes, savannas etc.). For example, how do different African coastal habitats compare in terms of incidences of marine resource exploitation, archaeological expressions of behavioral complexity, past population dynamics, and the extent to which each was interlinked or disconnected in the past? Neither do we have sufficiently representative coastal records even in southern and North Africa. Sealy (2016), for example, points out that sample size may be a notable problem, with more records dating to

MIS 6 needed in order to discern if anatomical and behavioral modernity were decoupled. Similarly, our ability to address questions of past population dynamics in southern Africa, one of the continent's better-studied regions, is still hampered by the small sample of well-resolved coastal sites (Carr et al. 2016).

With these limitations in mind, is it possible to link the behavioral complexity evident in the coastal contexts discussed here to these areas having acted as refugia (e.g., Marean 2011) and/or corridors for dispersal, both of which potentially encouraged cultural interconnectedness? In their paper, Carr et al. highlight previous studies on coasts as refugia both within and outside Africa, emphasizing the importance of marine resources as a reliable food source when inland settings were less favorable. Discussing climate-induced changes to the South African coastline, they suggest that population isolation may have occurred where regional geological structures prevented connections between the South African interior and its continental margins. Sealy assesses the validity of using evidence from Holocene Later Stone Age (LSA) sites (Wilton and post-Wilton) as a comparable model for demographic processes during the time of the Howiesons Poort and Still Bay. As a consequence, she makes some important arguments about population dynamics around the southern Cape coast, focusing on population size, density, expansions, contractions and connectedness during the MSA and LSA. By outlining the limitations of current datasets (e.g., problems with sample size, sampling bias, the temporal resolution of archaeological data, and the paucity of detailed paleoclimatic records), she challenges conventional interpretations of Late Pleistocene demographic processes in the region in relation to patterns of technological and behavioral complexity.

In North Africa, Jones and colleagues suggest that the Gebel Akhdar and its littoral margins in northern Cyrenaica preserved refugia for human populations during glacial periods, when populations appear to have been largely (but not entirely) confined to areas of the North African coast. During more humid episodes, on the other hand, the area appears to have been a critical fulcrum for connections via corridors with regions further to the south (e.g., trans-Sahara) and east (e.g., Nile valley). Also relevant are arguments for coastal migrations of MSA and LSA tool-makers out of Africa, and back into it, via the Sinai Peninsula and Levant (Jones et al. 2016), and of Nubian MSA tool-makers into Arabia and possibly back into Africa via the Horn (Van Peer 2016). The nature and timing of inland connections between the Maghreb and surrounding regions are addressed by Drake and Breeze, but more emphasis needs to be placed on assessing the role of corridors along coastlines themselves, particularly during the MSA. The evidence from northwest Africa (Hublin and McPherron 2012) points to a concentration of populations in areas of the Maghreb from MIS 6-2, but how isolated or connected was this region and where did connections take place? Are important clues located in coastal areas about which we have little archaeological and hominin fossil evidence (e.g., West Africa); in fact, how viable were other coastal habitats for supporting populations? While most coastal environments may present a richness of marine food resources, it is also important to emphasize the importance of freshwater sources (Jones et al. 2016; also Dewar and Stewart 2016) capable of supporting and sustaining populations. Freshwater location and availability must have played an important role in population size and migration corridors (e.g., Faure et al. 2002), and we need to investigate water source distribution in coastal settings together with the cultural ramifications of this (e.g., the use of ostrich eggshell or skin containers).

Apart from the over-representation of certain coastlines, whose remedy must await a major Equator-ward expansion of coastal archaeological research, there are two main challenges to reconstructing Stone Age population dynamics and paleoenvironments along Africa's shores. One is to move beyond a myopic focus on coastal cave sites and towards incorporating evidence from the wider landscape, including adjacent inland areas (e.g., Carr et al. 2016; Drake and Breeze 2016; Jones et al. 2016; Dewar and Stewart 2016; Stewart et al. 2016). With respect to the southern African record and the over-representation of evidence regarding human–environment interactions from coastal sites, Burrough (2016) notes the contrasting evidence from the Kalahari and other deep interior zones. Though genetic and archaeological evidence indicate a sustained human presence during the Late Pleistocene, substantially less is known about occupation in these vast inland areas. In northeast Libya, Jones et al. address this problem through a landscape survey of varied landforms within coastal, upland, and desert contexts, with the aim of contrasting MSA and LSA datasets from the rich coastal cave sequence of Haua Fteah against those from these different landscape settings.

The second challenge is to assess the predicaments posed by sea level fluctuations from MIS 6-2. These include mitigating the loss of evidence produced by submergence and marine erosion, and modeling human adaptive responses to sea level change. In areas of steep offshore topography, sea level changes do not appear to have significantly affected archaeological site preservation. An example of such an area is northern Cyrenaica. However, probable submergence of other sites east and west along North Africa's continental shelf may be masking important evidence of Pleistocene coastal migrations. Similarly, Dewar and Stewart note that the coastal plain of Namaqualand may have been an attractive area for human occupation during MIS 6, yet there are currently no known sites from this period, probably because they are now submerged. Marine transgressions, particularly during MIS 5e and the early Holocene, have also resulted in data loss via the erosion of sediments within some low altitude coastal cave sites (Carr et al. 2016; Sealy 2016). Carr et al. stress that the current submergence of the South African coastal plain (Agulhas Shelf) makes it difficult to model both landscape change in these areas and the resources available to human populations. The fact that MIS 6-2 sea level change had variable impacts depending on terrestrial geomorphology and ecology means that we need to carefully think about the consequences of this with respect to both human adaptive responses and migration patterns. Yet, they also note the employment of new geophysical surveying approaches in order to better understand submerged landscapes. We suspect that advances in underwater hunter-gatherer archaeology in other parts of the world (e.g., O'Shea et al. 2014) might be fruitfully applied to shallower stretches of Africa's continental shelf.

Deserts

Some 43% of the African continent is today classified as drylands (UNCCD, UNDP, and UNEP 2009). True desert biomes include the Sahara (Drake and Breeze 2016; Cancellieri et al. 2016; Jones et al. 2016; Van Peer 2016) and Namib Deserts. Semi-arid regions include the Kalahari (Burrough 2016; Robbins et al. 2016) and Namaqualand, a semi-arid southward extension of the Namib (Dewar and Stewart 2016). By far the most expansive arid zone of all is the Sahara, the Earth's largest tropical desert, stretching from the Atlantic Ocean to the Red Sea and bordered by semi-desert and Mediterranean biomes to the north and the grasslands of the Sahel to the south. Less-extreme conditions exist in some central Saharan mountainous regions (e.g., Ahaggar, Aïr and Tibesti massifs), and numerous oases interrupt the desert as does Africa's longest river, the Nile. Water supply to the Sahara varied significantly from MIS 6-2. Major fluctuations are also recorded for the Nile, the most important runoff source of which is the Ethiopian Highlands. Cycles of humidity and aridity, with corresponding contractions and expansion of the desert, had a clear impact on the distribution of freshwater, plant and animal resources throughout the Sahara. Glacial and stadial periods were generally associated with hyper-aridity and desert expansion, yet these were punctuated by Saharan pluvials (Drake and Breeze 2016; Jones et al. 2016), the environmental and human consequences of which remain poorly understood. Drake and Breeze argue that well-fed rivers would have existed in the Tibesti and Ahaggar mountains during humid episodes, feeding Saharan lakes (e.g., Lake Megachad and the Chotts Megalake) and supporting high biodiversity and varied resources.

Four chapters in this volume focus on human occupation of the Sahara from MIS 6-2. Concentrating on the Tadrart Acacus and Messak massifs in the Fezzan region of Libya, Cancellieri et al. report the results of their extensive landscape surveys, including the types and densities of lithic artifacts encountered. Similarly, Jones et al. discuss patterns of MSA and LSA human occupation at the northern limits of the Sahara desert in Cyrenaica. Other chapters take a much broader view of the Saharan record. Van Peer examines patterns of human occupation across North Africa and even into Eurasia, postulating demic connections on the basis of typo-technological similarities of regional lithic assemblages belonging in particular to the MSA. Using geoproxies and dated archaeological sites, Drake and Breeze propose episodes of enhanced humidity with paleohydrological networks across the Sahara and argue that occupations by Aterian tool-makers corresponded to humid periods. This is contra Cancellieri and colleagues who associate the Aterian with generally dry conditions, and caution against assumptions that wet equates with occupation and dry equates with abandonment.

In southern Africa, past and current climate and environments in the Kalahari are given detailed treatment by Burrough and Robbins and colleagues. Burrough remarks on the existence of extreme dry and extreme wet conditions in the Kalahari from MIS 6-2, the former marked by episodes of dune building, and the latter by variably sized lakes, some of which were massive and deep (e.g., Megalake Makgadikgadi, cf. Saharan Quaternary lakes such as lakes Megachad or Megafezzan). These lakes must have had a significant impact on human occupation, as argued by both Burrough and Robbins et al. Burrough notes a spatial correlation between extensive occurrences of Paleolithic stone tools and the rivers and lakes of the Okavango-Chobe-Zambezi system. Robbins et al. provide an account of Kalahari food resources during the MSA and LSA. Evidence of fish exploitation occurs in MSA and LSA contexts, but particularly in the latter where the archaeological record indicates intensive seasonal fishing. The chapter by Dewar and Stewart focuses on Namaqualand, describing current climatic systems that affect the region and their consequences for paleoenvironments and the composition and distribution of fauna. Against this ecological framework, they propose changes in human occupation and behavior in Namaqualand from MIS 6-2. Their chapter underscores the asynchronicity between desert responses to global climate change in northern and southern Africa. Contrasting with the Sahara, they argue that Namaqualand during glacial and stadial phases was colder but wetter than today, conditions that would have facilitated higher population densities. They highlight pulses of human occupation during MIS 4 in particular, indicated by the presence of technologies comparable to those of the Still Bay and Howiesons Poort at various sites (e.g., Apollo 11) . Similarly, periods of high water availability are documented during the Last Glacial Maximum (LGM) in MIS 2, a period that would have been hyper-arid in the Sahara.

Understanding deserts as fluctuating corridors and barriers is a key issue raised by every chapter focusing on MIS 6-2 population dynamics in this biome. The challenge lies in understanding the timing, characteristics and extent of these shifts, as well as the longevity of occupation in certain desert regions. In proposing peaks in humidity in North Africa from MIS 6-2, Drake and Breeze do specify that not all humid peaks corresponded with humid corridors across the Sahara (via a network of links between Saharan mountains, lakes and rivers, linking sub-Saharan Africa and the Maghreb in particular). Instead, they indicate that the viability of certain corridors would have been highly variable from MIS 6-2. For example, their data point to a humidity peak ~154 ka during MIS 6, broadly assumed to be a period of aridity in tropical desert regions, yet they argue that this was not sufficient to 'switch on' a corridor across the Sahara. On a smaller scale, Jones et al. propose a network of paleohydrological corridors that may have linked desert and coastal biomes of northern Cyrenaica during humid episodes, arguing that MSA occupation of desert habitats was closely tied to the distribution of water (and chert) sources and that fluctuating patterns of water availability was one of the factors underlying regional variability in MSA technologies. In contrast, Cancellieri et al. recommend looking for factors other than humidity to explain human occupation patterns, challenging the argument that 'green = life'. They suggest that humans migrated along both humid and arid corridors and were well-adapted to living in arid environments (cf. Garcea 2012).

Van Peer focuses on the North African MSA record as a whole, identifying technological similarities between some MSA sites within North Africa and beyond. He argues that different regional lithic taxonomies have disguised similarities between regional MSA records. While he does not refer to any specific corridors within Africa, he does propose phases and regions of population connection and size fluctuation. While the Sahara would have presented a large barrier to human occupation and dispersals during glacial phases in particular, its 'greening' during certain periods (e.g., MIS 5e) would have presented conduits for, but also impediments to, human dispersals. One such green barrier, he suggests, was the Nile Delta, where the presence of an interglacial forest could have prevented populations from expanding along a northern Mediterranean coastal route. Similarly, during humid episodes in the Kalahari, Robbins and colleagues argue that the great extent and depth of the Makgadikgadi Megalake may have acted as a barrier against dispersals, affecting mobility and migration patterns and isolating populations. Among other factors (e.g., small sample size), they remark that this may be one reason behind the lack of Still Bay or Howiesons Poort technologies in this region, with the paleolake having prevented their northward spread.

One important issue for consideration concerns population densities in arid zones from MIS 6-2; when and to what extent were the African deserts depopulated during this period? We may assume that there were episodes of depopulation in the Sahara and Namib deserts in particular, but perhaps there is insufficient evidence at present to confirm this. It is suggested that the Sahara was largely depopulated from MIS 4-2 (Cremaschi et al. 1998; Blome et al. 2012). Caution is recommended before accepting such broad-scale arguments, however, as it is possible that a persistence of small populations in some areas would leave barely visible, let alone dateable archaeological traces (Farr and Jones 2014). Although there are a large number of surface MSA and LSA sites from the Sahara (Cancellieri et al. 2016; Jones et al. 2016), there are very few dated sites from Saharan contexts, even those dating to humid interglacials (Drake and Breeze 2016). The absence of dated evidence from arid contexts remains a considerable problem.

During phases of greater desiccation from MIS 6-2, can we detect refugia within Africa's deserts (e.g., Larrasoaña 2012)? Cancellieri et al. argue that the Aterian assemblage at Uan Tabu in the Acacus dates to MIS 4, representative of residual populations that were adapted to mountainous and arid habitats. Assuming the accuracy of the relevant dates, their evidence suggests that this area was a refuge for populations during MIS 4. In fact, some of the mountain massifs of the Sahara – the Ahaggar, Aïr and Tibesti – today harbor relict forest and Mediterranean flora and fauna (Barham and Mitchell 2008) and experience relatively high precipitation levels (Drake and Breeze 2016). It is therefore possible that areas like these harbored human populations surviving in refugia during less favorable periods during MIS 6-2. Elsewhere in the Sahara, Jones et al. comment on the existence of Saharan pluvials from MIS 4-2; these would have provided some moisture but it is unknown if this would have been sufficient to sustain populations let alone encourage expansions of population sizes and distributions. The same could be said of other areas, such as the Kharga Oasis, with its aquifer-fed artesian springs that were active during MIS 4 and MIS 2. While Van Peer notes few sites in the Lower Nile Valley during the final MSA of MIS 4/3, population densities appear to have been high in those areas that were occupied, suggesting they acted as refugia during this period. Burrough also discusses the existence of refugia along fluvial networks in the southern African interior, where rivers fed from the tropics would have been critical water sources in otherwise water deficient regions. If hunter-gatherers aggregated near these wetter areas during Late Pleistocene dry periods, there should be increased archaeological visibility in these areas. Although MSA and

LSA sites (often surface occurrences) are numerous in the region, this has yet to be tested archaeologically, and very few of these sites are excavated and dated. The possibility of refugia existing within otherwise extreme arid environments (e.g., along river corridors, uplands or oases) is something that should be incorporated into models of past population dynamics within African deserts, as it already has been on other continents (e.g., Veth 1989, 1993), including their role in eliciting social, technological, and demographic changes (cf. Van Peer 2016).

Critical to discerning past population dynamics in deserts is the need to understand the complex relationships between deserts and adjacent biomes, in particular the impacts of spatiotemporal fluctuations in ecotones from MIS 6-2. In a number of localities in northern and southern Africa, for example, desert and coastal biomes are often directly adjacent or situated nearby. Some chapters have modeled population movements between coastal settings and their variably arid hinterlands (Dewar and Stewart 2016; Jones et al. 2016; Stewart et al. 2016). It is the interfaces between these different biomes that may have provided ecotonal advantages for hunter-gatherer groups, resulting in adaptive responses to variable resource types and distributions, with possible technological consequences. Jones et al. suggest that such adaptive responses to shifting habitats are reflected in the MSA and LSA records of Cyrenaica. Similarly, there are relationships between deserts and woodlands/rainforests (see below). Tropical rainfall over equatorial Africa feeds many rivers that terminate in or run through deserts. These include the Niger, Nile and Chari in North Africa, and, in southern Africa, the Okavango, which terminates as a delta in the northwest Kalahari. River systems probably acted as refugia zones or corridors at times from MIS 6-2 (e.g., Basell 2008; Stewart et al. 2016), and most notably during MIS 5e with the paleoriver and paleolake systems of the Sahara (e.g., Drake and Breeze 2016). In the Kalahari, Burrough argues that transitional periods during MIS 6-2 were marked by large-scale landscape and ecological changes, necessitating adaptive responses in the form of innovation and/or mobility. Transitional periods could have brought benefits as new ecotones were established. Understanding the implications of the shifts between biome boundaries adds an additional level of complexity, yet these shifts could be a driving force behind variable adaptive responses and evolutionary trajectories within and beyond such geographic regions.

In attempts to understand past population dynamics within and across desert biomes from MIS 6-2, we encounter three main challenges: managing the considerable variability within African desert environments; coping with preservational problems; and identifying drivers of hydrological change. Deserts are extraordinarily heterogeneous environments, and understanding past human adaptations to the great spatial and temporal variability within deserts, let alone their ecotonal habitats with adjacent biomes, constitutes a significant challenge. Dunefields, mountain massifs, coastlines, river systems, oases, and paleolakes within deserts presented hunter-gatherer populations with very varied opportunities and constraints. Although aridity is frequently emphasized as a driver of cultural change and innovation (e.g., Clark 1960), arguments for climatic *variability* as more influential are gaining traction (Burrough 2016; Jones et al. 2016; Stewart et al. 2016). For example, Burrough (2016) states that "In the context of past population dynamics, overarching climatic theories remain largely irrelevant, since it is regional environmental (rather than climatic) variability to which hominins are likely to have responded," where the spatial complexities of landscape and regional climate dynamics should be taken into greater account. For the Sahara, Cancellieri et al. make two important points of relevance. First, they emphasize that in central Saharan contexts, minor variations in rainfall can alter delicate equilibriums. Second, they emphasize that the Sahara possesses a considerable variety of ecological niches that can support human populations today. This suggests that viable habitats for hunter-gatherers can open and close within this vast desert with variable frequency, nuances that should be factored into demographic models.

In order to reveal how variability impacted past population dynamics, high-resolution dated paleoenvironmental records are essential, together with *in situ* archaeological contexts. This brings us to the second challenge – the problem of preservational issues in desert environments. Burrough remarks that the spatially diverse and fragmentary nature of geoproxy records has resulted in climate change models that are often contradictory and oversimplified. Similarly, it may not be possible to identify a Saharan corridor that corresponds to the MIS 6 humidity peak proposed by Drake and Breeze, simply due to an absence of preserved sediments and hence paleohydrological data. This, together with a lack of dated contexts from the Sahara, makes it very difficult to correlate human responses with climate change. Archaeological evidence in desert environments is most often encountered on the desert surface in erosive and deflated contexts, often as palimpsests (Cancellieri et al. 2016; Jones et al. 2016; Burrough 2016; Robbins et al. 2016; Dewar and Stewart 2016). Conversely, much cultural material is also inaccessible beneath deep sediment accumulations. Dating MIS 6-2 archaeological evidence in desert environments is thus highly problematic and often relies on subjective techno-typological correlations. This is a problem that recurs in each desert chapter. Although few in number, there are some notable excavated sites in desert contexts that have provided significant data; for example, Uan Tabu and Uan Afuda (Cancellieri et al. 2016), White Paintings Rockshelter (Robbins et al. 2016), Apollo 11 and Spitzkloof (Dewar and Stewart 2016). Furthermore, one benefit of desert archaeology is the high visibility of artifacts, albeit in surface contexts, meaning that broad landscape-scale patterns of artifact types, technologies, and densities are achievable that would otherwise be very difficult in other environments (e.g., woodlands, rainforests, or densely populated areas today).

The final challenge is to determine the drivers of hydrological changes within deserts. Again, understanding local variability in terms of landscape and regional climate dynamics is critical here. For example, do river and lake activations within desert biomes represent local amelioration or changes occurring in more proximal parts of hydrological systems, such as equatorial zones (e.g., Burrough)? Drake and Breeze comment that there is currently a poor understanding of the interplay between the African monsoon and the North Atlantic and Mediterranean westerlies with respect to the extent of Saharan 'greening'. Perhaps instead, enhanced humidity simply resulted in the reactivation of rivers that flowed through otherwise arid zones. In Namaqualand, Dewar and Stewart propose that colonization of the coastal plain by grasslands during marine regressions (during early MIS 2 in particular) was a consequence of two alternate mechanisms: increased precipitation with westward movement of the coastline, or reduced evaporation as a result of lower temperatures. These examples highlight that there were probably multiple potential drivers of paleohydrological change affecting African desert environments. These may have been proximal and/or distal, with specific regional landscape-driven and climate-driven idiosyncrasies that contributed to a complex yet delicate balance between wet and dry.

Grasslands, Woodlands, and Rainforests

Between ~20° S and ~20° N, Africa's drylands give way to the grasslands, woodlands, dry forests, and rainforests that dominate its core. Together, these biomes constitute well over half of the continent. The terms grassland and woodland as applied to modern African environments are nebulous and often subsume a spectrum of vegetation types, including shrubland, parkland, thicket, grassland savanna, woodland savanna, and dry forest. Africa's grasslands are predominantly savannas – that is, seasonally dry, edaphic tropical/subtropical grass-dominated plant communities with discontinuous tree cover (Osborne 2000; Timberlake et al. 2010). Savanna is the continent's largest single biome and Africa's are the world's most extensive (Osborne 2000). In contrast, temperate grasslands are much more restricted, found only in southern Africa's Highveld and Maloti-Drakensberg Mountains (Mucina and Rutherford 2006; Stewart et al. 2016) and highland Ethiopia. Unlike true savannas, these vegetation zones experience frost (Lewis and Berry 1988). Africa south of the rainforests is dominated by the enormous Miombo and Mopane humid savanna woodlands, which cover most of Angola, southern Democratic Republic of Congo (DRC), Zambia, southern Tanzania, Zimbabwe, Malawi and Mozambique. South of this are the dry savanna parklands of eastern Namibia, Botswana and northern South Africa that in places grade into semi-desert and desert. Moving north of the rainforests, three increasingly dry savanna belts (the Guinea, the Sudan and the Sahel) span nearly the continent's full width from Senegal to Sudan.

Where the southern and northern grasslands meet, to the east of the rainforests, are the savannas of East Africa, most famously exemplified by the Serengeti with its extraordinary mammalian, bird, and insect diversity and biomass. Two chapters deal with East Africa from MIS 6-2 (Faith et al. 2016; Lahr and Foley 2016). Both draw attention to the region's mosaic-like ecological structure underpinned by its complex climatic and geological history. Lahr and Foley synthesize a wide range of archaeological, fossil, and genetic evidence to explore patterning within East Africa and draw inter-regional comparisons. They suggest that the high habitat heterogeneity of East Africa would have promoted relative population stability through MIS 6-2 by providing a 'refugial network', or "set of independent, potentially asynchronous *refugia* linked by semi-contiguous geographic corridors" (Lahr 2013). By facilitating dispersals between relatively reliable and closely situated resource patches, such a network could have increased the options available to human groups adjusting to rapid climate changes or phases of reduced resource availability. A similar argument has been advanced by Basell (2008), who identifies the region's lake margins, rivers, and highlands as good candidates for Pleistocene human refugia, particularly through the prolonged aridity of MIS 6.

Regional linkages are also integral to Faith et al.'s chapter, which explores the affects of East Africa's environmental history on human paleodemography using insights from ungulate biogeography. Drawing on new paleontological data from Rusinga and Mfangano Islands in Lake Victoria, they argue that fluctuations in moisture availability, atmospheric CO_2, and topographic changes through the Middle and Late Pleistocene prompted major shifts in ungulate ranges. The presence in these bone assemblages of a diverse mix of ungulate taxa that today occupy discrete, often distant habitats indicates their deposition occurred during drier phases when Lake Victoria and others shrank and/or disappeared and grasslands expanded, interlinking the region via biogeographical corridors. That such corridors were also important for humans is something they suggest is reflected in the region's distribution of MSA lithic technologies. Small MSA bifacial points on the islands bear similarities to assemblages in the Rift Valley, which may signal movement from or interactions with regions further east during glacial or stadial phases. These point assemblages are consistently associated with proxy evidence for grassy environments. They contrast with woodland-associated assemblages of Lupemban affinity at other sites that hint at connections in the opposite direction (to the west) during warmer/wetter phases when Africa's forested core expanded.

Faith and colleagues thus emphasize glacial and stadial phases as promoting increased intra-regional connectivity. Interestingly, this is contrary to Basell (2008: 2496), who suggests that during these periods humans contracted into refugia that were "largely disconnected, separated by areas of open grassland, semi desert and desert" and "acted as barriers between hominin populations." Under more humid conditions, Basell (2008) argues that the distribution of human populations broadened with forest expansions, noting a tight correspondence in East Africa between MSA sites and woody settings. Conversely, Faith et al. (2016: 235) suggest that "during interglacials, the expansion of tropical forests across equatorial East Africa created a barrier that fragmented populations and restricted gene flow." Such interpretive differences underscore the desperate need throughout Africa for continuous, high-resolution regional paleoenvironmental archives that can provide data on *resource availability*, rather than just a broad sense of the predominant vegetation. Ideally, such archives should then be evaluated against regional archaeological records to identify the resource structures in which material cultures – and by extension people and/or information – either coalesce or fragment (cf. Mackay et al. 2014). Neither dry savannas nor dry forests typically impede foragers in the 'ethnographic present', and while rainforests and true deserts can (but by no means always do) act as barriers, the challenge is to understand whether, when, and the extent to which they may have done so in the past.

This is particularly pressing in East Africa, where habitat variability through time and space may have provided the selective context in which our species evolved. The location and timing of modern human origins within Africa are dealt with substantively by Lahr and Foley in their chapter. Dismissing recent molecular evidence favoring a southern African center of origin (Henn et al. 2011), they advocate East Africa as the most likely backdrop to both our biological and behavioral evolution. For support they cite the region's potential for harboring refugia (or 'refugial networks'), its uniquely ancient modern human fossil evidence (e.g., Omo Kibish, ~195 ka, McDougall et al. 2005; Herto, ~165 ka, White et al. 2003) and its early MSA archaeological record (e.g., Gademotta, >276 ka, Morgan and Renne 2008; Sahle et al. 2013). The timeframe they favor is gradual, with our phenotypic roots stretching into MIS 9-7 even if the first *sapiens*-specific morphological traits only appear in earliest MIS 6, and the largest strides toward cognitive complexity occurring ~300 ka with the advent of prepared core technologies (Foley and Lahr 1997). While their argument merits serious consideration, they air a debate that is beyond the scope of this introduction. Here we wish only to point out that the findings of Faith et al. suggest the region's patchwork ecology may be a largely interglacial phenomenon, with colder and drier glacial stages producing more homogeneous habitats with greatly diminished surface water. If so, this has implications for the extent to which the current ecological structure of East Africa can be used to support models of where *Homo sapiens* survived or speciated through a putative population bottleneck (Lahr and Foley 1998; Garrigan and Hammer 2006; Fagundes 2007; but see Sjödin et al. 2012), and re-emphasizes the need for crisper paleoenvironmental resolution at the local scale.

There is one landscape feature that would have always ensured African hunter-gatherers a degree of habitat heterogeneity, namely mountains. Widely spaced along the periphery of the continent's immense interior plateau, Africa's mountain systems are essentially biogeographical islands of cold-adapted, species-rich habitats punctuating the warmer, more homogeneous lowland 'sea'. Their broad geographic distribution, spanning the continent's full ~70° of latitude, with numerous geological substrates and multiple rainfall zones, makes them extremely diverse. Nevertheless, all the major mountain zones are stratified by distinct, altitudinally mediated vegetation belts, with numbers ranging from 3-4 in the Atlas and Maloti-Drakensberg and up to 6–8 in highland Ethiopia, the Rwnezori, Mount Kenya, and Kilimanjaro. While the highest afroalpine zones probably always discouraged prolonged or even sporadic human activity, mid-altitude afromontane belts would have offered foragers dividends for relatively minor increases in elevation. Broadly, these include greater resource diversity per unit area of terrain and reliable supplies of key resources, notably fuelwoods, plantfoods and medicines, seasonal hunting opportunities, rockshelters, high-quality volcanic or sedimentary toolstones and, perhaps most critically, freshwater. The latter – supplied by orographic rainfall and afroalpine snowmelt – may have been especially vital when seasonal or longer term changes in precipitation and/or evapotranspiration resulted in dwindling aquifers further downslope. In tropical Africa, highland settlement may also be expected to correlate with phases of marked humidity, conditions that encourage the spread of vector-borne parasitic diseases like trypanosomiasis and malaria that are absent or rare ≥1500 m a.s.l. (Ford 1971; Lindsay and Martens 1998).

Having long taken a backseat to other biomes, research into early human engagements with afromontane environments seems to be accelerating. Recent excavations by Stewart and colleagues in the Maloti-Drakensberg

Mountains of southern Africa are a case in point (Stewart et al. 2012, 2016). Their chapter presents the results of a multiproxy paleoenvironmental analysis of the ~85 kyr sequence at Melikane Rockshelter in the temperate grasslands of highland Lesotho (1800 m a.s.l.). Humans visiting the site through MIS 5-2 consistently encountered a landscape dominated by alpine grasses that today grow at higher altitude, suggesting a prevalence of cool conditions with subtle changes in temperature and moisture. The recurring presence of woody riparian taxa suggests precipitation was consistent enough to support water-loving vegetation along the deeply incised watercourses. Relating Melikane's record to others in the wider region, Stewart et al. posit a dual-source model of highland population influx. Regional warming, they suggest, would have encouraged mountain incursions from the adjacent coastal forelands because of either population increase there (push) or ameliorated conditions at altitude (pull). Phases of regional aridity or instability, however, more likely provoked influx originating in the Karoo Desert to the west, from which groups could follow the Orange-Senqu River to its stable headwater sources in the highlands. The latter appear to have been especially attractive during one such phase ~46–38 ka, when Melikane registers its most sustained occupation (Stewart et al. 2012). In contrast, intense cold – such as that experienced during the LGM and parts of MIS 4 – appears to have deterred highland incursions regardless of other environmental variables.

Recent work by Mercader and colleagues (2013) further north on Mozambique's Niassa Rift flank reveals a similar pattern. There, phases of intensified afromontane activity through MIS 5-3 occurred when paleoecological evidence suggests water availability was high in the highlands and low in the lowlands, echoing the situation in the Maloti-Drakensberg. Unlike Lesotho's temperate grasslands, Niassa's highlands show evidence of being densely forested at such times, prompting Mercader et al. (2013: 328) to wonder whether a 'montane woodland archipelago' linking southern and East Africa might have facilitated human dispersals across (and perhaps out of) the continent. Such a scenario corresponds well with Basell's (2008) findings and is tentatively consistent with recent work at Mochena Borago Rocksheter in highland Ethiopia (Brandt et al. 2012: 51). Nevertheless, all of these studies demonstrate the rewards of investigating human and environmental signatures in afromontane zones. These sites provide important counterparts to the better-known lowland hotbeds of research by filling hitherto unexplained 'gaps' in those records while enhancing our understanding of the range of selective pressures under which our species' behavioral repertoire evolved (Stewart et al. 2012; Dewar and Stewart 2012).

Returning to lower altitudes, Africa's largest lowland zone is dominated by the least understood biome of all: the rainforests. Africa's rainforests are better defined both ecologically and geographically than its grasslands and woodlands. These are the closed-canopy, stratified forests in equatorial areas humid enough for year-round growth. Annual precipitation averages 1600–2000 mm, annual humidity >80%, and the short dry season experienced in most areas is not severe enough to prevent plant growth as in grasslands and dry woodlands. Nevertheless, Africa's rainforests are significantly drier than those on other continents (Bonnefille 2011). Average monthly temperatures are high (≥17 °C) though not extreme, and diurnal and yearly temperature fluctuations are very low. Since nutrients in rainforests are locked up in plant biomass and recycle rapidly rather than enriching the soil, these ecosystems are more sensitive to climatic and anthropogenic changes than woodlands and grasslands (Lewis and Berry 1988). Africa's main rainforest zones center on the Congo Basin and West Africa's south coast. High levels of surplus runoff in the former generates Africa's largest river in terms of discharge (and second longest after the Nile). Yet neither the Congo, the Niger, nor any but the smallest central/West African rivers flow entirely within the present-day rainforest, passing instead through or into drier, more open woodland/grassland biomes (Lewis and Berry 1988; Burrough 2016). The Niger, for example, originates in well-watered highlands on the westernmost rainforest fringe, then flows northeast through the increasingly arid Sudanian and Sahelian belts to the southern Sahara, before returning to the rainforest to discharge at the Gulf of Guinea. Such rivers may have provided conduits for human movement and communication through MIS 6-2, both within the equatorial core and between it and adjacent biomes.

Our decision to group Africa's rainforests, woodlands, and grasslands into a single section of this book stems largely from the uneven geographic distribution of chapter topics. There are, however, also ecological grounds for considering them together as the distribution of these biomes was in flux from MIS 6-2 due to fluctuations in moisture balance. During drier climatic episodes a range of evidence suggests the rainforests fragmented into isolated refugia interspersed with open-canopy woodlands and grass savannas, with periods of enhanced moisture promoting widespread forest re-expansion (e.g., Van Noten 1982; Brook et al. 1990; Maley 1996; Jahns et al. 1998; Dupont et al. 2000; Runge 2000, 2001a, b; Bonnefille 2011). During the LGM, for example, it is estimated the African rainforests were ~15–30% of their current size, as compared to ~50% for those of the Amazon Basin (Anhuf 2000; Anhuf et al. 2006). The contrast between continents is due to tropical Africa's lower rainfall relative to other equatorial areas, which appears to have made the ecological consequences of Pleistocene climatic changes more extreme (Anhuf et al. 2006; Bonnefille 2011). Without question, such broad-brush scenarios oversimplify far more complex regional environmental histories (Mercader et al. 2000; Cornelissen 2002, 2016). The LGM refugia, for example, seem more likely to have been numerous, widely dispersed riverine and montane patches rather than large, relatively homogenous remnant rainforest cores as was previously thought (Moore 1998; Cornelissen 2002, 2013). Even today, Africa's rainforests are far from uniform, with variations in seasonality and topography both significantly influencing vegetation distributions (Bonnefille 2011). However, terrestrial paleoenvironmental archives in this huge region remain scarce, with fewer still reaching back before MIS 2 (e.g., Scholtz et al. 2011; Shanahan et al. 2012). Marine cores off West Africa are longer and more continuous (Dupont et al. 2000) but lack spatial precision, and complex transport processes of pollens and spores can impede even vegetation reconstructions of terrestrial source areas that are closely adjacent. Confounding matters further is the strong likelihood that today's rainforest 'remnants' represent inapposite analogues for those of the past due to long-term human modification (Mercader 2003: 2).

These and other issues hamper detailed reconstructions of the environments in which lowland tropical Africa was first peopled and subsequently inhabited. The subsistence challenges that confront hunter-gatherers occupying dense tropical forests are well known. Carbohydrate resources in rainforests are typically rare, inaccessible, and/or toxic, while prey taxa tend to be small, dispersed and often nocturnal. The extent to which such resource structures might have deterred prehistoric foragers from populating the world's rainforests has been widely debated (Headland 1987; Bailey et al. 1989; Townsend 1990; Bailey and Headland 1991; Colinvaux and Bush 1991; Endicott and Bellwood 1991; Mercader 2002, 2003; Barton 2005). The most extreme view holds that these environments were uninhabitable until the advent of farming when rainforest productivity increased through deforestation and exchange with agriculturalists ensured sufficient carbohydrates (Bailey et al. 1989). Africa has played a central role in contesting such views, with research conducted at various central African sites over the past two decades demonstrating that the human capacity to cope with dense closed-canopy forests pre-dates agriculture by a substantial margin of at least 20 kyr (Mercader et al. 2000; Mercader 2002, 2003). Emphasis has therefore shifted from the question of whether pre-agricultural foragers were able to inhabit Africa's rainforests towards fuller considerations of Pleistocene adaptations across the region. As with paleoenvironmental reconstruction, massive gaps in knowledge – in this case caused or exacerbated by severe taphonomic vagaries, logistical difficulties and political instability – make the task herculean.

In the final two chapters of this section, Cornelissen and Taylor take up the challenge. Both authors offer cogent syntheses of what is currently known about the Stone Age occupation of central Africa, drawing on a breadth of sources to maximize the data available. Cornelissen focuses on the rainforest's eastern zones in the northeastern DRC, an area that includes the Ituri forest. Until recently, little was known of this region's deep prehistory despite it being home to some of the world's best-studied hunter-gatherer groups (e.g., Turnbull 1961; Cavalli-Sforza 1986). Compiling data from published reports and unpublished museum collections, Cornelissen investigates the spatial distribution of virtually every recorded Stone Age occurrence in the region. As documented elsewhere in the central African rainforests, her research area's lithic assemblages are dominated by two very distinct traditions: (1) those lacking bifaces and typified by highly informal quartz microliths; and (2) others with bifaces, which can include foliates with clear Lupemban affiliation. These sharp technological contrasts reflect major temporal differences, as the two industries effectively bookend MIS 6-2.

At the few dated rockshelters where they occur, quartz microlithic assemblages span late MIS 3 to the late Holocene and have been described as LSA. Though more restricted in space than bifacial assemblages (Cornelissen 2016), their association with a range of paleoenvironmental proxies from forested to open suggests a versatile, expedient adaptation whose ~40 kyr duration testifies to its success (Cornelissen 2002). As Cornelissen points out, however, such diverse ecological associations and temporal continuity limits this technology's utility for demographic reconstruction. Moreover, it seems increasingly unlikely that successful rainforest colonization depended on innovations in lithic technology (Mercader 2003). At Ishango 11 in the Semliki Valley of the easternmost DRC, quartz microlithic tools have been recovered from LGM contexts alongside finely crafted bone harpoons and fish remains. The bone technology and species composition at Ishango 11 are nearly identical to those at the nearby site of Katanda with dates nearly four times as old (~80 ka; Feathers and Migliorini 2001). That fluvial fauna were being effectively captured in equatorial Africa by late MIS 5 may hint at the timing of permanent human settlement in the region, since such resources "may be important for exploring and exploiting an environment such as the rainforest with swamps and an abundant network of waterways" (Cornelissen 2016). But given the limited bone preservation in dense tropical forests, Cornelissen suggests the role of aquatic resources in early adaptations to such environments might be better explored via genetics.

Cornelissen's other main assemblage type – those with bifacial tools – appear to represent "a more ancient and more widely spread hunter-gatherer habitation of the region" (Cornelissen 2016). Though known from MIS 2 and Holocene contexts in west-central Africa (Cornelissen 2002), bifacial forms in the eastern forests are clearly older. Assemblages include handaxes, core-axes and picks, but the most intriguing are those classed as Lupemban, which can include blades and backed blades, prepared core and leaf-shaped points, and – most conspicuously – elongated, invasively flaked lanceolates that constitute the industry's *fossiles directeurs*. Discussion of the Lupemban brings us back to J.D. Clark's pioneering thoughts on population dynamics in Stone Age Africa. Clark envisaged the Lupemban as an adaptation to central Africa's dense forests whose innovation catalyzed the region's human colonization. This hypothesis, conceived over a half-century ago, remains untested. Yet recently obtained late Middle Pleistocene dates for Lupemban assemblages from key sites that include pigments and backed artifacts (Barham 2000, 2001, 2002a, b, 2012), taken together with the general African reorientation of behavioral origins research (McBrearty and Brooks 2000; Henshilwood and Marean 2003), render it today more tantalizing than ever.

In a chapter dealing exclusively with the Lupemban, Taylor provides a comprehensive and much-needed reappraisal of this enigmatic technocultural phenomenon. He generates a database of every reported Lupemban (or related) occurrence, which is then systematically interrogated to assess the data's robusticity by culling problematic cases. The process reveals the staggering extent of our ignorance; adequate paleoenvironmental proxy evidence, for example, exists for only 5% of Lupemban sites, a mere 17% have chronometric dates, and only 3% are dated using radiometric techniques capable of reaching beyond radiocarbon. Many occurrences lack the requisite documentation to establish provenance or affirm Lupemban affiliation, others are surface finds, and all have suffered some form of disturbance. In fact, at only two sites have Lupemban artifacts been recovered from coherent stratigraphic sequences – Twin Rivers and Kalambo Falls – both situated in Zambia's savanna woodlands. As Taylor notes, such extreme impoverishment "precludes the derivation of precise data pertaining to [the Lupemban's] techno-typological composition and spatiotemporal variability, and hampers attempts to test its potential behavioral significance" (Taylor 2016: 289). Nevertheless, the site distributions Taylor generates from his database – which serve as an independent, updated counterpoint to those in Clark (1967) – lend weight to Clark's original correlation of the Lupemban with wooded habitats and challenge arguments for its strict association with expanded glacial-phase grasslands (e.g., Banks et al. 2006). If a woodland correlation can be taken to indicate an emergence during the MIS 7 interglacial ($\sim$240–190 ka) (Barham 2001, 2012), the Lupemban clearly merits serious consideration as an archaeological residue of the first rainforest foragers and thus an early signature of behavioral complexity not unlike modern hunter-gatherers.

The Lupemban's seemingly long duration and continuity with the subsequent Tshitolian hints at high overall population stability in central Africa from MIS 6-2. Like the quartz microlithic industries that succeed it, viewing the Lupemban not as a biome-specific adaptation but rather as a flexible solution to foraging in shifting, variably wooded environments may help explain this longevity. Taylor points out that such a diverse, sophisticated technological repertoire probably performed well in both open and closed environments, a view supported by the limited proxy data available (van Zinderen Bakker 1969; Bishop and Reynolds 2000; Taylor et al. 2001). As with certain coastal and afromontane areas, intermediately wooded parts of central Africa probably acted as human refugia at times, though the locations of optimal areas would have changed as humid rainforests and dry grasslands waxed and waned with glacial cycles. As a corollary, high population stability also makes it plausible that the region contributed source populations to human dispersals both within and beyond Africa. Even if foragers only developed the adaptive tools necessary to thrive in the densest rainforests from MIS 3, the region may have always been more attractive when the canopies had thinned. As discussed by Cornelissen (2002), it is likely that the ecological reconfigurations and ecotones so created would have profoundly impacted the distributions of human groups through Africa's equatorial core from MIS 6-2 and perhaps before (see debate in Herries 2011; Barham 2012). For the moment, though, the Lupemban's spatiotemporal resolution is far too coarse to properly assess Clark's hypothesis, let alone assess the industry's broader evolutionary significance. The timing and ecological context of the earliest human incursions into central Africa will remain obscure until better chronostratigraphic sequences are discovered and developed. Likewise, suggestions that the Lupemban holds some sort of ancestral relationship to later, better-known bifacial MSA industries in North (the Aterian) and/or southern Africa (the Still Bay) must, in the absence of an improved central African record, remain speculative (Henshilwood 2008; Van Peer 2016).

Conclusion

This volume represents a first concerted effort to assess Stone Age population dynamics throughout Africa, and how these changed with fluctuations in global climate and more localized environments. The choice of MIS 6-2 is driven by multidisciplinary evidence that the last $\sim$200 kyr of the Pleistocene saw the consolidation of longer term trends toward the evolution of hominins physically and cognitively identical to ourselves. This was also the time interval during which modern human groups undertook multiple dispersals from the continent along various routes, culminating by late MIS 2 in the colonization of every habitable corner of the globe. Thus while earlier stages of the Middle Pleistocene were clearly fundamental in setting the stage for modern human evolution, the pronounced climatic and ecological volatility of MIS 6-2 hosted the final act. There are strong indications across multiple records that the distribution of human populations across Africa from MIS 6-2 was highly discontinuous. This is registered in the archaeological record by regional artifact traditions that alternately coalesce, fragment, or disappear entirely, and corroborated in the human fossil and genetic records by a high degree of morphological variability (Grine 2016) and evidence for deep time population structure (Soares et al. 2016), respectively.

However, to make real progress towards reconstructing population dynamics and understanding their evolutionary implications in Africa and beyond, we must build stronger bridges between archaeology, paleoenvironmental science, human paleontology and genetics (e.g., Lombard et al. 2013). We must work harder at disciplinary interfaces to integrate datasets that have inherently different resolutions and speeds of progression. How, for example, can we better assimilate insights from population genetics, which accrue at breakneck speeds but at relatively coarse historical resolution, into regional archaeological records, which are sharper but whose construction is painstakingly slow? Reconciling datasets *within* disciplines is also a major challenge, particularly when they represent different scales of phenomena. For example, the global-scale climate changes seen in the marine or ice core records must be interpreted in relation to their impact on regional- or local-scale paleoenvironments as reconstructed from discontinuous terrestrial archives. Similar challenges arise when attempting to compare chronologies derived from diverse dating techniques with different error margins, or interpreting variability in artifact form and frequency from a single site in relation to the full geographic and temporal span of that industry.

To examine population dynamics and their relationship to paleoenvironments is to examine change, and understanding the meaning of changes we detect is paramount. Disentangling demographic changes (e.g., population replacement) from the assimilation of external ideas (e.g., through diffusion) or *in situ* processes (e.g., cultural adaptation to environmental shifts) lay at the very core of inquiry into this topic. With such a shortage of human fossils throughout Africa from MIS 6-2, this will necessarily depend on more refined technological analyses of lithic industries and other material culture indicators. Similarly, translating coarse paleoenvironmental inferences into more explicit statements about resource abundance and structure is essential for establishing the human-scale ecological contexts needed for assessing adaptive change. On the other hand, reconciling our understanding of archaeological industries of seemingly extraordinary duration, including the Lupemban and Aterian, with those recently shown to be extremely short-lived, such as the Still Bay and Howiesons Poort, underscores the challenges of explaining continuity, and particularly of decoupling gross typological similarities across space and time from finer-scale shifts in technology that may prove more informative.

The ultimate goal towards which this book is merely a first step is to produce a diachronic, continental-scale understanding of human biogeography for Stone Age Africa. Similar efforts undertaken on other continents, particularly Australia, have proven highly fruitful (e.g., Veth 1989, 1993, 2005; O'Connor et al. 1993; Smith 1993, 2013; Hiscock and Wallis 2005; Hiscock 2008; Smith et al. 2008; Williams 2013). Perhaps our greatest hindrance in Africa is the extreme geographical unevenness and thus overall paucity of data. Though our knowledge of Stone Age Africa has grown immensely in the five decades since J.D. Clark (1960) crafted his wish list, vast tracts of the continent remain Paleolithic question marks. Poor coverage is particularly acute in West and central Africa, but also in substantial areas of the continent's north, east and south. While the chapters in this volume inevitably reflect this bias, they also sample the exciting array of datasets, approaches and ideas that are being brought to bear on understanding population dynamics in ancient Africa, their paleoenvironmental correlates and their evolutionary significance. Mapping and dating shifts in human biogeographical phenomena, including corridors, barriers, refugia, and abandonment zones, and how these articulate with regional archaeological, fossil and genetic records, should help unlock the secrets behind our species' biological and behavioral florescence.

Acknowledgments We thank Graeme Barker and Peter Mitchell for their thoughtful comments on an earlier draft of this introductory chapter.

References

Anhuf, D. (2000). Vegetation history and climate changes in Africa north and south of the equator (10°S to 10°N) during the last glacial maximum. In P. Smolka & W. Volkheimer (Eds.), *Southern hemisphere paleo- and neoclimates* (pp. 225–248). Berlin: Springer.

Anhuf, D., Ledru, M. P., Behling, H., Da Cruz, F. W., Cordeiro, R. C., Van der Hammen, T., et al. (2006). Paleo-environmental change in Amazonian and African rainforest during the LGM. *Palaeogeography, Palaeoclimatology, Palaeoecology, 239*(3–4), 510–527.

Avery, G., Halkett, D., Orton, J., Steele, T., Tusenius, M., & Klein, R. (2008). The Ysterfontein 1 Middle Stone Age rockshelter and the evolution of coastal foraging. *South African Archaeological Society Goodwin Series, 10*, 66–89.

Bailey, R. C., & Headland, T. N. (1991). The tropical rain forest: Is it a productive environment for human foragers? *Human Ecology, 19* (2), 261–285.

Bailey, R. C., Head, G., Jenike, M., Owen, B., Rechtman, R., & Zechenter, E. (1989). Hunting and gathering in tropical rain forest: Is it possible? *American Anthropologist, 91*(1), 59–82.

Banks, W. E., dErrico, F., Dibble, H. L., Krishtalka, L., West, D., & Olszewski, D. I., et al. (2006). Eco-cultural niche modeling: New tools for reconstructing the geography and ecology of past human populations. *PaleoAnthropology*, *4*, 68–83.

Barham, L. (1998). Possible early pigment use in south-central Africa. *Current Anthropology, 39*(5), 703–710.

Barham, L. (2000). *The Middle Stone Age of Zambia, south central Africa*. Bristol: CHERUB.

Barham, L. (2001). Central Africa and the emergence of regional identity in the Middle Pleistocene. In L. Barham & K. Robson-Brown (Eds.), *Human roots: Africa and Asia in the Middle Pleistocene* (pp. 65–80). Bristol: Western Academic and Specialist Press.

Barham, L. (2002a). Systematic pigment use in the Middle Pleistocene of south-central Africa. *Current Anthropology, 43*(1), 181–190.

Barham, L. (2002b). Backed tools in Middle Pleistocene central Africa and their evolutionary significance. *Journal of Human Evolution, 43* (5), 585–603.

Barham, L. (2012). Clarifying some fundamental errors in Herries' "A chronological perspective on the Acheulian and its transition to the Middle Stone Age in southern Africa: The question of the Fauresmith" (2011). *International Journal of Evolutionary Biology, 2012*, 1–5.

Barham, L. (2013). *From hand to handle: The first industrial revolution*. Oxford: Oxford University Press.

Barham, L., & Mitchell, P. (2008). *The first Africans: African archaeology from the earliest tool makers to most recent foragers*. Cambridge: Cambridge University Press.

Barton, R. N. E., Bouzouggar, A., Collcutt, S. N., Gale, R., Higham, T. F. G., Humphrey, L. T., et al. (2005). The late Upper Palaeolithic occupation of the Moroccan northwest during the Last Glacial Maximum. *African Archaeological Review, 22*(2), 77–100.

Basell, L. S. (2008). Middle Stone Age (MSA) site distributions in eastern Africa and their relationship to Quaternary environmental change, refugia and the evolution of *Homo sapiens*. *Quaternary Science Reviews, 27*(27–28), 2484–2498.

Bishop, L. C., & Reynolds, S. C. (2000). Fauna from Twin Rivers. In L. Barham & K. Robson-Brown (Eds.), *Human roots: Africa and Asia in the Middle Pleistocene* (pp. 217–222). Bristol: Western Academic and Specialist Press.

Blome, M. W., Cohen, A. S., Tryon, C. A., Brooks, A. S., & Russell, J. (2012). The environmental context for the origins of modern human diversity: A synthesis of regional variability in African climate 150,000–30,000 years ago. *Journal of Human Evolution, 62* (5), 563–592.

Bonnefille, R. (2011). Rainforest responses to past climatic changes in tropical Africa. In M. B. Bush, J. R. Flenley, & W. D. Gosling (Eds.), *Tropical rainforest responses to climatic change* (pp. 125–184). Berlin: Springer.

Bouzouggar, A., & Barton, R. N. E. (2012). The identity and timing of the Aterian in Morocco. In J. J. Hublin & S. McPherron (Eds.), *Modern origins: A North African perspective* (pp. 93–105). Dordrecht: Springer.

Bouzouggar, A., Barton, N., Vanhaeren, M., d'Errico, F., Collcutt, S., Higham, T., et al. (2007). 82,000-year-old shell beads from North Africa and implications for the origins of modern human behavior. *Proceedings of the National Academy of Sciences, 104*(24), 9964–9969.

Brandt, S. A., Fisher, E. C., Hildebrand, E. A., Vogelsang, R., Ambrose, S. H., Lesur, J., et al. (2012). Early MIS 3 occupation of Mochena Borago Rockshelter, southwest Ethiopian Highlands: Implications for Late Pleistocene archaeology, paleoenvironments and modern human dispersals. *Quaternary International, 274*, 38–54.

Brook, G. A., Burney, D. A., & Cowart, J. B. (1990). Paleoenvironmental data for Ituri, Zaire, from sediments in Matupi Cave, Mt. Hoyo. *Virginia Museum of Natural History Memoir, 1*, 49–70.

Brooks, C. E. P. (1914). The meteorological conditions of an ice-sheet and their bearing on the dessication of the globe. *Royal Meteorological Society Quarterly Journal, 40*, 53–70.

Brown, K. S., Marean, C. W., Herries, A. I., Jacobs, Z., Tribolo, C., Braun, D., et al. (2009). Fire as an engineering tool of early modern humans. *Science, 325*(5942), 859–862.

Burrough, S. L. (2016). Late Quaternary environmental change and human occupation of the southern African interior. In S. C. Jones & B. A. Stewart (Eds.), *Africa from MIS 6-2: Population dynamics and paleoenvironments* (pp. 161–174). Dordrecht: Springer.

Cancellieri, E., Cremaschi, M., Zerboni, A., & di Lernia, S. (2016). Climate, environment and population dynamics in Pleistocene Sahara. In S. C. Jones & B. A. Stewart (Eds.), *Africa from MIS 6-2: Population dynamics and paleoenvironments* (pp. 123–145). Dordrecht: Springer.

Cann, R. L. (1988). DNA and human origins. *Annual Review of Anthropology*, 127–143.

Carr, A. S., Chase, B. M., & Mackay, A. (2016). Mid to Late Quaternary landscape and environmental dynamics in the Middle Stone Age of southern South Africa. In S. C. Jones & B. A. Stewart (Eds.), *Africa from MIS 6-2: Population dynamics and paleoenvironments* (pp. 23–47). Dordrecht: Springer.

Cavalli-Sforza, L. L. (1986). *African pygmies*. Orlando, FL.: Academic Press.

Chamberlain, A. T. (2006). *Demography in archaeology. Cambridge manuals in archaeology*. Cambridge: Cambridge University Press.

Clark, J. D. (1959). *The prehistory of southern Africa*. Harmondsworth: Penguin Books.

Clark, J. D. (1960). Human ecology during Pleistocene and later times in Africa south of the Sahara. *Current Anthropology, 1*(4), 307–324.

Clark, J. D. (1967). *Atlas of African prehistory*. Chicago: University of Chicago Press.

Cochrane, G. W. G., Doelman, T., & Wadley, L. (2013). Another dating revolution for prehistoric archaeology? *Journal of Archaeological Method and Theory, 20*(1), 42–60.

Colinvaux, P. A., & Bush, M. B. (1991). The rain-forest ecosystem as a resource for hunting and gathering. *American Anthropologist, 93*(1), 153–160.

Collard, M., Ruttle, A., Buchanan, B., & O'Brien, M. J. (2013). Population size and cultural evolution in nonindustrial food-producing societies. *PLoS ONE, 8*(9), e72628.

Compton, J. S. (2011). Pleistocene sea-level fluctuations and human evolution on the southern coastal plain of South Africa. *Quaternary Science Reviews, 30*(5–6), 506–527.

Cornelissen, E. (2002). Human responses to changing environments in central africa between 40,000 and 12,000 BP. *Journal of World Prehistory, 16*(3), 197–235.

Cornelissen, E. (2013). Hunting and gathering in Africa's tropical forests at the end of the Pleistocene and in the Early Holocene. In P. Mitchell & P. Lane (Eds.), *The Oxford handbook of African archaeology* (pp. 403–417) Oxford: Oxford University Press.

Cornelissen, E. (2016). The later Pleistocene in the northeastern Central African rainforest. In S. C. Jones & B. A. Stewart (Eds.), *Africa from MIS 6-2: Population dynamics and paleoenvironments* (pp. 301–319). Dordrecht: Springer.

Cremaschi, M., Di Lernia, S., & Garcea, E. A. (1998). Some insights on the Aterian in the Libyan Sahara: Chronology, environment, and archaeology. *African Archaeological Review, 15*(4), 261–286.

d'Errico, F., Henshilwood, C., Vanhaeren, M., & van Niekerk, K. (2005). *Nassarius kraussianus* shell beads from Blombos Cave: Evidence for symbolic behaviour in the Middle Stone Age. *Journal of Human Evolution, 48*(1), 3–24.

d'Errico, F., Vanhaeren, M., & Wadley, L. (2008). Possible shell beads from the Middle Stone Age layers of Sibudu Cave, South Africa. *Journal of Archaeological Science, 35*(10), 2675–2685.

d'Errico, F., Vanhaeren, M., Barton, N., Bouzouggar, A., Mienis, H., Richter, D., et al. (2009). Additional evidence on the use of personal ornaments in the Middle Paleolithic of North Africa. *Proceedings of the National Academy of Sciences, 106*(38), 16051.

Dewar, G. (2008). *The archaeology of the coastal desert of Namaqualand, South Africa: A regional synthesis*. Oxford: BAR International Series 1761.

Dewar, G., & Stewart, B. A. (2016). Paleoenvironments, sea levels and land use in Namaqualand, South Africa, during MIS 6-2. In S. C. Jones & B. A. Stewart (Eds.), *Africa from MIS 6-2: Population dynamics and paleoenvironments* (pp. 195–212). Dordrecht: Springer.

Dewar, G., & Stewart, B. A. (2012). Preliminary results of excavations at Spitzkloof Rockshelter, Richtersveld, South Africa. *Quaternary International, 270*, 30–39.

Drake, N., & Breeze, P. (2016). Climate change and modern human occupation of the Sahara from MIS 6-2. In S. C. Jones & B. A. Stewart (Eds.), *Africa from MIS 6-2: Population dynamics and paleoenvironments* (pp. 103–122). Dordrecht: Springer.

Dupont, L. M., Jahns, S., Marret, F., & Ning, S. (2000). Vegetation change in equatorial West Africa: Time-slices for the last 150 ka. *Palaeogeography, Palaeoclimatology, Palaeoecology, 155*(1), 95–122.

Dusseldorp, G. L., & Langejans, G. H. J. (2013). Carry that weight: Coastal foraging and transport of marine resources during the South African Middle Stone Age. *Southern African Humanities, 25*, 105–135.

Endicott, K., & Bellwood, P. (1991). The possibility of independent foraging in the rain forest of peninsular Malaysia. *Human Ecology, 19*(2), 151–185.

Eriksson, A., & Manica, A. (2012). Effect of ancient population structure on the degree of polymorphism shared between modern human populations and ancient hominins. *Proceedings of the National Academy of Sciences, 109*(35), 13956–13960.

Erlandson, J. M. (2001). The archaeology of aquatic adaptations: Paradigms for a new millennium. *Journal of Archaeological Research, 9*(4), 287–350.

Faith, J. T., Tryon, C., & Peppe, D. J. (2016). Environmental change, ungulate biogeography, and their implications for early human dispersals in equatorial East Africa. In S. C. Jones & B. A. Stewart (Eds.), *Africa from MIS 6-2: Population dynamics and paleoenvironments* (pp. 233–245). Dordrecht: Springer.

Fagundes, N. J., Ray, N., Beaumont, M., Neuenschwander, S., Salzano, F. M., Bonatto, S. L., et al. (2007). Statistical evaluation of alternative models of human evolution. *Proceedings of the National Academy of Sciences, 104*(45), 17614–17619.

Farr, L., & Jones, S. (2014). Spatial and temporal variation in North African and Southwest Asian palaeo environmental and archaeological records during Marine Isotope Stage 4. In K. Boyle, R. Rabett, & C. Hunt (Eds.), *Living in the landscape: Essays in honour of Graeme Barker* (pp. 59–81). Cambridge: McDonald Institute for Archaeological Research.

Faure, H., Walter, R. C., & Grant, D. R. (2002). The coastal oasis: Ice age springs on emerged continental shelves. *Global and Planetary Change, 33*(1), 47–56.

Feathers, J. K., & Migliorini, E. (2001). Luminescence dating at Katanda – a reassessment. *Quaternary Science Reviews, 20*(5), 961–966.

Fischer, E. C., Bar-Matthews, M., Jerardino, A., & Marean, C. W. (2010). Middle and Late Pleistocene paleoscape modeling along the southern coast of South Africa. *Quaternary Science Reviews, 29*(11), 1382–1398.

Foley, R., & Lahr, M. M. (1997). Mode 3 technologies and the evolution of modern humans. *Cambridge Archaeological Journal, 7*, 3–36.

Ford, J. (1971). *The role of the trypanosomiases in African ecology. A study of the tsetse fly problem*. Oxford: Clarendon Press.

Garcea, E. A. A. (2012). Modern human desert adaptations: A Libyan perspective on the Aterian complex. In J.-J. Hublin & S. P. McPherron (Eds.), *Modern origins: A North African perspective* (pp. 127–142). Dordrecht: Springer.

Garrigan, D., & Hammer, M. F. (2006). Reconstructing human origins in the genomic era. *Nature Reviews Genetics, 7*(9), 669–680.

Green, R. E., Krause, J., Briggs, A. W., Maricic, T., Stenzel, U., Kircher, M., et al. (2010). A draft sequence of the Neandertal genome. *Science, 328*(5979), 710–722.

Grine, F. E. (2016). The Late Quaternary hominins of Africa: The skeletal evidence from MIS 6-2. In S. C. Jones & B. A. Stewart (Eds.), *Africa from MIS 6-2: Population dynamics and paleoenvironments* (pp. 323–381). Dordrecht: Springer.

Hammer, M. F., Woerner, A. E., Mendez, F. L., Watkins, J. C., & Wall, J. D. (2011). Genetic evidence for archaic admixture in Africa. *Proceedings of the National Academy of Sciences, 108*(37), 15123–15128.

Headland, T. N. (1987). The wild yam question: How well could independent hunter-gatherers live in a tropical rain forest ecosystem? *Human Ecology, 15*(4), 463–491.

Henn, B. M., Gignoux, C. R., Jobin, M., Granka, J. M., Macpherson, J. M., Kidd, J. M., et al. (2011). Hunter-gatherer genomic diversity suggests a southern African origin for modern humans. *Proceedings of the National Academy of Sciences, 108*(13), 5154–5162.

Henshilwood, C. S. (2008). Winds of change: Palaeoenvironments, material culture and human behaviour in the Late Pleistocene (~77 ka–48 ka ago) in the Western Cape Province, South Africa. *South African Archaeological Society Goodwin Series, 10*, 35–51.

Henshilwood, C. S., & Dubreuil, B. (2011). The Still Bay and Howiesons Poort, 77–59 ka: Symbolic material culture and the evolution of the mind during the African Middle Stone Age. *Current Anthropology, 52*(3), 361–400.

Henshilwood, C. S., & Marean, C. W. (2003). The origin of modern human behavior. *Current Anthropology, 44*(5), 627–651.

Henshilwood, C. S., d'Errico, F., Yates, R., Jacobs, Z., Tribolo, C., Duller, G. A., et al. (2002). Emergence of modern human behavior: Middle Stone Age engravings from South Africa. *Science, 295* (5558), 1278–1280.

Henshilwood, C. S., d'Errico, F., & Watts, I. (2009). Engraved ochres from the Middle Stone Age levels at Blombos Cave, South Africa. *Journal of Human Evolution, 57*(1), 27–47.

Henshilwood, C. S., d'Errico, F., van Niekerk, K. L., Coquinot, Y., Jacobs, Z., Lauritzen, S. E., et al. (2011). A 100,000-year-old ochre-processing workshop at Blombos Cave. *South Africa Science, 334*(6053), 219–222.

Herries, A. I. (2011). A chronological perspective on the Acheulian and its transition to the Middle Stone Age in southern Africa: The question of the Fauresmith. *International Journal of Evolutionary Biology, 2011*, 961401.

Hiscock, P. (2008). *Archaeology of ancient Australia*. London: Routledge.

Hiscock, P., & Wallis, L. A. (2005). Pleistocene settlement of deserts from an Australian perspective. In P. Veth, M. Smith, & P. Hiscock (Eds.), *Desert peoples: Archaeological perspectives* (pp. 34–57). Oxford: Blackwell.

Hublin, J. J., & McPherron, S. P. (2012). *Modern origins: A North African perspective*. Dordrecht: Springer.

Jacobs, Z., Roberts, R. G., Galbraith, R. F., Deacon, H. J., Grün, R., Mackay, A., et al. (2008). Ages for the Middle Stone Age of southern Africa: Implications for human behavior and dispersal. *Science, 322*(5902), 733–735.

Jahns, S., Hüls, M., & Sarnthein, M. (1998). Vegetation and climate history of west equatorial Africa based on a marine pollen record off Liberia (site GIK 16776) covering the last 400,000 years. *Review of Palaeobotany and Palynology, 102*(3), 277–288.

Jones, S., Antoniadou, A., Barton, H., Drake, N., Farr, L., Hunt, C., et al. (2016). Patterns of hominin occupation and cultural diversity across the Gebel Akhdar of northern Libya over the last ~200 kyr. In S. C. Jones & B. A. Stewart (Eds.), *Africa from MIS 6-2:*

Population dynamics and paleoenvironments (pp. 77–99). Dordrecht: Springer.

Klein, R. G., Avery, G., Cruz-Uribe, K., Halkett, D., Parkington, J. E., Steele, T., et al. (2004). The Ysterfontein 1 Middle Stone Age site, South Africa, and early human exploitation of coastal resources. *Proceedings of the National Academy of Sciences, 101*(16), 5708–5715.

Lahr, M. M. (2013). Why East Africa could be driving human evolution? http://www.in-africa.org/origins-2/why-east-africa/. Accessed 7 Nov 2014.

Lahr, M. M., & Foley, R. A. (1998). Towards a theory of modern human origins: Geography, demography, and diversity in recent human evolution. *Yearbook of Physical Anthropology, 41*, 137–176.

Lahr, M. M., & Foley, R. (2016). Human evolution in Late Quaternary eastern Africa. In S. C. Jones & B. A. Stewart (Eds.), *Africa from MIS 6-2: Population dynamics and paleoenvironments* (pp. 215–231). Dordrecht: Springer.

Langejans, G. H., van Niekerk, K. L., Dusseldorp, G. L., & Thackeray, J. F. (2012). Middle Stone Age shellfish exploitation: Potential indications for mass collecting and resource intensification at Blombos Cave and Klasies River, South Africa. *Quaternary International, 270*, 80–94.

Larrasoaña, J. C. (2012). A northeast Saharan perspective on environmental variability in North Africa and its implications for modern human origins. In J. J. Hublin & S. P. McPherron (2012). *Modern origins: A North African perspective* (pp. 19–34). Dordrecht: Springer.

Lee, R. B. (1963). The population ecology of man in the early Upper Pleistocene of southern Africa. *Proceedings of the Prehistoric Society, 29*, 235–257.

Leakey, L. S. B. (1951). *Olduvai Gorge*. London: Cambridge University Press.

Lewis, L. A., & Berry, L. (1988). *African environments and resources.* Boston: Unwin Hyman.

Lindsay, S. W., & Martens, W. J. (1998). Malaria in the African highlands: Past, present and future. *Bulletin of the World Health Organization, 76*(1), 33.

Lisiecki, L. E., & Raymo, M. E. (2005). A Pliocene-Pleistocene stack of 57 globally distributed benthic d18O records. *Paleoceanography, 20*, PA1003. doi:10.1029/2004PA001071.

Lombard, M., Schlebusch, C., & Soodyall, H. (2013). Bridging disciplines to better elucidate the evolution of early *Homo sapiens* in southern Africa. *South African Journal of Science, 109*(11/12), 1–8.

Mackay, A. (2016). Technological change and the importance of variability: The Western Cape of South Africa from MIS 6-2. In S. C. Jones & B. A. Stewart (Eds.), *Africa from MIS 6-2: Population dynamics and paleoenvironments* (pp. 49–63). Dordrecht: Springer.

Mackay, A., Stewart, B. A., & Chase, B. M. (2014). Coalescence and fragmentation in the Late Pleistocene archaeology of southernmost Africa. *Journal of Human Evolution, 72*, 26–51.

Maley, J. (1996). The African rain forest – main characteristics of changes in vegetation and climate from the Upper Cretaceous to the Quaternary. *Proceedings of the Royal Society of Edinburgh. Section B. Biological Sciences, 104*, 31–73.

Marean, C. W. (2010). Pinnacle Point Cave 13B (Western Cape Province, South Africa) in context: The Cape Floral kingdom, shellfish, and modern human origins. *Journal of Human Evolution, 59*(3–4), 425–443.

Marean, C. W. (2011). Coastal South Africa and the coevolution of the modern human lineage and the coastal adaptation. In N. F. Bicho, J. A. Haws, & L. G. Davis (Eds.), *Trekking the shore: Changing coastlines and the antiquity of coastal settlement* (pp. 421–440). New York Springer.

Marean, C. W. (2014). The origins and significance of coastal resource use in Africa and Western Eurasia. *Journal of Human Evolution, 77*, 17–40.

Marean, C. W., Bar-Matthews, M., Bernatchez, J., Fisher, E., Goldberg, P., Herries, A. I. R., et al. (2007). Early human use of marine resources and pigment in South Africa during the Middle Pleistocene. *Nature, 449*(7164), 905–908.

McBrearty, S. (2007). Down with the revolution. In K. Boyle, O. Bar-Yosef, & C. Stringer (Eds.), *Rethinking the human revolution: New behavioural and biological perspectives on the origin and dispersal of modern humans* (pp. 133–152). Cambridge: McDonald Institute for Archaeological Research Monographs.

McBrearty, S., & Brooks, A. S. (2000). The revolution that wasn't: A new interpretation of the origin of modern human behavior. *Journal of Human Evolution, 39*(5), 453–563.

McDougall, I., Brown, F. H., & Fleagle, J. G. (2005). Stratigraphic placement and age of modern humans from Kibish. *Ethiopia. Nature, 433*(7027), 733–736.

Mendez, F. L., Krahn, T., Schrack, B., Krahn, A. M., Veeramah, K. R., Woerner, A. E., et al. (2013). An African American paternal lineage adds an extremely ancient root to the human Y chromosome phylogenetic tree. *The American Journal of Human Genetics, 92*(3), 454–459.

Mercader, J. (2002). Forest people: The role of African rainforests in human evolution and dispersal. *Evolutionary Anthropology, 11*(3), 117–124.

Mercader, J. (2003). *Under the canopy: The archaeology of tropical rain forests.* New Brunswick: Rutgers University Press.

Mercader, J., Runge, F., Vrydaghs, L., Doutrelepont, H., Ewango, C. E., & Juan-Tresseras, J. (2000). Phytoliths from archaeological sites in the tropical forest of Ituri, Democratic Republic of Congo. *Quaternary Research, 54*(1), 102–112.

Mercader, J., Bennett, T., Esselmont, C., Simpson, S., & Walde, D. (2013). Phytoliths from Middle Stone Age habitats in the Mozambican Rift (105–29 ka). *Journal of Human Evolution, 64*(5), 328–336.

Mitchell, P. (2016). Africa from MIS 6-2: Where do we go from here? In S. C. Jones & B. A. Stewart (Eds.), *Africa from MIS 6-2: Population dynamics and paleoenvironments* (pp. 407–416). Dordrecht: Springer.

Moore, P. D. (1998). Palaeoecology: Did forests survive the cold in a hotspot? *Nature, 391*(6663), 124–127.

Morgan, L. E., & Renne, P. R. (2008). Diachronous dawn of Africa's Middle Stone Age: New ^{40}Ar/^{39}Ar ages from the Ethiopian Rift. *Geology, 36*(12), 967–970.

Mucina, L., & Rutherford, M. C. (2006). *The vegetation of South Africa, Lesotho and Swaziland.* Strelitzia 19. Pretoria: South African National Biodiversity Institute.

O'Connor, S., Veth, P., & Hubbard, N. (1993). Changing interpretations of postglacial human subsistence and demography in Sahul. In M. Smith, M. Spriggs, & B. Frankhauser (Eds.), *Sahul in review: Pleistocene archaeology in Australia, New Guinea and Island Melanesia*, (pp. 95–105). Canberra: Australian National University.

Orme, A. R. (2005). Africa, coastal geomorphology. In M. L. Schwartz (Ed.), *Encyclopedia of coastal science* (pp. 9–21). Dordrecht: Springer.

Osborne, P. L. (2000). *Tropical ecosystems and ecological concepts.* Cambridge: Cambridge University Press.

O'Shea, J. M., Lemke, A. K., Sonnenburg, E. P., Reynolds, R. G., & Abbott, B. D. (2014). A 9,000-year-old caribou hunting structure beneath Lake Huron. *Proceedings of the National Academy of Sciences, 111*(19), 6911–6915.

Parkington, J. (2001). Milestones: The impact of the systematic exploitation of marine foods on human evolution. In P. V. Tobias, M. A. Raath, J. Moggi-Cecchi, & H. Soodyall (Eds.), *Humanity from African naissance to coming millennia* (pp. 327–336). Firenze: Firenze University Press.

Parkington, J. (2003). Middens and moderns: Shellfishing and the Middle Stone Age of the western cape, South Africa. *South African Journal of Science, 99*(5–6), 243–247.

Parkington, J. (2006). *Shorelines, strandlopers, and shell middens*. Cape Town: Creda.

Parkington, J. (2010). Coastal diet, encephalization, and innovative behaviors in the late Middle Stone Age of southern Africa. In S. C. Cunnane & K. M. Stewart (Eds.), *Human brain evolution: The influence of freshwater and marine food resources* (pp. 189–202). New York: Wiley-Blackwell.

Powell, A., Shennan, S., & Thomas, M. G. (2009). Late Pleistocene demography and the appearance of modern human behavior. *Science, 324*(5932), 1298–1301.

Ramos, J., Bernal, D., Domínguez-Bella, S., Calado, D., Ruiz, B., Gil, M. J., et al. (2008). The Benzú Rockshelter: A Middle Palaeolithic site on the North African coast. *Quaternary Science Reviews, 27* (23), 2210–2218.

Reich, D., Green, R. E., Kircher, M., Krause, J., Patterson, N., Durand, E. Y., et al. (2010). Genetic history of an archaic hominin group from Denisova Cave in Siberia. *Nature, 468*(7327), 1053–1060.

Robbins, L. H., Brook, G. A., Murphy, M. L., Ivester, A. H., & Campbell, A. C. (2016). The Kalahari during MIS 6-2 (190–12 ka): Archaeology, paleoenvironment and population dynamics. In S. C. Jones & B. A. Stewart (Eds.), *Africa from MIS 6-2: Population dynamics and paleoenvironments* (pp. 175–193). Dordrecht: Springer.

Robertshaw, P. (1990). *A history of African archaeology*. London: James Currey.

Runge, J. (2000). Environmental and climatic history of the eastern Kivu area (D.R. Congo, ex Zaire) from 40 ka to the present. In P. Smolka & W. Volkheimer (Eds.), *Southern hemisphere paleo-and neoclimates* (pp. 249–262). Berlin: Springer.

Runge, J. (2001a). Central African palaeoclimates and palaeoenvironments since 40 ka – an overview. *Palaeoecology of Africa, 27*, 1–10.

Runge, J. (2001b). On the age of stone-lines and hillwash sediments in the eastern Congo basin: Palaeoenvironmental implications. *Palaeoecology of Africa, 27*, 19–36.

Sahle, Y., Hutchings, W. K., Braun, D. R., Sealy, J. C., Morgan, L. E., Negash, A., & Atnafu, B. (2013). Earliest stone-tipped projectiles from the Ethiopian rift date to >279,000 years ago. *PLoS ONE, 8* (11), e78092.

Scholz, C. A., Cohen, A. S., Johnson, T. C., King, J., Talbot, M. R., & Brown, E. T. (2011). Scientific drilling in the Great Rift Valley: The 2005 Lake Malawi Scientific Drilling Project – an overview of the past 145,000 years of climate variability in Southern Hemisphere East Africa. *Palaeogeography, Palaeoclimatology, Palaeoecology, 303*(1), 3–19.

Schwarcz, H. P. (2001). Chronometric dating of the Middle Pleistocene. In L. Barham & K. Robson-Brown (Eds.), *Human roots: Africa and Asia in the Middle Pleistocene* (pp. 41–53). Bristol: Western Academic and Specialist Press.

Sealy, J. (2016). Cultural change, demography, and the archaeology of the last 100 kyr in southern Africa. In S. C. Jones & B. A. Stewart (Eds.), *Africa from MIS 6-2: Population dynamics and paleoenvironments* (pp. 65–75). Dordrecht: Springer.

Sealy, J., & Galimberti, M. (2011). Shellfishing and the interpretation of shellfish sizes in the Middle and Later Stone Ages of South Africa. In N. F. Bicho, J. A. Haws, & L. G. Davis (Eds.), *Trekking the shore: Changing coastlines and the antiquity of coastal settlement* (pp. 405–419). New York: Springer.

Shanahan, T. M., Beck, J. W., Overpeck, J. T., McKay, N. P., Pigati, J. S., Peck, J. A., et al. (2012). Late Quaternary sedimentological and climate changes at Lake Bosumtwi, Ghana: New constraints from laminae analysis and radiocarbon age modeling. *Palaeogeography, Palaeoclimatology, Palaeoecology, 361*, 49–60.

Shea, J. J. (2011). *Homo sapiens* is as *Homo sapiens* was: Behavioral variability versus "behavioral modernity" in Paleolithic archaeology. *Current Anthropology, 52*(1), 1–35.

Sjödin, P., Sjöstrand, E. A., Jakobsson, M., & Blum, M. G. (2012). Resequencing data provide no evidence for a human bottleneck in Africa during the penultimate glacial period. *Molecular Biology and Evolution, 29*(7), 1851–1860.

Smith, M. A. (1993). Biogeography, human ecology and prehistory in the Sandridge Deserts. *Australian Archaeology, 37*, 35–50.

Smith, M. A. (2013). *The archaeology of Australia's deserts*. Cambridge: Cambridge University Press.

Smith, M. A., Williams, A. N., Turney, C. S. M., & Cupper, M. L. (2008). Human-environment interactions in Australian drylands: Exploratory time-series analysis of archaeological records. *The Holocene, 18*(3), 389–401.

Soares, P., Rito, T., Pereira, L., & Richards, M. B. (2016). A genetic perspective on African prehistory. In S. C. Jones & B. A. Stewart (Eds.), *Africa from MIS 6-2: Population dynamics and paleoenvironments* (pp. 383–405). Dordrecht: Springer.

Steele, T. E., & Álvarez-Fernández, E. (2011). Initial investigations into the exploitation of coastal resources in North Africa during the Late Pleistocene at Grotte des Contrebandiers, Morocco. In N. F. Bicho, J. A. Haws, & L. G. Davis (Eds.), *Trekking the shore: Changing coastlines and the antiquity of coastal settlement* (pp. 383–403). New York: Springer.

Steele, T. E., & Klein, R. G. (2008). Intertidal shellfish use during the Middle and Later Stone Age of South Africa. *Archaeofauna, 17*(6), 63–76.

Stewart, B. A., Dewar, G., Morley, M. W., Inglis, R. H., Wheeler, M., Jacobs, Z., et al. (2012). Afromontane foragers of the Late Pleistocene: Site formation, chronology and occupational pulsing at Melikane Rockshelter, Lesotho. *Quaternary International, 270*, 40–60.

Stewart, B. A., Dewar, G., Morley, M. W., Parker, A., & Allott, L. (2016). Follow the Senqu: Maloti-Drakensberg paleoenvironments and implications for modern human dispersals into mountain systems. In S. C. Jones & B. A. Stewart (Eds.), *Africa from MIS 6-2: Population dynamics and paleoenvironments* (pp. 247–271). Dordrecht: Springer.

Taylor, D. M., Marchant, R., & Hamilton, A. C. (2001). A reanalysis and interpretation of palynological data from the Kalambo Falls prehistoric site. In J. D. Clark (Ed.), *Kalambo Falls prehistoric site III, The Earlier Cultures: Middle and Early Stone Age* (pp. 66–81). Cambridge: Cambridge University Press.

Taylor, N. (2016). Across rainforests and woodlands: A systematic re-appraisal of the Lupemban Middle Stone Age in Central Africa. In S. C. Jones & B. A. Stewart (Eds.), *Africa from MIS 6-2: Population dynamics and paleoenvironments* (pp. 272–299). Dordrecht: Springer.

Texier, P. J., Porraz, G., Parkington, J., Rigaud, J. P., Poggenpoel, C., Miller, C., et al. (2010). A Howiesons Poort tradition of engraving ostrich eggshell containers dated to 60,000 years ago at Diepkloof Rock Shelter, South Africa. *Proceedings of the National Academy of Sciences, 107*(14), 6180.

Timberlake, J., Chidumayo, E., & Sawadogo, L. (2010). Distribution and characteristics of African dry forests and woodlands. In E. N. Chidumayo & D. J. Gumbo (Eds.), *The dry forests and woodlands of Africa: Managing for products and services* (pp. 11–41). London: Earthscan.

Tishkoff, S. A., Reed, F. A., Friedlaender, F. R., Ehret, C., Ranciaro, A., Froment, A., et al. (2009). The genetic structure and history of Africans and African Americans. *Science, 324*(5930), 1035–1044.

Townsend, P. K. (1990). On the possibility/impossibility of tropical forest hunting and gathering. *American Anthropologist, 92*, 745–747.
Turnbull, C. M. (1961). *The forest people*. New York: Simon & Schuster.
UNCCD, UNDP & UNEP (2009). *Climate change in the African drylands: Options and opportunities for adaptation and mitigation*. New York, Bonn, Nairobi: UNCCD, UNDP & UNEP.
van Andel, T. H. (1989). Late Pleistocene sea levels and the human exploitation of the shore and shelf of southern South Africa. *Journal of Field Archaeology, 16*(2), 133–155.
van Noten, F. L. (Ed.). (1982). *The archaeology of central Africa*. Graz: Akademische Druck-und-Verlagsanstalt.
Van Peer, P. (2016). Technological systems, population dynamics and historical process in the MSA of northern Africa. In S. C. Jones & B. A. Stewart (Eds.), *Africa from MIS 6-2: Population dynamics and paleoenvironments* (pp. 147–159). Dordrecht: Springer.
van Zinderen Bakker, E. M. (1969). The Pleistocene vegetation and climate of the basin. In J. D. Clark (Ed.), *Kalambo Falls prehistoric site. I: Geology, palaeoecology and detailed stratigraphy of the excavations* (pp. 57–84). Cambridge: Cambridge University Press.
Veth, P. (1989). Islands in the interior: A model for the colonization of Australia's arid zone. *Archaeology in Oceania, 24*(3), 81–92.
Veth, P. (1993). *Islands in the interior: The dynamics of prehistoric adaptations within the arid zone of Australia*. Ann Arbor: International Monographs in Prehistory.
Veth, P. (2005). Cycles of aridity and human mobility: Risk minimization among late Pleistocene foragers of the Western Desert, Australia. In P. Veth, M. Smith, & P. Hiscock (Eds.), *Desert peoples: Archaeological perspectives* (pp. 100–115). Oxford: Blackwell.
Vigilant, L., Stoneking, M., Harpending, H., Hawkes, K., & Wilson, A. C. (1991). African populations and the evolution of human mitochondrial DNA. *Science, 253*(5027), 1503–1507.
Volman, T. P. (1978). Early archeological evidence for shellfish collecting. *Science, 201*(4359), 911–913.
Wadley, L. (2013). Recognizing complex cognition through innovative technology in Stone Age and Paleolithic sites. *Cambridge Archaeological Journal, 23*(02), 163–183.
Walter, R. C., Buffler, R. T., Bruggemann, J. H., Guillaume, M. M. M., Berhe, S. M., Negassi, B., et al. (2000). Early human occupation of the Red Sea coast of Eritrea during the Last Interglacial. *Nature, 405*(6782), 65.
Wayland, E. J. (1934). Rifts, rivers, rains and early man in Uganda. *Journal of the Royal Anthropological Institute of London, 64*, 333–352.
White, T. D., Asfaw, B., DeGusta, D., Gilbert, H., Richards, G. D., Suwa, G., & Howell, F. C. (2003). Pleistocene *Homo sapiens* from Middle Awash, Ethiopia. *Nature, 423*(6941), 742–747.
Wilkins, J., Schoville, B. J., Brown, K. S., & Chazan, M. (2012). Evidence for early hafted hunting technology. *Science, 338*(6109), 942–946.
Will, M., Parkington, J. E., Kandel, A. W., & Conard, N. J. (2013). Coastal adaptations and the Middle Stone Age lithic assemblages from Hoedjiespunt 1 in the Western Cape, South Africa. *Journal of Human Evolution, 64*(6), 518–537.
Williams, A. N., Ulm, S., Cook, A. R., Langley, M. C., & Collard, M. (2013). Human refugia in Australia during the Last Glacial Maximum and terminal Pleistocene: A geospatial analysis of the 25–12 ka Australian archaeological record. *Journal of Archaeological Science, 40*(12), 4612–4625.
Wurz, S. (2013). Technological trends in the Middle Stone Age of South Africa between MIS 7 and MIS 3. *Current Anthropology, 54*, S305–S319.

Part I
Coasts

Chapter 2
Mid to Late Quaternary Landscape and Environmental Dynamics in the Middle Stone Age of Southern South Africa

Andrew S. Carr, Brian M. Chase, and Alex Mackay

Abstract The southern Cape of South Africa hosts a remarkably rich Middle Stone Age (MSA) archaeological record. Many of the associated caves and rock shelters are coastal sites, which contain evidence for varied occupational intensity and marine resource use, along with signs of notable landscape, environmental, and ecological change. Here, we review and synthesize evidence for Quaternary landscape and climatic change of relevance to the southern Cape MSA. We seek to highlight the available data of most relevance to the analysis and interpretation of the region's archaeological record, as well as critical data that are lacking. The southern Cape MSA occupation spans the full range of glacial-interglacial conditions (i.e., 170–55 ka). It witnessed marked changes in coastal landscape dynamics, which although driven largely by global eustatic sea level changes, were modulated by local-scale, often inherited, geological constraints. These prevent simple extrapolations and generalizations concerning paleolandscape change. Such changes, including pulses of coastal dune activity, will have directly influenced resource availability around the region's archaeological sites. Evidence for paleoclimatic change is apparent, but it is scarce and difficult to interpret. It is likely, however that due to the same diversity of rainfall sources influencing the region today, compared to parts of the continental interior, the southern Cape climate was relatively equable throughout the last 150 kyr. The region's paleoecology, particularly in relation to the coastal plains exposed during sea level lowstands, is a key element missing in attempts to synthesize and model the resources available to occupants of this region. Technology, settlement, and subsistence probably changed in response to these paleoclimate/landscape adjustments, but improvements in baseline archaeological and paleoenvironmental data are required to strengthen models of ecosystem variation and human behavioral response through the MSA.

Keywords Archaeology • Coast • Sea level • Eolian • Chronology • Paleoenvironment • Palaeoclimate

Preamble

Recent findings have placed increased emphasis on the role of southern Africa – notably the southern Cape coastal region (Fig. 2.1) – in the story of the emergence of modern humans (e.g., Marean 2010; Parkington 2010). A combination of the region's unique environment, ecology, and oceanographic setting, along with a series of artifact findings suggestive of cognitively modern behavior (Henshilwood et al. 2002, 2004, 2011; Marean et al. 2007; Brown et al. 2009), have led to the proposition that this region represents an early habitat of modern humans, perhaps as long as ~165 ka (Marean et al. 2007; Marean 2010, 2011). Recent interpretations of archaeological records emphasize the unique environmental context and history of the southern Cape (Marean 2011; Compton 2011).

The southern Cape's Middle Stone Age (MSA; c. 280–30 ka) archaeological record primarily comprises a series of coastal cave occupations. These include Die Kelders Cave (c. 90 km east of Cape Town), Blombos Cave (near to Still Bay), Pinnacle Point (west of Mossel Bay), Nelson Bay

A.S. Carr (✉)
Department of Geography, University of Leicester, Leicester, LE1 7RH, UK
e-mail: asc18@le.ac.uk

B.M. Chase
Centre National de la Recherche Scientifique, UMR 5554, Institut des Sciences de l'Evolution de Montpellier, Université de Montpellier, Bat.22, CC061, Place Eugène Bataillon, 34095 Montpellier Cedex 5, France

A. Mackay
Centre for Archaeological Science, School of Earth and Environmental Sciences, University of Wollongong, Northfields Avenue, Wollongong, NSW 2522, Australia

Sacha C. Jones and Brian A. Stewart (eds.), *Africa from MIS 6-2: Population Dynamics and Paleoenvironments*, Vertebrate Paleobiology and Paleoanthropology, DOI 10.1007/978-94-017-7520-5_2

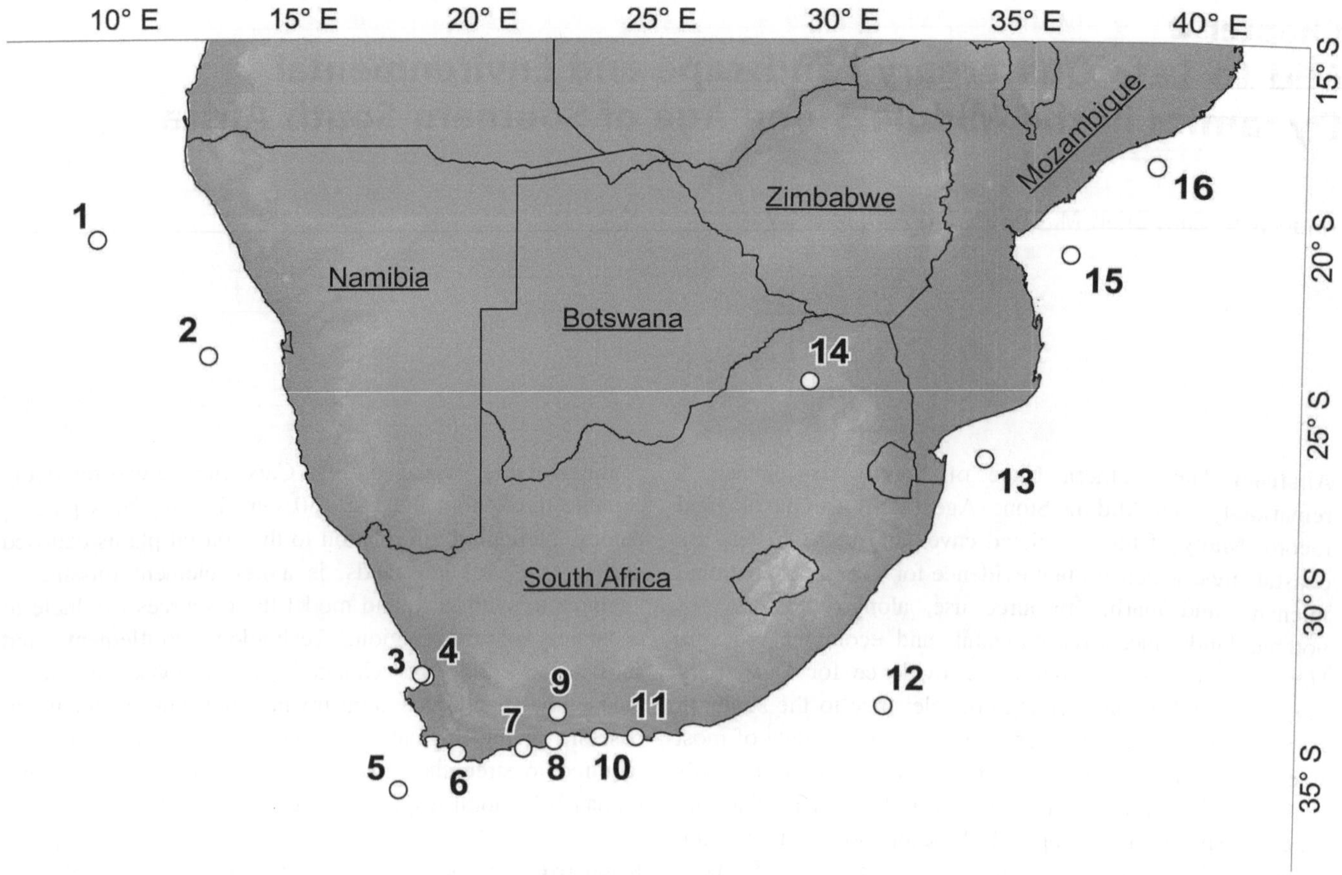

Fig. 2.1 Map of southern Africa with key marine and terrestrial paleoenvironmental sites and records. *Key 1* MD962094; *2* GeoB 1711-4; *3* Elands Bay Cave; *4* Diepkloof; *5* MD962081; *6* Die Kelders; *7* Blombos Cave; *8* Pinnacle Point-Crevice Cave; *9* Boomplaas Cave; *10* Nelson Bay Cave; *11* Klasies River; *12* MD962007; *13* MD962048; *14* Cold Air Cave; *15* MD79257; *16* MD79254

Cave (near Plettenberg Bay) and Klasies River (on the Tsitsikamma coast) (Fig. 2.1). A number of notable MSA sites on the west coast of South Africa are also situated on (e.g., Geelbeck [Kandel and Conard 2012] and Ysterfontein [Klein et al. 2004]) or relatively close (e.g., Diepkloof [Texier et al. 2010]) to the coastline. The significance of a dynamic coastal landscape for site formation, site preservation and in potentially explaining the occupational history of some sites has long been recognized (Tankard and Schweitzer 1976; Van Andel 1989). With the development of more robust chronologies for MSA occupations (Vogel et al. 1999; Feathers and Bush 2000; Feathers 2002; Jacobs et al. 2003, 2006, 2008, 2013; Tribolo et al. 2006, 2013; Jacobs and Roberts 2008; Jacobs 2010; Henshilwood et al. 2011) it is becoming possible to explore potential links between site occupation, human behavior, climate change, and landscape change in increasing detail (e.g., Cochrane et al. 2013).

This wider setting, comprising the regional landscape, climate, and ecology, their dynamism and influence on early human habitation, is the subject of this review. While the contemporary climatic and ecological conditions in the region are relatively well understood, evidence for paleoenvironmental change subsequent to the Last Interglacial (MIS 5e) remains limited and ambiguous (Chase and Meadows 2007; Chase 2010). New archaeological findings have made refining this record and unraveling its complexities a priority (Jacobs and Roberts 2008; Mackay 2011). An increasing pace of research, which includes several recent publications concerning local paleoclimate (Bar-Matthews et al. 2010; Chase 2010; Chase et al. 2013, 2015b; Quick et al. in press a, b) and landscape evolution (Bateman et al. 2011; Fisher et al. 2010; Cawthra et al. 2014) means that a regional-scale and holistic (re)consideration of paleoclimate and paleogeography is timely.

In this context, it is the purpose here to:

1. Provide an overview of the contemporary environmental setting of this region.
2. Synthesize newly published research concerning the geomorphic evolution of this coastal environment, highlighting the potential of these data to contribute to a holistic understanding of coastal landscape change.

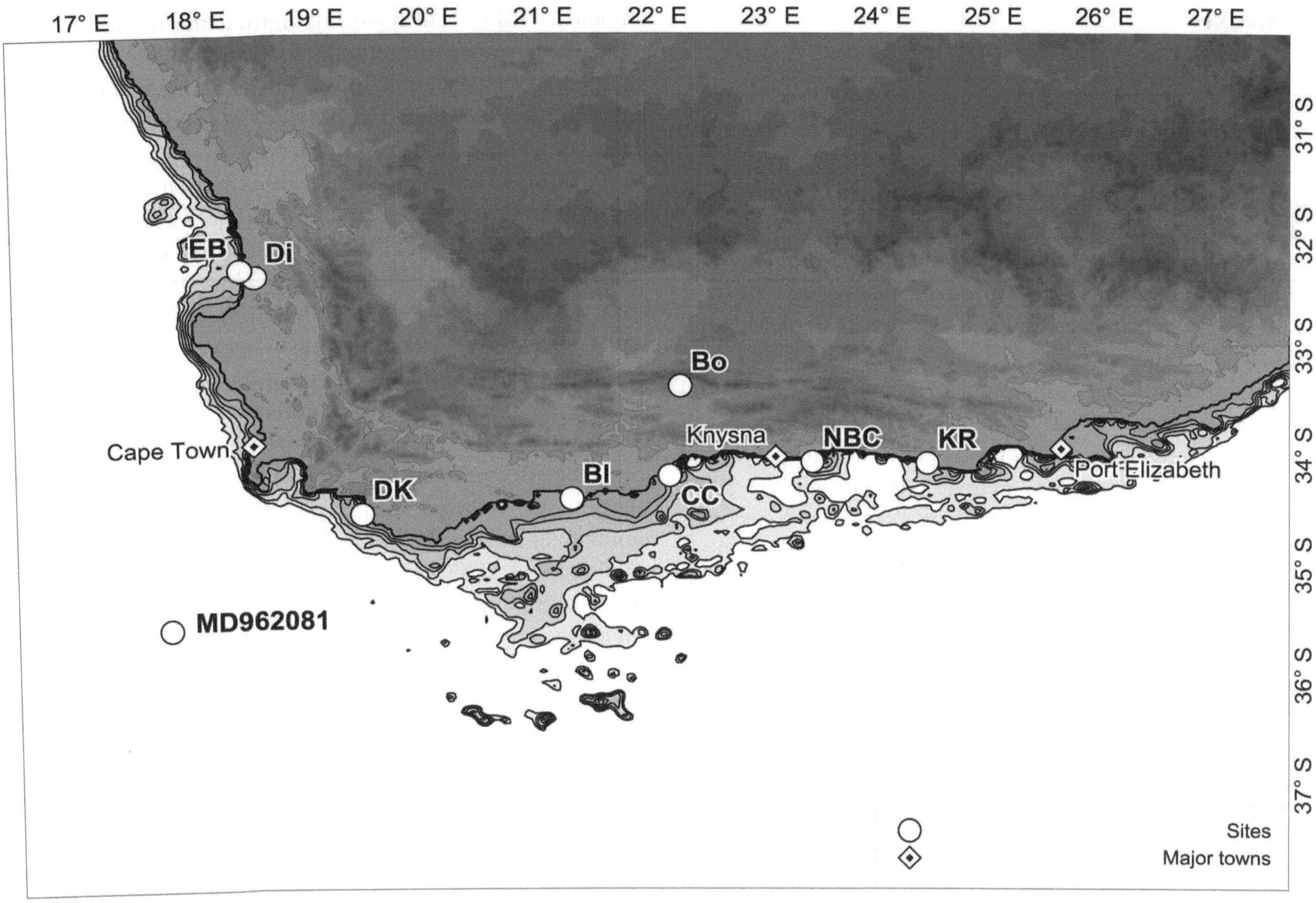

Fig. 2.2 Map of South Africa's southern Cape with regional paleoenvironmental and archaeological sites. Topographic contours are shown at 250 m intervals, while bathymetric contours are shown at 25 m intervals down to −125 m amsl, highlighting the extensive coastal plain that existed south of Cape Agulhas during the Last Glacial Maximum. EB = Elands Bay; DI = Diepkloof; CC = Crevice Cave; Bl = Blombos; Bo = Boomplaas; DK = Die Kelders; NBC = Nelson Bay Cave; KR = Klasies River

3. Analyze the currently available paleoclimatic data, and identify spatial, temporal, and interpretative gaps.
4. Consider, the relevance or potential relevance of points 2 and 3 for MSA archaeological research in this region.

Spatial and Temporal Scope

We limit our discussion largely to the temporal framework established for this volume, which spans MIS 6-2 (190–12 ka). This is broadly commensurate with the latter half of the MSA in southern Africa (Lombard 2012). It also includes the occurrences of the Still Bay and Howiesons Poort industries (dating to ~73.5–70.5 ka and ~66–58 ka, respectively; Jacobs et al. 2008), which have attracted particular interest (Jacobs et al. 2008; Compton 2011; McCall and Thomas 2012; Sealy 2016). Geographically (Fig. 2.2), we focus on the southern South African coastline from Cape Town to Port Elizabeth, which incorporates a number of important environmental facets: (1) South Africa's winter rainfall zone (WRZ), which presents a climatic gradient from the winter rainfall dominated environs of Cape Town, through the year-round rainfall zone (YRZ) of the south coast, to the interface with the summer rainfall zone (SRZ) north of Port Elizabeth (i.e., "Axis B" of Chase and Meadows [2007]); (2) the extensive offshore Agulhas Bank, which was variously exposed as eustatic sea level changed throughout MIS 6-2; (3) the peri-coastal region subject to the climatic influence of the Agulhas Current and localized near-coastal seasonal upwelling systems (Cohen and Tyson 1995); (4) the southern section of the diverse Cape Floristic Region (CFR, Linder 2003). It is also the location of major archaeological sites relevant to current debates surrounding modern human origins: Die Kelders Cave, Blombos Cave, the Pinnacle Point complex and Klasies River cave system.

Context

Given long-standing attempts to consider aspects of hominin evolution in the context of global-scale climatic forcing, there remains an impetus to provide contextual environmental information for fossil finds or sites of human occupation. These may be variously derived from local proxy data (perhaps from the same strata as the fossil material) or archives of "global-scale" change, epitomized by marine and ice core archives (Behrensmeyer 2006; Kingston 2007; deMenocal 2011). The latter approach presents some fundamental difficulties; primarily the assumption of direct and meaningful linkages between "global" or "hemispheric" scale signals, and "local scale" environmental information within archaeological sequences or single site archives (Behrensmeyer 2006; Kingston 2007; Chase 2010). This "scale-gap" becomes increasingly intractable deeper in the geological past, where proxy records are more fragmentary and/or equivocal (Kingston 2007).

In a region like South Africa, which lacks detailed terrestrial paleoenvironmental archives, this issue presents a significant problem, even for the middle and late Pleistocene. In fact, this scale issue has been a point of contention for many years (cf. Butzer 1984). Notwithstanding this, considerable progress has been made since Butzer's (1984) work; both in terms of the number of published proxy records and associated chronological control (Chase and Meadows 2007; Jacobs et al. 2008; Mitchell 2008).

Archaeology and the Coastal Landscape

An additional aim here is to consider information pertaining to wider Quaternary *landscape* change. Such data reflect our improved understanding of coastal geomorphic responses to long-term global climate forcing. The southern Cape landscape changed radically throughout MIS 6-2, primarily in response to global eustatic sea level changes (Van Andel 1989; Butzer 2004; Fisher et al. 2010; Cawthra et al. 2014). Our understanding of the nature and effects of such changes is increasingly supported by developments in geochronology and geophysical surveying. With such information we can begin to develop hypotheses on the nature of long-vanished landscapes on the submerged continental shelf.

Given, their potentially equable and resource-rich (ecotonal) settings, coastal environments have been variously highlighted as "refugia" and "migrational routes" for ancient human populations (*inter alia*: Stringer 2000; Bailey and Flemming 2008; Finlayson 2008; Compton 2011; Lambeck et al. 2011). Several workers in South Africa have also emphasized this, and have highlighted the marine environment as a source of reliable food resources through the vagaries of Pleistocene climate. Such marine resource usage has also been considered to provide clues concerning human behavior (Parkington 2003; Marean et al. 2007; Marean 2010, 2011). This interesting issue is not entirely unique to the southern Cape record, and has also been considered in the context of Neanderthal behavior (Stringer et al. 2008). Notwithstanding this, shell middens along the South African coast are widely distributed, both within rock shelters and the wider landscape. Marked contrasts between MSA and Late Stone Age (LSA) midden compositions have been reported, and the significance of these differences has been debated (Parkington 2003; Klein et al. 2004).

In the African context, the specific role of the continental margins as "refugia" during periods of aridity within the continental interior has been highlighted in various studies (Walter et al. 2000; Faure et al. 2002; Hetherington et al. 2008; Compton 2011). However, with the exception of the environmental archives provided by the East African lakes (e.g., Scholz et al. 2007; Castañeda et al. 2009) the continental Quaternary paleoclimatic record is sparse and geomorphic evidence of paleo-aridity in particular has proved difficult to interpret (Chase 2009; Thomas and Burroughs 2012; Burrough 2016). Although blanket claims of "Quaternary aridity" or "glacial aridity" should be treated with caution, phases of enhanced late Quaternary aridity can be identified within the southern African interior (e.g., Chase 2009, 2010; Chase et al. 2009, 2011; Chevalier and Chase 2015; Collins et al. 2014; Dupont et al. 2011; Lancaster 2002; Partridge et al. 1997; Scholz et al. 2007; Shi et al. 2001; Stager et al. 2011; Stuut et al. 2002; Thomas and Burrough 2012; Truc et al. 2013). Such periods of interior aridity are not necessarily restricted to, or specifically characteristic of "glacial" periods, but an emerging theme in southern African research has been the hypothesis that the coastal margins may have been of increased importance for human habitation during periods of interior aridity (e.g., Morris 2002; Hetherington 2008; Parkington 2010; Compton 2011; Blome et al. 2012). In the southern Cape this idea, perhaps to some extent, reflects the relatively mesic conditions we see in this region today (notably in the Knysna area). However, today's largely aseasonal and relatively humid rainfall regime along parts of the southern Cape owes its existence to the balanced influence of temperate and tropical rainfall systems, and this scenario is almost certainly sensitive to perturbations in global and regional circulation systems (Stuut et al. 2004; Chase and Meadows 2007; Chase 2010; Chase et al. 2013).

Understanding human occupation of the coastal zone is a challenging and inherently interdisciplinary task. Westley and Dix (2006) emphasize that coasts may: (1) represent equable and stable habitats; (2) offer uniformity in environment along-shore; (3) offer diversity and productivity in terms of resources; and (4) offer simplified landscapes for migration and navigation. Yet, they also note that these propositions should not be assumed. Inferences to this effect, based on archaeological investigations at a single site should be treated with caution. All modern coastal landscapes are geologically young, and the nature of paleocoasts must be inferred from preserved geological and geomorphic evidence. Most interpretations are derived from fragmentary evidence sampled largely in the subaerial landscape. Evidence for landscapes on the submerged continental shelf has, for obvious reasons, been largely lacking. Recent geophysical surveying approaches suggest that there is potential to resolve this issue (Cawthra et al. 2014).

Contemporary Setting and Drivers of Change

Landscape

The landscape of the southern Cape is today dominated by two key elements: (1) the Cape Fold Belt Mountains; and (2) a coastal platform. The Cape Fold Belt formed from the orogeny of the Ordovician Table Mountain Group (TMG) sandstones during the late Paleozoic, and today the eroded remnants of these mountains form a series of broadly coast-parallel ridges separating the southern Cape from the continental interior (Deacon et al. 1992; Compton 2011). The breakup of Gondwana had a fundamental influence on the southern Cape coastline; it created an initial platform seaward of the Cape Fold Mountains (Partridge and Maud 1987; Marker and Holmes 2010) (Fig. 2.2) and, due to a series of half-grabens formed during the fragmentation, divided the continental margin into distinct sedimentary basins (Broad et al. 2006). This structural control produced a series of resistant TMG sandstone headlands separated by basins (today these broadly correspond to coastal embayments) containing Late Mesozoic clastic sedimentary infills (e.g., Enon and Kirkwood Formations), as well as Neogene and Quaternary eolian and marginal-marine sedimentary deposits (Malan 1990; Marker and Holmes 2010) (Fig. 2.2).

Coastal Geomorphology

The varied geological, geomorphic, and marine settings of the southern Cape provide a diversity of environments and habitats, with a notable dichotomy between the rocky headlands and open sandy beaches. The South African continental shelf widens substantially on the south coast, forming the Agulhas Bank (Fig. 2.2). The southern Cape experiences high open water wave heights (median heights ~2.5 m at Knysna; Whitfield et al. 1983) and swell directions are predominantly from the southwest (Davies 1980). Long-shore sediment fluxes are typically in an easterly direction (Martin and Flemming 1986) and tidal ranges are generally low. The combination of the high wave energies and the low tidal ranges means that the southern Cape is classed as a wave-dominated coastline (Davis and Hayes 1984). Thus, with the exception of rocky headlands the coastal geomorphology is dominated by "barrier" landforms, specifically, wave-deposited sediments and associated landforms (e.g., Davis and Hayes 1984; Roy et al. 1994). The refraction of incoming swell waves means that wave energy is generally concentrated at headlands, and the embayments are characterized by wide intermediate to dissipative beaches. Many embayment beaches are backed by extensive coastal dunes, which in a number of locations are currently active (Tinley 1985). Seawards-younging sequences of eolian deposits imply that this situation has persisted since the Pliocene (e.g., Roberts et al. 2008).

East of Cape Agulhas, notably in the Still Bay and Knysna areas, landwards dune migration is limited by abundant vegetation and during the Pleistocene parabolic dunes stacked upon one another to form large composite barrier dune systems (Roberts et al. 2008, 2009; Bateman et al. 2011). Submerged barrier features are present on the continental shelf, notably off the Wilderness coast and bear testament to the close links between dune formation and relative sea level change in this region (Martin and Flemming 1986; Cawthra et al. 2014).

The headlands bounding the half-moon bays are frequently rocky and represent impediments to longshore sediment transport. In some locations this is facilitated by headland-bypass dune systems (Tinley 1985). These are long-established landscape elements (Bateman et al. 2004, 2008; Carr et al. 2006a; Carr and Botha 2012).

Climate

The climate of the southern Cape is a function of: (1) the interplay between the South Atlantic and Indian Ocean high pressure cells, and the cyclonic westerly systems; and (2) its position relative to the warm Agulhas Current, which, in conjunction with heating contrasts between the land and ocean, enhances moisture delivery to the coast (Tyson 1986; Jury et al. 1993; Lindesay 1998). For the majority of southern Africa, the austral summer is the wet season. This is driven by the southward migration of the ITCZ and the

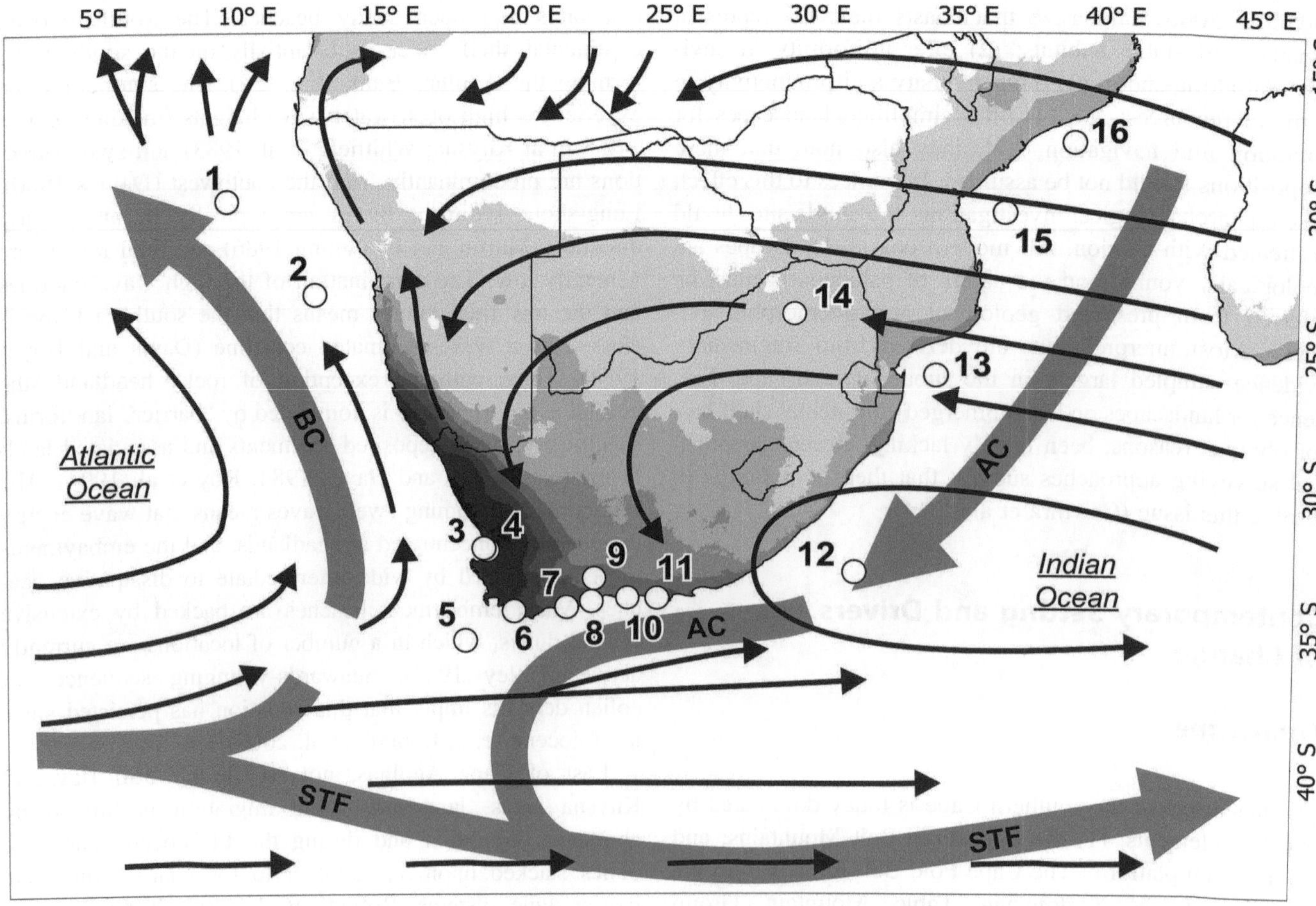

Fig. 2.3 Map of southern Africa with dominant atmospheric (*thin black arrows*) and oceanic circulation patterns (*thick gray arrows*). Major oceanic features include the Benguela Current (BC), the subtropical Front (STF), and the Agulhas Current (AC). Continental shading indicates the distribution of winter rain as a percentage of the annual total (darkest = 80%, lightest = 0%). Key marine and terrestrial paleoenvironmental sites and records are also shown: *1* MD962094; *2* GeoB 1711-4; *3* Elands Bay Cave; *4* Diepkloof; *5* MD962081; *6* Die Kelders; *7* Blombos Cave; *8* Pinnacle Point-Crevice Cave; *9* Boomplaas Cave; *10* Nelson Bay Cave; *11* Klasies River; *12* MD962007; *13* MD962048; *14* Cold Air Cave; *15* MD79257; *16* MD79254

influence of the tropical easterlies, producing a flux of moisture from the Indian Ocean. During the austral winter, the expansion of the circumpolar vortex causes the westerly cyclonic systems to track further north, bringing rain to the WRZ of the southwestern Cape (Fig. 2.3), while stable anticyclonic (dry) conditions prevail in the SRZ. The southern Cape represents a complex transitional zone between these regimes and large parts of it experience a year-round rainfall regime (YRZ) that derives moisture from both westerly systems and the Indian Ocean, along with more localized coastal disturbances (Tyson and Preston-Whyte 2000). Moving east of Cape Town, rainfall becomes less seasonal and shifts from a near semi-arid climate ($\sim$450 mm a^{-1} at Struisbaai) to a more temperate climate ($\sim$800 mm a^{-1} at Knysna).

It has long been argued that the positioning of the westerly cyclonic systems was a key driver of late Quaternary climatic variability in southwest Africa (van Zinderen Bakker 1967, 1976; Cockcroft et al. 1987). This has been linked to shifts in the position of the subtropical front (STF) (Fig. 2.3), which responds to the extent of Antarctic sea ice (van Zinderen Bakker 1976; Stuut et al. 2002, 2004; Chase and Meadows 2007; Chase 2010; Chase et al. 2013). These ideas are discussed further later in this review.

Ecology

The southern Cape lies largely within the Cape Floristic Region (CFR; Goldblatt and Manning 2002). The CFR comprises a number of separate biomes (Fig. 2.4), which are closely associated with specific soil types (Cowling and Holmes 1992; Low and Rebelo 1996). Fynbos ("*fine bush*") is characterized by small, needle-leafed ericoid shrubs, along with plants characterized by larger sclerophyllous leaves

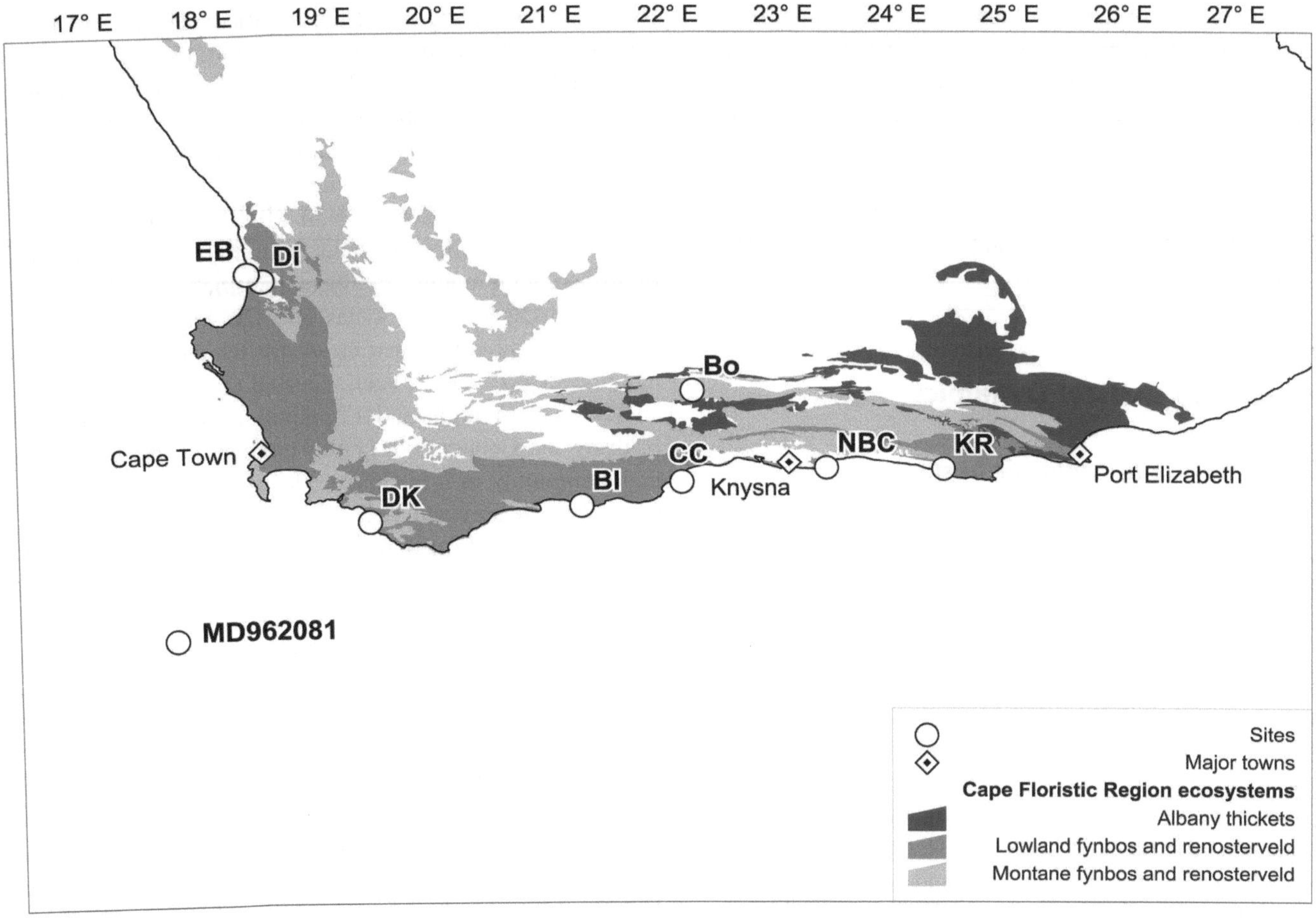

Fig. 2.4 Contemporary biomes of the southern Cape region. *EB* = Elands Bay; *DI* = Diepkloof; *CC* = Crevice Cave; *Bl* = Blombos; *Bo* = Boomplaas; *DK* = Die Kelders; *NBC* = Nelson Bay Cave; *KR* = Klasies River

(Cowling and Holmes 1992). Today it is dominated by plants using the C_3 photosynthetic pathway (Vogel et al. 1978).

Fynbos tends to be found as either upland Montane Fynbos, which is associated with the nutrient-poor sandy soils of the Cape Fold Belt or Lowland Fynbos, which is often associated with calcareous soils and dune sands. On finer-gained substrates (e.g., Bokkeveld Shale) *Renosterveld* is a common vegetation type, which, like fynbos, is broadly classed as a small-leaved shrub land (Cowling 1983; Cowling et al. 1988). It is often dominated by "Renosterbos" (*Elytropappus rhinocerotis*), and is associated with a higher proportion of grasses (including grasses using the C_4 photosynthetic pathway where more summer rains occur) and succulents (which will include CAM photosynthesis; *contra* Bar-Matthews et al. 2010). Its potential sensitivity to changing rainfall, in terms of the proportions of succulents and fynbos species has been noted (Cowling 1983; Low and Rebelo 1996), and possibly identified within the paleoecological record (Carr et al. 2006b; Quick et al. in press a, b). Thicket vegetation is associated with coastal dunes (Rebelo et al. 1991). A notable component of the southern Cape's vegetation is the afromontane forest of the Knysna area between Mossel Bay and Klasies River. This requires relatively humid year-round rainfall conditions and is relatively drought-sensitive (Cowling 1983), tending to be associated with ~800–1100 mm mean annual precipitation *and* low rainfall seasonality (a coefficient of variation of <17).

The responses of these vegetation communities during the Quaternary are poorly constrained. Evidence from the Cederberg in the Western Cape suggests that during the late Quaternary, Montane Fynbos was largely buffered from regional-scale climatic changes (Chase et al. 2015a, 2011) by the wide climatic tolerances of its plant genera, specific edaphic constraints (association with low nutrient sandstone substrates; Campbell and Werger 1988) and relatively reliable orographic rainfall (Cowling 1983; Meadows and Sugden 1993; Chase et al. 2011; Quick et al. 2011; Valsecchi et al. 2013). The paleoecology of the renosterveld (lowland fynbos, thicket and afromontane mosaic on the coastal lowlands), which is perhaps most relevant to the region's MSA archaeology, is discussed later, but is very poorly understood. Prior to the arrival of pastoralists and later, European colonists, a diverse range of fauna is thought to have occupied the southern Cape region (see Boshoff and Kerley 2001). In general, the large herbivores comprised of mixed feeders and browsers, including *inter alia*: African Elephant (*Loxodonta africana*), Cape Buffalo (*Syncerus caffer*), Red Hartebeest (*Alcephalus buselaphus*), Bontebok (*Damaliscus dorcas dorcas*), Quagga (*Equus quagga*), Blue Antelope (*Hippotragus leucophaeus*) and Cape Mountain Zebra (*Equus zebra zebra*).

Quaternary Coastal Dynamics

Quaternary Sea Level Change

Notable occupational hiatuses, particularly between MSA and LSA deposits, in the southern Cape coastal archaeological record have long been linked to eustatic sea level change(s) during MIS 5-2 (Van Andel 1989). Benthic oxygen isotope data imply 125–130 m of eustatic sea level change across glacial-interglacial cycles (e.g., Waelbroeck et al. 2002). Given the region's apparent tectonic stability (Roberts et al. 2012), this is probably a reasonable approximation for the southern Cape during the middle and late Quaternary (Ramsay and Cooper 2002; Compton 2011). Sea level change of this magnitude would have expanded the coastal platform by as much as ~50,000 km^2 during sea level lowstands (Fig. 2.2).

Sea Level Change, Site Formation and Site Occupation

Evidence from the MSA deposits of Pinnacle Point Cave 13b implies that fluctuating marine resource use during MIS 6 was correlated with eustatic sea level fluctuations (Marean et al. 2007). Compton (2011) considered the significance of relative sea level change from a broader perspective, emphasizing its role in moderating resource availability, competition, hunting practices and population density as the exposed continental shelf expanded and contracted. He also emphasized the importance of sea level change in controlling migration (of humans and other fauna) to and from the continental interior via Cape Hangklip and Plettenberg Bay. Such routes would have avoided the Cape Fold Belt and the presumed arid Karoo, but are now on the submerged continental shelf (Fig. 2.2). Periods of potentially easier access to the southern Cape lowlands via these east and west routes occurred only when sea levels were at least 75 m below present. Similarly, Parkington (2010) noted the "pulsing of landscape availability" on the continental shelf in the face of proposed periods of aridity in the continental interior 160–125 ka and 80–60 ka.

Sea level highstands would have significantly reduced the extent of the coastal lowlands and directly impinged on some MSA occupations, with obvious implications for site formation, occupation, and preservation (Hendey and Volman 1986). In the middle to late Quaternary two significant highstands exceeded contemporary sea level reaching ~13 m and 6–8 m above mean sea level (amsl) (Roberts et al. 2012). Recent age constraints for the former suggest that it relates to MIS 11 (391 ± 16 ka; 370 ± 14 ka to 388 ± 14 ka) (Jacobs et al. 2011; Roberts et al. 2012). The Pinnacle Point Cave 13b excavation contains potential evidence for this highstand, with rounded boulders identified at the base of the western excavation (Karkanas and Goldberg 2010). These are overlain by a laminated facies, for which an average OSL age of 385 ± 17 ka has been reported (Jacobs 2010).

The 6–8 m highstand (Tankard 1976a; Marker 1987; Ramsay and Cooper 2002) holds interest for the archaeological community as it is close to the altitude of some coastal cave sites (Hendey and Volman 1986). Luminescence dating has recently confirmed that deposits of this altitude most likely relate to MIS 5e (127 ± 6 to 116 ± 9 ka) (Jacobs and Roberts 2009; Carr et al. 2010a; Roberts et al. 2012) (Fig. 2.5).

In terms of its impact on coastal archaeological sites, a 6–8 m MIS 5e highstand would not have flooded cave 13b at Pinnacle Point, nor Blombos Cave (Henshilwood et al. 2001; Marean 2010). At present, however, PP13b (15 m amsl) is the only major southern Cape coastal site unequivocally occupied during MIS 5e, although the exact timing in relation to the highstand is unclear (Jacobs 2010). The densest occupation of the site seemingly occurred *after* MIS 5e, between 100 and 90 ka (Jacobs 2010). Pinnacle Point Cave 9 is much closer to sea level (c. 8–12 m amsl), but is protected by more recent rock fall debris (Marean et al. 2004). The exposure of this cave to the elements prior to the rock fall may explain the more limited archaeology it contains (Marean et al. 2004). The most recent ages from Blombos Cave (34 m amsl) constrain the MSA occupation to between 101 ± 4 ka and 68 ± 4 ka (Henshilwood et al. 2011), which therefore postdates the MIS 5e highstand. The reliability of a previously published age for the M3 phase of 143 ± 6 ka (Jacobs et al. 2006) is now questioned (Henshilwood et al. 2011).

There are as yet no published MSA records unequivocally relating to MIS 5e at Die Kelders or Klasies River. Die Kelders lies relatively close to the shore and the MIS 5e highstand would likely have significantly impacted the site. The base of cave sequence is only ~2 m above modern sea level (Tankard 1976b; Hendey and Volman 1986), implying that it was uninhabitable until MIS 5d. Currently published OSL ages provide a broad estimate of 80–60 ka for the MSA occupation, which as might be expected, postdates MIS 5e (Feathers and Bush 2000). The habitability of Klasies River during and after MIS 5e has been debated (Hendey and Volman 1986; Deacon and Lancaster 1988). The basal LBS Member overlies beach deposits at c. 8 m amsl (Deacon and Lancaster 1988; Deacon and Geleijnse 1988) and the faunal assemblages in the LBS are reportedly not dissimilar to

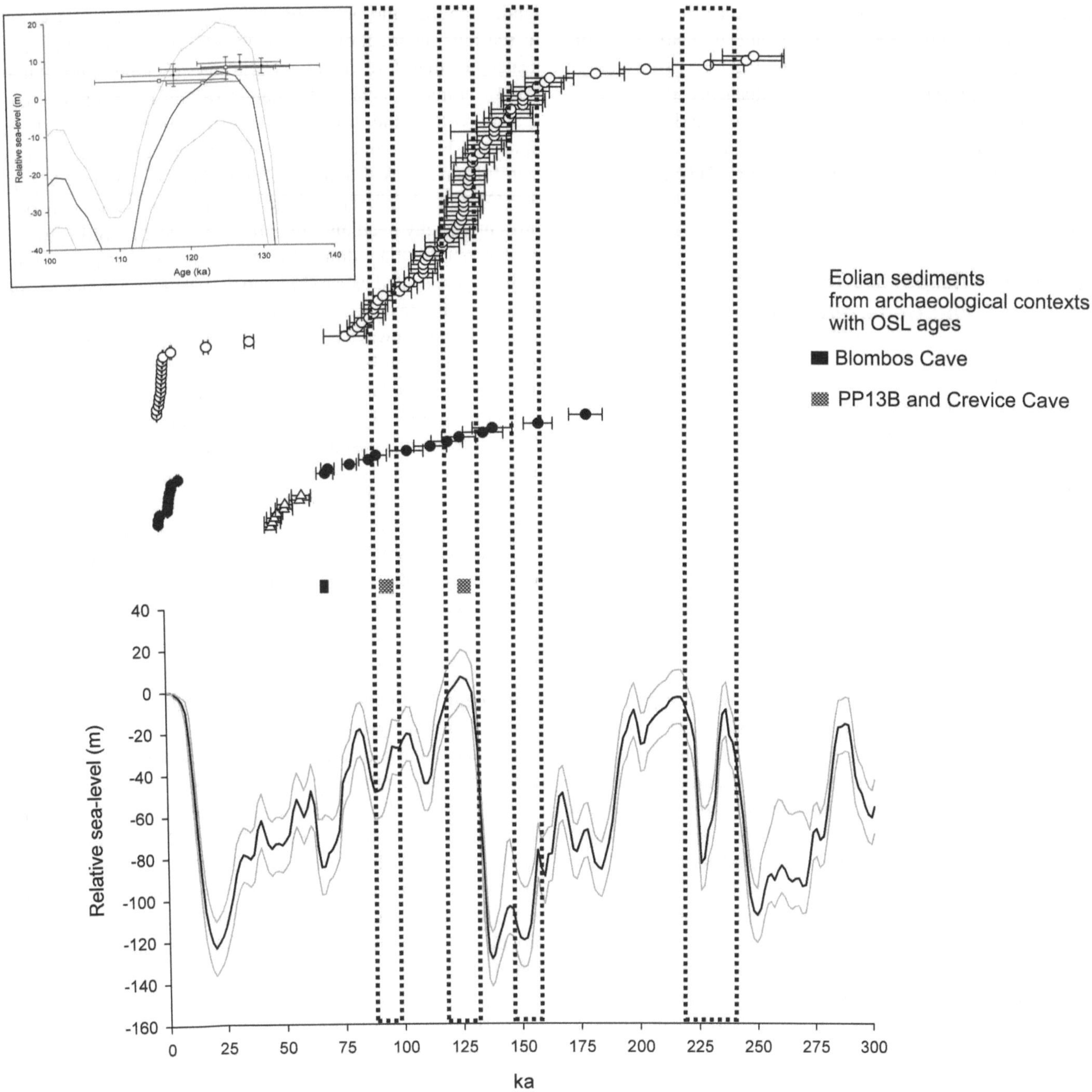

Fig. 2.5 Main Diagram: Distribution of optical luminescence ages (plotted in rank order) for eolian sediments along the southern Cape coastline, plotted relative to the eustatic sea level curve of Waelbroeck et al. (2002). *Open circles* are ages from the Wilderness Barrier dune systems (Bateman et al. 2011). *Filled circles* are OSL ages from the Agulhas Plain and Still Bay regions (Bateman et al. 2004, 2008; Carr et al. 2006a; Jacobs et al. 2003; Roberts et al. 2008). *Triangles* represent ages from pan-fringing lunette dunes on the Agulhas Plain (Carr et al. 2006a). The *dashed boxes* represent phases of eolian activity identified from statistical analysis of the Wilderness barrier dune OSL chronology (Bateman et al. 2011). Age ranges for eolian sediments within Blombos Cave (Jacobs et al. 2003), Pinnacle Point Cave 13b and Crevice Cave (Bar-Matthews et al. 2010; Jacobs 2010) are marked (*shaded boxes*). *Inset* The timing and magnitude of the southern Cape MIS 5e sea level highstand (Carr et al. 2010a; Roberts et al. 2012) relative to Waelbroeck et al.'s (2002) curve

Holocene ones (Deacon 1995) implying that the site was most likely occupied not long after MIS 5e – probably ~110 ka (Deacon 1995). More recent quartz OSL and feldspar IRSL ages of 110–115 ka (UW282) for the LBS in Cave 1 support this notion (Feathers 2002). The LBS is overlain by the Rock Fall Member, the SAS member and the Upper Member (which contains Howiesons Poort artifacts). Single grain OSL ages now suggest that these were deposited between c. 72 and c. 58 ka (Jacobs et al. 2008; Jacobs and Roberts 2008). Significant truncation of the Klasies River deposits was caused by the mid-Holocene sea level highstand (Deacon 1995), emphasizing the vulnerability of low altitude coastal sites to marine erosion. Mid-Holocene sea level probably peaked at 2–3 m amsl between 7500 and 6000 cal BP (Compton 2001).

Landscape Responses to Sea Level Change: Some Generalizations

While simple comparison between site altitude, global eustatic sea levels, and/or geological evidence for the magnitude of interglacial sea level highstands provides some basic context for the occupational records of specific sites, the response of wider sedimentary systems (and thus, the coastal *landscape*) to Quaternary sea level change is more complex, particularly within the embayed sections of the southern Cape coastline. Due to the region's tectonic stability, middle, and late Pleistocene sea level highstands reached similar points in the landscape, often reworking significant volumes of poorly lithified calcareous sediments within embayment fills (Roberts et al. 2008; Bateman et al. 2011).

Coastal responses to sea level change will comprise both the large-scale lateral translation of the shoreline across the continental shelf, as well as more subtle secondary effects, which confound simple interpretations of paleocoastal conditions and form. The lateral (onshore-offshore) response is a function of various factors; viz., sediment supply, accommodation space and continental shelf gradient (e.g., Storms et al. 2002; Cattaneo and Steel 2003). In addition, alongshore sediment fluxes, induced by waves approaching the shore obliquely, will respond to both changing wave energy and incident wave angle (Ashton et al. 2001). At large temporal and spatial scales the former will alter in response to wave attenuation (controlled by offshore bathymetry), while the angle of wave approach will respond to regional swell-wave conditions, as well as local wind systems and/or adjustment in the coastal plan-form itself (i.e., feedback response; Ashton et al. 2001). All are mediated by inherited geological characteristics (Roy et al. 1994).

The response of coastal locations at spatial/temporal scales relevant to the occupation of specific archaeological localities (scales of km and temporal scales of 10^2–10^3 years) is hard to assess. At this scale, noteworthy alterations of coastal environments need not occur in response to major environmental perturbations (e.g., Cooper et al. 2007). Variability in coastal environments reflects a subtle interplay of sediment supply, wave energy, and alongshore wave energy gradients, tidal currents, storm activity, and geological constraints. Understanding and predicting coastal sedimentary responses to sea level change, even over decadal and centennial timescales, is therefore a huge challenge (Cooper and Pilkey 2004). Hints of local-scale change (albeit over longer timescales) within the southern Cape geomorphic record are apparent. For example, at Cape Agulhas there was seemingly a shift from a rocky shore to a sandy shore setting *within* MIS 5e (Carr et al. 2010a). While at Pinnacle Point, which is presently rocky headland, the archaeological record contains evidence of eolian dune formation during MIS 5, which in-filled and sealed caves (Marean 2010). These dunes would undoubtedly have required a sandy beach as a sediment source; implying wave energy was less focused on this section of coast, allowing beach, and dune formation. Such changes imply that in headland locations accommodation space and/or sediment supply all respond to major sea level perturbations. Thus, models of coastal landscape based on modern bathymetry provide a useful means by which to consider the position of the coastline and the likely access to resources (i.e., site to shore distances) (Fisher et al. 2010), but the complexity of coastal sedimentary systems and their potential to respond to relatively subtle changes in climate, sea level, and sediment supply should be remembered, as should the uncertainties associated with eustatic sea level estimates derived from locations distant from southern Africa.

Landscape Responses to Sea Level Change: Coastal Eolian Systems

Our understanding of the timing and mechanisms of coastal dune formation on the southern Cape has grown considerably (Vogel et al. 1999; Shaw et al. 2001; Bateman et al. 2004, 2008, 2011; Carr et al. 2007, 2010a; Roberts et al. 2008, 2009). Some of this work was specifically motivated by reports of eolian sediments within cave sequences and shell middens within dune fields, but ultimately, it speaks more broadly to wider questions of coastal landscape adjustment and evolution.

A synthesis of some 104 coastal dune OSL ages from the southern Cape illustrates the drivers of eolian sediment accretion and preservation over glacial-interglacial timescales in what are presently subaerial environments (Fig. 2.5). The record suggests that dune formation was strongly mediated by eustatic sea level change, with dune activity broadly associated with periods of relatively high sea level (i.e., MIS 1, MIS 5e, and MIS 7). This is consistent with mechanisms of contemporary dune formation, whereby parabolic dunes do not migrate far from their primary sediment source, the modern shoreline. During sea level regressions eolian activity tracked the receding coastline across the Agulhas Bank, evidence for which is preserved in numerous coast-parallel dune ridges identified at depths of −40, −50, −65 to −70, and −80 to −90 m (Birch et al. 1978; Flemming et al. 1983; Martin and Flemming 1986) and at −33, −42, −77, −93, −97, −103, −108 and −115 m in the Wilderness embayment itself (Cawthra et al. 2014). The orientation of modern parabolic dunes and analysis of bedding in Quaternary dunes is consistent with south-westerly to north-westerly formative winds, implying an association with winter cyclonic systems (Flemming et al. 1983; Carr et al. 2006a; Roberts et al. 2008) and more specifically, that wind strength and sediment supply, rather than seasonal aridity, were/are the key factors mobilizing coastal dunes along the southern Cape.

In evaluating the response of coastal eolian systems over finer timescales, such as MIS 5e through to MIS 5a (Bateman et al. 2004: Fig. 9) identified separate phases of coastal eolian activity associated with MIS 5e, MIS 5c, and MIS 5a. The current synthesis (Fig. 2.5), comprising far more OSL ages from a longer stretch of coastline, suggests that such a separation is less clear at regional scales. Analysis of the OSL age distribution for the Wilderness Embayment barrier dunes reveals clusters of activity centered at c. 87–92 ka, 120–130 ka, 143–159 ka, and 221–241 ka (Bateman et al. 2011). Perhaps more important for the interpretation of single archaeological sites, the same study demonstrated that stratigraphic records at single sections of coastline (i.e., spatial scales of a few km) can differ significantly. This could be explained by recourse to the local bathymetry, which modulates the dune accumulation history of a particular locale by determining the duration over which the site was close to its beach sediment source (Bateman et al. 2011). The southern Cape geomorphic record thus demonstrates both large-scale/long-term drivers of coastal geomorphic change (glacial-interglacial sea level change), as well as local-scale variations in coastal response, driven by sediment supply and inherited geological constraints (e.g., bathymetry). These impart site-specific variation in the preserved coastal stratigraphic record.

The presence of eolian sediments within coastal rock shelters is abundantly clear at sites, such as Blombos Cave and Pinnacle Point. So far, published OSL ages for Pinnacle Point sites show good correspondence with the wider eolian geomorphic record (Fig. 2.5). For example, the LC-MSA (Upper) at Pinnacle Point 13b preserves evidence for a large dune that sealed the cave at 93 ± 4 ka (Jacobs 2010). Similarly, the Crevice Cave site indicates dune formation at 90 ± 2 ka (Bar-Matthews et al. 2010). These are both associated with the cluster of dune ages in the regional eolian record associated with MIS 5b (87–92 ka) (Bateman et al. 2011) (Fig. 2.5). At Blombos, new OSL ages, as well as previous TL ages and U-series dating constrain the M3 phase to 97.0 ± 2.7 ka (MIS 5c), the M2 phase to MIS 5a (weighted mean 82 ± 2 ka [MIS 5a]) and the M1 phase to 73 ± 3 ka (MIS 4/3; Henshilwood et al. 2011). Subsequent to the M1 Phase at 69 ± 4 ka, the cave was sealed by a coastal dune (BBC Hiatus) (Henshilwood et al. 2001; Jacobs et al. 2003; Henshilwood 2005), commensurate with the MIS 4 sea level regression. What happened at Blombos during and around MIS 5e is unclear at present, but to the east of Still Bay, a major phase of (eolian) barrier dune construction occurred between 140 ka and 121 ka, followed by later phases of barrier accretion at 114 and 90 ka (Roberts et al. 2008). The ages of the dune sands preserved in the cave (BBC Hiatus 69 ± 4 ka) and the eolianite remnants surrounding the cave (70 ± 4 ka and 71 ± 3 ka; Jacobs et al. 2003) are thus relatively young compared to both the eolianite east of Still Bay, and the synthesis from the Wilderness Embayment (Bateman et al. 2011), both of which imply that eolian activity was much reduced after c. 90 ka. Thus, although the occupations of Blombos and Pinnacle Point were seemingly influenced by coastal dune formation, there is apparent variation in the extent to which these phenomena reflect regional scale climatic/sea level changes (e.g., PP13b), as opposed to local geomorphic, geological, or preservational factors.

Sea Level Lowstands: Paleolandscapes and Paleohydrology

The bathymetry of the shelf is such that the area of landscape exposed off the southern Cape during MIS 5-3 was substantially less than that exposed relatively briefly during MIS 2 (Van Andel 1989) (Fig. 2.2). For instance, at Pinnacle Point the shoreline between MIS 5e and MIS 3 was between 0 and 37 km south of the modern coast, extending to more than 90 km at the Last Glacial Maximum (Fisher et al. 2010). This reflects the relatively steep shelf gradients close to the shore and a much flatter profile at depths below ~ −80 m (Dingle and Rogers 1972; Fisher et al. 2010).

The contemporary coastal platform, which lies between the Cape Fold Belt and the modern shore, is therefore relatively narrow compared to most of the Pleistocene. Today this lowland landscape represents an "unusual" topography and ecology (compared to the sandstone and fynbos vegetation of the Cape Fold Mountains), but it would have comprised a more substantial component of the landscape during the last glacial cycle. Dingle and Rogers (1972) note a distinct contrast in geological substrates between the western and the central/eastern areas of the Agulhas Bank. The coastal margin between Cape Town and Cape Agulhas is substantially rockier (pre-Mesozoic) to greater depths compared to areas east of Cape Agulhas. Recent sediment wedges are usually localized and the outer boundary of the rocky inner shelf extends to the 140 m isobath (Rogers 1985; Gentle 1987). Compton (2011) argued that this rocky terrain (especially between Cape Hangklip and Danger Point) would have been a major impediment for travel between the coastal lowlands and the interior of the Western Cape for all periods in which relative sea level was between 0 and 75 m below modern levels.

Given the strong edaphic controls on the region's vegetation communities, the ecology of the continental shelf during lowstands may have been strongly influenced by bedrock geology. The vegetation on rocky continental shelf regions around the Cape Peninsula and Cape Hangklip may not have been too dissimilar to modern fynbos vegetation. To the east, however, such quartzite exposures are limited more to headlands and few assumptions concerning substrates on the central and eastern Agulhas Bank can be made. It has been proposed that they may have included include finer grained, more nutrient-rich soil derived from Cretaceous clays (Dingle and Rogers 1972; Compton 2011). On the modern coastal platform today such finer grained soils are not associated with fynbos vegetation and potential modern analogues for such vegetation communities are more likely to lie (for example) in the Bokkeveld shale communities (e.g., the Agulhas Plain), which are associated with Renosterveld vegetation; Cowling et al. 1988) or Albany Thicket (Compton 2011). Renosterveld contains a grass component, which will contain greater or lesser proportions of C_3 and C_4 taxa depending on the annual distribution of rainfall (Mucina and Rutherford 2006), and its periodic occurrence in what are now offshore regions may offer some explanation for the fluctuating proportions of grazing and browsing fauna in the archaeological record (Deacon 1978; Klein 1976, 1983; Faith 2011a, b). At present such inferences are rather speculative. The complex sedimentary dynamics associated with sea level transgressions and regressions, this time across the continental shelf, will impart spatial variability in substrate composition, exposure, and preservation, irrespective of the temporal vegetation dynamics driven by climatic changes during the same periods. Notwithstanding, efforts are underway to model the "palaeoscapes" of the now submerged continental shelf (Marean et al. 2015).

Pleistocene Paleoenvironments and Paleoecology

Various attempts have been made to synthesize paleoenvironmental evidence from the southern Cape (Deacon and Lancaster 1988; Partridge 1990; Meadows and Baxter 1999; Partridge et al. 1999). Even more recent reviews are, however, significantly limited by a lack of reliable data (Chase 2010; Chase and Meadows 2007; Lewis 2008). There are, fortunately, a wide range of research initiatives underway, the findings of which (in contrast to many of the older records) will be independent of material within archaeological sequences. Here, we review the likely drivers of southern Cape climatic variability, which we consider in the context of recent terrestrial proxy records and marine proxies.

East of Still Bay, the southern Cape exists within what is largely a year-round rainfall zone (YRZ). This year-round precipitation is a function of moisture derived from temperate, tropical, and local storm systems (Fig. 2.3). Given these multiple controls, each of which influenced by different elements of the global oceanic-climate systems, it is likely that this is an ephemeral climatic regime. However, such diversity in potential moisture sources may have buffered the region from the more extreme climatic variability seen in the continental interior (e.g., Scholz et al. 2007).

Rainfall Regime Configurations During the Pleistocene

Colder conditions during the Pleistocene have often been closely linked with enhanced aridity throughout southern Africa, including the south coast (Partridge et al. 1999). A conceptual model of southern African climatic variability, in which there is an anti-phase relationship between the summer and winter rainfall systems of southern Africa, has been applied to interpretations of Pleistocene climatic variability (Tyson 1986; Cockcroft et al. 1987). From this it has been argued that when tropical systems intensify, the influence of the westerly storm systems declines, and vice versa (Tyson 1986; Cockcroft et al. 1987; Chase and Meadows 2007). These adjustments reflect latitudinal shifts in westerly storm tracks in response to changing hemispheric temperature gradients, Antarctic sea ice extent and the positioning of the subtropical front (STF), in addition to changes in the positioning and seasonal movements of the Atlantic and Indian Ocean subtropical high pressure systems.

Although – or perhaps because – these models were based on contemporary/historical climatic variability (Cockcroft et al. 1987) it has proved difficult to test such ideas in the proxy record. However, there are now a growing number of records derived from marine environments, which span the last 125 kyr or more and indicate significant, and often regionally unique, responses to global forcing mechanisms. Key among these records are: (1) marine records relating to position of the STF (Peeters et al. 2004; Bard and Rickaby 2009); (2) records indicative of changes in the character and flow of the Agulhas Current (Bard et al. 1997; Sonzogni et al. 1998; Peeters et al. 2004; Caley et al. 2011); and (3) records indicative of changes in the west coast Benguela upwelling system (Little et al. 1997a, b; Stuut et al. 2002; Pichevin et al. 2005). While these marine records do not necessarily directly reflect changes in terrestrial systems, they do indicate, at least in a general sense, variations in the underlying climate systems. Thus, they can be used to explore hypotheses of causation. For instance, Stuut et al. (2002) hypothesized that increased humidity on the western margins of Namibia during periods of relative global cold was the result of increased winter rainfall. This hypothesis finds support in the significant correlation between their record

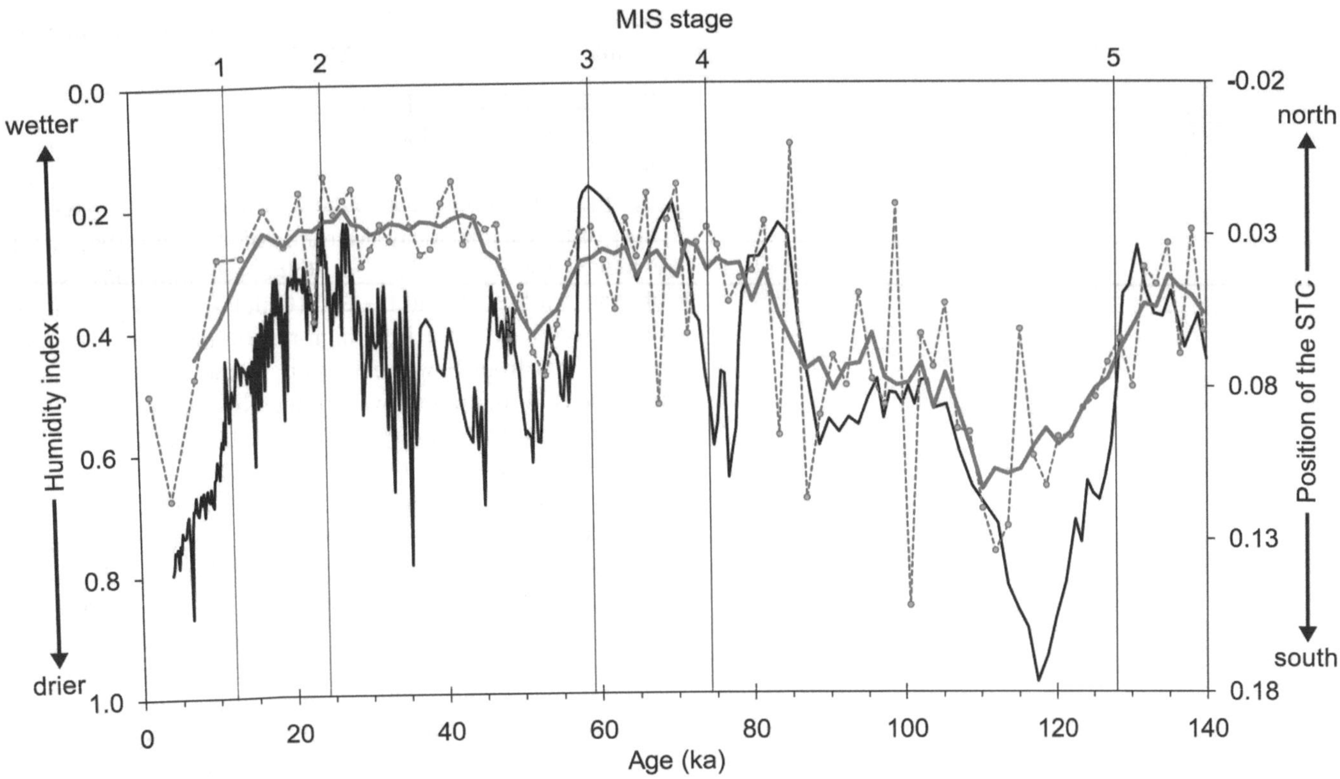

Fig. 2.6 Stuut et al.'s (2002) record from MD962094, which is interpreted as indicative of the significance of winter rainfall along the Namibian coastline (*Black solid line*). This is compared to Peeters et al.'s (2004) record (*dotted line*) indicative of the position of the STF (derived from the ratio: *Globorotalia truncatulinoides/(Neogloboquadrina pachyderma* (dex.) + *Globorotalia inflata* + *G. truncatulinoides*). The light *gray line* is the 7 point moving average for this dataset

and migrations of the STF (Peeters et al. 2004) (Fig. 2.6). Concerning tropical moisture systems, a pollen record reflecting conditions in the Limpopo Basin (Dupont et al. 2011) shows similarities with SSTs in the Mozambique Channel over the last 300 ka, as do several sites in northeast South Africa during the last 45,000 years (Chevalier and Chase 2015; Truc et al. 2013), implying a link between SSTs and continental humidity in the proximal summer rainfall zone. This link is seemingly less strong, however, in central South Africa, where variation in westerly systems also contributed to overall rainfall variability (Chevalier and Chase 2015).

Paleoclimatic Insights from MIS 2 and MIS 1

As previously reported (Chase and Meadows 2007), the MSA predates the majority of the region's terrestrial paleoenvironmental evidence (particularly evidence independent of archaeological/anthropic deposits). We can, however, consider evidence for terrestrial climate change during the last 25 kyr to illustrate potential paleoclimatic scenarios, and the responses of the driving systems. The late Pleistocene-Holocene transition (18–11 ka) and the mid- to late Holocene demonstrate the potential diversity of climatic configurations and the complexity of interpreting such changes based on proxy data (Figs. 2.3, 2.7).

The Late Pleistocene-Holocene Transition

As argued by Chase (2010) for MIS 4 it is possible to envisage scenarios within which both temperate and tropical moisture sources intensify at the same time (cf. Cockcroft et al. 1987). On the southern Cape the terminal Pleistocene, between 17.0 and 14.0 cal kBP, is potentially an example of this. Evidence from Boomplaas Cave (Fig. 2.1) indicates that this period witnessed the highest effective precipitation of the last c. 80 kyr (Scholtz 1986). In contrast, much of this period (including "Heinrich stadial 1" (HS-1)) was notably drier throughout parts of the Afro-Asian Monsoon region (cf. Stager et al. 2011, and reference therein). The early onset of warming in the Southern Hemisphere (Blunier et al. 1998; Pedro et al. 2011) becomes apparent in southwest Indian Ocean SSTs at ~17 ka (Sonzogni et al. 1998; Dupont et al. 2011). These warmer SSTs would have invigorated summer rainfall systems during the early parts of the glacial-interglacial transition, and evidence for this is seen in the increase of forest taxa in the Limpopo Basin (Dupont et al.

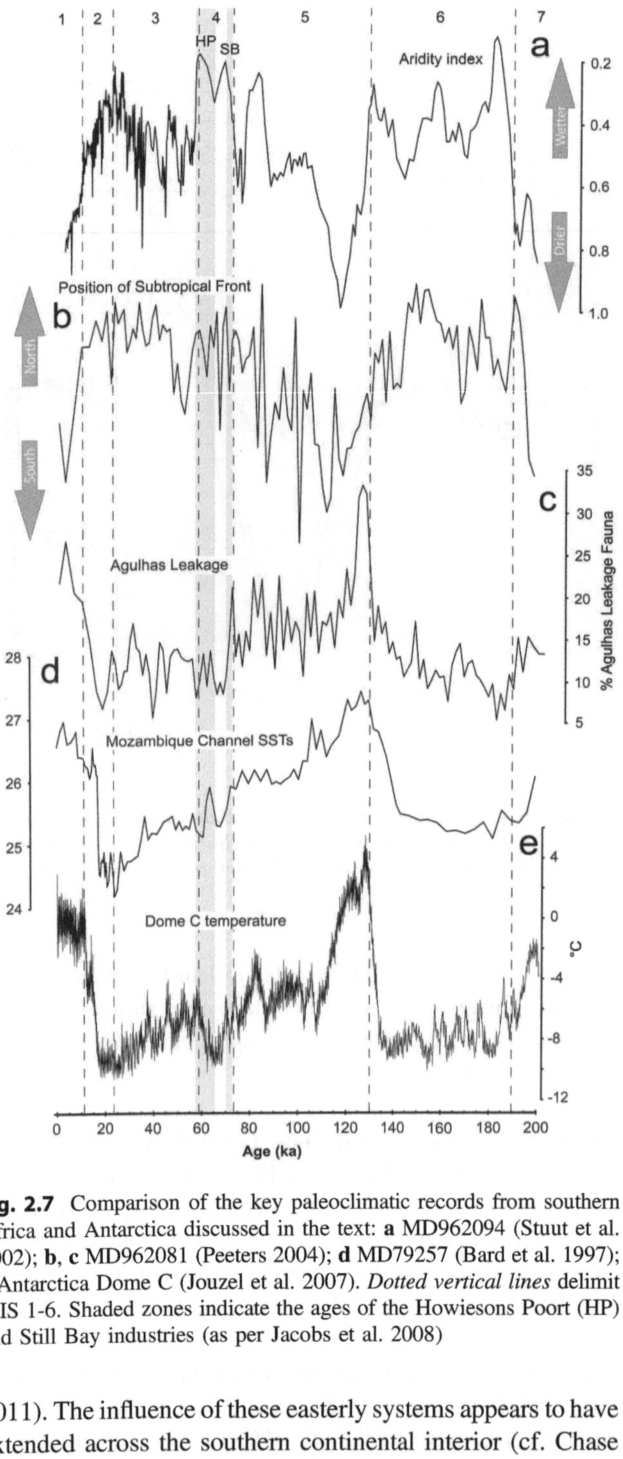

Fig. 2.7 Comparison of the key paleoclimatic records from southern Africa and Antarctica discussed in the text: **a** MD962094 (Stuut et al. 2002); **b**, **c** MD962081 (Peeters 2004); **d** MD79257 (Bard et al. 1997); **e** Antarctica Dome C (Jouzel et al. 2007). *Dotted vertical lines* delimit MIS 1-6. Shaded zones indicate the ages of the Howiesons Poort (HP) and Still Bay industries (as per Jacobs et al. 2008)

2011). The influence of these easterly systems appears to have extended across the southern continental interior (cf. Chase et al. 2015b), and more humid conditions in the Cederberg Mountains of the southwestern Cape during HS-1 have been attributed to increases in summer rainfall (Chase et al. 2015a). Furthermore, it would seem that there was no significant poleward shift in the STF (Peeters et al. 2004) until the beginning of the Holocene ($\sim$11 ka) indicating that little or no decline in winter rainfall had occurred. Thus, particularly humid conditions on the southern Cape from 17 to 14 ka can perhaps be explained as the combined influence of winter and summer rainfall (Chase and Meadows 2007).

The Mid- to Late Holocene

During the mid-Holocene (c. 7300–4500 cal BP) the drought-sensitive afromontane forest, which currently occupies the heart of the YRZ in the Knysna area (Fig. 2.4), was more restricted than today (Martin 1968; Scholtz 1986), while evidence for more arid mid Holocene conditions is also observed at Still Bay to the east (Quick et al. in press a). In detail, recent evidence from high-resolution rock hyrax midden records from Seweweekspoort in the Groot Swartberg mountains (170 km northwest of the Knysna area) suggest a distinct period of aridity 7–5 ka, which is coeval with anomalies in Antarctic sea ice extent. This is interpreted as good evidence for the role of the westerly systems (Chase et al. 2013) in driving variability in southern Cape climatic conditions. A subsequent period of drier climate and reduced forest cover also seems to have occurred between 2700 and 1300 cal BP (Scholtz 1986; Carr et al. 2006a, b; Quick et al. in press a). In this case, the underlying mechanisms are less straightforward, but a similar pattern of increased aridity at Cold Air Cave (Lee-Thorp et al. 2001) suggests that this could also reflect a reduction in summer rainfall (increasing southern Cape rainfall seasonality), perhaps due to lower Agulhas Current sea surface temperatures (Sonzogni et al. 1998), which would have promoted drier conditions on the south coast (Chase and Meadows 2007; Quick et al. in press a, b).

Paleoclimates During the MSA: MIS 5b-3 (95–60 Ka)

The waxing and waning influence of the major moisture-bearing systems was specifically considered by Chase (2010) in a review of MSA climates during MIS 4 and the Howiesons Poort (HP) and Still Bay (SB) industries ($\sim$74–58 ka [Jacobs et al. 2008; Jacobs and Roberts 2008; Bar-Matthews et al. 2010; McCall and Thomas 2012]). This review of proxy data from multiple southern African archaeological sites (e.g., Klein 1976, 1983; Tankard 1976b; Butzer et al. 1978; Avery 1982; Butzer 1984; Deacon et al. 1984; Klein and Cruz-Uribe 2000) concluded that during MIS 4 overall conditions were relatively cool and moist compared to the present (Chase 2010). Such findings imply that southern African climate systems did not follow apparent global trends during this period and it was suggested, based on records of variation in the westerly systems (Stuut et al. 2002) the STF (Peeters et al. 2004; Bard and Rickaby 2009), the Agulhas Current (van Campo et al. 1990; Peeters et al. 2004) and the influence of orbital obliquity on hemispheric temperature gradients, that MIS 4 was relatively humid as a result of increased contributions from both winter and summer rainfall systems (Chase 2010). New SST records from the southwest Indian Ocean (Caley et al. 2011) modify this hypothesis to some extent, but the basic premise still pertains, with periods of relatively elevated Indian Ocean SSTs combining with a more northerly position of the STF to enhance both summer rainfall and westerly derived rainfall, generating relatively wetter conditions across southern Africa. While more data will be required to verify this hypothesis, this scenario is comparable with that outlined for the late glacial and early Holocene. Recent findings from Sibudu Cave in KwaZulu-Natal (SRZ) imply summer rainfall comparable to the present during the HP, with the subsequent post HP period being relatively drier (Bruch et al. 2012).

More recently, there has been an attempt to directly consider high-resolution paleoenvironmental data for the southern Cape, although the link between the paleoenvironmental record and underlying processes is difficult to decipher. The Crevice Cave speleothem record from the Pinnacle Point excavations (Bar-Matthews et al. 2010) provides a high-resolution record of stable carbon and oxygen isotope variations for the period 90–53 ka (MIS 5b-MIS 3). Fluctuations in $\delta^{13}C$ within the record are interpreted as reflecting the relative abundance of C_3/C_4 grasses. In turn, this is interpreted as indicative of the degree of winter rainfall, which, gives the strong linkage between C_3 vegetation and the winter rainfall zone (i.e., growing season temperature), is how most records of this nature have also been interpreted (e.g., Lee-Thorp and Beaumont 1995; Scott and Vogel 2000). The $\delta^{13}C$ record shows some marked shifts, with periods of increased (open?) C_4-dominant vegetation inferred for 75–70 ka and 65–60 ka. Additionally, a period of rapid variation in the $\delta^{13}C$ signal at 65–70 ka is interpreted as a period of marked climatic/ecological instability between the SB and HP Industries. The mechanisms behind this are unclear.

Bar-Matthews et al.'s (2010) interpreted changes in vegetation type are not inconsistent with some long-standing ideas that suggest grassier environments probably existed on the continental shelf during periods of relative cold (and low sea level), explaining the increased prominence of grazing fauna in many MSA assemblages (e.g., Klein 1972). Although Rector and Read (2010) caution that it should not be assumed that such coastal plain grasses were C_4, proposing that complex mosaics of C_3 grasses and Fynbos, would also have been able to support grazing communities.

Despite a detailed analysis, the Crevice Cave record defies easy interpretation in some respects. The $\delta^{18}O$ record (interpreted as reflecting the seasonality of rainfall) shows little correlation with regional records of winter rainfall intensity, the position of the STF, or with data indicative of Agulhas Current flow and temperature (Stuut et al. 2002; Peeters et al. 2004; Caley et al. 2011). Counter to previous models (van Zinderen Bakker 1967, 1976; Cockcroft et al. 1987; Tyson 1999a, b; Chase and Meadows 2007) periods of *cooling* (through correlation with remote records from the EPICA ice core and an SST record from the Chatham Rise, New Zealand) are associated with *increases* in *summer* rainfall (lower $\delta^{18}O$) and the expansion of C_4 vegetation (higher $\delta^{13}C$). By way of perspective, it is important to note that the nearby (85 km north) Cango Cave speleothem record shows markedly different trends (de Wit et al. 2009). In the Cango record, variation in $\delta^{13}C$ shows strong correlations with changes in the strength of Agulhas flow along the south coast (Peeters et al. 2004), most notably including relatively enriched $\delta^{13}C$ values (more C_4) prior to MIS 4, and then a shift to more depleted values (more C_3) at ~70 ka implying increased winter rainfall during MIS 4. Similarly, the Holocene portion of the Cango Cave record (6000 cal BP to present) contrasts with the MIS 2 section of the record, with the former period exhibiting markedly higher $\delta^{13}C$ than the latter (Talma and Vogel 1992). Here, an interpretation concerning "interglacial" verses "glacial" vegetation, based on $\delta^{13}C$, would imply that the "glacial"/ "cooler" conditions were associated with *more* C_3 vegetation – the opposite of what is seen in the Crevice Cave scenario.

The discrepancies between the Cango and Crevice Cave records may relate to the fact that while the Cango Cave speleothem was recovered from a deep cave complex the Crevice Cave speleothem was recovered from a wave-cut crevice, which began to form after the hollow was sealed by coastal dunes c. 90 ka. The context for this speleothem record is therefore quite unusual. At present it is not entirely clear to what depth the cave was buried, and to what extent it was ventilated during formation. The latter aspect can have a substantial impact on isotopic equilibrium due to de-gassing effects in areas (or periods) of greater ventilation, creating variability unrelated to the inferred climatic parameters (Talma and Vogel 1992; Mickler et al. 2004; Tremaine et al. 2011). Some data (from two laminae) are presented concerning this nonequilibrium precipitation issue ("Hendy tests"), but these issues warrant further investigation. In the case of both the Cango and Crevice cave records, neither considers (or is easily able to consider) the influence of CAM plants, which are common at both sites. These may display a range of $\delta^{13}C$ values (Rundel et al. 1999), and may influence $\delta^{13}C$ signals in some paleoenvironmental archives in this region (Carr et al. 2010b).

Southern Cape Paleoecology

Aside from the aforementioned Crevice Cave record, paleoecological data for the MSA are largely restricted to faunal remains recovered from archaeological sites. Broadly speaking, there appears to be a significant correlation between glacial periods and increased numbers and diversity of grazing animals (Klein 1972, 1976, 1978, 1983; Klein and Cruz-Uribe 2000; Faith 2011a, b; Rector and Reed 2010). Based on these data, the inference has long been that glacial periods supported more open, grassier environments. Recent faunal evidence from PP13b, notably the Upper Roof Spall layer dating to 98–91 ka, has been interpreted as indicative of more open, mosaic habitats (Rector and Reed 2010), while data for the periods 134–94 ka and 102–91 ka were also thought to be suggestive of relatively open conditions, as well as moist ("vlei") conditions (Rector and Reed 2010). Nelson Bay Cave on the Robberg Peninsula is one of the few coastal archaeological sites with a faunal record crossing the Pleistocene-Holocene transition and a clear switch from dominantly grazing to dominantly browsing fauna seems to have occurred during the period 12,000–9,000 ^{14}C BP (Deacon 1978).

While the relationship between grazers and open, grassier environments is clear, it does not follow that this was the result of drier climates. Presently, the southern Cape coastal plain hosts a complex variety of vegetation types, and the shrubby renosterveld vegetation that would dominate parts of the coastal plain (were it not for modern land use practices) is on the drier end (250–550 mm a^{-1}) of the climatic continuum. It is only with increased humidity (500–750 mm a^{-1}) that grasses become a more important component of the vegetation (Cowling 1983).

Unfortunately, aside from the aforementioned speleothem records, there are very few data available that can assist in the interpretation of fluctuations within these faunal assemblages. Botanical remains are often poorly preserved in archaeological contexts, and there are few suitable and adequately studied wetlands in the region. At present only three lake sediment records extend beyond the Holocene (Fig. 2.1): (1) Voëlvlei and Soetendalsvlei from the Agulhas Plain (Carr et al. 2006b); (2) Rietvlei near to Still Bay (Carr et al. 2010c; Quick et al. in press a); and (3) Vankervelsvlei near to Knysna (Irving and Meadows 1997; Irving 1998; Quick et al. in press b). Voëlvlei and Soetendalsvlei have relatively coarse chronological control. However, Rietvlei and Vankervelsvlei are the subject of recent studies and provide detailed multi-proxy records spanning the last ~35 kyr (Quick et al. in press a) and 140 kyr (Quick et al. in press b), respectively.

At the Voëlvlei site modern vegetation is heavily modified by human activity, but the natural vegetation was probably renosterveld. The pollen records, derived from pan sediments, have limited chronological control. However, in conjunction with the surrounding geomorphic evidence they suggest a period of relative humidity, probably within MIS 3 (ages span >48,000–33,000 cal BP). The pollen spectra are rich in both fynbos pollen and characteristic renosterveld pollen. Grass pollen is also present, but not markedly abundant (Carr et al. 2006b). A core from the margins of nearby Soetendalsvlei dating to 14,400–13,300 cal BP produced pollen spectra similar to the modern limestone fynbos around the site, implying a comparable situation to the present (Carr et al. 2006b). Thus, evidence for significant reorganizations of the Pleistocene Agulhas Plain vegetation communities is rather equivocal, although coastal Fynbos was clearly present at Cape Agulhas from 14,000 cal BP. Further to the east, in the year-round rainfall zone, the Vankervelsvlei record spans ~140 kyr. The site is located within the drought-sensitive afromontane forest of the Knysna area (see section "Landscape Responses to Sea Level Change: Coastal Eolian Systems"), and like the Holocene records discussed above (e.g., Scholtz 1986) shows clear fluctuations in the extent of afromontane forest. MIS 2 is associated with increases in the relative significance of fynbos pollen, perhaps implying a decrease in humidity and/or increased rainfall seasonality (Irving 1998). Recent work has extended this record back to 140 kyr using luminescence dating (Quick et al. in press b). This study reveals distinctly warmer temperatures during MIS 5d compared to later MIS 5, MIS 4, and MIS 3. Evidence for increased summer rainfall during MIS 5d is also identified, but importantly although there is some evidence for increased rainfall seasonality from ~96 kyr onwards, significant reductions in overall humidity did not seemingly occur during MIS 4 and MIS 3, perhaps implying that reductions in (summer) rainfall were offset by lower evapotranspiration due to cooler temperatures. East of Still Bay, the Rietvlei wetland dates back to at least 35,000 cal BP (Carr et al. 2010c; Quick et al. in press a). Additional evidence suggests a persistent wetland of some form was present as early as MIS 5e (Roberts et al. 2008). Nearby MIS 5e eolianites contain trace fossils revealing a diversity of mammal fauna in the immediate area, suggesting that animals congregated here. The site is perhaps analogous to the coastal "vlei" environments inferred by Rector and Reed (2010). The Rietvlei record itself shows evidence for relatively humid conditions during MIS 3, but also distinct evidence for arid phases within this period. A clear contrast between relatively humid early Holocene and more arid mid Holocene conditions is also apparent.

Fundamentally, the nature of the vegetation on the continental shelf remains a critical unresolved element for interpretations of both the human and faunal records in this region (Rector and Reed 2010). Aside from the poorly defined nature and distribution of continental shelf substrates, the unknown hydrology of continental margin is a complicating factor. It is of specific relevance to models of coastal-zone habitability (Parkington 2003). The "coastal oasis" model argues that during sea level regressions steeper peri-coastal water table gradients increased the hydraulic head on continental aquifers, promoting spring activity and primary productivity on the continental margins (Faure et al. 2002). This potentially increased availability of water and biomass would have rendered the continental shelves more attractive environments for both grazing fauna and human occupation (e.g., Compton 2011). At present there is a little specific evidence to support this hypothesis on the southern Cape, although there is some limited geomorphic evidence for an adjustment in the southern Cape coastal hydrology in the Agulhas Plain salt pans (Carr et al. 2006a). These became inactive after c. 45 ka, as they ceased to intersect the water table, implying a response to sea level change consistent with that envisaged by Faure et al. (2002).

Summary: Integrating Human Occupation and Subsistence with Paleolandscape and Paleoclimates

A full understanding of the human story during the MSA requires an integration of the archaeological record with data pertaining to the landscapes and environments at that time. Linking specific aspects of human development such as stone tool industries or technocomplexes to environmental change is fundamentally difficult and contentious. For example, McCall (2007) argues for a direct correlation between the HP and colder temperatures during late MIS 4 (see also Ambrose and Lorenz 1990). Jacobs et al. (2008) argue that there can be no such association given that global temperatures at this time exhibit a *warming* trend. Hiscock et al. (2011) argue that the HP is a response to this warming. All of these arguments use global temperature data to model environmental change, and none is able to make significant use of local paleoenvironmental records that directly reflect subsistence conditions. As paleoenvironmental and paleogeographic evidence accrues, however, we can begin to offer a guide to some of the (potentially) most relevant facets of the environment for the region. We attempt to summarize these linkages in Fig. 2.8. Many elements of this diagram might be considered as generic, but based on the preceding we can highlight how they are uniquely manifested on the southern Cape.

Landscape Factors

Landscape factors are a function of geology and relative sea level change. Sea level change, mediated by local-scale geological control, has operated as an overarching driver of

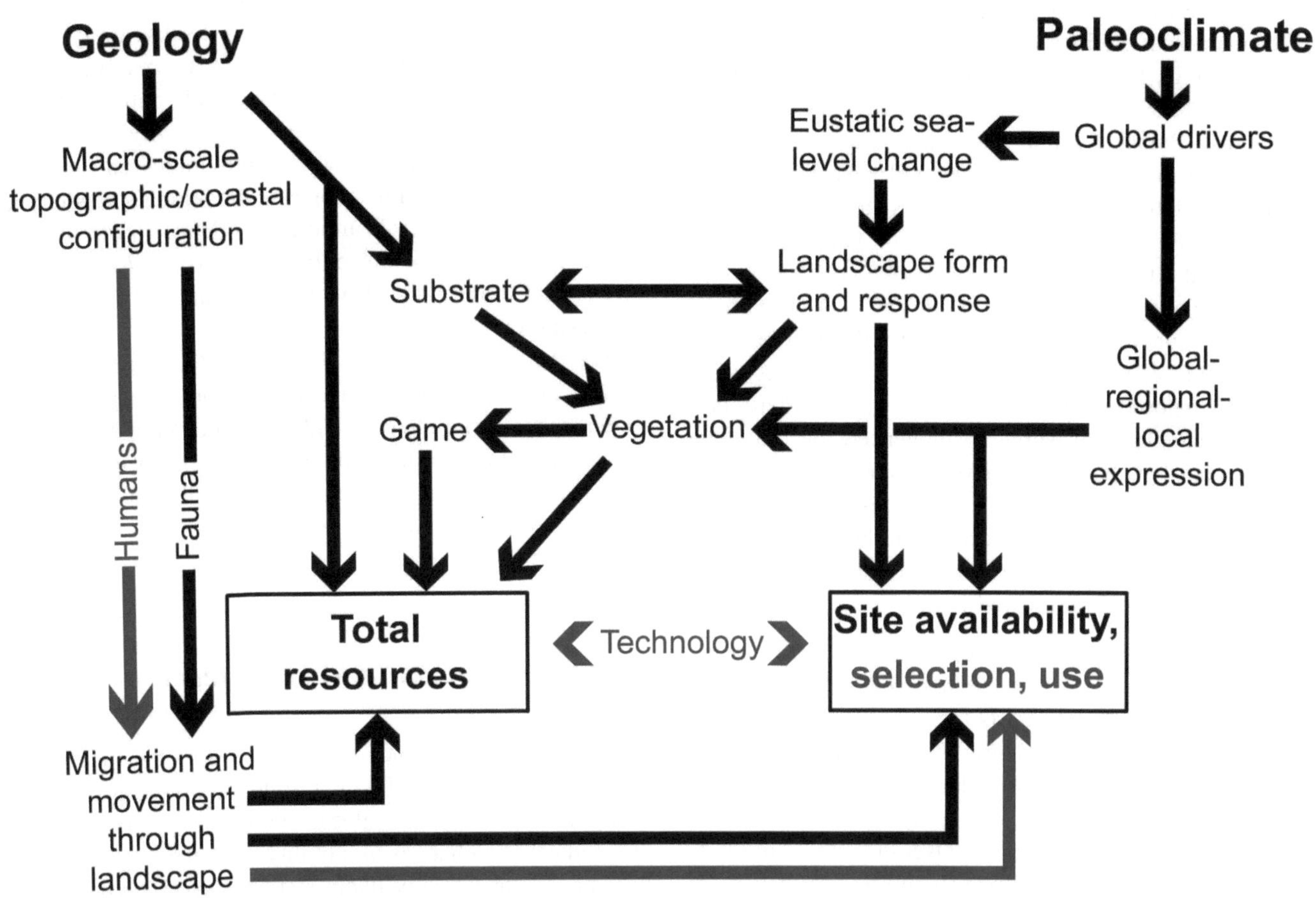

Fig. 2.8 A schematic representation of the interactions between landscape, palaeoclimate, resource availability (including fauna), and human behavior. The primary purpose is to summarize potential interconnections identified in this review, not to provide a prescriptive framework. The diagram highlights the role of (inherited) characteristics (e.g., geology) in mediating local responses to global-scale forcers of landscape and climate change

landscape dynamism and resource availability throughout the Pleistocene. It is a global-scale pacemaker (eustatic sea level change) mediated by (inherited) local geological control (e.g., Compton 2011). Landscape factors are somewhat interrelated, but are significant in terms of:

1. *Geological constraints act on landscape structure and topography.* They control coastal landscape configuration both locally (bays and headlands, sandy/rocky shores) and regionally (continental shelf topography). As a result they influence site to shore distance, marine resource availability, and site selection. Through the topography of the continental shelf they also determine local sensitivity to eustatic sea level changes. Regional geological structures might have presented impediments to migration between the interior and margins of South Africa. Periodic isolation of communities and fauna is a possible outcome (Compton 2011).
2. *Sea level constraints on coastal landscape, landscape dynamism and site occupation.* Eustatic sea level is a global signal but it is moderated at both local and regional scales by geological constraints. In coastal locations, sea level influences site access, habitability, and preference (i.e., occupational hiatuses), both directly through flooding/inaccessibility during sea level highstands (e.g., De Kelders during MIS 5e), and indirectly via local and regional-scale pulses of coastal eolian activity (Pinnacle Point, Blombos). The relative availability of marine food resources will have been directly influenced by most of these factors, as will the availability of fresh water (e.g., Avery 1974; Faure et al. 2002; Carr et al. 2006a).
3. *Soils, vegetation, and game resources.* Soil properties – largely a function of geology – control the distribution of the major vegetation types in this region. The vegetation on the exposed continental shelf during the Pleistocene was probably also strongly influenced by this phenomenon. It is potentially a key driver of habitat extent and heterogeneity, and thus, game resources (e.g., the extent of grazing). Habitat extent and heterogeneity will have been affected by relative sea level change. It remains a critical unresolved issue.

In considering these landscape factors the present evidence implies that caution is required when applying regional-scale trends at local (site) scales. Uncertainties in sediment supply/accommodation space and the absolute magnitude of relative sea level change remain (note that in southern Africa we are largely applying eustatic records derived from distant locations). Overinterpretation in the absence of stratigraphic evidence should be avoided.

Paleoclimatic and Paleoecological Factors

Paleoclimatic and paleoecological factors are not independent of our basic geological and sea level framework. They however are fundamentally related to *terrestrial and shore-margin ecosystem productivity.* In terrestrial environments, the availability, density, and type of water and food resources available are directly relevant to issues of population density, settlement organization, and technological change (Mackay 2009). They can be summarized as follows:

1. *The diversity of climatic drivers*: The southern Cape climate is a function of several components of the global climate circulation. This results in a diversity of moisture-bearing systems, which combine to create a variable, but resilient resource base. While the amount and seasonality of surface-available fresh water varied through the late Pleistocene, it is unlikely that the region was ever truly arid or "harsh".
2. *The complexity of vegetation response*: The available evidence suggests that the delivery of moisture to this region did vary, but the specific impacts on the region's vegetation communities, mediated by substrate type and availability (exposure), are difficult to resolve. Weak knowledge of the region's lowland and continental shelf paleoecology is a critical issue, particularly concerning the significance of "grassier" communities and the drivers of such structural changes in the region's vegetation. At present we have insufficient evidence to disentangle the role of paleoclimate (moisture source/seasonality) from substrate and as drivers of vegetation change on the coastal lowlands.
3. *Local mediating factors, which are difficult to predict*: As with landscape controls, local-scale controls will serve to buffer/mask the effects of global scale "climate deteriorations" through the influence of (for example) soil substrate patterns, marine resource availability and local hydrology (e.g., coastal springs).

Human Interaction and Subsistence

Human subsistence behavior is structured by the spatial and temporal distribution of key resources, principally water, food, and shelter (Kelly 1995). The effects of these factors are mediated by changes in mobility, settlement systems and prey choice; technology likely responds to all three.

Shelter selection on the south coast is likely to have been influenced by a combination of sea level and inherited (contingent) local responses. The appearance and disappearance of springs, shifts in beach ridges and dunes, and the reconfiguration of shore lines will all have influenced the ways in which sites were used and indeed whether they were used at all. That the operation of these factors is difficult to predict may provide some explanation for the limited temporal overlap or nonoverlap in relatively proximate south coast MSA sequences (e.g., Blombos, Klasies River, Nelson Bay Cave and Pinnacle Point). Such local controls, along with the small sample of well-resolved south coast sites, makes attempts to correlate periodicities of site usage with population fluxes problematic. Local resource availability and landscape configuration likely exert stronger control than absolute population size. As noted, proximity to shoreline will have influenced the viability of marine resource use, with implications for shelter use (e.g., Marean et al. 2007). Beyond this, however, shifts in the availability of marine resources may have affected patterns of mobility and technological systems, the former with potential impacts on duration of site occupancy. Sessile marine resources provide a reliable food source, which can be harvested with minimal technological constraint. Many marine resources can withstand longer and more intensive harvesting than can their terrestrial equivalents (Binford 2001; Kelly 1995). Proximity of marine resources may thus have allowed periods of extended occupancy within a regime of diminished residential mobility at near-shore sites during high-stand periods (though note Borrero and Barbarena 2006).

In a similar vein, shifts between seasonal and aseasonal moisture regimes resulting from changes in the relative strength of summer and winter rainfall systems can alter patterns in the organization of landscape use. Surface water availability has a structuring effect on mobility; with diminished surface water, a greater frequency of movements involving entire groups is expected (Kelly 1995; Read 2008). Conversely, extended residential occupation of sites becomes more viable with greater water availability (Mackay 2009). Due to attendant local resource suppression resulting from extended occupancy, a shift in the configuration of mobility from residential to logistical is plausible (cf., Binford 1980; Kelly 1983). Such a shift may have occurred during relatively humid phases, potentially explaining the large assemblage sizes in MIS 4 at many south coast sites (Mackay 2009). Greater incorporation of small game might be expected to follow local resource suppression under such circumstances.

A secondary effect of shifting seasonality may have been on the complexity of technological systems deployed. Ethnographic data suggest that length of growing season affects technological complexity (Bousman 1993; Collard et al. 2005; Read 2008; Torrence 1983). Shorter or less predictable rainy seasons would have operated to increase subsistence risk. A second controlling factor here, however, is effective temperature, which is difficult to model at the local scale with available data. Expanded grasslands may have supported large herds of grazers, potentially generating a stronger hunting-resource base, but fynbos is notably resilient and supports both browsing game and a rich suite of edible floral resources including tubers (Marean 2010; Parkington 1977). We might anticipate that changes in the past composition of floral communities in the southern Cape are likely to have influenced technological systems. Data from the LSA suggest that reductions in grasslands around the transition from MIS 2 to MIS 1 were associated with a shift from microlithic to macrolithic technologies (Deacon 1984). Without proposing a direct relationship we might expect to see technological changes of some kind tracking earlier shifts in southern Cape flora. As we discussed above, however, such vegetation responses are presently difficult to model, emphasizing the need for local archives.

Conclusions

The southern Cape hosts a remarkable archaeological record; the significance of which is steadily being revealed. Here we have sought to summarize the environmental facets most relevant to the interpretation of these archaeological findings. In doing so, we emphasize the legacy of geological controls in influencing both macro and meso-scale landscape responses to environmental change. The region's climate presents both challenges (complexity of interpretation) and opportunities (relevance to synoptic scale climatic controls), but it is likely that regional climates were never sufficiently "harsh" to fully prevent occupation of the southern Cape. Although the specific composition of south coast ecosystems during the late Quaternary remains unclear, the potential combination of mosaic-like vegetation communities and marine food resources implies a relatively diverse and resilient, if variable, resource base. Taken as a whole, the occupational record for the southern Cape probably spans much of the period 170–50 ka, the full range glacial-interglacial conditions. Systems of technology, settlement, and subsistence undoubtedly changed through this period and we have attempted to highlight some of the relevant factors and linkages, and how they may be operationalized. Improvements in baseline archaeological and paleoenvironmental data are now required to strengthen our modeling of ecosystem variation and human behavioral response through the MSA.

Acknowledgments Hayley Cawthra is thanked for providing some of the cited literature. John Compton, two anonymous reviewers, and the volume editor Brian Stewart are thanked for their helpful comments on an earlier version of this review.

References

Ashton, A., Murray, A. B., & Arnoult, O. (2001). Formation of coastline features by large-scale instabilities induced by high angle waves. *Nature, 414*, 296–300.

Avery, D. M. (1982). Micromammals as palaeoenvironmental indicators and an interpretation of the late Quaternary in the southern Cape Province, South Africa. *Annals of the South African Museum, 85*, 183–377.

Avery, G. (1974). Open station midden sites and associated features from the Pearly Beach area, southwestern Cape. *South African Archaeological Bulletin, 24*, 104–114.

Bailey, G. N., & Flemming, N. C. (2008). Archaeology of the continental shelf: Marine resources, submerged landscapes and underwater archaeology. *Quaternary Science Reviews, 27*, 2153–2165.

Bateman, M. D., Holmes, P. J., Carr, A. S., Horton, B. P., & Jaiswal, M. K. (2004). Aeolianite and barrier dune construction spanning the last two glacial-interglacial cycles from the southern Cape coast, South Africa. *Quaternary Science Reviews, 23*, 1681–1698.

Bateman, M. D., Carr, A. S., Murray-Wallace, C. V., Holmes, P. J., & Roberts, D. L. (2008). A dating inter-comparison study on Late Stone Age midden deposits, South Africa. *Geoarchaeology, 23*, 715–741.

Bateman, M. D., Carr, A. S., Dunajko, A. C., Holmes, P. J., Roberts, D. L., McLaren, S. J., et al. (2011). The evolution of coastal barrier systems: A case study of the Middle-Late Pleistocene wilderness barriers, South Africa. *Quaternary Science Reviews, 30*, 63–81.

Bard, E., & Rickaby, R. E. M. (2009). Migration of the subtropical front as a modulator of glacial climate. *Nature, 460*, 380–383.

Bard, E., Rostek, F., & Sonzogni, C. (1997). Interhemispheric synchrony of the last deglaciation inferred from alkenone palaeothermometry. *Nature, 385*, 707–710.

Bar-Matthews, M., Marean, C. W., Jacobs, Z., Karkanas, P., Fisher, E. C., Herries, A. I. R., et al. (2010). A high resolution and continuous isotopic speleothem record of palaeoclimate and palaeoenvironment from 90 to 53 ka from Pinnacle Point on the south coast of South Africa. *Quaternary Science Reviews, 29*, 2131–2145.

Behrensmeyer, A. K. (2006). Climate change and human evolution. *Science, 311*, 476–478.

Binford, L. R. (1980). Willow-smoke and dog's tails: Hunter-gatherer settlement systems and archaeological site formation. *American Antiquity, 45*, 4–20.

Binford, L. R. (2001). *Constructing frames of references: An analytical method for archaeological theory building using ethnographic and environmental data sets*. Berkeley: University of California Press.

Birch, G. F., De Plessis, A., & Willis, J. P. (1978). Offshore and on land geological and geophysical investigations in the Wilderness Lakes region. *Transactions of the Geological Society of South Africa, 81*, 339–352.

Blome, M. W., Cohen, A. S., Tryon, C. A., Brooks, A. S., Russell, J. (2012). The environmental context for the origins of modern human diversity: A synthesis of regional variability in African climate 150,000–30,000 years ago. *Journal of Human Evolution, 62* (5), 563–592.

Blunier, T., Chappellaz, J., Schwander, J., Daellenbach, A., Stauffer, B., Stocker, T. F., et al. (1998). Asynchrony of Antarctic and Greenland climate change during the last glacial period. *Nature, 394*, 739–743.

Borrero, L. A., & Barberena, R. (2006). Hunter gatherer home ranges and marine resource use: An archaeological case study from southern Patagonia. *Current Anthropology, 47*, 855–867.

Boshoff, A. F., & Kerley, G. I. H. (2001). Potential distributions of the medium- to large-sized mammals in the Cape Floristic Region, based on historical accounts and habitat requirements. *African Zoology, 36*, 245–273.

Bousman, C. B. (1993). Hunter-gatherer adaptations, economic risk and tool design. *Lithic Technology, 18*, 59–86.

Broad, D. S., Jungslager, E. H. A., McLachlan, I. R., & Roux, J. (2006). Offshore mesozoic basins. In M. R. Johnson, C. R. Anhaeusser, & R. J. Thomas (Eds.), *The geology of South Africa (pp. 553–571)*. Pretoria: The Geological Society of South Africa and Council for Geoscience.

Brown, K. S., Marean, C. W., Herries, A. I. R., Jacobs, Z., Tribolo, C., Braun, D., et al. (2009). Fire as an engineering tool of early modern humans. *Science, 325*, 859–862.

Bruch, A. A., Sievers, C., & Wadley, L. (2012). Quantification of climate and vegetation from southern African Middle Stone Age sites – an application using Late Pleistocene plant material from Sibudu, South Africa. *Quaternary Science Reviews, 45*, 7–17.

Burrough, S. L. (2016). Late Quaternary environmental change and human occupation of the southern African interior. In S.C. Jones & B.A. Stewart (Eds.), *Africa from MIS 6-2: Population dynamics and paleoenvironments* (pp. 161–174). Dordrecht: Springer.

Butzer, K. W. (1984). Archeogeology and Quaternary environment in the interior of southern Africa. In R. G. Klein (Ed.), *Southern African prehistory and palaeoenvironments* (pp. 1–64). Rotterdam: AA Balkema.

Butzer, K. W. (2004). Coastal eolian sands, paleosols, and Pleistocene geoarchaeology of the southwestern Cape, South Africa. *Journal of Archaeological Science, 31*, 1743–1781.

Butzer, K. W., Beaumont, P. B., & Vogel, J. C. (1978). Lithostratigraphy of Border Cave, KwaZulu, South Africa: A Middle Stone Age sequence beginning c. 195,000 BP. *Journal of Archaeological Science, 5*, 317–341.

Caley, T., Kim, J.-H., Malaize, B., Giraudeau, J., Laepple, T., Caillon, N., et al. (2011). High-latitude obliquity forcing drives the Agulhas leakage. *Climates of the Past, 7*, 2193–2215.

Campbell, B. M., & Werger, M. J. A. (1988). Plant form in the mountains of the Cape, South Africa. *Journal of Ecology, 76*, 637–653.

Cantaneo, A., & Steel, R. J. (2003). Transgressive deposits: A review of their variability. *Earth Science Reviews, 62*, 187–228.

Carr, A. S., & Botha, G. A. (2012). The coastal geomorphology of southern Africa. In P. J. Holmes & M. E. Meadows (Eds.), *Southern African geomorphology: Recent trends and new directions*. Sun Media: Bloemfontein.

Carr, A. S., Thomas, D. S. G., & Bateman, M. D. (2006a). Climatic and sea level controls on Late Quaternary eolian activity on the Agulhas Plain, South Africa. *Quaternary Research, 65*, 252–263.

Carr, A. S., Thomas, D. S. G., Bateman, M. D., Meadows, M. E., & Chase, B. M. (2006b). Late Quaternary paleoenvironments of the winter-rainfall zone of southern Africa: Palynological and sedimentological evidence from the Agulhas Plain. *Palaeogeography, Palaeoclimatology, Palaeoecology, 239*, 147–165.

Carr, A. S., Bateman, M. D., & Holmes, P. J. (2007). Developing a 150 ka luminescence chronology for the coastal dunes of the southern Cape, South Africa. *Quaternary Geochronology, 2*, 110–116.

Carr, A. S., Bateman, M. D., Roberts, D. L., Murray-Wallace, C. V., Jacobs, Z., & Holmes, P. J. (2010a). The last interglacial sea level high stand on the southern Cape coastline of South Africa. *Quaternary Research, 73*, 351–362.

Carr, A. S., Boom, A., & Chase, B. M. (2010b). The potential of plant biomarker evidence derived from rock hyrax middens as an indicator of palaeoenvironmental change. *Palaeogeography, Palaeoclimatology, Palaeoecology, 285*, 321–330.

Carr, A. S., Boom, A., Chase, B. M., Roberts, D. L., & Roberts, Z. E. (2010c). Molecular fingerprinting of wetland organic matter using pyrolysis-GC/MS: An example from the southern Cape coastline of South Africa. *Journal of Paleolimnology, 44*, 947–961.

Castañeda, I. S., Werne, J. P., Johnson, T. C., & Filley, T. R. (2009). Late Quaternary vegetation history of southeast Africa: The molecular isotope record from Lake Malawi. *Palaeogeography, Palaeoclimatology, Palaeoecology, 275*, 100–112.

Cattaneo, A., & Steel, R. J. (2003). Transgressive deposits: A review of their variability. *Earth Science Reviews, 62*, 187–228.

Cawthra, H. C., Bateman, M. D., Carr, A. S., Compton, J. S., & Holmes, P. J. (2014). Understanding Late Quaternary change at the land–ocean interface: A synthesis of the evolution of the Wilderness coastline, South Africa. *Quaternary Science Reviews, 99*, 210–223.

Chase, B. M. (2009). Evaluating the use of dune sediments as a proxy for paleo-aridity: A southern African case study. *Earth-Science Reviews, 93*, 31–45.

Chase, B. M. (2010). South African paleoenvironments during Marine Oxygen Isotope Stage 4: A context for the Howiesons Poort and Still Bay industries. *Journal of Archaeological Science, 37*, 1359–1366.

Chase, B. M., & Meadows, M. E. (2007). Late Quaternary dynamics of southern Africa's winter rainfall zone. *Earth-Science Reviews, 84*, 103–138.

Chase, B. M., Meadows, M. E., Scott, L., Thomas, D. S. G., Marais, E., Sealy, J., et al. (2009). A record of rapid Holocene climate change preserved in hyrax middens from southwestern Africa. *Geology, 37*, 703–706.

Chase, B. M., Quick, L. J., Meadows, M. E., Scott, L., Thomas, D. S. G., & Reimer, P. J. (2011). Late glacial interhemispheric climate dynamics revealed in South African hyrax middens. *Geology, 39*, 19–22.

Chase, B. M., Boom, A., Carr, A. S., Meadows, M. E., & Reimer, P. J. (2013). Holocene climate change in southernmost South Africa: rock hyrax middens record shifts in the southern westerlies. *Quaternary Science Reviews, 82*, 199–205.

Chase, B. M., Boom, A., Carr, A. S., Carré, M., Chevalier, M., Meadows, M. E., et al. (2015a). Evolving southwest African response to abrupt deglacial North Atlantic climate change events. *Quaternary Science Reviews, 121*, 132–136.

Chase, B. M., Lim, S., Chevalier, M., Boom, A., Carr, A. S., Meadows, M. E., & Reimer, P. J. (2015b). Influence of tropical easterlies in southern Africa's winter rainfall zone during the Holocene. *Quaternary Science Reviews, 107*, 138–148.

Chevalier, M., & Chase, B. M. (2015). Southeast African records reveal a coherent shift from high- to low-latitude forcing mechanisms along the east African margin across last glacial–interglacial transition. *Quaternary Science Reviews, 125*, 117–130.

Cochrane, G. W. G., Doelman, T., & Wadley, L. (2013). Another dating revolution prehistoric archaeology? *Journal of Archaeological Method and Theory, 20*, 42–60.

Cockcroft, M. J., Wilkinson, M. J., & Tyson, P. D. (1987). The application of a present-day climatic model to the late Quaternary in southern Africa. *Climatic Change, 10*, 161–181.

Cohen, A. L., & Tyson, P. D. (1995). Sea surface temperature fluctuations during the Holocene off the south coast of Africa: Implications for terrestrial climate and rainfall. *The Holocene, 5*, 304–312.

Collard, M., Kemery, M., & Banks, S. (2005). Causes of toolkit variation among hunter-gatherers: A test of four competing hypotheses. *Canadian Journal of Archaeology, 29*, 1–19.

Collins, J. A., Schefuß, E., Govin, A., Mulitza, S., & Tiedemann, R. (2014). Insolation and glacial–interglacial control on southwestern African hydroclimate over the past 140,000 years. *Earth and Planetary Science Letters, 398*, 1–10.

Compton, J. S. (2001). Holocene sea level fluctuations inferred from the evolution of depositional environments of the southern Langebaan Lagoon salt marsh, South Africa. *The Holocene, 11*, 395–405.

Compton, J. S. (2011). Pleistocene sea level fluctuations and human evolution on the southern coastal plain of South Africa. *Quaternary Science Reviews, 30*, 506–527.

Cooper, J. A. G., & Pilkey, O. H. (2004). Sea level rise and shoreline retreat: Time to abandon the Bruun Rule. *Global and Planetary Change, 43*, 157–171.

Cooper, J. A. G., McKenna, J., Jackson, D. W. T., & O'Connor, M. (2007). Meso-scale coastal behavior related to morphological self-adjustment. *Geology, 35*, 187–190.

Cowling, R. M. (1983). Phytochorology and vegetation history in the South-Eastern Cape, South Africa. *Journal of Biogeography, 10*, 393–419.

Cowling, R. M., & Holmes, P. M. (1992). Flora and vegetation. In R. M. Cowling (Ed.), *The ecology of the Fynbos: Nutrients, fire and diversity* (pp. 23–63). Cape Town: Oxford University Press.

Cowling, R. M., Campbell, B. M., Mustart, P., McDonald, D. J., Jarman, M. L., & Moll, E. J. (1988). Vegetation classification in a floristically complex area: The Agulhas Plain. *South African Journal of Botany, 54*, 290–300.

Davies, J. L. (1980). *Geographical variation in coastal development*. New York: Longman.

Davis, R. A., & Hayes, M. O. (1984). What is a wave dominated coast? *Marine Geology, 60*, 313–329.

Deacon, H. J. (1995). Two late Pleistocene-Holocene archaeological deposits from the southern cape, South Africa. *South African Archaeological Bulletin, 50*, 121–131.

Deacon, H. J., & Geleijnse, V. B. (1988). The stratigraphy and sedimentology of the main site sequence, Klasies River, South Africa. *South African Archaeological Bulletin, 43*, 5–14.

Deacon, H. J., Deacon, J., Scholtz, A., Thackeray, J. F., & Brink, J. S. (1984). Correlation of palaeoenvironmental data from the Late Pleistocene and Holocene deposits at Boomplaas Cave, southern Cape. In J. C. Vogel (Ed.), *Late Cainozoic Palaeoclimates of the Southern Hemisphere* (pp. 339–360). Rotterdam: AA Balkema.

Deacon, H. J., Jury, M. R., & Ellis, F. (1992). Selective regime and time. In R. M. Cowling (Ed.), *The ecology of Fynbos: Nutrients, fire and diversity* (pp. 6–22). Cape Town: Oxford University Press.

Deacon, J. (1978). Changing patterns in the Late Pleistocene/Early Holocene prehistory of southern Africa as seen from the Nelson Bay Cave stone artifact sequence. *Quaternary Research, 10*, 84–111.

Deacon, J., & Lancaster, N. (1988). *Late Quaternary palaeoenvironments of southern Africa*. Oxford: Clarendon Press.

deMenocal, P. B. (2011). Climate and human evolution. *Science, 311*, 540–541.

de Wit, M. J., Harris, C., Kramers, J., Doel, S., & Chase, B. M. (2009). Transient vegetation changes between 60 and 90 ka from South Africa's Cango Cave stalagmite record: A test for Indian Ocean monsoon variability and its relation to culture in Africa. Paper presented at the 5th EGU Alexander von Humboldt International Conference, Cape Town.

Dingle, R. V., & Rogers, J. (1972). Pleistocene palaeogeography of the Agulhas Bank. *Transactions of the Royal Society of South Africa, 40*, 155–165.

Dupont, L. M., Caley, T., Kim, J. H., Castañeda, I., Malaizé, B., & Giraudeau, J. (2011). Glacial-interglacial vegetation dynamics in south eastern Africa depend on sea surface temperature variations in the west Indian Ocean. *Climates of the Past Discussions, 7*, 2261–2296.

Faith, J. T. (2011a). Late Quaternary dietary shifts of the Cape grysbok (*Raphicerus melanotis*) in southern Africa. *Quaternary Research, 75*, 159–165.

Faith, J. T. (2011b). Ungulate community richness, grazer extinctions, and human subsistence behavior in southern Africa's Cape Floral Region. *Palaeogeography, Palaeoclimatology, Palaeoecology, 306*, 219–227.

Faure, H., Walter, R. C., & Grant, D. R. (2002). The coastal oasis: Ice age springs on emerged continental shelves. *Global and Planetary Change, 33*, 47–56.

Feathers, J. K. (2002). Luminescence dating in less than ideal conditions: Case studies from Klasies River Main Site and Duinefontein, South Africa. *Journal of Archaeological Science, 29*, 177–194.

Feathers, J. K., & Bush, D. A. (2000). Luminescence dating of Middle Stone Age deposits at Die Kelders. *Journal of Human Evolution, 38*, 91–119.

Finlayson, C. (2008). On the importance of coastal areas in the survival of Neanderthal populations during the Late Pleistocene. *Quaternary Science Reviews, 27*, 2246–2252.

Fisher, E. C., Bar-Matthews, M., Jerardino, A., & Marean, C. W. (2010). Middle and Late Pleistocene palaeoscape modelling along the southern coast of South Africa. *Quaternary Science Reviews, 29*, 1382–1398.

Flemming, B. W., Martin, A. K., & Rogers, J. (1983). *Onshore and offshore coastal eolianites between Mossel Bay and Knysna* (Technical Report No. 14). Cape Town: Marine Geosciences Group, University of Cape Town.

Gentle, R. I. (1987). *The geology of the inner continental shelf and Agulhas Arch: Cape Town to Port Elizabeth* (Council for Geoscience Bulletin No. 20). Cape Town: Marine Geoscience Unit, University of Cape Town.

Goldblatt, P., & Manning, J. C. (2002). Plant Diversity of the Cape Region of Southern Africa. *Annals of the Missouri Botanical Gardens, 89*, 281–302.

Hendey, Q. B., & Volman, T. P. (1986). Last Interglacial sea levels and coastal caves in the Cape Province, South Africa. *Quaternary Research, 25*, 189–198.

Henshilwood, C. S. (2005). Stratigraphic integrity of the Middle Stone Age levels at Blombos Cave. In F. d'Errico & L. Backwell (Eds.), *From tools to symbols: From early hominids to modern humans* (pp. 441–458). Johannesburg: Witwatersrand University Press.

Henshilwood, C. S., Sealy, J. C., Yates, R., Cruz-Uribe, K., Goldberg, P., Grine, F. E., et al. (2001). Blombos Cave, southern Cape, South Africa: Preliminary report on the 1992–1999 excavations of the Middle Stone Age levels. *Journal of Archaeological Science, 28*, 421–448.

Henshilwood, C. S., d'Errico, F., Yates, R., Jacobs, Z., Tribolo, C., Duller, G. A. T., et al. (2002). Emergence of modern human behaviour: Middle Stone Age engravings from South Africa. *Science, 295*, 1278–1290.

Henshilwood, C. S., d'Errico, F., Vanhaeren, M., van Niekerk, K., & Jacobs, Z. (2004). Middle Stone Age shell beads from South Africa. *Science, 303*, 404.

Henshilwood, C. S., d'Errico, F., van Niekerk, K. L., Coquinot, Y., Jacobs, Z., Lauritzen, S. E., et al. (2011). A 100,000-year-old ochre-processing workshop at Blombos Cave, South Africa. *Science, 334*, 219–222.

Hetherington, R., Wiebe, E., Weaver, A. J., Carto, S. L., Eby, M., & MacLeod, R. (2008). Climate, African and Beringian sub-aerial continental shelves, and migration of early people. *Quaternary International, 183*, 83–101.

Hiscock, P., Clarkson, C., & Mackay, A. (2011). Big debates over little tools: Ongoing disputes over microliths on three continents. *World Archaeology, 43*, 653–644.

Irving, S. J. (1998). *Late Quaternary Palaeoenvironments at Vankervelsvlei, near Knysna, South Africa*. M.Sc. Dissertation, University of Cape Town.

Irving, S. J. E., & Meadows, M. E. (1997). Radiocarbon chronology and organic matter accumulation at Vankervelsvlei, near Knysna, South Africa. *South African Geographical Journal, 79*, 101–105.

Jacobs, Z. (2010). An OSL chronology for the sedimentary deposits from Pinnacle Point Cave 13B – a punctuated presence. *Journal of Human Evolution, 59*, 289–305.

Jacobs, Z., & Roberts, R. G. (2008). Testing times: Old and new chronologies for the Howieson's Poort and Still Bay industries in environmental context. *South African Archaeological Society Godwin Series, 10*, 9–34.

Jacobs, Z., & Roberts, D. L. (2009). Last interglacial age for aeolian and marine deposits and the Nahoon fossil human footprints, southeast coast of South Africa. *Quaternary Geochronology, 4*, 160–169.

Jacobs, Z., Wintle, A. G., & Duller, G. A. T. (2003). Optical dating of dune sand from Blombos Cave, South Africa: I – multiple grain data. *Journal of Human Evolution, 44*, 599–612.

Jacobs, Z., Duller, G. A. T., Wintle, A. G., & Henshilwood, C. S. (2006). Extending the chronology of deposits at Blombos Cave, South Africa back to 140 ka using optical dating of single and multiple grains of quartz. *Journal of Human Evolution, 51*, 255–273.

Jacobs, Z., Roberts, R. G., Galbraith, R. F., Deacon, H. J., Grün, R., Mackay, A., et al. (2008). Ages for the Middle Stone Age of southern Africa: Implications for human behaviour and dispersal. *Science, 322*, 733–735.

Jacobs, Z., Roberts, R. G., Lachlan, T. J., Karkanas, K., Marean, C. W., & Roberts, D. L. (2011). Development of the SAR TT-OSL procedure for dating Middle Pleistocene dune and shallow marine deposits along the southern Cape coast of South Africa. *Quaternary Geochronology, 6*, 491–513.

Jacobs, Z., Hayes, E. H., Roberts, R. G., Galbraith, R. F., & Henshilwood, C. S. (2013). An improved OSL chronology for the Still Bay layers at Blombos Cave, South Africa: Further tests of single-grain dating procedures and a re-evaluation of the timing of the Still Bay industry across southern Africa. *Journal of Archaeological Science, 40*, 579–594.

Jouzel, J., Masson-Delmotte, V., Cattani, O., Dreyfus, G., Falourd, S., Hoffmann, G., et al. (2007). Orbital and millennial Antarctic climate variability over the past 800,000 years. *Science, 317*, 793–797.

Jury, M. R., Valentine, H. R., & Lutjeharms, J. R. (1993). Influence of the Agulhas Current on summer rainfall along the southeast coast of South Africa. *Journal of Applied Meteorology, 32*, 1282–1287.

Kandel, A. W., & Conard, N. J. (2012). Settlement patterns during the earlier and Middle Stone Age around Langebaan Lagoon, Western Cape (South Africa). *Quaternary International, 270*, 15–29.

Karkanas, P., & Goldberg, P. (2010). Site formation processes at Pinnacle Point Cave 13B (Mossel Bay, Western Cape Province, South Africa): Resolving stratigraphic and depositional complexities with micromorphology. *Journal of Human Evolution, 59*, 256–273.

Kelly, R. L. (1983). Hunter-gatherer mobility strategies. *Journal of Anthropological Research, 39*, 277–306.

Kelly, R. L. (1995). *The foraging spectrum: Diversity in hunter-gatherer lifeways*. Washington D.C.: Smithsonian Institute.

Kingston, J. D. (2007). Shifting adaptive landscapes: Progress and challenges in reconstructing early hominid environments. *Yearbook of Physical Anthropology, 50*, 20–58.

Klein, R. G. (1972). The Late Quaternary mammalian fauna of Nelson Bay Cave (Cape Province, South Africa): Its implications for megafaunal extinction and environment and cultural change. *Quaternary Research, 2*, 135–140.

Klein, R. G. (1976). The mammalian fauna of the Klasies River Mouth sites, southern Cape Province, South Africa. *South African Archaeological Bulletin, 31*, 74–98.

Klein, R. G. (1978). A preliminary report on the larger mammals from the Boomplaas stone age cave site, Cango Valley, Oudtshoorn District, South Africa. *The South African Archaeological Bulletin, 33*, 66–75.

Klein, R. G. (1983). Paleoenvironmental implications of Quaternary large mammals in the Fynbos Region. In H. J. Deacon, Q. B. Hendey, & J. J. N. Lambrechts (Eds.), *Fynbos paleoecology: A preliminary synthesis* (pp. 116–138). Pretoria: South African National Scientific Programmes.

Klein, R. G., & Cruz-Uribe, K. (2000). Middle and Later Stone Age large mammal and tortoise remains from Die Kelders Cave 1, Western Cape Province, South Africa. *Journal of Human Evolution, 38*, 169–195.

Klein, R. G., Avery, G., Cruz-Uribe, K., Halkett, D., Parkington, J. E., Steele, T., et al. (2004). The Ysterfontein 1 Middle Stone Age site, South Africa, and early human exploitation of coastal resources. *Proceedings of the National Academy of Sciences of the United States of America, 101*, 5708–5715.

Lambeck, K., Purcell, A., Flemming, N. C., Vita-Finzi, C., Alsharekh, A. B., & Bailey, G. N. (2011). Sea level and shoreline reconstructions for the Red Sea: Isostatic and tectonic considerations and implications for hominin migration out of Africa. *Quaternary Science Reviews, 30*, 3542–3574.

Lancaster, N. (2002). How dry was dry? Late Pleistocene palaeoclimates in the Namib Desert. *Quaternary Science Reviews, 21*, 769–782.

Lee-Thorp, J. A., & Beaumont, P. B. (1995). Vegetation and seasonality shifts during the late Quaternary deduced from $^{13}C/^{12}C$ ratios of grazers at Equus Cave. *South Africa. Quaternary Research, 43*, 426–432.

Lee-Thorp, J. A., Holmgren, K., Lauritzen, S.-E., Linge, H., Moberg, A., Partridge, T. C., et al. (2001). Rapid climate shifts in the southern African interior throughout the mid- to late Holocene. *Geophysical Research Letters, 28*, 4507–4510.

Lewis, C. A. (2008). Late Quaternary climatic changes, and associated human responses, during the last ~45,000 year in the Eastern and adjoining Western Cape, South Africa. *Earth-Science Reviews, 88*, 167–187.

Linder, H. P. (2003). The radiation of the Cape flora, southern Africa. *Biological Reviews, 78*, 597–638.

Lindesay, J. A. (1998). The present climates of southern Africa. In J. E. Hobbs, J. A. Lindesay, & H. A. Bridgman (Eds.), *Climates of the southern continents: Present, past and future* (pp. 5–52). Chichester: Wiley.

Little, M. G., Schneider, R. R., Kroon, D., Price, B., Bickert, T., & Wefer, G. (1997a). Rapid palaeoceanographic changes in the Benguela Upwelling System for the last 160,000 years as indicated by abundances of planktonic foraminifera. *Palaeogeography, Palaeoclimatology, Palaeoecology, 130*, 135–161.

Little, M. G., Schneider, R. R., Kroon, D., Price, B., Summerhayes, C. P., & Segl, M. (1997b). Trade wind forcing of upwelling, seasonality, and Heinrich events as a response to sub-Milankovitch climate variability. *Paleoceanography, 12*, 568–576.

Lombard, M. (2012). Thinking through the Middle Stone Age of sub-Saharan Africa. *Quaternary International, 270*, 140–155.

Low, A. B., & Rebelo, A. G. (1996). *Vegetation of southern Africa, Lesotho, and Swaziland*. Pretoria: Department of Environmental Affairs and Tourism.

Mackay, A. (2009). *History and selection in the Late Pleistocene archaeology of the Western Cape, South Africa*. Ph.D. Dissertation, Australian National University.

Mackay, A. (2011). Nature and significance of the Howiesons Poort to post-Howiesons Poort transition at Klein Kliphuis rockshelter, South Africa. *Journal of Archaeological Science, 38*, 1430–1440.

Malan, J. A. (1990). *The stratigraphy and sedimentology of the Bredasdorp Group, southern Cape Province*. M.Sc. Dissertation, University of Cape Town.

Marean, C. W. (2010). Pinnacle Point Cave 13B (Western Cape Province, South Africa) in context: The Cape Floral kingdom, shellfish, and modern human origins. *Journal of Human Evolution, 59*, 425–443.

Marean, C. W. (2011). Coastal South Africa and the coevolution of the modern human lineage and the coastal adaptation. In N. F. Bicho, J. A. Haws, & L. G. Davis (Eds.), *Trekking the shore: Changing coastlines and the antiquity of coastal settlement* (pp. 421–440). Dordrecht: Springer.

Marean, C. W., Nilssen, P. J., Brown, K., Jerardino, A., & Stynder, D. (2004). Paleoanthropological investigations of Middle Stone Age sites at Pinnacle Point, Mossel Bay (South Africa): Archaeology and hominid remains from the 2000 field season. *Paleoanthropology, 2*, 14–83.

Marean, C. W., Bar-Matthews, M., Bernatchez, J., Fisher, E., Goldberg, P., & Herries, A. I. R. (2007). Early human use of marine resources and pigment in South Africa during the Middle Pleistocene. *Nature, 449*, 905–909.

Marean, C. W., Anderson, R. J., Bar-Matthews, M., Braun, K., Cawthra, H. C., Cowling, R. M., et al. (2015). A new research strategy for integrating studies of paleoclimate, paleoenvironment, and paleoanthropology. *Evolutionary Anthropology: Issues, News, and Reviews, 24*, 62–72.

Marker, M. E. (1987). A note on marine benches of the southern Cape. *South African Journal of Geology, 90*, 120–124.

Marker, M. E., & Holmes, P. J. (2010). The geomorphology of the coastal platform in the southern Cape. *South African Geographical Journal, 92*, 105–116.

Martin, A. K., & Flemming, B. W. (1986). The Holocene shelf sediment wedge off the south and east coast of South Africa. In R. J. Knight & J. R. McLean (Eds.), *Shelf sands and sandstones* (Canadian journal of Petroleum Geologists Memoir No. 11) (pp. 27–44). Calgary: Canadian Society of Petroleum Geologists.

Martin, A. R. H. (1968). Pollen analysis of Groenvlei lake sediments, Knysna (South Africa). *Review of Palaeobotany and Palynology, 7*, 107–144.

McCall, G. S. (2007). Behavioral ecological models of lithic technological change during the later Middle Stone Age of South Africa. *Journal of Archaeological Science, 34*, 1738–1751.

McCall, G. S., & Thomas, J. T. (2012). Still Bay and Howiesons Poort foraging strategies: Recent research and models of cultural change. *African Archaeological Review, 29*, 7–50.

Meadows, M. E., & Baxter, A. J. (1999). Late Quaternary palaeoenvironments of the southwestern Cape, South Africa: A regional synthesis. *Quaternary International, 57–8*, 193–206.

Meadows, M. E., & Sugden, J. M. (1993). The Late Quaternary palaeoecology of a Floristic Kingdom – the southwestern Cape South Africa. *Palaeogeography, Palaeoclimatology, Palaeoecology, 101*, 271–281.

Mickler, P. J., Banner, J., Stern, L., Asmerom, Y., Edwards, R. L., & Ito, E. (2004). Stable isotope variations in modern tropical speleothems: Evaluating equilibrium versus kinetic isotope effects. *Geochimica et Cosmochimica Acta, 68*, 4381–4393.

Morris, A. G. (2002). Isolation and the origin of the Khoisan: Late Pleistocene and Early Holocene human evolution at the southern end of Africa. *Human Evolution, 17*, 231–240.

Mitchell, P. (2008). Developing the archaeology of Marine Isotope Stage 3. *South African Archaeological Society Goodwin Series, 10*, 52–65.

Mucina, L., Rutherford, M. C. (2006). *The vegetation of South Africa, Lesotho and Swaziland*. Strelitzia 19. Pretoria: South African National Biodiversity Institute.

Parkington, J. (1977). Soaqua: Hunter-fisher-gatherers of the Olifants River Valley, Western Cape. *South African Archaeological Bulletin, 35*, 73–83.

Parkington, J. (2003). Middens and moderns: Shell fishing and the Middle Stone Age of the Western Cape. *South African Journal of Science, 99*, 243–248.

Parkington, J. (2010). Coastal diet, encephalization, and innovative behaviours in the late Middle Stone Age of southern Africa. In C. Cunnane & K. M. Stewart (Eds.), *Human brain evolution* (pp. 189–201). Chichester: John Wiley.

Partridge, T. C., & Maud, R. R. (1987). Geomorphic evolution of southern Africa since the Mesozoic. *South African Journal of Geology, 90*, 179–208.

Partridge, T. C., Avery, D. M., Botha, G. A., Brink, J. S., Deacon, J., & Herbert, R. S. (1990). Late Pleistocene and Holocene climatic change in southern Africa. *South African Journal of Science, 86*, 302–306.

Partridge, T. C., deMenocal, P. B., Lorentz, S. A., Paiker, M. J., & Vogel, J. C. (1997). Orbital forcing of climate over South Africa: A 200,000-year rainfall record from the Pretoria Saltpan. *Quaternary Science Reviews, 16*, 1125–1133.

Partridge, T. C., Scott, L., & Hamilton, J. E. (1999). Synthetic reconstructions of southern African environments during the Last Glacial Maximum (21–18 kyr) and the Holocene Altithermal (8–6 kyr). *Quaternary International, 57–8*, 207–214.

Pedro, J. B., van Ommen, T. D., Rasmussen, S. O., Morgan, V. I., Chappellaz, J., & Moy, A. D. (2011). The last deglaciation: Timing the bipolar seesaw. *Climates of the Past, 7*, 671–683.

Peeters, F. J. C., Acheson, R., Brummer, G.-J. A., de Ruijter, W. P. M., Schneider, R. R., Ganssen, G. M., et al. (2004). Vigorous exchange between the Indian and Atlantic oceans at the end of the past five glacial periods. *Nature, 430*, 661–665.

Pichevin, L., Cremer, M., Giraudeau, J., & Bertrand, P. (2005). A 190 kyr record of lithogenic grain-size on the Namibian slope: Forging a tight link between past wind-strength and coastal upwelling dynamics. *Marine Geology, 218*, 81–96.

Quick, L. J., Chase, B. M., Meadows, M. E., Scott, L., & Reimer, P. J. (2011). A 19.5 kyr vegetation history from the central Cederberg Mountains, South Africa: Palynological evidence from rock hyrax middens. *Palaeogeography, Palaeoclimatology, Palaeoecology, 309*, 253–270.

Quick, L. J., Carr, A. S., Meadows, M. E., Boom, A., Bateman, M. D., Roberts, D. L., Reimer, P. J., & Chase, B. M. A late Pleistocene-Holocene multi-proxy record of palaeoenvironmental change from Still Bay, southern Cape coast, South Africa. *Journal of Quaternary Science*, in press a.

Quick, L. J., Meadows, M. E., Bateman, M. D., Kirsten, K. L., Mäusbacher, R., Haberzettl, T., & Chase, B. M. (2015). Vegetation and climate dynamics during the last glacial period in the fynbos-afrotemperate forest ecotone, southern Cape, South Africa. *Quaternary International*, in press b. doi:10.1016/j.quaint.2015.08.027.

Ramsay, P. J., & Cooper, J. A. G. (2002). Late Quaternary sea level change in South Africa. *Quaternary Research, 57*, 82–90.

Read, D. (2008). An interaction model for resource implement complexity based on risk and number of annual moves. *American Antiquity, 73*, 599–625.

Rebelo, A. G., Cowling, R. M., Campbell, B. M., & Meadows, M. E. (1991). Plant communities of the Riversdale Plain. *South African Journal of Botany, 57*, 10–28.

Rector, A. L., & Reed, K. E. (2010). Middle and late Pleistocene fauna of Pinnacle Point and their paleoecological implications. *Journal of Human Evolution, 59*, 340–357.

Roberts, D. L., Bateman, M. D., Murray-Wallace, C. V., Carr, A. S., & Holmes, P. J. (2008). Last Interglacial fossil elephant trackways dated by OSL/AAR in coastal aeolianites, Still Bay, South Africa. *Palaeogeography, Palaeoclimatology, Palaeoecology, 257*, 261–279.

Roberts, D. L., Bateman, M. D., Murray-Wallace, C. V., Carr, A. S., & Holmes, P. J. (2009). West coast dune plumes: Climate driven contrasts in dunefield morphogenesis along the western and southern South African coasts. *Palaeogeography, Palaeoclimatology, Palaeoecology, 271*, 34–48.

Roberts, D. L., Karkanas, P., Jacobs, Z., Marean, C. W., & Roberts, R. G. (2012). Melting ice sheets 400,000 year ago raised sea level by 13 m: Past analogue for future trends. *Earth and Planetary Science Letters, 357–358*, 226–237.

Rogers, J. (1985). Geomorphology, offshore bathymetry and Quaternary lithostratigraphy around the Bot River estuary. *Transactions of the Royal Society of South Africa, 45*, 211–227.

Roy, P. S., Cowell, P. J., Ferland, M. A., & Thom, B. G. (1994). Wave-dominated coasts. In R. W. G. Carter & C. D. Woodruffe (Eds.), *Coastal evolution: Late Quaternary shoreline morphodynamics* (pp. 33–86). Cambridge: Cambridge University Press.

Rundel, P. W., Esler, K. J., & Cowling, R. M. (1999). Ecology and phylogenetic patterns of carbon isotope discrimination in the winter-rainfall flora of the Richtersveld, South Africa. *Plant Ecology, 142*, 133–148.

Scholtz, A. (1986). *Palynological and palaeobotanical studies in the southern Cape*. M.Sc. Dissertation, University of Stellenbosch.

Scholz, C. A., Johnson, T. C., Cohen, A. S., King, J. W., Peck, J. A., Overpeck, J. T., et al. (2007). East African megadroughts between 135 and 75 thousand years ago and bearing on early-modern human origins. *Proceedings of the National Academy of Sciences of the United States of America, 104*, 16416–16421.

Scott, L., & Vogel, J. C. (2000). Evidence for environmental conditions during the last 20,000 years in southern Africa from ^{13}C in fossil hyrax dung. *Global and Planetary Change, 26*, 207–215.

Sealy, J. (2016). Cultural change, demography, and the archaeology of the last 100 kyr in southern Africa. In S. C. Jones & B. A. Stewart (Eds.), *Africa from MIS 6-2: Population dynamics and paleoenvironments* (pp. 65–75). Dordrecht: Springer.

Shaw, A. E., Holmes, P. J., & Rogers, J. (2001). Depositional landforms and environmental change in the headward vicinity of Dias Beach, Cape Point. *South African Journal of Geology, 104*, 101–114.

Shi, N., Schneider, R., Beug, H.-J., & Dupont, L. M. (2001). Southeast trade wind variations during the last 135 kyr: Evidence from pollen spectra in eastern South Atlantic sediments. *Earth and Planetary Science Letters, 187*, 311–321.

Sonzogni, C., Bard, E., & Rostek, F. (1998). Tropical sea-surface temperatures during the last glacial period: A view based on alkenones in Indian Ocean sediments. *Quaternary Science Reviews, 17*, 1185–1201.

Stager, J. C., Ryves, D. B., Chase, B. M., & Pausata, F. S. R. (2011). Catastrophic drought in the Afro-Asian monsoon region during Heinrich event 1. *Science, 331*, 1299–1302.

Stringer, C. (2000). Coasting out of Africa. *Nature, 405*, 24–27.

Stringer, C. B., Finlayson, J. C., Barton, R. N. E., Fernández-Jalvo, Y., Cáceres, I., Sabin, R. C., et al. (2008). Neanderthal exploitation of marine mammals in Gibraltar. *Proceedings of the National Academy of Sciences of the United States of America, 105*, 14319–14324.

Storms, J. E. A., Weltje, G. J., Van Dijke, J. J., Geel, C. R., & Kroonenberg, S. B. (2002). Process–response modeling of wave-dominated coastal systems: Simulating evolution and stratigraphy on geological timescales. *Journal of Sedimentary Research, 72*, 226–239.

Stuut, J.-B. W., Prins, M. A., Schneider, R. R., Weltje, G. J., Jansen, J. H. F., & Postma, G. (2002). A 300 kyr record of aridity and wind strength in southwestern Africa: Inferences from grain-size distributions of sediments on Walvis Ridge, SE Atlantic. *Marine Geology, 180*, 221–233.

Stuut, J.-B. W., Crosta, X., Van der Borg, K., & Schneider, R. R. (2004). On the relationship between Antarctic sea ice and southwestern African climate during the late Quaternary. *Geology, 32*, 909–912.

Talma, A. S., & Vogel, J. C. (1992). Late Quaternary paleotemperatures derived from a Speleothem from Cango Caves, Cape Province, South Africa. *Quaternary Research, 37*, 203–213.

Tankard, A. J. (1976a). The Pleistocene history and coastal morphology of the Ysterfontein-Elands Bay area, Cape Province. *Annals of the South African Museum, 69*, 73–119.

Tankard, A. J. (1976b). The stratigraphy of a coastal cave and its palaeoclimatic significance. *Palaeoecology of Africa, 9*, 151–159.

Tankard, A. J., & Schweitzer, F. R. (1976). Textural analysis of cave sediments: Die Kelders, Cape Province, South Africa. In D. A. Davidson & M. L. Shackley (Eds.), *Geoarchaeology: Earth science and the past* (pp. 289–316). Slough: Duckworth.

Texier, J.-P., Porraz, G., Parkington, J., Rigaud, J.-P., Poggenpoel, C., Miller, C., et al. (2010). A Howiesons Poort tradition of engraving ostrich eggshell containers dated to 60,000 years ago at Diepkloof Rock Shelter, South Africa. *Proceedings of the National Academy of Sciences of the United States of America, 107*, 6180–6185.

Thomas, D. S. G., & Burrough, S. L. (2012). Interpreting geoproxies of late Quaternary climate change in African drylands: Implications for understanding environmental change and early human behaviour. *Quaternary International, 253*, 5–17.

Tinley, K. L. (1985). *Coastal dunes of South Africa* (South African National Scientific Programmes Report No. 109). Pretoria: CSIR.

Torrence, R. (1983). Time budgeting and hunter-gatherer technology. In G. Bailey (Ed.), *Hunter-gatherer economy in prehistory* (pp. 11–22). Cambridge: Cambridge University Press.

Tremaine, D. M., Froelich, P. N., & Wang, Y. (2011). Speleothem calcite farmed *in situ*: Modern calibration of $\delta^{18}O$ and $\delta^{13}C$ palaeoclimate proxies in a continuously-monitored natural cave system. *Geochimica et Cosmochimica Acta, 75*, 4929–4950.

Tribolo, C., Mercier, N., Douville, E., Joron, J. L., Reyss, J. L., Rufer, D., et al. (2013). OSL and TL dating of the Middle Stone Age sequence at Diepkloof Rock Shelter (South Africa): a clarification. *Journal of Archaeological Science, 40*, 3401–3411.

Tribolo, C., Mercier, N., Selo, M., Valladas, H., Joron, J. L., & Reyss, J.-L. (2006). TL dating of burnt lithics from Blombos Cave (South Africa): Further evidence for the antiquity of modern human behaviour. *Archaeometry, 48*, 341–357.

Truc, L., Chevalier, M., Favier, C., Cheddadi, R., Meadows, M. E., Scott, L., et al. (2013). Quantification of climate change for the last 20,000 years from Wonderkrater, South Africa: Implications for the long-term dynamics of the Intertropical Convergence Zone. *Palaeogeography, Palaeoclimatology, Palaeoecology, 386*, 575–587.

Tyson, P. D. (1986). *Climatic change and variability in southern Africa*. Cape Town: Oxford University Press.

Tyson, P. D. (1999a). Atmospheric circulation changes and palaeoclimates of southern Africa. *South African Journal of Science, 95*, 194–201.

Tyson, P. D. (1999b). Late Quaternary and Holocene palaeoclimates of southern Africa: a synthesis. *South African Journal of Geology, 102*, 335–349.

Tyson, P. D., & Preston-Whyte, R. A. (2000). *The weather and climate of southern Africa*. Cape Town: Oxford University Press.

Valsecchi, V., Chase, B. M., Slingsby, J., Carr, A. S., Quick, L. J., Meadows, M. E., et al. (2013). A high resolution 15,600-year pollen and microcharcoal record from the Cederberg Mountains, South Africa. *Palaeogeography, Palaeoclimatology, Palaeoecology, 387*, 6–16.

van Campo, E., Duplessy, J. C., Prell, W. L., Barratt, N., & Sabatier, R. (1990). Comparison of terrestrial and marine temperature estimates for the past 135 kyr off southeast Africa: A test for GCM simulations of palaeoclimate. *Nature, 348*, 209–212.

Van Andel, T. H. (1989). Late Pleistocene sea levels and human exploitation of the shore and the shelf of southern South Africa. *Journal of Field Archaeology, 16*, 133–155.

van Zinderen Bakker, E. M. (1967). Upper Pleistocene stratigraphy and Holocene ecology on the basis of vegetation changes in sub-Saharan Africa. In W. W. Bishop & J. D. Clark (Eds.), *Background to evolution in Africa* (pp. 125–147). Chicago: University of Chicago Press.

van Zinderen Bakker, E. M. (1976). The evolution of late Quaternary palaeoclimates of southern Africa. *Palaeoecology of Africa, 9*, 160–202.

Vogel, J. C., Fuls, A., & Ellis, R. P. (1978). The geographical distribution of Kranz grasses in South Africa. *South African Journal of Science, 74*, 209–215.

Vogel, J. C., Wintle, A. G., & Woodbourne, S. M. (1999). Luminescence dating of coastal sands: Overcoming changes in environmental dose rate. *Journal of Archaeological Science, 26*, 729–733.

Waelbroeck, C., Labeyrie, L., Michel, E., Duplessy, J. C., McManus, J. F., Lambeck, K., et al. (2002). Sea level and deep water temperature changes derived from benthic foraminifera isotopic records. *Quaternary Science Reviews, 21*, 295–305.

Walter, R. C., Buffler, R. T., Bruggemann, J. H., Guillaume, M. M. M., Berhe, S. M., & Negassi, B. (2000). Early occupation of the Red Sea coast of Eritrea during the last interglacial. *Nature, 405*, 65–69.

Westley, K., Dix, J. (2006). Coastal environments and their role in prehistoric migrations. *Journal of Maritime Archaeology, 1*(1), 9–28.

Whitfield, A. K., Allanson, B. R., & Heinecken, T. J. E. (1983). Report 22. In A. E. F. Heydorn & J. R. Grindley (Eds.), *Estuaries of the Cape, part 2: Synopses of available information on individual systems* (CSIR Report No. 421). Stellenbosch: CSIR.

Chapter 3
Technological Change and the Importance of Variability: The Western Cape of South Africa from MIS 6-2

Alex Mackay

Abstract The South African Cape provides important evidence of behavioral and technological complexity in the period from MIS 6-2. Understanding the meaning of discontinuous temporal patterns in the distribution of technological systems is hampered by traditional culture historic approaches and culture evolutionary interpretations. These historical effects lead to depictions of the past as a series of stadial, progressive units. Evidence of variability is commonly suppressed and presumptions about what makes a technology advantageous go unquestioned. In this paper, key data used to generate existing stadial systems are considered from four sites in the Western Cape. Data are presented using the maximum available stratigraphic resolution within the constraints of the excavation systems used. Variability is shown to be a recurrent feature of technological systems. Rather than a series of discrete packages of innovation, technological change in this area is better understood in terms of the differential persistence of continually generated variation. The resulting picture is one of technologically flexible groups adapting rapidly and in some cases dramatically to changing circumstances through the Late Pleistocene.

Keywords South Africa • Western Cape • Lithic Technological Variability

Introduction

The Cape coastal and hinterland regions of South Africa feature prominently in discussions of human history and in particular in debates about the origins of behaviors distinctive of modern people. Rock shelter sites in these regions often preserve rich archives of archaeological material covering much of the period from MIS 6-2. Finds such as shell beads (Henshilwood et al. 2004; d'Errico et al. 2005, 2008), bone implements (Henshilwood and Sealy 1997; Henshilwood et al. 2001a; Backwell et al. 2008) and engraved fragments of ochre and ostrich eggshell (Henshilwood et al. 2002, 2009, 2014; Mackay and Welz 2008; Texier et al. 2010) have reshaped our understanding of human behaviors in these early periods, though such finds remain relatively infrequent components of pre-MIS 2 assemblages. Abundant, however, is evidence for the use of complex stone artifact technologies. These include the production of blades and bifacial points, the latter worked by soft hammer and possibly by pressure (Mourre et al. 2010; Hogberg and Larrson 2011), backed artifacts of a variety of forms, but most notably crescent- or segment-shaped pieces, and unifacial points. Stone selection patterns changed considerably, with some stone types preferentially sought despite increased acquisition costs (Mackay and Marwick 2011). There is also evidence for the use of heat treatment to alter the mechanical properties of flaked rocks at this time (Brown et al. 2009; Schmidt et al. 2013). Interestingly, the proliferation of implement types such as backed artifacts and bifacial points, along with the preferential selection of fine-grained rocks appears to be quite short-lived (Jacobs et al. 2008). This has led to the characterization of certain periods as unusually innovative, with attendant conjecture regarding the causes of both the appearance of technological complexity and its apparent subsequent demise (Jacobs and Roberts 2009; Powell et al. 2009; Chase 2010; Henshilwood and Dubreuil 2011).

Inhibiting our understanding of these issues are historically embedded assumptions about the factors underlying technological change. These affect the ways in which technological changes are presented and explained. This paper begins by considering the roles of culture history and culture

A. Mackay (✉)
Centre for Archaeological Science, School of Earth and Environmental Sciences, University of Wollongong, Northfields Avenue, Wollongong, NSW 2522, Australia
e-mail: mackay.ac@gmail.com

Sacha C. Jones and Brian A. Stewart (eds.), *Africa from MIS 6-2: Population Dynamics and Paleoenvironments*, Vertebrate Paleobiology and Paleoanthropology, DOI 10.1007/978-94-017-7520-5_3

evolutionism in structuring thought about technological change in southern Africa. The value of an alternative Darwinian approach, which centralizes changes in the through-time structure of variation, is then discussed. Technological sequences from four sites in the Western Cape region of South Africa are presented in light of this alternative view. It is shown that patterns of change are often fluid, while apparent innovations are shown rather to be recurrent aspects of technological expression, emphasized to a greater and lesser degree through time. These results highlight the adaptive and flexible nature of human technological systems in southern Africa during the Late Pleistocene, and make clear some important weaknesses in aspects of our present data and reasoning.

Culture Evolutionism and Culture History in Southern African Archaeology

Culture evolutionism and culture history have both had a considerable impact on archaeological research in southern Africa. Culture evolutionism can be defined as the idea that human socio-evolutionary history occurred in a sequence of stages, each leading to a more complex and effective way of life (Dunnell 1980). Culture history, on the other hand, is the idea that patterns in the archaeological record are reflections of past cultures, related to and differentiated from one another by a suite of shared traits (Lyman et al. 1997).

Culture evolutionism has its origins in nineteenth century social theory, wherein different forms of societal organization were placed in ranked systems according to their degree of advancement (Tylor 1929 [1871]; Morgan 1944 [1877]). Its application to archaeology, and specifically to the archaeology of technology, is underpinned by two assumptions. First, some technologies are inherently better than others. Second, better technologies will differentially persist over time. Quasilinear trends toward more complex technologies through time are logical consequences of these two assumptions.

Culture evolutionist thought is common in seminal early southern African archaeological work. Goodwin and van Riet Lowe (1929), for example, refer to waves of migrants bringing increasingly advanced technologies into southern Africa from the North, replacing less sophisticated groups. Gooch (as cited in Sanderson 1879) suggested that the degree of "adaptedness" of an artifact was inversely proportional to its age. More recently, Vishnyatsky (1994) has referred to the backed artifact-rich Howiesons Poort unit as "running ahead of time," on the basis that it antedates comparable systems in Europe by 20–30 ka. Numerous other researchers have referred to the "advanced" or "precocious" technologies of the Still Bay and Howiesons Poort (Butzer 1982; Foley and Lahr 2003; Henshilwood 2004; Mellars 2006a, b; Jacobs et al. 2008; Jacobs and Roberts 2009; Villa et al. 2010; Henshilwood and Dubreuil 2011); the inference being that there is a natural order to sequences of technological change, with these two units as early outliers.

The importance of culture evolutionism to archaeological research is that it provides a basis for inferring which technologies are most effective. The logic involved is both simple and circular. If we start with the two assumptions noted above, then it follows that technologies that occur more often in the recent past will be superior to those that occurred more often in the distant past. Thus, any given ancient technology, for example handaxes, will have been inferior to any given recent technology, for example microliths. These conclusions derive directly from the operational assumptions of the argument. It is important to note, however, that these assumptions are almost the only basis we presently have for inferring that microliths were more effective than handaxes.

The core problem with a culture evolutionist view of stone artifact technology is that the data required to test its conclusions are not readily available. Statements such as those to the effect that the Still Bay and Howiesons Poort technologies would have "substantially increased the efficiency and productivity of hunting activities" (Mellars 2006a: 9383), or "confer[red] a major adaptive advantage" (Foley and Lahr 2003: 117–118) lack the support of quantitative data.[1] Perhaps more problematically, the recent archaeological records of stone artifact using groups in Africa and around the world contain comparatively little evidence of technological complexity (White 1967; White and Thomas 1972; Brandt 1996; Shott and Sillitoe 2005; Weedman 2006; Conard 2007; Hiscock 2008), and instances of complexity loss are perhaps more common than often presumed (Torrence 1989; Sealy 2016). Idealized depictions of accumulating complexity through time (e.g., Foley and Lahr 1997, 2003; McBrearty and Brooks 2000;

[1]Accurately quantifying these effects would be extremely complex. Cost factors such as time/energy spent in stone acquisition (including replacement costs derived from implement use life), transport and implement production (including failure frequencies and learning costs) would need to be weighed against advantages of specific implements in terms of efficacy in pursuit of all relevant hunting/gathering and processing tasks. These factors would also have to be sensitive to, among other things, the effects of changes in mobility and prey choice. Moreover, the importance of within-system cost/benefit variability and absolute cost outlay would potentially be influenced by subsistence uncertainty and resource abundance (e.g., risk factors).

Henshilwood and Marean 2003) are thus inaccurate in important ways.[2]

Culture history has been another major theme of archaeological work in southern Africa. Primarily, early researchers were interested in using variation in stone artifact assemblages to identify past cultural groups (e.g., Penning 1887; Frames 1899; Kingston 1900; Peringuey 1911; Hewitt 1921; Burkitt 1928; Armstrong 1931; Stapleton and Hewitt 1927). In one example, Lewis-Abbott (1913) associated microliths with a group of "pygmy implement makers," The presence of microliths in Africa, Europe and Australia thus reflected a "great migration" (Lewis Abbott 1913: 147; see also Brown 1889; Mellars 2006b). Goodwin and Van Riet Lowe (1929) attributed almost all major archaeological variations to the arrival and passing of waves of immigrants. More recently, Volman's (1981) influential thesis on Late Pleistocene southern African technologies concluded that the form taken by a technology at any given point in time was largely the result of "fashion" (Volman 1981: 266; also note Kroeber 1909: 5, cited in O'Brien and Lyman 2007).

Along with some theoretical limitations (cf. O'Brien and Lyman 2007), culture history as employed in southern Africa has had the effect of reducing the archaeological record to a series of units, which are often temporally coarse. Caging history in such terms inevitably gives cultural/technological change a "blocky" appearance, in which characteristics that persist, sometimes for thousands of years, are replaced abruptly by a new set of characteristics (e.g., Singer and Wymer 1982; Thackeray 1992; Clark 1999; Wurz 2002; Minichillo 2005). These depictions also tend to be focused on a limited number of key markers that are most distinctive of a unit, rather than on all of the variation that a unit contains. In consequence, the difference between any two units invariably appears categorical (note Henshilwood and Dubreuil 2011). Combined, these factors make it impossible to discern change at levels finer than the culture historic unit itself, and long-term trends between units are necessarily obscured. This leads to perceptions of rapid and dramatic turnover, and to explanations phrased in terms of diffusion and population replacement (Singer and Wymer 1982; Bar-Yosef 2002; Jacobs et al. 2008).

Darwinian forms of archaeology provide a useful alternative to culture evolutionary and culture historic approaches (Leonard and Jones 1987; Jones et al. 1995; Lyman and O'Brien 1998; O'Brien and Lyman 2002; Mesoudi et al. 2006; Shennan 2008). From a Darwinian perspective, behavioral expression is viewed as inherently variable, with variability driven by a variety of random and targeted causes such as copying errors and problem solving, as well as social stimuli and population size (Boyd and Richerson 1985; Nieman 1995; Bettinger and Eerkens 1999; Shennan 2001, 2008; Eerkens and Lipo 2005). While variation is continually generated, the persistence of variants is shaped either by neutral processes such as drift or transmission biases, or by selective processes wherein more beneficial variants flourish while less beneficial variants tend to fail (Boyd and Richerson 1985; Shennan 2008). "Benefit" in this sense is context-specific rather than absolute, as it is in a culture evolutionist system. A variant that is beneficial in one context may be deleterious in another. Differentiating neutral effects from selective effects can be difficult given the complexity of transmission processes (Shennan 2008), though some insights are available. Drift will tend to produce variant homogeneity in small isolated populations or Markovian (battleship) structures in variant frequency where populations interact (Nieman 1995). Moreover, because they are neutral processes, drift, and transmission bias (including conformist and indirect bias [cf. Boyd and Richerson 1985]) need have no necessary relationship to environmental variation. Selection, on the other hand, is likely to be more strongly tied to environmental changes, particularly for items such as stone tools, which functioned as an interface between people and their subsistence environment (Kuhn 1995).

As the interest of Darwinian archaeology is in exploring and explaining the differential persistence of variation through time (Shennan 2008), the frequency distribution of variants becomes a central concern. As noted above, variation is often masked in existing approaches to southern African archaeology by the presentation of data in terms of previously defined culture historic units, such as the Howiesons Poort or Still Bay. Explanations for the causes of change are presumed with reference to culture evolutionism. In order to understand the differential persistence of variation, analytic units that maximize temporal resolution are required. The choice of variants is also of importance. In the present case, the variants chosen are the most basic of those that have been central to the key culture historic schemes deployed in the region. The objective is to use these same data to explore structural variation that is masked by traditional approaches, and through this, interrogate the validity of existing schemes, highlight any limitations, and ultimately to begin to approach underlying evolutionary factors.

[2]Some classes of technology do appear to become more prevalent and more varied through time, most notably organic assemblage components such as instruments, ornaments and implements made of wood and bone. Separating real increases in such items from increases resulting from taphonomic factors, however, is exceptionally problematic. As the Schoningen spears make clear (cf. Thieme 1997), the general absence of evidence for organic technologies in early assemblages cannot reasonably be taken as evidence of absence. As Surovell and Brantingham (2007) have demonstrated, preservation can create the appearance of directional change in cases where it almost certainly did not occur. Interpretations based on increases over time in the range and frequency of organic data must therefore be treated with considerable suspicion.

The Late Pleistocene Culture Historic Sequence in the Western Cape

The Western Cape of South Africa has a long history of Late Pleistocene research, with important sites such as Peers Cave having been excavated as early as the 1920s (Peers 1927). More recent work further to the north at Diepkloof (DRS) (Rigaud et al. 2006), Elands Bay Cave (EBC) (Parkington 1977; Volman 1981), Hollow Rock Shelter (HRS) (Evans 1994; Högberg and Larsson 2011), Klein Kliphuis (KKH) (Mackay 2006, 2010) and Klipfonteinrand (KFR) (Volman 1981; Mackay 2009) has resulted in a fairly well resolved and increasing well-dated culture historic sequence in the area (cf. Jacobs et al. 2008). It is on these sites, clustered in the Elands Bay and northern Cederberg Mountain regions that the present analysis is focused (Fig. 3.1).

The Late Pleistocene sequence in the Western Cape can roughly be divided into five broad culture historic units (Table 3.1). The earliest is the most heterogeneous and conflates all assemblages antedating ~75 ka and likely postdating 120 ka. Assemblages in this period include few morphologically regular retouched implements, a predominance of quartzite among materials used in artifact manufacture, and generally large flakes and cores (Volman 1981). After ~75 ka, the appearance of bifacial points marks the advent of the Still Bay. Silcrete has been noted to increase in prevalence at this time (Henshilwood et al. 2001b). The Still Bay ends ~70 ka, but the next known unit, Howiesons Poort, probably does not begin until ~65 ka (Tribolo et al. 2005, 2009; Valladas et al. 2005; Jacobs et al. 2008). The nature of any intervening assemblages is not well known at present, but recent work by Brown and colleagues (2012) suggests some

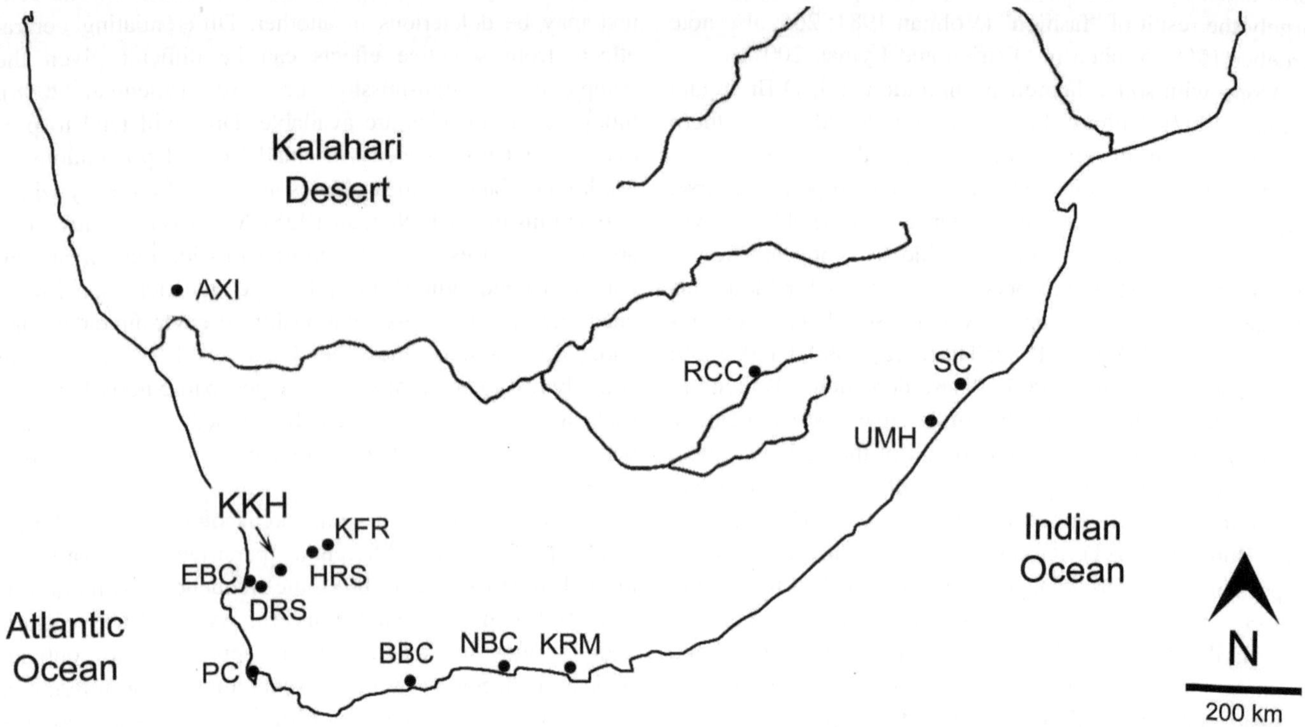

Fig. 3.1 Locations of sites discussed in the text

Table 3.1 Culture historic units in the Western Cape and their basic identifying characteristics

Culture historic unit	Age range (ka)	Dominant materials	Dominant implements
Late Pleistocene microlithic	12–30	Quartz	None
Later Middle Stone Age	30–50	Quartz	*Unknown*
Post-Howiesons Poort	50–59	Quartzite	Unifacial points
Howiesons Poort	59–65	Silcrete	Backed artifacts
Unnamed Howiesons Poort-like unit	65–70	*Unknown*	Backed artifacts
Still Bay	70–75	Silcrete/quartzite	Bifacial points
Earlier Middle Stone Age	75–120 (?)	Quartzite	None

continuity in the manufacture of backed artifacts through this period. This is discussed further below. The Howiesons Poort is most often associated with large numbers of backed artifacts, the production of laminar elements, and with even more dramatic increases in silcrete than are common in the Still Bay (Volman 1981; Thackeray 1992; Wurz 2002; Mackay 2010). At ~59 ka, the Howiesons Poort gives way to another heterogeneous unit referred to as the "post-Howiesons Poort", associated with the disappearance of backed artifacts and their replacement by unifacial points, diminished frequency of blades, and a return to prominence of quartzite. At some later point in this unit the frequency of quartz increases and unifacial points may disappear (Mackay 2010) – a unit I'll refer to here as the Later Middle Stone Age – though the timing and precise nature of changes in this period are again not well understood. At some point ~30 ka, prepared core reduction systems, which are common to all preceding units, disappear. At the sites of KKH and EBC quartz becomes dominant, and few implements are produced. Assemblages from this period might be assigned to the broad Late Pleistocene microlithic grouping (Deacon 1984; Orton 2006).

Technological Changes at Four Western Cape Sites

This section considers technological changes at four of the Western Cape sites noted above – DRS, HRS, KKH and KFR. The analytic focus is on the same kinds of data that have been used to generate the culture historic sequence discussed in the preceding section, and thus on material prevalence and implement types. This is done to draw out most strongly the contrasts between the two approaches.

Minimum analytic units differ between sites, as does the kind of the stone artifact data available. DRS and KKH are recently (re)excavated and provide the best resolution along with the most comprehensive analyses, including all technological classes (cores, implements, unretouched flakes, etc.). The samples from both sites derive from single 1 m column sequences, though the sample numbers are quite robust in both cases (DRS n = 6487; KKH n = 7604). Sequential data from HRS and KFR are more coarsely resolved and only implement-type data (e.g., morphologically regular retouched flakes) are available.

Changes in Materials and Implement Frequencies at Diepkloof

The Diepkloof sequence is known to include Still Bay, Howiesons Poort and post-Howiesons Poort units, along with assemblages antedating the Still Bay (Rigaud et al. 2006). While there are no published basal ages, if we accept that the Still Bay dates to early MIS 4 (*pace* Tribolo et al. 2009) then it follows that the oldest layers are likely to date to some part of MIS 5. The Pleistocene component of the site appears to terminate some time early in MIS 3 (Rigaud et al. 2006; Jacobs et al. 2008).

In some respects, Diepkloof is the type sequence for the region and is thus a particularly good candidate for exploring changes in aspects of stone artifact technology. The site is also very well stratified, making it possible to examine changes at high resolution. The convention used in the excavation of DRS was to name, rather than number layers. This allows the many discontinuous layers across the site to be reconciled more easily into a single stratigraphic sequence. As the present analysis only considers a single column from square L6, however, the layer names have been changed to sequential numbers. Layer numbers, layer names, artifact numbers, and ages are presented in Table 3.2. Note that the ages presented by Jacobs et al. (2008) are used here rather than those presented by Tribolo et al. (2009). The Still Bay ages presented by Tribolo et al. (2009) are inconsistent with ages for that industry in sites to the north (Vogelsang et al. 2010), east (e.g., Högberg and Larsson 2011) and south (e.g., Jacobs et al. 2013). This implies that: Diepkloof is out of phase with all others sites in the region, the ages presented by Tribolo et al. (2009) are inaccurate, or dating of the region generally is compromised by unresolved issues in luminescence assays. For now the match between ages at Diepkloof and Hollow Rock Shelter, ~60 km apart and dated by two different labs, lends weight to those ages.

Figure 3.2 presents changes in material prevalence through the DRS sequence. Only the three main material types – silcrete, quartzite, and quartz - are considered, though these in any case account for more than 80% of artifacts in all layers.

The early layers at DRS are dominated, as expected, by quartzite, though there is considerable variation in the prevalence of quartz. Silcrete, however, generally represents less than 5% of artifacts in these early layers. Around layer 38, the prevalence of silcrete increases dramatically to >20%, persisting at moderate to high levels through to layer 30 where it briefly becomes the dominant material. Immediately thereafter (layer 29), however, the prevalence of silcrete subsides, and though it continues in moderate frequencies in layers up to layer 26, it is quartz that dominates in this intervening period. Layers 22–12 show a sustained peak in silcrete, with the material accounting for a relative proportion of 40% or greater throughout. Layers 11–1 show alternating surges in quartz (11–7) and silcrete (6–1).

Figure 3.3 shows the occurrence of key implement types – backed artifacts, bifacial points, and unifacial points – through the sequence. Several things are worth noting. First,

Table 3.2 Diepkloof layers, artifact numbers and dates

Layer	Number of artifacts	Ages	Layer	Number of artifacts	Ages
1 (Burnt crust)	9	–	26 (Jess)	456	–
2 (Claude)	22	–	27 (Julia)	136	–
3 (Denzil)	68	–	28 (Kate)	139	–
4 (Danny)	79	–	29 (Kerry)	23	70.9 ± 2.3
5 (Debbie)	63	–	30 (Kenny)	142	–
6 (Darryl)	172	–	31 (Kegan)	199	–
7 (Deon)	28	–	32 (Keno)	105	–
8 (Eric)	153	–	33 (Kim)	26	–
9 (Ester)	168	–	34 (Larry)	177	–
10 (Edgar)	45	–	35 (Logan)	170	73.6 ± 2.5
11 (Eve)	203	–	36 (Liz)	72	–
12 (Eben)	62	–	37 (Leo)	125	–
13 (Fred)	114	–	38 (Lynn)	132	–
14 (Frank)	303	–	39 (Lauren)	223	–
15 (Frans)	85	–	40 (Mike)	46	–
16 (Fanie)	450	–	41 (Mark)	196	–
17 (Fiona)	80	–	42 (Moses)	5	–
18 (Fox)	92	–	43 (Maggie)	70	–
19 (Glen)	34	–	44 (Miles)	237	–
20 (Governor)	21	–	45 (Mary)	89	–
21 (John)	56	63.3 ± 2.2	46 (Noel)	96	–
22 (Jeff)	131	–	47 (Noah)	169	–
23 (Joy)	320	–	48 (Nina)	84	–
24 (Jack)	362	–	49 (Nel)	86	–
25 (Jude)	118	–	50 (Neva)	46	–

Includes artifacts >15 mm only. Ages are OSL determinations derived from Jacobs et al. (2008) and are presented as ka

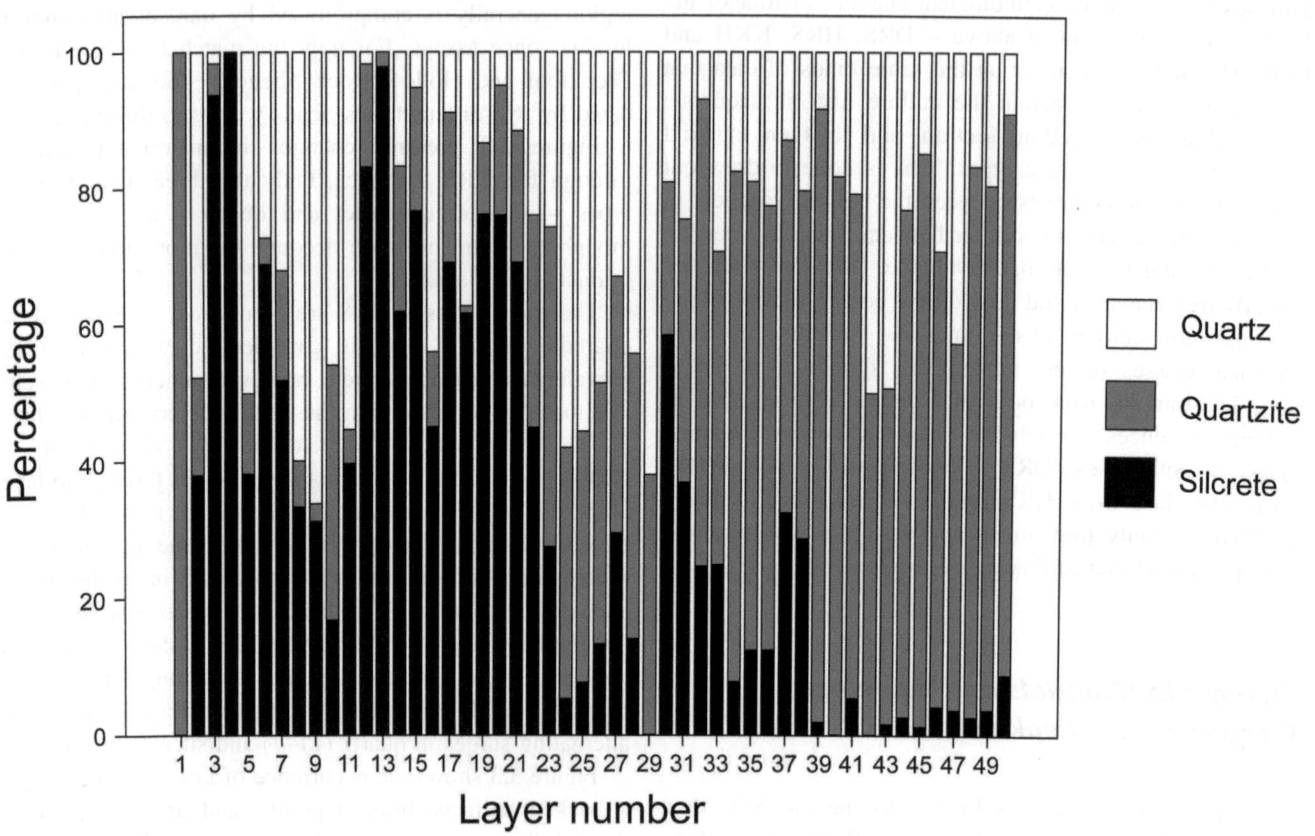

Fig. 3.2 Changes in material frequency at Diepkloof; relative proportions of silcrete, quartzite and quartz only

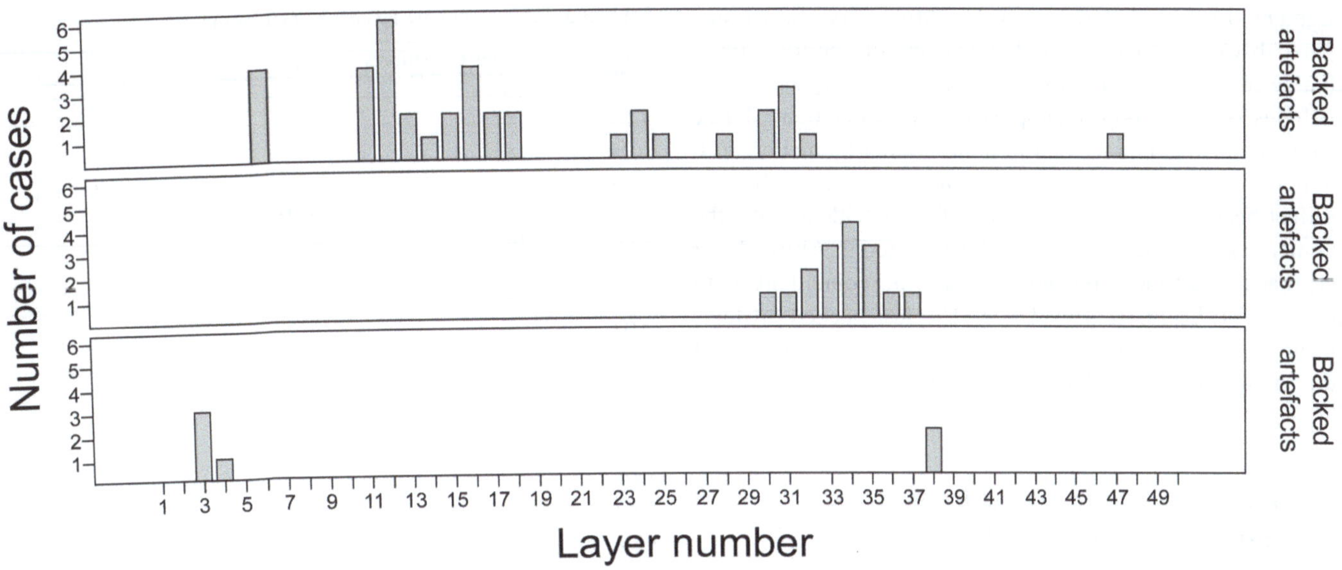

Fig. 3.3 Changes in implement type frequencies at Diepkloof; backed artifacts, bifacial points and unifacial points only

Fig. 3.4 Implements from Diepkloof. *Upper row* are points: **a**–**c** = bifacial points; **d**, **e** = unifacial points. *Lower row* are backed artifacts: **f**–**h** = truncated flakes; **i**, **j** = segments. The bifacial points all derive from the Still Bay. Unifacial point (**d**) derives from the immediately pre-Still Bay layers; (**e**) is post-Howiesons Poort. Truncated flake (**f**) is pre-Still Bay; (**g**) is Still Bay; (**h**) is Howiesons Poort. Segment (**i**) is Still Bay; (**j**) is Howiesons Poort

the oldest backed artifact, a truncated quartz flake (Fig. 3.4), occurs relatively deep in the sequence in layer 47. This is well before the implement type exhibits any sustained presence. Bifacial points are constrained to a single sequential cluster of eight layers from 37 to 30 inclusive, though their distribution through these layers shows an increase in frequency, a peak and a steady decline. Immediately preceding the appearance of bifacial points, however, is a pair of unifacial points in layer 38 (Fig. 3.4). This is also the layer where silcrete dramatically increases.

In the last few bifacial point-bearing layers backed artifacts recur, including several truncated pieces and a classic segment (Fig. 3.4). Rather than heralding a period of sustained backed artifact discard, however, the period immediately after the cessation of bifacial points (layers 29–19) includes only episodic examples of this implement. These layers are also generally rich in quartz rather than silcrete. It is not until layer 18 that a sustained cluster of backed artifact-bearing layers occurs, broadly coincident with the reemergence of silcrete as a dominant material. Above layer 11, backed artifacts become infrequent, coincident with another rise in quartz prevalence. The last backed artifacts occur in layer 6, with unifacial points reappearing in layer 4 after an absence of probably ~15 kyr.

Changes in Materials and Implement Frequencies at Klein Kliphuis

Klein Kliphuis was excavated in seven stratigraphic units divided into a series of ~25 mm spits. The sequence covers much of the period from the start of the Howiesons Poort through to the Late Pleistocene microlithic. Basal ages suggest that the sequence initiates around 66–68 ka, while the youngest Pleistocene layers are dated to around ~20 ka (Jacobs et al. 2008; Mackay 2010). Layer numbers, artifact numbers, and ages are presented in Table 3.3.

Figure 3.5 presents the prevalence of different material types through the KKH sequence. Again, only silcrete, quartzite, and quartz are used but again these account for >80% of artifacts in all layers. Stratigraphic units at KKH were originally allocated lower case roman numerals ("i" to "vii"), and the spits within them were allocated numbers. Here, the roman numerals have been changed to numbers (e.g., i = 1, ii = 2, iii = 3, etc.), and the labeling convention is as follows: layer 7.04 represents the fourth spit in the seventh stratigraphic unit; 6.12 is the 12th spit in the sixth unit, and so on (cf. Mackay 2010).

Silcrete dominates the oldest layers at the site, accounting for >80% of the sample in almost all layers below 6.05. The exception is layer 6.11 where the prevalence of silcrete briefly declines and the prevalence of quartz increases. However, this change is not random or unstructured. Quartz increases steadily from 6.15 to the peak in 6.11 and then decreases steadily again through to 6.07. This sequence of changes separates two distinct peaks in silcrete, suggesting an alternation in selection between the two materials, a pattern also noted at DRS.

Above its peak value in layer 6.07, silcrete prevalence gradually declines with quartzite becoming the dominant material above layer 6.04. Thereafter, silcrete is a relatively minor contributor to the assemblage with initially quartzite and subsequently quartz dominating.

Table 3.3 Klein Kliphuis layers, artifact numbers and dates

Layer	Number of artifacts	Ages
1.02	16	–
1.03	14	–
1.04	18	–
1.05	25	22.3 ± 0.3 cal kBP (Wk-20241) 33.0 ± 1.0 ka
1.06	10	–
2.01	3	–
2.02	6	–
2.03	5	–
3.01	21	–
4.01	55	>35.0 kBP
4.02	15	–
4.03	25	–
5.01	77	–
5.02	9	–
6.01	70	56.0 ± 3.0 ka
6.02	95	–
6.03	191	58.0 ± 2.0 ka
6.04	333	–
6.05	502	–
6.06	615	65.0 ± 3.0 ka
6.07	864	–
6.08	1077	–
6.09	945	–
6.10	517	60.0 ± 3.0 ka
6.11	202	–
6.12	558	–
6.13	749	64.0 ± 3.0 ka
6.14	336	–
6.15	116	–
7.01	63	–
7.02	30	–
7.03	10	–
7.04	3	–

Includes artifacts >15 mm only. Ages are OSL determinations derived from Jacobs et al. (2008), except where otherwise noted, and are presented as ka. Radiocarbon ages are presented as cal kBP

Figure 3.6 shows the distribution of backed artifacts and unifacial points through the sequence. Backed artifacts are common in the early layers but show a bimodal distribution that mirrors changes in silcrete prevalence. In layers 6.07 and 6.06 unifacial points appear and for this brief period in the sequence the two key implement types overlap. While this may be a consequence of stratigraphic mixing, as noted elsewhere (Mackay 2011) other implement types occur in these layers. These meld aspects of unifacial points and backed artifacts and thus a period where people either made or were aware of both is not implausible. Unifacial points were not observed above layer 5.01, where quartz becomes the dominant material, but a single backed artifact does occur in layer 1.04 (Fig. 3.7), dated to ~22.0 cal kBP (Table 3.3).

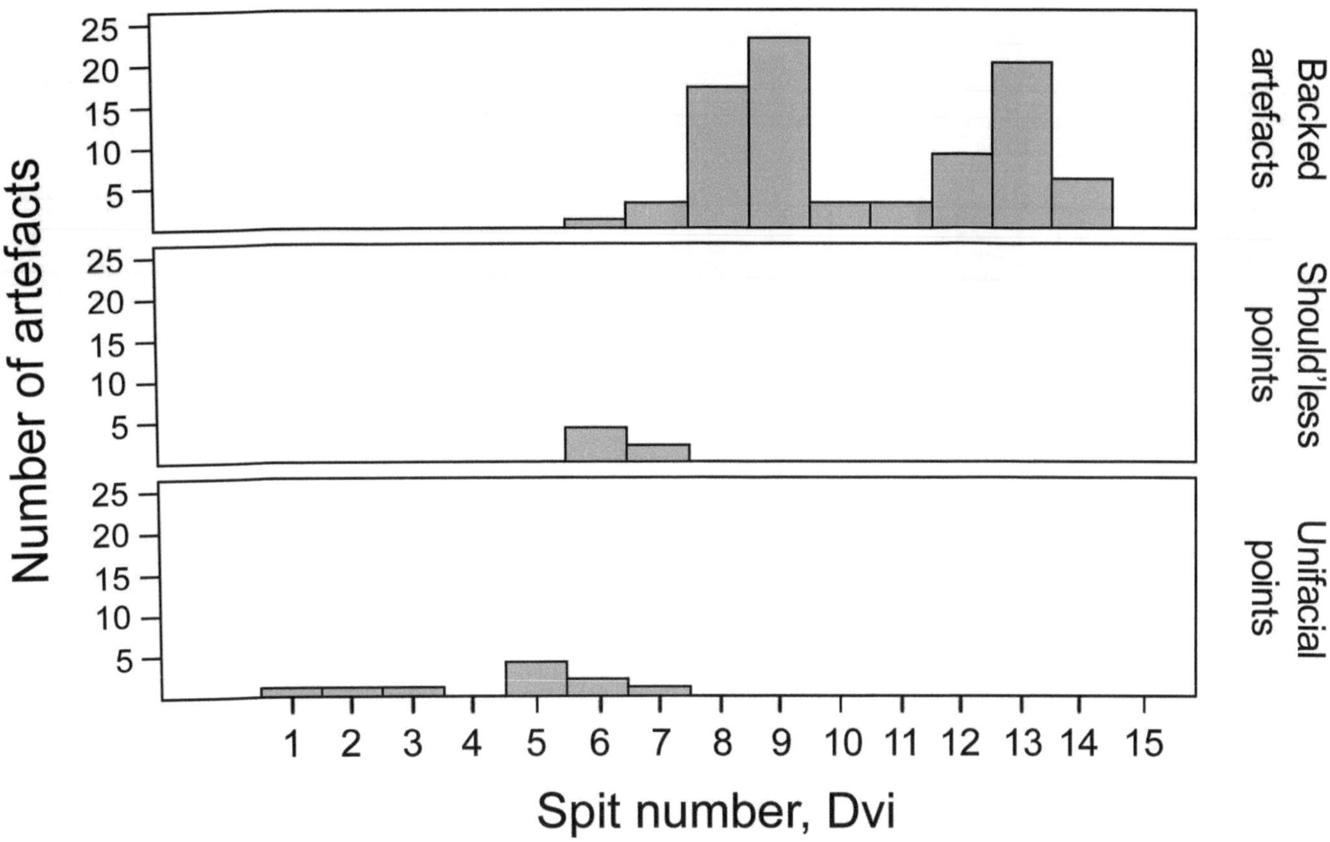

Fig. 3.5 Changes in material frequency at Klein Kliphuis; relative proportions of silcrete, quartzite and quartz only

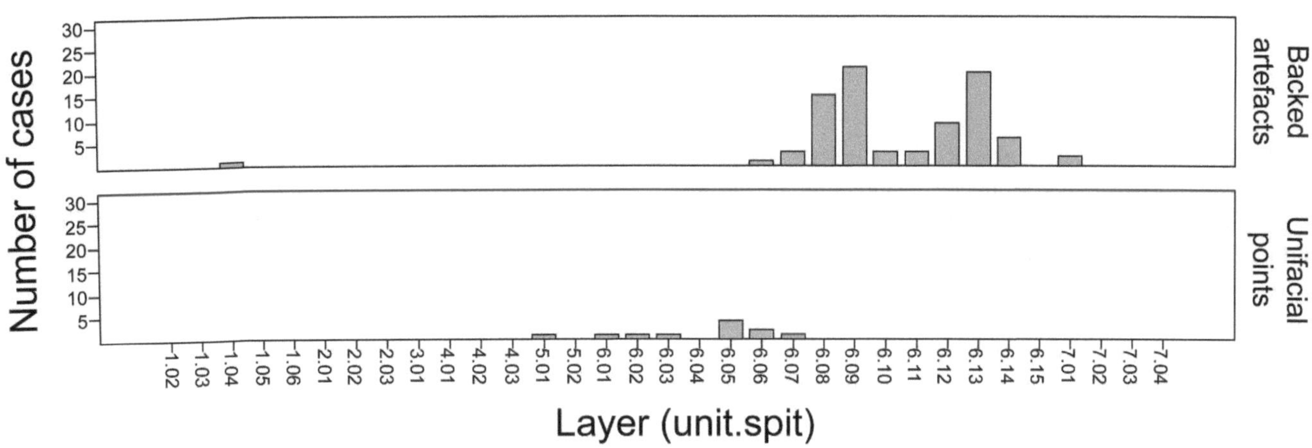

Fig. 3.6 Changes in implement type frequencies at Klein Kliphuis; backed artifacts and unifacial points only

Changes in Implement Frequencies at Hollow Rock Shelter and Klipfonteinrand

Hollow Rock Shelter and Klipfonteinrand provide more limited and/or coarsely resolved sequences than DRS and KKH, and only HRS has ages (Högberg and Larsson 2011). Previous analyses suggest the presence of a Still Bay and earlier technologies at HRS (Evans 1994; Högberg and Larsson 2011), and a Howiesons Poort was preceded by earlier technologies at KFR (Volman 1981; Mackay 2009). The interest here is on the occurrence of key implement types through these sequences. The HRS data presented derive from Evans (1994) published work while the KFR data derive from a reanalysis by the author of Parkington's excavated material from 1969.

Fig. 3.7 Implements from Klein Kliphuis. *Upper row* are backed artifacts: **a–c** = segments; **d**, **e** = truncated flakes; **f** = notched segment. *Lower row* are all unifacial points. All backed artifacts are from the Howiesons Poort except (**c**), which comes from the Late Pleistocene microlithic. All unifacial points are from the post-Howiesons Poort

Table 3.4 Numbers of implements by type in the Hollow Rock Shelter sequence

Layer	Backed artifacts	Unifacial points	Bifacial points
IA	0	1	25
IB	1	0	11
IIA	1	0	3
IIB	2	0	1

Table 3.5 Numbers of implements by type in the Klipfonteinrand sequence

Layer	Backed artifacts	Unifacial points
5	9	0
6	7	2
7	3	4
8	1	2
9	0	0

Table 3.4 shows the presence of implements in the sequences at HRS. This sequence includes all of the key types; and it's most interesting aspect is the co-occurrence of bifacial points and backed artifacts, and of bifacial points and unifacial points in different layers. Bifacial points occur in all layers but increase in frequency through the sequence, with the maximum number occurring in the uppermost layer. The backed artifacts, which include two truncated flakes and one other backed piece, are most common in the lower layers – that is, those preceding the peak in bifacial points. The single unifacial point occurs in the uppermost layer. Intriguingly, this sequence is the inverse of that observed at DRS, where unifacial points precede bifacial points and backed artifacts follow them.

The KFR sequence (Table 3.5) includes only backed artifacts and unifacial points. Backed artifacts occur in four of five layers but are most common in the upper two (layers 5 and 6).[3] Unifacial points occur in the middle three layers but are absent from the top and bottom of the sequence. Points are at their most common in the middle layer, layer 7.

Discussion

The technological changes presented above are characterized by two key patterns. First, changes in material prevalence usually occur as a series of alternating surges, some of which are gradual (e.g., 6.15–6.07 (quartz) and 6.07–6.04 (silcrete) at KKH; 24–20 (silcrete) and 9–4 (quartz) at DRS) and others of which are very abrupt (2.03–2.02 at KKH; 39–38 and 30–29 at DRS). Different rates of change and/or different mechanisms of change may underpin these alternatives. For example, the gradual phasing out of silcrete from 6.07 to 6.04 at KKH along with the change from backed artifacts to unifacial points is unlikely to reflect abandonment of the site or sudden population replacement (cf. Mackay 2011). The sudden switch from the silcrete-rich, bifacial point-bearing

[3]Layers 1–4 are divisions of the Holocene.

layer 30 to the quartz-rich and bifacial point-lacking layer 29 at DRS, on the other hand, might reflect an occupational discontinuity or the rapid abandonment of one technological system and uptake of another. The surge in quartz and decrease in backed artifacts from layers 6.12 to 6.09 at KKH is complemented by decreases in flake and core size and an increase in bipolar reduction (Mackay 2009) and as such almost certainly reflects a period of technological reorganization within what would otherwise be considered typical Howiesons Poort.

The second major pattern is that diagnostic implement types such as backed artifacts and unifacial points are recurrent assemblage components. Backed artifacts occur before bifacial points at DRS and with bifacial points at DRS and HRS. In the layers dated between ~70 and ~65 ka at DRS, backed artifacts occur but are infrequent, potentially matching the pattern at Pinnacle Point (Brown et al. 2012). Only with the surge in silcrete from layer 21 to layer 11 do backed artifacts become a sustained feature of the assemblage. Backed artifacts then recur some 35 ka after the Howiesons Poort, in the Late Pleistocene microlithic levels at KKH. They have also been noted in MIS 2 contexts at EBC (Mackay 2009).

Similarly recurrent, though perhaps less so, are unifacial points. These occur immediately before the appearance of bifacial points at DRS; with bifacial points at HRS; before backed artifacts at KFR; and both with and after backed artifacts at KKH. Even if we accept Volman's (1981) suggestion of stratigraphic problems at KFR, it seems unlikely that the excavators would have produced an assemblage that inverted the sequence, thus placing unifacial points before rather than after backed artifacts. In a similar vein, no argument has been made for a Howiesons Poort at HRS (cf. Hogeberg and Larrson 2011) and conflation of units would not have inverted the succession of backed artifacts and bifacial points.

Of the key markers considered here, only bifacial points are not recurrent in the assemblages examined. Their appearance in the Still Bay might well be considered the only clear technological "innovation" (or unprecedented variant) in these Late Pleistocene sequences. Even that should be weighed against the temporal and numerical limitations of the available samples.

While it might be tempting to view the patterns presented here as the result of stratigraphic problems or of unusually "noisy" assemblages, broader consideration of the southern African literature suggests that the recurrence of different implements through time is quite common. Volman (1981) for example, noted the presence of bifacial points in his early Middle Stone Age (MSA) 2a unit, while unifacial points and bifacial points occurred in almost all MSA 2b assemblages examined. Both of these units antedate the Howiesons Poort, though their relation to the Still Bay is unclear given that Volman did not formally recognize the unit. Unifacial points also occur with bifacial points in the Still Bay-ascribed layers at Apollo 11 (Jacobs et al. 2008). Consequently the co-occurrence of bifacial and unifacial points is not limited to HRS, and the presence of unifacial points both before and after the Howiesons Poort is not isolated to DRS. Further, with respect to MSA 2b, Volman (1981: 251) stated that: "[e]very site provided rare backed pieces which would not be out of place in Howiesons Poort assemblages," This includes Klasies River, whose sequence includes backed artifacts in its earlier MSA units (Singer and Wymer 1982). Both backed artifacts and bifacial points are also present in late post-Howiesons Poort assemblages at Rose Cottage Cave and Umhlatuzana in eastern South Africa (Wadley 2006; Kaplan 1990), while backed artifacts occur with bifacial points in the Still Bay at Sibudu (Wadley 2007). More recently, Wadley (2012) has documented the presence of bifacial points in contexts antedating the Still Bay at Sibudu.

In relation to issues of culture history, the data presented here suggest that far from occurring as a blocky sequence of discrete units, changes in implement type and material selection through the Late Pleistocene in the Western Cape were fluid and highly variable. Even where culture historic units are readily identified, such as in the Howiesons Poort at KKH and in the Still Bay at DRS, consideration of variation within units is informative. The middle of the Howiesons Poort at KKH witnesses a period where silcrete prevalence markedly dips and where the discard of backed artifacts also decreases in frequency. The Howiesons Poort thus appears multimodal, suggesting a sequence of on-going adaptive responses, rather than a single static entity. A bimodal Howiesons Poort can also be discerned at Klasies River (Singer and Wymer 1982) and at Nelson Bay Cave (Volman 1981).[4] As discussed above and elsewhere (Mackay 2011), the end of the Howiesons Poort at KKH is also gradual, effectively undercutting arguments for occupational discontinuities between the Howiesons Poort and post-Howiesons Poort.

At DRS, while bifacial points occur throughout the Still Bay, their frequency exhibits structured change. Bifacial points occur first in small numbers before later increasing. The same trend occurs at HRS (and also Blombos [cf. Villa et al. (2010)]. At HRS, the occurrence of bifacial points is truncated at its maximum frequency by the apparent cessation of occupation at the site. At DRS, however, bifacial point frequency declines with the appearance of backed artifacts. Yet the appearance of backed artifacts at the end of the Still Bay does not herald their sustained presence but

[4]Multimodal Howiesons Poort assemblages have been published for Diepkloof (Porraz et al. 2013) and Klipdrift (Henshilwood et al. 2014) since this paper was written.

rather initiates a quartz-rich period in which they occur episodically. It is thus possible that the late Still Bay at DRS witnessed a period of technological instability, potentially involving concurrent or alternating use of quite different technological systems, neither of which was sustained to any great degree past 70 ka.

The time range from 70 to 65 ka covers the unnamed unit between the Still Bay and Howiesons Poort. Whether this period at DRS is ultimately ascribed to the Howiesons Poort or some variant of it is a culture historic issue (note Brown et al. 2012). However, it is clear that there are marked differences between material selection and implement manufacture at this time and in layers above 21. Thus, at the very least, consideration of the data at layer-by-layer resolution suggests more variability than had previously been noted.

In relation to issues of cultural evolution the notion that technologies such as backed artifacts or bifacial points were "advanced" or of unusually great adaptive benefit cannot be sustained. Backed artifacts are repeatedly deployed and abandoned through the Late Pleistocene. This does not fit easily with the idea that these tools were a universal panacea (e.g., Mellars 2006a, b). Bifacial points occur only once in the sequences considered and persist for <5 ka, possibly for as little as 2 ka (Jacobs et al. 2008). If such tools were so advantageous, why is their appearance so singular and so brief? An associated question is why the replacement of backed artifacts by unifacial points at the end of the Howiesons Poort is seen as a loss of complexity (e.g., Powell et al. 2009), when no robust evidence demonstrates the superiority of the former over the latter. Technological complexity did not necessarily disappear at the end of the Howiesons Poort; more likely it simply changed form.

The technological changes witnessed in the Western Cape through the Late Pleistocene are readily reconciled with a Darwinian view of cultural change. Variants such as backed artifacts and unifacial points appear continually to have been thrown up, with their frequency increasing under some circumstances and decreasing under others. Both neutral and selective processes may have played a part in these changes. The frequency distribution of bifacial points, for example, displays what might be interpreted as Markovian structures, potentially implying the operation of drift, and their demise with the introduction of a selectively equivalent variant (e.g., backed artifacts). Similar arguments might be made with respect to material selection patterns. In general terms, however, drift probably had a limited impact on temporal patterning in this study. Drift is expected to lead to homogenization of variants most readily in small isolated populations (Nieman 1995), yet the implement-type markers of the Howiesons Poort most notably, but also of the Still Bay and post-Howiesons Poort, are spread across southern Africa making the premise of isolated populations unlikely. If drift was solely to account for technological change in this period across southern Africa we might reasonably expect more differentiation of variant-preference in disparate locations. A combination of drift and selection provides a more plausible explanation (e.g., Powell et al. 2009). While material selection patterns do differ across southern Africa in the Howiesons Poort and Still Bay, the associated inference of functional equivalence between material types is questionable (Mackay 2008). Moreover, even if material selection itself was neutral, it may have functioned as an indicator trait for changes in mobility that were under selection (e.g., Ambrose and Lorenz 1990).

While there may be a role for other selectively neutral processes in these patterns (such as indirect bias, cf. Boyd and Richerson 1985), the broad correlation between environmental and technological change more readily implies active selective pressures (for extended discussion, see Mackay and Marwick 2011). For example, the frequency of bifacial points only begins to surge with the onset of glacial MIS 4 (Jacobs et al. 2008). The coldest part of that period, as attested by Southern Hemisphere ice cores (Jouzel et al. 2007), broadly correlates with the abandonment of bifacial points and a surge in quartz. Backed artifacts flourish in the warmer later parts of MIS 4 and give way to unifacial points with the transition to MIS 3 (Jacobs et al. 2008). MIS 2, much like mid-MIS 4, includes occasional backed pieces and is rich in quartz at KKH and also EBC (Mackay 2009). While the paucity of local terrestrial environmental data in the study area (Carr et al. 2016) and the complexity of environment/technology relationships precludes making strong statements about the nature of causality, correlation with global scale environmental change strongly implies some role for active selective pressures in technological change. The resulting picture is one of fluid technological adaptation to rapidly changing circumstances through the Late Pleistocene.

Concluding Remarks

In effect, Darwin took an orderly world and made it messy, replacing immutable categories selected and consciously emplaced by divine force with the differential persistence of randomly generated heritable variation. There is much to be gained from taking a similarly Darwinian view of technological change. Southern African people in the Late Pleistocene were not members of unchanging groups brought into existence only to be replaced. Nor were they early points on a developmental continuum. As the data presented here make clear, people in the Late Pleistocene were technologically and behaviorally complex, varying, and adapting rapidly through the Late Pleistocene. This complexity can only be appreciated when the messy grandeur of variability is considered.

Acknowledgments Thanks to Sacha Jones and Brian Stewart for organizing the African MIS 6-2 Conference and for putting together this volume. Professor Judith Sealy provided helpful advice on an earlier draft of this paper. Three anonymous reviewers provided excellent and helpful critique and are thanked for their time and effort.

References

Ambrose, S., & Lorenz, K. G. (1990). Social and ecological models for the Middle Stone Age in southern Africa. In P. A. Mellars (Ed.), *The emergence of modern humans* (pp. 3–33). Edinburgh: Edinburgh University Press.

Armstrong, A. L. (1931). Rhodesian archaeological expedition (1929): Excavations in Bambata Cave and researches on prehistoric sites in southern Rhodesia. *The Journal of the Royal Anthropological Institute of Great Britain and Ireland, 61*, 239–276.

Backwell, L., d'Errico, F., & Wadley, L. (2008). Middle Stone Age bone tools from the Howiesons Poort layers, Sibudu Cave, South Africa. *Journal of Archaeological Science, 35*, 1566–1580.

Bar-Yosef, O. (2002). The Upper Paleolithic revolution. *Annual Review of Anthropology, 31*, 363–393.

Bettinger, R. L., & Eerkens, J. W. (1999). Point typologies, cultural transmission, and the spread of bow-and-arrow technology in the prehistoric Great Basin. *American Antiquity, 64*, 231–242.

Boyd, R., & Richerson, P. J. (1985). *Culture and the evolutionary process*. Chicago: University of Chicago Press.

Brandt, S. A. (1996). The ethnoarchaeology of flaked stone tool use in southern Ethiopia. In G. Pwiti & R. Soper (Eds.), *Aspects of African archaeology* (pp. 733–738). Harare: University of Zimbabwe Publications.

Brown, J. A. (1889). On some small, highly specialized forms of stone implements, found in Asia, North Africa, and Europe. *The Journal of the Royal Anthropological Institute of Great Britain and Ireland, 18*, 134–139.

Brown, K., Marean, C., Herries, A., Jacobs, Z., Tribolo, C., Braun, D., et al. (2009). Fire as an engineering tool of early modern humans. *Science, 325*, 859–862.

Brown, K., Marean, C., Jacobs, Z., Schoville, B., Oestmo, S., Fisher, E., et al. (2012). An early and enduring advanced technology originating 71,000 years ago in South Africa. *Nature, 491*, 590–594.

Burkitt, M. C. (1928). *South Africa's past in stone and paint*. Cambridge: Cambridge University Press.

Butzer, K. W. (1982). Geomorphology and sediment stratigraphy. In R. J. Singer & J. Wymer (Eds.), *The Middle Stone Age at Klasies River Mouth in South Africa* (pp. 33–42). Chicago: University of Chicago Press.

Carr, A. S., Chase, B. M., & Mackay, A. (2016). Mid to Late Quaternary landscape and environmental dynamics in the Middle Stone Age of southern South Africa. In S. C. Jones & B. A. Stewart (Eds.), *Africa from MIS 6-2: Population dynamics and paleoenvironments* (pp. 23–47). Dordrecht: Springer.

Chase, B. M. (2010). South African palaeoenvironments during Marine Oxygen Isotope Stage 4: A context for the Howiesons Poort and Still Bay industries. *Journal of Archaeological Science, 37*, 1359–1366.

Clark, A. M. B. (1999). Late Pleistocene technology at Rose Cottage Cave: A search for modern behavior in an MSA context. *African Archaeological Review, 16*, 93–119.

Conard, N. J. (2007). Cultural evolution in Africa and Eurasia during the Middle and Late Pleistocene. In W. Henke & I. Tattersall (Eds.), *Handbook of paleoanthropology* (pp. 2001–2037). Berlin: Springer.

d'Errico, F., Henshilwood, C., Vanharen, M., & van Niekerk, K. (2005). *Nassarius kraussianus* shell beads from Blombos Cave: Evidence for symbolic behaviour in the Middle Stone Age. *Journal of Human Evolution, 48*, 3–24.

d'Errico, F., Vanhaeren, M., & Wadley, L. (2008). Possible shell beads from the Middle Stone Age layers of Sibudu Cave, South Africa. *Journal of Archaeological Science, 35*, 2675–2685.

Deacon, J. (1984). Later Stone Age people and their descendants in southern Africa. In R. G. Klein (Ed.), *Southern African prehistory and palaeoenvironments* (pp. 221–328). Rotterdam: Balkema.

Dunnell, R. C. (1980). Evolutionary theory and archaeology. In M. B. Schiffer (Ed.), *Advances in archaeological method and theory* (Vol. 3, pp. 38–99). New York: Academic Press.

Eerkens, J., & Lipo, C. (2005). Cultural transmission, copying errors, and the generation of variation in material culture and the archaeological record. *Journal of Archaeological Science, 24*, 316–334.

Evans, U. (1994). Hollow Rock Shelter, a Middle Stone Age site in the Cederberg. *Southern African Field Archaeology, 3*, 63–73.

Foley, R. A., & Lahr, M. M. (1997). Mode 3 technologies and the evolution of modern humans. *Cambridge Journal of Archaeology, 7*, 3–36.

Foley, R., & Lahr, M. M. (2003). On stony ground: Lithic technology, human evolution, and the emergence of culture. *Evolutionary Anthropology, 12*, 109–122.

Frames, E. M. (1899). On some stone implements found in a cave in Griqualand-East, Cape Colony. *Journal of the Anthropological Institute of Great Britain and Ireland, 28*, 251–257.

Goodwin, A. J. H., & Van Riet Lowe, C. (1929). The Stone Age cultures of South Africa. *Annals of the South African Museum, 27*, 1–146.

Henshilwood, C. S. (2004). The origins of modern human behaviour—exploring the African evidence. In T. Oestigaard, N. Anfinset & T. Saetersdal (Eds.), *Combining the past and the present: Archaeological perspectives on society* (BAR International Series No. 1210) (pp. 95–106). Oxford: Archaeopress.

Henshilwood, C. S., & Dubreuil, B. (2011). The Still Bay and Howiesons Poort, 77–59 ka: Symbolic material culture and the evolution of the mind during the African Middle Stone Age. *Current Anthropology, 52*, 361–400.

Henshilwood, C. S., & Marean, C. (2003). The origin of modern human behavior: A review and critique of models and test implications. *Current Anthropology, 44*, 627–651.

Henshilwood, C. S., & Sealy, J. (1997). Bone artifacts from the Middle Stone Age at Blombos Cave, Southern Cape, South Africa. *Current Anthropology, 38*, 890–895.

Henshilwood, C. S., d'Errico, F., Marean, C., Milo, R., & Yates, R. (2001a). An early bone tool industry from the Middle Stone Age at Blombos Cave, South Africa: Implications for the origins of modern human behaviour, symbolism and language. *Journal of Human Evolution, 41*, 631–678.

Henshilwood, C. S., Sealy, J., Yates, R., Cruz-Uribe, K., Goldberg, P., Grine, F., et al. (2001b). Blombos Cave, Southern Cape, South Africa: Preliminary report on the 1992–1999 excavations of the Middle Stone Age levels. *Journal of Archaeological Science, 28*, 421–448.

Henshilwood, C. S., d'Errico, F., Yates, R., Jacobs, Z., Tribolo, C., Duller, G., et al. (2002). Emergence of modern human behaviour: Middle Stone Age engravings from South Africa. *Science, 295*, 1278–1280.

Henshilwood, C. S., d'Errico, F., Vanhaeren, M., van Niekerk, K. L., & Jacobs, Z. (2004). Middle Stone Age shell beads from South Africa. *Science, 304*, 404.

Henshilwood, C. S., d'Errico, F., & Watts, I. (2009). Engraved ochres from the Middle Stone Age levels at Blombos Cave, South Africa. *Journal of Human Evolution, 57*, 27–47.

Henshilwood, C. S., van Niekerk, K. L., Wurz, S., Delagnes, A., Armitage, S. J., Rifkin, R. F., et al. (2014). Klipdrift Shelter, southern Cape, South Africa: Preliminary report on the Howiesons Poort layers. *Journal of Archaeological Science, 45*, 284–303.

Hewitt, J. (1921). On several implements and ornaments from Strandloper sites in the Eastern Cape Province. *South African Journal of Science, 18*, 454–467.

Hiscock, P. (2008). *The archaeology of ancient Australia*. London: Routledge.

Högberg, A., & Larsson, L. (2011). Lithic technology and behavioural modernity: New results from the Still Bay site, Hollow Rock Shelter, Western Cape Province, South Africa. *Journal of Human Evolution, 61*, 133–155.

Jacobs, Z., Hayes, E. H., Roberts, R. G., Galbraith, R. F., & Henshilwood, C. S. (2013). An improved OSL chronology for the Still Bay layers at Blombos Cave, South Africa: Further tests of single-grain dating procedures and a re-evaluation of the timing of the Still Bay industry across southern Africa. *Journal of Archaeological Science, 40*(1), 579–594.

Jacobs, Z., & Roberts, R. G. (2009). Catalysts for Stone Age innovations. What might have triggered two short-lived bursts of technological and behavioral innovation in southern Africa during the Middle Stone Age? *Communicative and Integrative Biology, 2*, 191–193.

Jacobs, Z., Roberts, R. G., Galbraith, R. F., Deacon, H. J., Grün, R., Mackay, A., et al. (2008). Ages for Middle Stone Age innovations in southern Africa: Implications for modern human behavior and dispersal. *Science, 322*, 733–735.

Jones, T., Leonard, D., & Abbott, A. L. (1995). The structure of selectionist applications in archaeology. In P. A. Teltser (Ed.), *Evolutionary archaeology: Methodological issues* (pp. 13–32). Tuscon: University of Arizona Press.

Jouzel, J., Masson-Delmotte, V., Cattani, O., Dreyfus, G., Falourd, S., Hoffmann, G., et al. (2007). Orbital and millennial Antarctic climate variability over the past 800,000 years. *Science, 317*, 793–795.

Kaplan, J. (1990). The Umhlatuzana Rock Shelter sequence: 100 000 years of Stone Age history. *Natal Museum Journal of Humanities, 2*, 1–94.

Kingston, H. D. R. (1900). Notes on some caves in the Tzitzikama or Outeniqua District, near Knysna, South Africa and the objects found therein. *Journal of the Anthropological Institute of Great Britain and Ireland, 30*, 45–49.

Kuhn, S. L. (1995). *Mousterian lithic technology*. Princeton: Princeton University Press.

Leonard, R. D., & Jones, G. T. (1987). Elements of an inclusive evolutionary model for archaeology. *Journal of Anthropological Archaeology, 6*, 199–219.

Lewis-Abbott, W. J. (1913). Pygmy implements from Cape Colony. *Man, 13*, 143–147.

Lyman, R. L., & O'Brien, M. J. (1998). The goals of evolutionary archaeology. *Current Anthropology, 39*, 615–652.

Lyman, R. L., O'Brien, M. J., & Dunnell, R. C. (1997). *The rise and fall of culture history*. New York: Plenum Press.

Mackay, A. (2008). A method for estimating edge length from flake dimensions: Use and implications for technological change in the southern African MSA. *Journal of Archaeological Science, 35*, 614–622.

Mackay, A. (2009). History and selection in the Late Pleistocene archaeology of the Western Cape, South Africa. Ph.D. Dissertation, Australian National University, Canberra.

Mackay, A. (2010). The Late Pleistocene archaeology of Klein Kliphuis rock shelter, Western Cape: 2006 excavations. *South African Archaeological Bulletin, 65*, 134–147.

Mackay, A. (2011). The nature and significance of the transition from the Howiesons Poort to the post-Howiesons Poort at Klein Kliphuis rockshelter, South Africa. *Journal of Archaeological Science, 38*, 1430–1440.

Mackay, A., & Marwick, B. (2011). Costs and benefits in technological decision making under variable conditions: Examples from the Late Pleistocene in southern Africa. In B. Marwick & A. Mackay (Eds.), *Keeping your edge: Recent approaches to the organisation of stone artefact technology* (BAR International Series No. 2273) (pp. 119–134). Oxford: Archaeopress.

Mackay, A., & Welz, A. (2008). Engraved ochre from a Middle Stone Age context at Klein Kliphuis in the Western Cape of South Africa. *Journal of Archaeological Science, 35*, 1521–1532.

McBrearty, S., & Brooks, A. (2000). The revolution that wasn't: A new interpretation of the origin of modern human behaviour. *Journal of Human Evolution, 39*, 453–563.

Mellars, P. A. (2006a). Why did modern human populations disperse from Africa ca. 60,000 years ago? A new model. *Proceedings of the National Academy of Sciences, 103*, 9381–9386.

Mellars, P. A. (2006b). Going east: New genetic and archaeological perspectives on the modern human colonization of Eurasia. *Science, 313*, 796–800.

Mesoudi, A., Whiten, A., & Laland, K. N. (2006). Towards a unified science of cultural evolution. *Brain and Behavioral Sciences, 29*, 329–383.

Minichillo, T. (2005). Middle Stone Age lithic study, South Africa: An examination of modern human origins. Ph.D. Dissertation, University of Washington.

Morgan, L. H. (1944) [1877]. *Ancient society*. First Indian Edition. Calcutta: N.G. Sinha Roy.

Mourre, V., Villa, P., & Henshilwood, C. S. (2010). Early use of pressure flaking on lithic artifacts at Blombos Cave, South Africa. *Science, 330*, 659–662.

Nieman, F. D. (1995). Stylistic variation in evolutionary perspective: Inferences from decorative diversity and interassemblage distance in Illinois Woodland ceramic assemblages. *American Antiquity, 60*, 7–36.

O'Brien, M. J., & Lyman, R. L. (2002). Evolutionary archaeology: Current status and future prospects. *Evolutionary Anthropology, 11*, 26–36.

O'Brien, M. J., & Lyman, R. L. (2007). The epistemological nature of archaeological units. *Anthropological Theory, 2*, 37–56.

Orton, J. (2006). The Later Stone Age lithic sequence at Elands Bay, Western Cape, South Africa: Raw materials, artifacts and sporadic change. *Southern African Humanities, 18*, 1–28.

Parkington, J. E. (1977). Follow the San. Ph.D. Dissertation, University of Cambridge.

Peers, B. (1927). Excavation report of 1927 on Peers Cave. Unpublished Manuscript, South African Museum.

Penning, W. H. (1887). Notes on a few stone implements found in South Africa. *Journal of the Anthropological Institute, 16*, 68–70.

Peringuey, L. (1911). The Stone Ages of South Africa. *Annals of the South African Museum, 8*, 1–218.

Porraz, G., Texier, P.-J., Archer, W., Piboule, M., Rigaud, J.-P., & Tribolo, C. (2013). Technological successions in the Middle Stone Age sequence of Diepkloof Rock Shelter, Western Cape, South Africa. *Journal of Archaeological Science, 40*(9), 3376–3400.

Powell, A., Shennan, S., & Thomas, M. G. (2009). Late Pleistocene demography and the appearance of modern human behavior. *Science, 324*, 1298–1301.

Rigaud, J.-P., Texier, P.-J., Parkington, J., & Poggenpoel, C. (2006). Le mobilier Stillbay et Howiesons Poort de l'abri Diepkloof. La chronologie du Middle Stone Age sud-africain et ses implication. *Comptes Rendus Palevol, 5*, 839–849.

Sanderson, J. (1879). Notes in connection with stone implements from Natal. *The Journal of the Anthropological Institute of Great Britain and Ireland, 8*, 15–21.

Schmidt, P., Porraz, G., Slodczyk, A., Bellot-gurlet, L., Archer, W., & Miller, C. E. (2013). Heat treatment in the South African Middle

Stone Age: Temperature induced transformations of silcrete and their technological implications. *Journal of Archaeological Science, 40*(9), 3519–3531.

Sealy, J. (2016). Cultural change, demography, and the archaeology of the last 100 kyr in southern Africa. In S. C. Jones & B. A. Stewart (Eds.), *Africa from MIS 6-2: Population dynamics and paleoenvironments* (pp. 65–75). Dordrecht: Springer.

Shennan, S. (2008). Evolution in archaeology. *Annual Review of Anthropology, 37*(1), 75–91.

Shennan, S. J. (2001). Demography and cultural innovation: A model and its implications for the emergence of modern human culture. *Cambridge Archaeological Journal, 11*, 5–16.

Shott, M. J., & Sillitoe, P. (2005). Use life and curation in New Guinea experimental used flakes. *Journal of Archaeological Science, 32*, 653–663.

Singer, R., & Wymer, J. (1982). *The Middle Stone Age at Klasies River Mouth in South Africa.* Chicago: University of Chicago Press.

Stapleton, P., & Hewitt, J. (1927). Stone implements from a Rock-Shelter at Howieson's Poort near Grahamstown. *South African Journal of Science, 24*, 574–587.

Surovell, T. A., & Brantingham, P. J. (2007). A note on the use of temporal frequency distributions in studies of prehistoric demography. *Journal of Archaeological Science, 34*, 1868–1877.

Texier, P.-J., Porraz, G., Parkington, J., Rigaud, J.-P., Poggenpoel, C., Miller, C., et al. (2010). A Howiesons Poort tradition of engraving ostrich eggshell containers dated to 60,000 years ago at Diepkloof Rock Shelter, South Africa. *Proceedings of the National Academy of Science, 107*, 6180–6185.

Thackeray, A. I. (1992). The Middle Stone Age south of the Limpopo River. *Journal of World Prehistory, 6*, 385–440.

Thieme, H. (1997). Lower Palaeolithic hunting spears from Germany. *Nature, 385*, 807–810.

Tribolo, C., Mercier, N., & Valladas, H. (2005). Chronology of the Howiesons Poort and Still Bay techno-complexes: Assessment and new data from luminescence. In F. d'Errico & L. Blackwell (Eds.), *From tools to symbols: From early hominids to modern humans* (pp. 493–511). Johannesburg: Wits University Press.

Tribolo, C., Mercier, N., Valladas, H., Joron, J.-L., Guibert, P., Lefrais, Y., et al. (2009). Thermoluminescence dating of a Stillbay-Howiesons Poort sequence at Diepkloof Rock Shelter (Western Cape, South Africa). *Journal of Archaeological Science, 36*, 730–739.

Torrence, R. (1989). Re-tooling: Towards a behavioral theory of stone tools. In R. Torrence (Ed.), *Time, energy and stone tools* (pp. 57–66). Cambridge: Cambridge University Press.

Tylor, E. B. (1929) [1871]. *Primitive culture: Researches into the development of mythology, philosophy, religion, language, art, and custom.* Fifth Edition (2nd reprint). London: John Murray.

Valladas, H., Wadley, L., Mercier, N., Tribolo, C., Reyss, J.-L., & Joron, J.-L. (2005). Thermoluminescence dating on burnt lithics from Middle Stone Age layers at Rose Cottage Cave. *South African Journal of Science, 101*, 169–174.

Villa, P., Soriano, S., Teyssandier, N., & Wurz, S. (2010). The Howiesons Poort and MSA III at Klasies River main site, Cave 1A. *Journal of Archaeological Science, 37*, 630–655.

Vogelsang, R., Richter, J., Jacobs, Z., Eichhorn, B., Linseele, V., & Roberts, R. G. (2010). New excavations of Middle Stone Age deposits at Apollo 11 Rockshelter, Namibia: Stratigraphy, archaeology, chronology and past environments. *Journal of African Archaeology, 8*(2), 185–218.

Vishnyatsky, L. B. (1994). "Running ahead of time" in the development of Palaeolithic industries. *Antiquity, 68*, 134–140.

Volman, T. (1981). The Middle Stone Age in the Southern Cape. Ph.D. Dissertation, University of Chicago.

Wadley, L. (2006). A typological study of the final Middle Stone Age stone tools from Sibudu Cave, Kwazulu-Natal. *South African Archaeological Bulletin, 60*, 51–63.

Wadley, L. (2007). Announcing a Still Bay industry at Sibudu Cave, South Africa. *Journal of Human Evolution, 52*, 681–689.

Wadley, L. (2012). Two "moments in time" during the Middle Stone Age occupation of Sibudu, South Africa. *Southern African Humanities, 24*, 79–97.

Weedman, K. J. (2006). An ethnoarchaeological study of hafting and stone tool diversity among the Game of Ethiopia. *Journal of Archaeological Method and Theory, 13*, 189–238.

White, J. P. (1967). Ethnoarchaeology in New Guinea: Two examples. *Man, 6*, 409–414.

White, J. P., & Thomas, D. H. (1972). What mean these stones? Ethnotaxonomic models and archaeological interpretations in the New Guinea Highlands. In D. L. Clarke (Ed.), *Models in archaeology* (pp. 275–308). London: Methuen.

Wurz, S. (2002). Variability in the Middle Stone Age lithic sequence, 115,000-60,000 years ago at Klasies River, South Africa. *Journal of Archaeological Science, 29*, 1001–1015.

Chapter 4
Cultural Change, Demography, and the Archaeology of the Last 100 kyr in Southern Africa

Judith Sealy

Abstract Recent research has highlighted the role of demography in cultural innovation and transmission. Some authors have suggested that changes in population size may be associated with the appearance and disappearance of the Still Bay and Howiesons Poort, c. 75–60 ka, and that the advanced cognitive and behavioral skills attested by these industries underpinned the expansion of early modern humans out of Africa. Our ability to test demographic hypotheses in the Middle Stone Age (MSA) is limited, because we still have only a few well-excavated sites from the relevant periods. Turning to the much better documented record of Holocene hunter-gatherers in southern Africa, I explore this line of argument using the appearance and disappearance of the mid-Holocene microlithic "Wilton" as a model. Estimates of relative population sizes based on archaeological evidence from the Holocene do not fit the predictions of the model, calling into question its utility in the MSA.

Keywords Still Bay • Howiesons Poort • Middle Stone Age • Technological innovation • Demographic history • Modern human behavior

Introduction

Archaeologists are good at documenting patterns. The distributions of artifacts, settlement layouts, burial styles, and other aspects of material culture through time and across space are the cornerstones of our reconstructions of life in the past. It is, however, much more difficult to explain why these patterns existed, and what caused changes to occur. A number of recent papers invoke demographic change as a driver of cultural innovation and transmission, building on work by Boyd and Richerson (1985, 2000). The core of the argument is that technological and cultural complexity and change depend partly on population size and density (including the degree of interconnectedness of social groups). Many of the skills that humans depend upon to survive and reproduce are learned. Larger populations are more likely to yield occasional innovative individuals who develop new and better solutions to a problem, and will also contain a greater number of able pupils to learn these improved ways of doing things. Cultural innovations are therefore more likely to be passed on and retained as part of the technological repertoire in larger populations. Henrich (2004) argues that this is true even if transmission of new behaviors is imperfect and incomplete. The opposite is also true: small populations may lose adaptive behaviors (Henrich 2004, 2006; Kline and Boyd 2010).

Powell et al. (2009) apply this model to the appearance of "modern human behavior." Developing work by Shennan (2001), they explore the idea that demographic factors may account for the appearance of personal ornaments, abstract art, systematic production of standardized bone artifacts, finely flaked bifacial stone points, backed microliths, and other "complex" items of material culture often taken as indicators of modernity. This set of innovations is currently best known from the Still Bay and Howiesons Poort industries of southern Africa (Henshilwood et al. 2002, 2009; d'Errico et al. 2005, 2008; Mackay and Welz 2008; Texier et al. 2010; Henshilwood and Dubreuil 2011), although several of its elements have also been reported from elsewhere in Africa, the Levant, and parts of Europe, in some cases at earlier dates (e.g., d'Errico 2003; Vanhaeren

J. Sealy (✉)
Department of Archaeology, University of Cape Town,
Private Bag X3, Rondebosch, 7701, South Africa
e-mail: Judith.Sealy@uct.ac.za

Sacha C. Jones and Brian A. Stewart (eds.), *Africa from MIS 6-2: Population Dynamics and Paleoenvironments*,
Vertebrate Paleobiology and Paleoanthropology, DOI 10.1007/978-94-017-7520-5_4

et al. 2006; Bouzouggar et al. 2007; d'Errico et al. 2009; Zilhao et al. 2010). In a synthesis of many dates from multiple sites, Jacobs et al. (2008) reported the Still Bay to date to 72–70 ka and the Howiesons Poort to 65–60 ka. This picture has recently been challenged by Tribolo et al. (2013) who propose a longer chronology for the Still Bay and Howiesons Poort, with more gradual transitions from one phase of the MSA to another. These studies have yet to be reconciled. At Katanda, in the Democratic Republic of Congo, barbed bone spear-points and other bone artifacts have been recovered from riverside sites that have yielded dates between 90 and 70–60 ka (Brooks et al. 1995; Yellen et al. 1995; Yellen 1998). Since modern human anatomy dates back to at least 200 ka (White et al. 2003; McDougal et al. 2005), why is there such a long gap before we see significant numbers of these artifactual markers of more complex behavior? We have very few sites dating to MIS 6, so part of the problem may simply be sample size. One school of thought argues that before 50 ka, although overall human morphology (including that of the cranium) may have been modern, the brain was not yet fully modern. These scholars suggest that cognitive modernity was the result of a genetic mutation at ~50 ka (Mithen 1996; Klein 2008), rapidly followed by expansion of fully modern human populations out of Africa. This view is, however, losing support in the face of an increasing body of evidence from the MSA for decoration (Henshilwood et al. 2002; d'Errico et al. 2005, 2008; Mackay and Welz 2008; Texier et al. 2010), the ability to plan and execute complex multistage tasks (Brown et al. 2009; Wadley et al. 2009; Henshilwood et al. 2011) and other sophisticated behaviors. If these archaeological traces are reliable indicators of modern behavior, then such behavior significantly predates 50 ka. The very limited amount of evidence available from MIS 6 hints that it is perhaps as old as modern anatomy (Marean et al. 2007; Brown et al. 2009). We need more evidence from MIS 6, but even if we look only at MIS 5-3, it is clear that indicators of "advanced" behavior are *intermittently* present in the MSA archaeological record. Why did these appear and disappear? What caused the emergence of the Still Bay and Howiesons Poort, and their subsequent disappearance? If these were "advanced" technologies that conferred some advantage on their makers (Mellars 2005, 2006), why did people stop making them?

One idea is that climatic and environmental changes were causal or contributing factors (H.J. Deacon 1989; Lombard 2005; Minichillo 2005; Thackeray 2007; Henshilwood 2008; Ziegler et al. 2013). Climates in eastern Africa during late MIS 5-4 were at least intermittently harsh (Ambrose 1998, 2003; Cohen et al. 2007; Scholz et al. 2007), but conditions in southern Africa were probably more equable. Chase (2010) and Carr et al. (2016) suggest that in southern Africa, MIS 4 was, on the whole, cool and moist. Bar-Matthews et al. (2010), in their analyses of speleothems from Pinnacle Point, report more positive $\delta^{13}C$ values between 66 and 55 ka. They interpret these as indicative of a larger proportion of summer rainfall, resulting in more C_4 grasses at that time. They also note a short-lived negative spike in speleothem $\delta^{13}C$ values at 72 ka, attributed to a brief increase in winter rains. This is, of course, the time of the Still Bay, which these authors suggest may have been a technological response to rapid environmental change.

Jacobs et al. (2008) point out that the Howiesons Poort spans a period of substantial temperature change as recorded in ice cores, and is therefore unlikely to be highly constrained by climate or environment. Cochrane (2008) argues that at Sibudu, the Howiesons Poort to post-Howiesons Poort transition does not coincide with significant climatic change. All in all, changing environmental conditions may play some part, but do not seem to be an adequate explanation for the appearance and disappearance of the Still Bay and Howiesons Poort – two very different lithic industries – in a relatively short time.

Powell et al. (2009) propose that demographic factors were important. Using Henrich's transmission model as a starting point, they model late Pleistocene population dynamics and find that differences in population density at a small-scale (regional) level, and/or differences in mobility, i.e., the extent to which groups were in contact with one another, can indeed lead to geographical variation in the accumulation of knowledge and skills. Drawing on genetic studies, they estimate the likely sizes of ancestral populations and suggest that population densities in MIS 4 in sub-Saharan Africa were comparable to those in early Upper Paleolithic Europe. They argue that their results support the hypothesis that population growth stimulated the emergence of the Still Bay and Howiesons Poort (although they do not, as might be expected, see evidence for population reduction when these industries disappear). This appears to fit well with the finding by Atkinson et al. (2009) of a major expansion between 86 and 61 ka in populations with mitochondrial DNA haplogroup L3 – the lineage that expanded out of Africa.

This is an intriguing idea, but there are some difficulties. As Powell et al. (2009) point out complex patterns of demographic change – such as repeated bottlenecks and expansions – may be undetectable on the basis of the evidence available at present. Jacobs et al. (2008) reported a gap of approximately 7 ka between the Still Bay and the Howiesons Poort, and variations in population size on this time-scale so far back in the past would be difficult to resolve using genetic evidence. Obtaining a sufficiently close chronological fit between population estimates based on Bayesian coalescent inference of mtDNA haplogroups and the appearance and disappearance of artifactual assemblages in the archaeological sequences is a problem (Jacobs and Roberts 2008). Nevertheless, the proposal that demographic change drove (or enabled?) innovation in the MSA is plausible and attractive in the way that it links evidence from archaeology and genetics (Jacobs and Roberts 2009a, b). What other evidence do we have for population size during the MSA?

Population Size During the MSA

It is notoriously difficult to estimate population size on the basis of archaeological evidence, especially for hunter-gatherer societies: we do not know what proportion of the material originally deposited in a particular archaeological site is preserved, and we do not know how frequently people moved camp. Robust landscape-scale patterning in numbers of sites and quantities of deposit may, however, offer clues to *relative* population size (cf. Deacon and Thackeray 1984). Most MSA sites (especially open sites) are undated, but if we look only at Still Bay and Howiesons Poort occurrences, it is clear that the Still Bay is known from only a handful of sites (Apollo 11, Hollow Rock Shelter, Diepkloof Rock Shelter, Peers Cave, and nearby caves in Fish Hoek, Blombos Cave, Paardeberg, Sibudu, Umhlatuzana), as well as a few open sites (Blombos Sands, Klein Jongensfontein, Hangklip, Maitland, Soutfontein 1) (Fig. 4.1). Sites with Howiesons Poort deposits are much more common: we currently have more than 30 excavated caves, as well as some open-air sites. This pattern has persisted over several decades, remaining unchanged by the recent resurgence of interest in the MSA when archaeologists have been targeting the Still Bay and Howiesons Poort for further investigation. It is therefore a robust pattern, and it certainly raises the possibility that population sizes may have been different during these two periods. Jacobs et al. (2008), in their wide-ranging study of MSA sites, also reported more occupations dating to immediately post-Howiesons Poort times ($\sim$58 ka) than to Still Bay times.

Only four sites (Apollo 11, Diepkloof, Sibudu, and Umhlatuzana) contain both Still Bay and Howiesons Poort deposits and have been excavated to currently accepted standards. The relevant areas of all four excavations are at present limited in extent, so it is difficult to tell whether greater volumes of material were deposited during Howiesons Poort

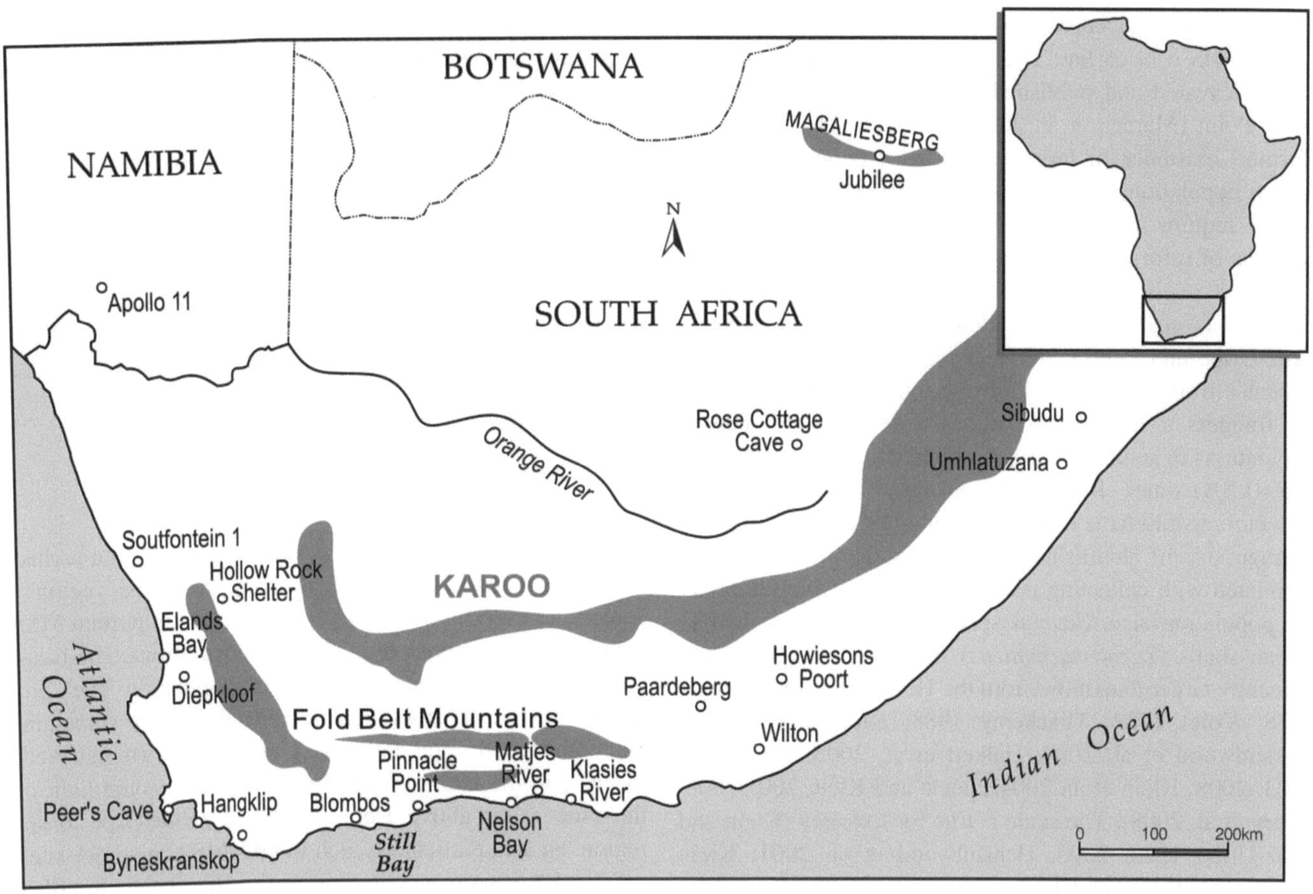

Fig. 4.1 Map showing sites mentioned in the text

compared with Still Bay times, as one would expect if populations were bigger. At Apollo 11, the Howiesons Poort assemblage is particularly large—more than half of the total stone assemblage from the original excavations (Vogelsang et al. 2010). At Sibudu, Howiesons Poort deposits are more substantial than Still Bay, and have yielded many more artifacts (Wadley 2006, 2008). Post-Howiesons Poort levels dated to ~58 ka show strong evidence of more intensive use of the shelter: the rate of deposition of archaeological sediments increased and the density of stone flakes rose markedly. The excavators argue that populations were larger, and/or more time was spent in the shelter. Evidence for more regular site maintenance at this time is consistent with this suggestion (Wadley et al. 2011). It is also worth noting that at Klasies River, Nelson Bay Cave and Klein Kliphuis, Howiesons Poort assemblages are large relative to those from preceding and succeeding periods (Volman 1981; Singer and Wymer 1982; Mackay 2010, 2011a).

If we look at MSA sites in general, some show evidence of very large quantities of deposit. A number of caves along the south coast contain small quantities of brecciated MSA material cemented to the rock walls, sometimes meters above the present ground surface (personal observation). These sites were once filled with MSA deposits that have disappeared, probably washed out by the c. +5 m sea levels of the last interglacial, which means that these occupations date to MIS 6 or earlier. The only site of this kind that has been excavated and published to date is Cave 13B at Pinnacle Point (Marean et al. 2007), but there are others. The (former) existence of large quantities of deposit hints that MIS 6 populations sometimes attained substantial sizes, but it will require a great deal more work to develop these snippets of information into a more coherent picture.

Impact on resources may be one way of investigating population size. Richard Klein, together with Kathy Cruz-Uribe and Teresa Steele, has amassed a great deal of evidence from shellfish and tortoise remains. These are easy for foragers to collect and were staple foods for coastal populations in southern Africa in both MSA and Later Stone Age (LSA) times. Klein and colleagues have argued that collectors would have taken larger animals first, and that the average size of shellfish or tortoises is therefore inversely correlated with collecting pressure, itself effectively a proxy for population size. Several species of limpets as well as turban shells (*Turbo sarmaticus*) from MSA sites are significantly larger than those from the Holocene LSA (Volman 1978; Voigt 1982; Thackeray 1988; Klein 1998, 2008; Henshilwood et al. 2001; Halkett et al. 2003; Parkington 2003, 2008; Klein et al. 2004; Steele and Klein 2005/2006; Avery et al. 2008). The same is true for tortoises (Klein and Cruz-Uribe 1983, 2000; Henshilwood et al. 2001; Klein et al. 2004; Steele and Klein 2005/2006). Klein and others interpret this to mean that populations in the MSA were smaller than in the LSA. Klein has linked this to his argument that MSA people were not yet cognitively modern, and therefore not as efficient at extracting food from their environment as LSA people—hence MSA populations were unable to grow to the same size as LSA populations (Klein 2008). The underlying assumption here is that mollusks and tortoises grew at the same rates in MSA (glacial) and Holocene (interglacial) times, a premise that has yet to be investigated in detail. Environmental factors may indeed play a greater role in these size differences than previously thought (Teske et al. 2007; Sealy and Galimberti 2011), which would mean that differences in population size between the MSA and the LSA have been overestimated.

Since our interest here is in possible variation in population sizes *within* the MSA, are there any significant shifts in shellfish or tortoise sizes during this time? In answering this question, we are limited to evidence from securely dated assemblages, or to long sequence sites where we can use relative dating (position within the stratigraphic column) to investigate possible change over time. Very few such sites have sufficiently large numbers of well-preserved (i.e., measurable) shellfish. Information has been published only for Klasies River and Blombos Cave. At Klasies, the sizes of *Turbo* opercula do not differ significantly between the Howiesons Poort and other phases of the MSA (Voigt 1982). Blombos does not have Howiesons Poort, but it does have Still Bay; the sizes of *Turbo* opercula from Still Bay levels are not significantly different from underlying older levels (Henshilwood et al. 2001).

Unfortunately, the sum total of archaeological evidence from the MSA is too sparse to be able to make confident statements about population sizes at that time. Work is in progress at several localities, but at present we have only a few well-excavated and well-dated MSA sites. We therefore cannot use archaeological evidence to critically evaluate hypotheses about demographic change during the MSA.

The LSA Archaeological Sequence

If, however, we turn to the LSA, we have far more archaeological evidence, much of it from the same region – sometimes even the same sites – that contain important MSA deposits. It remains difficult to move from site-based archaeology to an understanding of population-level processes, but we have enough information to say something about changes in population size during the terminal Pleistocene and the Holocene, and this may shed some light on the issues raised above. Here I will focus on the Cape coastal region, an area particularly rich in both MSA and LSA sites.

The LSA sequence in this region has been described extensively (Louw 1960; H.J. Deacon 1976; J. Deacon 1972,

1978, 1984; Parkington 1981, 1986a, b, 1988; Schweitzer and Wilson 1982; Inskeep 1987; Binneman 1995; Döckel 1998; Jerardino et al. 2008; Jerardino 2010). Briefly, the earliest LSA comprised rather unstandardized lithic assemblages, succeeded around 22 ka by the Robberg Industry, which was dominated by small bladelets and bladelet cores. Rare retouched forms included scrapers and occasional backed pieces. Robberg assemblages continued to be made until about 12 ka, although at a few sites they persisted until the early Holocene (Opperman 1987; Wadley 1997). Across southern Africa, we have relatively few sites with Robberg assemblages, perhaps because populations were small at the Last Glacial Maximum and immediately thereafter (H. J. Deacon 1976; J. Deacon 1984).

Around 12 ka, Robberg assemblages were replaced by macrolithic assemblages assigned to the Oakhurst Complex. These are found throughout southern Africa, in both caves and open-air sites. Along the Cape coast, early Oakhurst Complex assemblages had few diagnostic stone artifacts, but later large scrapers appeared; bone artifacts were a feature of both phases. People very likely also made items from perishable materials such as wood. Oakhurst deposits are much more common and far more substantial than those of the Robberg, with sites such as Elands Bay Cave (Parkington 2006), Byneskranskop (Schweitzer and Wilson 1982), Matjes River Rock Shelter (Louw 1960; Döckel 1998) and others all containing large volumes of material: parts of Layer D at Matjes River were 16-feet deep. In the interior Karoo, Bousman (2005) noted that the presence of large quantities of Oakhurst Complex material make it reasonable to infer that these populations were more densely packed on the landscape than Robberg ones.

At ~8 ka, the Oakhurst Complex gave way to a microlithic industry first reported from Wilton Large Rock Shelter in the Eastern Cape. Today, the name "Wilton" is usually restricted to material from the Cape coastal area, although very similar assemblages were made over a large area of the subcontinent, including Zimbabwe and southern Namibia. They are characterized by small artifact size and a wide range of standardized retouched ("formal") tool types, especially small scrapers and backed elements. Fine-grained raw materials were sought out; along the Cape coast this contrasts with the preceding and succeeding assemblages, which were dominated by quartzite artifacts. Backed items included segments. In the south, including at the Wilton name site, segments were limited to the mid-Holocene and were replaced by backed blades in the later part of the microlithic period. There were also other changes within the Wilton (J. Deacon 1972; Inskeep 1987), but the microlithic character and relatively high frequencies of formal tools, especially small scrapers, are identifying features. Non-lithic artifacts such as bone points, awls, and linkshafts also occur with Wilton assemblages, as do decorative items made from bone, shell, and ostrich egg-shell.

Wilton assemblages are best known from the coastal regions of South Africa, extending inland as far as the Cape Fold Belt Mountains. They are absent from the central Karoo region, but occur infrequently in the Northern Cape, north of the Orange River (Humphreys and Thackeray 1983; Bousman 2005), at Rose Cottage Cave in the eastern Free State (Wadley 1997), and at Jubilee Shelter in the Magaliesberg (Wadley 1986), also further north in Zimbabwe and southern Namibia (Wendt 1976; Cooke 1979; Walker 1995). The Karoo was effectively depopulated during Wilton times (8–4 ka), probably because warmer conditions made the already hot and dry environment unsuitable for occupation (J. Deacon 1974, 1984; Sampson 1985; Hart 1989; Mitchell 2002; Bousman 2005). A number of sites elsewhere in the country also show gaps in occupation between 8 and 4 ka (e.g., Parkington 1981; Opperman 1987; Parkington et al. 1988). Even coastal and near-coastal sites usually contain only limited volumes of Wilton deposit. Several authors have noted that in general, Wilton remains are less prominent on the landscape than those of the Oakhurst Complex (Wadley 1989; Mitchell 2002; Bousman 2005).

Up to this point, very similar artifact assemblages were made across large areas of South Africa, but after 4 ka there was more regional variation. In the northwest, microlithic assemblages persisted into the last few hundred years (Webley 2002; Orton 2002, 2006; Dewar 2008). Along most of the southern Cape coast, the Wilton was replaced by an informal macrolithic industry consisting almost entirely of crudely struck flakes and chunks, usually made from quartzite. There are almost no retouched pieces, although in the east one finds occasional large quartzite segments (Binneman 1995). Bone artifacts were a prominent feature of post-Wilton assemblages and probably replaced stone artifacts in precision tools, e.g., as arrow points. On the basis of the evidence we have at present, the Wilton to post-Wilton transition in the southern Cape did not involve a progressive loss of artifact types, it was an abrupt replacement of one type of assemblage by another. Interestingly, radiocarbon dates show that this was not necessarily synchronous even at nearby sites (Inskeep 1987; Binneman 1995; Kyriacou 2009).

Post-Wilton sites along the Cape coast contain very large volumes of deposit. There are also many open sites dating between 4 and 2 ka, including the massive "megamiddens" of the Elands Bay area (Parkington et al. 1988; Jerardino 2010). There can be no doubt that post-Wilton populations were large and densities were high. Expanding populations reoccupied the Karoo after approximately 4 ka (J. Deacon 1974, 1984). In at least some already inhabited areas, groups settled on the landscape became much more sedentary (Sealy 2006; Hall 1990). Exploitation of marine resources peaked, and there is abundant evidence of collecting plant foods, hunting and snaring of small animals, and other indicators of "intensification".

In the last 2 ka, food production was introduced in South Africa, as iron-using mixed farmers moved into the eastern and northern parts of the country, and pastoralism replaced hunting and gathering in much of the west and south. Hunter-gatherers continued to live in mountainous areas and dry regions that lack good grazing. The last two millennia are not relevant to this chapter. Questions about cultural change in relation to population expansion and contraction are explored within an entirely hunter-gatherer world.

Howiesons Poort and Wilton

As a number of researchers have noted (Goodwin and van Riet Lowe 1929; H.J. Deacon 1989, 1992; Ambrose and Lorenz 1990; Wurz 1999) Howiesons Poort and Wilton assemblages are similar in a number of ways. Both are characterized by a high frequency of retouched pieces with very standardized backed elements, including segments (Singer and Wymer 1982; H.J. Deacon 1995; J. Deacon 1995; Wurz 1999, 2002; Soriano et al. 2007; Wadley 2008; Mackay 2011b). At least some Holocene backed segments were used as arrow points, as perhaps were some Howiesons Poort specimens (Lombard and Pargeter 2008; Lombard and Phillipson 2010), although these were probably used in a variety of ways (Wadley and Mohapi 2008). At sites in the Cape, both the Wilton and the Howiesons Poort show a higher incidence of fine-grained raw materials compared with preceding and succeeding assemblages (Singer and Wymer 1982; Wurz 2002; Porraz et al. 2008; Mackay 2011a). In other parts of southern Africa, where fine-grained raw materials were more commonly used during the MSA, this shift does not occur (Soriano et al. 2007). Both the Howiesons Poort and the Wilton were widespread, occurring over large parts of the southern African subcontinent. Thanks to the recent dating program of Jacobs and Roberts (2008), we now know that the Howiesons Poort lasted about 5 ka (but longer in the Tribolo et al. [2013] chronology), the Wilton about 4 ka. The similarities are sufficiently compelling that researchers interested in the Howiesons Poort have used comparisons with the Wilton to explore a range of issues including technology, social relations, landscape use, and behavioral modernity (H.J. Deacon 1989, 1995; Ambrose and Lorenz 1990; Wurz 1999). Is there evidence for demographic increase associated with the appearance of the Wilton, in the same way that Powell et al. (2009) have proposed for the Howiesons Poort?

As the brief outline above indicates, the Holocene LSA sequence along the southern Cape coast includes two periods (Oakhurst Complex and post-Wilton) in which the numbers of sites and volumes of deposit are consistent with the suggestion that population densities were high. During these times, people made informal macrolithic stone artifact assemblages. In coastal and near-coastal localities, tools were made mainly from quartzite available in the immediate vicinity of the sites. Few recognizable tool "types" occurred during these periods, although bone artifacts were a regular and no doubt important part of the toolkit. The mid-Holocene Wilton provides a sharp contrast. Wilton artifactual assemblages were microlithic with many standardized formal tool types made from a range of fine-grained raw materials. Non-lithic artifacts continued to be made. Population distribution was discontinuous, with the central Karoo region of South Africa inhabited very sparsely, if at all. Occupation was concentrated on the coastal plain and Fold Belt Mountains, and areas to the north of the Karoo. Even here, archaeological traces are sparser than in Oakhurst Complex and post-Wilton times. Wilton material culture is undoubtedly more complex than that of the Oakhurst Complex or post-Wilton, but the population appears to have been smaller.

Interpretations of the Wilton have varied from early views of it as a florescence of LSA tool-making to more recent and more informed assessments by Deacon (1984) and Wadley (1989), who argued that it was a way of coping with hard times. The standardization of stone artifact assemblages over long distances clearly indicates a degree of contact and communication between distant groups. Wadley (1989) suggested that people might have formed networks of alliances maintained through mechanisms such as gift giving. Using the Ju'/hoansi system of *hxaro* as a model, such networks could have served as a kind of insurance system. If conditions in one area became too harsh, people could make extended visits to exchange partners, staying with them until conditions improved. H.J. Deacon (1992) has made the same argument for the Howiesons Poort. Regardless of the details, Wilton communities must have been in contact with one another over long distances. Perhaps this degree of interconnectedness means that the *effective* population size was large, even if people were thinly scattered across the landscape? This may be true – but in earlier (Oakhurst complex) times there must also have been a degree of long-distance communication, since macrolithic informal artifact assemblages were then widespread across the subcontinent.

All in all, the rich and relatively well-known Holocene archaeological record of southern Africa does not provide a good correlation between evidence for larger and denser populations and more complex tool-kits. We should not expect a perfect correlation, and there will no doubt be case studies that do not fit the model. Obtaining sufficiently detailed data to test these hypotheses is difficult. Even the use of historical ethnographic evidence is contentious (Henrich 2004, 2006; Read 2006); much older archaeological evidence is likely to be even more so. There are also real problems of scale and resolution, given that processes such as population increase may occur over a much shorter time scale than can be distinguished in many archaeological deposits.

It is unclear just how far one can take the parallels between the Howiesons Poort and the Wilton. There are certainly many points of similarity, but there are also likely to be differences in the reasons why they developed and why they both disappeared after a few thousand years. In the Holocene, we have enough evidence to see that, at least at the Wilton to post-Wilton transition in the southern Cape, an increase in population occurred at a time of simplification of technology, as far as we can tell from the artifactual assemblages that are preserved. Complexity in stone artifacts does not necessarily equate with complexity in the tool-kit as a whole, and much is not preserved: we rarely find artifacts made from leather, wood, or plant fiber, although there are some examples from very dry caves. Bone artifacts preserve well in many sites (especially coastal shell middens), and occur in both Wilton and post-Wilton contexts. Powell et al. (2009) find those components of Still Bay and Howiesons Poort tool-kits that are preserved adequate to infer complexity. By the same argument, the change in artifact assemblages that we see at the Wilton to post-Wilton transition is the inverse of what the demographic model predicts. It would require very special pleading to make a case for Wilton population densities – even effective densities – having been greater than those of the post-Wilton, in which we can demonstrate massive population growth, expansion into previously unoccupied areas, and simplification of tool-kits. The Holocene case study does not rule out the idea that population increase contributed to the development of the Howiesons Poort (and/or the Still Bay), but it does not inspire confidence.

What about other causes of technological complexity? There is an extensive literature on the role of risk in shaping artifactual assemblages, with a number of authors contending that shifts toward small, standardized, highly retouched artifacts, including backed elements, have to do with risk reduction strategies (e.g., Hiscock 1994; Kuhn 1994; Fitzhugh 2001; Bousman 2005; Hiscock et al. 2011). The social networking implied by widespread artifact standardization may have contributed toward the same goal (Wadley 1989; Ambrose and Lorenz 1990; Deacon 1992, 1995; Wurz 1999). The mid-Holocene was a time of hotter and drier climates, especially in the interior of South Africa, but conditions on the coastal forelands would have been buffered by proximity to the ocean. Our current understanding is that climates during late MIS 4, the time of the Howiesons Poort, were relatively mild, as outlined above (Cochrane 2008; Jacobs et al. 2008; Chase 2010; Carr et al. 2016). There is no evidence that environmental or resource-based risks were especially high at that time.

Several researchers have recently argued that the nature of shifts in artifact assemblages between 75 and 60 ka was relatively gradual, indicating a greater degree of technological continuity than previously thought. It is also suggested that this is inconsistent with explanations based on population expansion and contraction (Villa et al. 2010; Mackay 2011a). Newly excavated assemblages together with more detailed examination of older collections show that the transitions into and out of the Howiesons Poort involve considerable complexity (Cochrane 2008; Porraz et al. 2008; Villa et al. 2010; Mackay 2011a, 2016). We need, however, to document patterns of technological change more fully in the Late Pleistocene and Holocene of southern Africa, supplemented by more detailed quantitative palaeoclimatic data. A clearer understanding of how such transitions occurred is surely a prerequisite for answering the more difficult "why" questions.

Acknowledgments I am grateful to Alex Mackay for helpful comments on this chapter, and Sally Adam for drawing the map.

References

Ambrose, S. H. (1998). Late Pleistocene human population bottlenecks, volcanic winter, and differentiation of modern humans. *Journal of Human Evolution, 34*, 623–651.

Ambrose, S. H. (2003). Did the super-eruption of Toba cause a human population bottleneck? Reply to Gathorne-Hardy and Harcourt-Smith. *Journal of Human Evolution, 45*, 231–237.

Ambrose, S. H., & Lorenz, K. G. (1990). Social and ecological models for the Middle Stone Age in southern Africa. In P. Mellars (Ed.), *The Emergence of Modern Humans: An Archaeological Perspective* (pp. 3–33). Edinburgh: Edinburgh University Press.

Atkinson, Q. D., Gray, R. D., & Drummond, A. J. (2009). Bayesian coalescent inference of major human mitochondrial DNA haplogroup expansions in Africa. *Proceedings of the Royal Society B, 276*, 367–373.

Avery, G., Halkett, D., Orton, J., Steele, T., Tusenius, M., & Klein, R. (2008). The Ysterfontein 1 Middle Stone Age rock shelter and the evolution of coastal foraging. *South African Archaeological Society Goodwin Series, 10*, 66–89.

Bar-Matthews, M., Marean, C. W., Jacobs, Z., Karkanas, P., Fisher, E. C., Herries, A. I. R., et al. (2010). A high resolution and continuous isotopic speleothem record of paleoclimate and paleoenvironment from 90 to 53 ka from Pinnacle Point on the south coast of South Africa. *Quaternary Science Reviews, 29*, 2131–2145.

Binneman, J. N. F. (1995). Symbolic construction of communities during the Holocene Later Stone Age in the south-eastern Cape. Ph. D. Dissertation, University of the Witwatersrand, Johannesburg.

Bousman, C. B. (2005). Coping with risk: Later Stone Age technological strategies at Blydefontein Rock Shelter, South Africa. *Journal of Anthropological Archaeology, 24*, 193–226.

Bouzouggar, A., Barton, N., Vanhaeren, M., d'Errico, F., Collcutt, S., Higham, T., et al. (2007). 82,000-year-old shell beads from North Africa and implications for the origins of modern human behaviour. *Proceedings of the National Academy of Sciences, 104*, 9964–9969.

Boyd, R., & Richerson, P. J. (1985). *Culture and the evolutionary process*. Chicago: University of Chicago Press.

Boyd, R., & Richerson, P. J. (2000). Memes: Universal acid or a better mousetrap? In R. Aunger (Ed.), *Darwinizing culture: The status of memetics as a science* (pp. 143–162). Oxford: Oxford University Press.

Brooks, A. S., Helgren, D. M., Cramer, J. S., Franklin, A., Hornyak, W., Keating, J. M., et al. (1995). Dating and context of three Middle Stone Age sites with bone points in the Upper Semliki Valley, Zaire. *Science, 268*, 548–553.

Brown, K. S., Marean, C. W., Herries, A. I. R., Jacobs, Z., Tribolo, C., Braun, D. R., et al. (2009). Fire as an engineering tool of early modern humans. *Science, 325*, 859–862.

Carr, A. S., Chase, B. M., & Mackay, A. (2016). Mid to Late Quaternary landscape and environmental dynamics in the Middle Stone Age of southern South Africa. In S. C. Jones & B. A. Stewart (Eds.), *Africa from MIS 6-2: Population dynamics and paleoenvironments* (pp. 23–47). Dordrecht: Springer.

Chase, B. M. (2010). South African palaeoenvironments during Marine Oxygen Isotope Stage 4: A context for the Howiesons Poort and Still Bay industries. *Journal of Archaeological Science, 37*, 1359–1366.

Cochrane, G. W. G. (2008). The transition from Howiesons Poort to post Howiesons Poort industries in southern Africa. *South African Archaeological Society Goodwin Series, 10*, 157–167.

Cohen, A. S., Stone, J. R., Beuning, K. R. M., Park, L. E., Reinthal, P. N., Dettman, D., et al. (2007). Ecological consequences of early Late Pleistocene megadroughts in tropical Africa. *Proceedings of the National Academy of Sciences, 104*, 16422–16427.

Cooke, C. K. (1979). Excavations at Diana's Vow Rock Shelter, Maroni District, Zimbabwe, Rhodesia. *Occasional Papers of the National Museum of Rhodesia A, 4*, 115–148.

Deacon, H. J. (1976). *Where hunters gathered: A study of Holocene Stone Age people in the Eastern Cape* (South African Archaeological Society Monograph No. 1). Cape Town: South African Archaeological Society.

Deacon, H. J. (1989). Late Pleistocene palaeoecology and archaeology in the Southern Cape, South Africa. In P. Mellars & C. Stringer (Eds.), *The human revolution – behavioural and biological perspectives on the origins of modern humans* (pp. 547–564). Edinburgh: Edinburgh University Press.

Deacon, H. J. (1992). Southern Africa and modern human origins. *Philosophical Transactions of the Royal Society of London B, 337*, 177–183.

Deacon, H. J. (1995a). Two Late Pleistocene-Holocene archaeological depositories from the Southern Cape, South Africa. *South African Archaeological Bulletin, 50*, 121–131.

Deacon, H. J., & Thackeray, J. F. (1984). Late Pleistocene environmental changes and implications for the archaeological record in southern Africa. In J. C. Vogel (Ed.), *Late Cainozoic palaeoclimates of the southern hemisphere* (pp. 375–390). Rotterdam: Balkema.

Deacon, J. (1972). Wilton: An assessment after fifty years. *South African Archaeological Bulletin, 27*, 10–48.

Deacon, J. (1974). Patterning in the radiocarbon dates for the Wilton/Smithfield complex in southern Africa. *South African Archaeological Bulletin, 29*, 3–18.

Deacon, J. (1978). Changing patterns in the Late Pleistocene/early Holocene prehistory of southern Africa as seen from the Nelson Bay Cave stone artifact sequence. *Quaternary Research, 10*, 84–111.

Deacon, J. (1984). *The Later Stone Age of southernmost Africa* (BAR International Series No. 213). Oxford: Archaeopress.

Deacon, J. (1995b). An unsolved mystery at the Howiesons Poort name site. *South African Archaeological Bulletin, 50*, 110–120.

d'Errico, F. (2003). The invisible frontier. A multiple species model for the origin of behavioural modernity. *Evolutionary Anthropology, 12*, 188–202.

d'Errico, F., Henshilwood, C., Vanhaeren, M., & van Niekerk, K. (2005). Nassarius kraussianus shell beads from Blombos Cave: Evidence for symbolic behaviour in the Middle Stone Age. *Journal of Human Evolution, 48*, 3–24.

d'Errico, F., Vanhaeren, M., & Wadley, L. (2008). Possible shell beads from the Middle Stone Age layers of Sibudu Cave, South Africa. *Journal of Archaeological Science, 35*, 2675–2685.

d'Errico, F., Vanhaeren, M., Barton, N., Bouzouggar, A., Mienis, H., Richter, D., et al. (2009). Additional evidence on the use of personal ornaments in the Middle Paleolithic of North Africa. *Proceedings of the National Academy of Sciences, 106*, 16051–16056.

Dewar, G. I. (2008). *Archaeology of the coastal desert of Namaqualand, South Africa* (BAR International Series No. 1761). Oxford: Archaeopress.

Döckel, W. (1998). Re-investigation of the Matjes River Rock Shelter. M.A. Dissertation, University of Stellenbosch.

Fitzhugh, B. (2001). Risk and invention in human technological evolution. *Journal of Anthropological Archaeology, 20*, 125–167.

Goodwin, A. J. H., & van Riet Lowe, C. (1929). The Stone Age cultures of South Africa. *Annals of the South African Museum, 27*, 1–289.

Halkett, D., Hart, T., Yates, R., Volman, T. P., Parkington, J. E., Orton, J., et al. (2003). First excavation of intact Middle Stone Age layers at Ysterfontein, Western Cape Province, South Africa: Implications for Middle Stone Age ecology. *Journal of Archaeological Science, 30*, 955–971.

Hall, S. L. (1990). Hunter-fisher-gatherers of the Fish River Basin: A contribution to the Holocene prehistory of the Eastern Cape. D.Phil. Dissertation, University of Stellenbosch.

Hart, T. J. G. (1989). Haaskraal and Volstruisfontein, Later Stone Age events at two rockshelters in the Zeekoe Valley, Great Karoo, South Africa. M.A. Dissertation, University of Cape Town.

Henrich, J. (2004). Demography and cultural evolution: How adaptive cultural processes can produce maladaptive losses – the Tasmanian case. *American Antiquity, 69*, 197–214.

Henrich, J. (2006). Understanding cultural evolutionary models: A reply to Read's critique. *American Antiquity, 71*, 771–782.

Henshilwood, C. S. (2008). Winds of change: Palaeoenvironments, material culture and human behaviour in the Late Pleistocene (~77 ka–48 ka ago) in the Western Cape Province, South Africa. *South African Archaeological Society Goodwin Series, 10*, 35–51.

Henshilwood, C. S., & Dubreuil, B. (2011). The Still Bay and Howiesons Poort, 77–59 ka. Symbolic material culture and the evolution of the mind during the African Middle Stone Age. *Current Anthropology, 52*, 361–400.

Henshilwood, C. S., Sealy, J. C., Yates, R., Cruz-Uribe, K., Goldberg, P., Grine, F. E., et al. (2001). Blombos Cave, southern Cape, South Africa: Preliminary report on the 1992–1999 excavations of the Middle Stone Age levels. *Journal of Archaeological Science, 28*, 421–448.

Henshilwood, C. S., d'Errico, F., Yates, R., Jacobs, Z., Tribolo, C., Duller, G. A. T., et al. (2002). Emergence of modern human behaviour: Middle Stone Age engravings from South Africa. *Science, 295*, 1278–1280.

Henshilwood, C. S., d'Errico, F., & Watts, I. (2009). Engraved ochres from the Middle Stone Age levels at Blombos Cave, South Africa. *Journal of Human Evolution, 57*, 27–47.

Henshilwood, C. S., d'Errico, F., van Niekerk, K. L., Coquinot, Y., Jacobs, Z., Lauritzen, S.-E., et al. (2011). A 100,000-year-old ochre-processing workshop at Blombos Cave, South Africa. *Science, 334*, 219–222.

Hiscock, P. (1994). Technological responses to risk in Holocene Australia. *Journal of World Prehistory, 8*, 267–292.

Hiscock, P., Clarkson, C., & Mackay, A. (2011). Big debates over little tools: Ongoing disputes over microliths on three continents. *World Archaeology, 43*, 653–664.

Humphreys, A. J. B., & Thackeray, A. I. (1983). *Ghaap and Gariep: Later Stone Age studies in the Northern Cape* (South African

Archaeological Society Monograph No. 2). Cape Town: South African Archaeological Society.
Inskeep, R. R. (1987). *Nelson Bay Cave, Cape Province, South Africa. The Holocene levels* (BAR International Series No. 357). Oxford: Archaeopress.
Jacobs, Z., & Roberts, R. G. (2008). Testing times: Old and new chronologies for the Howiesons Poort and Still Bay industries in environmental context. *South African Archaeological Society Goodwin Series, 10*, 9–34.
Jacobs, Z., & Roberts, R. G. (2009a). Human history written in stone and blood. *American Scientist, 97*, 302–309.
Jacobs, Z., & Roberts, R. G. (2009b). Catalysts for Stone Age innovations. What might have triggered two short-lived bursts of technological and behavioral innovation in southern Africa during the Middle Stone Age? *Communicative and Integrative Biology, 2*, 191–193.
Jacobs, Z., Roberts, R. G., Galbraith, R. F., Deacon, H. J., Grün, R., Mackay, A., et al. (2008). Ages for the Middle Stone Age of southern Africa: Implications for human behaviour and dispersal. *Science, 322*, 733–735.
Jerardino, A. (2010). Large shell middens in Lamberts Bay, South Africa: A case of hunter-gatherer resource intensification. *Journal of Archaeological Science, 37*, 2291–2302.
Jerardino, A., Branch, G. M., & Navarro, R. (2008). Human impact on precolonial West Coast marine environments of South Africa. In J. M. Erlandson & T. C. Rick (Eds.), *Human impacts on marine environments* (pp. 279–296). Berkeley: University of California Press.
Klein, R. G. (1998). Why anatomically modern people did not disperse from Africa 100,000 years ago. In T. Akazawa, K. Aoki, & O. Bar-Yosef (Eds.), *Neanderthals and modern humans in Western Asia* (pp. 509–521). New York: Plenum Press.
Klein, R. G. (2008). Out of Africa and the evolution of human behaviour. *Evolutionary Anthropology, 17*, 267–281.
Klein, R. G., & Cruz-Uribe, K. (1983). Stone Age population numbers and average tortoise size at Byneskranskop Cave 1 and Die Kelders Cave 1, Southern Cape Province, South Africa. *South African Archaeological Bulletin, 38*, 26–30.
Klein, R. G., & Cruz-Uribe, K. (2000). Middle and Later Stone Age large mammal and tortoise remains from Die Kelders Cave 1, Western Cape Province, South Africa. *Journal of Human Evolution, 38*, 169–195.
Klein, R. G., Avery, G., Cruz-Uribe, K., Halkett, D., Parkington, J. E., Steele, T., et al. (2004). The Ysterfontein 1 Middle Stone Age site, South Africa, and early human exploitation of coastal resources. *Proceedings of the National Academy of Sciences, 101*(16), 5708–5715.
Kline, M. A., & Boyd, R. (2010). Population size predicts technological complexity in Oceania. *Proceedings of the Royal Society B, 277*, 2559–2564.
Kuhn, S. L. (1994). A formal approach to the design and assembly of mobile toolkits. *American Antiquity, 59*, 426–442.
Kyriacou, K. (2009). The reinvestigation of Hoffman's/Robberg Cave – the artefactual and shellfish assemblages, 2007. M.Phil. Dissertation, University of Cape Town.
Lombard, M. (2005). The Howiesons Poort of South Africa: What we know, what we think we know, what we need to know. *Southern African Humanities, 17*, 33–55.
Lombard, M., & Pargeter, J. (2008). Hunting with Howiesons Poort segments: Pilot experimental study and the functional interpretation of archaeological tools. *Journal of Archaeological Science, 35*, 2523–2531.
Lombard, M., & Phillipson, L. (2010). Indications of bow and stone-tipped arrow use 64,000 years ago in KwaZulu-Natal, South Africa. *Antiquity, 84*, 1–14.
Louw, J. T. (1960). *Prehistory of the Matjes River Rock Shelter* (National Museum Memoir No. 1). Bloemfontein: The National Museum.
Mackay, A. (2010). The Late Pleistocene archaeology of Klein Kliphuis Rock Shelter, Western Cape, South Africa: 2006 excavations. *South African Archaeological Bulletin, 65*, 132–147.
Mackay, A. (2011a). Nature and significance of the Howiesons Poort to post-Howiesons Poort transition at Klein Kliphuis rockshelter, South Africa. *Journal of Archaeological Science, 38*, 1430–1440.
Mackay, A. (2011b). Potentially stylistic differences between backed artefacts from two nearby sites occupied ~60,000 years before present in South Africa. *Journal of Anthropological Archaeology, 30*, 235–245.
Mackay, A. (2016). Technological change and the importance of variability: The Western Cape of South Africa from MIS 6-2. In S. C. Jones & B. A. Stewart (Eds.), *Africa from MIS 6-2: Population dynamics and paleoenvironments* (pp. 49–63). Dordrecht: Springer.
Mackay, A., & Welz, A. (2008). Engraved ochre from a Middle Stone Age context at Klein Kliphuis in the Western Cape of South Africa. *Journal of Archaeological Science, 35*, 1521–1532.
Marean, C. W., Bar-Matthews, M., Bernatchez, J., Fisher, E., Goldberg, P., Herries, A. I. R., et al. (2007). Early human use of marine resources and pigment in South Africa during the Middle Pleistocene. *Nature, 449*, 905–908.
McDougal, I., Brown, F. H., & Fleagle, J. G. (2005). Stratigraphic placement and age of modern humans from Kibish, Ethiopia. *Nature, 433*, 733–736.
Mellars, P. (2005). The impossible coincidence. A single-species model for the origins of modern human behaviour in Europe. *Evolutionary Anthropology, 14*, 12–27.
Mellars, P. (2006). Going east: New genetic and archaeological perspectives on the modern human colonization of Eurasia. *Science, 313*, 796–800.
Minichillo, T.J. (2005). Middle Stone Age lithic study, South Africa: An examination of modern human origins. Ph.D. Dissertation, University of Washington.
Mitchell, P. (2002). *The archaeology of southern Africa*. Cambridge: Cambridge University Press.
Mithen, S. (1996). *The prehistory of the mind. A search for the origins of art, religion and science*. London: Thames & Hudson.
Opperman, H. (1987). *The Later Stone Age of the Drakensberg range and its foothills* (BAR International Series No. 339). Oxford: Archaeopress.
Orton, J. (2002). Patterns in stone: The lithic assemblage from Dunefield Midden, Western Cape, South Africa. *South African Archaeological Bulletin, 57*, 31–37.
Orton, J. (2006). The Later Stone Age lithic sequence at Elands Bay, Western Cape, South Africa: Raw materials, artefacts and sporadic change. *Southern African Humanities, 18*(2), 1–28.
Parkington, J. E. (1981). The effects of environmental change on the scheduling of visits to the Elands Bay Cave, Cape Province, S.A. In I. Hodder, G. Isaac, & N. Hammond (Eds.), *Pattern of the past: Studies in honour of David Clarke* (pp. 341–359). Cambridge: Cambridge University Press.
Parkington, J. E. (1986a). Stone-tool assemblages, raw material distributions and prehistoric subsistence activities: The Late Stone Age in South Africa. In G. N. Bailey & P. Callow (Eds.), *Stone Age prehistory* (pp. 181–194). Cambridge: Cambridge University Press.
Parkington, J. (1986b). Landscape and subsistence changes since the Last Glacial Maximum along the western Cape coast. In L. G. Straus (Ed.), *The end of the Paleolithic in the Old World* (BAR International Series No. 284) (pp. 201–227). Oxford: Archaeopress.
Parkington, J. (1988). The Pleistocene/Holocene transition in the Western Cape, South Africa: Observations from Verlorenvlei. In J. Bower & D. Lubell (Eds.), *Prehistoric cultures and the*

environments in the Late Quaternary of Africa (BAR International Series No. 405) (pp. 197–205). Oxford: Archaeopress.

Parkington, J. (2003). Middens and moderns: Shellfishing and the Middle Stone Age of the Western Cape, South Africa. *South African Journal of Science, 99*, 243–247.

Parkington, J. (2006). *Shorelines, strandlopers and shell middens*. Cape Town: Krakadouw Trust.

Parkington, J. (2008). Limpet sizes in Stone Age archaeological contexts at the Cape, South Africa: Changing environment or human impact? In A. Antczak & R. Cipriani (Eds.), *Early human impact on megamolluscs* (BAR International Series No. 266) (pp. 175–184). Oxford: Archaeopress.

Parkington, J., Poggenpoel, C., Buchanan, B., Robey, T., Manhire, T., & Sealy, J. (1988). Holocene coastal settlement patterns in the western Cape. In G. Bailey & J. E. Parkington (Eds.), *The archaeology of prehistoric coastlines* (pp. 22–41). Cambridge: Cambridge University Press.

Porraz, G., Texier, P.-J., Rigaud, J.-P., Parkington, J., Poggenpoel, C., & Roberts, D. L. (2008). Preliminary characterization of a Middle Stone Age lithic assemblage preceding the "classic" Howiesons Poort complex at Diepkloof Rock-Shelter, Western Cape Province, South Africa. *South African Archaeological Society Goodwin Series, 10*, 105–121.

Powell, A., Shennan, S., & Thomas, M. G. (2009). Late Pleistocene demography and the appearance of modern human behaviour. *Science, 324*, 1298–1301.

Read, D. (2006). Tasmanian knowledge and skill: Maladaptive imitation or adequate technology? *American Antiquity, 71*, 164–184.

Sampson, C. G. (1985). Atlas of Stone Age settlement in the Central and Upper Seacow Valley (Memoirs of the National Museum, No. 20). Bloemfontein: The National Museum.

Scholz, C. A., Johnson, T. C., Cohen, A. S., King, J. W., Peck, J. A., Overpeck, J. T., et al. (2007). East African megadroughts between 135 and 75 thousand years ago and bearing on early-modern human origins. *Proceedings of the National Academy of Sciences, 104*, 16416–16421.

Schweitzer, F. R., & Wilson, M. L. (1982). Byneskranskop 1: A Late Quaternary living site in the southern Cape Province, South Africa. *Annals of the South African Museum, 88*, 1–203.

Sealy, J. (2006). Diet, mobility and settlement pattern among Holocene hunter-gatherers in southernmost Africa. *Current Anthropology, 47*, 569–595.

Sealy, J., & Galimberti, M. (2011). Shellfishing and the interpretation of shellfish sizes in the Middle and Later Stone Ages of South Africa. In N. F. Bicho, J. A. Haws, & L. G. Davis (Eds.), *Trekking the shore: Changing coastlines and the antiquity of coastal settlement* (pp. 405–419). New York: Springer.

Shennan, S. (2001). Demography and cultural innovation: A model and its implications for the emergence of modern human culture. *Cambridge Archaeological Journal, 11*, 5–16.

Singer, R., & Wymer, J. (1982). *The Middle Stone Age at Klasies River Mouth in South Africa*. Chicago: University of Chicago Press.

Soriano, S., Villa, P., & Wadley, L. (2007). Blade technology and tool forms in the Middle Stone Age of South Africa: The Howiesons Poort and post-Howiesons Poort at Rose Cottage Cave. *Journal of Archaeological Science, 34*, 681–703.

Steele, T. E. & Klein, R. G. (2005/2006). Mollusk and tortoise size as proxies for Stone Age population density in South Africa: Implications for the evolution of human cultural capacity. *Munibe (Antropologia-Arkeologia), 57*(1), 221–237.

Teske, P. R., Papadopoulos, I., McQuaid, C. D., Newman, B. K., & Barker, N. P. (2007). Climate change, genetics or human choice: Why were the shells of mankind's earliest ornament larger in the Pleistocene than in the Holocene? *PLoS ONE, 2*(7), e614. doi:10.1371/journal.pone.0000614.

Texier, P.-J., Porraz, G., Parkington, J., Rigaud, J.-P., Poggenpoel, C., Miller, C., et al. (2010). A Howiesons Poort tradition of engraving ostrich eggshell containers dated to 60,000 years ago at Diepkloof Rock Shelter, South Africa. *Proceedings of the National Academy of Sciences, 107*, 6180–6185.

Thackeray, J. F. (1988). Molluscan fauna from Klasies River, South Africa. *South African Archaeological Bulletin, 43*, 27–32.

Thackeray, J. F. (2007). Sea levels and chronology of Late Pleistocene coastal cave deposits at Klasies River in South Africa. *Annals of the Transvaal Museum, 44*, 219–220.

Tribolo, C., Mercier, N., Douville, E., Joron, J.-L., Reyss, J.-L., Rufer, D., et al. (2013). OSL and TL dating of the Middle Stone Age sequence at Diepkloof Rock Shelter (South Africa): A clarification. *Journal of Archaeological Science, 40*, 3401–3411.

Vanhaeren, M., d'Errico, F., Stringer, C., James, S. L., Todd, J. A., & Mienis, H. K. (2006). Middle Paleolithic shell beads in Israel and Algeria. *Science, 312*, 1785–1788.

Villa, P., Soriano, S., Teyssandier, N., & Wurz, S. (2010). The Howiesons Poort and MSA III at Klasies River main site, Cave 1A. *Journal of Archaeological Science, 37*, 630–655.

Vogelsang, R., Richter, J., Jacobs, Z., Eichhorn, B., Linseele, V., & Roberts, R. G. (2010). New excavations of Middle Stone Age deposits at Apollo 11 rockshelter, Namibia: Stratigraphy, archaeology, chronology and past environments. *Journal of African Archaeology, 8*, 185–218.

Voigt, E. A. (1982). The molluscan fauna. In R. Singer & J. Wymer (Eds.), *The Middle Stone Age at Klasies River Mouth in South Africa* (pp. 155–186). Chicago: University of Chicago Press.

Volman, T. P. (1978). Early archaeological evidence for shellfish collecting. *Science, 210*, 911–913.

Volman, T. P. (1981). The Middle Stone Age in the southern Cape. Ph. D. Dissertation, University of Chicago.

Wadley, L. (1986). Segments of time: A mid-Holocene Wilton site in the Transvaal. *South African Archaeological Bulletin, 41*, 54–62.

Wadley, L. (1989). Legacies from the Later Stone Age. *South African Archaeological Society Goodwin Series, 6*, 42–53.

Wadley, L. (1997). Rose Cottage Cave: Archaeological work 1987–1997. *South African Journal of Science, 93*, 439–444.

Wadley, L. (2006). Partners in grime: Results of multi-disciplinary archaeology at Sibudu Cave. *Southern African Humanities, 18*, 315–341.

Wadley, L. (2008). The Howieson's Poort industry of Sibudu Cave. *South African Archaeological Society Goodwin Series, 10*, 122–132.

Wadley, L., & Mohapi, M. (2008). A segment is not a monolith: Evidence from the Howiesons Poort of Sibudu, South Africa. *Journal of Archaeological Science, 35*, 2594–2605.

Wadley, L., Hodgskiss, T., & Grant, M. (2009). Implications for complex cognition from the hafting of tools with compound in the Middle Stone Age, South Africa. *Proceedings of the National Academy of Sciences, 106*, 9590–9594.

Wadley, L., Sievers, C., Bamford, M., Goldberg, P., Berna, F., & Miller, C. (2011). Middle Stone Age bedding construction and settlement patterns at Sibudu, South Africa. *Science, 334*, 1388–1391.

Walker, N. J. (1995). *Late Pleistocene and Holocene hunter-gatherers of the Matopos*. Uppsala: Societas Archaeologica Upsaliensis.

Webley, L. (2002). The re-excavation of Spoegrivier Cave on the west coast of South Africa. *Annals of the Eastern Cape Museums, 2*, 19–49.

Wendt, W. E. (1976). "Art mobilier" from the Apollo 11 cave, South West Africa: Africa's oldest dated works of art. *South African Archaeological Bulletin, 31*, 5–11.

White, T. D., Asfaw, B., DeGusta, D., Gilbert, H., Richards, D. G., Suwa, G., & Howell, F. C. (2003). Pleistocene Homo sapiens from Middle Awash, Ethiopia. *Nature, 423*, 742–747.

Wurz, S. (1999). The Howiesons Poort backed artefacts from Klasies River: An argument for symbolic behaviour. *South African Archaeological Bulletin, 54*, 38–50.

Wurz, S. (2002). Variability in the Middle Stone Age lithic sequence, 115,000–60,000 years ago at Klasies River, South Africa. *Journal of Archaeological Science, 29*, 1001–1015.

Yellen, J. E. (1998). Barbed bone points: Tradition and continuity in Saharan and sub-Saharan Africa. *African Archaeological Review, 15*, 173–198.

Yellen, J. E., Brooks, A. S., Cornelissen, E., Mehlman, M. J., & Stewart, K. (1995). A Middle Stone Age worked bone industry from Katanda, Upper Semliki Valley, Zaire. *Science, 268*, 553–556.

Ziegler, M., Simon, M. H., Hall, I. R., Barker, S., Stringer, C., & Zahn, R. (2013). Development of Middle Stone Age innovation linked to rapid climate change. *Nature Communications, 4*, 1905.

Zilhão, J., Angelucci, D. E., Badal-Garcia, E., d'Errico, F., Daniel, F., Dayet, L., et al. (2010). Symbolic use of marine shells and mineral pigments by Iberian Neandertals. *Proceedings of the National Academy of Sciences, 107*, 1023–1028.

Chapter 5

Patterns of Hominin Occupation and Cultural Diversity Across the Gebel Akhdar of Northern Libya Over the Last ~200 kyr

Sacha Jones, Annita Antoniadou, Huw Barton, Nick Drake, Lucy Farr, Chris Hunt, Robyn Inglis, Tim Reynolds, Kevin White, and Graeme Barker

Abstract Excavations at Haua Fteah cave in Cyrenaica, Libya, have revealed a cultural sequence that may span the last glacial–interglacial-glacial cycle. The TRANS-NAP project has been re-excavating Haua Fteah and conducting geoarchaeological survey of an ecologically diverse landscape that includes the fertile Gebel Akhdar and littoral, pre-desert, and desert biomes. A major aim of this project is to characterize cultural and environmental changes across the region and correlate the surface archaeology with that from Haua Fteah. To date, 181 sites have been recorded, ranging from the Middle Stone Age (MSA) to Late Stone Age (LSA). Their geographic distribution suggests temporal variation in patterns of hominin habitat preference, with significantly more LSA than MSA sites at higher elevations. The surface archaeology also points to substantial spatiotemporal technological variation within the MSA. These patterns may be explained by both paleoenvironmental change and paleodemographic shifts in the region, resulting in a variety of hominin adaptive responses.

Keywords Cyrenaica • Middle Stone Age • Late Stone Age • Haua Fteah • Mediterranean littoral • Sahara • Dispersal routes

S. Jones (✉) · L. Farr · G. Barker
McDonald Institute for Archaeological Research, University of Cambridge, Downing Street, Cambridge, CB2 3ER, UK
e-mail: scj23@cam.ac.uk

A. Antoniadou
School of Geography, Archaeology and Palaeoecology, Queen's University, Belfast, Northern Ireland, BT7 1NN, UK

H. Barton
School of Archaeology and Ancient History, University of Leicester, Leicester, LE1 7RH, UK

N. Drake
Department of Geography, King's College London, London, WC2R 2LS, UK

C. Hunt
School of Natural Sciences and Psychology, Liverpool John Moores University, Byrom Street, Liverpool, L3 3AF, UK

R. Inglis
Department of Archaeology, University of York, The King's Manor, York, YO1 7EP, UK

T. Reynolds
Department of History, Classics & Archaeology, Birkbeck, University of London, Russell Square, Bloomsbury, London, WC1B 5DQ, UK

K. White
Department of Geography, University of Reading, Whiteknights, Reading, RG6 6AB, UK

Introduction

North Africa during the last glacial–interglacial–glacial cycle witnessed a number of complex demographic and behavioral changes. The oldest *Homo sapiens* fossils in North Africa, from Jebel Irhoud in Morocco, currently point to colonization of the region during MIS 6 (Hublin 2001: 110; Smith et al. 2007; Stringer and Barton 2008; Balter 2011; Hublin and McPherron 2012). The first MSA technologies in North Africa also date to MIS 6 and to even earlier in MIS 7 (e.g., McBrearty and Brooks 2000; Garcea 2010a, 2012a, b; Hawkins 2012), where the latter raises the question of which hominin populations and species were responsible for the early MSA across the region. From MIS 6 onwards, other key demographic and behavioral processes occurred in North Africa. For example, modern humans are argued to have dispersed across the Sahara and out of Africa during the climatic optimum of MIS 5e (Drake et al. 2008, 2011; Osborne et al. 2008; Hublin and Klein 2011). It was during MIS 5 that classic markers of symbolic behavior or "behavioral modernity" (e.g., shell beads, ochre use) start appearing in MIS 5 Aterian contexts, rivaling in nature and age similar cultural material from southern Africa

Sacha C. Jones and Brian A. Stewart (eds.), *Africa from MIS 6-2: Population Dynamics and Paleoenvironments*, Vertebrate Paleobiology and Paleoanthropology, DOI 10.1007/978-94-017-7520-5_5

(McBrearty and Brooks 2000; Bouzouggar et al. 2007; Barton et al. 2009; d'Errico et al. 2009). From late MIS 4 until the Holocene, a paucity of archaeological evidence from the Sahara suggests that populations became confined to coastal areas as the Sahara became increasingly arid (Cremaschi et al. 1998; Ambrose 1998; Barham and Mitchell 2008: 265). Yet, it is questionable that the Sahara was uninhabitable throughout this period, as suggested by recent evidence for pluvial episodes across the Sahara from MIS 4-2, albeit variable in timing, location and frequency (Smith 2012). Evidence for past water sources derived from aquifer-fed artesian springs and dating to MIS 4 and MIS 2 in the Kharga Oasis further point to greater water availability than perhaps previously assumed (Blackwell et al. 2012). Using the coastline in mid-MIS 3, populations are argued to have dispersed into North Africa from southwest Asia (McBurney 1967; Olivieri et al. 2006; Pereira et al. 2010). Possible markers of this include the appearance of early LSA ("Upper Paleolithic" or "Dabban") industries restricted mainly to northern Cyrenaica in Libya (McBurney and Hey 1955; McBurney 1967), although this may have been an autochthonous development (e.g., Iovita 2009; Garcea 2010b, c). Few comparable early LSA sites are known from the Nile Valley (Vermeersch 2010), northeast Libya (Barich et al. 2006; Garcea and Giraudi 2006), or the Maghreb (Barton et al. 2007). At the end of this period, around the time of the Last Glacial Maximum (LGM), an explosion of backed bladelet industries appears in archaeological contexts across North Africa, clearly marking a change in the region's population dynamics (e.g., Close 2002).

These broad-scale changes from MIS 6-2 are only the highlights of what would have been a far more complex period in recent human evolution, the details of which are becoming apparent only recently as North Africa gains more prominent status in paleoanthropological debates (e.g., Stringer and Barton 2008; Balter 2011; Hublin and McPherron 2012). Focusing on this critical period in recent human evolution, this study explores new archaeological evidence from northern Cyrenaica in Libya and addresses several key questions relating to past behaviors and population dynamics. For example, what were the relationships between changing resource distribution (e.g., water, plants, animals) from MIS 6-2, and raw material location, patterns of occupation, and behavior during the MSA and LSA? As this study demonstrates that the MSA of the region, as with the rest of North Africa (Van Peer 2016), exhibits substantial technological variability, to what extent was this due to climate change and consequential demographic changes, and to what extent was it a result of behavioral adaptation of pre-existing populations to habitat and societal changes? Finally, MIS 6-2 was characteristically a period of modern human dispersals within Africa and beyond, yet at various points in time certain geographic regions (e.g., the Sahara) were barriers to population movements. When and how were demographic connections most plausible between northern Cyrenaica and regions beyond during this period? Furthermore, what was the degree of population isolation and what were the possible cultural manifestations of this? The evidence presented in this chapter, albeit hindered by limited fossil evidence and a lack of chronometric dates for the surface archaeology, goes some way towards answering these questions. These issues, however, continue to form part of the research agenda of the ERC-funded TRANS-NAP project (*Cultural Transformations and Environmental Transitions in North African Prehistory*) of which this study is a part.

The TRANS-NAP project began in 2009 and one of its key aims is to characterize the cultural and environmental changes that occurred across an ecologically diverse North African landscape over the last ~200 kyr, from the earliest MSA to the transition from hunting and gathering subsistence strategies to Neolithic farming. The project's study region is centered on the Gebel Akhdar ("Green Mountain") in northeast Libya, an area of relatively high rainfall that forms a fertile zone within an otherwise arid east–west belt stretching from the eastern edge of the Maghreb to the Nile Valley, with the Sahara Desert lying directly to the south. The project's ~150 by 150 km study region traverses four biomes; from north to south these are: the littoral, upland (Gebel Akhdar), pre-desert, and desert zones. Of particular interest is the relationship between climatic and environmental fluctuations, and changing human behaviors and population histories from the end of the Middle Pleistocene to the Holocene. This involves the use of archaeological and multi-proxy environmental data from multiple sites to: (1) reconstruct past paleoenvironments in the region over the last ~200 kyr; (2) determine when and how *H. sapiens* colonized the region; (3) establish if certain behaviors enabled successful colonization by *H. sapiens*; and (4) ascertain if occupation was continuous through glacial periods and if the Gebel Akhdar may have acted as a refugium. These issues are being addressed through the reinvestigation of Haua Fteah, a cave first excavated in the 1950s (McBurney 1967), and geoarchaeological survey of the four biomes within the study region. This chapter concentrates on the results of these geoarchaeological surveys, where patterns of occupation and changing technologies across the four biomes during the MSA and LSA are described and interpreted with respect to spatiotemporal changes in hominin adaptive behaviors. Additional evidence is drawn upon to posit theories about past population dynamics in northern Cyrenaica as well as connections with regions beyond from MIS 6-2.

Geoarchaeological Surveys in the TRANS-NAP Project Study Region

Previous geoarchaeological surveys in northern Cyrenaica concentrated on the coast and its immediate hinterland (Watson 1949; McBurney and Hey 1955; McBurney 1947, 1967, 1968; Hey 1968; Barker et al. 2007, 2008), but did not venture southward across the Gebel Akhdar and into the arid zones. McBurney and Hey (1955) pioneered this research, reporting evidence from several open-air and cave sites. Typical MSA lithic types (referred to as "Middle Paleolithic" or "Levalloiso-Mousterian") were recovered from raised beach deposits (e.g., at Wadi Haula, El Atrun, and Ras Aamer). Other key finds from Wadi Salib, near Tocra, include discoidal and flake-blade cores as well as a rare bifacial tool (a possible handaxe). They report more extensive investigations at one open-air site (Hajj Creiem) and two cave sites (Hagfet ed Dabba and Hagfet et Tera). Hajj Creiem, in Wadi Derna, produced *in situ* MSA lithics (e.g., Levallois cores, points, and scrapers) and excavations at Hagfet ed Dabba, in the Wadi Kuf, produced large collections of fauna and blade-based assemblages. The latter represents the type-site of the "Upper Paleolithic" of the region, termed the "Dabban". Excavations at Hagfet et Tera, near Benghazi, produced some MSA but predominantly LSA ("Oranian") artifacts (McBurney and Hey 1955). An Aterian presence in northern Cyrenaica is not compelling, being weakly represented at Hagfet et Tera, Ras Aamer, and Haua Fteah (McBurney and Hey 1955; McBurney 1967: 129; Reynolds 2013).

Haua Fteah, a ~80 m-wide and ~20 m-high cave, was first excavated by McBurney and colleagues over three field seasons in the 1950s (McBurney 1967). The cave's deposits were excavated as a series of trenches to a depth of 14 m, revealing tens of thousands of lithic artefacts and faunal fragments, including two *H. sapiens* mandibles in MSA ("Levalloiso-Mousterian") levels. Excavations produced a complex cultural succession, from the idiosyncratic and inappropriately named "pre-Aurignacian" (an early MSA lithic collection from the cave's deepest sediments) to the Neolithic and Historic periods, intervened by particularly diverse and lengthy MSA and LSA sequences. The "pre-Aurignacian" was estimated by McBurney (1967) to date to ~80–60 ka but is now thought to be significantly older, possibly extending into MIS 6 (Moyer 2003; Barker et al. 2007) (an ongoing dating program is addressing this). Although McBurney argued for relatively continuous occupation of the cave from the "pre-Aurignacian" to the present day, gaps are present in the sequence and the density of material is highly variable throughout the profile (Klein and Scott 1986), particularly in the "pre-Aurignacian" and later MSA layers. Since McBurney's excavations, several other researchers have analyzed different aspects of both the faunal (Klein and Scott 1986; MacDonald 1997; Wall-Scheffler 2007) and lithic (Close 1986; Chazan 1995; Hiscock 1996; Moyer 2003) collections. In 2007, the Cyrenaican Prehistory Project (later becoming the TRANS-NAP project) embarked on renewed investigations of the cave (Barker et al. 2007, 2008, 2009, 2010; Farr et al. 2014; Hunt et al. 2010, 2011; Rabett et al. 2013), with the aim to re-excavate new trenches alongside McBurney's old section faces and conduct high-resolution sampling for dating, paleoenvironmental, and lithic analyses. Research at Haua Fteah is ongoing, involving multiple project members and analytical techniques.

This chapter reports new data collected during geoarchaeological surveys of an area of ~150 by 150 km that traverses four different ecological zones. From north to south, these are the littoral, upland, pre-desert, and desert zones (Fig. 5.1). The littoral zone is a narrow coastal strip, dominated by Mediterranean vegetation and bordered by the steep northerly escarpments of the Gebel Akhdar. Haua Fteah lies within the littoral zone, ~1 km south of the shoreline. Quaternary deposits are common here, where north-flowing wadis incise the Gebel, depositing alluvium at their terminus. Raised beach deposits also exist, some revealing surface and *in situ* lithic artifacts (McBurney and Hey 1955). The Gebel Akhdar, a ~200 by 50 km (McBurney 1967: 324) upland zone, is an area of relatively high precipitation (approximately 200–550 mm annual rainfall [McBurney and Hey 1955; Elfadli 2009]) and ecological productivity within an otherwise semi-arid coastal strip and arid hinterland of North Africa. Reaching a maximum height of ~879 m above mean sea level (amsl), the Gebel dramatically terminates in the north and west as two escarpments, yet gradually diminishes in elevation to the south and east. The uplands and escarpments are dissected by numerous wadi systems and their tributaries, forming steep-sided gorges in certain areas. Vegetation coverage is typical Mediterranean maquis scrub and some areas are heavily forested, receiving snow coverage in winter. As the Gebel gradually descends southwards, there is a sharp decline in vegetation, partly caused by climate (particularly a southwards decrease in precipitation) but also by overgrazing by livestock. This undulating and relatively dry and devegetated landscape marks the beginning of the pre-desert, with a gradual southward decrease in plant cover until reaching the desert proper. These zones are also characterized by river channels that have their headwaters in the uplands, run through the pre-desert, and terminate in fans feeding paleolakes in the desert. Divisions between pre-desert and desert are somewhat arbitrary, and the boundaries between them would have fluctuated during MIS 6-2 as the rivers and lakes oscillated between being perennial and ephemeral (Fig. 5.1).

The first geoarchaeological investigations were conducted in the southern Gebel Akhdar, pre-desert, and desert during

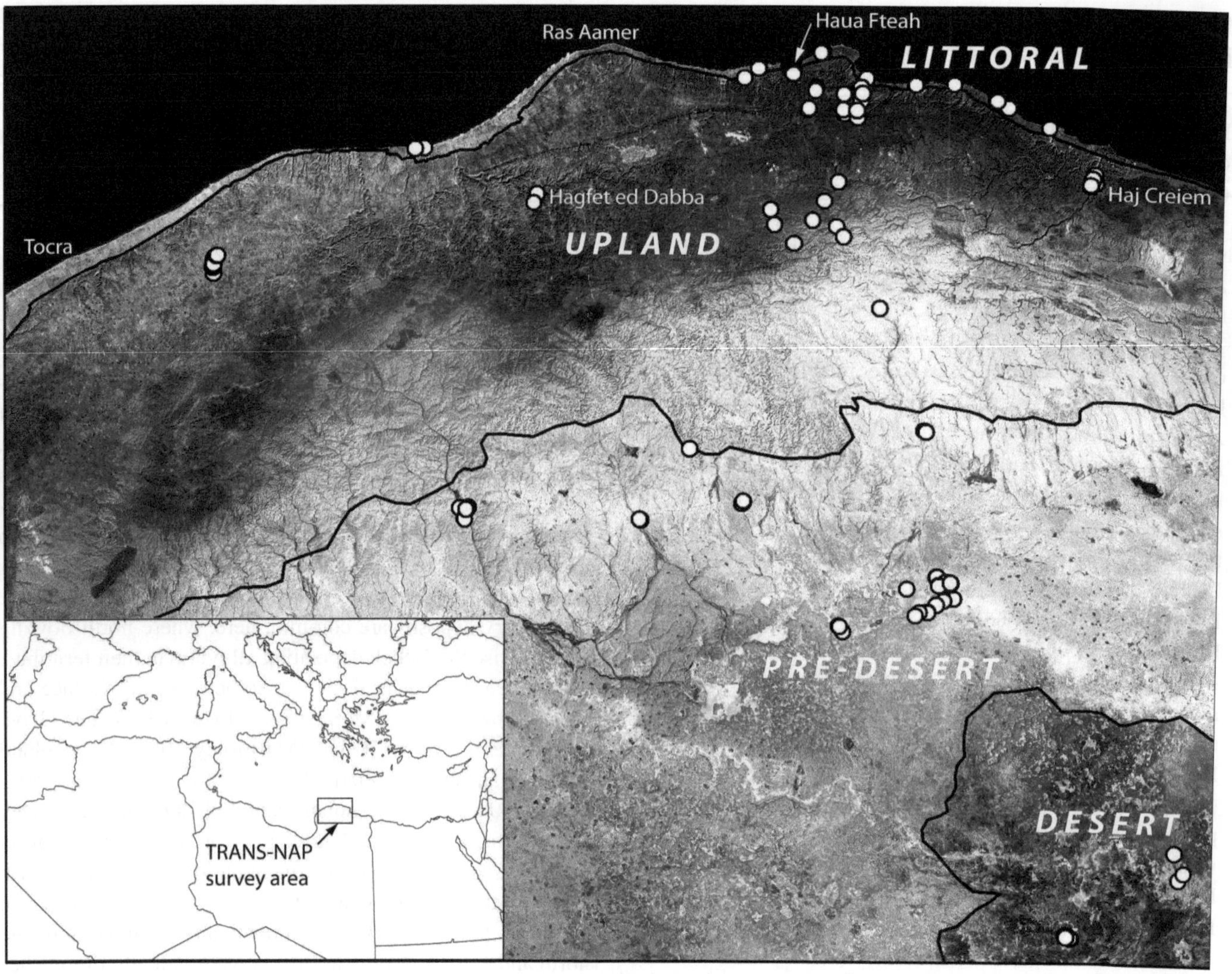

Fig. 5.1 Landsat satellite map showing the distribution of Paleolithic sites in the TRANS-NAP study area, divided into the four ecological zones: littoral, upland Gebel Akhdar, pre-desert, and desert. Key sites mentioned in the text are indicated and the inset shows where the surveyed area is situated in North Africa

two field seasons in 2009 and 2010 (Barker et al. 2009, 2010; Jones et al. 2011) (authors: SJ, LF, ND, KW, HB). In the same years, a second team surveyed the coastal zone and Wadi Derna, building on previous work conducted by McBurney and Hey (1955) (authors: AA, CH). Additional sites were recorded in the area immediately surrounding Haua Fteah during surveys in 2007 and 2008; however, less archaeological information is available due to a different recording system. Surveys of the Gebel, pre-desert, and desert employed a different methodology from those conducted along the coast and Wadi Derna. In the former, analysis of remote sensing imagery followed by targeted ground survey proved to be the most valuable starting point, given the large size of the study region and the unexplored nature of territories to the south of the littoral zone. Landsat TM and ETM+ false-color composite images were used to locate particular areas of interest (Barker et al. 2009, 2010), from the enclosed topographic depression at Al-Marj to headwaters, wadi channels, plateaus, stony desert plains, alluvial fans, and paleolakes. An objective was to explore the archaeological record of a variety of landforms with the aim to discern relationships between a site's geomorphological context and the archaeology therein (Jones et al. 2011). While the vast majority of sites discovered in the Gebel, pre-desert, and desert consisted of surface lithic scatters, artifacts at Al-Marj are exposed in the sections, spoil heaps, and floor of an ~8.5 km-long and ~3 m-deep canal that was dug into the northern end of the basin in the late 1970s. An ~29 m-deep core was drilled in this former enclosed lake basin, nestled within the western Gebel Akhdar, in order to collect samples for paleoenvironmental reconstruction (Barker et al. 2010). A principal objective of the coastal and Wadi Derna surveys was to locate sites with *in situ* Paleolithic artifacts and establish the relationship between paleoenvironmental and cultural change along the coast (Antoniadou 2012). At these sites, the lithostratigraphy and paleoenvironmental signatures of sedimentary sequences were established using gross sedimentary characteristics, thin-section analysis, palynology, and mollusks, and samples were collected for radiocarbon, OSL, and Uranium-series dating. With the exception of data recorded in 2007 and 2008, all sites were recorded using the same forms, enabling direct comparison of data. Artifact descriptions were included in these forms and small collections of surface lithics were often made. Where possible, material was assigned to an appropriate period of the Paleolithic. As the majority of sites consist of surface lithic scatters, dating the artifacts remains a considerable problem. As a means to partially resolve this, one of the ongoing objectives of the TRANS-NAP project is to use detailed technological comparisons of lithics from Haua Fteah and the landscape sites, together with an ongoing and extensive program of dating the cave's sediments, to narrow down the age of many of the region's surface lithic scatters. As part of this objective, it is important to account for potential technological variation caused by spatial as well as temporal differences in behavior (e.g., those caused by cave-based versus wider landscape-based activities).

Archaeological Patterns Across Biomes in Northern Cyrenaica

Four seasons of archaeological surveys across the region from 2007 to 2010 resulted in the identification of 178 new archaeological sites. Tables 5.1 and 5.2 present data from all of these sites, as well as Haua Fteah, Hagfet ed Dabba (CPP1647), and Hajj Creiem (CPP1825a), which have been documented previously (McBurney and Hey 1955; McBurney 1967). Sites within each biome have been grouped into different geographic clusters (Fig. 5.2). The number of sites in each cluster is not a measure of occupation density but is a function of survey bias. Some areas have been surveyed more intensively than others; for example, the littoral and hinterland areas to the north and south of Haua Fteah, respectively, the Baltat ar Ramlah (east) paleolake and the

Table 5.1 Site location data recorded during geoarchaeological survey of the different geographic clusters within the four biomes

Geographic cluster	Number of sites (elevation range[a])	Landforms (no. of sites)	Chert source[b]
Desert			
Baltat lawlah al Halq (west)	5 (120–140 m)	Paleolake (3); hill slope (1); hill top (1)	1
Baltat lawlah al Halq (east)	3 (120–127 m)	Paleolake (2); wadi channel (1)	0
Pre-desert			
Baltat ar Ramlah (east)	20 (128–144 m)	Alluvial fan (9); hill slope (1); paleolake (8); paleolake island (1); plateau (1)	2
Baltat ar Ramlah (west)	3 (128–138 m)	Hill slope (1); hill top (1); paleolake island (1)	1
Wadi al Hammam	2 (231–233 m)	Alluvial fan (2)	1
Wadi Ramlah	3 (277–284 m)	Plain (2); wadi terrace (1)	0
Wadi Samalus	12 (262–281 m)	Hill slope (8); hill top (3); slope base (1)	6
Wadi Tanamlu	4 (217–328 m)	Alluvial fan (3); plateau (1)	0
Gebel Akhdar			
Al-Marj canal	37 (249–289 m)	Basin (37)	1
Haua Fteah hinterland	13 (108–225 m)	Hill slope (7); hill top (4); plateau (1); wadi channel (1)	6
North Gebel (Lamludah north)	17 (241–653 m)	Hill slope (5); hill top (4); plateau (3); tufa dam (1); wadi channel (2); wadi terrace (1); slope base (1)	2
South Gebel	1 (459 m)	Plain (1)	0
Upper Gebel (Lamludah south)	8 (624–802 m)	Hill slope (4); wadi channel (3); wadi terrace (1)	0
Wadi Derna	6 (90–146 m)	Hill slope (2); wadi channel (3); wadi terrace (1)	0
Wadi Kuf Gebel	3 (341–367 m)	Cave (3)	0
Littoral			
Chersa	10 (0–18 m)	Coastal terrace (10)	0
El Atrun	7 (5–6 m)	Coastal terrace (7)	0
Haua Fteah littoral	19 (10–75 m)	Cave (1); coastal terrace (10); hill slope (2); hill top (1); slope base (2); wadi channel (1); wadi terrace (2)	0
Ras al Hillal	3 (8–13 m)	Coastal terrace (3)	0
Wadi Gladia littoral	3 (7–87 m)	Coastal terrace (2); hill slope (1)	0
Wadi Kuf littoral	2 (1–14 m)	Cave (1); coastal terrace (1)	0

Data includes site elevation, landforms upon which sites are located and frequency of chert sources recorded

[a]Elevation range derived from ASTER 30 m DEM

[b]Number of chert sources is under-reported in all littoral zone clusters and in Wadi Derna and Wadi Kuf. This variable was only recorded during surveys in 2009 and 2010 in the pre-desert, desert, and Gebel Akhdar

Table 5.2 Archaeological data collected during geoarchaeological survey of the different geographic clusters within the four biomes, including the frequency of MSA sites, LSA sites, sites preserving both MSA and LSA material, and sites preserving undiagnostic lithics

Geographic cluster	Paleolithic phase(s)	Number of sites	Site Identification numbers	Examples of lithic technologies present
Desert				
Baltat lawlah al Halq (west)	MSA	3	8049, 8050, 8053	Levallois, discoidal
	MSA & LSA	1	8054	Discoidal, blade, microblade
	LSA	1	8051	Blade
Baltat lawlah al Halq (east)	MSA	2	8056, 8057	Discoidal
	LSA	1	8055	
Pre-desert				
Baltat ar Ramlah (east)	MSA	13	8007, 8009, 8010, 8011, 8012, 8013, 8014, 8025, 8059, 8060, 8061, 8062, 8063	Levallois, discoidal, bifacial, tanged
	MSA & LSA	1	8021	Levallois
	LSA	2	8015, 8024	Blade, bladelet
	Unknown	4	8008, 8017, 8023, 8058	
Baltat ar Ramlah (west)	MSA & LSA	2	8026, 8027	Discoidal, bladelet
	LSA	1	8028	Blade, bladelet, backed
Wadi al Hammam	MSA	2	8043, 8044	Discoidal
Wadi Ramlah	MSA	1	8003	Bifacial
	MSA & LSA	1	8004	
	Unknown	1	8005	
Wadi Samalus	MSA	3	8029, 8031, 8034	Levallois, discoidal
	MSA & LSA	4	8032, 8033, 8035, 8038	Levallois, discoidal, bifacial, blade, bladelet, backed
	LSA	4	8037, 8040, 8041, 8042	Blade, bladelet
	Unknown	1	8030	
Wadi Tanamlu	MSA	4	8045, 8046, 8047, 8048	Levallois, discoidal
Gebel Akhdar				
Al-Marj canal	MSA	33	EM: 3, 4, 9, 10, 11, 17, 18, 20, 21, 101, 102, 103, 105, 106, 107, 109, 110, 113, 114, 117, 119, 120, 121–131	Levallois, discoidal, handaxe, bifacial, stemmed, tanged, blade, bladelet
	Unknown	4	EM: 111, 112, 116, 118	
Haua Fteah hinterland	MSA	1	2020	
	LSA	1	8069	Blade, bladelet
	Unknown	11	1013, 1016, 2004, 2005, 2006, 2007, 2009, 2011, 2012, 2013, 2016	
North Gebel	MSA	1	8120	Levallois
	MSA & LSA	1	8119	Levallois blade, truncated-faceted, carinated end scraper, bladelet, backed
	LSA	9	8100, 8102, 8103, 8104, 8107, 8111, 8116, 8117, 8118	Bifacial point, chamfered blade, bladelet, backed
	Unknown	6	8101, 8105, 8106, 8108, 8109, 8110	
South Gebel	Unknown	1	8000	
Upper Gebel	LSA	6	8064, 8065, 8066, 8068, 8112, 8113	Discoidal, blade, bladelet, backed, bifacial point, carinated endscraper
	Unknown	2	8114, 8115	
Wadi Derna	MSA	2	1825a (Hajj Creiem), 1846	See McBurney and Hey (1955) for Hajj Creiem technologies
	LSA	1	1817	Bladelet
	Unknown	3	1815, 1816, 2021	

(continued)

Table 5.2 (continued)

Geographic cluster	Paleolithic phase(s)	Number of sites	Site Identification numbers	Examples of lithic technologies present
Wadi Kuf Gebel	LSA	1	1647 (Hagfet ed Dabba)	See, e.g., McBurney and Hey (1955) for Hagfet ed Dabba technologies
	Unknown	2	1639, 1642	
Littoral				
Chersa	MSA	8	1830, 1831, 1832b, 1835, 1837, 1838, 1848, 1851	Levallois
	LSA	1	1834	Backed
	Unknown	1	1849	Blade
El Atrun	MSA	5	1801, 1802, 1803, 1805, 1823	Levallois
	LSA	1	1800	Bladelet
	Unknown	1	1798	
Haua Fteah littoral	MSA	2	1008, 2002	Levallois, bifacial
	MSA & LSA	2	Haua Fteah, 1826	See, e.g., McBurney (1967) for Haua Fteah technologies
	LSA	2	1006, 1529	Blade, bladelet
	Unknown	13	1000, 1001, 1002, 1003, 1004, 1005, 1007, 1009, 1010, 1011, 2017, 2018, 2019	
Ras al Hillal	MSA & LSA	1	1818	Blade, bifacial point
	Unknown	2	1792, 1794	
Wadi Gladia	MSA & LSA	1	1571a	
	LSA	1	1827	Bladelet, backed
	Unknown	1	1571b	
Wadi Kuf littoral	LSA	1	2008 (Hagfet al-Gama)	Bladelet, backed (Capsian) (Barker et al. 2008)
	Unknown	1	1813	

Particular types of lithic artifacts encountered in each geographic cluster are provided. Further details regarding the lithic technologies found in some of these geographic clusters are available elsewhere (Barker et al. 2009, 2010; Jones et al. 2011)

[a]"Unknown" refers to sites preserving lithic artifacts that cannot be assigned to either the Middle Stone Age or Late Stone Age (i.e., at sites where only a small number of lithics are present and/or there are no artifacts present that are typologically or technologically diagnostic)

Al-Marj canal. The range of site elevations for each geographic cluster, derived from 30-m-resolution ASTER DEM using a GIS, is provided to express the wide spectrum of site elevations within the Gebel Akhdar in particular. Geoarchaeological data is given for sites in each geographic cluster, including: (1) the type and frequency of different landforms on which the sites are located; (2) the number of sites revealing MSA, LSA, and both MSA and LSA technologies, as well as those that preserve lithics which cannot be assigned reliably to either the MSA or LSA; and (3) typical lithic technologies encountered in each geographic cluster (Fig. 5.3). The presence of certain lithic technologies as well as other traits has enabled some sites to be assigned to different phases within the MSA, suggesting considerable MSA variation across the region. Table 5.1 also gives the number of those archaeological sites where sources of chert were encountered. Depending on the year and location of the survey, this site attribute was not always sought or recorded, and the chert source figures are therefore an underestimate. Chert sources were always recorded during surveys of the desert, pre-desert, and most of the Gebel. Figures 5.4 and 5.5 illustrate the location of MSA and LSA sites encountered in each geographic cluster, including those sites with sources of chert.

Littoral

Two sets of deposits are present in the littoral zone, to at least 90 m amsl. These include marine, lagoonal, and shoreline deposits, interbedded with highly lithified alluvial fan and colluvial sediments. The younger coastal sequences extend to at least MIS 7 and deposits of at least two and possibly as many as four sea-level highstands are represented. These are interbedded with complex and laterally variable deposits including lagoonal clays, aeolian sands, alluvial fan gravels, paleosols, and colluvial diamicts and screes. Also present are deposits of a major paleolake system in the Wadi Derna, which persisted from MIS 5 into the Late Glacial. The deposits at Hajj Creiem (McBurney and Hey 1955) are part of this lacustrine depositional complex.

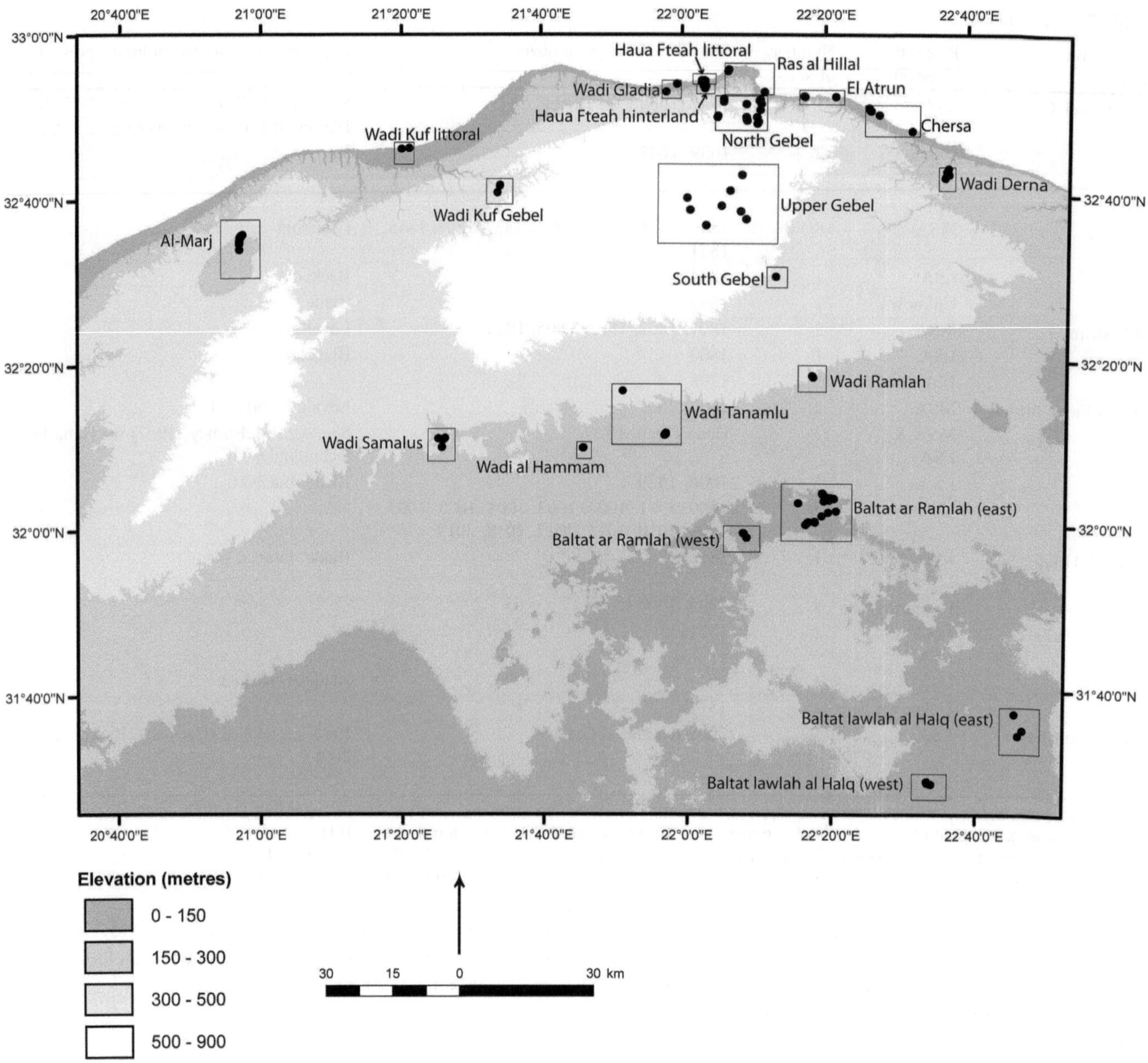

Fig. 5.2 Geographic clusters of Paleolithic sites in the TRANS-NAP study area according to elevation

Archaeological materials are widely distributed throughout the coastal zone; 44 sites have been recorded, although few have yielded large collections. These are divided into six different zones based on geographic location (Table 5.1; Fig. 5.2), yet all are similar in terms of elevation and the types of landform present. All are found at elevations <87 m amsl, where most sites are located below 49 m amsl. The majority of sites (75%) are situated on coastal terraces (including raised beach deposits), where lithics are either on the surface or *in situ*. Of the coastal sites, 34% are MSA, 14% LSA, and 9% preserve both MSA and LSA technologies (the remaining 43% of sites preserve lithics that cannot be assigned to a particular period). Typical artifact types found in these open coastal sites include Levallois cores and flakes, as well as blades and bladelets (including backed varieties). Sites showing *in situ* MSA core reduction are associated with MIS 5 and possibly MIS 7 raised beaches. An Aterian presence is not evident in any of the coastal sites documented as part of this study; however, tanged artifacts are reported from Haua Fteah (McBurney 1967) and at Ras Aamer (McBurney and Hey 1955). Later LSA material with "Dabban" affinities is not widely distributed in the littoral zone; however, one site

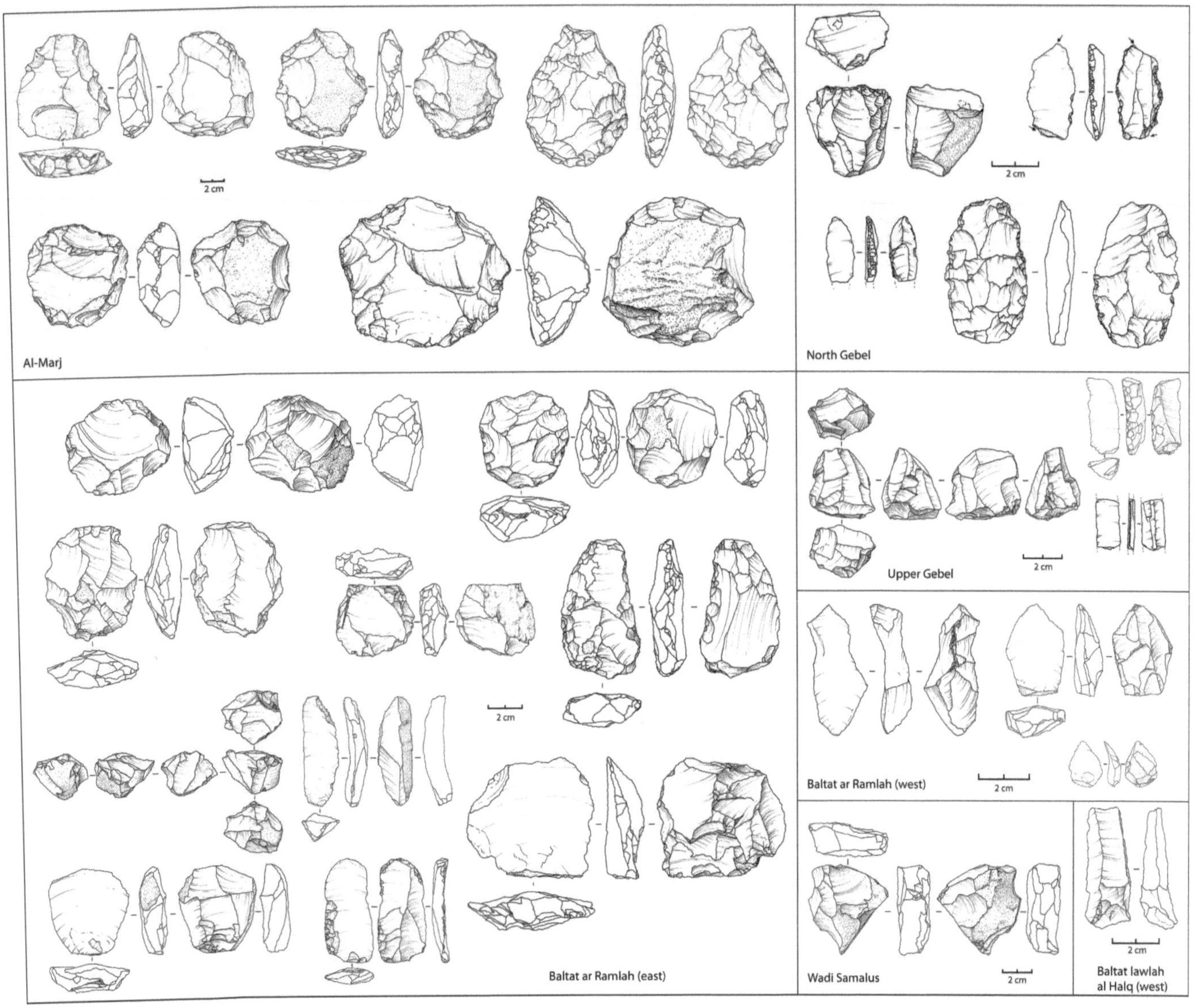

Fig. 5.3 Examples of MSA and LSA lithic artifacts recovered during surveys in seven of the geographic clusters depicted in Fig. 5.2 (clockwise from *top left*): *1* MSA artifacts from Al-Marj (Levallois and discoidal cores, handaxe); *2* LSA artifacts from the North Gebel cluster (bladelet core, chamfered blade, bifacial foliate, backed bladelet fragment); *3* LSA artifacts from the Upper Gebel (bidirectional core, side-retouched blade, bladelet fragment); *4* artifacts from MSA and LSA sites in Baltat ar Ramlah (west) (crested blade, flakes); *5* blade from LSA site in Baltat lawlah al Halq (west); *6* multiplatform core from probable LSA site in Wadi Samalus; *7* predominantly MSA artifacts from Baltat ar Ramlah (east) (Levallois, discoidal and bifacial cores, flakes, blades)

near Haua Fteah consists of a ~28.0 cal kBP fireplace (IntCal09 calibration curve) associated with a long blade industry in a skeletal paleosol interbedded in valley-fill deposits. Older, higher level marine, lagoonal, and riverine deposits are also present, which were laid down during one or more marine transgressions rising to ~90 m amsl. A single U/Th date suggests that these deposits are more than 350 ka. Very rare cores and flakes are present within the basal conglomerate of the transgressive marine deposits, making these the oldest known artifacts in the region so far. Significantly, this is consistent with a pre-modern human presence in the region during the Middle Pleistocene.

Gebel Akhdar

The 85 sites recorded in the Gebel Akhdar fall into 7 different zones, based on geographical location and topographic and landform differences. The majority (n = 37) of these are located in the Al-Marj Basin, an area that is archaeologically unique in terms of the types and densities of artifacts present. The 1970s' canal cutting through the basin has exposed MSA and possibly late Acheulean artifacts within alluvial and lacustrine sediments. Lithic types include numerous core forms (e.g., Levallois, discoidal, bifacial,

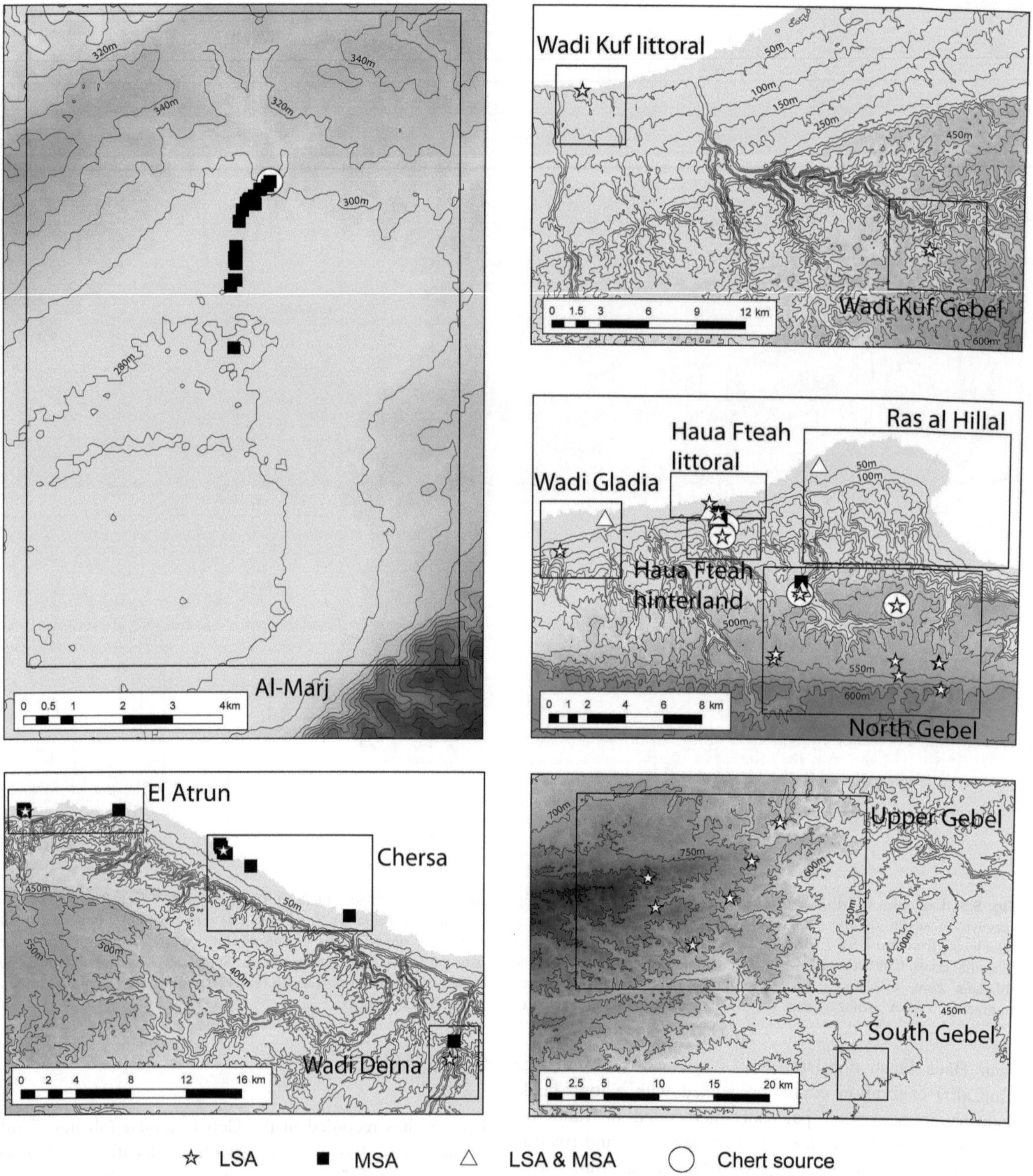

Fig. 5.4 Distribution of chert sources, MSA sites, LSA sites, and sites preserving both MSA and LSA artifacts in geographic clusters (see Fig. 5.2) located in the northern half of the study region

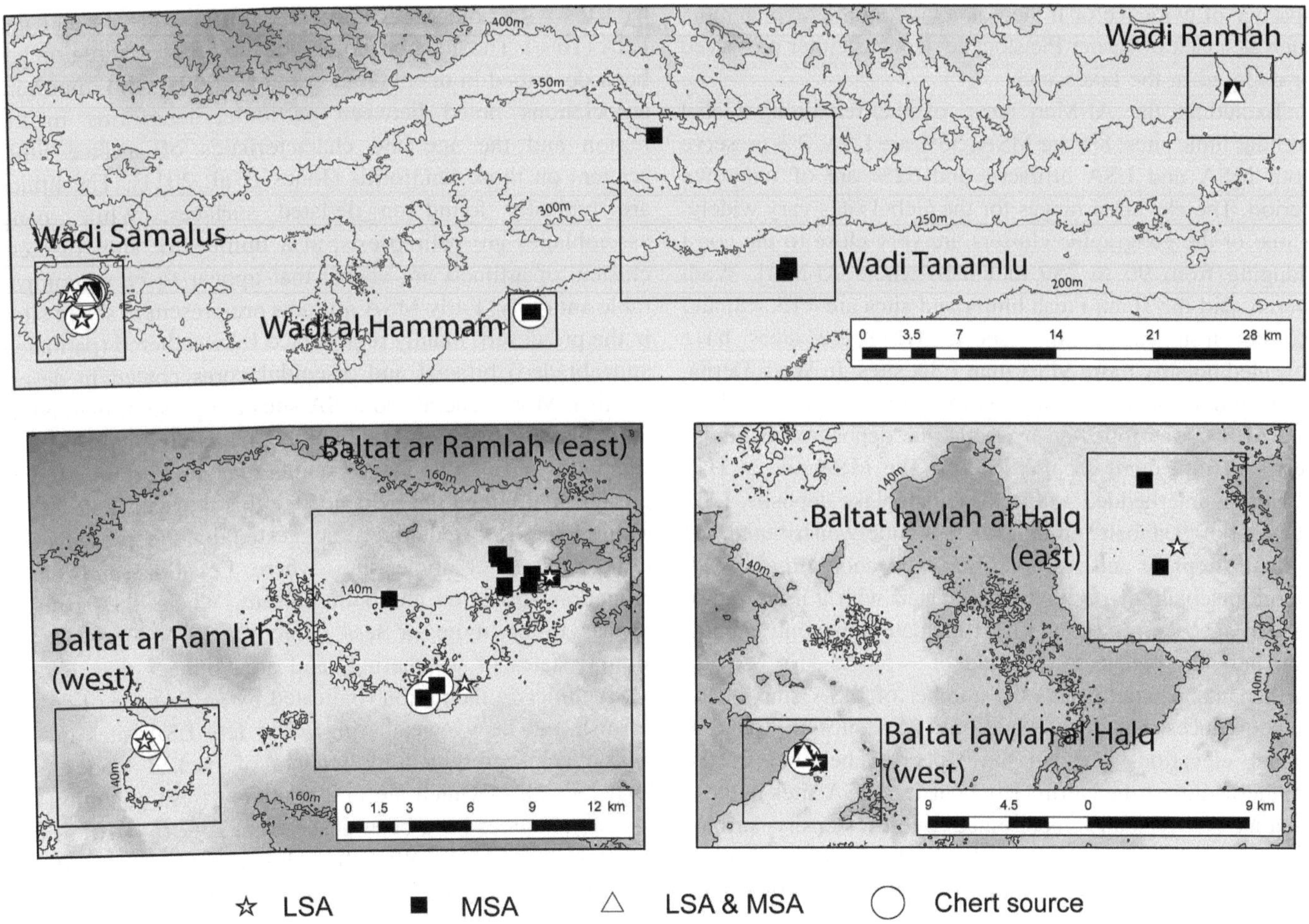

Fig. 5.5 Distribution of chert sources, MSA sites, LSA sites, and sites preserving both MSA and LSA artifacts in geographic clusters (see Fig. 5.2) located in the southern half of the study region

multi-platform), tanged and stemmed artifacts, and numerous handaxes of various shapes and sizes. These artifacts are found within the canal's sections, predominantly in fluvial gravels and sands, and exhibit different states of patination and abrasion. Currently, it is not possible to establish a chronology for the different artifact types as the majority of artifacts recovered so far are either in reworked fluvial contexts or are surface finds. Al-Marj potentially preserves artifacts from the Late Acheulean, early MSA, and Aterian; however, the presence of handaxes could equally be part of the region's MSA tradition. It is significant that the discovery of true Acheulean bifaces was extremely rare during the landscape surveys undertaken by McBurney (McBurney et al. 1948; McBurney and Hey 1955). An earlier surface collection of seven bifaces from Tocra (McBurney 1947) and a broken butt from a putative "amygdaloid" biface from Lete (Petrocchi 1940), both near Benghazi, are the nearest Acheulean material, albeit broken and heavily worn. Three handaxes, however, were described as possessing a smooth patina and sharp edges suggesting they had eroded out relatively recently (McBurney 1947: 78). Of these three pieces, one is discoidal and another is also more likely to be a core (McBurney 1947: Fig. 15, No. 3). This material was considered to be an "evolved Acheulian-Middle Paleolithic type" (McBurney 1947: 78). A small number of other bifacial pieces have been recorded but these are thought to be Neolithic (McBurney 1947; McBurney and Hey 1955). Given this paucity of evidence, *H. sapiens* may have been the first hominin species to inhabit Cyrenaica, using an industry that was either late Acheulean or part of the MSA. The presence of rare bifaces in the "Levalloiso-Mousterian" layers of Haua Fteah might suggest the latter. The apparent absence of earlier Middle Pleistocene and Early Pleistocene technologies is surprising given that the Gebel Akhdar was relatively resource-rich for hominins, when compared to other regions of Libya, and that earlier technologies (handaxes and pebble tools) have been found at sites in other areas along the northern fringe of North Africa (e.g., Ternifine and Ain Hanech in Algeria) (McLaren and Reynolds 2009: 778). It remains to be confirmed if this reflects a true absence of evidence or if there is a taphonomic explanation, such as a lack of earlier Pleistocene deposits either preserved or exposed in the landscape.

Excluding the Al-Marj sites, of the remaining Gebel Akhdar lithic sites, 8% are MSA, 38% are LSA, 2% preserve both MSA and LSA artifacts, and 52% are of unknown period. The elevation ranges for the Gebel sites vary widely. Three of the geographic clusters are very close to the coast (ranging from 90 to 289 m amsl), where Al-Marj, Wadi Derna, and the Haua Fteah hinterland sites are <15, <8, and <2 km from the coast, respectively. These areas have revealed notably more MSA than LSA sites. In Wadi Derna, ~40 m-thick sequences suggest that fluvial deposition from late MIS 6 was followed by lacustrine deposition behind a series of tufa dams from MIS 5-2. The MSA site at Hajj Creiem is interbedded within these paleolake deposits. LSA material of "Dabban" affinities is less widely distributed, yet a small fireplace and two long blades in cool-climate Wadi Derna paleolake deposits are associated with a finite radiocarbon age of ~52.0 ^{14}C kBP (this date is beyond calibration but may equate to a period of ~58–55 ka). Moving further inland (southward) the number of MSA sites in the Gebel reduces dramatically and only LSA sites are found at higher elevations (i.e., in the Upper Gebel and in the southern sites of the North Gebel cluster). LSA sites in these regions are typically represented by low-density surface scatters of flakes, blades, and bladelets, including backed forms, geometric microliths, and bladelet cores. Artifacts are mostly either lightly patinated or unpatinated, with and without abraded edges, and commonly manufactured on a high-quality blue-gray chert, sources of which have yet to be discovered. A lack of visible sources, the low density of the lithic scatters, and an absence of *in situ* flaking products all suggest that these were the products of mobile hunter-gatherer populations that carried out chert exploitation and the early stages of lithic reduction elsewhere. These LSA sites are likely to be terminal Pleistocene or early Holocene in date (Barker et al. 2009). The scatters of backed blades and blade and microblade cores, encountered during surveys along the Wadi el Qlaa in the North Gebel cluster, have been likened to the Dabban and Oranian layers at Haua Fteah (Barker et al. 2010).

Pre-desert

The 44 sites in the pre-desert region have been divided into six geographic clusters; four are located in the lower reaches of wadi systems that originate in the Gebel and drain southward across the pre-desert. Two clusters are paleolake regions, both at the terminus of the Wadi ar Ramlah. The majority of sites in the pre-desert are MSA (53%), followed by sites with both MSA and LSA artifacts (18%) and just LSA (16%). The geoarchaeology of the pre-desert region has been described in detail (Barker et al. 2009), with important associations noted between particular landforms in the region and the age and characteristics of surface lithic scatters on those landforms (Jones et al. 2011). The lithics are typically found on deflated surfaces. While many assemblages are palimpsests, at a number of sites discrete clusters of artifacts are visible that appear to be of comparable antiquity. Early MSA artifacts are present at seven sites in the pre-desert, mainly represented by weathered (patinated and abraded) bifacial and discoidal cores coated in desert varnish. More generalized MSA sites are present, marked in particular by the presence of typical Levallois cores. Two fragments of possible tanged points may suggest an Aterian presence, although the evidence for this is weak. Late MSA technologies are tentatively suggested by the presence of lithic scatters at some sites where Levallois and blade reduction strategies are both present, where the artifacts appear to be more or less contemporary (suggested by similar states of weathering) and are younger than earlier MSA lithics in the region (indicated by an absence of desert varnish and heavy weathering). The few LSA sites in the region are represented by scatters of backed blades and bladelets, all of which are in a relatively fresh condition.

Chert sources have been observed at 23% (n = 10) of the pre-desert sites. The Wadi Samalus area revealed nodular chert sources within limestone bedrock at six sites and evidence of primary flaking activity at three of these. The Baltat ar Ramlah (east) paleolake has been surveyed relatively intensively, revealing one particularly important site, CPP 8009, where there is a high-density scatter of MSA lithics (possibly later MSA), located on an area of raised ground (a "paleolake island", ~400 by 250 m) within the paleolake. Here, chert has been prized from outcrops within the limestone bedrock. In the immediate vicinity, there is evidence of high-density lithic scatters and large rotated cores attesting to the early stages of core reduction. Refined Levallois cores, blades, retouched flakes, and other artifact types are present throughout the island and there is also evidence of artifact recycling, suggesting that the area was repeatedly visited by hominin groups.

Desert

The Baltat lawlah al Halq paleolakes were identified using remote sensing imagery and subjected to a small-scale geoarchaeological survey. Eight sites were recorded in eastern and western paleolakes in the Baltat lawlah at Halq system: MSA (n = 5), MSA and LSA (n = 1), and LSA (n = 2). A source of chert was recorded at one of these sites.

MSA artifacts include heavily weathered discoidal cores and flakes with desert varnish as well as relatively fresh Levallois cores. The differences in weathering patterns suggest that these artifacts may belong to different phases of the MSA. The LSA is represented by blade and microblade technologies; however, the presence of a Neolithic bifacial point means that these may be early Holocene rather than terminal Pleistocene in age.

Three main patterns are evident from the results of geoarchaeological surveys across all four biomes. First, MSA sites are mainly concentrated around the paleolakes of the desert and pre-desert, along the southern flanks of the Gebel where chert sources are available, and along coastal Cyrenaica and a relatively short distance inland into the Gebel Akhdar. In contrast, there is a dramatic reduction in the number of MSA sites at higher elevations in the Gebel; instead, late Pleistocene and possibly early Holocene LSA sites prevail. After excluding the Al-Marj sites, located in the Gebel but within a topographic depression and only ~15 km from the coast, as well as the desert sites as they are too few in number for statistical testing, there exists a statistically significant relationship between biome and the frequency of MSA and LSA sites present in those biomes (χ^2 (2) = 21.573, $p < 0.001$). According to this chi-square test, both the littoral and pre-desert zones have significantly more MSA and fewer LSA sites than expected, whereas the Gebel Akhdar has more LSA sites and fewer MSA sites than expected. Second, confirming this observation but including all 181 sites located in the four biomes and 21 geographic clusters, a comparison of site elevations (using ASTER 30 m resolution elevation data) for MSA sites, LSA sites, and sites where both MSA and LSA lithics are present reveals that the only statistically significant relationship exists between LSA sites and MSA sites. Here, significantly more LSA sites are located at higher elevations than MSA sites (Kruskal-Wallis test, $p = 0.017$). Pairwise comparisons reveal no statistically significant differences between the elevations at which LSA versus MSA and LSA sites are found, or between those at which MSA versus MSA and LSA sites are located (Fig. 5.6). Third, there is substantial typo-technological variation within the MSA across the study region. There are at least three different types of MSA assemblages that probably belong to earlier phases of the MSA: (1) the "pre-Aurignacian", known from Haua Fteah; (2) the handaxe

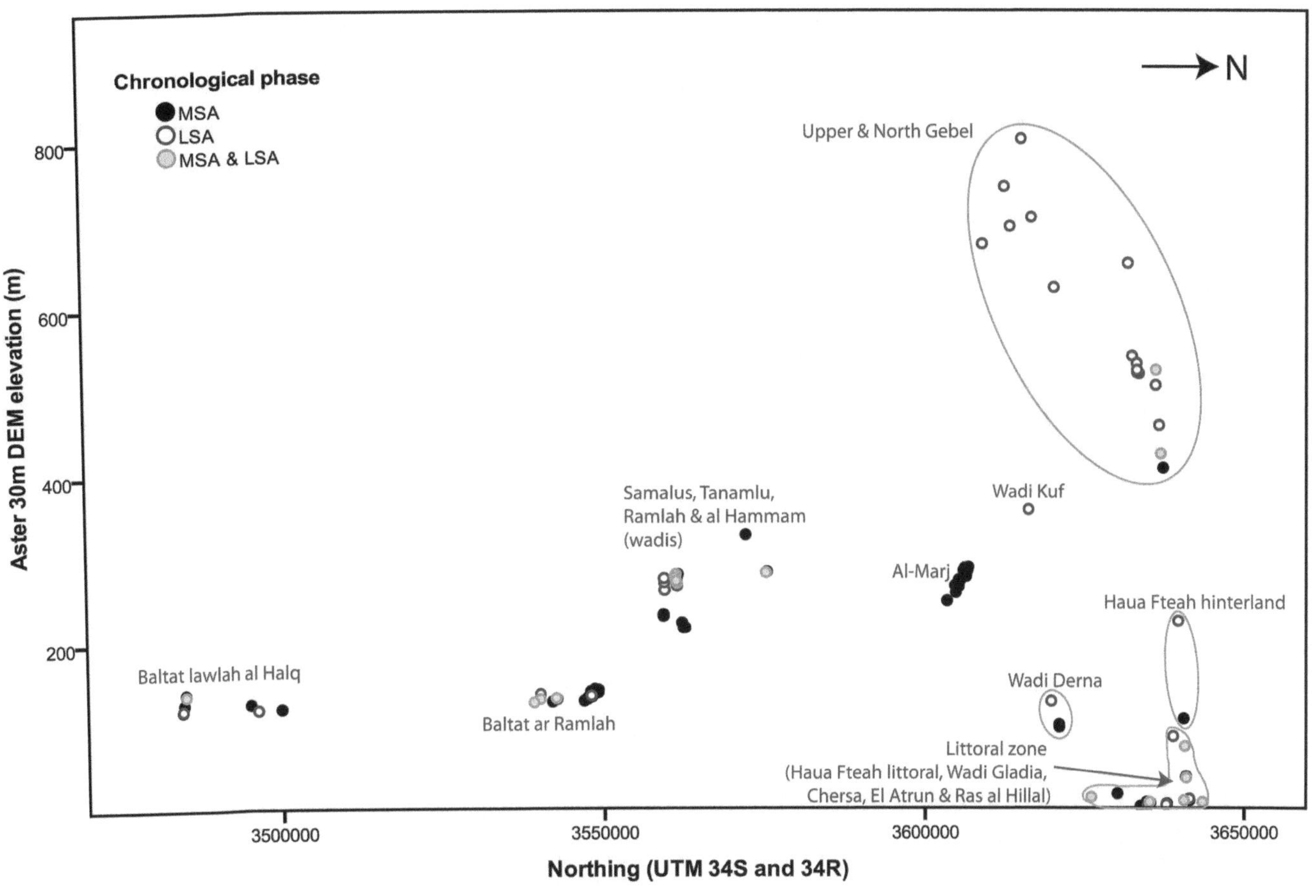

Fig. 5.6 Scatterplot depicting the distribution of sites with MSA, LSA, and both MSA and LSA artifacts according to site elevation and geographic cluster

and Levallois core technologies at Al-Marj, although it remains to be determined if these were contemporary; and (3) the heavily weathered discoidal and bifacial chert cores with desert varnish from the pre-desert and desert. In addition, other phases of the MSA are also suggested: (1) a general MSA with prepared core technologies, including classic Levallois cores and discoidal cores exhibiting centripetal preparation; (2) the Aterian on the coast and possibly at Al-Marj and Baltat ar Ramlah, although an Aterian presence in the region as a whole is relatively weak compared to other regions of North Africa; and (3) a later MSA phase represented by a combination of Levallois and blade reduction strategies.

Discussion

Surveys of the TRANS-NAP study region have revealed the presence of Paleolithic artifacts in surface and *in situ* contexts across all four ecological zones and associated with varied landforms. Although a clear chronology for these sites is currently lacking, dates have already been obtained for some coastal sites with *in situ* archaeology that extend back into the Middle Pleistocene. Without direct dates for the surface archaeology, we have to rely on information garnered from the techno-typological criteria of the lithic scatters. Using this approach, 80 of the sites across the region preserve lithics with MSA affinities, 33 with LSA, 14 with both MSA and LSA, and 54 of unknown affiliation. Although MSA and LSA sites are distributed throughout the surveyed area, two main patterns in the archaeological record are apparent: (1) a difference in MSA and LSA site location according to elevation; and (2) notable technological diversity within the MSA in particular. Explanations are provided for why this dichotomy between the distribution of MSA and LSA sites may exist. For example, why were paleolakes a focus of MSA occupation and why is there a lack of MSA sites at higher elevations in the Gebel Akhdar? Causes of the diversity seen in the MSA are proposed, emphasizing possible underlying microevolutionary processes. The surveyed area is considered in terms of the resources available to human populations and how changing resource distribution may have governed hominin behavioral adaptations as well as population dynamics in the region.

MSA and LSA Patterns of Occupation in Northern Cyrenaica

Water supply would have been a driving force behind occupation patterns and population densities across the region, where water availability dictated the type and abundance of plant and animal resources. In the coastal zone and Gebel Akhdar, the presence of numerous springs and relatively high precipitation rates offered sufficient water for hunter-gatherer groups, even during glacial periods when large areas elsewhere in North Africa were experiencing hyper-arid conditions. These regions were comparatively rich in plant and animal foods, from marine resources and terrestrial land snails to game of varying sizes. In fact, archaeological remains from Haua Fteah indicate that adaptations such as marine resource exploitation were present from as early as the "pre-Aurignacian" (Klein and Scott 1986). Today, vegetation cover in the pre-desert and desert is substantially less, albeit not entirely absent in the former. Here, arid-adapted animal species would have represented suitable prey, perhaps constituting a more reliable source of energy than plant foods. Contrasting with water supplies to the coast and Gebel, those to the pre-desert and desert zones were unpredictable but not absent. Since 1985, Landsat TM Quicklooks imagery reveals that the Baltat ar Ramlah paleolake has contained water approximately every 5 years, providing a source of water for livestock for more than a year in man-made trenches dug into the paleolake (Jones et al. 2011). Modern wells in the vicinity also attest to water sources in the area. Moving further south into the desert paleolakes, water becomes increasingly unavailable. Drawing on evidence from other studies (e.g., Osborne et al. 2008; Drake et al. 2008, 2011), the paleolakes and wadis in the pre-desert and desert would have been relatively well-watered environments during interglacial periods, exhibiting greater plant cover and attracting animals and human groups. Increased water supply during humid episodes in MIS 5, in particular, is argued to have been one of the main factors determining MSA occupation in these biomes, with the most viable habitats and reasonably long-term and high-density occupation being most likely during MIS 5e. In contrast, arid conditions of the later Pleistocene explain a relative lack of terminal Pleistocene LSA sites in the area. This has been noted in other arid areas of North Africa from MIS 4-2 where populations existed at very low densities (Ambrose 1998) (e.g., in the Nile Valley [Van Peer and Vermeersch 2007; Barham and Mitchell 2008: 266]) or may have abandoned areas such as the Libyan Sahara after 60 ka until the Holocene (Cremaschi et al. 1998). Microlithic surface sites do exist in the pre-desert and desert, yet some or all of these may be early Holocene in age.

In addition to providing water and a diverse array of food resources during humid periods, the pre-desert and desert paleolakes would have attracted MSA hominins for two further reasons: (1) their location along potential dispersal routes that followed waterways located to the south, east, and west, active during humid interglacials; and (2) the sources of chert on the "paleo-islands" and southern escarpments of the Gebel, as is suggested by the presence of *in situ* flaking scatters, evidence of chert extraction from limestone boulders and large cores in the early stages of reduction. In fact, the location of chert sources across all four biomes suggests that proximity to raw materials was a driving force behind MSA site location and may explain the lack of MSA sites at higher elevations in the Gebel. Chert sources are documented in the pre-desert and desert paleolakes, are readily available both along the coast and at lower elevations in the northernmost parts of the Gebel, and are present in abundance and are of high quality in the Al-Marj Basin. These are all places where MSA sites are located, whereas chert sources are far more elusive in the Gebel at higher elevations. At these elevations, LSA artifacts were commonly made from a high-quality blue-gray chert, the source(s) of which remain unknown. As all lithic scatters in these areas are of low density, the LSA populations whose traces exist in the Upper Gebel were possibly not as tethered to raw material sources as MSA groups; instead, these LSA sites are the remnants of small mobile hunter-gatherer groups with portable tool kits, exploiting the Gebel for its plant and animal resources. The LSA sites in the Gebel and coastal biomes could represent the terminal Pleistocene backed bladelet producing populations that spread across coastal North Africa after ~20 ka, from the Near East to the Maghreb (Close 2002; Barton et al. 2007) but were either absent or existing at very low population densities in the more arid regions of North Africa. Arguably, the relatively high number of LSA sites along the coast and in the Gebel was a result of population compaction into these areas during MIS 2, following a contraction of viable habitats. As seen in Haua Fteah, evidence of LSA technologies continues into the early Holocene with the Capsian (McBurney 1967; Barker et al. 2009) and some of these LSA sites could also date to this period.

MSA Diversity in Northern Cyrenaica

The MSA of North Africa exhibits considerable technological diversity. This was probably a result of both diachronic change and contemporary regional variation in the cultural record. In order to qualify and quantify this technological diversity, further chronological resolution of the archaeological record is required, and different regional research approaches, terminologies, and typologies need to be taken into account. Currently, various early MSA records are documented across North Africa, including those from the: (1) Western Desert in Egypt, at Bir Tarfawi and Bir Sahara East from ~230 ka (McBrearty and Brooks 2000) and at Kharga Oasis 220 ± 20 ka (Churcher et al. 1999); (2) Nile Valley from early MIS 5 (early Nubian complex) (van Peer 1998); (3) Jebel Gharbi in northwest Libya during MIS 5 ("generalized" early MSA, lacking tanged artifacts) (Barich et al. 2006); (4) the "pre-Aurignacian" at Haua Fteah, dating to MIS 5 and possibly MIS 6 (Moyer 2003; Barker et al. 2007); and (5) ~171 ka levels (lacking tanged artifacts) in Ifri n'Ammar cave in Morocco (Richter et al. 2010). The presence of tanged points is regarded as marking the North African Aterian culture, one of the many African MSA variants (e.g., Clark 1992), and assumed by some to succeed the early MSA. Garcea (2010a), for example, argues that the Aterian in northwest Libya follows the early MSA after a hiatus and population discontinuity. Others argue that the Aterian may not represent a chronologically distinct cultural complex, where "Middle Paleolithic" layers at Ifri n'Ammar (~171 and ~130 ka) alternate with Aterian layers (~145 and ~83 ka), based on a respective absence and presence of tanged points (Richter et al. 2010). This appears to be the case at other sites in North Africa, including Haua Fteah, where the putative Aterian layers are succeeded by "Mousterian" layers (McBurney 1967; Wendorf and Schild 1992; Richter et al. 2010). An absence of tanged points is argued to not necessarily denote the absence of the Aterian (Garcea 2012a), as other lithic technologies are also included in the Aterian, such as bifacial foliates, small Levallois and discoidal cores, an increase in flake laminarity (Stringer and Barton 2008) and flakes with bulbar basal thinning and bifacial retouch (Garcea 2010a). Yet, all of these can be part of non-Aterian MSA assemblages in North Africa, elsewhere in Africa and outside Africa. The definition of the Aterian remains problematic (e.g., Kleindienst 2001) and circular arguments regarding what is and what is not "Aterian" are commonplace. In contrast, the Nile Valley MSA record appears to be more diverse than that further to the west; however, to some extent this could be a result of different regional research strategies and ways of describing the lithic evidence. Technological differences between Nile Valley assemblages reveal an early Nubian complex (early MIS 5), later Nubian complex (later MIS 5), Lower Nile Valley complex and Khormusan (MIS 4 and early MIS 3), Taramsan (MIS 3) (Van Peer 1998; Vermeersch 2010), and absence of the Aterian. Leaving definitions of the latter aside, this brief summary of the North African evidence reveals a complex MSA record (see Van Peer 2016), perhaps unsurprising given the paleoclimatic and demographic shifts that influenced the region from MIS 6 onwards.

Geoarchaeological surveys of the TRANS-NAP study region have revealed a comparable level of diversity within the MSA of northern Cyrenaica. In Haua Fteah, this diversity can be arranged chronologically, whereas only a relative chronology can be applied tentatively to the surface archaeology. Temporal as well as spatial factors may explain the variety of MSA technologies evident across the landscape. For example, three different forms of early MSA are discernible that may or may not overlap chronologically. These include the "pre-Aurignacian" at Haua Fteah, heavily weathered discoidal cores with desert varnish at pre-desert and desert surface sites, and the Levallois cores and handaxes found throughout the Al-Marj canal; however, the stratigraphic relationship between the latter technologies at Al-Marj remains unresolved. Other forms of MSA are evident, including a more general MSA, characterized by reduction techniques such as Levallois, centripetal and radial, and a possible late MSA that features both Levallois flake and blade cores. If defined on the basis of the presence of tanged or foliate points, an Aterian signature in the region is weak, particularly in the pre-desert and desert biomes. This is in conflict with evidence from elsewhere in North Africa where Aterian technologies are argued to represent an adaptive response by hunter-gatherers inhabiting arid environments with ephemeral water sources (Garcea 2012b; cf. Drake and Breeze 2016). The lack of Aterian lithic types in the pre-desert and desert may be a case of absence of evidence rather than evidence of absence, or it may suggest that technological adaptations were different in this region of North Africa when compared to similar environments elsewhere. More controversially, this pattern may reflect a true demographic divide between Cyrenaica and other regions where Aterian technologies are clearly present.

To explain the possible causes behind diversity in the MSA of northern Cyrenaica without a clear chronology for the surface assemblages does present a challenge. There are a number of factors, not necessarily mutually exclusive, which may underlie this diversity, for example: (1) late Middle and Late Pleistocene paleoenvironmental change; (2) fluctuating population densities and structure within the region; (3) migration of populations into and out of the region; and (4) behavioral adaptation to different habitats and resources therein (e.g., food, water and raw materials). Reconstruction of past paleoenvironments in the region is part of the TRANS-NAP project's ongoing research, particularly with respect to the Haua Fteah sediments, Al-Marj core, raised beach deposits, and Wadi Derna sections. McBurney (1967) and others identified alternate cold and warm phases in Haua Fteah, by analyzing gross sedimentology (Sampson 1967), faunal profiles (Higgs 1967; Klein and Scott 1986), and conducting pioneering oxygen isotope studies of shell (McBurney 1967: 54–59). While ages for climatic phases older than ~46 ka were not based on chronometric dates, the age and characteristics of these phases are being revised and refined using various dating techniques (Douka et al. 2014). New evidence from Haua Fteah reveals a major phase of cooling in the MSA levels as indicated by layers of limestone roof-spall that correlate broadly with MIS 4 (Barker et al. 2008, 2009; Inglis 2012) and a drop in occupation density. A further decline in MSA occupation occurred with the deposition of sediments that mark warmer but fluctuating environments, corresponding to the global climatic oscillations of MIS 3.

Unlike in the littoral zone, explorations of the pre-desert and desert zones have produced few sections suitable for paleoenvironmental studies. Here, a key issue concerns the relationship between paleohydrological fluctuations in these biomes and cultural diversity. For example, there is clear evidence of humid and wet conditions across the Sahara in MIS 5e (Osborne et al. 2008; Drake et al. 2011). During MIS 5e and Middle Pleistocene humid interglacials, large lakes (e.g., Lake Megafazzan) and river systems (e.g., Kufra) are argued to have covered ~10% of Libya. Although there were several humid episodes after ~120 ka, these resulted in smaller and more fragmented lake systems, marking a trend towards an overall reduction in water availability during the Late Pleistocene (Drake et al. 2008). Such humid episodes are documented in the Fazzan Basin at 100, 74, 47, 30 and 14–13 ka (Mattingly et al. 2007), marking the activation of lakes (White et al. 2006). At Bir Tarfawi in the Eastern Sahara, different phases of the MSA are associated with different phases of lake activity in the late Middle and Late Pleistocene (Wendorf et al. 1987). Substantial lakes are absent from this region from MIS 4-2 when humid periods would have been short-lived and low intensity, yet it remains unclear how habitable the area would have been during such periods (Smith 2010). Although the eastern Sahara is ~1100 km south of northern Cyrenaica, similar humid cycles may have occurred in the latter from MIS 6 onward that were probably greater in intensity given the relatively high precipitation rates in the Gebel Akhdar. Of particular relevance to northern Cyrenaica is the identification of cold and dry phases in the alluvial fan sequences derived from the Wadi Zuiana (~90 km west of Haua Fteah) at various times from MIS 6-2 (Rowan et al. 2000; Macklin et al. 2002). Phases of alluviation occurred during arid periods, with the transition from forested steppe to steppe vegetation and subsequent land surface destabilization. Two alluviation episodes occurred during MIS 6 at ~183–179 ka and ~157–138 ka, correlating with periods of substantial global cooling. Other periods of alluviation occurred ~111–109 ka (MIS 5d), ~88 ka (MIS 5b/5a boundary), ~69 ka (MIS 4), ~42 ka (MIS 3), and three times during MIS 2. In contrast, periods of incision took place during humid peaks in MIS 5e and the early Holocene (Macklin et al. 2002). These alternating arid and humid phases from MIS 6-2 would have resulted in fluctuating environmental conditions and been one of the factors underlying shifts in human behavior and the diversity of lithic technologies from the MSA onwards.

Various microevolutionary processes, driven by paleoenvironmental changes from MIS 6-2, influenced past population dynamics in the region. During humid phases, an increase in cultural change and diversity in the region is predicted, where both demographic expansions within the region and the arrival of new populations resulted in new social structures, cultural exchange and transmission, and consequently, technological change and an increase in cultural diversity. In the pre-desert and desert zones, pulses of occupation during humid phases are hypothesized, with long periods of no occupation. This ebb and flow of new populations into the region at different points in the past could explain some of the region's MSA technological diversity. Habitats of the northern Gebel Akhdar and littoral zone contrast starkly with those of the pre-desert and desert, essentially representing an island of more Mediterranean-like vegetation and fertile landscape than the arid landscapes to the south, east and west. This area likely acted as a refugium during cold and dry phases across North Africa. Although population extinctions were unlikely in both the Gebel Akhdar and coastal biomes as a result of climatic downturns, smaller scale group extinctions were possible. During such periods, populations would have compacted into smaller areas, resulting in an increase in population density and possibly stimulating technological change through both increased competition and behavioral adaptation to altered population structure and habitats. Specifically long-term isolation of populations in the region could have taken place during extended climatic perturbations, potentially causing both a loss of cultural diversity through cultural drift (e.g., Henrich 2004) and technological differences between northern Cyrenaica and surrounding regions (e.g., the Nile Valley or Maghreb). Fluctuating population density is supported by evidence from Haua Fteah. This shows notable shifts in occupation intensity from the "pre-Aurignacian" to Dabban, with hiatuses documented in the "pre-Aurignacian" levels. Occupation densities are particularly low in the late MSA and early LSA layers, correlating broadly with the marked environmental fluctuations of MIS 3 (Inglis 2012). A human presence is, however, still documented in the cave during glacial periods in both the MSA and LSA (Higgs 1967; Klein and Scott 1986; Inglis 2012). In fact, the Late Dabban layers record an explosion in artifact numbers when compared to the Early Dabban, a jump that occurred as the area cooled (but did not necessarily become hyper-arid) towards MIS 2. This potentially indicates a contraction of populations into certain locations that remained habitable, due to adequate local rainfall as well as the possible presence of springs (Inglis 2012). Littoral zone sites with bladelet-based LSA industries, presumably dating to during and after the LGM, are widely distributed in contexts that, if *in situ*, suggest both climatic amelioration (e.g., paleosols) and climatic severity (e.g., fluvial and colluvial deposits). Continuity of activity at relatively high densities from the Dabban through the LGM to the later LSA "Oranian" and "Capsian" is supported by the new excavations in Haua Fteah.

This technological diversity could be a result of environmental changes that caused demographic changes, for example, as a consequence of new populations that expanded into the region during pluvial periods, along the coast from areas to the west and east, or across the Sahara. Yet, a degree of this diversity would have been a consequence of behavioral adaptation by pre-existing populations to altered habitats as well as social structures and practices. This may explain some of the spatial as well as temporal cultural differences within the study area. For example, shifts in hunter-gatherer subsistence strategies, caused by changes in resource type and availability, are argued to have varied in each biome because of differences in resource base. In short, hunting and gathering practices employed along the coast (e.g., for the exploitation of marine resources) would have been different from those required in the pre-desert and desert (e.g., to successfully hunt arid-adapted mammals), and these differences would have had technological manifestations. In addition, a degree of technological diversity can be explained by a site's proximity to stone materials. For example, many of the pre-desert and desert sites are close to raw material sources, yet the lower quality chert evident thus far in these biomes, when compared to that in coastal areas or at Al-Marj, may explain some regional behavioral and technological differences. In sum, the different factors that may be contributing to the patterns seen in the archaeological record are numerous and complex. It is very difficult to discern the extent to which the archaeological record marks indigenous cultural developments or signifies new populations in the region, or more plausibly, to what extent both processes were in operation. This is a problem further confounded by the absence of a decent sample of human remains from northern Cyrenaica and surrounding regions. At this stage, therefore, it is important to look beyond the study area and examine the geographic links that may have existed with other regions and how these might have influenced past population dynamics.

Demographic Connections from MIS 6-2

Considering the impact of population isolations and migrations on the Paleolithic record of the region raises questions about when and where connections existed between northern Cyrenaica and regions beyond. For example, was northern Cyrenaica at a "cross-roads" between the Maghreb, Sahara, Nile Valley and Levant, and if so, how did this manifest itself in the archaeological record? Unfortunately, paleoclimatic and paleoenvironmental evidence is sparse and it is not possible to draw firm conclusions (Drake et al. 2013; Drake and Breeze 2016). The most likely timing of dispersals across the Sahara was during MIS 5, when the desert contracted and was replaced by grasslands and extensive lake and river systems (Osborne et al. 2008; Drake et al. 2011). Fossil river channels have been identified to the south of the TRANS-NAP study region, including the Kufra river system and the Serir Tibesti and Sahabi rivers (Drake et al. 2008, 2011), as well as an ~800 km-long uninterrupted drainage network running from the Tibesti Mountains to the Sirte Basin (Osborne et al. 2008). While these "corridors" represented plausible hominin migration routes, the archaeological signatures of these exact routes have yet to be discovered and it is unknown how frequently they were active in the past. Discoveries of artifacts from the Oldowan to Neolithic in "megalake" basins, the catchments of which form a corridor across the Sahara, do show that the desert was regularly, albeit periodically, occupied by people during climatic optima (Drake et al. 2008, 2011).

A trans-Saharan migration route to northern Cyrenaica that was active during the last humid phases is suggested by the presence of a diverse array of Sub-Saharan savannah fauna in the Late Pleistocene and Holocene deposits of Haua Fteah (Klein and Scott 1986). Other migration routes may have also been active during periods from MIS 6-2. For example, Drake and Breeze (2016) show that the northern Sahara and the Levant show brief periods of humidity between 129 and 92 ka and at approximately 76, 44, 37 and 29 ka. Consequently, population movements between Cyrenaica and these regions possibly occurred at these times. Low sea levels during glacial humid periods (e.g., 44, 37 and 29 ka) would have exposed a larger, littoral plain along the north coast of North Africa; however, at the Gebel Akhdar a coastal plain would have been relatively limited given its steep offshore topography (Lambeck and Purcell 2005). If populations, avoiding hyper-aridity in the center of North Africa, were utilizing sites along this lower coastline, potentially as far as the Levant, then much of the evidence for occupation during such periods would now be submerged, hampering assessment of such linkages (Bailey and Fleming 2008; Garcea 2010b). It is possible that early LSA industries such as the Dabban were the result of such connections, with technological diffusion, not necessarily population diffusion (e.g., McBurney 1967), occurring within a broad southern Mediterranean littoral zone (Garcea 2010c). McBurney and Hey (1955: 7–8) favored a route along the Mediterranean littoral but also suggested an additional route that ran from northwest Libya to the Augila (Awjilah) oasis, along the northern extent of the Sand Sea and connecting with Siwa Oasis in Egypt. No paleoclimate research has been conducted in these regions to verify this route though there is evidence for humidity in the eastern Sahara between 129 and 92 ka and at about 76 ka (Drake and Breeze 2016); therefore, at least the eastern part of this route would have been viable during these periods.

Several previous studies have looked at the paleohydrology of the Sahara, particularly from MIS 5e to the Holocene (Osborne et al. 2008; Drake et al. 2008, 2011; Smith 2010), yet northern Cyrenaica has been excluded from these models. Establishing the paleohydrological links between northern Cyrenaica and regions further to the south is crucial for identifying the potential routes that hominins may have taken into and out of our region, and when these routes were active. Figure 5.7 provides an evaluation of Saharan paleohydrology, developed by Drake et al. (2011) who demonstrate how such links can be made. This shows the simplified Holocene paleohydrology of North Africa, illustrating a dense river network in most areas as well as numerous large paleolakes. Dating of the paleolake sediments in these basins shows that many of them were active during MIS 5 (Drake et al. 2011) and probably during previous humid periods. The Sahabi, Serir Tibesti, and Kufra rivers link with fluvio-lacustrine systems in the southern Sahara, and form corridors across the desert that could have allowed hominin dispersals to the region just south of Cyrenaica. The Kufra River terminates in a giant alluvial fan >500 km from Cyrenaica, while the Serir Tibesti and Sahabi rivers feed a giant paleolake. This paleolake, Sabkhat al Qenien, is only 65 km from the nearest hydrological system draining the Gebel Akhdar and therefore offers a more attractive route than the Kufra. Using methods outlined in Drake et al. (2011), Fig. 5.8 provides a map of the paleohydrology of the Cyrenaica/Gulf of Sirte region. This shows the existence of a number of routes into and out of the study region via the Sabkhat al Qenien. Hominin groups could have crossed the 65 km between the Sabkhat and the nearest Cyrenaican fluvio-lacustrine system. They could have followed the river that feeds into the Sabkhat from the east and drains towards the eastern coastal margin of Cyrenaica. Alternatively, they could have traversed the field of numerous small paleolakes found on the platform between the Sabkhat and Cyrenaica. Together, this shows the existence of several viable routes that hominins may have followed between the Sahara and northern Cyrenaica during climatic optima. Some of the surface archaeology reported

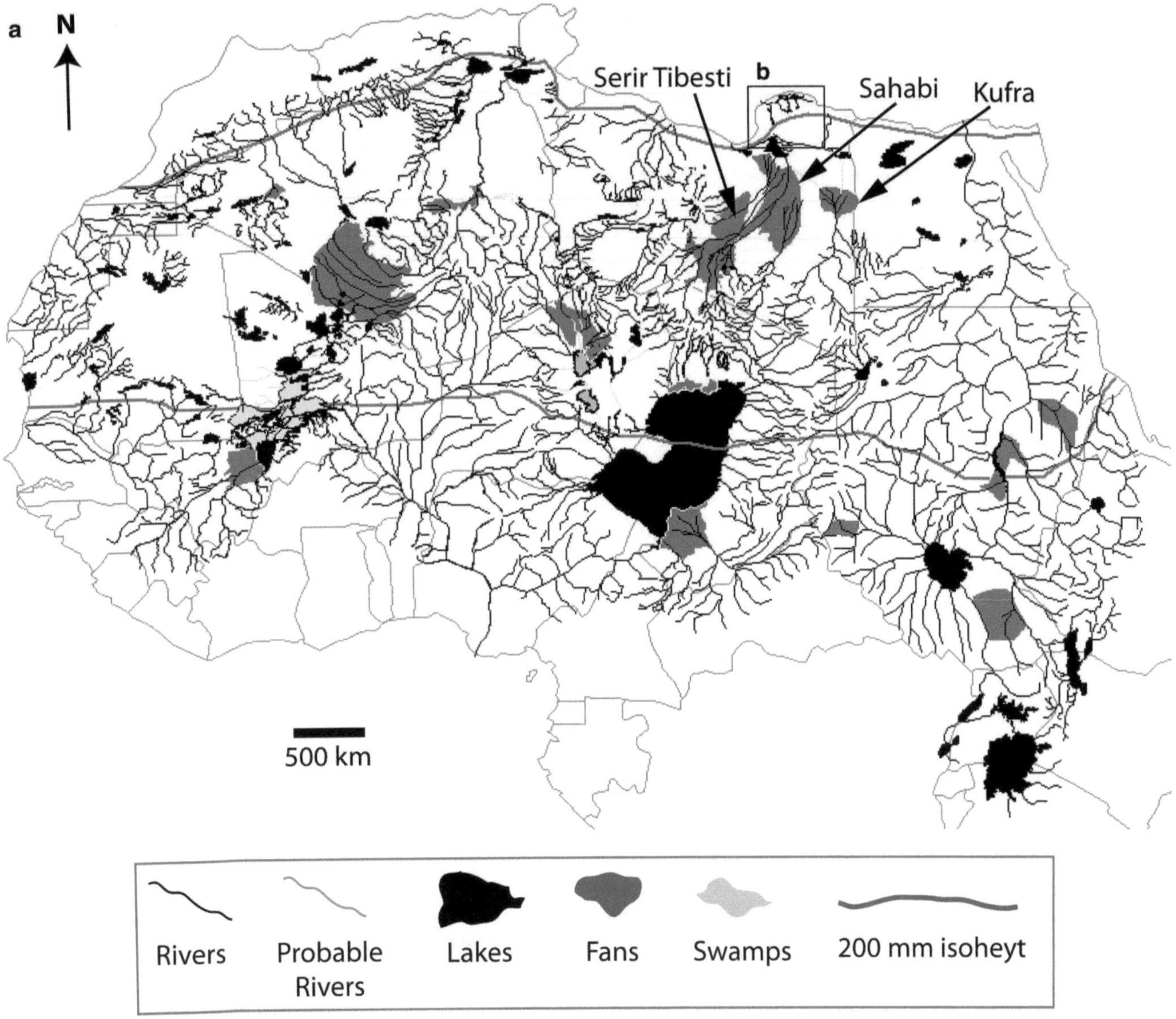

Fig. 5.7 **a** Late Pleistocene and early Holocene paleohydrology of the Sahara (c. 11–8 ka) developed from Drake et al. (2011). The 200-mm isohyet indicates the current limit of the Sahara Desert. **b** Area enlarged in Fig. 5.8

here, such as that from the pre-desert and desert paleolakes and wadis, may represent the first recorded signatures of people using such routes into and out of the study area.

Concluding Remarks

The distribution across the surveyed area of sites preserving typical MSA and LSA technologies reveals different patterns of occupation from MIS 6-2; for example, a lack of MSA sites at higher elevations and relatively few LSA sites in the pre-desert and desert. This pattern most plausibly reflects different adaptive responses to environmental change, shifting modes of resource exploitation (food, water and raw materials), dissimilar patterns of mobility and, arguably, differences in demographic structure and social behaviors. Despite this dichotomy between the MSA and LSA, the former itself was highly variable, both regionally and chronologically. This is unsurprising given that the MSA of the region lasted from at least MIS 6 until the middle of MIS 3, a period of significance in terms of climate change as well as modern human anatomical and behavioral evolution and dispersals. Population densities in the region fluctuated from MIS 6-2, evident from the cultural sequence at Haua Fteah and littoral sites with stratified archaeology. This is also apparent from the type and distribution of sites in the pre-desert and desert, areas that were not continuously

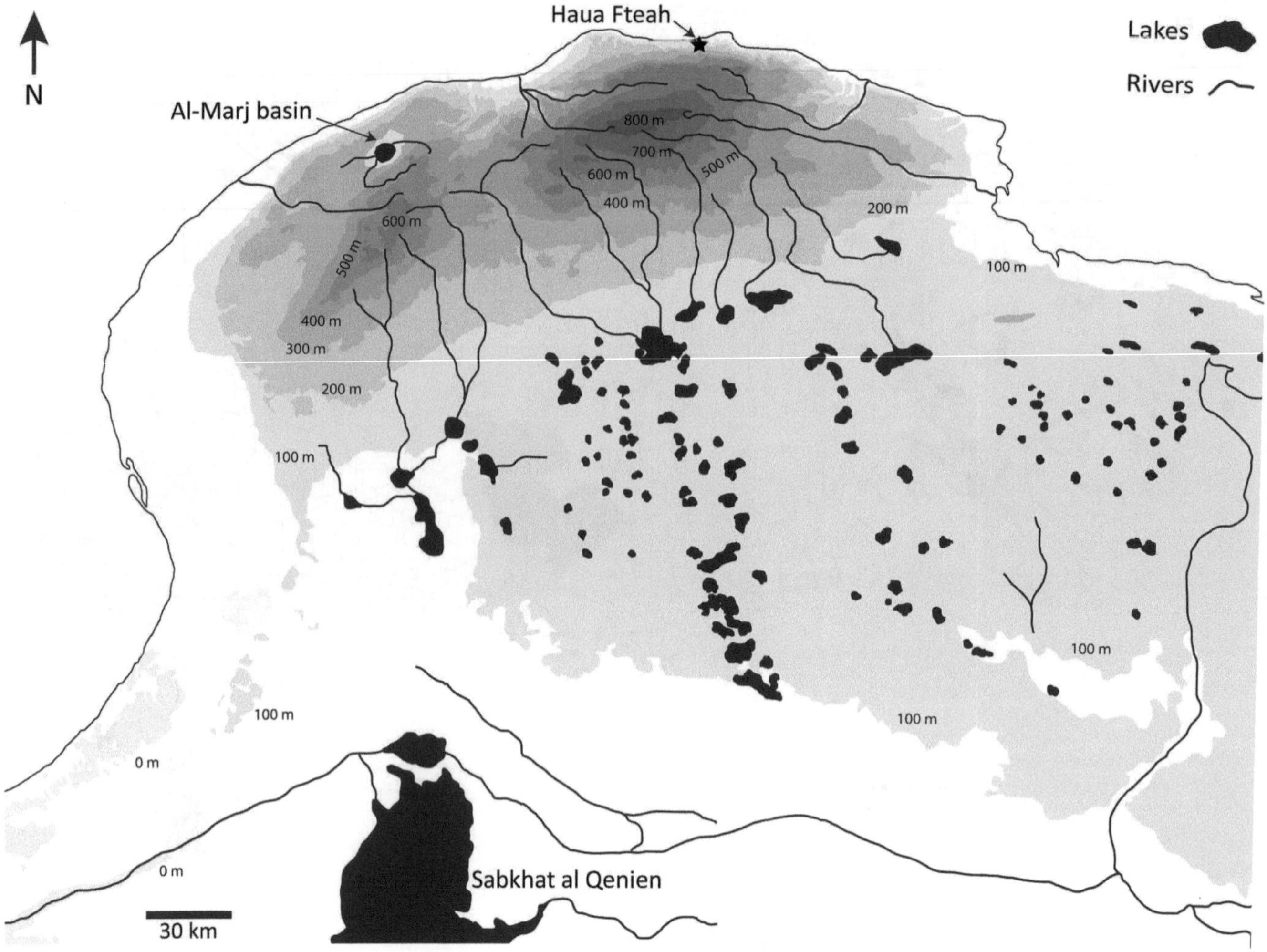

Fig. 5.8 Late Pleistocene and early Holocene (c. 11–8 ka) paleohydrology of Cyrenaica and the northern Sahara developed using the methods described by Drake et al. (2011). See Fig. 5.7 for the location of the mapped area in North Africa

occupied. Here, MSA and LSA sites are the result of an ebb and flow of populations that contracted and expanded across the landscape at various points in the past and probably in various directions. In contrast, areas of refugia likely existed in parts of the Gebel Akhdar and littoral zone, providing adequate resources of food, water and raw materials during climatic downturns. Several characteristics of the region's Paleolithic record are somewhat idiosyncratic when compared to the rest of North Africa (e.g., the "pre-Aurignacian", possible lack of Aterian, and Dabban). It is possible that the Gebel Akhdar's unique fertile geography in an otherwise arid area of North Africa provided extraordinary conditions for technological, behavioral, and demographic change. Yet, to what extent these are truly idiosyncratic requires both precise chronologies for archaeological data and inter-regional lithic comparisons with assemblages from elsewhere in North Africa and the Levant. Only then will it be possible to refine our interpretations of the archaeological record, to test some of the aforementioned interpretations, and to reconstruct more precisely the complex processes behind hominin behavioral and demographic shifts in the region from MIS 6-2.

Acknowledgments Data presented in this paper were collected during four field seasons from 2007 to 2010 and involved numerous team members to whom we are most grateful. We thank the Department of Antiquities of Libya for permission to undertake this work and the European Research Council and the Society for Libyan Studies for financial support. We are indebted to our colleagues in Libya for their support, including Giuma Anag, Salah Aghab, Ahmed Buzaian, Ahmed Saber, Mohammed Twati, and the late Abdulgader Mzeine. We thank the three reviewers for their important and helpful comments.

References

Ambrose, S. H. (1998). Chronology of the Later Stone Age and food production in East Africa. *Journal of Archaeological Science, 25*, 377–392.

Antoniadou, A. (2012). The geoarchaeology of Cyrenaica, Libya. Ph.D. Dissertation, Queens University Belfast.

Bailey, G. N., & Fleming, N. C. (2008). Archaeology of the continental shelf: Marine resources, submerged landscapes and underwater archaeology. *Quaternary Science Reviews, 27*(23–24), 2153–2165.

Balter, M. (2011). Was North Africa the launch pad for modern human migrations? *Science, 331*, 20–23.

Barham, L., & Mitchell, P. (2008). *The first Africans: African archaeology from the earliest toolmakers to most recent foragers*. Cambridge: Cambridge University Press.

Barich, B. E., Garcea, E. A. A., & Giraudi, C. (2006). Between the Mediterranean and the Sahara: Geoarchaeological reconnaissance in the Jebel Gharbi, Libya. *Antiquity, 80*, 567–582.

Barker, G., Hunt, C., & Reynolds, T. (2007). The Haua Fteah, Cyrenaica (northeast Libya): Renewed investigations of the cave and its landscape, 2007. *Libyan Studies, 38*, 93–114.

Barker, G., Basell, L., Brooks, I., Burn, L., Cartwright, C., Cole, F., et al. (2008). The Cyrenaican Prehistory Project 2008: The second season of investigations of the Haua Fteah cave and its landscape, and further results from the initial (2007) fieldwork. *Libyan Studies, 39*, 175–221.

Barker, G., Antoniadou, A., Barton, H., Brooks, I., Candy, I., Drake, N., et al. (2009). The Cyrenaican Prehistory Project 2009: The third season of investigations of the Haua Fteah cave and its landscape, and further results from the 2007 to 2008 fieldwork. *Libyan Studies, 40*, 55–94.

Barker, G., Antoniadou, A., Armitage, S., Brooks, I., Candy, I., Connell, K., et al. (2010). The Cyrenaican Prehistory Project 2010: The fourth season of investigations of the Haua Fteah cave and its landscape, and further results from the 2007 to 2009 fieldwork. *Libyan Studies, 41*, 63–88.

Barton, N. E., Bouzouggar, A., Ramsey, C. B., Collcutt, S., Higham, T. F. G., Humphrey, L. T., et al. (2007). Abrupt climatic change and chronology of the Upper Palaeolithic in Northern and Eastern Morocco. In P. Mellars, K. Boyle, O. Bar-Yosef, & C. Stringer (Eds.), *Rethinking the human revolution: New behavioural and biological perspectives on the origin and dispersal of modern humans* (pp. 177–186). Cambridge: McDonald Institute for Archaeological Research.

Barton, R. N. E., Bouzouggar, A., Collcutt, S. N., Schwenninger, J. L., & Clark-Balzan, L. (2009). OSL dating of the Aterian levels at Dar es-Soltan I (Rabat, Morocco) and implications for the dispersal of modern *Homo sapiens*. *Quaternary Science Reviews, 28*, 1914–1931.

Blackwell, B. A. B., Skinner, A. R., Mashriqi, F., Deely, A. E., Long, R. A., Gong, J. J. J., et al. (2012). Challenges in constraining pluvial events and hominin activity: Examples of ESR dating molluscs from the Western Desert. *Egypt. Quaternary Geochronology, 10*, 430–435.

Bouzouggar, A., Barton, N., Vanhaeren, M., d'Errico, F., Collcutt, S., Higham, T., et al. (2007). 82,000-year-old shell beads from North Africa and implications for the origins of modern human behavior. *Proceedings of the National Academy of Sciences, 104*(24), 9964–9969.

Chazan, M. (1995). Variability in the Middle Paleolithic of Haua Fteah, Libya. *Paléorient, 21*(1), 65–73.

Churcher, C. S., Kleindienst, M. R., & Schwarcz, H. P. (1999). Faunal remains from a Middle Pleistocene lacustrine marl in Dakhleh Oasis, Egypt: palaeoenvironmental reconstructions. *Palaeogeography, Palaeoclimatology, Palaeoecology, 154*, 301–312.

Clark, J. D. (1992). African and Asian perspectives on the origins of modern humans. *Philosophical Transactions: Biological Sciences, 337*(1280), 201–215.

Close, A. E. (1986). The place of the Haua Fteah in the Late Palaeolithic of North Africa. In G. N. Bailey & P. Callow (Eds.), *Stone Age prehistory: Studies in memory of Charles McBurney* (pp. 169–180). Cambridge: Cambridge University Press.

Close, A. E. (2002). Backed bladelets are a foreign country. *Archeological Papers of the American Anthropological Association, 12* (1), 31–44.

Cremaschi, M., Di Lernia, S., & Garcea, E. A. A. (1998). Some insights on the Aterian in the Libyan Sahara: Chronology, environment, and archaeology. *African Archaeological Review, 15*(4), 261–286.

d'Errico, F., Vanhaeren, M., Barton, N., Bouzouggar, A., Mienis, H., Richter, D., et al. (2009). Additional evidence on the use of personal ornaments in the Middle Paleolithic of North Africa. *Proceedings of the National Academy of Sciences, 106*(38), 16051–16056.

Douka, K., Jacobs, Z., Lane, C., Grün, R., Farr, L., Hunt, C., et al. (2014). The chronostratigraphy of the Haua Fteah cave (Cyrenaica, northeast Libya). *Journal of Human Evolution, 66*, 39–63.

Drake, N. & Breeze, P. (2016). Climate change and modern human occupation of the Sahara from MIS 6-2. In S. C. Jones & B. A. Stewart (Eds.), *Africa from MIS 6-2: Population dynamics and paleoenvironments* (pp. 103–122). Dordrecht: Springer.

Drake, N. A., El-Hawat, A. S., Turner, P., Armitage, S. J., Salem, M. J., White, K. H., et al. (2008). Palaeohydrology of the Fazzan Basin and surrounding regions: The last 7 million years. *Palaeogeography, Palaeoclimatology, Palaeoecology, 263*, 131–145.

Drake, N. A., Blench, R. M., Armitage, S. J., Bristow, C. S., & White, K. H. (2011). Ancient watercourses and biogeography of the Sahara explain the peopling of the desert. *Proceedings of the National Academy of Sciences, 108*, 458–462.

Drake, N. A., Breeze, P., & Parker, A. (2013). Paleoclimate in the Saharan and Arabian deserts during the Middle Paleolithic and the potential for hominin dispersals. *Quaternary International, 300*, 48–61.

Elfadli, K. I. (2009). *Precipitation data of Libya*. Tripoli: Climate Department, Libyan National Meteorological Center (LNMC).

Farr, L., Lane, R., Abdulazeez, F., Bennett, P., Holman, J., Marasi, A., et al. (2014). The cyrenaican prehistory project 2013: The seventh season of excavations in the Haua Fteah cave. *Libyan Studies, 45*, 163–173.

Garcea, E. A. A. (2010a). The spread of Aterian peoples in North Africa. In E. A. A. Garcea (Ed.), *South-eastern Mediterranean peoples between 130,000 and 10,000 years ago* (pp. 37–53). Oxford: Oxbow Books.

Garcea, E. A. A. (2010b). The Lower and Upper Late Stone Age of North Africa. In E. A. A. Garcea (Ed.), *South-eastern Mediterranean peoples between 130,000 and 10,000 years ago* (pp. 54–65). Oxford: Oxbow Books.

Garcea, E. A. A. (2010c). Bridging the gap between in and out of Africa. In E. A. A. Garcea (Ed.), *South-eastern Mediterranean peoples between 130,000 and 10,000 years ago* (pp. 174–181). Oxford: Oxbow Books.

Garcea, E. A. A. (2012a). Successes and failures of human dispersals from North Africa. *Quaternary International, 270*, 119–128.

Garcea, E. A. A. (2012b). Modern human desert adaptations: A Libyan perspective on the Aterian complex. In J.-J. Hublin & S. P. McPherron (Eds.), *Modern origins: A North African perspective* (pp. 127–142). Dordrecht: Springer.

Garcea, E. A. A., & Giraudi, C. (2006). Late Quaternary human settlement patterning in the Jebel Gharbi. *Journal of Human Evolution, 51*, 411–421.

Hawkins, A. L. (2012). The Aterian of the oases of the Western Desert of Egypt: Adaptation to changing climatic conditions? In J.-J. Hublin & S. P. McPherron (Eds.), *Modern origins: A North African perspective* (pp. 157–175). Dordrecht: Springer.

Henrich, J. (2004). Demography and cultural evolution: How adaptive cultural processes can produce maladaptive losses – the Tasmanian case. *American Antiquity, 69*(2), 197–214.

Hey, R. W. (1968). The Quaternary geology of the Jabal al Akhdar coast. In F. T. Barr (Ed.), *Geology and archaeology of northern*

Cyrenaica, Libya (pp. 159–165). Amsterdam: Petroleum Exploration Society of Libya.

Higgs, E. (1967). Environment and chronology: The evidence from mammalian fauna. In C. B. M. McBurney (Ed.), *The Haua Fteah (Cyrenaica) and the Stone Age of the south-east Mediterranean* (pp. 16–44). Cambridge: Cambridge University Press.

Hiscock, P. (1996). Transformations of Upper Palaeolithic implements in the Dabba industry from Haua Fteah (Libya). *Antiquity, 70*, 657–664.

Hublin, J.-J. (2001). Northwestern African Middle Pleistocene hominids and their bearing on the emergence of *Homo sapiens*. In L. Barham & K. Robson-Brown (Eds.), *Human roots. Africa and Asia in the Middle Pleistocene* (Vol. 99–121). Bristol: CHERUB, Western Academic & Specialist Press Ltd.

Hublin, J.-J., & Klein, R. G. (2011). Northern Africa could also have housed the source population for living humans. *Proceedings of the National Academy of Science, 108*(28), E277.

Hublin, J.-J., & McPherron, S. P. (Eds.). (2012). *Modern origins: A North African perspective*. Dordrecht: Springer.

Hunt, C., Davison, J., Inglis, R., Farr, L., Reynolds, T., Simpson, D., et al. (2010). Site formation processes in caves: The Holocene sediments of the Haua Fteah, Cyrenaica. *Libya. Journal of Archaeological Science, 37*(7), 1600–1611.

Hunt, C. O., Reynolds, T. G., El-Rishi, H. A., Buzaian, A., Hill, E., & Barker, G. W. (2011). Resource pressure and environmental change on the North African littoral: Epipalaeolithic to Roman gastropods from Cyrenaica. *Libya. Quaternary International, 244*(1), 15–26.

Inglis, R. H. (2012). Human occupation and changing environments during the Middle to Later Stone Age: Soil micromorphology at the Haua Fteah, Libya. Ph.D. Dissertation, University of Cambridge.

Iovita, R. P. (2009). Reevaluating Connections between the early Upper Paleolithic of northeast Africa and the Levant: Technological differences between the Dab-ban and the Emiran. In J. J. Shea & D. E. Lieberman (Eds.), *Transitions in prehistory: Essays in honor of Ofer Bar-Yosef* (pp. 125–142). Oxford: Oxbow Books.

Jones, S., Farr, L., Barton, H., Drake, N., White, K., & Barker, G. (2011). Geoarchaeological patterns in the pre-desert and desert ecozones of northern Cyrenaica. *Libyan Studies, 41*, 89–97.

Klein, R. G., & Scott, K. (1986). Re-analysis of faunal assemblages from the Haua Fteah and other Late Quaternary archaeological sites in Cyrenaican Libya. *Journal of Archaeological Science, 13*, 514–542.

Kleindienst, M. R. (2001). What is the Aterian? The view from Dakhleh Oasis and the Western Desert, Egypt. In C. A. Marlow & A. J. Mills (Eds.), *The Oasis Papers 1: The proceedings of the first conference of the Dakhleh Oasis Project* (pp. 1–14). Oxford: Oxbow Books.

Lambeck, K., & Purcell, A. (2005). Sea-level change in the Mediterranean Sea since the LGM: Model predictions for tectonically stable areas. *Quaternary Science Reviews, 24*(18–19), 1969–1988.

MacDonald, K. C. (1997). The avifauna of the Haua Fteah (Libya). *Archaeozoologia, 9*, 83–102.

Macklin, M. G., Fuller, I. C., Lewin, J., Maas, G. S., Passmore, D. G., Rose, J., et al. (2002). Correlation of fluvial sequences in the Mediterranean basin over the last 200 ka and their relationship to climate change. *Quaternary Science Reviews, 21*, 1633–1641.

Mattingly, D., Lahr, M., Armitage, S., Barton, H., Dore, J., Drake, N., et al. (2007). Desert migrations: People, environment and culture in the Libyan Sahara. *Libyan Studies, 38*, 1–42.

McBrearty, S., & Brooks, A. (2000). The revolution that wasn't: A new interpretation of the origin of modern human behaviour. *Journal of Human Evolution, 39*, 453–563.

McBurney, C. B. M. (1947). The Stone Age of the Libyan littoral: The results of a war-time reconnaissance. *Proceedings of the Prehistoric Society, 13*, 56–84.

McBurney, C. B. M. (1967). *The Haua Fteah (Cyrenaica) and the Stone Age of the south-east Mediterranean*. Cambridge: Cambridge University Press.

McBurney, C. B. M. (1968). Pleistocene and early post-Pleistocene archaeology of Libya. In F. T. Barr (Ed.), *Geology and archaeology of northern Cyrenaica, Libya* (pp. 13–21). Amsterdam: The Petroleum Exploration Society of Libya.

McBurney, C. B. M., & Hey, R. W. (1955). *Prehistory and Pleistocene geology in Cyrenaican Libya*. Cambridge: Cambridge University Press.

McBurney, C. B. M., Hey, R. W., & Watson, W. (1948). First report of the Cambridge archaeological expedition to Cyrenaica (1947). *Proceedings of the Prehistoric Society, 14*, 33–45.

McLaren, S. J., & Reynolds, T. (2009). Early humans in dryland environments: A geoarchaeological perspective. In A. J. Parsons & A. Abrahams (Eds.), *Geomorphology of desert environments* (pp. 773–798). Springer.

Moyer, C. (2003). *The organisation of lithic technology in the Middle and early Upper Palaeolithic industries at the Haua Fteah, Libya*. Ph.D. Dissertation, University of Cambridge.

Olivieri, A., Achilli, A., Pala, M., Battaglia, V., Fornarino, S., Al-Zahery, N., et al. (2006). The mtDNA legacy of the Levantine early Upper Palaeolithic in Africa. *Science, 314*(5806), 1767–1770.

Osborne, A. H., Vance, D., Rohling, E. J., Barton, N., Rogerson, M., & Fello, N. (2008). A humid corridor across the Sahara for the migration of early modern humans out of Africa 120,000 years ago. *Proceedings of the National Academy of Science, 105*(43), 16444–16447.

Pereira, L., Silva, N., Franco-Duarte, R., Fernandes, V., Pereira, J., Costa, M., et al. (2010). Population expansion in the North African Late Pleistocene signalled by mitochondrial DNA haplogroup U6. *BMC Evolutionary Biology, 10*(1), 390.

Petrocchi, G. (1940). *Ricerche preistoriche in Cirenaica. Africa Italiana, 7*, 1–33.

Rabett, R., Farr, L., Hill, E., Hunt, C., Lane, R., Moseley, H., et al. (2013). The Cyrenaican Prehistory Project 2012: The sixth season of excavations in the Haua Fteah cave. *Libyan Studies, 44*, 113–125.

Reynolds, T. (2013). The Middle Palaeolithic of Cyrenaica: Is there an Aterian at the Haua Fteah and does it matter? *Quaternary International, 300*, 171–181.

Richter, D., Moser, J., Nami, M., Eiwanger, J., & Mikdad, A. (2010). New chronometric data from Ifri n'Ammar (Morocco) and the chronostratigraphy of the Middle Palaeolithic in the Western Maghreb. *Journal of Human Evolution, 59*(6), 672–679.

Rowan, J. S., Black, S., Macklin, M. G., Tabner, B. J., & Dore, J. (2000). Quaternary environmental change in Cyrenaica evidenced by UTh, ESR and OSL of coastal alluvial fan sequences. *Libyan Studies, 31*, 5–16.

Sampson, C. G. (1967). Coarse granulometric analysis of the Haua Fteah deposits. In C. B. M. McBurney (Ed.), *The Haua Fteah (Cyrenaica) and the Stone Age of the south-east Mediterranean* (pp. 50–54). Cambridge: Cambridge University Press.

Smith, J. R. (2010). Palaeoenvironments of eastern North Africa and the Levant in the late Pleistocene. In E. A. A. Garcea (Ed.), *South-eastern Mediterranean peoples between 130,000 and 10,000 years ago* (pp. 6–17). Oxford: Oxbow Books.

Smith, J. R. (2012). Spatial and temporal variation in the nature of Pleistocene pluvial phase environments across North Africa. In J.-J. Hublin & S. P. McPherron (Eds.), *Modern origins: A North African perspective* (pp. 35–47). Dordrecht: Springer.

Smith, T. M., Tafforeau, P., Reid, D. J., Grün, R., Eggins, S., Boutakiout, M., et al. (2007). Earliest evidence of modern human life history in North African early *Homo sapiens*. *Proceedings of the National Academy of Science, 104*(15), 6128–6133.

Stringer, C., & Barton, N. (2008). Putting North Africa on the map of modern human origins. *Evolutionary Anthropology, 17*, 5–7.

Van Peer, P. (1998). The Nile Corridor and the out of Africa model: An examination of the archaeological record. *Current Anthropology, 39*, S115–S140.

Van Peer, P. (2016). Technological systems, population dynamics and historical process in the MSA of northern Africa. In S. C. Jones & B. A. Stewart (Eds.), *Africa from MIS 6-2: Population dynamics and paleoenvironments* (pp. 147–159). Dordrecht: Springer.

Van Peer, P., & Vermeersch, P. M. (2007). The place of northeast Africa in the early history of modern humans: New data and interpretations on the Middle Stone Age. In P. Mellars, K. Boyle, O. Bar-Yosef, & C. Stringer (Eds.), *Rethinking the human revolution* (pp. 187–198). Cambridge: McDonald Institute for Archaeological Research.

Vermeersch, P. M. (2010). Middle and Upper Palaeolithic in the Egyptian Nile Valley. In E. A. A. Garcea (Ed.), *South-eastern Mediterranean peoples between 130,000 and 10,000 years ago* (pp. 66–88). Oxford: Oxbow Books.

Wall-Scheffler, C. M. (2007). Digital cementum luminance analysis and the Haua Fteah hominins: How seasonality and season of use changed through time. *Archaeometry, 49*(4), 815–826.

Watson, W. (1949). The surface flint implements of Cyrenaica. *Man, 49*, 100–104.

Wendorf, F., & Schild, R. (1992). The Middle Palaeolithic of North Africa: A status report. In F. Klees & R. Kuper (Eds.), *New light on the Northeast African past* (pp. 39–80). Köln: Heinrich Barth-Institut.

Wendorf, F., Close, A. E., & Schild, R. (1987). Recent work on the Middle Palaeolithic of the Eastern Sahara. *African Archaeological Review, 5*, 49–63.

White, K., Charlton, M., Drake, N., McLaren, S., Mattingly, D., & Brooks, N. (2006). Lakes of the Edeyen Ubari and the Wadi al-Hayat. In D. Mattingly, S. McLaren, E. Savage, Y. al-Fasatwi, & K. Gadgood (Eds.), *The Libyan Desert: Natural resources and cultural heritage* (pp. 123–130). London: The Society for Libyan Studies.

Part II
Deserts

Chapter 6
Climate Change and Modern Human Occupation of the Sahara from MIS 6-2

Nick Drake and Paul Breeze

Abstract We have implemented probability density function (PDF) analysis of optically stimulated luminescence (OSL) and uranium series (U/Th) dates from humid sites in North Africa from 20 to 200 ka to identify humid periods. We then combine this with maps of Saharan paleohydrology to identify humid corridors across the Sahara that could have provided dispersal routes for anatomically modern humans (AMH). We then apply a similar analysis to the Aterian with a map of the spatial distribution of Aterian sites and a database of dated locations. Results suggest humid periods centered on 76 ka and between 92 and 129 ka provide green corridors across the Sahara that could have allowed AMH to cross it. Aterian sites are found preferentially in the 92 and 129 ka humid corridor, indicating that they may have used it. This hypothesis is supported by the fact that Aterian PDF curve peaks at the time of this corridor. These results suggest that Aterians occupied a "green Sahara," and this is supported by an analysis of the fauna associated with Aterian sites. However, most of the dated Aterian sites and faunal locations are found in the Maghreb, with few sites from the Sahara. Thus the evidence is not conclusive and further research is needed into Aterian chronology and environmental preferences within the Sahara.

Keywords Sahara • Paleoclimate • Paleohydrological mapping • Aterian • Dispersals

N. Drake (✉) · P. Breeze
Department of Geography, King's College London, London, WC2R 2LS, UK
e-mail: nick.drake@kcl.ac.uk

Introduction

In the last decade, genetic (e.g., Tishkoff et al. 2009) and archeological (e.g., McBrearty and Brooks 2000) evidence has converged to show that sub-Saharan Africa is at the center of human evolution, and the issue of dispersal out of this region is thus becoming increasingly important. The first "out of Africa" dispersals may have occurred as early as 2.5 Ma (Dennell 2009). This could have been followed by three waves of hominin dispersals out of Africa between 1.9 and 0.7 Ma (Bar-Yosef and Belfer-Cohen 2001) with some suggesting a further migration around 600 ka (e.g., Lahr and Foley 1998). These migrations were followed by those of modern humans sometime between 130 and 50 ka, with the possibility of multiple migrations in both directions during this period (Gunz et al. 2009).

Many different migration routes have been proposed, either crossing the Sahara or following the coast. A Red Sea coastal route out of Africa has been proposed for modern humans, once they had begun to utilize marine resources (Stringer 2000; Walter et al. 2000). It has also been suggested that the Bab el Mandeb provided a route "out of Africa" that could have avoided the Sahara (Kingdon 1993). Initially human genetic research (Macaulay et al. 2005) indicated that early modern humans used this route, despite the fact that a land bridge has not been present for the last 5 Ma (Fernandes et al. 2006), requiring a water crossing. A subsequent comprehensive reassessment of the archaeological record (Crassard 2010; Maher 2010; Rose and Usik 2010) alongside animal (Fernandes 2010) and human genetics (Cabrera et al. 2010; Rídl et al. 2010) supported an "out of Africa" exodus via the Levant rather than the Bab el Mandeb. However, recent research has provided some archaeological evidence supporting the southern route (Armitage et al. 2011), thus this route cannot be dismissed. Notwithstanding this, no direct evidence for coastal migrations has yet been found, possibly because they occurred at times of low sea level, consequently the routes are now

Sacha C. Jones and Brian A. Stewart (eds.), *Africa from MIS 6-2: Population Dynamics and Paleoenvironments*, Vertebrate Paleobiology and Paleoanthropology, DOI 10.1007/978-94-017-7520-5_6

underwater and archaeological sites are difficult to locate and survey. Thus it is proving hard to substantiate the viability of these potential migration corridors.

The other likely dispersal route is across the Sahara and then out of Africa. Traditionally, it has been assumed that the Nile was the primary corridor for the dispersal of hominins across the Sahara desert. Alternatively a "Green Sahara" route has been proposed whereby this now hyperarid region could have been crossed during "green" periods of greater humidity (Lahr 1996; Drake and Bristow 2006; Drake et al. 2011). This was first suggested by Duveyrier (1864) to explain the existence of the Nile crocodiles in isolated Saharan oases and has periodically been promoted since, notably by Dumont (1982) who recognized the trans-Saharan distribution of numerous aquatic animal species. Recent evidence for this is provided by Osborne et al. (2008) who used isotope analysis of a marine core from the southern Mediterranean and aquatic snails from the adjacent North African landmass to suggest a river corridor across the Sahara during MIS 5 and by Tjallingii et al. (2008) who used paleoclimate modeling and variations in dust concentrations in a marine core to infer humid conditions during MIS 1 and MIS 5. Drake et al. (2011) further develop these ideas. They demonstrate that more animals, including aquatic species, colonized North Africa via the Sahara than via the Nile corridor. This was possible because during the Holocene humid period the region contained a series of linked lakes, rivers, and giant alluvial fans, which facilitated these migrations. Holocene human occupation appears to have been influenced by the resultant faunal distributions, with different hunter-gatherer groups occupying different ecological niches. Furthermore, green corridors across the Sahara also appear to have existed during MIS 5, providing viable dispersal routes at a time when the first migration of modern humans out of Africa probably occurred (Drake et al. 2011).

The details of the nature of the Saharan climate in these different humid episodes are poorly understood. Today the Sahara receives two distinctly different sources of rainfall, the African Monsoon in the south and North Atlantic and Mediterranean westerlies in the north. The monsoon moves further north at times when precession produces higher solar insolation over the desert, causing it to contract. The climatic control on the location of westerlies is less well understood. However, there is a suggestion that their position varies with global temperatures (Toggweiler and Russell 2008), strengthening and contracting towards the poles during warm periods (interglacials), or reducing in strength and moving towards the equator during colder periods (glacials). Drake et al. (2013) advocate that the southward movement of the northern margin of the Sahara can be expected when the global climate is cool enough to keep the westerlies in a southern position, but not too cool to make them so weak they cannot penetrate far inland. This situation might be expected to occur at the beginning of interglacial periods, a time when the northward migration of the monsoon is causing the southern margin of the Sahara to move north due to increased insolation. It is not clear if these two rainfall systems ever migrate far enough to merge and form a truly green Sahara. Or if they simply reactivate rivers that flow across the arid zone that separates them, allowing the migration of animals and hominins across the now much contracted Sahara desert via these riparian corridors.

The "green Sahara" hypothesis is an attractive one for explaining dispersals from sub-Saharan Africa as the desert is scattered with lithics associated with hominins. However, our understanding of the relationship between "out of Africa" events, the Saharan archaeological record and its paleoclimate is fragmentary. Early hominin fossils have been found on the northern and southern margins of the Sahara (Brunet et al. 1995; Raynal et al. 2001; Brunet et al. 2002), but not in the center. In contrast, stone tools from periods ranging from the Acheulean to Neolithic have been found throughout much of the desert. Significantly they have been located in the green corridors that Drake et al. (2011) show form potential dispersal routes across the Sahara (Tillet 1985; Aumassip 2004; Mattingly et al. 2007; Lahr et al. 2008), suggesting that these regions were not only periodically habitable for hominins, but that they were also regularly occupied by them. However, uncertainties abound because direct chronological evidence from the desert is in short supply; all but a handful of these lithic sites being undated. Thus, insecure archaeological chronologies mean that it is difficult to relate climate change to hominin response.

The Aterian, a distinctive late Middle Stone Age (MSA) industry found throughout much of North Africa, provides an important example of the potentially important role played by the Sahara in hominin dispersals, but also of the problems created by this chronological uncertainty. Discovery at Aterian sites of structured living space, symbolic ornaments, ochre and bone tools (Bouzouggar et al. 2007) and modern human skeletal remains (Smith et al. 2007) indicate that the Aterians exhibited many aspects of behavioral modernity (Barton et al. 2009). Aterian sites are found throughout much of the Sahara, suggesting that the desert played an important role in modern human dispersal, but raising the question of how and when this was achieved, and if it formed part of a wider "out of Africa" migration. However, dating the Aterian has proved problematic. Radiocarbon dating initially indicated the period to be between 20 and 40 ka, but recent luminescence dating shows that it is much older, ranging from 30 to 140 ka (Barton et al. 2009; Richter et al. 2010), making the Aterian technocomplex among the oldest manifestations of modern humans.

However, the vast majority of the sites subjected to luminescence dating are in the Maghreb, and the paucity of securely dated paleoclimatic and archaeological sites within the Sahara during the Aterian period means that there is little consensus on when exactly occupation occurred or indeed what the Saharan climate regime was like at the time.

This lack of understanding has led some authors to associate Aterian occupation with a humid Saharan climate (Débenath et al. 1986) while others suggest they were desert adapted (Clark 1980). The latter is currently the prevailing view (Cremaschi et al. 1998; Garcea 2012; Hawkins 2012; Larrasoaña 2012; Cancellieri et al. 2016), though there are detractors (Scerri 2013). If the proponents of the view that Aterian populations were desert adapted are correct, then not only was the Sahara a preferential route for modern human dispersal, it was also a region of preferred occupation for modern humans. Thus rather than representing a barrier to their dispersal it might have provided a favored environment.

Aims and Objectives

In order to shed light on potential associations between the Aterian and the green Sahara, a spatial and temporal evaluation of dated Saharan "humid deposits" and Aterian sites has been conducted from MIS 6-2 (186–20 ka). The aim of the paleoclimate analysis is to further our understanding of the timing of Saharan climate change and the spatial distribution of the evidence. When and where are the humid corridors across the Sahara, and when does it form a barrier to modern human dispersal? Following this, an identical spatial and temporal evaluation of the Aterian archaeology of North Africa is conducted, with the aim of evaluating the similarities between the two datasets and to see if the Saharan paleoclimate can inform our understanding of the Aterian occupation of the desert.

Finally, the faunal remains from Aterian sites are assessed for their climate affinity in order to further our understanding as to whether Aterians were desert adapted or occupied the Sahara during humid periods. If the Aterian is an arid-adapted technology then faunal remains found at Aterian sites would consist of animals such as gazelles, hare, jackal, lizard, rodents, and fox; animals found today in the Sahara where there is 50–100 mm rainfall. However, if they were living in a more humid climate, sites should contain elements from a more humid-adapted species. If species crossed the Sahara then there would be more humid savanna animals (e.g., zebra, elephant, hippopotamus, etc.). However, if animals came "into Africa" from Europe or Asia via the Sinai Peninsula they would have a more Mediterranean flavor (e.g., pig, deer, etc.).

Methods

Temporal Evaluation of Saharan Paleoclimate

In order to evaluate the terrestrial record of former humid episodes within the Sahara, published data regarding dated deposits formed under humid conditions were reviewed, using sum probability density function (PDF) analyses (Drake et al. 2013). PDF analyses permit the identification of clustering among radiometric dates with associated uncertainties, and have formerly been applied in several previous midlatitude desert belt Quaternary paleoenvironmental studies. On a localized scale PDF analyses have been used to define chronologies of alluvial fan, fluvial and dune deposition (Zielhofer et al. 2008; Blechschmidt et al. 2009; Preusser 2009), while on a broader spatiotemporal scale the method has been used to produce syntheses of regional paleohumidity data (Glennie and Singhvi 2002; Parker and Rose 2008; Parker 2009; Smith 2012; Drake et al. 2013).

Data for this study came from published deposits dated using the optically stimulated luminescence (OSL) and uranium/thorium (U/Th) methods, but excluded radiocarbon dates, which have been suggested to be unreliable after $\sim$20 ka (Fontes and Gasse 1989) and to influence the accuracy of PDF analyses through effects related to radiocarbon calibration (Chiverrell et al. 2011). The U/Th and OSL dates reviewed are from paleolake sediments, fluvial deposits, spring deposits (tufa), and speleothems. A conservative approach was taken during the selection of dates, dates that had been reported as being potentially erroneous were not used, and for sequences dated with several methods, only the dates produced by the dating method showing the greatest stratigraphic consistency were selected. If dates were available from several studies at the same site then the most recent were used (assuming these met the stratigraphic consistency criteria), and in cases where an individual stratigraphic unit had been dated multiple times, the guidelines proposed by Geyh (2008) were followed to produce weighted averages for dates that grouped statistically, with those that did not retained as individual events. This was necessary since including large numbers of dates for a single deposit would erroneously amplify the PDF curve and bias the results, and this method allowed reduction of the dates to the minimum statistically acceptable number. These selection criteria and treatments produced 144 dates falling within the period ranging from 20 to 200 ka, when considering the dating errors to the 2σ level.

The PDF analysis of these dates allowed a temporal examination of when the dates clustered in time while at the same time considering their associated errors. This was performed by calculating the individual probability distribution for each date based on the mean and error, before summing the resultant distributions in order to produce a plot of sum probability over time (Fig. 6.1a). Within the curve produced by this analysis peaks reflect clustering of the input dates, interpreted to represent humid episodes. The relative height of the peaks is a product of both the number of dates contributing to the curve at that point, and the size of the errors, with small errors producing distinct peaks and large errors

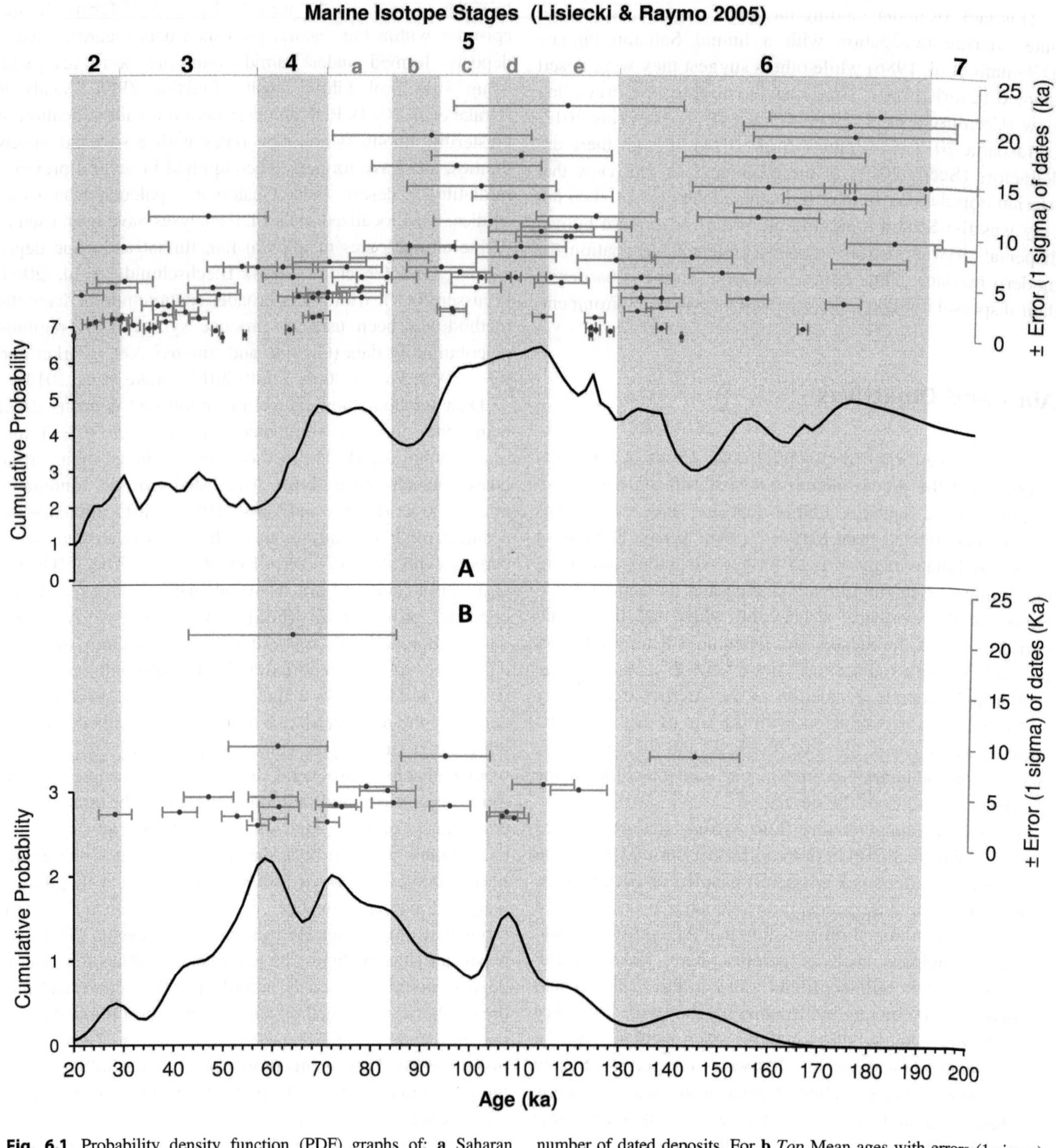

Fig. 6.1 Probability density function (PDF) graphs of: **a** Saharan climate; and **b** Dated Aterian sites. For **a** *Top* Mean ages with errors (1 sigma) used as inputs to PDF analysis. Data has been separated vertically according to size of errors for display. Bottom: PDF curve displaying sum probability fluctuations over time. N indicates the number of dated deposits. For **b** *Top* Mean ages with errors (1 sigma) used as inputs to PDF analysis. Data have been separated vertically according to size of errors for display. *Bottom* PDF curve displaying sum probability fluctuations over time. N indicates the number of dated deposits

producing a smoother curve spread over a large age range. Thus the size of each peak is not a direct measure of how intense a particular humid episode may have been. Further detail regarding the method, and the details of each date employed in this study can be found in Drake et al. (2013).

Spatial Evaluation of Saharan Paleoclimate

Dated sample sites were analyzed spatially by overlaying them on a paleohydrological map of the Sahara developed by Drake et al. (2011) (Fig. 6.2). The map was constructed using a combination of remote sensing and geographical information systems (GIS) methods. GIS analysis of the Shuttle Radar Topography Mission (SRTM) 90 m resolution digital elevation model (DEM) was employed to map river channels that exhibited a significant topographic expression, and closed basins, likely locations for paleolakes to develop during past humid periods. These closed basins were investigated for the presence of lake shoreline landforms that, if detected, were used to map lake shorelines using the methods outlined by Drake and Bristow (2006). Landsat Thematic Mapper (TM) was used in parallel with the DEM analysis to map river channels with little topography, such as those found on alluvial fans, and to analyze the closed basins defined by the DEM to see if they contained paleolake sediments that were identified by interpretation of Landsat TM false color composites. Most lakes were identified by the presence of lake sediments rather than shoreline landforms, as only a few of the larger lakes expressed well-developed shorelines. Finally, visual interpretation of the rivers and lakes in conjunction with the DEM was used to determine the catchment areas of the identified fluvio-lacustrine systems. This spatial analysis was checked at selected sites using fieldwork (e.g., Drake et al. 2008, 2011), and more comprehensively using an extensive literature review of past of past studies of Saharan paleolakes and rivers (see Drake et al. (2011) for further details and references).

The dated sites associated with each peak in probability identified by the temporal analysis (Fig. 6.1a) were then overlain on the paleohydrological map (Fig. 6.2). However, some of these samples represent more than simply points on

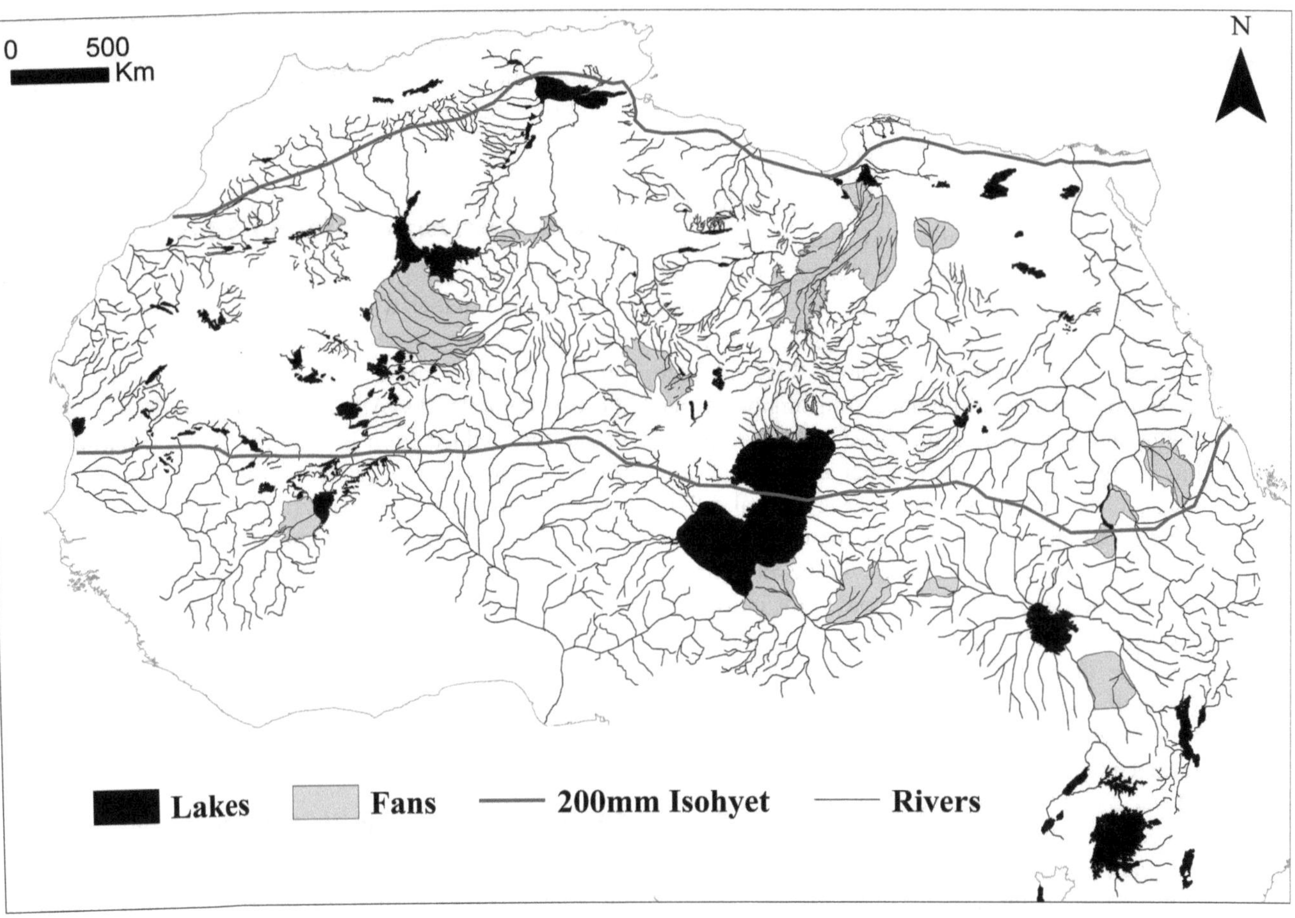

Fig. 6.2 The paleohydrology of Northern Africa. The boundary of the Sahara is shown by the 200 mm rainfall isohyet. Only those North African rivers that drain into the Sahara are marked. Developed from Drake et al. (2011)

the map. Dated paleolake and river sites represent evidence of humidity in their entire catchment areas, thus it is reasonable to assume that they show when the entire catchments they are within were wet. Therefore dated paleolakes and rivers were linked to their catchments in order to determine the regions of humidity each sample represents. These catchments were then displayed (hatched areas) in Figs. 6.3, 6.4, 6.5 and 6.6 to see if there was evidence for humidity within them during the humid period defined by the probability peak being displayed. This allowed an assessment of how spatially representative each peak in humid probability is for the Sahara as a whole, and more importantly if the dated sites linked to provide corridors that could have been used by hominins and animals to disperse across the Sahara. The other dated sites were speleothems and spring tufas that are formed by point water sources and consequently only provide paleoenvironmental information for that point, and could not be spatially analyzed in this manner. However, such samples do suggest that the vicinity of the dated deposit was humid at the time. The paleohydrological maps associated with each PDF peak in humidity are presented in Figs. 6.3, 6.4, 6.5 and 6.6.

Temporal Evaluation of the Aterian

A temporal review through PDF analysis was also applied to dated Aterian sites in North Africa. As with the paleoclimate analysis, only securely stratified sites dated using OSL, TL, or U/Th methods were used (Table 6.1), and site locations were recorded for spatial examination. Although significant debate has been raised regarding the characteristics that can be used to define the Aterian as a techno-typological complex (for reviews, cf. Bouzouggar and Barton 2012; Richter et al. 2012: 62) here we have refrained from imposing our own interpretation upon the data, and have used all deposits suggested as being Aterian by the original authors.

Some Aterian sites have been dated using several different dating methods, and where this was the case the most recent reassessments were chosen for use in the analysis. In some cases, multiple dates had been recorded for individual Aterian deposits. If the original authors had produced weighted average dates for these deposits (Richter et al. 2010; Jacobs et al. 2011, 2012), these dates were used. Data without pre-existing averages came from the Grotte des Pigeons lower laminated group E (Bouzouggar et al. 2007) and from level 3a of Rhafas cave (Mercier et al. 2007), which we averaged to produce single dates for the statistically coeval groupings within each deposit, once again following the procedures defined by Geyh (2008).

For some sites the available chronology for the Aterian was only relative to adjacent absolutely dated levels (Dar es-Soltan II and Ain Zharga). For Ain Zargha, U/Th dates were reported from calcretes above and below the Aterian level (Barich and Garcia 2008; Garcea 2012). Although the reported means are identical, no error is reported for the level above the Aterian (preventing the use of this date in a PDF analysis or weighted averaging), so here we have used the date of the lower calcrete (64 ± 21 ka) to represent the Aterian. For Dar es-Soltan II we decided to produce a weighted average of the dates for deposits overlying and underlying the Aterian (101 ± 11 ka, and 129 ± 6 ka, respectively) (Schwenninger et al. 2010) in order to produce a date representing the intervening Aterian (level 6) (Débenath 1976). While analysis showed that these deposits cannot be considered coeval and averaging is not statistically viable, we decided to produce an average regardless, in order to have a date that roughly represents the Aterian at this site. As a result, this particular date (121.8 ka ± 5.7 ka, Table 6.1) should be treated with caution, however, the Aterian finds were from the base of level 6 (Débenath 1976), and the averaged date trends towards that of the underlying layer 7 (129 ± 6 ka), providing some loose stratigraphic support for the averaged figure. Finally, dates from deposits below level 6 of the Mugharet el 'Aliya sequence were not included, based on suggestions the Aterian materials in these levels may be intrusive (Wrinn and Rink 2003). The compiled data, presented in Table 6.1, was used as the input dates for the PDF analysis (25 dates in total), the result of which is shown in Fig. 6.1b.

Spatial Analysis of Aterian Sites

A literature review was also conducted of all dated and undated Aterian sites with locations recorded in order to produce a map of these sites. The map was largely based on the one compiled by Aumassip (2004) but with sites published since added from Lahr et al. (2008), Mattingly et al. (2007) and Jones et al. (2011). In order to see if there was a spatial association between Aterian sites and any of the humid episodes identified by the PDF analysis this map was then overlaid on the paleohydrological maps outlining the spatial distribution of the evidence for different Saharan humid periods identified by the PDF analysis (Fig. 6.6b). If an association can be found it might suggest the Aterian technocomplex was present in the region in question.

Fauna Associated with Aterian Sites

In order to provide ancillary information on the environmental context of the Aterian technocomplex a literature review was conducted to determine the different animals

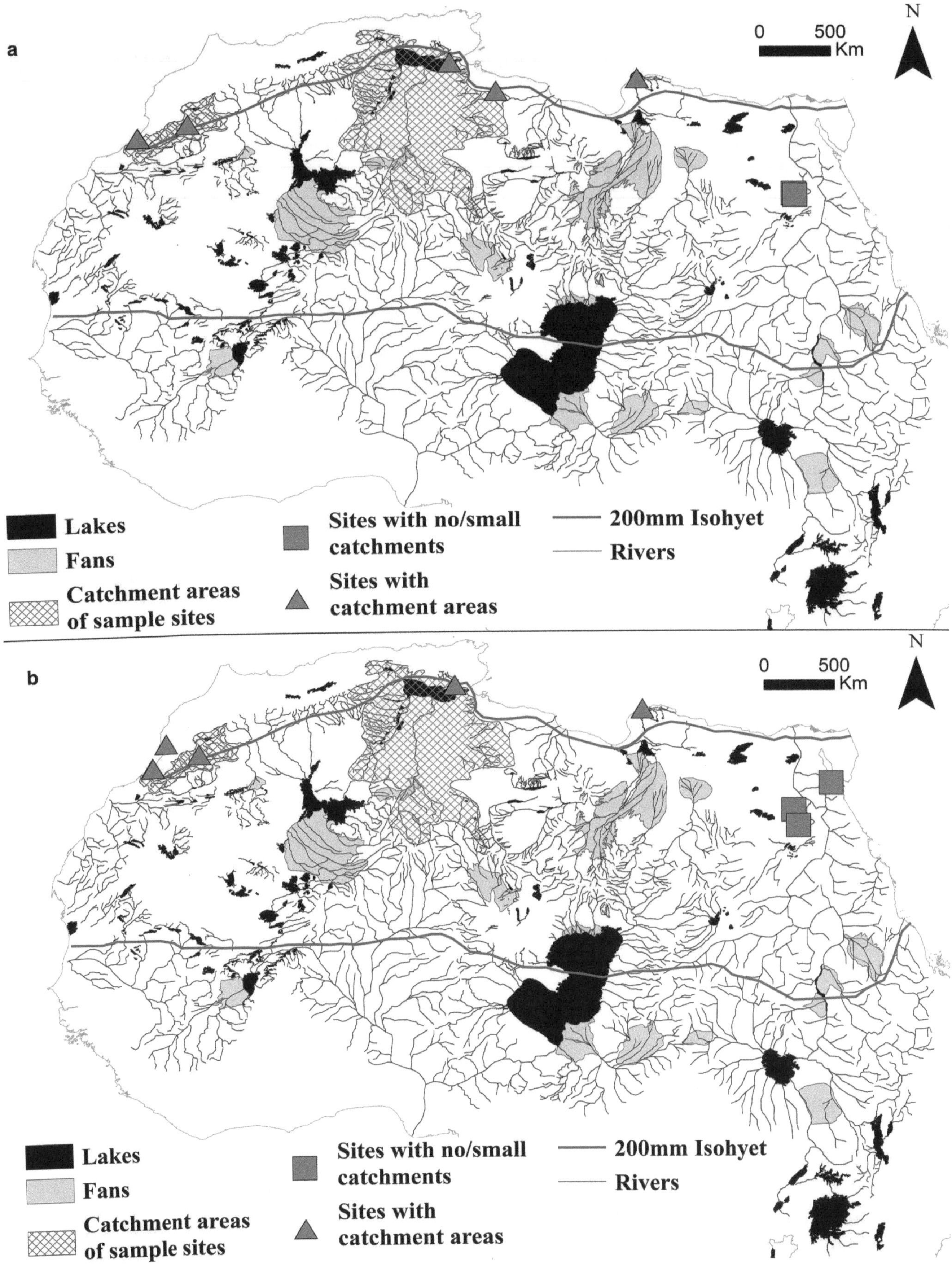

Fig. 6.3 The spatial distribution for sites and associated catchment areas for locations that are associated with: **a** the 29 ka; and **b** the 44 ka PDF peaks

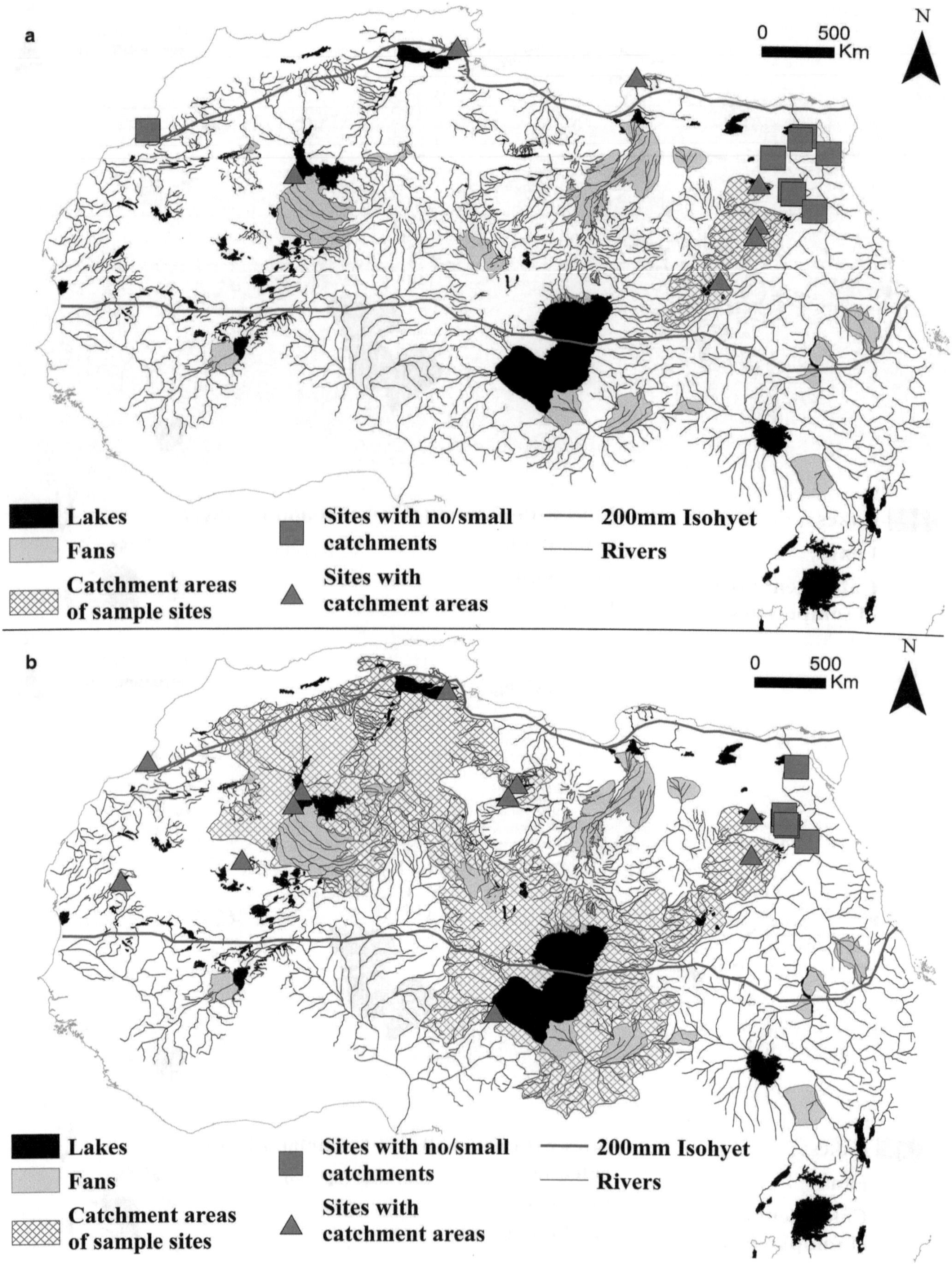

Fig. 6.4 The spatial distribution of sites and associated catchment areas for locations associated with: **a** a PDF peak from 62 to 84 ka; and **b** a PDF peak between 92 and 129 ka

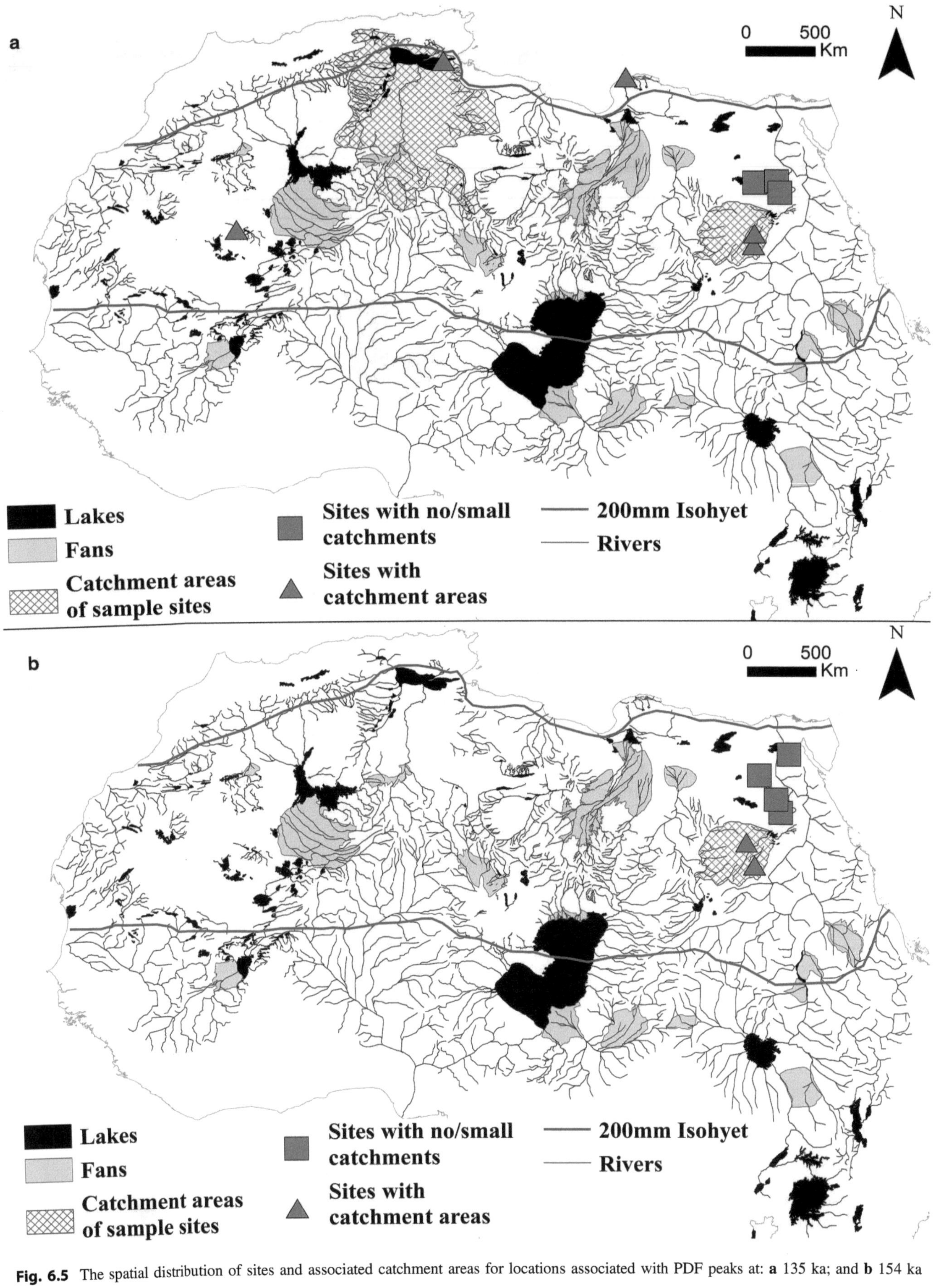

Fig. 6.5 The spatial distribution of sites and associated catchment areas for locations associated with PDF peaks at: **a** 135 ka; and **b** 154 ka

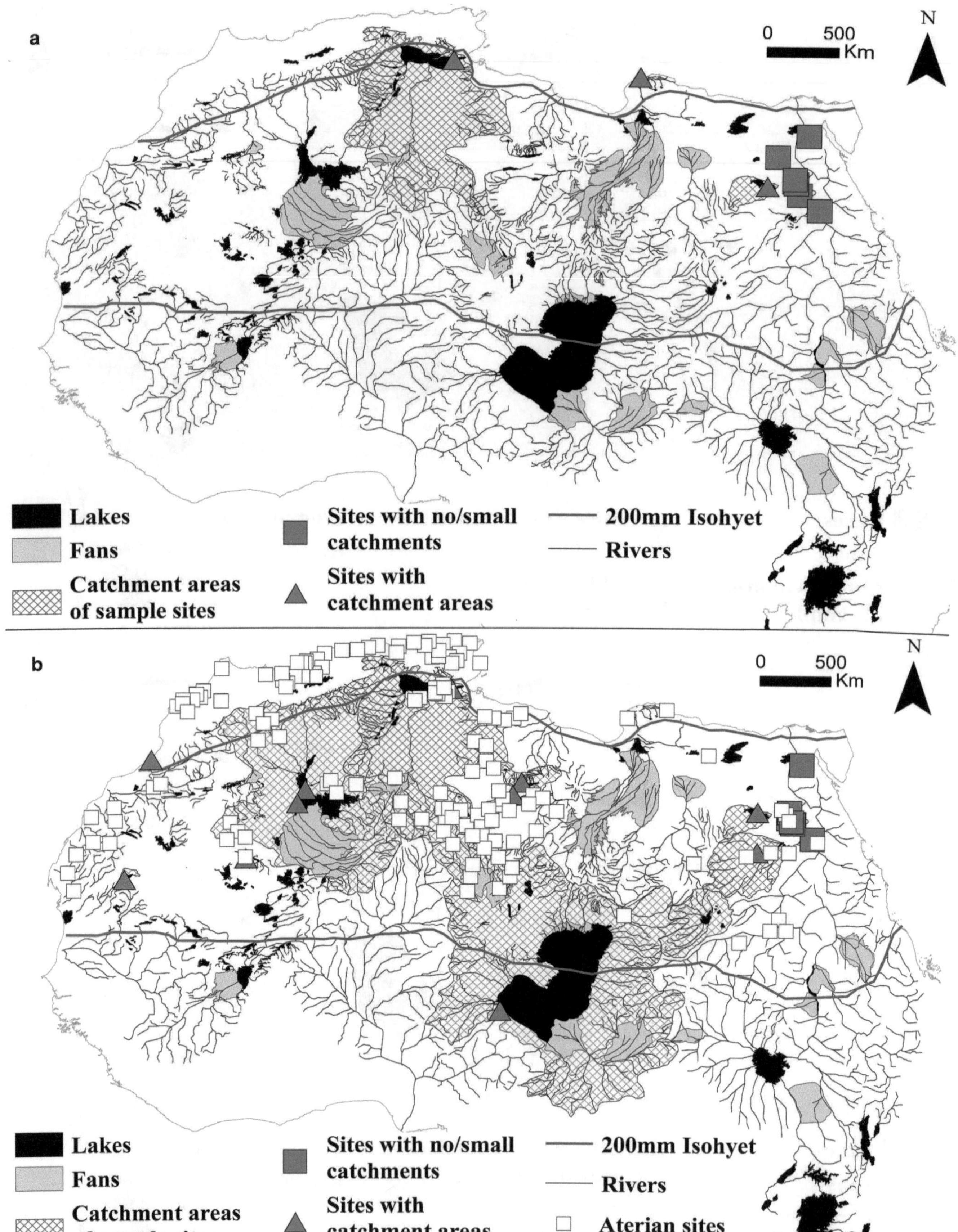

Fig. 6.6 **a** The spatial distribution of sites and associated catchment areas for locations that are associated with the 180 ka, and **b** the 92 and 129 ka, humid corridor across the Sahara with Aterian sites overlain

Table 6.1 Data used in Aterian PDF analysis ($N = 24$)

Reference	Location	Context	Method	Sample and notes	Age (ka)	1σ error (ka)
Texier et al. (1988)	Chaperon-Rouge 1	Cave habitation layer	TL	OX TL 724 g	28.20	3.30
Debenath et al. (1986)	El Harhoura 1	Cave habitation layer	TL	BOR 56	41.16	3.5
Wrinn and Rink (2003)	Mugharet el 'Aliya	Level 5/6	ESR	97122a	47.00	5.00
Barton et al. (2009)	Dar es-Soltan I	OSL 13 (Aterian level c1) Bayesian date	OSL	X2389	52.70	3.10
Nespoulet et al. (2008), Jacobs et al. (2012)	El Harhoura 2 (Temara)	Level 3	OSL (single grain)	Weighted average (produced by original authors)	56.90	2.20
Mercier et al. (2007)	Rhafas Cave	Level 3a	OSL	GR 73	60.00	5.00
Bouzougger et al. (2007)	Grotte des Pigeons, Taforalt	Lower laminated group E	OSL	Weighted average (X2639 and X2643)	*60.20*	*2.83*
Martini et al. (1998), Cremaschi et al. (1998)	Uan Tabu	Layer 22, habitation layer	OSL	UT1 (c-g)	61.00	10.00
Barton et al. (2009)	Dar es-Soltan I	OSL 12 (Aterian level c2) Bayesian date	OSL	X2388	61.20	4.00
Barich and Garcea (2008), Garcea (2012)	Ain Zharga	27E N-trench (Calcrete below Aterian)	UTH	n/a	64.00	21.00
Bouzougger et al. (2007)	Grotte des Pigeons, Taforalt	Lower laminated group E (upper)	OSL	Weighted average (X2399 and X2400)	70.92	2.51
Mercier et al. (2007)	Rhafas Cave	Level 3a	OSL	Weighted average (GR 57, 58 and 73)	72.75	4.21
Nespoulet et al. (2008), Jacobs et al. (2012)	El Harhoura 2 (Temara)	Level 4a	OSL (single grain)	EH08-9	74.00	4.00
Wengler (2006)	Wadi El Hay	Wadi el Hay section	UTH	Proto-Aterian base of Aterian sequence	79.00	6.00
Richter et al. (2010)	Ifri n'Ammar	Upper OS	TL	Weighted average (produced by original authors)	83.30	5.60
Bouzougger et al. (2007)	Grotte des Pigeons, Taforalt	Lower laminated group E	OSL	X2640 (statistically individual)	84.50	4.40
Nespoulet et al. (2008), Jacobs et al. (2011)	El Mnasra	Upper Level 4 (Aterian)	OSL (single grain)	Uppermost level 4	95.00	9.00
Jacobs et al. (2011)	Contrebandiers	Roche trench and layer IV-2	OSL (single grain)	Weighted average (produced by original authors)	95.90	4.10
Nespoulet et al. (2008), Jacobs et al. (2012)	El Harhoura 2 (Temara)	Levels 9-4b	OSL (single grain)	Weighted average (produced by original authors)	106.50	3.10
Jacobs et al. (2011)	Contrebandiers	Layer 4/V-1b & V-2	OSL (single grain)	Weighted average (produced by original authors)	107.40	3.50
Nespoulet et al. (2008), Jacobs et al. (2012)	El Mnasra	Levels 7,11, and lower 4	OSL (single grain)	Weighted average (produced by original authors)	108.90	2.90
Barton et al. (2009)	Dar es-Soltan I	OSL 6 (Aterian level I) Bayesian date	OSL	X2382	114.70	6.20
Schwenninger et al. (2010)	Dar es-Soltan II	Layer 7 & Layer 5 averaged to get date for Aterian level 6	OSL	Weighted average (X1854 and K0314) (not statistically viable, but used to define undated Aterian unit)	121.79	5.67
Richter et al. (2010)	Ifri n'Ammar	Upper OI	TL	Weighted average (produced by original authors)	145.00	9.00

Italics mark weighted averages produced during this study for statistically coeval dates within individual deposits (original sample names are given in brackets)

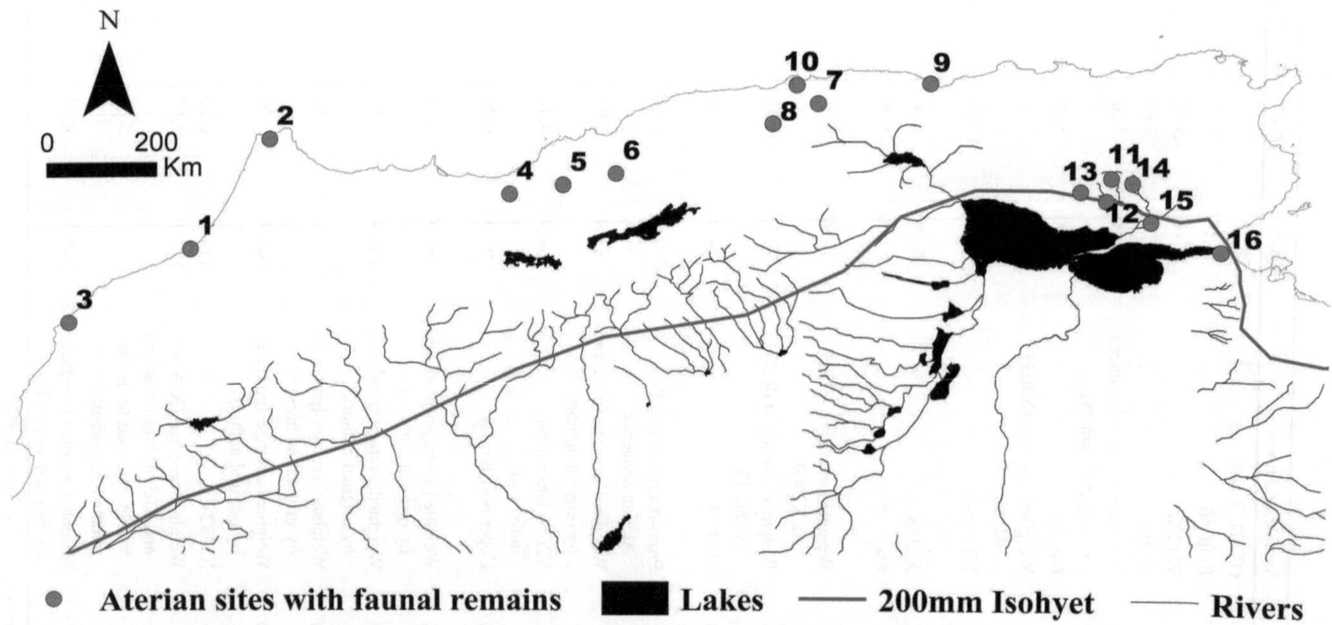

Fig. 6.7 Location of Aterian sites that contain faunal remains overlain on an extract of the paleohydrological map shown in Fig. 6.2. Only those North African rivers that drain into the Sahara are marked. All sites are in the Maghreb. Sites close together are displayed under the same number as outlined below. Morocco: *1* Dar el Soltan (Ruhlmann 1951), Doukkala II (Michel and Wengler 1993; Geraads 2008), El Harhoura 1 (Zouhrah Cave) (Debénath and Sbihi-Alaoui 1979; Aouraghe 2000; Aouraghe and Abbassi 2002; Bailon and Aouraghe 2002; Aouraghe 2004), El Harhoura 2 (Aouraghe and Debénath 1999; Debénath and Sbihi-Alaoui 1979; Nespoulet et al. 2008); El Mnasra (Nespoulet et al. 2008); *2* Mugharet el Aliya (Vaufrey 1955; Arambourg 1967; Howe et al. 1967); *3* El Khenzira 1 (Ruhlmann 1936; Vaufrey 1955), El Khenzira 2-B (Ruhlmann 1936; Vaufrey 1955); *4* Taforalt (Vaufrey 1955). Algeria: *5* Grotte de Polygone (Vaufrey 1955); *6* Oued Saida (Doumergue and Poirier 1894); *7* Les Allobroges (Bagtache et al. 1984; Balout and Roubet 1980; Hadjouis 1986); *8* Berard (Vaufrey 1955); *9* Ali Bacha – Bougie (Romer 1938), Ali Bacha-4 – Bougie (Vaufrey 1955); *10* Ain Taya (Vaufrey 1955); *11* Oglats de Chachaas (Reygasse 1921–1922; Balout 1955; Vaufrey 1955); *12* Oued Djebbana (Bir el Ater) (Vaufrey 1955; Morel 1974a); *13* Oued Djouf el Djemel (Le Du 1934; Morel 1978). Tunisia: *14* Ain Metherchem (Vaufrey 1936, 1955); *15* El Guettar (Vaufrey 1954, 1955); *16* Oued el Akarit (Roset 2005). The location of Oued Gorea (Antoine 1934) could not be determined, so is not included on the map

found at Aterian sites and their frequency of occurrence among the sites. Twenty-six sites that contained animal remains were identified (Fig. 6.7). Most of these excavations were conducted a long time ago and they usually only recorded a list of animals found in the Aterian level or levels, rather than the number of animals present in each level. To consider this problem a table was constructed outlining the number of Aterian sites each animal was found at, in order to provide an indication of the animals most commonly associated with Aterian peoples (Table 6.2).

Results

Spatial and Temporal Analysis of the Paleohydrology

The paleohydrology of the Sahara is shown in Fig. 6.2. It is clear that a dense paleoriver network with many very large alluvial fans and numerous closed basins that supported many lakes, some of which were enormous, formerly covered the region. The largest of these is Lake Megachad, a lake of 361,000 km^2 during the Holocene (Drake and Bristow 2006) and probably slightly larger during MIS 5 (Drake et al. 2011). To its north the Lake Megachad catchment adjoins a large catchment that drains northwards from the Tassili n-Ajjer and Hoggar Mountains. This catchment terminates in the basin of the Chotts, where a large lake (that we call the Chotts Megalake) existed at similar times to Lake Megachad (Causse et al. 2003). Drake et al. (2011) estimate the lake to have been about 30,000 km^2 (Fig. 6.2). To the west of the catchment of the Chotts there is a further vast closed basin in Algeria (Fig. 6.2). Conrad and Lapparient (1991) report a large paleolake in this basin that that has an estimated area of at least 32,000 km^2 (Drake et al. 2011) (Fig. 6.2). This lake has also been dated to a similar time to those outlined above (Causse et al. 1988).

It is important to note that the large river systems in the southern Sahara, such as the Nile, Chari, and Niger, have their headwaters outside the desert in Sub-Saharan Africa (Fig. 6.2). Drake et al. (2011) showed that the alluvial fan distributary channels and paleolake overspills provide links between the majority of the Saharan paleohydrological systems. Thus the Nile, Chari, and Niger River headwaters form sub-Saharan refugia for Saharan biota during arid periods, and subsequently provide the dispersal routes back into the Sahara upon the onset of humid phases (Drake et al. 2011). As the Nile River has its headwaters in East Africa, where anatomically modern humans (AMH) are thought to have evolved at about 200 ka (McDougall et al. 2005), it would have formed a primary migration route for hominins into the green Sahara, with the interconnected nature of the Saharan paleohydrology promoting subsequent dispersal throughout the region.

The analyses of Drake et al. (2011) and Lézine et al. (2011) show that the Saharan paleohydrological systems outlined in Fig. 6.2 were active during the early to middle Holocene. The humid period PDF analysis is used here to evaluate the evidence for this paleohydrological activity during MIS 6-2. Peaks in the PDF curve represent times when the Sahara contained more humid sites (Fig. 6.1a), while troughs characterize a reduced number of humid dates, and thus represent drier periods. There are, however, two important artifacts of the samples and their dating that also affect the PDF curve and complicate this interpretation. First, as the age increases, there is a tendency for the error to get larger, whereupon the curve gets smoother, with both the peaks and troughs declining. In order to be able to recognize when different sized errors are contributing to the PDF curve in this manner, we have plotted each date and its associated error above the PDF curve (Fig. 6.1a). Second, sediments are highly susceptible to erosion during arid periods and thus preservation of older humid period deposits becomes progressively less likely the older they get because they experience more arid phases during which they may be eroded. The severity of this erosion problem can be illustrated by the fact that for the period between MIS 6-2 we have identified 144 dated paleohydrological sedimentary units from 32 localities in the Sahara. Yet for the Holocene there are 1500 dated records (Lézine et al. 2011).

Notwithstanding these problems a number of probability peaks and troughs are evident in the PDF curve. The first rise in probability is near the boundary between MIS 3 and MIS 2, peaking at ~29 ka. The evidence producing this humidity peak consists of fluvial sediments from northern Libya (Wadi Zewana: Rowan et al. 2000) and Morocco (Oued Noun: Weisrock et al. 2006; Tassint: Thorp et al. 2002), lacustrine sediments from southern Tunisia (Chott el Djerid: Causse et al. 2003), tufa from Egypt (Kharga Oasis: Blackwell et al. 2012; Wadi Abu Had-Dib: Hamdan 2000), calcrete from northern Libya (Ras el Wadi: Giraudi 2005), and travertine from Morocco (Oued Noun: Weisrock et al. 2006, 2008). These deposits provide evidence for humidity in the eastern Sahara and its northern margin during the end of MIS 3 and the start of MIS 2 (Fig. 6.3a), but no evidence for humidity further south in the Sahara proper, apart from in the catchment of the Chotts Megalake that could indicate some humidity in the Hoggar Mountains of Southern Algeria but could also represent more local input from Mediterranean weather systems.

During MIS 3 there are further humidity peaks at 37 and 44 ka, with the 37 ka peak relating solely to travertines and fluvial sediments from the northern margins of the Moroccan Sahara (Oued Noun and Imouzzer: Weisrock et al. 2008). Evidence for humidity at 44 ka however, comes from both the northern Saharan semi-arid margin and from the east-central Sahara, in the form of fluvial deposits from Libya (Rowan et al. 2000), southern Tunisia (White et al. 1996) and Morocco (Thorp et al. 2002; Weisrock et al. 2008), and spring tufa from the Kharga Oasis (Sultan et al. 1997) and Wadi Abu Had-Dib (Hamdan 2000) in Egypt (Fig. 6.3b). While the vast majority of these sites are found along the northern margins of the Sahara, humidity is also indicated on the southern periphery of the desert at ~44 ka by eastern Atlantic marine records off the Sahelian coast (Lezine and Casanova 1991; Castañeda et al. 2009). Additionally, speleothem (Lisker et al. 2010; Frumkin et al. 2011) and tufa (Waldmann et al. 2010) records from the Dead Sea region provide evidence for concurrent humidity in the Southern Mediterranean. Thus, evidence for humidity at 44 ka is widespread, and humid conditions in the northern, eastern, and southern Sahara are all indicated at this time, though there is a paucity of central and western Saharan evidence.

The end of MIS 4 appears to be arid, and, as we go further back in time, the humid probability rises to a peak from 62 to 84 ka, centered on 76 ka during late MIS 5. This is not the only humid period in MIS 5. There is a second between 92 and 129 ka, with the peak at about 115 ka. These peaks are separated by a period when there is less evidence for humidity, at around 87 ka. The two MIS 5 peaks are the most pronounced in the PDF record, being characterized by 34 dated humidity proxies in the form of speleothem, tufa, fluvial, and lacustrine deposits. Thus, the Sahara exhibits more evidence for humidity during MIS 5 than for any other period. The 76 ka humid period is predominantly contributed to by speleothems (Wadi Sannar and Djara Caves: Osmond and Dubois 2004) and tufa (Kharga Oasis: Osmond and Dabous 2004; Kurkur Oasis: Crombie et al. 1997) from Egypt, travertine from Morocco (Oued Noun: Weisrock et al. 2008), and lacustrine sediments identified in Algeria (Kadda: Causse et al. 1988) and the Egyptian and Sudanese Western Desert (Bir Sahara East and Bir Tarfawi: Wendorf et al. 1993, Selima Sandsheet, Wadi Hussein and Wadi Arid: Szabo et al. 1995). Fluvial sediments from Libya (Wadi Zewana: Rowan et al. 2000) compliment this evidence. When this PDF peak is mapped using the paleohydrology (Fig. 6.4a) this data provides evidence for a corridor of enhanced humidity across the eastern Sahara, but there is little evidence in the central Sahara, and nothing in the west. Thus it may have been possible for hominins to disperse across the eastern Sahara during MIS 5a, and this would have been facilitated by the fact that the Nile was flowing at this time (Revel et al. 2010).

The 92–129 ka humid period shows evidence for widely distributed Saharan humidity. Furthermore, this evidence lends itself to the GIS mapping approach as it is dominated by paleolakes with very large catchments, such as Lake Megachad in the central and southern Sahara (Drake et al. 2011) and the Ahnet-Mouydir Megalake and the Chotts Megalake in north-central regions (Causse et al. 1988, 2003) (Fig. 6.4b). These giant lakes and associated catchments link to form corridors across the Sahara that could have allowed AMH dispersal across the region. The other paleolake catchments, though smaller, are dispersed across much of the Sahara, in Mauritania (Sebkah Chemchane: Lezine and Casanova 1991), Mali (Toudenni Depression: Petit-Mare et al. 1994), Algeria (Kadda: Causse et al. 1988), Libya (Wadi el-Agial and Wadi ash Shati: Armitage et al. 2007), and Egypt (Dakhla Oasis, Bir Tarfawi and Bir Sahara East: Wendorf et al. 1993; Brookes 2010). This widespread evidence for lacustrine activity is supplemented by additional dates from Egypt for the deposition of tufas (Kharga and Kurkur Oases: Osmond and Dabous 2004; Crombie et al. 1997) and a speleothem (Wadi Sannar Cave: Osmond and Dabous 2004). In summary, there is strong evidence for a humid corridor across the central Sahara at this time (Fig. 6.4b), and there is a wide distribution of dates in other regions that point to the possibility of this corridor representing a wider "green Sahara" across much, if not the entire desert during this period of time.

There is evidence for both humidity and aridity during MIS 6. It starts with a small humid peak centered on 135 ka, followed by a significant falloff in the probability to a minima at 145 ka, with other peaks at 154 and 180 ka. Interpretation of these PDF peaks is complicated by the fact that the errors on some dates are very large (Fig. 6.1a), spreading from peaks into troughs and in extreme cases contributing to two peaks and the troughs in between. This is particularly evident for the 154 and 180 ka PDF maxima, where dates from speleothems from Wadi Sannar and Djara Cave (Osmond and Dabous 2004), and from tufas from Kharga Oasis (Sultan et al. 1997), Wadi Abu Had-Dib (Hamdan 2000) and Dakhla Oasis (Brookes 1993), all contribute to both peaks, explaining why the decline in probability between them is so small. Dates contributing solely to the 154 ka peak come from lacustrine sediments from Tunisia (Chott el Djerid: Causse et al. 2003) and Egypt (Bir Sahara and Selima Oasis: Szabo et al. 1995) and from an Egyptian tufa (Kharga Oasis: Smith et al. 2004). Dates contributing solely to the 180 ka peak derive from lacustrine deposits from Tunisia (Chott el Djerid: Causse et al. 2003), fluvial sediments from Libya (Wadi Zewana: Rowan et al. 2000) and Morocco (Oued Noun: Weisrock et al. 2008), and tufas from the oases of Kharga (Sultan et al. 1997) and Kurkur (Crombie et al. 1997) in Egypt.

Only the dates that contribute to a single peak have been used in the paleohydrological mapping of these MIS 6 humid episodes (Figs. 6.5a, b, and 6.6a). None of these peaks provides a corridor across the desert, with dates being concentrated in the north and east of the Sahara. Yet the existence of any evidence for humidity in MIS 6 is surprising given the fact that this is a glacial period and they are generally thought to be cold and dry in tropical deserts. Furthermore, the 154 ka peak is tantalizingly close to the date for the presence of the first AMHs north of the Sahara, at Jebel Irhoud in Morocco at about 160 ka (Smith et al. 2007). As AMHs are thought to have evolved in East Africa at about 200 ka (McDougall et al. 2005), MIS 6 is a likely time for their crossing the Sahara, though a late MIS 7 dispersal could also explain the Jebel Irhoud hominins and cannot be discounted. Clearly, with current data no MIS 6 corridor can be demonstrated (Figs. 6.5a, b and 6.6a). Yet there is archaeological evidence that can be interpreted to indicate that AMHs have crossed the Sahara, and the fact that existing paleohydrological data is sparse may be due to the sediment preservation problems outlined above, and thus this record may be incomplete. Further work is needed to understand this issue, perhaps targeting these elusive older sediment outcrops.

Spatial and Temporal Analysis of the Aterian

When plotted on the paleohydrological map, the spatial distribution of Aterian sites in North Africa appears to show some correspondence with the 92–129 ka Saharan green corridor (Fig. 6.6b). Indeed the sites exhibit two predominant clusters, one along the coast of the Maghreb and the other along this green corridor across the Sahara, supporting the view that the Sahara did not form an effective barrier to the dispersal of hominins across the desert. This corridor encompasses the Ahaggar and Tibesti Mountains in the central Sahara; regions that today receive more rainfall than the surrounding plains, and that can be expected to have received significantly more during past humid periods. Indeed, the rivers that originate in these mountains feed an extremely well developed paleohydrology, commonly forming giant alluvial fans, which when active may have been much like the Okavango delta is today. They also feed numerous lakes, some of which were huge, including Lake Megachad in the southern Sahara, the Ahnet-Mouydir Megalake and the Chotts Megalake in the north-central Sahara (Drake et al. 2011). Thus the corridor contained a diverse array of environments that are likely to have provided a varied selection of resources that may have proved attractive to the people responsible for the Aterian technocomplex. Furthermore, the major river systems in the southern end of this corridor were linked to the Nile during humid phases (Drake et al. 2011; Bezault et al. 2011), thus there were direct riparian and lacustrine links from East Africa to this corridor, providing potential pathways for hominin dispersal from East Africa, across the Sahara, and to the Maghreb. Such routes were likely to be attractive, given the fact that the majority of East African archaeological sites that can be attributed to this period are found in either riverine or lacustrine environments (Basell 2008).

Though there is no unequivocal hominin fossil evidence in the Sahara demonstrating the migration of AMHs from East Africa to the Maghreb at 92–129 ka, mammals such as the giant buffalo *Pelorovis antiquus,* and the hartebeest *Alcelaphus buselaphus* appear to have done so during MIS 5 (Geraads 2010), providing evidence for an effective dispersal route across the Sahara at around this time. Furthermore, this interpretation fits with the vast majority of the dated Aterian sites in the Maghreb. The 92–129 ka green corridor corresponds with the bulk of the evidence for the earliest Aterian occupation of the region, with the earliest peak in the Aterian PDF plot at this time (Fig. 6.1b) and this is roughly coincident with the first appearance of AMHs in the Levant (Grün et al. 2005). However, this explanation of the evidence is complicated by the fact that the earliest evidence for Aterians in the Maghreb is found at Ifri n'Ammar at about ~140 ka (Richter et al. 2010). If it were possible to cross a "green Sahara" during MIS 6 then this would help to explain the start of the Aterian archaeological record in the Maghreb. However, there is only a small amount of evidence for humidity in the Sahara at this time and the errors associated with the Richter et al. (2010) dates are so large they do not preclude an early MIS 5 date for the earliest Aterian. Thus archaeological and paleoclimatological evidence sustains the view of a green Sahara open to dispersal of modern human groups during MIS 5, though the earliest dates for the Aterian technocomplex provide an alternative interpretation; that Aterians could have been present in the Maghreb prior to this, during late MIS 6 (Richter et al. 2010). It is clear that other AMH groups were in the Maghreb during MIS 6 (Smith et al. 2007), and it is even possible that the Aterians could have evolved there from them (Balter 2011). If this were the case then any Aterian occupation of the Sahara during MIS 5 would have been from the Maghreb southwards. Clearly more work is needed to clarify these issues, and dating of Aterian sites in the Sahara is a priority as it could help determine if Aterians arrived in the Sahara before the Maghreb, or *vice versa*.

Changes in the number of dated Aterian sites over time can provide a measure of their demography, with a larger number of sites suggesting larger populations (Gamble et al. 2005; Blome et al. 2012). Cumulative PDFs can be used to illustrate these population changes while taking into account the errors on the dates. The Aterian PDF curve (Fig. 6.1b) provides many similarities to that of the Saharan paleoclimate (Fig. 6.1a), indicating that climate change could have played a role in Aterian demography. This is rather surprising, given that the paleoclimate information comes from throughout the arid and semi-arid zone of North Africa, while the dated Aterian sites are largely restricted to the Maghreb, with only a couple of sites from the Sahara proper (Fig. 6.6b). The PDF of Aterian sites starts to rise significantly at the beginning of MIS 5, reaching a peak at approximately the same time as the Saharan humidity PDF (~109 ka) (Fig. 6.1a). The Aterian PDF curve then falls, coinciding with a fall in the humidity PDF, before rising to a maxima in MIS 5a, again, roughly coinciding with the Saharan MIS 5a humidity PDF maxima. Both PDFs fall at the start of MIS 4, however, they then diverge, with a further peak in the Aterian PDF curve during late MIS 4 while the evidence for Saharan humidity continues to decline. At the start of MIS 3 the Aterian PDF declines, but at about 42 ka there is some evidence for a brief arrest to this drop, at a similar time to the ~45 ka northern Saharan climate amelioration, before a further decline to extinction by 20 ka. In summary, these results suggest that in general Aterian populations rose during North African humid periods and declined during arid ones. These results therefore imply that the Aterian technocomplex was not an arid adaption as has been claimed by some (Garcea 2012; Larrasoaña 2012; Hawkins 2012; Cremaschi et al. 1998; Debénath et al. 1986). It is likely that populations expanded and they dispersed into the Sahara during humid periods and subsequently declined during arid phases. It is also possible that any Aterians in the Sahara at the end of MIS 5 were "pumped" out of the Sahara into the Maghreb when the Sahara became hyper-arid during early MIS 4 and late MIS 3. Thus, it is possible that the MIS 4 peak in Aterian probability is linked to an environmentally forced migration from the now arid Sahara into the more humid Maghreb, where the vast majority of dated sites come from.

Fauna Associated with the Aterian

To investigate further the possibility that Aterian populations were adapted to humid climates we have investigated the animals associated with Aterian sites, to see if they are either arid or humid adapted (Table 6.2). The results show that Aterians appear to be associated with savanna animals that mostly originate in sub-Saharan Africa, with a smattering of paleoarctic elements originating outside Africa from more northern latitudes. This finding indicates not only an Aterian humid environment adaption, but also that many of the animals

Table 6.2 The fauna found at the Aterian sites that contain faunal remains and their frequency of occurrence

Common name	Species	Number of sites
Zebra (extinct)	*Equus mauritanicus*	18
Hartebeest	*Alcelaphus buselaphus*	17
Wild Boar	*Sus scrofa*	12
Aurochs (extinct)	*Bos primigenius*	13
Rhinoceros	*Rhinocerotidae*	11
Warthog	*Phacochoerus aethiopicus*	9
Blue Wildebeest	*Connochaetes taurinus*	9
Golden Jackal	*Canis aureus*	8
Gazelle	*Gazella* sp.	8
Red Fox	*Vulpes vulpes*	8
Crested Porcupine	*Hystrix cristata*	8
Barbary Sheep	*Ammotragus lervia*	7
Gazelle (extinct)	*Gazella atlantica*	6
Spotted Hyena	*Crocuta crocuta*	6
Striped Hyena	*Hyaena hyaena*	5
Deer	*Cerus* sp.	5
Dorcas Gazelle	*Gazella dorcas*	5
Hippopotamus	*Hippopotamus amphibious*	5
Leopard	*Felis Pardus*	5
Spur-thighed Tortoise	*Testudo graeca*	5
Giant Buffalo (extinct)	*Pelorovis antiquus*	4
Lion	*Panthera leo*	4
Elephant	*Elephantidae*	4
Bovid	*Bovidae* sp.	4
Equid	*Equus* sp.	4
Cuvier's Gazelle	*Gazella cuvieri*	4
Ostrich	*Struthio camelus*	4
Hare	*Lepus* sp.	4
European Rabbit	*Oryctolagus* sp.	3
Hedgehog	*Erinaceus ep.*	3
Camel	*Camelidae*	3
Oryx	*Oryx* sp.	2
Reedbuck	*Redunca* sp.	2

A total of 26 sites contain both faunal remains and Aterian artifacts. The location of these sites is shown in Fig. 6.7

that they were found with migrated across a green Sahara, just as the Aterian populations could have done during MIS 5.

These findings must be considered in context with the spatial location of the sites. This is shown in Fig. 6.7, which demonstrates that all the sites come from the Maghreb, three from the edge of the Sahara and seven from the semi-arid region to the north. This paper has suggested two main spatial distributions of Aterian sites, the first along the coast of the Maghreb and the second along the MIS 5 Saharan humid corridor (Fig. 6.6b). These distributions yield considerable information about the coastal Aterian populations, and less about those in what is now the Sahara Desert (Fig. 6.7). Only the three sites in eastern Algeria and southern Tunisia on the edge of the Sahara tell us about Aterian desert adaptations. When we look at the fauna associated with these sites they appear to form a subset of the savanna fauna found throughout the region, with three zebra (*Equus mauritanicus*), two rhinocerotidae (*Ceratotherium mauritanicum?*), two aurochs (*Bos primigenius*), two hartebeest (*Alcelaphus buselaphus*), one striped hyena (*Hyaena hyaena*), one giant buffalo (*Pelorovis antiquus*), one roan antelope (*Hippotragus equinus*), and a camel (*Camelidae*). Thus, faunal evidence does not suggest that this region has a distinctly different climate from elsewhere in the Maghreb during Aterian times. In summary, clearly in the Maghreb there are Aterians who are not desert adapted, while on the edge of what is now the Sahara the evidence suggests occupation of an environment considerably more humid than today.

Conclusions

Combining PDF analysis, paleohydrological mapping and GIS analysis of sites provides an effective way of analyzing archaeological and paleoenvironmental data. The method reveals times when it was possible for AMH to disperse across the Sahara. However, the data presented here is far from complete, and more paleoclimate information, site excavation, and chronological data are needed to help understand the paleoclimate of the Sahara, the nature and development of the Aterian industry, Aterian environmental preferences, and their role in the dispersal of modern humans out of Africa. Notwithstanding this, the research presented here sheds some light on these issues.

The combined PDF and paleohydrological mapping approach suggests that during MIS 3 and the start of MIS 2 there are three brief periods of enhanced humidity in North Africa, with much evidence along the northern margins of the Sahara, some in the east, a bit in the south, but little in the center. MIS 5 provides considerably more evidence for humidity across the Sahara. There is a humidity peak centered on 76 ka that provides evidence for enhanced environmental conditions across the eastern Sahara, but little evidence in the center, and nothing in the west. There is a decline in humidity at around 87 ka but the PDF then rises to a broad peak between 92 and 129 ka, providing evidence for a broad humid corridor across the central Sahara and further evidence for pluvial conditions scattered across much of the desert, suggesting the possibility of a wider "green Sahara." Given the evidence, it appears that it would have been possible for AMH to disperse across the eastern Sahara during MIS 5a, and to occupy much, if not all, of the Sahara during MIS 5c, 5d, and 5e. There is evidence for humidity during MIS 6 at 135, 154, and 180 ka, however, dated sites are concentrated in the north and east of the Sahara and do not form corridors across the desert.

There is some evidence for an Aterian occupation of the Sahara during the 92 and 129 ka humid phase as Aterian sites are found preferentially in the MIS 5 humid corridor across the Sahara that is evident at this time. Furthermore, both the Aterian and Saharan humidity PDF curves peak at the time this corridor was active. However, none of the dated Aterian sites are from the Sahara (Table 6.1), and thus the evidence is not conclusive.

The investigation into the question of whether the fauna associated with Aterian sites were either arid or humid adapted showed that they were predominantly humid adapted, and were principally savanna animals that appear to have migrated across a "green Sahara," just like the Aterians could have done during MIS 5. However, as with the dating, most sites come from the Maghreb, and there is a lack of sites from the Sahara proper. Thus the results presented here, though indicating an Aterian preference for a savanna environment, do not unequivocally demonstrate it, and further chronometric and archaeological research is required in the Sahara proper to put this issue to rest.

References

Antoine, M. (1934). Notes de prehistoire marocaine VIII: Un gisement atérien en place dans 1es alluvions de l'Oued Goréa, près de Casablanca. *Bulletin de la Sociéte de Prehistoire du Maroc, 8*, 7–34.

Aouraghe, H. (2000). Les carnivores fossiles d'El Harhoura 1, Temara, Maroc. *L'Anthropologie, 104*, 147–171.

Aouraghe, H. (2004). Les populations de mammifères atériens d'El Harhoura 1 (Témara, Maroc). *Bulletin d'Archéologie Marocaine, 20*, 83–104.

Aouraghe, H., & Abbassi, M. (2002). Les rongeurs du site Atérien d'El Harhoura 1 (Témara, Maroc). *Quaternaire, 13*, 125–136.

Aouraghe, H., & Debénath, A. (1999). Les équidés du Pléistocène supérieur de la Grotte Zouhrah á El Harhoura, Maroc. *Quaternaire, 10*, 283–292.

Arambourg, C. (1967). Appendix A. Observations sur la faune des Grottes d'Hercule près de Tanger, Maroc. In B. Howe (Ed.), *The Paleolithic of Tangier, Morocco: Excavations at Cape Ashakar, 1939–1947* (pp. 181–186). Cambridge, Massachusetts: The Peabody Museum.

Armitage, S. J., Drake, N. A., Stokes, S., El-Hawat, A., Salem, M. J., White, K., et al. (2007). Multiple phases of North African humidity recorded in lacustrine sediments from the Fazzan Basin. *Libyan Sahara, Quaternary Geochronology, 2*, 181–186.

Armitage, S. J., Jasim, S. A., Marks, A. E., Parker, A. G., Usik, V. I., & Uerpmann, H. (2011). The southern route "Out of Africa": Evidence for an early expansion of modern humans into Arabia. *Science, 331*, 453–456.

Aumassip, G. (2004). *Prehistoire du Sahara et de ses abords*. Paris: Maisonneuve and Larose.

Bagtache, B., Hadjouis, D., & Eisenmann, V. (1984). Présence d'un Equus caballin (*E. algericus* n. sp.) et d'une autre espècen nouvelle d'Equus (*E. melkiensis* n. sp.) dans l'Atérien des Allobroges, Algérie. *Comptes Rendus de l'Academie des Sciences de Paris, 298*, 609–612.

Bailon, S., Aouraghe, H. (2002). Amphibiens, chéloniens et squamates du pléistocène supérieur d'El Harhoura 1 (Témara, Maroc). *Geodiversitas, 24*, 821–830.

Balout, L. (1955). *Préhistoire de l'Afrique du Nord*. Paris: Arts et Métiers Graphiques.

Balout, L., & Roubet, C. (1980). Apports de la préhistoire à la stratigraphie et la cronologie au Sahara et au Maghreb. In M. A. Williams & H. Faure (Eds.), *The Sahara and the Nile* (pp. 163–172). Rotterdam: Balkema.

Balter, M. (2011). Was North Africa the launch pad for modern human migrations? *Science, 331*, 20–23.

Barich, B. E., & Garcea, E. A. A. (2008). Ecological patterns in the Upper Pleistocene and Holocene in the Jebel Gharbi, northern Libya: Chronology, climate and human occupation. *African Archaeological Review, 25*, 87–97.

Basell, L. S. (2008). Middle Stone Age (MSA) distributions in eastern Africa and their relationship to Quaternary environmental change, refugia and the evolution of *Homo sapiens*. *Quaternary Science Reviews, 27*, 2484–2498.

Barton, R. N. E., Bouzouggar, A., Collcutt, S. N., Schwenninger, J.-L., & Clark-Balzan, L. (2009). OSL dating of the Aterian levels at Dar es-Soltan I (Rabat, Morocco) and implications for the dispersal of modern *Homo sapiens*. *Quaternary Science Reviews, 28*, 1914–1931.

Bar-Yosef, O., & Belfer-Cohen, A. (2001). From Africa to Eurasia — early dispersals. *Quaternary International, 75*, 19–28.

Bezault, E., Balaresque, P., Toguyeni, A., Fermon, Y., Araki, H., Baroiller, J.-F., & Rognon, X. (2011). Spatial and temporal variation in population genetic structure of wild Nile tilapia (*Oreochromis niloticus*) across Africa. *BMC Genetics, 102*, 1–16.

Blackwell, B., Skinner, A., Mashriqi, F., Deely, A. E., Long, R. A., Gong, J. J. J., et al. (2012). Challenges in constraining pluvial events and hominin activity: Examples of ESR dating molluscs from the Western Desert. *Egypt. Quaternary Geochronology, 10*, 430–435.

Blechschmidt, I., Matter, A., Preusser, F., & Rieke-zapp, D. (2009). Monsoon triggered formation of Quaternary alluvial megafans in the interior of Oman. *Geomorphology, 110*, 128–139.

Blome, M. W., Cohen, A. S., Tryon, C. A., Brooks, A. S., & Russell, J. (2012). The environmental context for the origins of modern human diversity: A synthesis of regional variability in African climate 150,000–30,000 years ago. *Journal of Human Evolution, 62*, 563–592.

Bouzouggar, A., & Barton, R. N. E. (2012). The identity and timing of the Aterian in Morocco. In J. J. Hublin & S. P. McPherron (Eds.), *Modern origins: A North African perspective* (pp. 93–105). Dordrecht: Springer.

Bouzouggar, A., Barton, R. N. E, Vanhaeren, M., d'Errico, F., Collcutt, S., Higham, T., et al. (2007). 82,000-year-old shell beads from North Africa and implications for the origins of modern human behaviour. *Proceedings of the National Academy of Sciences of America, 104*, 9964–9969.

Brookes, I. (1993). Geomorphology and Quaternary geology of the Dakhla Oasis region, Egypt. *Quaternary Science Reviews, 12*, 529–552.

Brookes, I. (2010). Spatially variable sedimentary responses to orbitally driven pluvial climate during Marine Oxygen Isotope Stage 5.1, Dakhla Oasis. *Egypt. Quaternary Research, 74*, 252–264.

Brunet, M., Beauvilain, A., Coppens, Y., Heintz, E., Moutaye, A. H. E., & Pilbeam, D. (1995). The first australopithecine 2,500 kilometres west of the Rift Valley (Chad). *Nature, 378*, 273–275.

Brunet, M., Guy, F., Pilbeam, D., Mackaye, H. T., Likius, A., Ahounta, D., et al. (2002). A new hominid from the Upper Miocene of Chad, Central Africa. *Nature, 418*, 145–151.

Cabrera, V. M., Abu-Amero, K. K., Larruga, J. M., & González, A. M. (2010). The Arabian Peninsula: Gate for human migrations out of

Africa or cul-de-sac? A mitochondrial DNA phylogeographic perspective. In M. D. Petraglia & J. I. Rose (Eds.), *The evolution of human populations in Arabia: Paleoenvironments, prehistory and genetics* (pp. 279–289). London: Springer.

Cancellieri, E., Cremaschi, M., Zerboni, A., & di Lernia, S. (2016). Climate, environment and population dynamics in Pleistocene Sahara. In S. C. Jones & B. A. Stewart (Eds.), *Africa from MIS 6-2: Population dynamics and paleoenvironments* (pp. 123–145). Dordrecht: Springer.

Castañeda, I. S., Mulitza, S., Schefuss, E., Lopes dos Santos, R. A., Sinninghe Damsté, J. S., & Schouten, S. (2009). Wet phases in the Sahara/Sahel region and human dispersal patterns in North Africa. *Proceedings of the National Academy of Sciences, 106*, 20159–20163.

Causse, C., Conrad, G., Fontes, J., Gasse, F., Gibert, E., & Kassir, A. (1988). The last Pleistocene humid phase in the western Sahara, dated to about 80-100 ka. *Comptes Rendus – Academie des Sciences, Serie II, 306*, 1459–1464.

Causse, C., Ghaleb, B., Chkir, N., Zouari, K., Ben Ouezdou, H., & Mamou, A. (2003). Humidity changes in southern Tunisia during the Late Pleistocene inferred from U-Th dating of mollusc shells. *Applied Geochemistry, 18*, 1691–1703.

Chiverrell, R. C., Thorndycraft, V. R., & Hoffmann, T. O. (2011). Cumulative probability functions and their role in evaluating the chronology of geomorphological events during the Holocene. *Journal of Quaternary Science, 26*, 76–85.

Clark, J. D. (1980). Human populations and cultural adaptations in the Sahara and Nile during prehistoric times. In M. A. J. Williams & H. Faure (Eds.), *The Sahara and the Nile* (pp. 527–582). Rotterdam: Balkema.

Conrad, G., & Lapparient, J. R. (1991). The appearance of Cardium fauna and foraminifers in the great lakes of the early Quaternary period in the Algerian Central Sahara Desert. *Journal of African Earth Sciences, 12*, 375–382.

Crassard, R. (2010). The Middle Paleolithic of Arabia: The view from the Hadramawt region, Yemen. In M. D. Petraglia & J. I. Rose (Eds.), *The evolution of human populations in Arabia: Paleoenvironments, prehistory and genetics* (pp. 151–168). London: Springer.

Cremaschi, M., Di Lernia, S., & Garcea, E. A. A. (1998). Some Insights on the Aterian in the Libyan Sahara: Chronology, environment and archaeology. *African Archaeological Review, 15*, 261–286.

Crombie, M., Arvidson, R., Sturchio, N., El Alfy, Z., & Abu Zeid, K. (1997). Age and Isotopic constraints on Pleistocene pluvial episodes in Western Desert, Egypt. *Paleogeography, Paleoclimatology, Paleoecology, 130*, 337–355.

Debénath, A. (1976). La grotte de Dar-es-Soltane II à Rabat (Maroc). *Bulletins et Mémoires de la Société d'Anthropologie de Paris, XIII série, 3*, 181–182.

Debénath, A., & Sbihi-Alaoui, F. (1979). Découverte de deux nouveaux gisements préhistoriques prés de Rabat (Maroc). *Bulletin de la Société Préhistorique Française, 76*, 11–12.

Debénath, A., Raynal, J.-P., Roche, J., Texier, J.-P., & Ferembach, D. (1986). Stratigraphie, habitat, typologie et devenir de l'Aterien marocain: Donnees recentes. *L'Anthropologie, 90*, 233–246.

Dennell, R. (2009). *The Paleolithic settlement of Asia*. Cambridge: Cambridge University Press.

Doumergue, F., & Poirier, E. (1894). *La grotte préhistorique de l'Oued Saïda*. Oran: Paul Perrier.

Drake, N. A., & Bristow, C. (2006). Shorelines in the Sahara: Geomorphological evidence for an enhanced monsoon from paleolake Megachad. *The Holocene, 16*, 901–911.

Drake, N. A., El-Hawat, A. S., Turner, P., Armitage, S. J., Salem, M. J., White, K. H., et al. (2008). Paleohydrology of the Fazzan Basin and surrounding regions: The Last 7 million years. *Paleogeography, Paleoclimatology, Paleoecology, 263*, 131–145.

Drake, N. A., Blench, R. M., Armitage, S. J., Bristow, C. S., & White, K. H. (2011). Ancient watercourses and biogeography of the Sahara explain the peopling of the desert. *Proceedings of the National Academy of Sciences of America, 108*, 458–462.

Drake, N. A., Breeze, P. S., & Parker, A. G. (2013). Paleoclimate in the Saharan and Arabian deserts during the Middle Paleolithic and the potential for hominin dispersals. *Quaternary International, 300*, 48–61.

Dumont, H. J. (1982). Relict distribution patterns of aquatic animals: Another tool in evaluating late Pleistocene climate changes in the Sahara and Sahel. *Paleoecology of Africa, 14*, 1–24.

Duveyrier, H. (1864). *Exploration du Sahara: Les Touâreg du nord*. Paris: Challamel Aine.

Fernandes, C. A. (2010). Bayesian coalescent inference from mitochondrial DNA variation of the colonization time of Arabia by the hamadryas baboon (*Papio hamadryas hamadryas*). In M. D. Petraglia & J. I. Rose (Eds.), *The evolution of human Populations in Arabia: Paleoenvironments, prehistory and genetics* (pp. 89–103). London: Springer.

Fernandes, C., Rohling, E. J., & Siddall, M. (2006). Absence of Quaternary Red Sea land bridges: Biogeographic implications. *Journal of Biogeography, 33*, 961–966.

Fontes, J. C., & Gasse, F. (1989). On the ages of humid Holocene and late Pleistocene phases in North Africa – remarks on "late Quaternary climatic reconstruction for the Maghreb (North Africa)" by P. Rognon. *Paleogeography, Paleoclimatology, Paleoecology, 70*, 393–398.

Frumkin, A., Bar-Yosef, O., & Schwarcz, H. P. (2011). Possible paleohydrologic and paleoclimatic effects on hominin migration and occupation of the Levantine Middle Paleolithic. *Journal of Human Evolution, 60*, 437–451.

Gamble, C., Davies, W., Pettitt, P., Hazelwood, L., & Richards, M. (2005). The archaeological and genetic foundations of the European population during the late glacial: Implications for "agricultural thinking". *Cambridge Archaeological Journal, 15*, 193–223.

Garcea, E. A. A. (2012). Modern human desert adaptations: A Libyan perspective on the Aterian complex. In J.-J. Hublin & S. P. McPherron (Eds.), *Modern origins: A North African perspective* (pp. 127–142). Dordrecht: Springer.

Geraads, D. (2008). Plio-Pleistocene carnivora of northwestern Africa: A short review. *Comptes Rendus Palevol, 7*, 591–599.

Geraads, D. (2010). Biogeographic relationships of Pliocene and Pleistocene north-western African mammals. *Quaternary International, 212*, 159–168.

Geyh, M. (2008). The handling of numerical ages and their random uncertainties. *Eiszeitalter und Gegenwart, 57*, 239–252.

Giraudi, C. (2005). Eolian sand in peridesert northwestern Libya and implications for Late Pleistocene and Holocene Sahara expansions. *Paleogeography, Paleoclimatology, Paleoecology, 218*, 161–173.

Glennie, K., & Singhvi, A. (2002). Event stratigraphy, paleoenvironment and chronology of SE Arabian deserts. *Quaternary Science Reviews, 21*, 853–869.

Grün, R., Stringer, C., McDermott, F., Nathan, R., Porat, N., Robertson, S., et al. (2005). U-series and ESR analyses of bones and teeth relating to the human burials from Skhul. *Journal of Human Evolution, 49*, 316–334.

Gunz, P., Bookstein, F. L., Mitteroecker, P., Stadlmayr, A., Seidler, H., & Weber, G. W. (2009). Early modern human diversity suggests subdivided population structure and a complex out-of-Africa scenario. *Proceedings of the National Academy of Sciences of America, 106*, 6094–6098.

Hadjouis, D. (1986). Présence du genre Kobus dans le Pléistocène supérieur d'Algérie. *L'Anthropologie, 90*, 317–320.

Hamdan, M. A. E. (2000). Quaternary travertines of Wadis Abu Had-Dib Area, eastern desert, Egypt: Paleoenvironment through field, sedimentology, age and isotopic study. *Sedimentology of Egypt, 8*, 49–62.

Hawkins, A. (2012). The Aterian of the Western Desert oases of Egypt: Adaptation to changing climatic conditions? In J.-J. Hublin & S. McPherron (Eds.), *Modern origins: a North African perspective* (pp. 157–175). Dordrecht: Springer.

Howe, B., Arambourg, C., & Briggs L. C. (1967). *The Paleolithic of Tangier, Morocco* (Peabody Museum, Harvard University Bulletin No. 22). Cambridge, Massachusetts: American School of Prehistoric Research.

Jacobs, Z., Meyer, M. C., Roberts, R. G., Aldeias, V., Dibble, H., & El Hajraoui, M. A. (2011). Single-grain OSL dating at La Grotte des Contrebandiers ("Smugglers' Cave"), Morocco: Improved age constraints for the Middle Paleolithic levels. *Journal of Archaeological Science, 38*, 3631–3643.

Jacobs, Z., Roberts, R. G., Nespoulet, R., El Hajraoui, M. A., & Debénath, A. (2012). Single-grain OSL chronologies for Middle Paleolithic deposits at El Mnasra and El Harhoura 2, Morocco: Implications for Late Pleistocene human-environment interactions along the Atlantic coast of northwest Africa. *Journal of Human Evolution, 62*, 377–394.

Jones, S., Farr, L., Barton, H., Drake, N. A., White, K. H., & Barker, G. (2011). Geoarchaeological patterns in the pre-desert and desert ecozones of northern Cyrenaica. *Libyan Studies, 42*, 11–19.

Kingdon, J. (1993). *Self-made man: Human evolution from Eden to extinction*. New York: Wiley.

Lahr, M. M. (1996). *The evolution of modern human cranial diversity: A study in cranial variation*. Cambridge: Cambridge University Press.

Lahr, M. M., & Foley, R. A. (1998). Towards a theory of modern human origins: Geography, demography and diversity in recent human evolution. *Yearbook of Physical Anthropology, 41*, 137–176.

Lahr, M. M., Armitage, S., Barton, H., Crivellaro, F., Drake, N., Foley, R., et al. (2008). DMP III: Pleistocene and Holocene paleoenvironments and prehistoric occupation of Fazzan, Libyan Sahara. *Libyan Studies, 39*, 263–294.

Larrasoaña, J. C. (2012). A northeast Saharan perspective on environmental variability in North Africa and its implications for modern human origins. In J.-J. Hublin & S. McPherron (Eds.), *Modern origins: A North African perspective* (pp. 19–34). Dordrecht: Springer.

Le Du, R. (1934). Station atérienne de l'Oued Djouf-el-Djebel. Région de Tébessa-Chéria. *Recueil des Notices et Memoires de la Societe Archeologique de la Province de Constantine, 62*, 201–217.

Lézine, A., & Casanova, J. (1991). Correlated oceanic and continental records demonstrate past climate and hydrology of North Africa (0-140 ka). *Geology, 19*, 307–310.

Lézine, A.-M., Hély, C., Grenier, C., Braconnot, P., & Krinner, G. (2011). Sahara and Sahel vulnerability to climate changes, lessons from Holocene hydrological data. *Quaternary Science Reviews, 30*, 3001–3012.

Lisker, S., Vaks, A., Bar-Matthews, M., Porat, R., & Frumkin, A. (2010). Late Pleistocene paleoclimatic and paleoenvironmental reconstruction of the Dead Sea area (Israel), based on speleothems and cave stromatolites. *Quaternary Science Reviews, 29*, 1201–1211.

Macaulay, V., Hill, C., Achilli, A., Rengo, C., Clarke, D., Meehan, W., et al. (2005). Single, rapid coastal settlement of Asia revealed by analysis of complete mitochondrial genomes. *Science, 308*, 1034–1036.

Maher, L. A. (2010). The Late Pleistocene of Arabia in relation to the Levant. In M. D. Petraglia & J. I. Rose (Eds.), *The evolution of human populations in Arabia: Paleoenvironments, prehistory and genetics* (pp. 187–204). London: Springer.

Martini, M., Sibilia, E., Zelaschi, C., Troja, S. O., Forzese, R., Gueli, A. M., et al. (1998). TL and OSL dating of fossil dune sand in the Uan Afuda and Uan Tabu rockshelters, Tadrart Acacus (Libyan Sahara). In M. Cremaschi & S. Di Lernia (Eds.), *Wadi Teshuinat – paleoenvironment and prehistory in south-western Fezzan* (pp. 67–73). Firenze: All'Insegna del Giglio.

Mattingly, D., Lahr, M., Armitage, S., Barton, H., Dore, J., Drake, N., et al. (2007). Desert migrations: People, environment and culture in the Libyan Sahara. *Libyan Studies, 38*, 115–156.

McBrearty, S., & Brooks, A. S. J. (2000). The revolution that wasn't: A new interpretation of the origin of modern human behavior. *Journal of Human Evolution, 39*, 453–563.

McDougall, I., Brown, F. H., & Fleagle, J. G. (2005). Stratigraphic placement and age of modern humans from Kibish, Ethiopia. *Nature, 433*, 733–736.

Mercier, N., Wengler, L., Valladas, H., Joron, J.-L., Froget, L., & Reyss, J.-L. (2007). The Rhafas cave (Morocco): Chronology of the Mousterian and Aterian archaeological occupations and their implications for Quaternary geochronology based on luminescence (TL/OSL) age determinations. *Quaternary Geochronology, 2*, 309–313.

Michel, P., & Wengler, L. (1993). Le site paléontologique et archéologique de Doukkala II (Maroc, Pléistocène moyen et supérieur): Premier jalon en Afrique du Nord d'un comportement humain assimilable à "charognage contrôlé et actif". *Comptes Rendus de l'Académie des Sciences, Paris, 317, Série II*, 557–562.

Morel, J. (1974). La Station éponyme de L'oued Djebbana à Bir-El-Ater - est Algérien. *L'Anthropologie, 78*, 53–81.

Morel, J. (1978). L'industrie atérienne de l'Oued Djouf El Djemel. Comparaison avec l'industrie de l'Oued Djebbana. Le complexe atérien du Maghreb Oriental. *Bulletin de la Société Préhistorique Française, 75*, 487–500.

Nespoulet, R., Hajraoui, M. A., Amani, F., Ben Ncer, A., Debénath, A., El Idrissi, A., et al. (2008). Paleolithic and Neolithic occupations in the Témara Region (Rabat, Morocco): recent data on hominin contexts and behavior. *African Archaeological Review, 25*, 21–39.

Osborne, A. H., Vance, D., Rohling, E. J., Barton, N., Rogerson, M., & Fello, N. (2008). A humid corridor across the Sahara for the migration "Out of Africa" of early modern humans 120,000 years ago. *Proceedings of the National Academy of Sciences of America, 105*, 16444–16447.

Osmond, J., & Dabous, A. A. (2004). Timing and intensity of groundwater movement during Egyptian Sahara pluvial periods by U-series analysis of secondary U in ores and carbonates. *Quaternary Research, 61*, 85–94.

Parker, A. (2009). Pleistocene climate change in Arabia: Developing a framework for Hominin dispersal over the last 350 ka. In M. D. Petraglia & J. I. Rose (Eds.), *The evolution of human populations in Arabia: Paleoenvironments, prehistory and genetics* (pp. 39–49). New York: Springer.

Parker, A., & Rose, J. (2008). Climate change and human origins in southern Arabia. *Proceedings of the Seminar for Arabian Studies, 38*, 25–42.

Petit-Maire, N., Reyss, J., & Fabre, J. (1994). A last interglacial paleolake in the hyperarid Sahara of Mali (23 N). *Comptes Rendus – Academie des Sciences, Serie II, 319*, 805–809.

Preusser, F. (2009). Chronology of the impact of Quaternary climate change on continental environments in the Arabian Peninsula. *Comptes Rendus Geoscience, 341*, 621–632.

Raynal, J. P. (2001). The earliest occupation of North-Africa: The Moroccan perspective. *Quaternary International, 75*, 65–75.

Revel, M., Ducassou, E., Grousset, F. E., Bernasconi, S. M., Migeon, S., Revillon, S., et al. (2010). 100,000 Years of African monsoon variability recorded in sediments of the Nile margin. *Quaternary Science Reviews, 29*, 1342–1362.

Reygasse, M. (1921–1922). Etudes de Palethnologie Maghrébine (deuxième série). Recueil des *Notices et Mémoires de la Société Archéologique, Historique et Géographique de Constantine, 53*, 159–204.

Richter, D., Moser, J., Nami, M., Eiwanger, J., & Mikdad, A. (2010). New chronometric data from Ifri n'Ammar (Morocco) and the chronostratigraphy of the Middle Paleolithic in the Western Maghreb. *Journal of Human Evolution, 59*, 672–679.

Richter, D., Moser, J., & Nami, M. (2012). New data from the site of Ifri n'Ammar (Morocco) and some remarks on the chronometric status of the Middle Paleolithic in the Maghreb. In J.-J. Hublin & S. P. McPherron (Eds.), *Modern origins: A North African perspective* (pp. 61–78). Dordrecht: Springer.

Rídl, J., Edens, C. M., & Cerný, V. (2010). Mitochondrial DNA structure of Yemeni population: Regional differences and the implications for different migratory contributions. In M. D. Petraglia & J. I. Rose (Eds.), *The evolution of human populations in Arabia: Paleoenvironments, prehistory and genetics* (pp. 269–279). London: Springer.

Romer, A. S. (1938). Mammalian remains from some Paleolithic stations in Algeria. In A. W. Pond, L. Chapuis, A. S. Romer, & F. C. Baker (Eds.), Prehistoric habitation sites in the Sahara and North Africa (pp. 165–184). Beloit College, Beloit WI: Logan Museum Bulletin, 5.

Rose, J. I., & Usik, V. I. (2010). The "Upper Paleolithic" of South Arabia. In M. D. Petraglia & J. I. Rose (Eds.), *The evolution of human populations in Arabia: Paleoenvironments, prehistory and genetics* (pp. 169–187). London: Springer.

Roset, J.-P. (2005). El-Akarit et le Paleolithique moyen en Tunisie. In Ministère des Affaires etrangeres (Ed.), *Archeologies, 20 ans de recherches françaises dans le monde* (pp. 225–226). Paris: ADPF.

Rowan, J. S., Black, S., Macklin, M. G., Tabner, B. J., & Dore, J. (2000). Quaternary environmental change in Cyrenaica evidenced by U-Th, ESR and OSL dating of coastal alluvial fan sequences. *Libyan Studies, 31*, 5–16.

Ruhlman, A. (1936). *Les grottes prehistoriques d'El Khenzira (region de Mazagan)* Publications du Service des antiquites du Maroc No. 2). Morocco: Service des antiquites du Maroc.

Ruhlmann, A. (1951). *La grotte préhistorique de Dar es-Soltan* (Hesperis No. 11). Paris: Institut de Hautes Etudes Marocaines.

Scerri, E. M. L. (2013). The Aterian and its place in the North African Middle Stone Age. *Quaternary International, 300*, 111–130.

Schwenninger, J., Collcutt, S., Barton, N., Bouzouggar, A., Clark-Balzan, L., Abdeljalil, M., et al. (2010). A new luminescence chronology for Aterian cave sites on the Atlantic coast of Morocco. In E. A. A. Garcea (Ed.), *South-eastern Mediterranean peoples between 130,000 and 10,000 years ago* (pp. 18–36). Oxford: Oxbow.

Smith, J. R. (2012). Spatial and temporal variation in the nature of Pleistocene pluvial phase environments across North Africa. In J.-J. Hublin & S. P. McPherron (Eds.), *Modern origins: A North African perspective* (pp. 35–47). Dordrecht: Springer.

Smith, J. R., Giegengack, R., Schwarcz, H. P., McDonald, M., Kleindientst, M. R., Hawkins, A. L., et al. (2004). A reconstruction of Quaternary pluvial environments and human occupations using stratigraphy and geochronology of fossil-spring tufas, Kharga Oasis. *Egypt. Geoarchaeology, 19*, 407–439.

Smith, T. M., Tafforeau, P., Reid, D. J., Grün, R., Eggins, S., Boutakiout, M., et al. (2007). Earliest evidence of modern human life history in North African early *Homo sapiens*. *Proceedings of the National Academy of Sciences of America, 104*, 6128–6133.

Stringer, C. (2000). Coasting out of Africa. *Nature, 405*, 24–27.

Sultan, M., Sturchio, N., Hassan, F., Hamdan, M., Mahmood, A. M., Alfy, Z. E., et al. (1997). Precipitation source inferred from stable isotopic composition of Pleistocene groundwater and carbonate deposits in the Western desert of Egypt. *Quaternary Research, 48*, 29–37.

Szabo, B. I., Haynes, C. V., & Maxwell, T. A. (1995). Ages of Quaternary pluvial episodes determined by uranium-series and radiocarbon dating of lacustrine deposits of eastern Sahara. *Paleogeography, Paleoclimatology, Paleoecology, 113*, 227–242.

Texier, J., Huxtable, J., Rhodes, E., Miallier, D., & Ousmoi, M. (1988). Nouvelles données sur la situation chronologique de l'Atérien du Maroc et leurs implications. *Comptes Rendus – Academie des Sciences, Serie II, 307*, 827–832.

Thorp, M., Glanville, P., Stokes, S., & Bailey, R. (2002). Preliminary optical and radiocarbon age determinations for Upper Pleistocene alluvial sediments in the southern Anti Atlas Mountains, Morocco. *Comptes Rendus- Academie des Sciences Geoscience, 334*, 903–908.

Tishkoff, S. A., Reed, F. A., Friedlaender, F. R., Ehret, C., Ranciaro, A., et al. (2009). The genetic structure and history of Africans and African Americans. *Science, 324*, 1035–1044.

Tillet, T. (1985). The Paleolithic and its environment in the northern part of the Chad basin. *African Archaeological Review, 3*, 163–177.

Tjallingii, R., Claussen, M., Stuut, J.-B. W., Fohlmeister, J., Jahn, A., Bickert, J. T., et al. (2008). Coherent high- and low-latitude control of the northwest African hydrological balance. *Nature Geoscience, 1*, 670–675.

Toggweiler, J. R., & Russell, J. (2008). Ocean circulation in a warming climate. *Nature, 451*, 286–288.

Vaufrey, R. (1936). Stratigraphie Capsienne, Swiatowit, Vol. XVI, 1934–1935.

Vaufrey, R. (1954). La Faune d'El Guettar. *Karthago, 5*, 83–87.

Vaufrey, R. (1955). *Préhistoire de l'Afrique: Maghreb*. Paris: Masson.

Waldmann, N., Torfstein, A., & Stein, M. (2010). Northward intrusions of low- and mid-latitude storms across the Saharo-Arabian belt during past interglacials. *Geology, 38*, 567–570.

Walter, R. C., Buffler, R. T., Bruggemann, J. H., Guillaume, M. M. M., Berhe, S. M., Negassi, B., et al. (2000). Early human occupation of the Red Sea coast of Eritrea during the last interglacial. *Nature, 405*, 65–69.

Weisrock, A., Wengler, L., Mathieu, J., Ouammou, A., Fontugne, M., Mercier, N., et al. (2006). Upper Pleistocene comparative OSL, U/Th and 14C datings of sedimentary sequences and correlative morphodynamical implications in the south-western Anti-Atlas (Oued Noun, 29° N, Morocco). *Quaternaire, 17*, 45–59.

Weisrock, A., Rousseau, L., Reyss, J. L., Falguères, C., Ghaleb, B., Bahain, J.-J., et al. (2008). Travertins de la bordure nord du Sahara marocain: Dispositifs morphologiques, datations U/Th et indications paléoclimatiques. *Géomorphologie: Relief. Processus, Environnement, 3*, 153–168.

Wendorf, F., Schild, R., & Close, A. E. (1993). Summary and conclusions. In F. Wendorf, R. Schild, & A. Close (Eds.), *Egypt during the last Interglacial, the Middle Paleolithic of Bir Tarfawi and Bir Sahara East* (pp. 552–573). New York: Plenum Press.

Wengler, L. (2006). Innovations et normes techniques dans le Paléolithique moyen et supérieur du Maghreb: Une alternative aux migrations? In L. Astruc, F. Bon, V. Léa, P.-Y. Milcent, & S. Philbert (Eds.), *Normes techniques et pratiques sociales, de la simplicité des outillages pré- et protohistoriques* (pp. 93–105). Antibes: APDCA.

White, K., Drake, N., Millington, A., & Stokes, S. (1996). Constraining the timing of alluvial fan response to Late Quaternary climatic changes, southern Tunisia. *Geomorphology, 17*, 295–304.

Wrinn, P. J., & Rink, W. J. (2003). ESR dating of tooth enamel from Aterian levels at Mugharet el 'Aliya (Tangier, Morocco). *Journal of Archaeological Science, 30*, 123–133.

Zielhofer, C., Faust, D., & Linstädter, J. (2008). Late Pleistocene and Holocene alluvial archives in the southwestern Mediterranean: Changes in fluvial dynamics and past human response. *Quaternary International, 181*, 39–54.

Chapter 7
Climate, Environment, and Population Dynamics in Pleistocene Sahara

Emanuele Cancellieri, Mauro Cremaschi, Andrea Zerboni, and Savino di Lernia

Abstract The study area witnessed alternate paleoenvironmental and population events influenced by glacial/interglacial conditions. Paleosols, relict fluvial bodies, lacustrine carbonatic deposits, sand dunes, and other features underline the severely fluctuating activity of water resources. The study region (SW Libya) provides two different data sets: (1) two stratified, dated, Middle Stone Age/Aterian sites; and (2) hundreds of surface lithic scatters rarely associated with paleoenvironmental proxies. Early/Middle Pleistocene human occupation is presumable, but the bulk of evidence is from the late Middle/Late Pleistocene. Productive environments possibly housed human groups with a Late Acheulean technology during MIS 7. Most of the MSA evidences are barely diagnostic from a techno-typological point of view. Exceptions are made for scanty but precise similarities with sub-Saharan early MSA finds, suggesting the presence of modern humans in MIS 6, and for the Aterian, an example of MIS 4 arid landscape adaptation. Although MIS 3/2 post-Aterian human presence is not demonstrable, signs of a generalized LSA technology are recognizable in the Messak, where stony raw materials could have attracted task-specific temporary occupants.

Keywords Acacus and Messak • Arid landscapes adaptation • Central Sahara • Large-scale survey • Lithic technology • Quaternary environments and paleoclimate • Southwest Libya

Background

The long chronicle of Pleistocene geological and archaeological research in the Fezzan, southwestern Libya, started in the 1930s with Italian teams (Graziosi 1937), followed by French groups (Petit-Maire 1982) and by independent or isolated researchers, who significantly increased the available dataset (e.g., Le Quellec 1998).

However, the beginning of systematic research in our study region (Fig. 7.1) – the Tadrart Acacus, Messak and surrounding areas – started in the early 1990s, thanks to the survey and excavations carried out by the Italian-Libyan Mission of Sapienza University of Rome (e.g., Cremaschi and di Lernia 1998b). This research expanded in the following years to areas that had already been explored (Cremaschi and di Lernia 2000; Anag et al. 2002), or were about to be explored (Anag and di Lernia 2007), for oil prospection and extraction by oil companies. Major foci of these missions were actions related to cultural heritage rescue, and the assessment of the archaeological risks. Into this framework should also be inserted the surveys carried out by British teams in different locations of the Fezzan (Lahr et al. 2010).

Specific, question-oriented research programs have considered the analysis of Aterian occupation (Cremaschi et al. 1998; Garcea 2001) and the study of human dispersal across the Sahara (Osborne et al. 2008; Drake et al. 2011; Drake

E. Cancellieri · S. di Lernia (✉)
Dipartimento di Scienze dell'Antichità, Sapienza Università di Roma, Via dei Volsci 122, 00185 Rome, Italy
e-mail: savino.dilernia@uniroma1.it

E. Cancellieri
Istituto Italiano di Paleontologia Umana, c/o Museo Civico di Zoologia, via Ulisse Aldrovandi 18, 00197 Rome, Italy
e-mail: emanuele.cancellieri@uniroma1.it

M. Cremaschi · A. Zerboni
Dipartimento di Scienze della Terra "A. Desio", Università degli Studi di Milano, Via L. Mangiagalli 34, 20133 Milan, Italy
e-mail: mauro.cremaschi@unimi.it

A. Zerboni
e-mail: andrea.zerboni@unimi.it

S. di Lernia
School of Geography, Archaeology and Environmental Studies, University of the Witwatersrand, Private Bag 3, Johannesburg, 2050, South Africa

Sacha C. Jones and Brian A. Stewart (eds.), *Africa from MIS 6-2: Population Dynamics and Paleoenvironments*, Vertebrate Paleobiology and Paleoanthropology, DOI 10.1007/978-94-017-7520-5_7

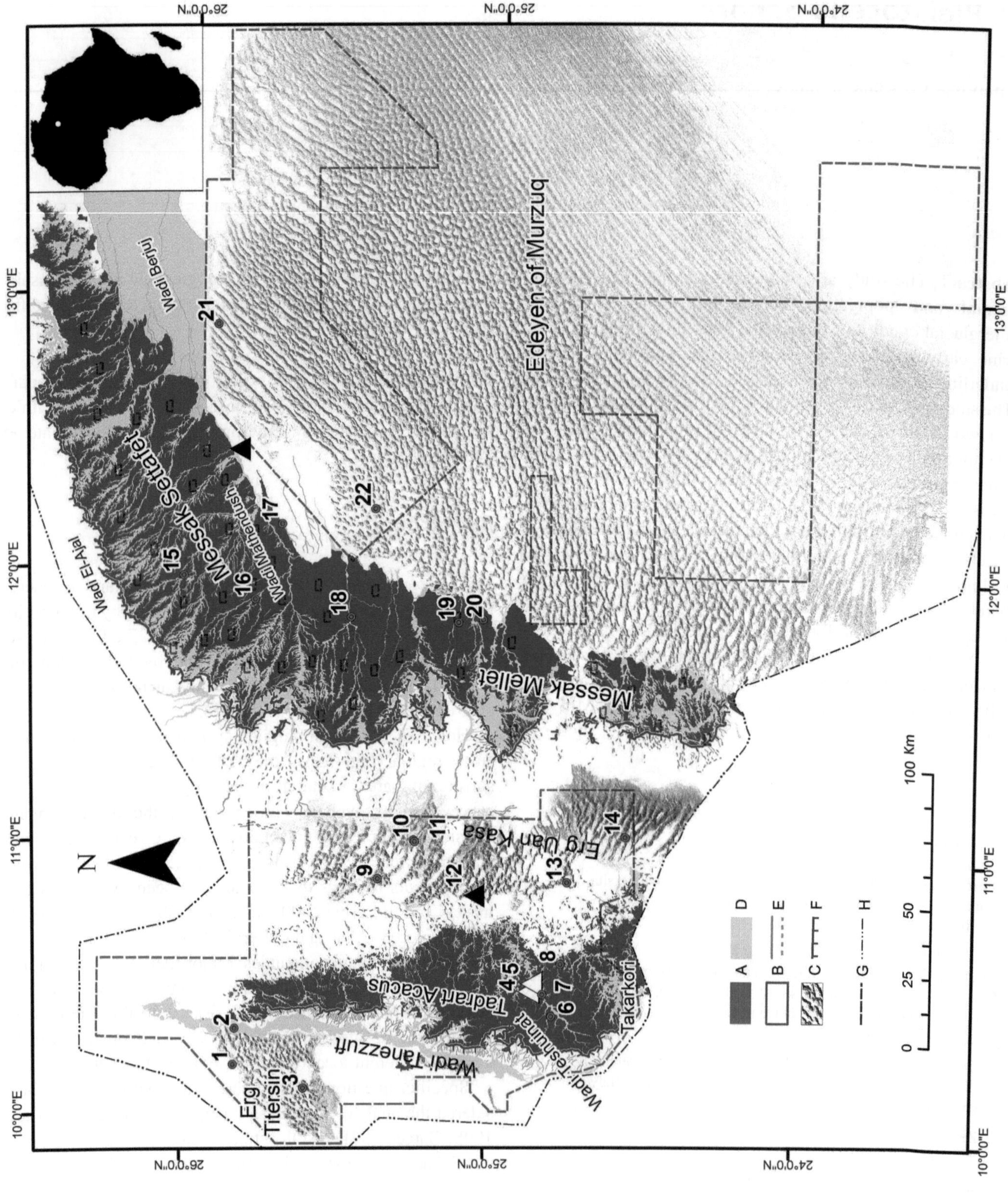

Messak Settafet
Messak Mellet
Edeyen of Murzuq
Wadi Berjuj
Wadi El-Ajal
Wadi Mathendush
Erg Uan Kasa
Tadrart Acacus
Wadi Tanezzuft
Wadi Teshuinat
Takarkori
Erg Titersin
N
A
B
C
D
E
F
G
H
0 25 50 100 Km
10°0'0"E
11°0'0"E
12°0'0"E
13°0'0"E
24°0'0"N
25°0'0"N
26°0'0"N

Fig. 7.1 Simplified geomorphological map of the study area. The localities and archaeological units cited in the text (*numbers*), origin of U/Th (*black triangles*) and luminescence dating (*white triangles*) are indicated. *Key A* sandstone massifs; *B* pediments and residual surfaces; *C* dune fields; *D* main valleys and fluvial deposits; *E* fluvial net; *F* escarpment; *G* main survey areas; *H* limits of the map. *1* 96/295; *2* 96/221; *3* 01/134; *4* Uan Afuda; *5* UanTabu; *6* AT8; *7* TH126; *8* Uan Telocat; *9* 03/92; *10* 03/104; *11* 03/105; *12* TH101; *13* 94/84; *14* 97/71; *15* MRSR2000/231; *16* MT114; *17* MT22; *18* Wadi Imrawen; *19* 10/578; *20* Wadi Tidwa; *21* 94/6; *22* M4a/114

and Breeze 2016), undertaken by different teams until the beginning of the conflict in Libya (in March 2011).

The research on Pleistocene archaeology is strongly connected to paleoclimatic and paleoenvironmental studies, given the peculiar nature of Saharan contexts (e.g., Petit-Maire 1988; Castañeda et al. 2009). At these latitudes, minor variations in rainfall can durably alter delicate equilibriums; however, the idea of a "green Sahara" should most likely be challenged, and the equation "wet = occupation versus dry = abandonment" should be revised. In fact, Saharan topographical contexts reveal an incredible variety of econiches that allow human frequentation even today.

Although hampered by the scarceness of stratified contexts and datable materials, we argue that a research strategy involving fieldwork on a large geographical scale is useful. This allows for the incorporation of diverse topographical and morphological contexts, and identification of the macro-dynamics of land use, lines of human penetration, and mechanisms of human dispersal.

The Study Area: Geologic and Physiographic Elements

Located in Southwest Libya between the latitudes 26° and 24° N (Fig. 7.1), the area of research geologically constitutes the wide geosyncline of the Murzuq Basin, composed of Paleozoic and Mesozoic sandstone, marls, and limestone lying upon the intrusive formation of the Tassili (Desio 1936; Hallet 2002; El-ghali 2005).

The Tadrart Acacus, at the western reach of the syncline, is an elongated massif that runs north to south and is bisected by a fossil drainage network (Zerboni et al. 2015). The massif is delimited to the west by a scarp, and grades through a pediment to the east to the dunes of Erg Uan Kasa. Due to solutional processes, the walls flanking the wadis (dried river valleys) in the massif are characterized by the presence of cavities and rock shelters. The Messak is a plateau gently tilted eastward and delimited to the NW by an abrupt scarp up to 160 m high (Perego et al. 2011). The plateau is interpreted as a relict of a Tertiary peneplain with Inselberg-type relief (e.g., Busche 1980; Cremaschi 1998; Zerboni 2008), bisected by a dense net of sinuous wadis with a dendritic pattern. Hamada surfaces, a particular type of desert landscape consisting of high rocky plateaus, are typical of the Messak region. Its clasts (also including lithic artifacts) are coated by a continuous layer of dark rock varnish (Cremaschi 1996; Zerboni 2008).

Wide dune fields (edeyen or erg) cover most of the lowlands surrounding the Acacus and Messak massifs. Consisting of different types of coalescent dunes, the sand seas present a complex morphology. The largest sandy domes rise up to 100 m, with a spacing of several hundred meters. The Edeyen of Murzuq lies in the center of the homonymous basin. The Erg Uan Kasa is located between the Acacus and the Messak Mountains, and the Erg Titersin is located in a depression that is bounded to the north and west by the rock terraces of the Algerian Tassili. Today, there is no evidence for water table outcropping within the limits of the dune fields.

Climate and Paleoclimate

The present climate is hyperarid. The mean annual temperature is 30 °C, and the mean annual rainfall is between 0 and 20 mm (Fantoli 1937). The Holocene and Middle–Late Pleistocene interglacials were characterized by environmental conditions that were wetter than today, and the main hydrological changes were triggered by the expansion and withdrawal of the southwest African monsoon (e.g., Gasse 2000; Drake et al. 2011).

Interdisciplinary studies of the Pleistocene in SW Libya started with paleoenvironmental research in the Wadi Shati, at the northern margin of the Murzuq Basin. This research revealed the existence of an extensive lake formation dating to the Middle and Late Pleistocene, with associated Middle Stone Age (MSA) artifacts (Petit-Maire 1982). The highstand of the lake was U/Th dated between 140 and 130 ka (MIS 5d and 5e), while subsequent reactivations of the basins at 90 and 40 ka are uncertain (Gaven 1982).

More recently, research has focused on the lacustrine terraces distributed throughout the Wadi El-Ajal and the area surrounding Murzuq. Here, several carbonatic terraces accumulated during humid episodes at >420, 380–290, 260–205, and 140–125 ka, roughly coincident with the MIS 9, 7, and 5e interglacial periods. Moreover, Drake et al. (2011) obtained OSL results indicating a wet phase related to the last interglacial MIS 5 (119–98 ka). After the termination of the Last Glacial Maximum, the most recent recharge of the aquifers occurred during the Holocene African Humid Period (e.g., Cremaschi et al. 2010).

A further consideration of the late Quaternary paleoclimate of the central Sahara concerns the variation in the intensity of precipitation. Analyzing the rate of accumulation of subsequent members from the lacustrine deposits in the Murzuq-Shati area (Al Mahruqah Formation), Geyh and Thiedig (2008) noticed a trend of shrinking lake size across the Quaternary that was even more pronounced in the Late Pleistocene and Early Holocene. This phenomenon is interpreted as a reduction of the intensity of humid episodes throughout the Pleistocene. Other authors (Szabo et al. 1995; Smith et al. 2004; Brookes 2010) confirmed a decreasing trend of the lacustrine sedimentation in northeastern Africa over time during the late Quaternary.

Methods of Fieldwork in Desert Environments: Potentials and Limits

Our systematic survey has been characterized since its start by the need to make available a basic archaeological assessment of the area (for example, virtually no sites were known at the beginning of the 1990s in the Murzuq and Uan Kasa sand seas), and to address the progressive impoverishment of the surface documentation. The sampling of lithic artifacts has always been kept to a minimum.

Two distinct methodologies of fieldwork have been applied: *extensive* reconnaissance and *intensive* survey: the former covering large areas and aiming at screening the geoarchaeological potential, and the latter normally carried out on foot within specific research-oriented programs (Cremaschi and di Lernia 1998b, 2000).

In the sand seas, the extensive surveys have generally been carried out by vehicle, with walking control, where lacustrine sediments and stone concentrations are perfectly visible from a great distance. Isolated findings or lightly buried palimpsests are, of course, less visible and therefore less documented. In specific topographic contexts, such as the wadis of the Acacus Mountains, we performed intensive surveys systematically targeting all of the rock shelters and caves (Fig. 7.2d); here, the Holocene fill hides the rare Pleistocene deposits, and only deep soundings revealed the presence of *in situ* Pleistocene contexts. Intensive surveys have also been conducted as an assessment of the archaeological potential of areas to be prospected and/or exploited for oil drilling, as was the case in the Murzuq sand sea (Anag and di Lernia 2007) and the Messak plateau (Cremaschi and di Lernia 2000; di Lernia 2011). The intensive surveys have normally been carried out on foot by teams following a sampling procedure that allows for the statistical management of collected data. The sampling principles were always defined by spatial criteria, whether arbitrary or natural. Examples include the adoption of linear survey strips within one out of three interdune basins in the Edeyen of Murzuq (Anag and di Lernia 2007), or within 4 × 2 km transects in the Messak (Gallinaro et al. 2012; Biagetti et al. 2013). In the latter case, we added the field documentation of a square area of 10 × 10 m every 100 m along selected strips in response to the very high density of artifacts that were encountered (Cancellieri and di Lernia 2013).

The integration of data from both of these field activities is used to map the distribution of Pleistocene contexts, which can be matched against their topographic features. Although the scale of resolution is often low, the extent of territorial research is interesting (Table 7.1).

From a methodological viewpoint, however, the principal issue to be addressed in this type of environment is the definition of "site" (e.g., Foley 1981; Olszewski et al. 2010). The clusters of artifacts in the sand seas are often easily identifiable discrete entities based on criteria such as size, shape, contiguity, and abundance (Fig. 7.2f). The identification of sites is not so easy on the continuous stony surfaces, generated by dismantling and selective wind erosion, of the Messak and Acacus Mountains. Here, artifacts related to distinct occupations mostly lie within palimpsests that exhibit virtually no discontinuity (Fig. 7.2a). In response to this, one of the best methodological approaches is to split the surface record according to regular and arbitrary grids.

Table 7.1 Extent estimates (km^2) of extensive and intensive surveys carried out in the study area

Region	Size	%	Extensive		Intensive		Total extent	%
			Extent	%	Extent	%		
Erg Titersin	1200	*1.3*	700	*58.3*	–	–	700	*58.3*
Wadi Tanezzuft	1200	*1.3*	1000	*83.3*	–	–	1000	*83.3*
Tadrart Acacus	5300	*5.9*	–	–	1600	*30.2*	1600	*30.2*
Erg Uan Kasa	4400	*4.9*	3200	*72.7*	–	–	3200	*72.7*
Edeyen of Murzuq	59200	*66.1*	4600	*7.8*	3200	*5.4*	7800	*13.2*
Messak	18200	*20.4*	–	–	700	3.8	700	3.8
Total	89500	*100.0*	9500	*10.6*	5500	*6.1*	15000	*16.7*

Fig. 7.2 Diverse physiographic units found throughout the study area. *Key* **a** Messak, plateau with desert pavement; **b** Messak, endorheic depression; **c** Messak, wadi system; **d** Acacus, Uan Afuda Cave; **e** Acacus, escarpment, viewed from Wadi Tanezzuft; **f** Murzuq, interdune basin; **g** Tanezzuft, dune system

For these reasons, the Pleistocene evidence in our study area is uneven, ranging from isolated handaxes to stratified *in situ* assemblages. As a consequence, we prefer to avoid the term "site," and conservatively adopt the term "Archaeological Unit" (AU). Different AUs have been identified, and when possible, their techno-typological features have been recorded and described following different sources (e.g., Bordes 1961; Tixier 1963; Inizan et al. 1999).

The long time span over which the data was collected (upon which this review builds) witnessed substantial changes in research priorities and methods. Large portions of our data set have characteristics of homogeneity and comparability, especially when dealing with single macro-areas (i.e., southern Murzuq, Messak, or central Acacus). If taken as a whole, these data can also be reduced to a small number of categories. Doing so results in a significant loss of data, especially in the case of site categorization. Nevertheless, we feel that such categorization is crucial to this regional review, and is conducted according to the existence of recognizable spatial limits, either natural or artificial. This categorization lumps together heterogeneous aspects, and does not aim at a classification of "site" attributes. Rather, this approach aims to categorize "data" attributes, and splits these attributes into three broad blocks. To summarize, we have used the following categories:

1. "Isolated AU": isolated recorded artifact, usually bearing a strong chrono-cultural connotation. In most cases, these artifacts are *absolutely* isolated (namely: within an almost empty area); but this aspect is not accounted for here.
2. "Ephemeral AU": the spatial limits of the dispersion are not recognizable or unknown. An important distinguishing feature of ephemeral AUs is that the artifacts are highly scattered, and the density is very low overall.
3. "Solid AU": the spatial limits of the dispersion are recognizable or known and are defined by (i) natural features (e.g., caves, rock shelters, or endorheic depressions); (ii) a well-defined spatial distribution (e.g., a dense scatter with a recognizable size and shape); or (iii) artificial criteria (e.g., excavation or sampling spot).

Table 7.2 Number and type of located AUs, arranged by region

	Isolated		Ephemeral		Solid		Total	
	n	%	n	%	n	%	n	%
Erg Titersin	3	*1.0*	9	*3.6*	12	*1.1*	24	*1.5*
Wadi Tanezzuft	3	*1.0*	9	*3.6*	2	*0.2*	14	*0.9*
Tadrart Acacus	2	*0.7*	5	*2.0*	34	*3.1*	41	*2.5*
Erg Uan Kasa	6	*2.1*	28	*11.1*	22	*2.0*	56	*3.4*
Edeyen of Murzuq	205	*70.4*	201	*79.8*	53	*4.9*	459	*28.3*
Messak	72	24.7	–	–	958	*88.6*	1030	63.5
Total	291	*100.0*	252	*100.0*	1081	*100.0*	1624	*100.0*

In this sense, solid AU constitutes the best approximation of the common concept of "site," while isolated and ephemeral AUs may be considered similar to the common concept of "find-spot." Based on these criteria, some basic differences are evident (Table 7.2).

The Acacus and Messak samples are characterized exclusively by intensive surveys. This choice of survey methods is also due to their specific archaeological record: the Acacus is characterized by rock shelters and caves (and partially by Late Tertiary–Pleistocene paleosurfaces), and the Messak plateau is characterized by the typical hamada surface, which has low archaeological visibility and is very difficult to access by car. The incidence of surveyed areas is very low for the Messak (3.8%), whereas in the Acacus, approximately 30% of the mountain range has been surveyed.

The sand seas of Uan Kasa and Titersin, where extensive surveys by vehicle are especially remunerative due to both ease of accessibility and high visibility of surface remains, have been surveyed in high percentages (>50%). The Edeyen of Murzuq, the largest sand sea in Libya, has been subject to a combination of extensive and intensive surveys (13.2%). Finally, the Tanezzuft valley has been largely surveyed, together with other adjacent regions.

Altogether, nearly 17% of the entire study area has been visited and surveyed with different degrees of resolution. Unsurprisingly, due to the physical and logistical constraints outlined above, large parts of the study area have not been visited at all.

The number of located AUs (Table 7.2) obviously reflects the type of field strategy employed, which is adapted to the specific physiographical context. As an example, isolated finds are very common in the Murzuq, where intensive survey, high visibility, and excellent accessibility allowed the collection of single items. The number of solid AUs increases in the areas where intensive survey has been adopted, with a significant difference between the Murzuq sand sea and the Messak Plateau. The former shows apparently discrete assemblages, whereas most of the solid AUs of the latter should be considered as contexts arbitrarily defined at fixed intervals along survey strips or, far more rarely, in different locations of one general area.

Pleistocene Paleoenvironments in the SW Fezzan

Landforms and Stratigraphic Contexts

In comparison with the evidence of Holocene wet periods (e.g., Cremaschi and Zerboni 2009), the geological indications of Pleistocene morphogenesis are rare, poorly preserved, and rather discontinuous. The evidence consists of remnants of weathered paleosurfaces (often associated with paleosols), slope deposits, weathered dunes, and fluvial and lacustrine deposits, which originated under contrasting (wet and dry) paleoenvironmental conditions. Notwithstanding their rare occurrence, the deposits and landforms dating to the Pleistocene are, without exception, associated with stone artifacts, indicating an extensive presence of humans in the area throughout the Quaternary.

In the mountain areas, most of the Stone Age AUs were found on paleosurfaces. These paleosurfaces originated at, or after, the beginning of the Pleistocene (Busche 2010) and are the outcome of geomorphological processes under alternating climatic conditions. Etching processes promoted by pluvial conditions led to the formation of thick paleosols of the tropical types (Zerboni et al. 2011), such as *sols ferralitiques* (sensu Duchaufour 1983), whereas subsequent arid periods favored wind erosion and the dismantling of pedological covers. A single U/Th measure of 236 +93.8/–46.6 ka (roughly MIS 7) was obtained on pedogenic carbonatic concretions (pedogenic calcrete) from a ferralitic paleosol. Despite the large degree of uncertainty, the result is of some interest because it might indicate that these paleosols were already developed and subjected to processes of aridification during the Middle Pleistocene.

On the top of the plateau, the ferralitic paleosol is discontinuously preserved and covered by weathered aeolian sand related to a Middle Pleistocene desert expansion. At the interface between the paleosol and superposed weathered aeolian sand, Acheulean artifacts were found (MT114). The situation appears to be rather different inside the valleys cut into the Messak Plateau (Perego et al. 2011). Here, gravel megabars

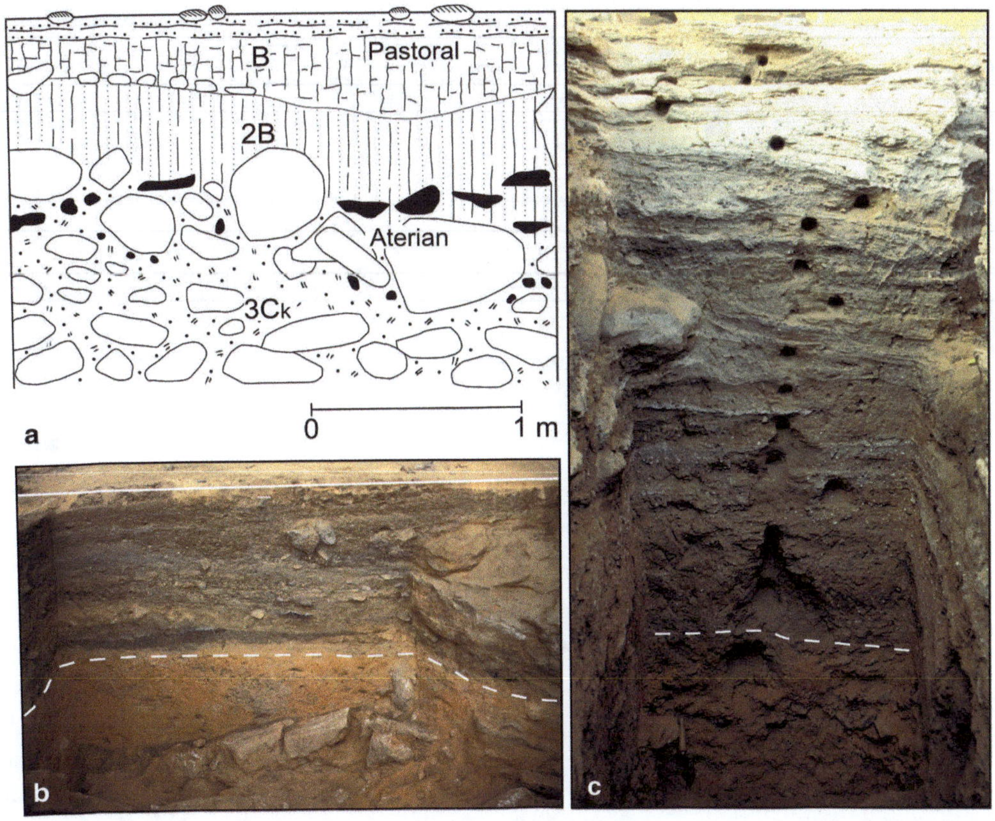

Fig. 7.3 Some of the Aterian/MSA contexts described in the text. *Key* **a** The stratigraphic section at MT22 (modified from Cremaschi 1998). The complex sequence of soil horizons developed on sand, B and 2B, and gravel, 3Ck. The upper soil contains Pastoral Neolithic artifacts (Holocene). The 3Ck horizon developed on Pleistocene fluviatile gravel, and at its top Aterian artifacts (in *black*) are present. **b** The stratigraphic section at Uan Afuda. **c** The stratigraphic section at Uan Tabu. *Note* in (**b**) and (**c**) the change in texture and color at the transition between the Holocene and the Late Pleistocene parts of the sequences (indicated by *dashed lines*)

generally include allochthonous rolled Acheulean and MSA artifacts. At MT22 (Fig. 7.3a), along Wadi Mathendush, the gravel that included such artifacts was superposed by a dense cluster, interpreted as a workshop, of unweathered Aterian artifacts, subsequently buried by a Late Pleistocene aeolian sand (Cremaschi 1998; Cremaschi et al. 1998).

At the base of the Holocene stratigraphic sequence of Uan Afuda, Uan Tabu, and Uan Telocat in the central Acacus, Pleistocene aeolian sand associated with Stone Age tools was unearthed. At Uan Afuda, the sandy sequence is two meters thick and is intercalated by two levels of fallen blocks (Fig. 7.3b). At its top, a reddish paleosol is present. The base of the upper level of collapsed blocks included MSA/Aterian artifacts (di Lernia 1999).

Measures of the sand above the upper block layer provided TL dates of 70 ± 9.5 BP and 73.5 ± 10 ka, and an OSL date of 69 ± 7 ka; a further OSL date obtained from the sand covering the lower level of fallen blocks gave an age of 90 ± 10 ka (Cremaschi et al. 1998; Martini et al. 1998). In the Uan Tabu rock shelter, Pleistocene aeolian sand is less developed (Fig. 7.3c), but it includes a high concentration of Aterian artifacts (Garcea 2001). The aeolian sand was OSL dated to 61 ± 10 ka (Martini et al. 1998). Possible Aterian artifacts included in the aeolian sand were also observed at the base of the Uan Telocat sequence, but were not studied in detail.

The dating of the lower sand deposit at Uan Afuda refers to the beginning of dune aggradation inside the cave at the very end of MIS 5a. The measure of Uan Tabu and the upper layers of Uan Afuda date the Aterian frequentation of the area to MIS 4.

Remnants of rubified aeolian dunes associated with MSA and Acheulean artifacts were found at several locations (areas of Wadi Teshuinat and Takarkori). Residual fossil dunes associated with Acheulean artifacts also occur in the sand seas (e.g., 94/84 at Uan Kasa). Yet, most of the geological remains of the Pleistocene period consist of lacustrine marls and caliche, and, to a minor extent, of limited outcrops of deeply weathered paleosols and fluvial deposits, all of which stem from the pluvial periods of the Pleistocene (Zerboni et al. 2015). In the absence of direct dating, their approximate age is inferred from their geological context and the lithic assemblages to which they are related.

Fig. 7.4 Some examples of Pleistocene paleoenvironmental contexts. *Key* **a** buried and deeply weathered B horizon in Messak Settafet; **b** terrace of Pleistocene lacustrine deposit with a handaxe; **c** a caliche crust in the Edeyen of Murzuq displaying a possibly *in situ* handaxe; **d** rubified soil buried by fluvial deposits; **e** remnant of a rubified fossil dune emerging from the sand of the Erg Uan Kasa. Scale = 15 cm

In a much-eroded outcrop of lacustrine marls covered by a caliche crust (03/105), located in an interdune corridor of the Erg Uan Kasa, a rich Mode 1 lithic assemblage associated with well-preserved fauna, including turtles and large mammals, was discovered. A few similar artifacts were found in an outcrop of a deeply weathered paleosol emerging from a dune ridge in Erg Uan Kasa.

MSA artifacts associated with fauna (mainly small ungulates) of uncertain age were recovered in the Edeyen of Murzuq at M4a/114. These were found lying at the top of a rubified soil, cemented in an iron oxide crust and buried by fluvial deposits (Fig. 7.4d). While no direct dating is available, this AU must be regarded as quite old from a geological point of view, and it has to be referred to at least the pluvial stages of MIS 5, if not to a previous stage.

The most remarkable Pleistocene lacustrine deposits were observed in the interdune corridors at the northwestern edge of the Edeyen of Murzuq. These deposits wrap the upper terraces existing along these corridors (e.g., Zerboni et al. 2011). In one case, an Acheulean handaxe was collected *in situ*. Large lacustrine deposits associated with scatters of Acheulean lithic assemblages were also discovered at the northwest fringe of Erg Titersin.

Several concentrations of Acheulean artifacts also occur along the interdune corridors. These lie on bleached hydromorphic sand and small gravel (*serir*), and are highly reflective on satellite photographs. These areas are interpreted as deeply eroded paleosurfaces. The lithics are in a generally fresh state, and primary distributions are preserved at some locations. In some cases (03/92, 97/71, 03/104, and 94/6), the clusters of artifacts are associated with faunal remains consisting almost exclusively of laminae of elephant teeth – the only remains that have survived the process of weathering.

U/Th dating of a caliche crust at the top of a Pleistocene lacustrine deposit in the Erg Uan Kasa (TH 101) resulted in a date of 68.7 +4.4/–4.2 ka. This date is linked not to the lacustrine sedimentation (related to a wet phase), but instead to a subsequent dry period of high evapotranspiration fostering the caliche's formation. In this interpretation, the date fits well with those related to the sand ingression at Uan Afuda cave and Uan Tabu. However, the limited number of dates and their large standard deviation may mask a deeper complexity in wet/dry fluctuations during MIS 4.

Environmental Changes Throughout the Pleistocene

While rather fragmentary, evidences of "pluvial" climatic contexts, including deeply developed paleosols and lacustrine deposits, indicate the oldest Pleistocene formations recorded in the area and also include the earliest Stone Age industry (possibly Oldowan).

Pedogenic concretions at the top of the ferralitic paleosol at the western margin of the Messak are indicative of a phase of enhanced evaporation rate during MIS 7. However, the earliest well-preserved evidence of severe desert conditions consists of the widespread aeolian sand occurring at the top of the paleosols on the Messak Plateau, and the dune remnants inside the ergs and in the Acacus Mountains. This phase may be assigned to the late Middle Pleistocene, tentatively to MIS 6, and it was weathered in MIS 5. Most of the lacustrine deposits of the area may be attributed to this same pluvial period. Although in the absence of dating, their possible correlation with the limnites of the Al Mahruqah formation described by Petit-Maire (1982), and Geyh and Thiedig (2008) in the Murzuq-Shati areas, must remain speculative at the moment. Occurring in all of these geological contexts, the Acheulean complexes seem to span a long period including different climatic phases: wet conditions during which lake sedimentation was widespread over the whole area, and possibly the dry environment of MIS 6.

The gravel deposits inside the Messak wadis testify to strong rainfall during MIS 5. At MT22, the existence of an unweathered Aterian workshop indicates that at that time, fluvial aggradation ceased. Dry conditions indicated by generalized dune aggradation dominated the late MIS 4. While minor climatic fluctuations may not be excluded, generally dry conditions accompanied the Aterian frequentation of the area (cf. Drake and Breeze 2016). Except for limited outcrops of aeolian sand covering Aterian and MSA AUs, geological evidence that may be assigned to MIS 3 and MIS 2 is almost absent in the studied area. During these periods, it is likely that wind erosion largely prevailed (Swezey 2001).

Pleistocene Human Occupation: Technological Framework and Regional Distribution

Context and Technological Traits

The archaeological data are grouped and labeled following the "Mode" nomenclature (Clark 1977), adopted here to provide a synthetic techno-typological characterization. Our groups are defined as follows:

- Mode 1, assemblages with artifacts typologically classifiable as choppers/chopping tools, and flakes/flake tools that are clearly far more worn than the "younger" diagnostic items that are also found within a very few meters of them (Fig. 7.5a–c).
- Mode 2, assemblages with handaxes.
- Mode 2 + 3, assemblages with handaxes and Levallois artifacts.
- Mode 3, assemblages with Levallois artifacts
- Mode 3-Aterian, assemblages with tanged pieces, and/or foliate points, and possibly (though not necessarily) Levallois artifacts.
- Mode 4, assemblages with non-Levallois laminar artifacts, supposedly Pleistocene in age.

Approximately, 16% of the records lack diagnostic traits, and are classified as Pleistocene *sensu lato*. These records are included only in Tables 7.2 and 7.3.

Table 7.3 Number and incidence of AUs, arranged by technology and region

	Erg Titersin		Wadi Tanezzuft		Tadrart Acacus		Erg Uan Kasa		Edeyen of Murzuq		Messak		Total	
	n	%	n	%	n	%	n	%	n	%	n	%	n	%
Mode1	–	–	–	–	–	–	1	*1.8*	–	–	8	*0.8*	9	*0.6*
Mode 2	13	*54.2*	5	*35.7*	13	*31.7*	27	*48.2*	322	*70.2*	54	*5.2*	434	*26.7*
Mode 2 + 3	4	*16.7*	–	–	3	*7.3*	5	*8.9*	30	*6.5*	125	*12.1*	167	*10.3*
Mode 3	–	–	8	*57.1*	9	*22.0*	8	*14.3*	28	*6.1*	535	*51.9*	588	*36.2*
Mode3-Aterian	1	*4.2*	1	*7.1*	8	*19.5*	6	*10.7*	19	*4.1*	30	*2.9*	65	*4.0*
Mode 4	–	–	–	–	–	–	–	–	–	–	105	*10.2*	105	*6.5*
Pleistocene *sensu lato*	6	*25.0*	–	–	8	*19.5*	9	*16.1*	60	*13.1*	173	*16.8*	256	*15.8*
Total	24	*100*	14	*100*	41	*100*	56	*100*	459	*100*	1030	*100*	1624	*100*

Mode 1

There is increasing concern surrounding the recognition of early technology within surface scatters. This is for two fundamental reasons. First, choppers, chopping tools, and ordinary flakes refer to a simple technology, which is also the most widespread way to modify stones. Second, complex patterns of raw material management and reduction are present in the technological history of mankind for at least 3.3 Ma (Harmand et al. 2015); these are also carried out by flaking methods (i.e., discoid) that suggest a high degree of predetermination (De Lumley 2006), and are also attributed to more recent phases of human evolution. Due to the fact that a simple/archaic flaking process alone does not necessarily indicate early technology, when attempting an early attribution (possibly Oldowan), we adopted the cautionary principle of assessing weathering degree introduced above. According to this principle, nine AUs (Fig. 7.6) have been recognized, eight in the Messak (Fig. 7.5a, d), and one in the Erg Uan Kasa (Fig. 7.5e). Notably, the latter (03/105) also contained faunal remains of large mammals (Fig. 7.5f) and turtles in the same context. This might be assigned to a much stronger "pluvial environment," but the association with the lithics remains to be fully assessed.

Mode 2

Nearly half of this group (n. 200, 46%; 176 are from Murzuq) is composed of isolated AUs (Table 7.4; Fig. 7.6). Approximately, 21% (91) of the findings are solid AUs. These are mostly found in the sand seas and on the Messak Plateau. Two Acacus shelters yielded isolated artifacts. In the same area, open-air AUs are located within wadi alluvia, fossil dunes, and on structural surfaces. In the Messak, Mode 2 artifacts have been recovered in stratigraphic positions at MT22 unit 1 (Fig. 7.3a) and Wadi Tidwa (Cremaschi and di Lernia 1998a; Cremaschi et al. 1998).

Also, in the Messak, a series of AUs provided information about technology and typology. The AUs mostly returned bifaces. The tools are notches, denticulates, and scrapers on undifferentiated blanks. Quartzarenite is the only raw material used. The cores are both unpatterned and discoid (*sensu* Boëda 1993). The evidence provided by flakes attests to the use of the Kombewa-type method of lithic production. Moreover, we identified possible evidence for the application of the Tabelbala-Tachengit method, used in the Saharan Acheulean to produce large cleaver blanks.

Table 7.4 Technological features of AUs, arranged by region and type of AU

Group	Region	Type of archaeological unit							
		Isolated		Ephemeral		Solid		Total	
		n	%	n	%	n	%	n	%
Mode 2	Erg Titersin	3	*1.5*	5	*3.5*	5	*5.5*	13	*3.0*
	Wadi Tanezzuft	1	*0.5*	3	*2.1*	1	*1.1*	5	*1.2*
	Tadrart Acacus	1	*0.5*	1	*0.7*	11	*12.1*	13	*3.0*
	Erg Uan Kasa	2	*1.0*	11	*7.7*	14	*15.4*	27	*6.2*
	Edeyen of Murzuq	176	*88.0*	123	*86.0*	23	*25.3*	322	*74.2*
	Messak	17	*8.5*	–	–	37	*40.7*	54	*12.4*
	Total Mode 2	200	*100.0*	143	*100.0*	91	*100.0*	434	*100.0*
Mode 2 + 3	Erg Titersin	–	–	–	–	4	*2.7*	4	*2.4*
	Tadrart Acacus	–	–	–	–	3	*2.0*	3	*1.8*
	Erg Uan Kasa	–	–	2	*11.1*	3	*2.0*	5	*3.0*
	Edeyen of Murzuq	–	–	16	*88.9*	14	*9.4*	30	*18.0*
	Messak	–	–	–	–	125	*83.9*	125	*74.9*
	Total Mode 2 + 3	–	–	18	*100.0*	149	*100.0*	167	*100.0*
Mode 3	Erg Titersin	–	–	–	–	–	–	–	–
	Wadi Tanezzuft	2	*7.4*	5	*23.8*	1	*0.2*	8	*1.4*
	Tadrart Acacus	1	*3.7*	3	*14.3*	5	*0.9*	9	*1.5*
	Erg Uan Kasa	–	–	5	*23.8*	3	*0.6*	8	*1.4*
	Edeyen of Murzuq	10	*37.0*	8	*38.1*	10	*1.9*	28	*4.8*
	Messak	14	*51.9*	–	–	521	*96.5*	535	*91.0*
	Total Mode 3	27	*100.0*	21	*100.0*	540	*100.0*	588	*100.0*
Mode 3-Aterian	Erg Titersin	–	–	–	–	1	*3.8*	1	*1.5*
	Wadi Tanezzuft	–	–	1	*8.3*	–	–	1	*1.5*
	Tadrart Acacus	–	–	1	*8.3*	7	*26.9*	8	*12.3*
	Erg Uan Kasa	2	*7.4*	4	*33.3*	–	–	6	*9.2*
	Edeyen of Murzuq	10	*37.0*	6	*50.0*	3	*11.5*	19	*29.2*
	Messak	15	*55.6*	–	–	15	*57.7*	30	*46.2*
	Total Mode 3 - Aterian	27	*100.0*	12	*100.0*	26	*100.0*	65	*100.0*

Fig. 7.5 Findings from Mode 1, Mode 2, and Mode 2 + 3 archaeological units (AUs). Box (*top left*) examples of different weatherings and patinas on artifacts ascribable to different techno-typological classes, found within the same sample area (10 × 10 m) in the Messak. *Key* **a** very eolized chopping tool with a grayish patina; **b** denticulate with a reddish patina (the natural stone color is visible on the fresh fracture on the distal end on the *right*); **c** flake with fresh edges and almost no patina; **d**, **e** chopping tools; **f** faunal element; **g** end-struck Tabelbala-Tachengit large flake; **h** cleaver; **i**, **n**, **q** bifaces; **l** core on flake; **m**, **o** Levallois flakes; **f** "giant" Levallois core. Artifacts **a–d**, **g**, **i–f** are from the Messak; *E*, **f** are from Erg Uan Kasa; **h**, and **q** from Edeyen of Murzuq. Artifacts **a–c**, **e**, **f**, **i–m**, and **n**, **f** are from the same archaeological units, respectively (scale bar 5 cm)

An example is the large flake (c. 17 cm long) (Fig. 7.5g), which was apparently produced using the variant of the method described by Clark (1994), and was found within an AU that also comprised bifaces. The flake shows a centripetal preparation, followed by large removals affecting the left side. Therefore, once removed by an end-struck blow, the flake shows the characteristic *débordant* side of the Tabelbala-Tachengit cleaver blanks, to which it is assimilated.

Mode 2 + 3

Approximately, 90% (149) of this group is composed of solid AUs (Table 7.4; Fig. 7.6). Large concentrations have been recognized in the Erg Titersin and Erg Uan Kasa (i.e., 96/295 and 94/84), comprising dozens of bifaces, cleavers, Levallois products, and by-products (Cremaschi and di Lernia 1998a), and in the Edeyen of Murzuq. Here, the frequency and distribution of the AUs suggests a diffuse occupation (Anag and di Lernia 2007).

On the Messak Plateau, significant concentrations of artifacts, many appearing to be workshops, have been identified. These often occur in conjunction with large quartzarenite outcrops. The variability in the application of the Levallois method (based on field observations of 50 AUs including Levallois cores) suggests a slightly higher incidence of centripetal and lineal schemes (Fig. 7.5p) relative to the recurrent uni–bidirectional ones (Table 7.5). The tool inventory includes notches, denticulates, and scrapers, also made on Levallois flakes. Bifaces are the most common tools, while cleavers are rare. Quartzarenite is the only raw material used.

Mode 3

The rare solid AUs from the lowlands were found in the Erg Uan Kasa, Edeyen of Murzuq, and Wadi Tanezzuft, while Erg Titersin lacks any evidence (Fig. 7.7).

In the Acacus, two sheltered AUs have been identified on structural surfaces. In Wadi Teshuinat, TH 126 was filled by

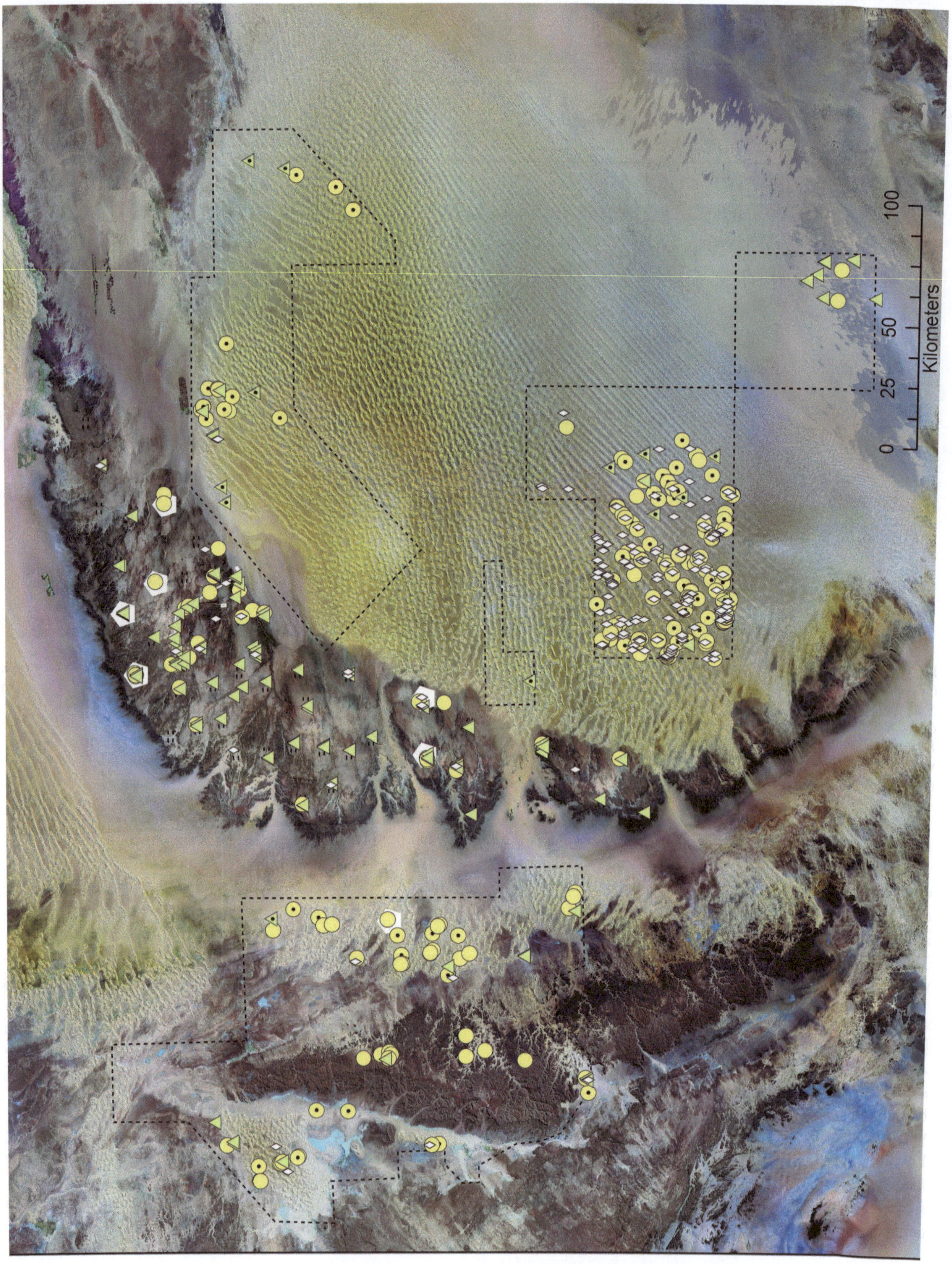
0
25
50
100
Kilometers

Fig. 7.6 Distribution of archaeological units (AUs). *Key* Mode 1 (pentagon); Mode 2 (circle = solid AU; circle-dot = ephemeral AU; diamond = isolated AU); and Mode 2 + 3 (triangle = solid AU; triangle-dot = ephemeral AU). Dashed line = approximated simplified maximum areas of the survey. Messak polygons are not all visible at this scale

a fossil dune containing several Levallois cores and flakes. Open-air AUs are located on both the alluvia and the terraces.

The samples from the Messak Plateau provide approximately 91% of the evidence in this category (Table 7.4). Excluding isolated AUs, more than 90% of the Messak record is represented by workshops. These are frequently located in the vicinity of quartzarenite outcrops. More than half of these findings exhibit nearly complete reduction sequences, whereas approximately 30% lack retouched blanks. The AUs not classified as workshops (approximately 10%) are characterized by the exclusive presence of retouched implements, and are sometimes associated with unretouched Levallois flakes.

In comparison to the Mode 2 + 3 group, the variability in the application of the Levallois method (based on field observations of 182 AUs including Levallois cores) highlights a decrease in the incidence of AUs with recurrent centripetal cores (52–49%); an increase in the frequency of AUs including uni-bidirectional cores (38–41%); and a substantial decrease in the occurrence of AUs with lineal cores (54–42%: Table 7.5b).

The technological variability within the same AU seems to be lower than that of the Mode 2 + 3 group, and the exclusive presence of recurrent uni-bidirectional cores shows an increase relative to the Mode 2 + 3 group (14–20%: Table 7.5a).

The scrapers, notches, and denticulates are frequently made on Levallois blanks. Levallois point production, although barely represented, is accomplished using both "classic" (Fig. 7.8g) and Nubian schemes (Fig. 7.8a, h). Quartzarenite is almost the only raw material used. Quartz and fossil wood artifacts have been occasionally observed.

AU 10/578, recently discovered, deserves a more detailed description. The site is located in the southeast Messak, in an open-air surface context that corresponds with a fine-grained quartzarenite outcrop. The assemblage appears to be internally coherent, and all of the artifacts share the same degree of aeolization and patina. Together with a series of Levallois cores (one of the Nubian Type 2, Fig. 7.8a) and Levallois blades and flakes, three large backed pieces and one mesial fragment of a flat foliate (Fig. 7.8b–e) have been recovered within an area of 10 × 10 m, indicating close affinities with the Lupemban (Barham 2000, 2002; Barham and Mitchell 2008).

Table 7.5 Number and incidence of AUs with Levallois cores recognized in the "Mode 2 + 3" and "Mode 3" groups in the Messak (fieldwork 2010–2011)

Levallois core type	A				B			
	Mode 2 + 3		Mode 3		Mode 2 + 3		Mode 3	
	n	%	n	%	n	%	n	%
Lineal (flakes)	10	*20.0*	41	*22.5*	27	*54.0*	77	*42.3*
Lineal (points)	–	–	2	*1.1*	–	–	6	*3.3*
Recurrent centripetal	12	*24.0*	47	*25.8*	26	*52.0*	89	*48.9*
Recurrent uni-bidirectional	7	*14.0*	37	*20.3*	19	*38.0*	75	*41.2*
Lineal (flakes) Recurrent centripetal	9	*18.0*	16	*8.8*				
Lineal (flakes) Recurrent uni-bidirectional	7	*14*	12	*6.6*				
Recurrent centripetal Recurrent uni-bidirectional	4	*8.0*	15	*8.2*				
Lineal (flakes) Recurrent centripetal Recurrent uni-bidirectional	1	*2.0*	8	*4.4*				
Lineal (points) Recurrent centripetal	–	–	1	*0.5*				
Lineal (points) Recurrent uni-bidirectional	–	–	1	*0.5*				
Lineal (points) Recurrent centripetal Recurrent uni-bidirectional	–	–	2	*1.1*				
Total observed AUs	50	*100.0*	182	*100.0*	50	*100.0*	182	*100.0*

A co-occurrence of different types of core within the same AU; *B* occurrence of the single types of core

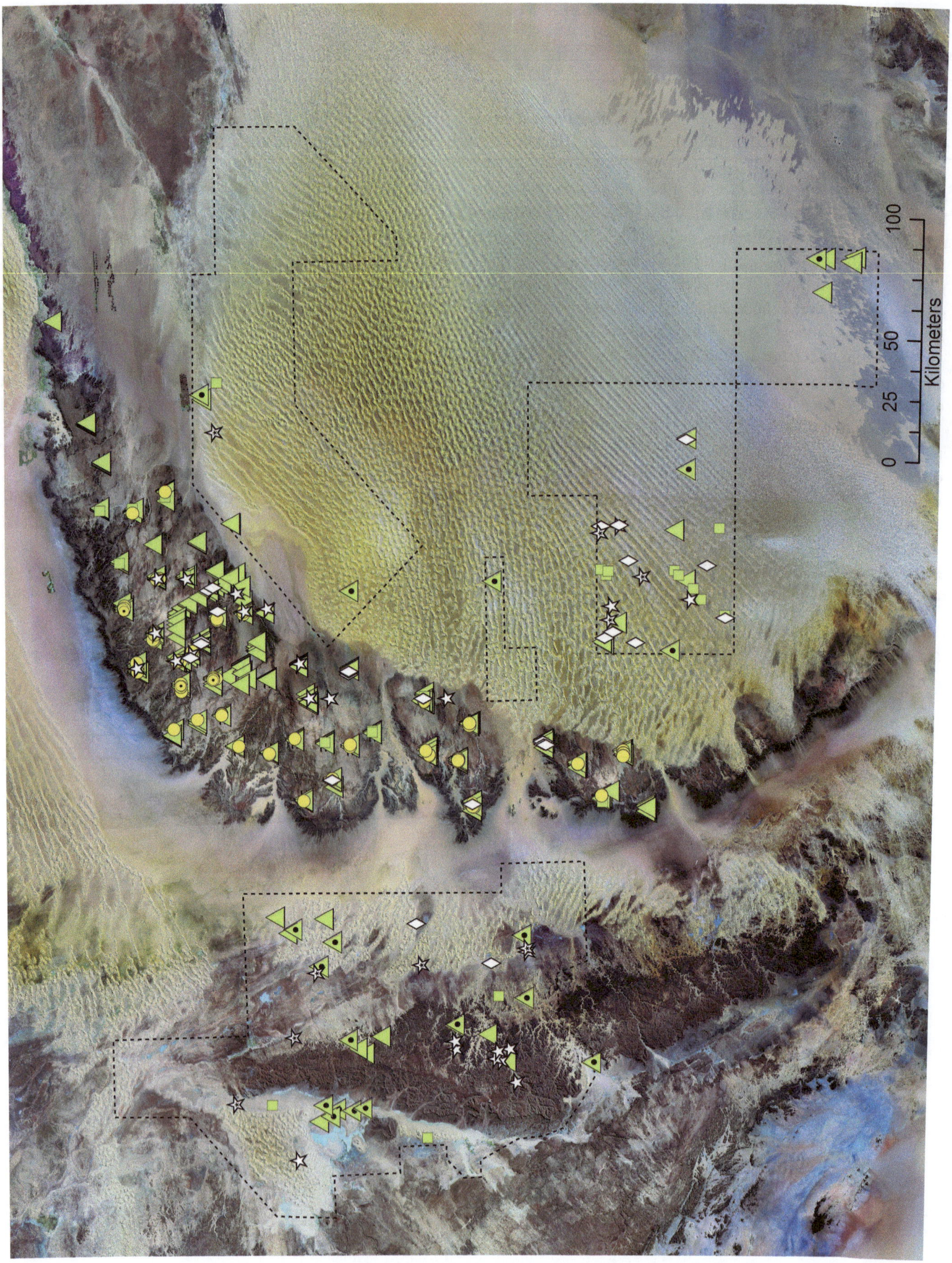
0
25
50
100
Kilometers

◀ **Fig. 7.7** Distribution of archaeological units (AUs). *Key* Mode 3 (triangle = solid AU; triangle-dot = ephemeral AU; square = isolated AU): Mode 3-Aterian (star = solid AU; star–star = ephemeral AU; diamond = isolated AU); and Mode 4 (circle = solid AU; circle-dot = isolated AU). *Dashed line* approximated simplified maximum areas of the survey. Messak polygons are not all visible at this scale

Mode 3-Aterian

Erg Titersin contains a single Aterian occurrence (01/134), partly covered by lacustrine sediments (Fig. 7.7). AU 96/221 constitutes the only evidence from the Wadi Tanezzuft. This AU returned a handful of artifacts, including two tanged points. Erg Uan Kasa contains a slightly larger amount of Aterian contexts. These are mostly isolated and ephemeral AUs. The Edeyen of Murzuq parallels these findings, with the majority of AUs comprising isolated artifacts mainly recovered from areas of serir (Cremaschi et al. 1998; Anag and di Lernia 2007).

The Acacus Aterian occupation developed inside caves, shelters, and open-air contexts on structural surfaces (Table 7.4). At Uan Tabu, approximately 3,000 artifacts testify to almost complete Levallois (including the Nubian variant) and non-Levallois reduction sequences. The raw materials inventory is quite diversified. The tools include Levallois flakes and points, side-scrapers, and rare tanged tools. The nature of the assemblage may indicate a "residential" character of occupation (Cremaschi et al. 1998; Garcea 2001). The limited sample of Uan Afuda is the result of a brief occupation, but the assemblage lacks diagnostic traits. The Aterian attribution of this site is based on stratigraphic and chronological correlations with Uan Tabu, although an earlier MSA phase cannot be discounted (di Lernia 1999). At AT 8, located on a saddle on the highest structural surface, a well-preserved spatial configuration of the artifacts, including flakes, cores, and tanged tools, is possible. Given its topographic position, this AU was tentatively interpreted as a hunting-related station (Cremaschi et al. 1998).

The Messak returned 30 AUs, mostly on the surface of the hamada, but also in a stratigraphic position at Wadi Imrawen, Wadi Tidwa, and MT 22, Unit 2 (Cremaschi et al. 1998). "Late Paleolithic" traits, however, have also been noted at MT 22, Unit 2 by Garcea (2001).

Mode 4

This group, comprising 105 AUs, is only found in the Messak. An overestimation of Mode 4 AUs is likely, as similar blades can in fact be obtained with different methods, and it is often difficult to provide a solid diagnosis. Moreover, many scatters lack cores or denote an expedient laminar reduction from poorly structured cores.

The AUs are spread over almost the entire massif, largely on the plateau (approximately 90%), and are mostly workshops. Techno-typological field observations on a sample of 70 AUs revealed that (i) blade cores (noted at 43 AUs) (Fig. 7.8o–q) are much more frequent than bladelet cores (noted at 3 AUs), and include both single platform (including pyramidal) and opposed platform variants; (ii) the use of thick blanks as cores is also attested; (iii) the start and management of production is indicated by some crests and neo-crests (Fig. 7.8r); and (iv) full-production blades are triangular, trapezoidal, or polygonal in section, and were extracted from single and opposed platform cores. The tools include retouched blades, truncations, and end-scrapers (Fig. 7.8s). Quartzarenite is the most used raw material. Quartz and fossil wood artifacts have been occasionally observed.

A group of findings, mainly in central Messak Settafet, includes very large blades (up to 40 cm in length; Fig. 7.8t). Mega-blades are not new to Messak archaeology, as different research teams have recognized them at different locales in the past. As also noted by Reynolds (2007), these blades are the outcome of a true serial laminar technology. They also show a clear link to the raw material of which they are mostly made: flat slabs of fine-grained quartzarenite, whose natural geometry allows a perfect management of the laminar reduction with little or no effort required to shape the raw blocks.

Large size, a vague geographical clustering, the exploitation of raw materials with similar morphologies, and (usually) a low degree of aeolization suggest that the mega-blade workshops were a rather homogeneous entity. At MSRS2000/231, a test excavation indicated that such artifacts lay on the top of a relict soil, but not inside it (Cremaschi and di Lernia 2000).

Regional Synthesis

One of the aims of our regional perspective is the identification of gross occupation patterns over large geographic areas. In this sense, Fig. 7.9 shows the correlation between the number and type of AU, technology (Mode 1 is deliberately omitted), physiography, and archaeological remains. These remains have been categorized according to the occurrence or co-occurrence of two basic classes of lithic artifacts: by-products (Bp) and formal tools (Ft). "Bp" units have an exclusive or predominant presence of unretouched flakes and/or cores, "Ft" units have an exclusive or predominant presence of retouched tools, and "BpFt" units have both ("na" = not assessed). Admittedly, this assessment is very basic, but it is useful to broadly correlate environmental settings and archaeological contexts. We also acknowledge that the observed data patterning strongly reflects our

Fig. 7.8 Findings from Mode 3, Mode 3-Aterian, and Mode 4 archaeological units (AUs). *Key* **a** Nubian core; **b** foliate fragment; **c–e** backed tools; **f** refitting Levallois core and shaping flake; **g** Levallois point; **h** Nubian point; **i–n** Aterian tools; **o–q** blade cores; **r** neo-crest; **s** end-scraper; **t** mega-blade. All artifacts are from Messak, except: **m** from Erg Uan Kasa; and **h**, **n** from Edeyen of Murzuq. Artifacts **a–e** are from the same archaeological unit (scale bar 5 cm)

methodological choices; an example is the fact that Mode 2, Mode 2 + 3, and Mode 3-Aterian do not contain "Bp."

Mode 2 and Mode 2 + 3

It is worth looking at these two groups together. The lowland findings indicate a strong emphasis on "Ft." Large clusters of handaxes, but also high numbers of isolated finds, mostly within interdune basins, can be related to diffuse subsistence activities paralleled by limited contextual toolkit production and maintenance. The Messak AUs show a higher incidence of cases with inferably complete reduction sequences, starting from the procurement of the local raw material. In the Acacus, the AUs on alluvium are principally "Ft," while on structural surfaces and dune margins, the AUs are exclusively "BpFt."

Mode 3

"Ft" AUs are almost absent in the sand seas. "BpFt" and "Bp" AUs are present on both interdune basins and dunes margins, and are also recognizable in the dune margins of Wadi Tanezzuft. Mode 3 AUs on the Messak are clearly dominated by workshops. Unlike the previous groups, a much higher number of contexts are located in alluvial deposits and endorheic depressions. In the Acacus, AUs are located at the alluvium level and on structural surfaces.

Mode 3-Aterian

These AUs occur in almost all of the physiographic contexts in the study area. "Ft" AUs occur exclusively in the dune margins of the sand seas, while "BpFt" AUs are also present within interdune basins. In the Acacus, the archaeological contexts at the alluvium level are mainly shelters or caves, while open-air contexts are almost always confined to higher structural surfaces. This topographic distinction is also thought to be functional, with the former (Uan Tabu *in primis*) interpreted as residential sites, and the latter (AT8) as possible specialized locations (Cremaschi et al. 1998).

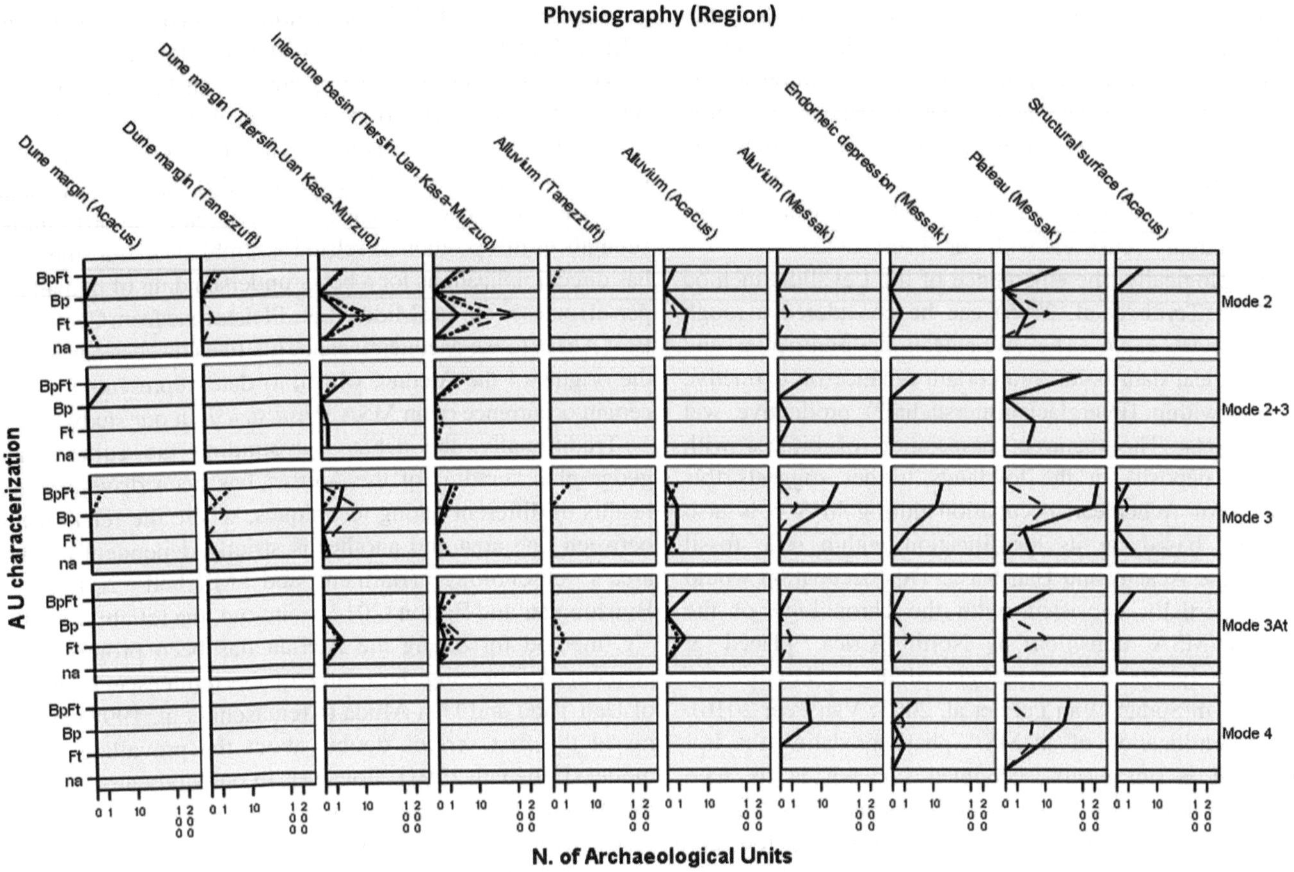

Fig. 7.9 Number and type of archaeological units according to technology (Mode), regional/physiographic location (Physiography region) and archaeological content (see text for details). *Key* solid line = solid AU; dashed line = ephemeral AU; dotted line = isolated AU. Log scale

Additionally, given the projectile nature of some of the tanged points, hunting practices in the Messak may have taken place along side lithic raw material procurement and processing.

Mode 4

The Mode 4 data, all from Messak, suggest occupation events aimed at the procurement and processing of raw material. The presence of "BpFt" and "Ft" AUs also indicates the existence of other types of activities, but the absence of armatures precludes hunting.

Discussion

In those cases where the choppers and chopping tools (Mode 1) exhibit a greater degree of surface weathering than other Stone Age objects in the same location, they have been tentatively attributed to the Oldowan. However, their chronological position is only speculative, and has yet to be defined. Although equally conjectural, occurrences of Oldowan artifacts that have recently been reported from Wadi Shati, the sand sea of Ubari, and northern Messak Settafet by Lahr et al. (2010) lend weight to the idea that "something very early" may exist. While tempting, the evidence as it stands is not enough to postulate the presence of hominins in this part of the Sahara during the Early Pleistocene. Therefore, we are still unable to contribute to the discussion of the earliest peopling of North Africa, which, to date, has only been documented at Aïn Hanech, in Algeria. Here, Oldowan artifacts have been dated to approximately 1.8 Ma (Sahnouni et al. 2002). However, severe criticisms of both the chronology and archaeological bases of that finding suggest that a date of 1.2 Ma would be more reliable (Geraads et al. 2004).

An Acheulean industry is well attested throughout the study area, but its potentially very long duration and the lack of dated contexts prevent us from better defining its evolution and trajectories. An hypothetical earlier phase is indicated by robust and deeply worn handaxes and traces of "large-flake" technologies (i.e., Tabelbala-Tachengit), which possibly predate the Levallois method (e.g. Rolland 1995) and that were adopted in the Saharan Acheulean to produce large cleaver blanks (Clark 1994), assumed to have disappeared from the Levant and North Africa before 0.5 Ma (Sharon 2010). Evidence for all of these elements is sparse, and we cannot substantially build upon them. Later Acheulean contexts are well represented, often signaled by flat symmetrical bifaces and the extensive use of "true" Levallois reduction strategies (Fig. 7.5o). Notwithstanding notable and recent achievements in the study of African Levallois technology, its origins and evolution are still poorly understood. Nevertheless, later Acheulean sites from South and East Africa, as well as from the northwestern Sahara (i.e., Mourre 2003; Tryon et al. 2006; Villa 2009; Wilkins et al. 2010), more frequently indicate lineal and centripetal variants, which is roughly consistent with our observations.

Chronologically, the emergence of the Levallois method in MIS 8 (Tryon et al. 2006) can be considered a rough *terminus post quem*. The Acheulean environmental and archaeological data point to a certain reliance on extractive activities within (interglacial/interstadial?) productive wet environments. The recurrent association of bifaces with lacustrine deposits in the lowlands further suggests this reliance. An Acheulean occupation during MIS 6 is also suggested, based on its identification within relic fossil dunes in the Acacus and Uan Kasa. This occupation would also be partially consistent with the chronology of the Acheulean/MSA transition in North Africa, placed at approximately 200 ka, as recognized at the key site of Sai8-B-11 in Sudan (Van Peer et al. 2003; Van Peer 2016).

The identification of an MSA phase predating the last interglacial is not easily accounted for, nor is its Late Pleistocene development. As recently discussed by Cancellieri and di Lernia (2013), most of the surface assemblages included in our Mode 3 data set include un-diagnostic material (i.e., "classic" Levallois artifacts, scrapers, notches, and denticulates) and cannot, therefore, automatically be included within a narrow definition of the MSA. Excluding the Aterian, archaeological occurrences with MSA diagnostics are generally rare. Point production, which is among the formal distinguishing features of the MSA (McBrearty and Brooks 2000; Brooks et al. 2006), is represented by sparse evidence. To date, Nubian point-reduction schemes have been recognized as having been associated with the Aterian industry of Uan Tabu (Garcea 2001); and, in the Messak and the Edeyen of Murzuq, within open-air undated contexts that do not contain Aterian artifacts (Fig. 7.8a, h).

Given the similarities of its assemblage with Lupemban industries elsewhere, one of the Messak AUs (AU 10/578) represents a possible early occurrence of the MSA. At Twin Rivers, in Zambia, this occurrence is bracketed between 240 and 170 kyr (Barham 2000). Lupemban industries have also been recorded at sites in North Africa, i.e., at Sai 8-B-11 in Sudan. Here, the Lupemban is overlain by sands dated to 152 ± 10 ka, and is stratified above Sangoan levels overlain by sands dated to 182 ± 20 ka (Van Peer et al. 2003; Van Peer 2004). Hence, on a purely techno-typological ground, the MSA in our study area can be speculatively rooted in MIS 6, which is largely consistent with the timing of the spread of early modern humans across North Africa as a consequence of MIS 7 population dispersals from Central Africa (Van Peer et al. 2008; Van Peer 2016).

The suggestion that the Lupemban can be considered the origins of the Aterian has often also been discussed on the basis of its peculiar bifacial technology (e.g., Caton-Thomson 1946; Clark 1993; Kleindienst 2001; Garcea 2012a). The occasional presence of tanged implements (Barham 2000) further strengthens this position. A related assemblage in our study area has direct implications for a better understanding of human dispersal routes in the late Middle/Late Pleistocene from Central and East Africa to the Mediterranean. This fuels discussions around the origins of the Aterian, which, to date, represents the only certain occurrence of an MSA *sensu stricto* in our study area.

Traditionally, the debate surrounding the cultural and geographic "cradle" of the Aterian has been driven by the results of different dating techniques, where the relationship between one area and another is strictly dependent on that area's chronology (Barham and Mitchell 2008). As Bouzouggar and Barton (2012) point out, the reliability of the ^{14}C method for dating the Aterian has been progressively challenged following the results of the luminescence dating of Uan Tabu and Uan Afuda (Cremaschi et al. 1998), which raised the first serious doubts about the prevailing 1990s model (Debénath 1994). However, in our study area, outside the Acacus, the Aterian is still a regionally undated palimpsest. Given the early appearance of the Aterian in the Maghreb in late MIS 6/early MIS 5 (Barton et al. 2009; Richter et al. 2010; Schwenninger et al. 2010), and assuming it had a sub-Saharan origin, the occurrence of the Aterian in the Sahara should be contemporary, if not earlier (Garcea 2012b). However, the available chronology of the conclusively Aterian assemblage of Uan Tabu confines it to within hyperarid MIS 4. As noted by Garcea (2012a), there is a meaningful relationship between settlement choices and geographic gradients (latitude and altitude), with societies following varying ecotones and organizing the subsistence settlement pattern around water points. The hypothesis of a mountainous adaptation to drying environments is also consistent with the often-claimed nature of refugia in the central massifs (e.g., Clark 1998). As today, these were characterized by residual water availability during the driest phases. However, extending the Acacus model to the entire study area is less convincing, particularly when the lowlands are included. Here, the presence of Aterian contexts is especially important. Yet, in MIS 4, these lowlands must have more closely resembled the extreme desert environment of today.

To postulate the existence of a Late Stone Age (LSA) occupation during the Late Pleistocene is rather problematic. Blade technology, the main aspect upon which our reasoning builds, no longer has any intrinsic chronological or cultural value, given its early occurrence (0.5 Ma) and routine adoption in the MSA and LSA (Bar-Yosef and

Kuhn 1999; McBrearty and Brooks 2000; Johnsons and McBrearty 2010). The rarity of artifacts with chrono-cultural information, and the assumption that the lack of evidence after the Aterian is due to severe climatic conditions (e.g., Clark 1998), make the issue even more problematic.

Well-developed laminar industries of the early Holocene are present throughout most of the regions, except in the Messak. These industries are related to the first Early Acacus groups that repopulated the region after wetter conditions were fully re-established in the Late Pleistocene (Cremaschi and di Lernia 1999; Cancellieri and di Lernia 2014). These industries could represent a natural and obvious comparison for the laminar assemblages of the Messak. However, laminar production here is mostly bladelet oriented, and the toolkit comprises both common backed tools (i.e., straight-backed points) and more peculiar artifacts, such as Ounan points. None of these artifacts have been found in the Messak, which precludes our ability to make any firm comparisons. A blade component is present, for example, in the Epipaleolithic industries of Adrar Bous and Greboun, in Niger; and is related to the production of Ounan points and other projectile armatures (Smith 1993; Clark 2008). Once again, comparisons are difficult.

Turning back to the Messak and the Pleistocene, "Late Paleolithic" assemblages have been previously identified at Tidwa surface and Imrawen 1A, 1B, and 2A on the basis of either technological (e.g., crest technique, laminar reduction, and special butt preparation) or typological traits. The chronological position is conjecturally Late Pleistocene (Garcea 2001; Van Peer 2001).

Due to the absence of armatures and the almost non-diagnostic typology of the tools (i.e., end-scrapers, truncations, and retouched blades), comparisons with the North African LSA (e.g., Garcea 2010) are few. Considering the nature of the frequentation, which was supposedly aimed at raw material procurement, some of the features that we observed could be comparable with the Upper Paleolithic/LSA of the Nile valley (Vermeersch 2010). Here, similar core reduction strategies, and consequent specific blanks and by-products, seem to be comparable from both a dimensional and technological perspective. Comparative factors include the non-microlithic character of the tools, the use of the crest technique, and the volumetric concept of both laminar reduction and the maintenance of core convexities (i.e., *semi-tournante* and *tournante* modalities, bipolar reduction, neo-crests, and flank extractions).

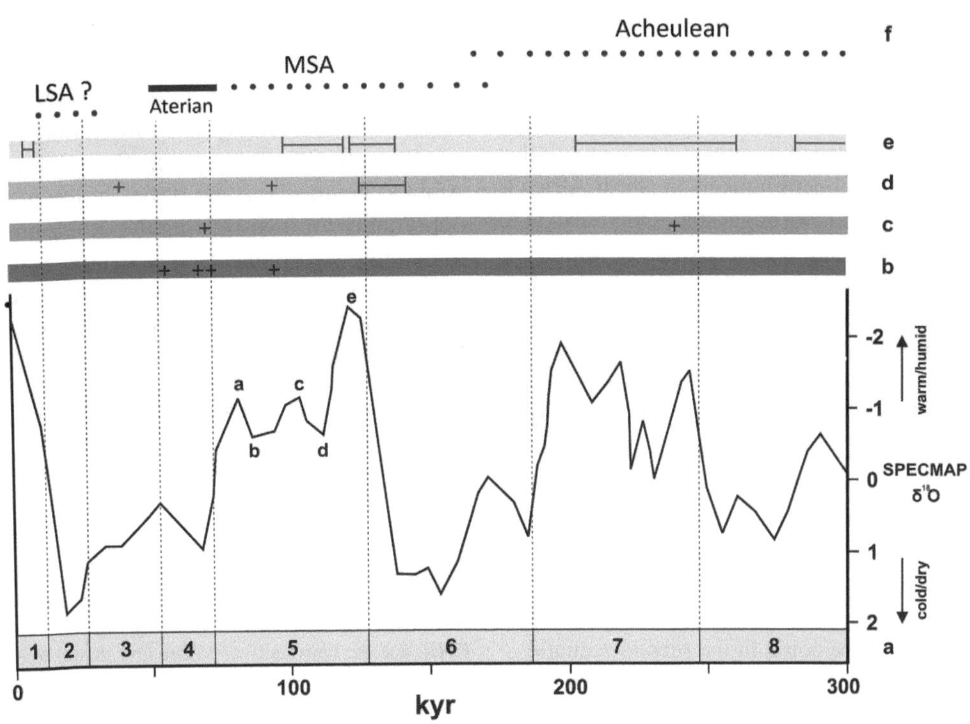

Fig. 7.10 Definition of last glacial/interglacial cycles using the SPECMAP foraminifera $\delta^{18}O$ time series (Imbrie et al. 1984); MIS stages and substages are also indicated. *Key* **a** MIS; **b** luminescence dating for wind activity (Cremaschi 1998; Cremaschi et al. 1998; di Lernia 1999; Martini et al. 1998; Garcea 2001); **c** deposition of carbonatic concretions (Zerboni et al. 2011; this study); **d** U/Th dating of the lake terraces in the Wadi Shati (Gaven 1982); **e** radiometric dating for the lake deposits in the Murzuq and El-Ajal regions (for the Pleistocene: Armitage et al. 2007; Drake et al. 2008 2011; Geyh and Thiedig 2008; for the Holocene: Cremaschi and Zerboni 2009); **f** suggested chronology for the Stone Age techno-complexes of the study area hypothetically comprised between MIS 8 and MIS 2: *bold line* represents chronometric dated interval; *dotted lines* represent hypothetical intervals

Considering that some of the technological characteristics that we observed are consistent with "Upper Paleolithic" *sensu lato* concepts of core reduction, we do not exclude the possibility of an MIS 3/2 post-Aterian human presence.

Finally, the mega-blade workshops are a distinct entity within the *corpus* of the laminar industries of the Messak, but their chronological attribution is also controversial (Reynolds 2007). We favor a late chronological position, possibly at the end of the Late Pleistocene/Early Holocene, but it is important to remember that in Libya, large blades that may be similar are present at Haua Fteah in the "pre-Aurignacian" (McBurney 1967; Moyer 2003; Douka et al. 2014).

Concluding Remarks

The decline in humidity recognized throughout the Quaternary severely affected the ecological productivity of certain areas throughout time. This change contributed to the creation and disappearance of specific econiches and human adaptations. It is possible that from the last interglacial onward, wet interglacial environments became increasingly less productive than during previous ones, when a diffuse Acheulean presence relied heavily on the availability and distribution of large freshwater environments. In this respect, the distribution of MSA (*sensu lato*) artifacts, in contrast to that of the Acheulean, reveals that they were strikingly less prevalent in lowland areas (Figs. 7.7, 7.8).

The paleoclimatic and paleoenvironmental framework that witnessed the dispersal of modern humans in North Africa is almost unknown. Despite this, we suggest that their arrival to our study area from Central and/or East Africa was influenced by conditions of environmental variability and fragmentation; and that this could have taken place during late MIS 6. Their technology comprised composite tool "spare parts," the backed pieces, which occur at different rates throughout the MSA (Hiscock and O'Connor 2005). The use of maintainable composite tools might be seen as a solution to, and a way of reducing the risk of, being under-equipped (Hiscock 1994). On the one hand, they allow for the exploitation of environments that were becoming unfamiliar due to rapid climate change. Yet, equally, their use also facilitates explorative pioneering movements across unknown geographic ranges (Barham and Mitchell 2008). Our evidence could fit the second scenario.

The Aterian of the hyperarid MIS 4 Acacus is the second piece of evidence that we use to make our argument. Currently, these are the only occurrence of dated MSA artifacts in the entire study area, and therefore constitute our only certain evidence. The age of the Aterian outside the Acacus is unknown, but is likely to have had an earlier southern origin. Given this, we suspect that the mountain adaptations evident in this region, although successful, reflect a residual occupation of humans who were skilled in "arid survival."

Finally, although yet not demonstrable, signs of a generalized LSA technology are found in the Messak. Here, the large quantities of high-quality stony raw materials could have attracted task-specific temporary occupants, and activated arid "stony" corridors from nearby areas. It is possible, in fact, that during late MIS 3 and MIS 2, areas rich in raw materials also witnessed human frequentation, as observed in other parts of North Africa. For example, the Jebel Gharbi in northern Libya (Garcea and Giraudi 2006) and Nazlet Khater 4 in Egypt (Leplongeon and Pleurdeau 2011).

To conclude, the present paleoenvironmental and archaeological reconstruction generates challenging questions surrounding the issue of whether "green = life." As we have seen, the opposite is neither necessarily nor automatically true. At least three population dynamics (explorative occupation, residual occupation, and special-task occupation) corresponding to three different chrono-cultural phases occurred (Aterian), or are presumed to have occurred (early MSA and LSA *sensu lato*), within conditions of environmental stress during arid phases at the end of the Middle Pleistocene and in the Late Pleistocene (Fig. 7.10). During the time period in question, humans moved along both "green" and "brown" corridors, and dwelled in arid landscapes. It is therefore necessary to consider evidence for factors other than humidity to account for unexpected relations with the environment. Extending research beyond the orthodox link between humans and humidity can lead to exciting results. As Groucutt and Petraglia (2012) have recently noted with reference to the Arabian Peninsula, some places in the world provide a rich opportunity for investigating the development of adaptations to extreme environments. The central Sahara must certainly be counted among this number. Although the evidence from this region is not yet able to provide good answers, it seems that it allows us, for the moment, to ask good questions.

Acknowledgments The present chapter is the outcome of a series of fieldwork and laboratory analyses over more than 20 years. We thank all of the colleagues of the Libyan Department of Archaeology for their help and support. The research has been carried out under the aegis of the Italian-Libyan Archaeological Mission in the Acacus and Messak, directed since 2002 by SDL, and it has been funded by several bodies: Sapienza University of Rome (Grandi Scavi), which holds the permit of survey and excavations in the Acacus and Messak; the Ministry of Foreign Affairs (DGPCC); the Ministry of University and Research (Cofin, PRIN); the CNR; and the University of Milan. Our thanks to Alessandro Perego for drawing Fig. 7.1. We thank Stefano Biagetti and Marina Gallinaro for their comments on the manuscript draft. We thank Sacha Jones and Brian Stewart for inviting SDL to attend the conference and submit a paper, and we thank three anonymous reviewers for their careful review and valuable comments and suggestions. The paper was designed and written by SDL, with contributions by EC (archaeological analysis), MC, and AZ (geoarchaeological and paleoenvironmental data). MC and SDL provided the original and unpublished dataset of the geoarchaeological contexts discussed herein. This work is dedicated to *all* of our Libyan friends.

References

Anag, G., & di Lernia, S. (2007). The archaeological survey: Aims, methodology and results. In G. Anag, L. Cosentino, & S. Di Lernia (Eds.), *Edeyen of Murzuq. Archaeological survey in the Libyan Sahara*. Firenze: All'Insegna del Giglio.

Anag, G., Cremaschi, M., di Lernia, S., & Liverani, M. (2002). Environment, archaeology, and oil: The Messak Settafet rescue operation (Libyan Sahara). *African Archaeological Review, 19*, 67–73.

Armitage, S. J., Drake, N. A., Stokes, S., El-Hawatc A., Salem, M. J., White, K., et al. (2007). Multiple phases of North African humidity recorded in lacustrine sediments from the Fazzan Basin, Libyan Sahara. *Quaternary Geochronology, 2*(2), 181–186.

Bar-Yosef, O., & Kuhn, S. L. (1999). The big deal about blades: Laminar technologies and human evolution. *American Anthropologist, 101*, 322–338.

Barham, L. (2000). *The Middle Stone Age of Zambia, south-central Africa*. Bristol: Western Academic and Specialist Press.

Barham, L. (2002). Backed tools in Middle Pleistocene central Africa and their evolutionary significance. *Journal of Human Evolution, 43*, 585–603.

Barham, L., & Mitchell, P. (2008). *The first Africans*. Cambridge: Cambridge University Press.

Barton, R. N. E., Bouzouggar, A., Collcutt, S., Schwenninger, J.-L., & Clark-Balzan, L. (2009). OSL dating of the Aterian levels at Dar es-Soltan I (Rabat, Morocco) and implications for the dispersal of modern *Homo sapiens*. *Quaternary Science Reviews, 28*, 1914–1931.

Biagetti, S., Cancellieri, E., Cremaschi, M., Gauthier, C., Gauthier, Y., Zerboni, A., et al. (2013). The 'Messak Project': Archaeological research for cultural heritage management in SW Libya. *Journal of African Archaeology, 11*(1), 55–74.

Boëda, E. (1993). Le débitage discoïde et le débitage Levallois recurrent centripède. *Bulletin de la Société Préhistorique Française, 90*, 392–404.

Bordes, F. (1961). *Typologie du Paléolithique ancien et moyen*. Bordeaux: Publications de l'Institut de Préhistoire de l'Université de Bordeaux, Mémoire n° 1, Delmas.

Bouzouggar, A., & Barton, N. (2012). The identity and timing of the Aterian in Morocco. In J. J. Hublin & S. P. McPherron (Eds.), *Modern origins: A North African perspective*. Dordrecht: Springer.

Brookes, I. A. (2010). Spatially variable sedimentary responses to orbitally driven pluvial climate during Marine Oxygen Isotope Stage 5.1, Dakhla Oasis region. *Egypt. Quaternary Research, 74*, 252–264.

Brooks, A., Nevell, L., Yellen, J. E., & Hartman, G. (2006). Projectile technologies of the African MSA. Implications for modern human origins. In E. Hovers & S. Kuhn (Eds.), *Transitions before the transition. Evolution and stability in the Middle Paleolithic and Middle Stone Age*. New York: Springer.

Busche, D. (1980). On the origin of the Msak Mallat and Hamadat Manghini escarpment. In J. Salem & S. Busrewil (Eds.), *The geology of Libya* (Vol. 3, pp. 837–848). Amsterdam: Elsevier.

Busche, D. (2010). North-eastern Niger: Sandstone landscape of the Sahara. In P. Mignon (Ed.), *Geomorphological landscapes of the world* (pp. 171–182). Dordrecht: Springer.

Cancellieri, E., & di Lernia, S. (2013). Middle Stone Age human occupation and dispersals. New data from the Messak plateau (SW Libya, central Sahara). *Quaternary International, 300*, 142–152.

Cancellieri, E., & di Lernia, S. (2014). Re-entering the central Sahara at the onset of the Holocene: A territorial approach to Early Acacus hunter–gatherers (SW Libya). *Quaternary International, 320*, 43–62.

Castañeda, I. S., Mulitza, S., Schefuß, E., Lopes dos Santos, R. A., Sinninghe Damsté, J. S., & Schouten, S. (2009). Wet phases in the Sahara/Sahel region and human migration patterns in North Africa. *Proceedings of the National Academy of Sciences, 106*, 20159–20163.

Caton-Thompson, G. (1946). The Aterian industry: Its place and significance in the Palaeolithic World. *The Journal of the Royal Anthropological Institute of Great Britain and Ireland, 76*, 87–130.

Clark, G. (1977). *World prehistory: A new perspective*. Cambridge: Cambridge University Press.

Clark, J. D. (1993). The Aterian of the central Sahara. In L. Krzyzaniak, M. Kobusiewicz, & J. Alexander (Eds.), *Environmental change and human culture in the Nile Basin and northern Africa until the second millennium BC* (pp. 49–67). Poznan: Poznan Archaeological Museum.

Clark, J. D. (1994). The Acheulian industrial complex in Africa and elsewhere. In R. S. Corruccini & R. L. Ciochon (Eds.), *Integrative paths to the past: Paleoanthropological advances in honor of F. Clark Howell* (pp. 451–469). Englewood Cliffs: Prentice Hall.

Clark, J. D. (1998). Before food production in North Africa: Overview and discussion. In S. di Lernia & G. Manzi (Eds.), *Before food production in North Africa: Questions and tools dealing with resource exploitation and population dynamics at 12,000–7,000 bp* (pp. 163–177). Forlì: ABACO.

Clark, J. D. (2008). Epipalaeolithic aggregates from Greboun and Adrar Bous. In J. D. Clark & D. Gifford-Gonzalez (Eds.), *Adrar Bous: Archaeology of a central Saharan granitic ring complex in Niger* (pp. 163–178). Tervuren: Royal Museum for Central Africa.

Cremaschi, M. (1996). The rock varnish in the Messak Settafet (Fezzan, Libyan Sahara), age, archaeological context, and paleo-environmental implication. *Geoarchaeology, 11*, 393–421.

Cremaschi, M. (1998). Late Quaternary geological evidence for environmental changes in south-western Fezzan (Libyan Sahara). In M. Cremaschi & S. di Lernia (Eds.), *Wadi Teshuinat – palaeoenvironment and prehistory in south-western Fezzan (Libyan Sahara)* (pp. 13–47). Roma-Milano: CNR.

Cremaschi, M., & di Lernia, S. (1998a). The geoarchaeological survey in the central Tadrart Acacus and surroundings (Libyan Sahara). Environment and cultures. In M. Cremaschi & S. Di Lernia (Eds.), *Wadi Teshuinat. Palaeoenvironment and prehistory in south-western Fezzan (Libyan Sahara)*. Milano: CNR.

Cremaschi, M., & di Lernia, S. (Eds.). (1998b). *Wadi Teshuinat. Palaeoenvironment and prehistory in south-western Fezzan (Libyan Sahara)*. Milano: CNR.

Cremaschi, M., & di Lernia, S. (1999). Holocene climatic changes and cultural dynamics in the Libyan Sahara. *African Archaeological Review, 15*, 211–238.

Cremaschi, M., & di Lernia, S. (Eds.). (2000). *Lasmo N-FC 174 concession Area. The Messak Settafet rescue operation (Libyan Sahara): Preliminary report*. Tripoli and Rome: Department of Antiquities and CIRSA.

Cremaschi, M., & Zerboni, A. (2009). Early to Middle Holocene landscape exploitation in a drying environment: Two case studies compared from the central Sahara (SW Fezzan, Libya). *C. R. Geosci, 341*, 689–702.

Cremaschi, M., di Lernia, S., & Garcea, E. A. A. (1998). Some insights on the Aterian in the Libyan Sahara: Chronology, environment, and archaeology. *African Archaeological Review, 15*, 261–286.

Cremaschi, M., Zerboni, A., Spötl, C., & Felletti, F. (2010). The calcareous tufa in the Tadrart Acacus Mt. (SW Fezzan, Libya). An early Holocene palaeoclimate archive in the central Sahara. *Palaeogeography, Palaeoclimatology, Palaeoecology, 287*, 81–94.

Debénath, A. (1994). L'Aterian du nord de l'Afrique et du Sahara. *Sahara, 6*.

De Lumley, E. (2006). Il y a 2.5 millions d'années… un seuil majeur de l'hominisation. L'émergence de la pensée conceptuelle et des strategies maîtrisées du débitage de la pierre. *Comptes Rendus Palevol, 5*, 119–126.

Desio, A. (1936). Riassunto sulla costituzione geologica del Fezzan. *Bollettino della Società Geologica Italiana, 55*, 319–355.

di Lernia, S. (Ed.). (1999). *The Uan Afuda Cave. Hunter gatherer societies of central Sahara*. Firenze: All'Insegna del Giglio.

di Lernia, S. (Ed.). (2011). *The Messak project: Cultural and natural heritage and sustainable tourism. The geo-database of the Messak (deliverable D1)*. Roma: Università di Roma Sapienza.

Douka, K., Jacobs, Z., Lane, C., Grün, R., Farr, L., Hunt, C., et al. (2014). The chronostratigraphy of the Haua Fteah cave (Cyrenaica, northeast Libya). *Journal of Human Evolution, 66*(0), 39–63.

Drake, N., & Breeze, P. (2016). Climate change and modern human occupation of the Sahara from MIS 6-2. In S. C. Jones & B. A. Stewart (Eds.), *Africa from MIS 6-2: Population dynamics and paleoenvironments* (pp. 103–122). Dordrecht: Springer.

Drake, N. A., El-Hawat, A. S., Turner, P., Armitage, S. J., Salem, M. J., White, K. H., et al. (2008). Palaeohydrology of the Fazzan Basin and surrounding regions: The last 7 million years. *Palaeogeography, Palaeoclimatology, Palaeoecology, 263*, 131–145.

Drake, N. A., Blench, R. M., Armitage, S. J., Bristow, C. S., & White, K. H. (2011). Ancient watercourses and biogeography of the Sahara explain the peopling of the desert. *Proceedings of the National Academy of Sciences, 108*, 458–462.

Duchaufour, P. (1983). *Pedologie: 1. Pedogenese et classification*. Paris: Masson.

El-ghali, M. A. K. (2005). Depositional environments and sequence stratigraphy of paralic glacial, paraglacial and postglacial Upper Ordovician siliciclastic deposits in the Murzuq basin, SW Libya. *Sedimentary Geology, 177*, 145–173.

Fantoli, A. (1937). Clima. In Regia società geografica Italiana (Ed.), *Il Sahara Italiano* (pp. 97–119). Roma: Società Italiana Arti Grafiche.

Foley, R. (1981). *Off-site archaeology and human adaptation in eastern Africa* (BAR International Series No. 97). Oxford: Archaeopress.

Gallinaro, M., Gauthier, C., Gauthier, Y., Le Quellec, J. L., Abdel Aziz, S., Biagetti, S., et al. (2012). The Messak project. Cultural and natural preservation and sustainable tourism (south-western Libya). *Antiquity, 086*(331). http://www.antiquity.ac.uk/projgall/gallinaro331/. Accessed 6 June 2014.

Garcea, E. A. A. (Ed.). (2001). *Uan Tabu in the settlement history of Libyan Sahara*. Firenze: All'Insegna del Giglio.

Garcea, E. A. A. (2010). The lower and upper later Stone Age of North Africa. In E. A. A. Garcea (Ed.), *South-eastern Mediterranean peoples between 130,000 and 10,000 years ago* (pp. 54–65). Oxford: Oxbow Books.

Garcea, E. A. A. (2012a). Modern human desert adaptation: A Libyan perspective on the Aterian complex. In J. J. Hublin & S. P. McPherron (Eds.), *Modern origins: A North African perspective*. Dobrecht: Springer.

Garcea, E. A. A. (2012b). Successes and failures of human dispersals from North Africa. *Quaternary International, 270*, 119–128.

Garcea, E. A. A., & Giraudi, C. (2006). Late Quaternary human settlement patterning in the Jebel Gharbi. *Journal of Human Evolution, 51*, 411–421.

Gasse, F. (2000). Hydrological changes in the African tropics since the Last Glacial Maximum. *Quaternary Science Reviews, 19*, 189–211.

Gaven, C. (1982). Radiochronologie isotopique Ionium-Uranium. In N. Petit-Maire (Ed.), *Le Shati. Lac Pleistocene du Fezzan (Libye)* (pp. 44–54). Paris: CNRS.

Geraads, D., Raynal, J. P., & Eisenmann, V. (2004). The earliest human occupation of North Africa: A reply to Sahnouni et al. (2002). *Journal of Human Evolution, 46*, 751–761.

Geyh, M. A., & Thiedig, F. (2008). The Middle Pleistocene Al Mahrúqah Formation in the Murzuq basin, northern Sahara, Libya evidence for orbitally-forced humid episodes during the last 500,000 years. *Palaeogeography, Palaeoclimatology, Palaeoecology, 257*, 1–21.

Graziosi, P. (1937). Preistoria. In reale società geografica Italiana (Ed.), *Il Sahara Italiano* (pp. 243–274). Roma: Società Italiana Arti Grafiche.

Groucutt, H., & Petraglia, M. (2012). The prehistory of the Arabian Peninsula: Deserts, dispersals, and demography. *Evolutionary Anthropology, 21*, 113–125.

Hallet, D. (2002). *Petroleum geology of Libya*. Amsterdam: Elsevier Science.

Harmand, S., Lewis, J. E., Feibel, C. S., Lepre, C. J., Prat, S., Lenoble, A., et al. (2015). 3.3-million-year-old stone tools from Lomekwi 3, West Turkana, Kenya. *Nature, 521*(7552), 310–315.

Hiscock, P. (1994). Technological responses to risk in Holocene Australia. *Journal of World Prehistory, 8*, 267–292.

Hiscock, P., & O'Connor, S., (2005). Arid paradises or dangerous landscapes. A review of explanations for Paleolithic assemblage change in arid Australia and Africa. In P. Veth, M. Smith, & P. Hiscock (Eds.), *Desert peoples. Archaeological perspectives* (pp. 58–77). Oxford: Blackwell.

Imbrie, J., Hays, J. D., McIntyre, A., Mix, A. C., Morley, J. J., Pisias, N. G., et al. (1984). The orbital theory of Pleistocene climate: Support from a revised chronology of the marine d^{18}O record. In A. Berger, J. Imbrie, J. Hays, G. Kukla, & B. Saltzman (Eds.), *Milankovitch and climate, part 1* (pp. 269–305). Boston: Reidel.

Inizan, M. L., Reduron-Ballinger, M., Roche, H., & Tixier, J. (1999). *Préhistoire de la pierretaillée, 5. Technology and terminology of knapped stone*. Nanterre: CREP.

Johnsons, C. R., & McBrearty, S. (2010). 500,000 year old blades from the Kapthurin Formation, Kenya. *Journal of Human Evolution, 58*, 193–200.

Kleindienst, M. R. (2001). What is the Aterian? The view from Dakhleh Oasis and the western desert, Egypt. In M. Marlow & A. J. Mills (Eds.), *The Oasis Paper 1: The proceedings of the first conference of the Dakhleh Oasis Project* (pp. 1–14). Oxford: Oxbow Books.

Lahr, M., Foley, R., Crivellaro, F., Maillo Fernandez, J., Wilshaw, A., Purdon, A., et al. (2010). DMP XI: Preliminary results from 2010 fieldwork on the human prehistory of the Libyan Sahara. *Libyan Studies, 41*, 137–158.

Le Quellec, J. L. (1998). *Art rupestre et préhistorie du Sahara. Le Messak Libyen*. Paris: Bibliotheque Scientifique Payot.

Leplongeon, A., & Pleurdeau, D. (2011). The Upper Palaeolithic lithic industry of Nazlet Khater 4 (Egypt): Implications for the Stone Age/Palaeolithic of northeastern Africa. *African Archaeological Review, 28*, 213–236.

Martini, M., Sibilia, E., Zelaschi, C., Troja, S. O., Forzese, R., Caputa, C., et al. (1998). TL and OSL dating of fossil dune sand in the Uan Afuda and Uan Tabu sites (Tadrart Acacus, Libyan Sahara). In M. Cremaschi & S. di Lernia (Eds.), *Wadi Teshuinat. Palaeoenvironment and prehistory in south-western Fezzan (Libyan Sahara)* (pp. 67–72). Roma-Milano: CNR.

McBrearty, S., & Brooks, A. (2000). The revolution that wasn't: A new interpretation of the origin of modern human behavior. *Journal of Human Evolution, 39*, 453–563.

McBurney, C. B. M. (1967). *The Haua Fteah (Cyrenaica) and the Stone Age of the south-east Mediterranean*. Cambridge: Cambridge University Press.

Mourre, V. (2003). *Implications culturelles de la technologie des hachereaux*. Ph.D. Dissertation, Université de Paris X, Nanterre.

Moyer, C.C. (2003). *The organisation of lithic technology in the Middle and early Upper Palaeolithic industries at the Haua Fteah, Libya*. Ph.D Dissertation, University of Cambridge.

Olszewski, D. I., Dibble, H. L., McPherron, S. P., Schurmans, U. A., Chiotti, L., & Smith, J. R. (2010). Nubian Complex strategies in the Egyptian high desert. *Journal of Human Evolution, 59*, 188–201.

Osborne, A. H., Vance, D., Rohling, E. J., Barton, N., Rogerson, M., & Fello, N. (2008). A humid corridor across the Sahara for the migration of early modern humans out of Africa 120,000 years ago. *Proceedings of the National Academy of Sciences, 105*, 16444–16447.

Perego, A., Zerboni, A., & Cremaschi, M. (2011). The geomorphological map of the Messak Settafet and Mellet (Central Sahara, SW Libya). *Journal of Maps, v2011*, 464–475. doi: 10.4113/jom.2011.1207.

Petit-Maire, N. (1982). *Le Shati. Lac Pleistocene du Fezzan (Libye)*. Paris: CNRS.

Petit-Maire, N. (1988). Climatic change and man in the Sahara. In J. Bower & D. Lubell (Eds.), *Prehistoric cultures and environments in the late Quaternary of Africa* (BAR International Series No. 405) (pp. 19–42). Oxford: Archaeopress.

Reynolds, T. (2007). Section 42. The lithics. In D. Mattingly (Ed.), *The archaeology of Fazzan. Volume 2, site gazetteer, pottery and other survey finds* (Vol. 2, pp. 432–447). Tripoli and London: Department of Antiquities and Society for Libyan Studies.

Richter, D., Moser, J., Nami, M., Eiwanger, J., & Mikdad, A. (2010). New chronometric data from Ifri n'Ammar (Morocco) and the chronostratigraphy of the Middle Palaeolithic in the western Maghreb. *Journal of Human Evolution, 59*, 672–679.

Rolland, N. (1995). Levallois technique emergence: Single or multiple? A review of the Euro-African record. In H. L. Dibble & O. Bar-Yosef (Eds.), *The definition and interpretation of Levallois technology* (pp. 361–379). Madison: Prehistory Press.

Sahnouni, M., Hadjouis, D., van der Made, J., Derradji, A.-E.-K., Canals, A., Medig, M., et al. (2002). Further research at the Oldowan site of Ain Hanech, north-eastern Algeria. *Journal of Human Evolution, 43*, 925–937.

Schwenninger, J.-L., Collcutt, S. N., Barton, N., Bouzouggar, A., Clark-Balzan, L., El Hajraoui, M. A., et al. (2010). A new luminescence chronology for Aterian cave sites on the Atlantic coast of Morocco. In E. A. A. Garcea (Ed.), *South-eastern Mediterranean peoples between 130,000 and 10,000 years ago* (pp. 18–36). Oxford: Oxbow Books.

Sharon, G. (2010). Large flake Acheulian. *Quaternary International, 223–224*, 226–233.

Smith, A.B. (1993). Terminal Palaeolithic industries of Sahara: A discussion of new data. In L. Krzyzaniak (Ed.), *Environmental change and human culture in the Nile basin and northern Africa until 2nd Millennium B.C.* (pp. 69–75). Poznan: Muzeum Archeologiczne.

Smith, J. R., Giegengack, R., Schwarcz, H. P., McDonald, M. M. A., Kleindienst, M. R., Hawkins, A. L., et al. (2004). A reconstruction of Quaternary pluvial environments and human occupations using stratigraphy and geochronology of fossil-spring tufas, Kharga Oasis. *Egypt. Geoarchaeology, 19*, 407–439.

Swezey, C. (2001). Eolian sediment responses to late Quaternary climate changes: Temporal and spatial patterns in the Sahara. *Palaeogeography, Palaeoclimatology, Palaeoecology, 167*, 119–155.

Szabo, B. J., Haynes, C. V, Jr, & Maxwell, T. A. (1995). Ages of Quaternary pluvial episodes determined by Uranium-series and radiocarbon dating of lacustrine deposits of eastern Sahara. *Palaeogeography, Palaeoclimatology, Palaeoecology, 113*, 227–242.

Tixier, J. (1963). *Typologie de l'Epipaléolithique du Maghreb. Memoires du centre de recherches anthropologiques, préhistoriques et ethnographiques*. Alger, Paris: A.M.G.

Tryon, C. A., McBrearty, S., & Texier, P. J. (2006). Levallois lithic technology from the Kapthurin Formation, Kenya: Acheulian origin and Middle Stone Age diversity. *African Archaeological Review, 22*, 199–299.

Van Peer, P. (2001). Observations on the Palaeolithic of the south-western Fezzan and thoughts on the origin of the Aterian. In E. A. A. Garcea (Ed.), *Uan Tabu in the settlement history of Libyan Sahara*. All'Insegna del Giglio: Firenze.

Van Peer, P. (2004). Did Middle Stone Age moderns of sub-Saharan Africa descent trigger on Upper Palaeolithic revolution in the Lower Nile Valley? *Anthropologie, 43*, 215–226.

Van Peer, P. (2016). Technological systems, population dynamics and historical process in the MSA of northern Africa. In S.C. Jones & B. A. Stewart (Eds.), *Africa from MIS 6-2: Population dynamics and paleoenvironments* (pp. 147–159). Dordrecht: Springer.

Van Peer, P., Fullagar, R., Stokes, S., Bailey, R. M., Moeyersons, J., Steenhoudt, F., et al. (2003). The Early to Middle Stone Age transition and the emergence of modern human behaviour at site 8-B-11, Sai Island, Sudan. *Journal of Human Evolution, 45*, 187–193.

Van Peer, P., Roots, V., & Vermeersch, P. M. (2008). A wasted effort at the quarry: Wear analysis and interpretation of an MSA lanceolate point from Taramsa-8. *Egypt. PaleoAnthropology, 2008*, 234–250.

Vermeersch, P. M. (2010). Middle and Upper Palaeolithic in the Egyptian Nile Valley. In E. A. A. Garcea (Ed.), *South-eastern Mediterranean peoples between 130,000 and 10,000 years ago* (pp. 66–88). Oxford: Oxbow Books.

Villa, P. (2009). Discussion 3: The Lower to Middle Paleolithic transition. In M. Camps & P. Chauhan (Eds.), *Sourcebook of Paleolithic transitions. Methods, theories, and interpretations* (pp. 265–270). New York: Springer.

Wilkins, J., Pollarolo, L., & Kuman, K. (2010). Prepared core reduction at the site of Kudu Koppie in northern South Africa: Temporal patterns across the earlier and Middle Stone Age boundary. *Journal of Archaeological Science, 37*, 1279–1292.

Zerboni, A. (2008). Holocene rock varnish on the Messak plateau (Libyan Sahara): Chronology of weathering processes. *Geomorphology, 102*, 640–651.

Zerboni, A., Trombino, L., & Cremaschi, M. (2011). Micromorphological approach to polycyclic pedogenesis on the Messak Settafet plateau (central Sahara): Formative processes and palaeoenvironmental significance. *Geomorphology, 125*, 319–335.

Zerboni, A., Perego, A., & Cremaschi, M. (2015). Geomorphological map of the Tadrart Acacus massif and the Erg Uan Kasa (Libyan Central Sahara). *Journal of Maps, 11*, 772–787. doi:10.1080/17445647.2014.955891.

Chapter 8
Technological Systems, Population Dynamics, and Historical Process in the MSA of Northern Africa

Philip Van Peer

Abstract This paper presents an alternative historical interpretation of regional patterns in MSA lithic industries of northern Africa, based on the observation that our current systematics often disguise real similarities and differences in the archaeological record. With regard to the early MSA, it is argued that the geographical distribution of the Sangoan is much wider than previously acknowledged and that it is present up to the Mediterranean coast. It may be the archaeological signature of an early expansion of anatomically modern humans. From the Last Interglacial onward, the distribution of the Nubian Complex records population influx in the eastern Sahara and in regions east of the Nile, including the Red Sea mountains and the Arabian peninsula. During the middle part of MIS 5, human populations in many parts of northern Africa may have been small or even absent. It is only in its final phase that this tendency is reversed again and that Late Nubian Complex sites frequently occur in the entire eastern range of the Saharan-Arabian belt. To the west, the Late Nubian Complex has a marked boundary and it is argued that the Aterian of the Central Sahara records a phenomenon of cultural assimilation between western and eastern populations. All these demographic processes are triggered by the aridification of northern Africa, ultimately leading to a phase of profound cultural and social change of which an Upper Palaeolithic mode of production is the outcome.

Keywords Middle Stone Age • Nile Valley • Sahara • Maghreb • Sangoan • Lupemban • Nubian Complex • Aterian

Introduction

In the last 20 years, the incorporation of evolutionary genetics in the study of recent human evolution has reestablished demography as a key driver of cultural change. With haplotype distributions serving as proxies, migrations become an almost tangible concept, with expected articulations in material cultures. The latter may even stand in a causal relationship with the demographic phenomenon, as reflections of behavioral change preceding demic diffusion.

The analytical upscaling demanded by historical models of population processes confronts us with the limitations of our taxonomic systems and the intellectual traditions in which they are inscribed. The latter, preoccupied with behavioral systems of local human groups, are simply pre-programmed for analytical "splitting" of large biogeographical zones. Even though casual supra-regional comparisons usually form a part of regional syntheses, "lumping" has not been the epistemological rule for many decades. Now that the paradigmatic climate is turning, it appears that comparative studies at the appropriate scale are seriously hampered by the lack of analytical concordance. Further, without an adequate grammar, to paraphrase A. Whittle (1996), historical description and evolutionary explanation cannot be achieved.

The North African Middle Stone Age (MSA) / Middle Paleolithic in particular is an example of how archaeological grammars can shape different historical "realities" according to the research tradition from which they originate. The aim of this paper is to point out how some technological similarities among MSA sites at the scale of the whole of North Africa, and even beyond, have been disguised by regional taxonomies and nomenclatures. Paradigmatic frontiers often are more difficult to breach than geographic boundaries, but once the shift is achieved, unexpected patterns of similarity become obvious. Taking a cursory review of the late Middle and early Late Pleistocene record from northeast Africa as its starting point, some supra-regional analogies are identified

P. Van Peer (✉)
Prehistoric Archaeology Unit, Institute of Geo-Sciences, Katholieke Universiteit Leuven, Celestijnenlaan 200E, 3001 Louvain, Belgium
e-mail: philip.vanpeer@ees.kuleuven.be

Sacha C. Jones and Brian A. Stewart (eds.), *Africa from MIS 6-2: Population Dynamics and Paleoenvironments*, Vertebrate Paleobiology and Paleoanthropology, DOI 10.1007/978-94-017-7520-5_8

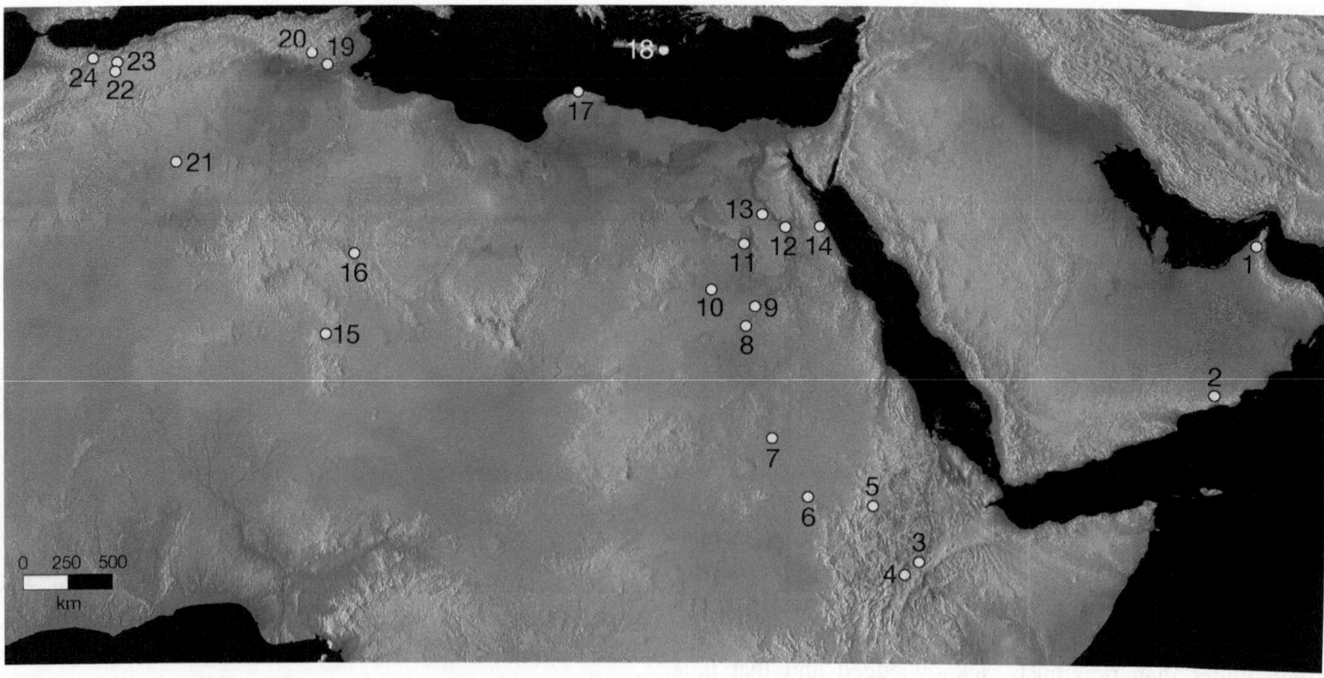

Fig. 8.1 Map of sites mentioned in the text. *1* Jebel Faya; *2* Dhofar; *3* Herto; *4* K'one; *5* Gorgora; *6* Abu Hagar; *7* Khor Abu Anga; *8* Sai 8-B-11; *9* Arkin 5/8; *10* Bir Tarfawi; *11* Kharga; *12* Taramsa 1; *13* Nazlet Khater 4; *14* Sodmein Cave; *15* Adrar Bous; *16* Uan Tabu; *17* Haua Fteah; *18* Preveli; *19* El Guettar; *20* 'S Baikia; *21* Zaouia El Kebira; *22* Shab al Ghar; *23* Rhafas Cave; *24* Ifri N' Ammar

and interpreted in the framework of the early dispersal history of anatomically modern humans (Fig. 8.1).

The Middle Pleistocene MSA

Geographic Patterns in Technological Systems

The Northeast African Sangoan and Lupemban

Elsewhere, I have argued that the MSA in northeast Africa is a comparatively short stage, beginning abruptly at ~200 ka (Van Peer 2004). In terms of technological constellations, the earliest assemblages are characterized by discoidal debitage strategies (Fig. 8.2:2) in combination with bifacial technology in the form of core-axes (Fig. 8.2:1). The latter are a distinct hallmark of the Sangoan, and technically, they are very different from handaxes. Following Clark and Kleindienst (2001:50), core-axes are defined as bifacial tools produced by hard hammer technique, with a retouched blunt distal end. In cross-section they are thick. These authors distinguish between early and late forms, indicating that the latter are more symmetrical, usually with two core-axe ends on opposite sides. In section, these ends may appear as symmetrical (axe), asymmetric (adze) or concavo-convex (gouge). In addition to those traits, I might add the scalariform appearance of the bifacial retouch on the lateral sides of a core-axe. In numerous specimens, the transversal symmetry is shaped by different generations of retouch, resulting in characteristically thick sections. Due to the repetitive retouch, the edges often have acquired a 'battered' aspect. While both technically and morphometrically core-axes are very different from handaxes, the two technological systems have often been conflated and this is most certainly part of the confusion surrounding the Sangoan, next to site formation histories.

In the region of concern here, Sangoan technology is found at Middle Nile Valley sites such as Khor Abu Anga (Arkell 1949), Sai 8-B-11 (Van Peer et al. 2003), and Arkin 8 (Chmielewski 1968), and further to the south at Abu Hagar (Lacaille 1951) and Herto (Clark et al. 2003).[1] At Sai 8-B-11, spatial analysis of a primary context occupation level from the base of a paleochannel, combined with a detailed technological study, suggest the presence of two behavioral systems in the same landscapes. At the larger scale, this pattern of site formation may be in line with

[1]This identification represents the present author's opinion, based on the data presented in the cited reference and on a cursory inspection of some artefacts during a visit to the National Museum in Addis Abeba, facilitated by Tim White on the occasion of *The Middle Stone Age of East Africa and Modern Human Origins* conference, National Museums of Kenya and National Museum of Ethiopia, 2005.

population influx in the northern reaches of the Nile Valley during MIS 7.

Behaviorally speaking, the Sangoan material culture is the consequence of a subsistence system with an emphasis on sub-surface exploitation of resources, both foodstuffs and mineral materials. During the period of MIS 6 it changes into a Lupemban facies, with the addition of lanceolate foliates and volumetric blade production (Fig. 8.2:3).

Other Northern Sangoan and Lupemban Occurrences

There are several documented instances of Sangoan material constellations in North Africa. With all caution that is due, Sangoan occurrences can even be claimed for the island of Crete as I shall discuss below.

The Maghreb

Northwest of the Tunisian city of Gafsa, at the Algerian border, there is a cluster of sites that were discovered by M. Reygasse early in the 20th century, including the eponymous site for the Aterian, Bir-el-Ater. At the locality of S'baikia, a surface industry was collected containing numerous bifacial tools ranging from thick symmetrical forms to bifacial foliates. This constellation was coined as the Sbaikian (Balout 1955). Another Sbaikian assemblage comes from Oued Mahrouguet (Anonymous 1956) and its bifacials show very striking morphological and, as far as it can be judged from the drawings, technical similarities with the core-axes from Sai 8-B-11.

From Shab Al Ghar, in the high peneplain of central Morocco, comes an assemblage (Wengler 1993: 649–653) with very obvious Sangoan traits. It does contain a number of quite typical core-axes, which Wengler (1993) identifies alternatively as *bifaces amygdaloïdes* (Wengler 1993: Fig. 268,1) or *bifaces ovalaires* (Wengler 1993: Fig. 268,2). The entire technological and typological make-up of this small assemblage, which includes a very distinct tranchet on a large flake (1993: Fig. 269,3), is remarkably similar to the Sangoan at Sai 8-B-11.

The Sahara

At Adrar Bous, in Niger, several Acheulean and Aterian surface sites were recorded (Clark et al. 2008). One of the claimed Acheulean occurrences—Main Wadi S120—may in fact be Sangoan, at least partially. At any rate, the morphologies of many bifacial tools here are different from those at the other Acheulean locations, such as Diatomite Dune Terrace S144. The latter contains typical Late Acheulean lanceolate forms together with large flake cleavers. At S120, the symmetrical *limande* is much more common and many of these, in fact, are probably core-axes. Again, it is difficult to infer technical features from conventional drawings but some of these tools seem to exhibit the Sangoan technique of superimposing generations of lateral retouch on thick, stocky volumes. However, it would be surprising that a Sangoan connotation for S120 would have gone unnoticed by J.D. Clark, who was familiar with the sub-Saharan record, even if it is not clear to what degree he was involved in the actual lithic analysis.

Leaving S120 aside, Clark does identify core-axes in the so-called Aterian assemblage from Yellowstone Hill S140 (Clark et al. 2008: Fig. 5.7). In this part of Adrar Bous, artifacts were collected from the surface of yellow-brown aeolian sands from which they were seemingly eroding. These are likely dunes associated with the edge of a paleolake. S140 was collected from an area of 100 m^2 (Clark et al. 2008: 110). Unfortunately, there is no section from this locale but using the one from nearby Lookout Hill as a reference (Clark et al. 2008: Fig. 3.8a [b]), it is clear that artifacts occur in different stratigraphic positions within lacustrine and aeolian sands. Clearly and predictably, this lacustrine area has been occupied multiple times, both at historical and geological timescales. Surface collections even when restricted in surface, must be expected to be palimpsests and this applies to S140 as well, as acknowledged by Clark et al. (2008: 110). The core-axes undoubtedly are included in what is considered the Lower Paleolithic portion of the collection: "some artefacts of apparent Lower Palaeolithic provenance were encountered" (Clark et al. 2008: 110). In fact, they are early MSA Lupemban, as most of the S140 material might actually be. The debitage technology comprises discoidal reductions and a clear blade component from volumetric cores (Clark et al. 2008: Fig. 5.10, 6). The Aterian tanged pieces are only an admixture and it is not appropriate to identify this as an Aterian assemblage.

The Mediterranean

Very recently, Lower Paleolithic occurrences have been reported from nine sites on the southeastern coast of Crete (Strasser et al. 2010). At the surface of pediments in front of the raised beach conglomerates at different elevations, quartz artifacts were collected, some of which are bifacial tools.

Strasser et al. (2010: 173) assert that at the sites of Preveli 2 and 3 "the artefacts appear to be derived from these

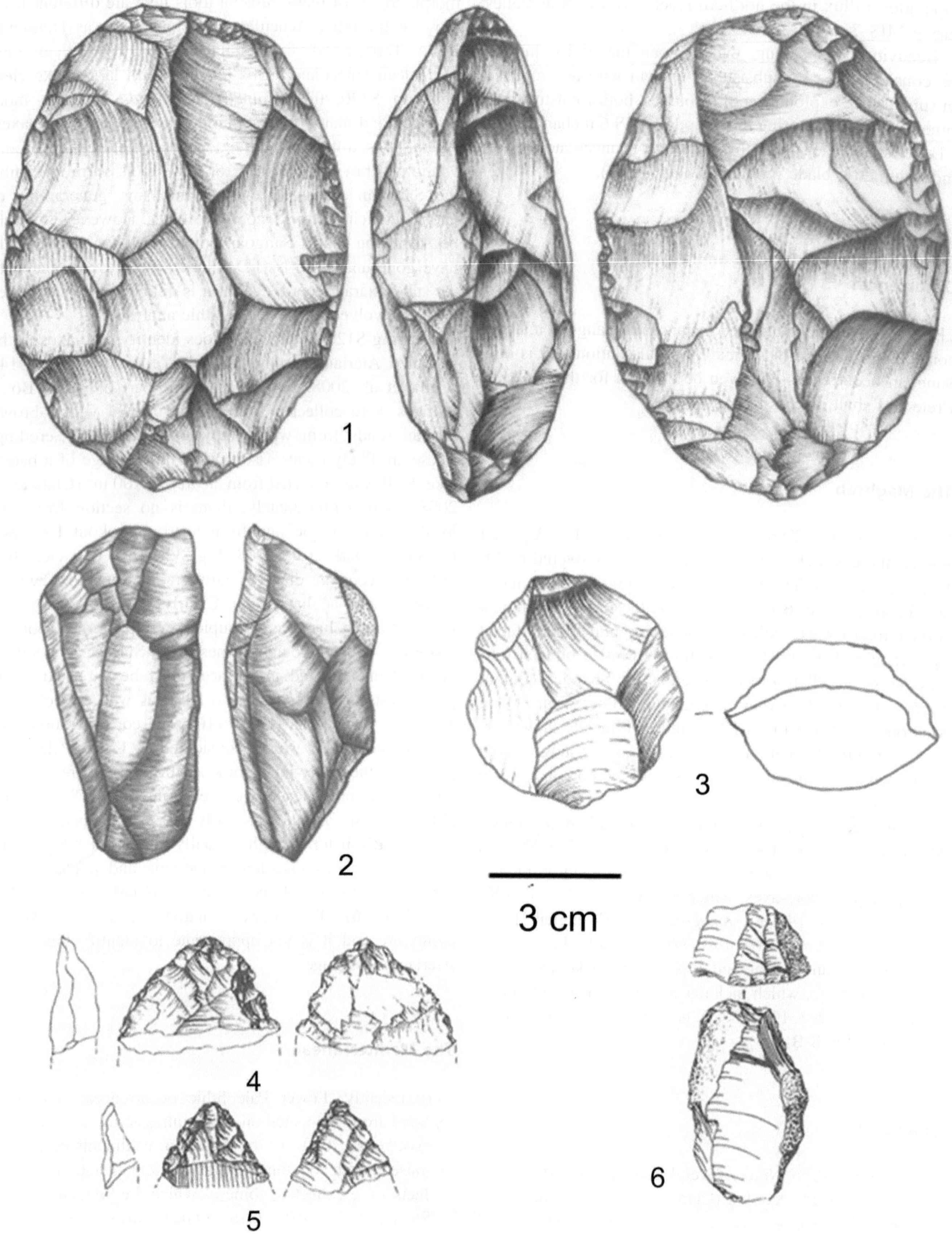

Fig. 8.2 Selected lithic items from early MSA and Early Nubian Complex assemblages. *1* Core-axe (Middle Sangoan level, Sai 8-B11); *2* Volumetric blade core (Lupemban level, Sai 8-B-11); *3* Discoidal core (Lower Sangoan, Sai 8-B-11); *4–5* Foliate tips (early Nubian Complex, Bir Tarfawi BT-14 Grey Phase I); *6* Nubian end scraper (early Nubian Complex, Bir Tarfawi BT-14 Grey Phase I). *4–6* are reproduced from Wendorf et al. (1993: Figs. 22.20, 22.25)

conglomerates," but as long as they are not unequivocally observed within the beach deposits, there is no stratigraphic argument as to the age of those artifacts. At the site of Preveli 7, Paleolithic artifacts are claimed to be in stratigraphic context within reworked paleosol deposits on which another "well-preserved paleosol Bt horizon exhibiting the characteristics of a highly mature soil" (Strasser et al. 2010: 174) is supposedly developed. Even if artifacts were contained within these deposits, their Pleistocene age is certainly not beyond doubt. It is most likely very difficult to distinguish an *in situ* Bt from reworked pedogenetic material, especially in a sedimentological environment as shown in the published picture of the profile (Strasser et al. 2010: Fig. 29). As the time of reworking is unknown, nothing can be said about the dating of artifacts that might eventually be comprised within them.

While at this point there does not seem to be good, independent chronological confirmation for the Paleolithic of Crete, the artifacts do suggest that this can be expected in the future. Based on technological considerations, Strasser et al. (2010: 178) suggest that the "lithic artifacts may belong to more than one Palaeolithic industry, including in traditional terms the Lower and Middle Palaeolithic." I do not believe that there is any evidence of Lower Paleolithic technologies in the current assemblages. Obviously, it is the bifacial tools that give way to this assertion but all these specimens in fact evoke the Sangoan/Lupemban core-axe technology (e.g., Strasser et al. 2010: Figs. 32–34). They are not bifaces in the morphological nor in the technical sense.

Human Biogeography

Sound chronological evidence for any of the regions or sites mentioned above is very rare indeed. Likewise, it is difficult to infer the historical process underlying the presence of this early MSA technological system from the present archaeological evidence, both in terms of technological phylogenies and site formation processes.

Yet, a few things are clear. Despite the absence of detailed technical lithic analysis in most of the concerned cases, we do observe one and the same homologous technological system. This is the Sangoan/Lupemban system that uses debitage techniques for symmetrical core-axe manufacture. The latter typically occurs in assemblage constellations where discoidal and volumetric blade reductions are also practiced. Further, it seems unlikely that the core-axe technical system could be part of a technological lineage originating in local Late Acheulean *façonnage*. How the core-axe technical process might be related to earlier systems is unknown, but scanty evidence on site formation processes in the region of concern here suggests that the former is intrusive. At Sai 8-B-11, for instance, interstratified Acheulean and Sangoan occupation events occur over a relatively short period of time. Sometimes these spatial scatters form palimpsests at the same land surface. They are always stratigraphically associated with fluvial gravels suggesting active channels and humid conditions. The present chronological evidence suggests that this occurred in MIS 7 (Van Peer et al. 2003). As for the human anatomies associated with the Sangoan, the evidence is again scanty. There is no basis for associating the Singa fossil with the small Sangoan assemblage from Abu Hagar, other than its proximity and the likelihood that they come from the same fossiliferous context. At Herto, the spatial association between the human fossils and lithic industry is better but the Sangoan nature of (part of) the latter awaits demonstration.

While it will take much more field evidence to draw the precise historical contours of the phenomenon, there is much reason to believe that the archaeological record testifies to a very significant range expansion of Middle Pleistocene *Homo sapiens*, including regions outside of Africa. If Crete is included, it is most likely that it was reached through island hopping from the Turkish landmass in the northeast. In this area, north of the Hellenic Arc (Taymaz et al. 2007), distances across the open sea would have been minimal.

The Middle Stone Age During MIS 5

Geographic Patterns in Technological Systems

The Northeast African Nubian Complex

By the last interglacial (MIS 5e), the early MSA industries have changed into the Nubian Complex, an MSA variant specific to this part of Africa (Marks 1968a). Its hallmark is the Nubian reduction strategy, a specific manifestation of the Levallois system designed to produce pointed forms (Fig. 8.3:1,3,4). Its origin can be found in Lupemban foliate *façonnage*. This Nubian production system is a very distinct association of discrete technical and morphological traits and it shows an evolutionary sequence throughout MIS 5. The latter is well demonstrated at the site of Taramsa 1, in the Lower Nile Valley (Van Peer et al. 2010).

In its spatial distribution across sites, this new technological system of Levallois blank production is employed in the context of specific logistical tactics involving, among other things, the anticipated production of points at workshops. In the Nile Valley, it has been argued that the Nubian Complex settlement system is founded on logistical tactics involving residential areas at the floodplain edges, and special activity locations elsewhere in the landscape (Van Peer 1991, 2001; but see also Olszewski et al. 2010).

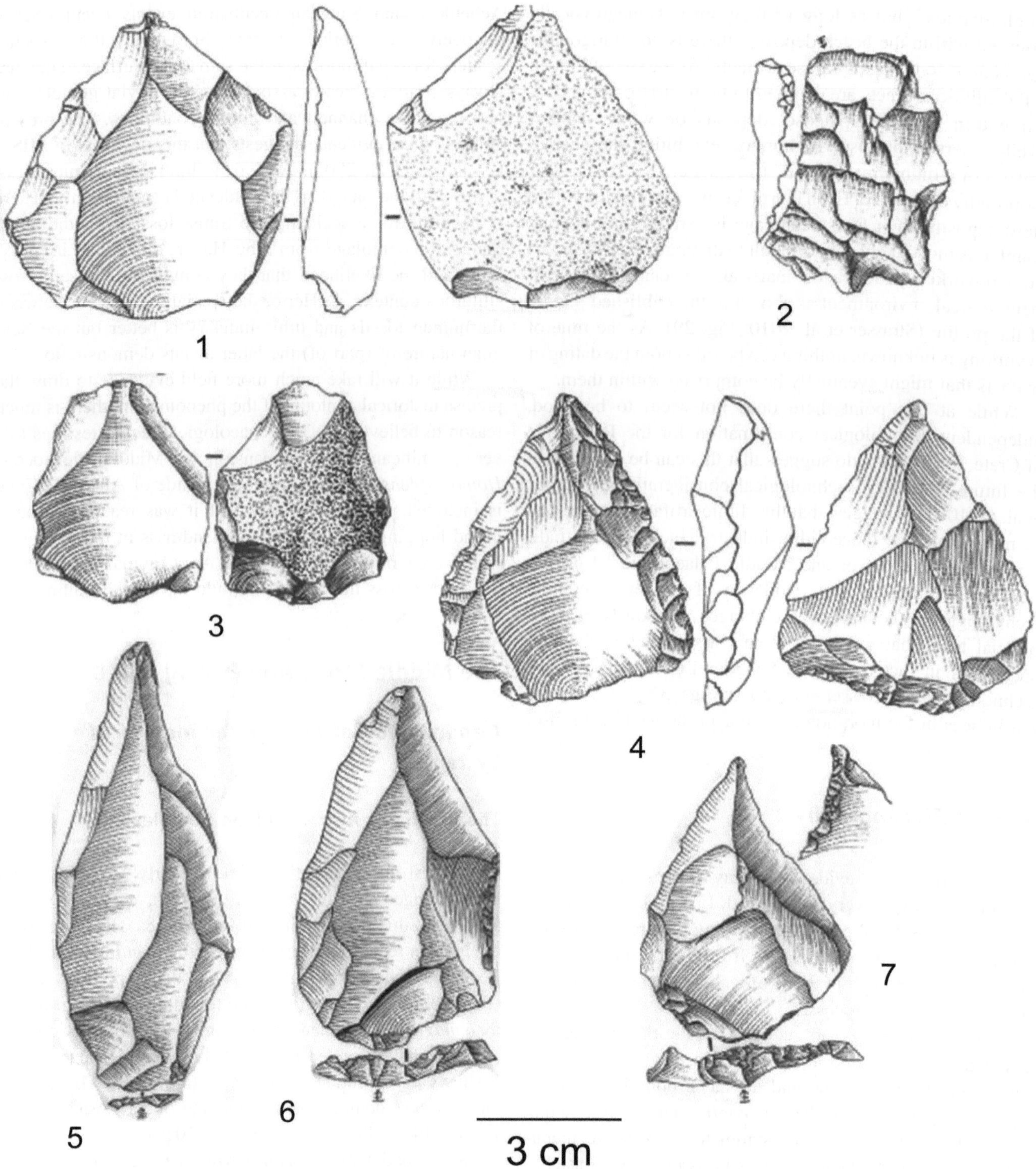

Fig. 8.3 Selected lithic items from the Late Nubian Complex assemblage of Nazlet Khater 1. *1* Nubian type 1 core (*Middle layer*); *2* Nubian ridge preparation element (*Upper layer*); *3* Nubian type 1 core (*Upper layer*); *4* Nubian type 2 core (Middle layer); *5–6* Nubian points (*Middle layer*); *7* Nazlet Khater point (*Upper layer 1*)

There is little chronological and stratigraphic evidence to assess the internal variability within the Nubian Complex both at a regional scale and at a satisfactory level of resolution. At the few sites with a stratigraphic succession of Nubian Complex assemblages, an early phase with thin leaf-shaped foliates (Fig. 8.2:3,4) and so-called "Nubian end scrapers" (Guichard and Guichard 1965) (Fig. 8.2:5) appears to be present. On the other hand, it is not unlikely that bifacial foliate frequencies vary across assemblages according to the same logistical tactics that underlie the patterning in the Nubian point production systems. In terms of its assemblage features, the later Nubian Complex differs from the earlier in several respects. The differences are most evident in the absence of bifacial foliates and in the much more frequent presence of the Nubian type 1 method of point production (Fig. 8.3:2,5,6). These points or pointed flakes served as blanks for tools such as the Nazlet Khater point (Fig. 8.3:7). A few absolute dates suggest that such Late Nubian Complex (LNC or, alternatively, Nubian Complex *sensu stricto*) assemblages date to the end of MIS 5. For example, site E-87-3 is associated with the 5a Green Lake phase at Bir Tarfawi (Hill 1993). At Taramsa 1, aeolian sands filling in a late Nubian Complex exploitation pit have an OSL age determination of 79 ka (Van Peer et al. 2010). At that same site, the child burial represents a firm reaffirmation of the association of the late Nubian Complex with anatomically modern humans.

The Nubian Complex in the South

The distribution of Nubian Complex sites in Africa is much wider than northeast Africa *sensu stricto* (i.e., the Lower Nile Valley and its immediately neighboring deserts). It occurs in the Horn of Africa (Clark 1954) and in Ethiopia, at sites such as K'one (Kurashina 1978) and Gorgora (Clark 1988). At K'one, different concentrations were excavated and there is quite some lithic variability among them. The largest collection comes from K'one 5E and it is overwhelmingly dominated by the Nubian reduction strategy. However, it is not a "classic" MIS 5a assemblage and the Nubian cores show many idiosyncratic technical features.[2] There are no dates for K'one and it is possible that sector 5E is post MIS 5a.

The Nubian Complex in the East

At rock shelter FAY-NE1, in the Jebel Faya Mountains in southeastern Arabia, assemblage C from the lowermost layer contains bifacial implements in association with Levallois and volumetric blade reduction strategies (Armitage et al. 2011: 454). The assemblage is not yet published in great detail, but a Lupemban/early Nubian Complex is clearly apparent (Armitage et al. 2011: Fig. 2,1). The authors themselves note that, "technological patterns at FAY-NE1 show greater similarities with East and northeast Africa than with other sites in Arabia" (Armitage et al. 2011: 455). OSL dates place assemblage C at the end of the Middle Pleistocene or in the first part of MIS 5.

Even more compelling evidence for the southwest Asian occurrence of the Nubian Complex comes from Dhofar in Oman (Rose et al. 2011; Usik et al. 2013). This region is literally strewn with Nubian Complex sites, which, on the one hand, all share the Nubian 1 technological system, and, on the other, display quite some variability in terms of derived technical features (Usik et al. 2013). So far, there are two OSL ages that suggest a position in the middle of MIS 5 for some of these Dhofar sites (Rose et al. 2011).

The Nubian Complex in the West

An early MIS 5 manifestation of the Nubian Complex is encountered at Haua Fteah Cave, in Cyrenaica, Libya. Here, the Pre-Aurignacian occurs in Phase A of the sequence, underneath the so-called Levallois-Mousterian (McBurney 1967; Wendorf and Schild 1992; Clark 1982; Reynolds 2013). New fieldwork at Haua Fteah is being undertaken in the framework of the TRANS-NAP project led by Graeme Barker (Jones et al. 2016), and, in that context, a reassessment of these assemblages was recently published (Reynolds 2013). Based on the latter (e.g., Reynolds 2013: Fig. 5) and on a personal evaluation of a selection of Pre-Aurignacian artifacts during a short research visit in early 2013, it is clear that there are very obvious similarities between the Pre-Aurignacian and the Taramsa Phase II technology (Van Peer et al. 2010). The latter has been identified as an incipient stage in the technological development of the Nubian production system and the relevant assemblage at Taramsa 1 was sandwiched between aeolian sands dating to 88.8 ± 9.5 ka (above) and 117 ± 10.5 ka (below), respectively (Van Peer et al. 2010). As the artifacts display a quite strongly developed surface patina, the underlying date is probably closer to deposition time. Significantly, the tips of three bifacially worked tools were reported for this Haua Fteah Pre-Aurignacian (McBurney 1967), two of which are definitely the distal ends of lanceolate foliates. This is another distinct similarity with the Lupemban/early Nubian Complex.

Some of the key sites in Gertrude's Caton-Thompson's (1952) sequence for Kharga Oasis have recently been revisited in the context of a U-series dating program (Smith et al. 2007). One of the original sections at Bulaq Wadi 3 was exactly relocated (Locus 1) and the tufa above a small lithic assemblage was dated to 114.4 ± 4 ka. A number of the artifacts from this "Levalloiso-Khargan" assemblage are

[2]Personal observation during a research visit at the National Museum in Addis Abeba. June 2012.

illustrated (Caton-Thompson 1952: Plates 74–75) and again, the similarities with the Taramsa Phase II technology seem clear.

Whereas the Nubian Complex in its late MIS 5 form is present in the regions east and south of the Lower Nile Valley, such is not the case further to the west. Beyond the western border of present-day Egypt, Late Nubian Complex sites do not occur anymore. However, in the Aterian of the Sahara there are very clear traces of the Nubian blank production system. Of utmost significance for the close association between Nubian Levallois and tang technology is the assemblage from Uan Tabu in the Acacus region of the Central Sahara. Here, aeolian sand unit IV contains an Aterian assemblage with classical tanged points elaborated on Nubian blanks and is OSL-dated to 61 ± 10 ka (Garcea 2001). In the assemblage from Zaouia El Kebira in the Saoura basin of the northwestern Sahara, the Nubian 1 method of production is very extensively used (Chavaillon 1971). It is likewise attested in Aterian assemblages at Adrar Bous (Clark 1993; Clark et al. 2008).

Human Biogeography

MIS 5 in Africa is characterized by enhanced climatic variability at various temporal scales (Carto et al. 2009). It incorporates the Last Interglacial (MIS 5e) and a stadial/interstadial sequence (MIS 5d-a). At the millennial scale it includes two Heinrich events (H7 and H9) during which cooling of the southern Atlantic occurred. Unfortunately, the archaeological chronostratigraphic record is generally not at a resolution that permits the isolation of these climatic stages and this is obviously a fundamental obstacle to biogeographic modeling. Yet, some patterns can be cautiously identified at substage resolution.

The Last Interglacial

It is a logical expectation that in the Last Interglacial massive new territories became available to humans within northeast Africa and that there was population influx into the eastern Sahara and the Red Sea Mountains (Derricourt 2005). Empirical evidence for this comes from early Nubian Complex sites in the depressions at Bir Tarfawi and Bir Sahara (Wendorf and Schild 1992; Wendorf et al. 1993), and from Sodmein Cave in the Eastern Desert. The homology of the Pre-Aurignacian at Haua Fteah and the early Nubian Complex at Taramsa 1 suggests that, as a result of this range expansion, northeast Africa and the eastern Sahara up to the Mediterranean coast formed one large biogeographic zone with a common technological system. This zone may be bounded in the west by the North African watershed and the river system that discharged into the Gulf of Sirte (Osborne et al. 2008).

On the other hand, traditional "refuge" areas such as the Nile Valley may have become less favorable for occupation and if so, this must have been another stimulus for dispersal. The full interglacial valley may have been too densely forested to have been a suitable environment for human occupation (Kleindienst 2000). Some evidence in line with this is found at site 440 in Nubia (Shiner 1968). Here, a true non-calcic brown soil was developed on local sandy sediments and covered by Nilotic silts. The two archaeological levels at this site are associated with sediments that predate the period of pedogenesis and not with the soil formation itself. The presence of an Interglacial forest in the Nile Valley would also have made the Nile delta unsuitable for occupation and an unlikely corridor for dispersal. Thus, a northern route along the Mediterranean coast probably did not exist under full interglacial conditions.

The MIS 5d and 5b Stadials

Mid-MIS 5 comprises two stadials, each ending with a Heinrich event and separated by the short interstadial of MIS 5c. Climatic modeling for H9 at 105 ka and for the period immediately preceding it (Carto et al. 2009) has shown that severe arid conditions during MIS 5d prevailed throughout the whole of northern Africa – much more so than in central and southern parts of the continent. Expansion of barren soil environments would have been particularly pronounced in West and North Africa (Carto et al. 2009: 146). Carto et al. (2009) conclude that the same pattern would apply to other Heinrich events, including H7 at 85 ka.

With all the reservations that are in place, our own work in northeast Africa does seem to indicate that human occupation is very restricted in the middle portion of MIS 5. A thick sterile layer in the Sodmein Cave sequence (Van Peer et al. 1996), most likely deposited in MIS 5b, separates early and late Nubian Complex assemblages and represents evidence of human absence. A similar hiatus is suggested by the OSL-chronology for the Taramsa 1 workshop.

In the Maghrebian zone, the sequence at Rhafas cave (Wengler 1993) may be instructive in the same sense. Here the aeolian sands of level 6 have an OSL age determination of 107 ± 12 ka (Mercier et al. 2007), and level 3a containing an Aterian assemblage has a TL age of ~70 ka. In stark contrast to these two important occupation events, the levels in between have only produced very small assemblages.

At least a number of the Nubian Complex sites in Dhofar in the south Arabian Peninsula would appear to date from mid-MIS 5 precisely, according to the OSL age estimates available (Rose et al. 2011). The technological similarity of the Dhofar Nubian Complex and the Nile Valley sites is stunning. According to normal taxonomic rule, all these material culture constellations should go under the same techno-complex designation regardless of the geographic presence of the Red Sea. The only possible historic explanation for this pattern seems to be demic diffusion into Arabia during arid MIS 5, when population evidence in Africa is scarce. The conventional dating of the earliest modern humans in the Levantine area, based primarily on averaged values obtained by different methods and of different layers at Qafzeh (Hovers 2009), is between 119 and 85 ka. Chronologically, therefore, this is perfectly concordant with the expansion into Arabia.

The MIS 5a Interstadial

After the reduced population levels during mid-MIS 5, Nubian Complex sites increase in number in MIS 5a, and the technological constellations now assume their classic form. Numerous workshops are known on the terraces of the Lower Nile Valley and on the fringes of the high desert (Olszewski et al. 2010), though very few of them are actually dated. It would seem that the techno-complex reaches its maximal geographic distribution, at least to the south and east. According to the present geographical and chronological patterning in Nubian Complex assemblages, it is quite possible that the technological adaptations giving way to the Late Nubian Complex occurred in Arabia and that they were reintroduced in northeast Africa during MIS 5a. In any case, recent genetic work has demonstrated that there have been migrations of modern humans back into Africa already during the Pleistocene (Sánchez-Quinto et al. 2012).

Strangely enough, the distribution in western North Africa is limited. In fact, it seems more limited at this point than earlier on in MIS 5. At Haua Fteah, for instance, there is no evidence at all of a Late Nubian Complex occurrence. However, it is imperative in this context to consider the Aterian.

I have earlier advanced the idea (Van Peer 2001, 2004; Van Peer et al. 2010) that the Aterian is really the Saharan manifestation of the Nubian Complex. Its origin should be situated in the context of populations who had reestablished themselves in the eastern Sahara coping with its final aridification following H7 (Carto et al. 2009). While this interpretation got some support (Clark et al. 2008), it might now seem to be contradicted by the most recent chronological evidence from the Maghreb, and from the site of Ifri n'Ammar in particular (Nami and Moser 2010). Yet, this is the case if that evidence is only taken at superficial face value. A more precise consideration of site formation processes and technologies, in fact, provides additional support for the late emergence of the Aterian as an adaptation to aridity. I do believe that in its essence, the hypothesis of a (eastern) Saharan origin of the Aterian still stands even in the face of the new empirical evidence from the Maghreb, as I will argue below.

The Question of the Aterian

As far as the definition of the Aterian is concerned, Wengler (1993) makes a distinction between proto-Aterian, which is a Mousterian with tanged elements, and Aterian, which is defined in Bordesian terms (i.e., a Mousterian facies with specific typological and technological indices). Although this terminology accommodates the perception of a classic culture-historic developmental sequence in the Maghreb, it is a useful distinction as it at least helps to solve some of the definitional ambiguities surrounding the Aterian.

Let us, for the moment, conceive of the Aterian in the first sense, that is, as a Mousterian or Middle Stone Age industry with some tanged elements, and assess the pattern of its spatiotemporal distribution. First, it should be pointed out that not even under this simple definition is complete coherence achieved since some authors have also considered bifacial foliates as an Aterian "guide-fossil" (Guichard and Guichard 1965; Wendorf and Schild 1992). Leaving these aside, it is still clear that such an Aterian is present in eastern North Africa and that it is at least as old if not older than in the Maghreb, even accounting for the OSL-estimate of 145 ± 9 for the proto-Aterian at Ifri n'Ammar (Nami and Moser 2010). A few tanged elements occur in a number of Nubian Complex assemblages from the Nile Valley, such as E-78-11 (Singleton and Close 1980) and Arkin 5 (Chmielewski 1968). There are some tangs present as well in MIS 5 assemblages at Bir Sahara and Bir Tarfawi (Wendorf et al. 1987, 1993).

Under this definition, El Guettar-lowest level (Gruet 1954) is an Aterian assemblage as well as the one from nearby Ain Meterchem (Balout 1955). Interestingly, a comparative attribute analysis of Levallois flakes using principal components analysis not only places El Guettar near Nubian Complex assemblages from the Lower Nile, but also next to "unquestionably" Aterian assemblages such as Bir-el-Ater and Ain Mansourah that come from the same region as El Guettar (Van Peer 1991).

Based on all this evidence, the interpretation can be proposed that both the proto-Aterian and Nubian Complex alike are inscribed in lineages with the earlier Sangoan/ Lupemban of sub-Saharan origin. This gives credit to

G. Caton-Thompson's (1946): 28 suggestion that "...this is the possibility that Central Equatorial Africa was the shaping ground [of the Aterian] and the control centre of Africa's later Palaeolithic industries." In any case, tanged pieces are part of the late Middle Pleistocene technological array here, for instance in the Lupemban of the Twin Rivers site in Zambia (Barham et al. 2000).

As a further consequence of the demographic crisis during mid-MIS 5 in North Africa, and as something very different from the proto-Aterian, the Aterian in its second definition above has emerged. Using technical traits that had been part of the technological package for a long time, it was an adaptation to the constraints of arid Saharan environments (Van Peer 2001, 2004). This would imply that it is intrusive in the Maghreb, *contra* the interpretation of its gradual local emergence via the proto-Aterian (Debénath 1996; Wengler 2001). Perhaps, the gradual influx of groups from the Sahara can provide the same archaeological pattern. Based on a few raw material provenance studies, it has been argued that these Aterian groups were highly mobile foragers. In the Aterian of Adrar Bous, for instance, silicified vitric tuff has been introduced from a distance of 280 km (Clark 1993; Clark et al. 2008).

Elaborating on the hypothesis that the Late Nubian Complex represents a reflux of populations into Africa, we may see the "Nubian Aterian" as at Uan Tabu (Cancellieri et al. 2016) and Zaouia El Kebira (Chavallion 1971) as the result of a process of cultural assimilation among the Late Pleistocene Saharan populations.

The End of the MSA in Northeast Africa

In terms of occupational density, the present archaeological evidence from the Lower Nile Valley is ambiguous. On the one hand, there are very few MIS 4/3 sites. On the other hand, in those few areas where final MSA occupations are attested, some evidence suggests dense population. This is particularly the case in Upper Egypt, at sites such as Taramsa 1. The enormous quantities of raw material that have been exploited here during MIS 4/3, and the technological change that occurs, have been interpreted as indicating population pressure in this area (Van Peer et al. 2010). Several explanations for this apparent contradiction are possible. Taphonomic conditions may have played a part in eroding sites alongside the deposits of reduced floodplains from this time (Schild et al. 1992). Second, with its reduced floodplains and perhaps increased seasonality, only certain portions of the Valley may have provided the conditions to sustain large populations (Schild and Wendorf 1986: 31).

The lithic technology characteristic of the Nubian Complex is no longer present in such assemblages. Technological analyses of assemblages from Taramsa 1 (Van Peer et al. 2010), Nazlet Safaha (Van Peer et al. 2002) and Khormusan sites in Nubia (Marks 1968b) show various trajectories of lithic technological change among populations with complex social behavior patterns.

Under these conditions, social change could have come about in groups who were competing over resources, ultimately leading to the acquisition of a new social system that provided the foundation for an Upper Paleolithic way of life. One of the future research challenges is to find out what that specific new social form was. Internal competition, inequality, property acquisition, population increase, and enhanced territorial expansion may all be part of that answer. One Nile Valley site hints at the persistence of the Taramsan blade technology in an early Upper Paleolithic context. The blade assemblage from the site of Nazlet Khater 4 in Middle Egypt evokes some Taramsan features (Vermeersch et al. 2002). In underground mine shafts, chert pebbles were exploited here in a period between 40 and 35 ka. The use of bifacial axes as exploitation tools is also reminiscent of the Taramsan. One axe was found as a burial gift in one of the two burials that are associated with the chert mine of Nazlet Khater 4. This may be the individual's personal item and an indication of social status or economic specialization. It testifies to the acknowledgement of individuality and the existence of social/economic distinction in this society. Was there, in this respect, a qualitative difference with other contemporaneous societies?

In the southern Levant, the Emiran level I at Boker Tachtit has transitional features leading to a fully Upper Paleolithic Ahmarian technology in level IV. The beginning of this transitional sequence is usually placed at $\sim$47 ka, but it may be slightly prior to 50 ka (Marks 2003). A Levantine Late Middle Paleolithic antecedent for the Emiran is presently unknown (Marks 2003) and as a consequence, some have proposed that the stimulus for this process of change came from outside, and from the Nile Valley in particular (Bar-Yosef 2000, 2003; Tostevin 2003). The Taramsan and Emiran reduction systems, both studied by means of completely refitted sequences, show a common adaptation of the classic Levallois surface exploitation to volume exploitation (Van Peer et al. 2010). Whether they actually represent the same technical system composed of the same technical modules will have to be established through a first-hand comparative analysis. The possibility of homology and of cultural diffusion is at least entertained by the present evidence and, as the Taramsan is older than the Emiran, that diffusion of an incipient Upper Paleolithic technology must be from Africa to the Levant.

Conclusion

Over the last 25 years, knowledge of the evolution of modern humans and the historic processes of their establishment across the Old World is gradually taking shape. It is archaeology's role to critically develop models based on genetic diversity patterns into historic and explanatory scenarios. Almost by definition, the latter are bound to reveal a more complex pattern than the history recorded in the genetic make-up of present humans. The pattern of linked technologies across Late Middle and early Late Pleistocene sites in northern Africa and adjacent regions suggests that this is exactly the case.

The early MSA Sangoan/Lupemban complex most likely represents the signature of the earliest expansion of modern humans, genetically recorded in the distribution of haplogroup L1 (Watson et al. 1997). Interestingly, at frequencies of less than 1%, L1 haplotypes survive in present Mediterranean and Arabian populations (Forster 2004). This early technological system contains, in an ancestral form, all the technical traits that then reappear throughout the later MSA, at various times and in different places.

Episodes of extreme arid conditions occurred in tropical Africa between 135 and 90 ka, while the North experienced humid interstadial conditions during this time. After 90 ka, the pattern reversed and final aridification of the Sahara set in (Carto et al. 2009). It is most probable that Heinrich Event H9 in particular had serious demographic consequences at the continental scale, especially since high population densities seem to have been reached in sub-Saharan Africa at ~100 ka (Powell et al. 2009). In the region northeast of the North African watershed, a network through which technological knowledge flowed during mid-MIS 5 is observed in the distribution of Taramsa Phase II technology. Possibly (but not necessarily), this network represents demic diffusion. More compelling evidence for mid-MIS 5 demic diffusion, this time Out of Africa, comes from the Levant and from southern Arabia.

It is possible that the Late Nubian Complex emerged in Arabia and that it was reintroduced in Africa at the end of MIS 5. In Saharan and Maghrebian industries, tanged forms reappear, this time within classical Aterian constellations. We may witness a rare example of acculturation in Paleolithic times in the overlay of tangs on Nubian points. This would just be another indication of the dramatic effects of the final aridification of the Sahara on human populations, which were using greatly extended territories in the first place. The much-debated L3 expansion out of Africa is most likely just another iteration of this process of severe cultural change.

In my final remark, I want to return to the methodological issue touched upon in the introduction. This scenario, however plausible it might be, draws necessarily on debatable interpretations of the archaeological record. Most importantly, there is an urgent need for an analytical, comparative study of early Late Pleistocene North Africa that clarifies similarity patterns in lithic technology and assesses them in terms of the alternative historic processes that can produce this record. Only then can the true significance of this part of the Old World with regard to the global issue of modern human origins and dispersals appear.

Acknowledgments I thank the editors for inviting me to write a contribution to this volume. I express my gratitude to institutions and colleagues who have invited me to take part in their projects or who have permitted me to look at collections: Graeme Barker (McDonald Institute for Archaeological Research, Cambridge) and the TRANS-NAP project for the Haua Fteah collection; François Bon (Université de Toulouse) and the direction of the National Museum of Ethiopia; Nena Galanidou (University of Crete) for introducing me to the southeast coast of Crete; Anthony E. Marks for receiving me in Santa Fe and conversations on Paleolithic Arabia; Jeff Rose (University of Birmingham) for my participation in the 2012 mission in Dhofar, Oman. The research underlying this paper was funded by *Fund for Scientific Research—Flanders* (Project Ref. K8.007.12N).

References

Anonymous (1956). *Collections préhistoriques, Planches Album n°1.* Paris: Musée d'Ethnographie et de Préhistoire du Bardo, Arts et Métiers Graphiques.

Arkell, A. J. (1949). *The Old Stone Age in the Anglo-Egyptian Sudan* (Sudan Antiquities Service Occasional Papers No. 1). Khartoum: Sudan Antiquities Service.

Armitage, S. J., Jasim, S. A., Marks, A. E., Parker, A. G., Usik, V. I., & Uerpmann, H.-P. (2011). The southern route 'Out of Africa': Evidence for an early expansion of modern humans into Arabia. *Science, 331*, 453–456.

Balout, L. (1955). *Préhistoire de l'Afrique du Nord. Essai de chronologie*. Paris: Arts et Métiers Graphiques.

Barham, L., Simms, M. J., Gilmour, M., & Debenham, N. (2000). Twin Rivers, excavation and behavioural record. In L. Barham (Ed.), *The Middle Stone Age of Zambia, South Central Africa* (pp. 165–216). Bristol: Western Academic & Specialist Press.

Bar-Yosef, O. (2000). The Middle and Early Upper Paleolithic in Southwest Asia and neighboring regions. In O. Bar-Yosef & D. Pilbeam (Eds.), *The geography of Neandertals and modern humans in Europe and the greater Mediterranean* (Peabody Museum Bulletin No. 8.) (pp. 107–156). Cambridge: Harvard University Press.

Bar-Yosef, O. (2003). Reflections on the selected issues of the Upper Palaeolithic. In A. N. Goring-Morris & A. Belfer-Cohen (Eds.), *More than meets the eye. Studies on Upper Palaeolithic diversity in the Levant* (pp. 265–273). Oxford: Oxbow Books.

Cancellieri, E., Cremaschi, M., Zerboni, A., & di Lernia, S. (2016). Climate, environment and population dynamics in Pleistocene Sahara. In S. C. Jones & B. A. Stewart (Eds.), *Africa from MIS 6-2: Population dynamics and paleoenvironments* (pp. 123–145). Dordrecht: Springer.

Carto, S. L., Andrew, J., Weaver, A. J., Hetherington, R., Lam, Y., & Wiebe, E. C. (2009). Out of Africa and into an ice age: On the role of global climate change in the late Pleistocene migration of early modern humans out of Africa. *Journal of Human Evolution, 56*, 139–151.

Caton-Thompson, G. (1946). The Aterian industry: Its place and significance in the Palaeolithic world. The Royal Anthropological Institute, London: Huxley Memorial Lecture.

Caton-Thompson, G. (1952). *Kharga Oasis in prehistory*. London: Athlone Press.

Chavaillon, N. (1971). L'Atérien de la Zaouia el Kebira au Sahara nord-occidental (République Algérienne). *Lybica, 19*, 9–52.

Chmielewski, W. (1968). Early and Middle Paleolithic sites near Arkin, Sudan. In F. Wendorf (Ed.), *The prehistory of Nubia* (pp. 110–147). Dallas: Fort Burgwin Research Center and Southern Methodist University Press.

Clark, J. D. (1954). *The prehistoric cultures of the Horn of Africa*. Cambridge: Cambridge University Press.

Clark, J. D. (1982). The transition from Lower to Middle Palaeolithic in the African continent. In A. Ronen (Ed.), *The transition from Lower to Middle Palaeolithic and the origin of modern man* (BAR International Series No. 151) (pp. 235–256). Oxford: Archaeopress.

Clark, J. D. (1988). The Middle Stone Age of East Africa and the beginnings of regional identity. *Journal of World Prehistory, 2*, 235–306.

Clark, J. D. (1993). The Aterian of the Central Sahara. In L. Krzyzaniak, M. Kobusiewicz, & J. Alexander (Eds.), *Environmental and cultural change in North Africa through the 2nd mill. BC* (pp. 129–143). Poznan, Poznan Archaeological Museum.

Clark, J. D., & Kleindienst, M. (2001). The Stone Age cultural sequence: Terminology, typology and raw material. In J. D. Clark (Ed.), *Kalambo Falls prehistoric site* (Vol. III, pp. 34–65). Cambridge: Cambridge University Press.

Clark, J. D., Beyene, Y., Woldegabriel, G., Hart, W. K., Renne, P. R., Gilbert, H., et al. (2003). Stratigraphic, chronological and behavioural contexts of Pleistocene *Homo sapiens* from Middle Awash, Ethiopia. *Nature, 423*, 747–752.

Clark, J. D., Agrilla, E. J., Galloway, A., Garcea, E. A. A., Gifford-Gonzalez, D., Hall, D. N., et al. (2008). *Adrar Bous: Archaeology of a Central Saharan granitic complex in Niger* (Studies in Human Sciences No. 170) (p. 403). Tervuren: Royal Museum for Central Africa.

Debénath, A. (1996). L'Atérien du nord de l'Afrique et du Sahara. *Sahara, 6*, 21–30.

Derricourt, R. (2005). Getting "Out of Africa": Sea crossings, land crossings and culture in the hominin migrations. *Journal of World Prehistory, 19*, 119–132.

Forster, P. (2004). Ice Ages and the mitochondrial DNA chronology of human dispersals: A review. *Philosophical Transactions of the Royal Society of London B, 359*, 255–264.

Garcea, E. A. A. (2001). A reconsideration of the Middle Palaeolithic/Middle Stone Age in North Africa after the evidence from the Libyan Sahara. In E. A. A. Garcea (Ed.), *Uan Tabu in the settlement history of the Libyan Sahara* (Arid Zone Archaeology Monographs No. 2.) (pp. 25–50). Rome: Edizioni all'insegna del giglio.

Gruet, M. (1954). Le gisement Moustérien d'el Guettar. *Karthago, 5*, 3–79.

Guichard, J., & Guichard, G. (1965). The Early and Middle Paleolithic of Nubia: A preliminary report. In F. Wendorf (Ed.), *Contributions to the prehistory of Nubia* (pp. 57–116). Dallas: Fort Burgwin Research Center and Southern Methodist University Press.

Hill, C. L. (1993). E-87-3. A small, dry-season occupation at the onset of the green phase. In F. Wendorf, R. Schild, A. E. Close, & Associates (Eds.), *Egypt during the Last Interglacial: The Middle Paleolithic of Bir Tarfawi and Bir Sahara East* (pp. 459–470). New York: Plenum Press.

Hovers, E. (2009). *The lithic assemblages of Qafzeh Cave*. Oxford: Oxford University Press.

Jones, S., Antoniadou, A., Barton, H., Drake, N., Farr, L., Hunt, C., et al. (2016). Patterns of hominin occupation and cultural diversity across the Gebel Akhdar of northern Libya over the last ~200 kyr. In S. C. Jones & B. A. Stewart (Eds.), *Africa from MIS 6-2: Population dynamics and paleoenvironments* (pp. 77–99). Dordrecht: Springer.

Kleindienst, M. R. (2000). On the Nile corridor and the Out-of-Africa model. *Current Anthropology, 41*, 107–109.

Kurashina, H. (1978). An examination of prehistoric lithic technology in east-central Ethiopia. Ph.D. Dissertation, University of California, Berkeley.

Lacaille, A. D. (1951). The stone industry of Singa-Abu Hugar. In *Fossil mammals of Africa, No. 2, the Pleistocene fauna of two Blue Nile sites* (pp. 43–51). London: The British Museum.

Marks, A. E. (1968a). The Mousterian industries of Nubia. In F. Wendorf (Ed.), *The prehistory of Nubia* (Vol. I, pp. 194–314). Dallas: Fort Burgwin Research Center and Southern Methodist University Press.

Marks, A. E. (1968b). The Khormusan: An Upper Pleistocene industry in Sudanese Nubia. In F. Wendorf (Ed.), *The prehistory of Nubia* (Vol. I, pp. 315–391). Dallas: Fort Burgwin Research Center and Southern Methodist University Press.

Marks, A. E. (2003). Reflections on Levantine Upper Palaeolithic studies: Past and present. In A. N. Goring-Morris & A. Belfer-Cohen (Eds.), *More than meets the eye. Studies on Upper Palaeolithic diversity in the Levant* (pp. 249–264). Oxford: Oxbow Books.

McBurney, C. B. M. (1967). *The Haua Fteah (Cyrenaica) and the Stone Age of the southeast Mediterranean*. Cambridge: Cambridge University Press.

Mercier, N., Wengler, L., Valladas, H., Joron, J.-L., Frogeta, L., & Reyssa, J.-L. (2007). The Rhafas Cave (Morocco): Chronology of the Mousterian and Aterian archaeological occupations and their implications for Quaternary geochronology based on luminescence (TL/OSL) age determinations. *Quaternary Geochronology, 2*, 309–313.

Nami, M., & Moser, J. (2010). *La Grotte d'Ifri N'Ammar. Tome 2, Le Paléolithique Moyen*. Wiesbaden: Reichert Verlag.

Olszewski, D. I., Dibble, H. L., McPherron, S. P., Schurmans, U. A., Chiotti, L. V., & Smith, J. R. (2010). Nubian Complex strategies in the Egyptian high desert. *Journal of Human Evolution, 59*, 188–201.

Osborne, A. H., Vance, D., Rohling, E. J., Barton, N., Rogerson, M., & Fello, N. (2008). A humid corridor across the Sahara for the migration of early modern humans out of Africa 120,000 years ago. *PNAS, 105*, 16444–16447.

Powell, A., Shennan, S. & Thomas, M.G. (2009). Late Pleistocene demography and the appearance of modern human behavior. *Science, 324*, 1298–1301.

Reynolds, T. (2013). The Middle Palaeolithic of Cyrenaica: Is there an Aterian at the Haua Fteah and does it matter? *Quaternary International, 300*, 171–181.

Rose, J. I., Usik, V. I., Marks, A. E., Hilbert, Y. H., Galletti, C. S., Parton, A., et al. (2011). The Nubian Complex of Dhofar, Oman: An African Middle Stone Age industry in southern Arabia. *PLoS ONE, 6*(11), e28239. doi:10.1371/journal.pone/0028239.

Sánchez-Quinto, F., Botigué, L. R., Civit, S., Arenas, C., Ávila-Arcos, M. C., Bustamante, C. D., et al. (2012). North African populations carry the signature of admixture with Neandertals. *PLoS ONE, 7*, e47765. doi:10.1371/journal.pone/0047765.

Schild, R., & Wendorf, F. (1986). The geological setting. In F. Wendorf, R. Schild, & A. E. Close (Eds.), *The prehistory of Wadi Kubbaniya. The Wadi Kubbanyia skeleton: A Late Paleolithic burial from southern Egypt* (Vol. I, pp. 7–32). Dallas: Southern Methodist University Press.

Schild, R., Wendorf, F., & Close, A. E. (1992). Northern and eastern Africa climate changes between 140 and 12 thousand years ago. In F. Klees & R. Kuper (Eds.), *New light on the northeast African past* (pp. 81–98). Köln: Heinrich Barth Institut.

Singleton, W. L., & Close, A. E. (1980). Report on Site E-78-11. In F. Wendorf, R. Schild, & A. E. Close (Eds.), *Loaves and fishes: The prehistory of Wadi Kubbaniya* (pp. 229–237). Dallas: Southern Methodist University, Department of Anthropology.

Shiner, J. (1968). Miscellaneous sites. In F. Wendorf (Ed.), *The prehistory of Nubia* (Vol. II, pp. 630–650). Dallas: Fort Burgwin Research Center and Southern Methodist University Press.

Smith, J. R., Hawkins, A. L., Asmerom, Y., Polyak, V., & Giegengack, R. (2007). New age constraints on the Middle Stone Age occupations of Kharga Oasis, Western Desert, Egypt. *Journal of Human Evolution, 52*, 690–701.

Strasser, T. F., Thompson, F., Panagopoulou, E., Karkanas, P., Runnels, C. N., McCoy, F. W., et al. (2010). Stone Age seafaring in the Mediterranean. Evidence from the Plakias region for Lower Palaeolithic and Mesolithic habitation of Crete. *Hesperia, 79*, 145–190.

Taymaz, T., Yilmaz, Y., & Dilek, Y. (2007). *The geodynamics of the Aegean and Anatolia: Introduction*. London: The Geological Society.

Tostevin, G. B. (2003). A quest for antecedents: A comparison of the terminal Middle Palaeolithic and the early Upper Palaeolithic of the Levant. In A. N. Goring-Morris & A. Belfer-Cohen (Eds.), *More than meets the eye. Studies on Upper Palaeolithic diversity in the Levant* (pp. 54–67). Oxford: Oxbow Books.

Usik, V. I., Rose, J. I., Hilbert, Y. H., Van Peer, P., & Marks, A. E. (2013). Nubian Complex reduction strategies in Dhofar, southern Oman. *Quaternary International, 300*, 244–266.

Van Peer, P. (1991). Interassemblage variability and Levallois styles: The case of the northern African Middle Palaeolithic. *Journal of Anthropological Archaeology, 10*, 107–151.

Van Peer, P. (2001). The Nubian Complex settlement system in northeast Africa. In N. J. Conard (Ed.), *Settlement dynamics of the Middle Paleolithic and Middle Stone Age* (pp. 45–64). Tübingen: Kerns Verlag.

Van Peer, P. (2004). Did Middle Stone Age moderns of sub-Saharan African descent trigger an Upper Paleolithic revolution in the Lower Nile Valley? *Anthropologie: International Journal of the Science of Man, XLII*, 215–226.

Van Peer, P., Vermeersch, P. M., Moeyersons, J., & Van Neer, W. (1996). Palaeolithic sequence of Sodmein Cave, Red Sea Mountains, Egypt. In G. Pwiti & R. Soper (Eds.), *Aspects of African archaeology. Papers from the 10th congress of the pan-African association for prehistory and related studies* (pp. 149–156). Harare: University of Zimbabwe Publications.

Van Peer, P., Vermeersch, P. M., & Paulissen, E. (2002). Chert quarrying in the Nazlet Safaha area. In P. M. Vermeersch (Ed.), *Palaeolithic quarrying sites in Upper and Middle Egypt* (Egyptian Prehistory Monographs No. 3) (pp. 139–209). Leuven: Leuven University Press.

Van Peer, P., Fullagar, R., Stokes, S., Bailey, R. M., Moeyersons, J., Steenhoudt, F., et al. (2003). The Early to Middle Stone Age transition and the emergence of modern human behaviour at site 8-B-11, Sai Island, Sudan. *Journal of Human Evolution, 45*, 187–193.

Van Peer, P., Vermeersch, P. M., & Paulissen, E. (2010). *Chert quarrying, lithic technology and a modern human burial at the Palaeolithic site of Taramsa 1, Upper Egypt* (Egyptian Prehistory Monographs No. 5). Leuven: Leuven University Press.

Vermeersch, P. M., Paulissen, E., & Vanderbeken, T. (2002). Nazlet Khater 4, an Upper Palaeolithic underground chert mine. In P. M. Vermeersch (Ed.), *Palaeolithic quarrying sites in Upper and Middle Egypt* (pp. 211–271). Leuven: Leuven University Press.

Watson, E., Forster, P., Richards, M., & Bandelt, H.-J. (1997). Mitochondrial footprints of human expansions in Africa. *American Journal of Human Genetics, 61*, 691–704.

Wendorf, F., & Schild, R. (1992). The Middle Paleolithic of North Africa: A status report. In F. Klees & R. Kuper (Eds.), *New light on the northeast African past* (pp. 39–78). Köln: Heinrich Barth Institut.

Wendorf, F., Close, A. E., & Schild, R. (1987). Recent work on the Middle Palaeolithic of the Eastern Sahara. *The African Archaeological Review, 5*, 49–63.

Wendorf, F., Schild, R., & Close, A. E. (Eds.). (1993). *Egypt during the last interglacial: The Middle Paleolithic of Bir Tarfawi and Bir Sahara East*. New York: Plenum Press.

Wengler, L. (1993). Cultures préhistoriques et formations quaternaires du Maroc oriental. Relations entre comportements et paléoenvironnements au Paléolithique moyen. Ph.D. Dissertation, Université de Bordeaux I, Bordeaux.

Wengler, L. (2001). Settlements during the Middle Palaeolithic of the Maghreb. In N. J. Conard (Ed.), *Settlement dynamics of the Middle Paleolithic and Middle Stone Age* (pp. 65–90). Tübingen: Kerns Verlag.

Whittle, A. (1996). *Europe in the Neolithic. The creation of new worlds*. Cambridge: Cambridge University Press.

Chapter 9
Late Quaternary Environmental Change and Human Occupation of the Southern African Interior

Sallie L. Burrough

Abstract The interior southern African basin (Kalahari) is a remarkable region, with a complex and dynamic environmental history and a long record of utilization by human populations during the late Quaternary. Paleoenvironmental reconstructions are beginning to provide a spatially detailed record of landscape and hydrological dynamics in the Kalahari, with a strong chronometric underpinning for records of environmental extremes. Theories concerning the distribution of early people in the landscape place great importance on the temporal dynamics of water availability, and may be particularly relevant in the Kalahari where there is significant evidence of hydrologic/climatic-driven landscape change. High amplitude environmental variability during MIS 6-2 is evidenced by periods of dune building within currently stabilized dunefields and the intermittent existence of large lacustrine systems such as Megalake Makgadikgadi that remain all but ephemerally dry under present-day conditions. That the wider Kalahari was, at times, a key resource for Stone Age populations is evident from the extensive occurrence of stone tools, most notably in association with the fluvial networks and lake basins of the Okavango-Chobe-Zambezi system. Today, these riparian corridors link the semiarid desert region to the southern subtropics and, in the past, drove environmental change in the Kalahari, potentially impacting the occupation and dispersal of hominins within the interior southern African basin.

Keywords Kalahari • Paleohydrology • Paleoenvironmental change • Dunes • Lakes • Stone Age archaeology

S.L. Burrough (✉)
School of Geography and the Environment, Oxford University, South Parks Road, Oxford, OX1 3QY, UK
e-mail: sallie.burrough@ouce.ox.ac.uk

Introduction

Attempts to draw correlations between climatic events and trends and the African Stone Age archeological record are increasingly prevalent within the Quaternary literature (e.g., Scholz et al. 2007; Jacobs and Roberts 2009; Stager et al. 2011). These attempt to establish the causes of both behavioral changes by, and the geographical distribution of, early modern humans. Environmental contributions to hominin dispersal out of Africa are a particular point of interest (Maslin and Christiensen 2007). The influence of aridity on human resource use and mobility is often emphasized, though the spatial complexities of both landscape and climate dynamics are frequently overlooked (Thomas and Burrough 2012).

Within southern Africa, such human environment associations have been strongly focused on Middle Stone Age (MSA) technological change within coastal sites (e.g., Jacobs et al. 2008). Significantly less is known about early people in the southern African interior, despite the genetic and archeological evidence for sustained periods of regional occupation during the late Quaternary. The deficit of research is largely a consequence of the nature of the environment of the interior, which, by consequence of its geological history, offers few closed sites and, due to relatively arid conditions, preserves little organic material. Until recently, difficulties of paleoenvironmental reconstruction in present-day deserts have hindered our understanding of their long-term landscape evolution and environmental change. Technological and methodological advances, however, are now facilitating the development of robust records of Quaternary landscape dynamics: in the Kalahari, these records reveal environmental changes of significant amplitude and frequency during MIS 6-2. This chapter examines the potential consequences of this dynamic environmental history for early humans, the evidence for which is abundant but remains poorly investigated. Critically, however, no consideration of the dynamics of past climatic systems can be effectively conducted without assessing the nature and

Sacha C. Jones and Brian A. Stewart (eds.), *Africa from MIS 6-2: Population Dynamics and Paleoenvironments*, Vertebrate Paleobiology and Paleoanthropology, DOI 10.1007/978-94-017-7520-5_9

behavior of the climatic systems dominant in an area today, since past conditions are conventionally presented in terms of changes from the present. This chapter begins with a consideration of the controls on central southern African climate today, before reviewing the paleoenvironmental records derived from within the region itself.

Climate and Environment of the Kalahari Basin

The Kalahari basin, a region in excess of 2.5 million km^2 and comprising the largest continuous sand sea on earth (Thomas and Shaw 1991), dominates the southern African interior. The area north and west of the Okavango Delta is sometimes called the Northern Kalahari; the heart of the region, including the area occupied by the Okavango Delta and Makgadikgadi basin, the Middle Kalahari; and the remainder of the basin, the Southern Kalahari, following Passarge (1904). Geologically, it is characterized by Late Cretaceous and Early Tertiary sedimentary rocks, and overlain by unconsolidated sands, weathered and reworked from the underlying lithology during the Pliocene and Quaternary (Haddon 2005). This sand mantled landscape has been subsequently molded into a significant suite of depositional landforms dynamic over Quaternary timescales. These Kalahari sediments straddle a significant latitudinal climatic and environmental gradient ranging from the miombo and montane forests of Zambia and Angola (where mean annual precipitation [MAP] exceeds 1000 mm/year) to the arid deserts of southwest Botswana and the Northern Cape of South Africa (MAP < 200 mm/yr)

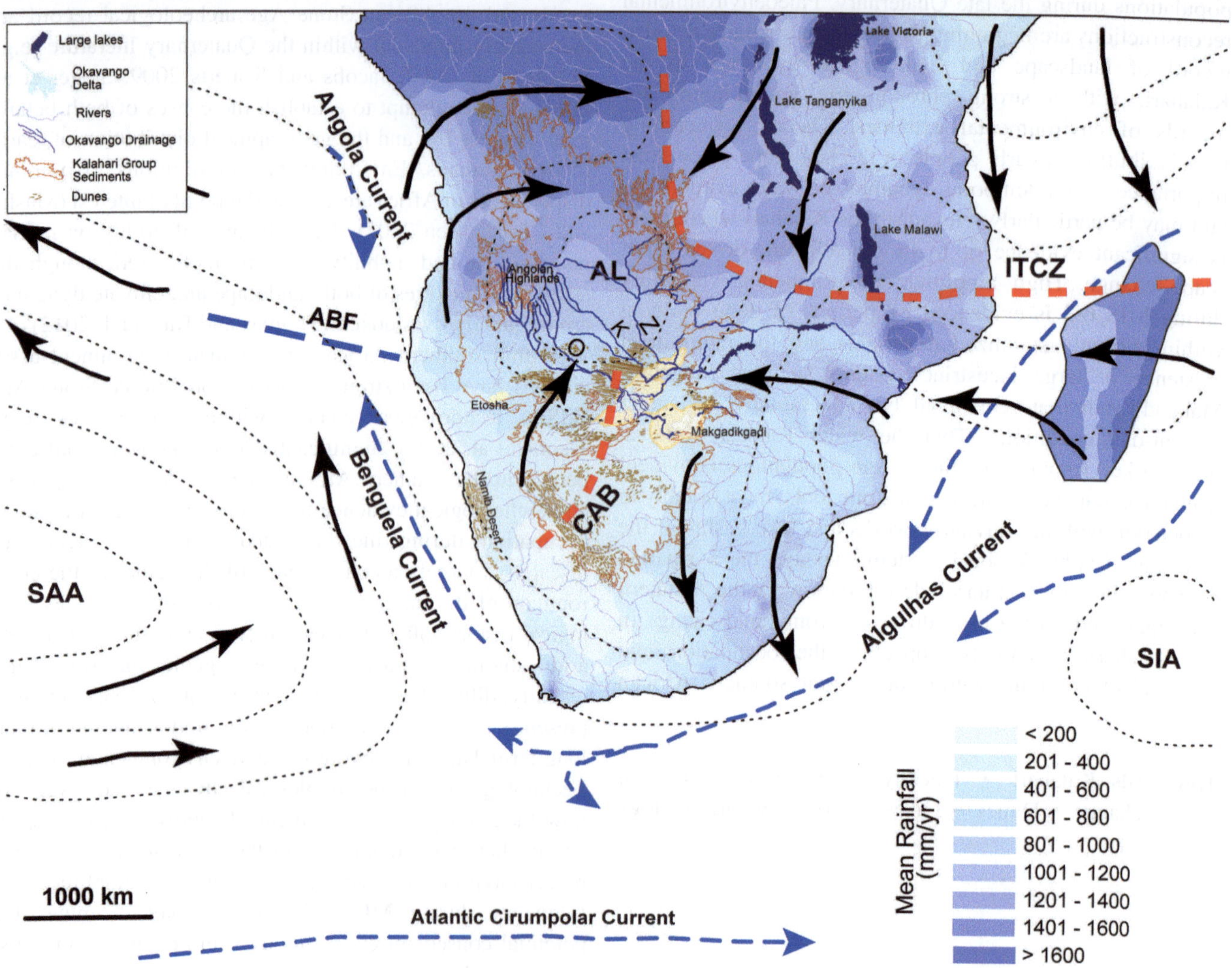

Fig. 9.1 Hydrological systems and dunefields of the Kalahari and the circulation systems affecting southern Africa (modified after Nicholson 1996; Gasse et al. 2008). *O* Okavango River; *K* Kwando River; *Z* Zambezi River; *SAA* South Atlantic Anticyclone; *SIA* South Indian Anticyclone; *AL* Angolan Low; *ABF* Angola-Benguela Front. The mean austral summer position of the ITCZ (Intertropical Convergence Zone) and CAB (Congo Air boundary) and major climatic regions (from Gasse et al. 2008) are indicated

(Fig. 9.1). This gradient can be measured not only in terms of mean precipitation amount but also in the length of the wet season (decreasing from northeast to southwest) and the degree of interannual variability (increasing from northeast to southwest). The Kalahari lies within southern Africa's semi-arid subtropical high-pressure belt, and is dominated by the Southern Hemisphere summer migration of the Inter Tropical Convergence Zone (ITCZ). This brings with it heavy convective rainfall between November and April, and migrates north of the equator in the austral winter allowing the dominance of high-pressure systems and dry conditions over the southern African interior.

In addition, synoptic conditions such as the presence of an eastern equatorial trough (Taljaard 1981) are responsible for low-level easterly moisture fluxes, originating from the Indian Ocean and facilitating the presence of unstable uplifting air via tropical-extratropical interactions. In the northern parts of the Kalahari, summer rainfall can occur in conjunction with recurved South Atlantic air, which enters the region from the Congo and Angola; while in the Middle and Southern Kalahari, easterlies blowing around the Indian Ocean high are a more likely source of rainfall (Thomas and Shaw 1991). In the southwestern Kalahari, the occasional passage of frontal systems associated with Atlantic westerlies that dominate the winter rainfall zone also makes a contribution to the annual rainfall during the winter season (Tyson 1986). The Indian Ocean is the major moisture source for most of the summer rainfall zone driving an additional zonal rainfall pattern that causes the eastern region of southern Africa to be wetter than the west. Within the Middle and Southern Kalahari, low annual precipitation and high rates of evaporation result in a region wide moisture deficit and a notable lack of surface water outside of the Okavango delta and its northerly perennial fluvial channels. It is the reorganization of these circulation systems that are frequently invoked to explain Quaternary climatic and environmental change in the Kalahari during MIS 6-2, typically characterized as regional transitions between humid and arid conditions, and, by implication, with important consequences for early human populations.

Wetlands in a Dryland: The Importance of Hydrology to Interior Southern Africa

The presence of surface water, in particular the hydrological systems fed by tropical rainfall that terminate in the Middle Kalahari, play a critical role in the survival of many species in the Kalahari, driving the migratory patterns of many large mammal and bird species. This area is primarily fed by the tropically-sourced Cubango and Cuito Rivers. These are the major tributaries of the endoreic Okavango River, the Kwando/Chobe drainage system and, prior to its Pliocene (Lister 1979) or Early-Mid Pleistocene (Bond 1975; Shaw and Thomas 1988) capture and coastal reorientation, the Upper Zambezi. Some surface water reaches the eastern part of the Middle Kalahari via small rivers sourced in western Zimbabwe, notably the Nata, which flows seasonally into Sua Pan, in the east of the Makgadikgadi system.

The northern rivers share a common source in the Angolan highlands, ~1000 km from the Middle Kalahari itself. The Lunda Divide forms a watershed on the Angola plateau and separates rivers, such as the Cuango, that flow north into the Congo River from those flowing in a southerly direction. A few tributaries join the Zambezi, but most, other than the Kwando, constitute the active catchment of the Okavango River. The Kunene also has its source in this region but takes a more westerly course toward the Atlantic. Other smaller rivers (*oshanas*) flow south forming the internal drainage of the Etosha Basin. Many of these rivers and streams are highly seasonal. For example, the tributaries of the Cubango remain completely dry for much of the year.

At 18° S, the Okavango, Kwando, and Zambezi Rivers reach the tectonically active zones of the Gwembe Trough and Okavango Depression (Thomas and Shaw 1991). Here, the Okavango terminates in an alluvial fan, known as the "the Okavango Delta" – a "wet zone" in a dryland region, now internationally renowned for its rich biodiversity. The morphometry and hydrology of this alluvial fan is dominated by two major conjugate faults: a larger north easterly striking "half-graben," which holds the Delta; and a smaller north westerly striking faultline (Gumbricht et al. 2001) occupied by the Panhandle (Scholz et al. 1976; McCarthy and Hancox 2000). The total catchment of the Okavango River basin is approximately 530,000 km^2. Under present-day conditions, however, 95% of the Okavango's flow is contributed from an area of only 135,000 km^2, situated within Angola (Andersson et al. 2003). The easterly catchment (Cuito River) responds to instabilities in tropical lows in the Indian Ocean easterlies, while the western catchment (Cubango River) is more strongly influenced by variability in the Atlantic equatorial westerlies (McCarthy et al. 2000). Approximately 450 mm/yr (or 5×10^9 m^3) of local precipitation falls annually over the delta itself (McCarthy and Hancox 2000). This is usually confined to the period between December and February compared to a mean annual inflow into the fan of 11×10^9 m^3 (Ellery and McCarthy 1998). The majority of surface water takes months to arrive from its northerly source and peaks as a flood wave during the winter dry season. Critically then, the hydrological regime of the Middle Kalahari, and the functioning of much of its biodiversity, is principally controlled by the climatology of the tropics at 12° S rather than locally in the region of the Okavango fan (~18°–21° S).

The common catchment region of the Zambezi, Kwando, and Okavango Rivers means that changes in their hydrological regimes are likely to co-vary over time. Mazvimavi and Wolski (2006) demonstrated that the average annual discharges of the Okavango and Zambezi Rivers are closely related, with a correlation coefficient of 0.70. This has important implications for flood regimes of the southern African interior, both under contemporary conditions and in relation to past Quaternary environmental change. The Kwando is connected to the Okavango system via the Makwegana (Selinda Spillway), but today diverts east to the Zambezi via the Linyanti and Chobe Rivers, forming ephemeral connections between these three substantial river systems (Fig. 9.3). The Mambova Falls marks the Chobe-Zambezi confluence where the rivers flow east across a Karoo basalt ridge along a north-south-oriented fault line. During peak flows, the impediment of the ridge results in an annual backflood up to 20 km westwards along the Chobe (Thomas and Shaw 1991). Evidence for periods of extreme flooding during the Quaternary is derived from erosional notches cut along the hill slope at 935 m above mean sea level (amsl), from calcretized alluvial terraces at 932–934 m amsl (Shaw and Thomas 1988), and from shorelines bounding the backflooding zone (Burrough and Thomas 2008). It is hypothesized that Zambezi flow made a significant contribution to the surface water of the Middle Kalahari during these periods of high discharge and back-ponding (Shaw and Thomas 1988; Burrough and Thomas 2008), providing an important link between these two sizeable hydrological systems during the Quaternary.

Quaternary Environmental Dynamics in the Kalahari

The magnitude of past environmental changes within the Kalahari has left a strong signature in the landscape that has been recognized and documented for more than 100 years (Livingstone 1858; Passarge 1904; Grove 1969; Thomas and Shaw 2002). Shifts in landscape processes in response to prevailing climatic events and trends have been encoded into a range of archives including extensive now largely inactive dunefields, speleothems, fluvial channels, and today's predominantly dry lake systems. Such juxtaposed signals of extreme "wet" conditions and extreme "dry" conditions testify to the amplitude of environmental change in the southern African interior. The drivers of large-scale landscape and environmental shifts remain poorly understood, but hypotheses regarding spatial reorganizations of climate and their causal mechanisms have been put forward by a number of researchers. These can be summarized as: (1) the latitudinal displacement of major circulation features; (2) intensification or weakening of major circulation features; and (3) longitudinal displacement of circulation features (c.f. Tyson 1999; Chase and Meadows 2007; Singarayer and Burrough 2015).

Dryland landscapes are particularly challenging environments in which to assess the nature of past climate changes. Regions such as the Kalahari, one of the driest parts of Africa, are often devoid of, or have only limited, biological and geochemical proxies or closed sites. This restricts their potential for paleoenvironmental reconstructions. The reliance on spatially diverse and fragmentary records has resulted in contradictory, and often oversimplified, explanations of climatic change that fall short of providing the precision with which to adequately test the climatic hypotheses described above. More recently, marked advances in the ability to establish, and date, suitable proxy records from such areas, have enabled significant steps to be made toward the reconstruction of a more detailed picture of environmental dynamics (Drake and Bristow 2006; Burrough et al. 2009a) and a necessary reevaluation of the interpretation of these data. These geoproxy records constitute an important facet in the understanding of Late Pleistocene dynamics of the Kalahari. Their interpretation is considered below. Following this, the relationships between archaeology and paleoenvironments are examined. This examination integrates other, limited-occurrence paleoenvironmental proxies from specific site associations, though it is the picture of past landscape dynamics as a whole, including the spatial variance of environmental conditions, that generates the perspective that is necessary to understand hominin usage of the Kalahari environment.

Kalahari Dune Archives

Dunefields dominate large areas of the Kalahari. These in turn are primarily composed of linear sand dunes that, unlike more mobile transverse and barchans dunes, have a greater propensity to accumulate and store sand over time, due to their formative mechanism (see, for example, Thomas 1992; Lancaster 2011). They are therefore potentially important archives of past environmental conditions, and have been analyzed in this way in the Kalahari since Flint and Bond (1968). Following technological and methodological developments, more than 200 luminescence ages for dunefield activity have been produced in the last 20 years providing information on dune activity that extends back to 185 ka. However, far from increasing the certainty with which we are able to interpret the environmental change occurring in MIS 6-2, the growing body of luminescence ages has destabilized initial interpretations (e.g., Stokes et al. 1997) and is difficult to interpret when synthesized as a regional body of data (Thomas and Burrough 2013). Recent analyses

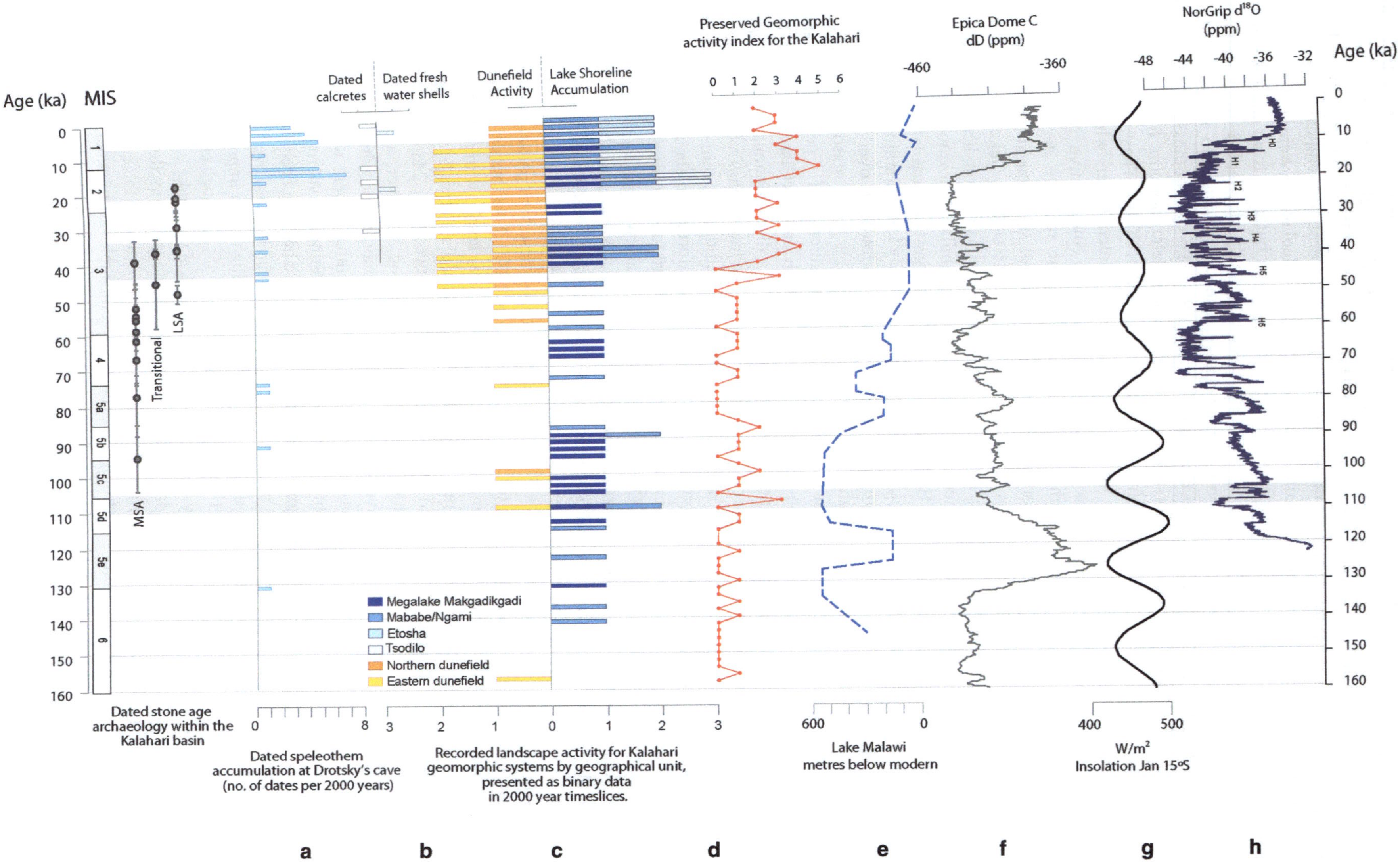

Fig. 9.2 Paleoenvironmental records from the Kalahari in the context of regional and global records. **a** Dated speleothem accumulation shown as number of dates per 2000 year time slice at Drotsky's cave (interpreted as indicative of wetter periods) (Cooke and Verhagen 1977; Shaw and Cooke 1986; Brook et al. 1996); **b** Dated calcretes and fresh water shells from the Middle Kalahari (Shaw 1985; Shaw and Thomas 1993); **c** Summary of geoproxy records (active dunefields and lake highstands) within the Kalahari region; **d** Preserved geomorphic activity index, i.e., total number of geographical units preserving a record in each 2 ka time slice (Thomas and Burrough 2012); **e** Lake Malawi lake level status derived from the synthesis of sediment core paleoclimate indicators with observations from seismic reflection records (Scholz et al. 2011); **f** EPICA Dome C Ice Cores Deuterium Data (Jouzel et al. 2007); **g** Insolation at 15° S (Berger 1992); **h** NorGRIP d^{18}O record (NorGRIP members 2004). Archaeological dates relating to the MSA/LSA transition (Feathers et al. 1997; Barham 2000; Robbins et al. 2000; Brook et al. 2008; Ivester et al. 2010) are shown for comparison

suggest that in the driest parts of the interior, particularly in the southwest Kalahari, the landscape lies close to mobility thresholds such that for dune ages used as proxies, short-term multiyear drought events may be indistinguishable from significant arid episodes (Thomas and Burrough 2012; Bailey and Thomas 2014). Combined with the issue of discontinuous preservation, current analyses propose that these records should be viewed in the context of regional landscape activation and transition rather than in simple terms as evidence of "arid" (Stokes et al. 1997) or "windy" (Chase and Thomas 2006) periods.

By contrast, dunefields from more northerly or easterly areas of the Kalahari, which lie wholly within the range of the annual ITCZ-induced wet season and that are often heavily wooded today, are likely only to become mobile during extremely prolonged periods of dryness sufficient to see significant die-off of vegetation. Even allowing for the potential for incomplete preservation of sediments from past dune building events (due to their potential for reactivation during later periods of dune dynamics), these features are paleoenvironmentally more significant than their more southerly counterparts. While they are less sensitive to short-term episodic activation, they provide a more significant proxy for Precipitation-Evaporation (P-E) deficits in the context of large-scale Quaternary environmental change. Theoretical studies suggest that the greatest preservation potential for accumulation phases of linear dune proxies is the transition period out of a mean dry or windy phase (Kocurek 1998; Telfer et al. 2010), though not all transitions may be preserved if further periods of aeolian activity are severe enough to remobilize entire dunes. The effect of preservation potential on the production of a robust record of past dune activity can also be enhanced by the intrinsically unsystematic sampling of dunefields. This arises from sampling individual dunes as representative landforms of a large area and can to some extent be removed by combining data from individual sample locations and reducing a data set to a binary record of "switched on" or "switched off" state at the dunefield-scale (Fig. 9.2).

Kalahari Lake Archives

In addition to dune mobilization, another important regional proxy is the suite of shoreline deposits associated with the, now-dry, paleolake Makgadikgadi system at the terminal sump of the Okavango. These landforms have undergone detailed mapping, surveying, and investigation for many decades, and the recognition of the former presence of a large lake system was the subject of numerous publications (e.g., Cooke and Verstappen 1984; Shaw 1985; Shaw and Cooke 1986; Shaw and Thomas 1993). It is only recently that a systematic dating program has enabled a detailed analysis of the timing of lake highstand events to be established (Shaw et al. 2003; Burrough et al. 2007, 2009a; Burrough and Thomas 2008). Like Kalahari dunes, these landform proxies should be interpreted with care. Two of the basins that make up the paleolake system (Mababe and Ngami) are fed directly by tributaries from the Okavango delta and can become partially decoupled from regional environmental change, responding instead to avulsion within the Okavango fan (Fig. 9.3). To the southeast of the fan, a horst feature separates the delta domain from the Makgadikgadi Depression (Haddon 2005). These are connected by the down-cutting of the Boteti River, which acts much like an overspill valve. The conservative record of lake highstand phases is thus more robustly derived from the shorelines of the Makgadikgadi sump basin. Highstands in this basin are usually referred to as "megalake" events, since the implication of shoreline construction here is the existence of a 66,000 km^2 lake, similar in size to Lake Victoria in East Africa, the world's largest tropical lake.

Like linear dunes, these sandy lake shorelines appear to be compound features and accumulated over multiple lake highstand events punctuated by more seasonal conditions that facilitated the formation of pedogenic and non-pedogenic calcretes (Nash and McClaren 2003). Like dunes, these geoproxies offer a discontinuous snapshot into the past enabling us to place the transitions from the extremes of regional environmental conditions within the context of global change, but offer little information about the subtleties of intervening gaps in the record (c.f. Burrough et al. 2009a).

Dating suggests that megalake Makgadikgadi highstands were centered on 131 ± 11 ka; 105 ± 4 ka and 92 ± 2 ka with subsequent paleolake transgression phases occurring between 66 ± 5 ka and 62 ± 8 ka; 40 ± 4 ka–37 ± 5 ka and 28 ± 2 ka–25 ± 3 ka (weighted means of these high phases are centered on 64 ± 2 ka; 39 ± 2 ka and 27 ± 1 ka, but do not preclude significant fluctuation of lake levels within each phase) (Burrough et al. 2009a). A very distinct high lake level phase also occurred at 17 ± 1.5 ka (between 19 ± 2 ka and 15 ± 2 ka). What remain poorly understood, however, are the specific drivers of these hydrological changes and whether high lake phases occurred when *local* climates were wetter, drier, or comparable to those of today. Simply invoking local climate change alone is oversimplistic: it has long been suggested that the magnitude of sustained rainfall increase over the lake itself required to fill the basin is so high to be improbable (Cooke and Verstappen 1984; Shaw 1985; Burrough et al. 2007). The key to filling the extensive Kalahari lake systems lies in understanding its relationship to the rivers flowing from the north with tropical sources. The driving force of these major hydrological excursions is thus relevant to understanding the occupation and mobility of hominins in the Kalahari, in as much as it determines the wider geographical pattern of water availability in the southern African interior.

The persistent dominance of easterly wind systems during the Quaternary is evidenced by the orientation of shorelines (wave-built shorelines are found on the western boundaries of Kalahari lake basins). However, both Nash et al. (2006) and Burrough et al. (2009b) note the possible role of recurved Atlantic moisture in driving hydrological change via increases in P-E balance in the upper western catchment. The long latitudinal extent of the hydrological system that feeds the paleolake could, theoretically, enable a tropical wet pulse to be imported into a dry Kalahari. While this implies a possible decoupling from local environmental conditions and the feasibility of large lakes in a dry environment, HADCM3 coupled ocean atmosphere model simulations suggest that the very large lake surface area of a full Makgadikgadi system (66,000 km^2) would have had a significant feedback effect on the local and regional climate (Burrough et al. 2009b). These simulations predict a resultant increase in the local P-E balance (i.e., in effective rainfall) by up to 15%, with associated changes in vegetation (Burrough et al. 2009, 2012). Biophysical feedbacks, which are often insufficiently considered in paleoenvironmental reconstructions, render our ability to distinguish between the drivers and responses of environmental change very challenging. In this case, they imply that even if large lake systems could exist under arid conditions in the Kalahari, an important consequence would be the localized environmental and ecosystem modification toward conditions more amenable to both humans and animals alike.

Together, these geoproxy records suggest that, far from the largely geomorphologically inactive landscape that characterizes much of the Kalahari under contemporary conditions, the region is, in fact, extraordinarily dynamic over long timescales and has repeatedly experienced both extreme water deficits and excesses with significant implications for both hominins and other taxa. The resolution of these geoproxies is highly influenced by preservation and is currently often insufficiently fine-scale to allow unpicking of discrete wet or dry "events". Vitally, however, in the absence of high-resolution biological proxies, these geoproxies do capture a major, and often ignored, component of environmental dynamics, namely *variability* (Thomas and Burrough 2012). This would have played a crucial role in driving innovation, survival, and the capacity of hominins to move successfully (Maslin and Christensen 2007). Detectable periods of extreme climatic variability/change resulting in significant landscape activity are highlighted in Fig. 9.2. These suggest that the relative amplitude and frequency of preserved change was particularly marked on the cusp of the transitions into and out of the last glacial period at ~110 ka and 19–12 ka, respectively, and during MIS 3 at 40–36 ka. Further evidence from the Stampriet Aquifer in the southwest Kalahari (Stute and Talma 1998) suggests that at least the later transition, a period for which we have a suite of paleoenvironmental records, was also a time of marked temperature change. This additionally drove competition between plant species (Scott et al. 2008, 2012), and resulted in significant changes to prevailing vegetation assemblages. Transitional periods in the Kalahari during MIS 6-2, characterized by large-scale landscape and ecosystem change, are thus likely to have been testing times for humans and animals alike requiring a significant level of adaptive capacity in the form of innovation and/or mobility.

Populations and Paleoenvironmental Change in the Kalahari

Regional Records

Whether seasonal or on longer timescales, current geoarchaeological theory attaches great importance to the role of climatic and environmental change in providing environmental constraints and opportunities for movement, dispersal, and adaptive resource use by humans. Like modern populations in southern Africa, Late Pleistocene groups would have been strongly dependent on surface water availability, particularly during times of increased aridity, and, as today, would have been vulnerable both to extreme floods and droughts (Flint 2006). The migratory patterns of many species, particularly ungulates, within the continental interior are highly dependent on the dynamics of the Okavango flood pulse. Historically, animals have moved from dry season permanent water bodies to temporary water holes in the wet season grasslands including millions of springbok and wildebeest that moved from the central Kalahari to the southwest and hundreds of thousands of zebra and wildebeest that seasonally migrated from the Boteti River to wet season pools in the Makgadikgadi grasslands (Bartlam Brooks et al. 2011). Until the development of modern economies in the Kalahari from the mid-twentieth century onwards, mobility, both by hunter-gatherers and pastoralists, had also been a widespread practice (e.g., Sporton and Twyman 2002; Hitchcock 2002) as a legitimate resilience strategy to resource fluctuations.

Brooks (1984) investigated the varying seasonal landscape use by modern Juc'hoansi (!Kung) San in the northwest Kalahari and found that dispersed and highly mobile wet season groups aggregate near water during the dry season. It is these aggregation camps, at resource-rich sites, that are occupied for longer periods and leave a visible record of usage. If these patterns hold for prehistory, archeological visibility in the landscape should strongly correlate with areas where water is available during Late Pleistocene dry phases. In Zambia, investigations at sites such as Kabwe, Kalambo Falls, Twin Rivers and in particular at Mumbwa Caves have demonstrated marked changes in archeological visibility through time (Barham 2000). This has led to hypotheses that fluvial systems may have formed key refugia at times of regionally dry conditions during the Pleistocene, when occupation of sites distant from water sources was absent or limited (Barham 2001; Avery 2003). Taking these interconnected systems as a whole, the North-South fluvial corridors of the interior that bring flood pulses from the tropics, together with the basin sumps in which these systems terminate, are likely to have been critical as the last vestiges of perennial food and water at times of regional water deficit.

Whether or not such mobility is visible in the archeological record of the drier Kalahari zone has yet to be systematically tested. Despite the depth of Kalahari prehistory, a general consensus that the region was "a backwater of archaeological interest" (Walker 1998: 67), has led to a dearth of archeological investigations. The lack of closed, stratified sites has greatly contributed to the continued deficit of regional Stone Age archeological research. As a consequence, with the exception of the sites of Tsodilo Hills and ≠Gi, little is known about the archaeology of the African subcontinental interior, particularly when compared to the major advances made in the last two decades in regions to the south and north.

Stone Age archeological assemblages, however, are numerous and are frequently noted within the gray literature. The site register of the National Museum of Botswana records 15 sites with tool typologies designated as Early Stone Age (ESA), more than 50 MSA archeological occurrences and more than 100 Late Stone Age (LSA) sites (*National Sites Register*, National Museum of Botswana). Many of these are based on observed surface finds related to impact assessments for infrastructure development, and few receive any formal excavation. Figure 9.3 illustrates the location of some of these sites (where their specific locations are recorded and published) and their relationship to the hydrological system of the Middle Kalahari.

The greatest concentration of reported ESA sites is found in the hardveld fluvial channels in southeast Botswana. Here, Wayland identified a total of 36 sites in the vicinity of the Taung, Ngotwane, and Limpopo Rivers (Cooke 1979; Thomas and Shaw 1991), and Ebert et al. (1976) reported the location of more than 400 Acheulean artifacts at Serowe. Further north, ESA sites maintain a strong fluvial association. *In situ* sites are reported by Aldiss (1987) from the

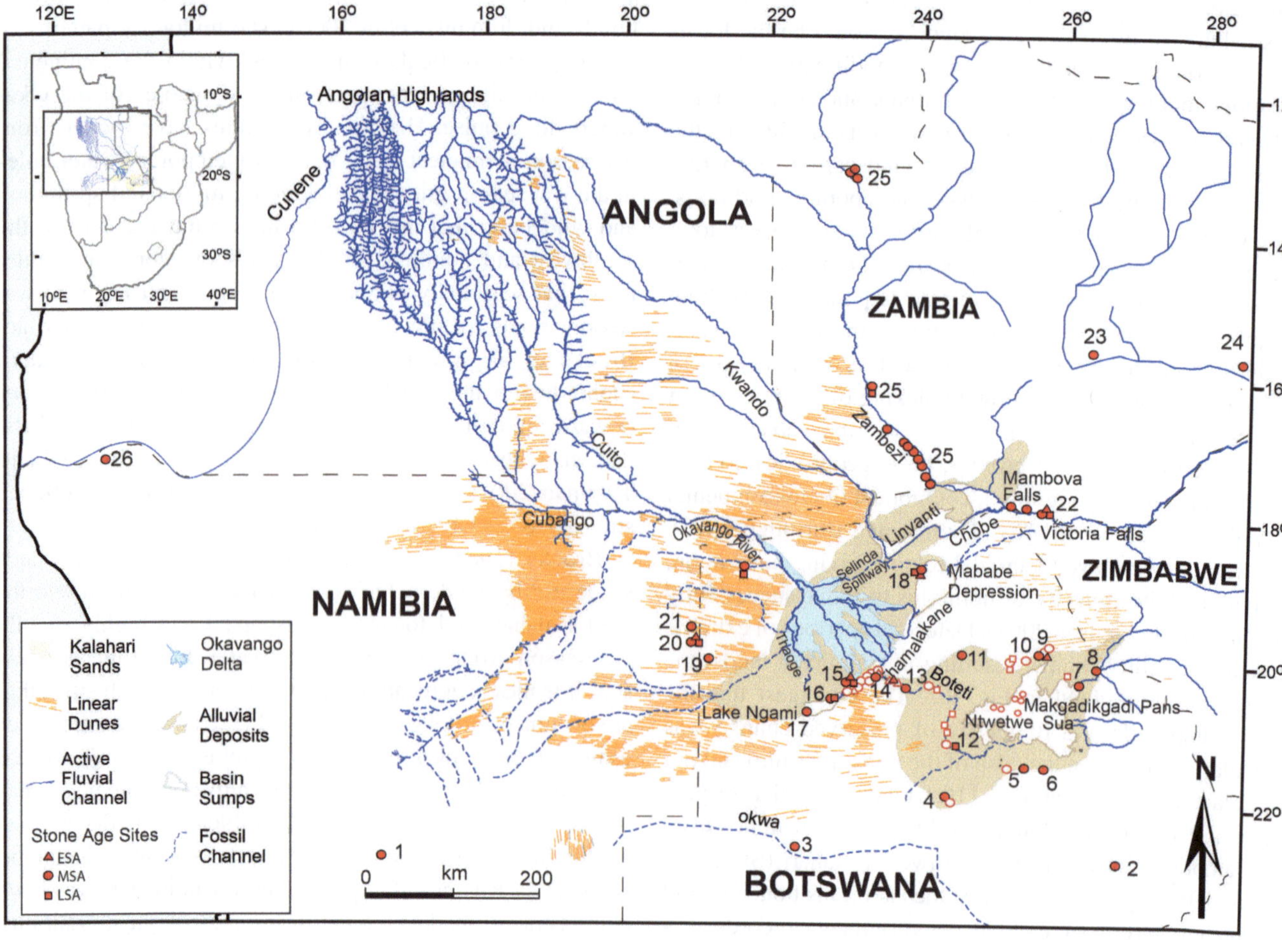

Fig. 9.3 Kalahari Stone Age sites reported in the literature (see Table 9.1). Closed symbols are those sites published in academic journals. Open symbols include sites reported in gray literature including archeological impact assessments where specific locations are given

Okwa Valley, and by Yellen (1971) in the Quangwa Valley. In addition to the numerous ESA handaxes found with the Nchabe, Thamalakane and Boteti Rivers (Campbell 1988), a transitional ESA/MSA factory site remains intact at Samedupe Drift (Wayland 1950), one of several localities where silcrete outcrops along the Boteti River. None of these sites have been dated, and their integration with the paleoenvironmental record from MIS 6-2 remains unknown.

MSA sites are more numerous, though most are still characteristically open in nature. Specifically, the region surrounding the Thamalakane, the Boteti, and the Makgadikgadi basin seems to have been an important locality for early humans. The basin floor of Makgadikgadi itself, in particular that of Ntwetwe Pan, is also yielding extensive suites of MSA artifacts. These tools are generally made from locally formed silcrete and are similar in nature to those noted from Kudiakam Pan, also within the Makgadikgadi basin. Robbins (1987) notes that these artifacts are relatively fresh and characterized by well-made unifacial and bifacial points, notched pieces, denticulates, and a number of small handaxes, some of which were found embedded on their edges suggesting erosion *in situ*. Levallois cores at both pans are a significant component of the debitage. Within Ntwetwe, an additional component of the archeological assemblage includes some extraordinarily large and unusual artifacts (Fig. 9.4). Like the ESA sites, these assemblages remain unsystematically investigated and without chronological control, making their integration into the record of paleoenvironmental change currently unattainable.

There are very few well-dated archeological assemblages within the interior of southern Africa. ≠Gi, an open site associated with permanent water and a well-developed MSA industry characterized by a variety of retouched points, is associated with extinct large *bovids* dated to 77 ± 11 TL kBP (Brooks et al. 1990). A bracketing ostrich eggshell, dating to 34 ± 1 ^{14}C kBP, has also been obtained from the transition pan margin sediments that separate the MSA from the LSA blade industries (Brooks et al. 1990). At Tsodilo, the MSA spans the range from ~90–50 ka at White Paintings rock shelter (Robbins et al. 2000, 2016; Ivester et al. 2010), with a dramatic increase in stone artifacts during the MSA levels at c.35 ka, and indications of long-term reoccupation of the shelter (Robbins and Murphy 1998). Recent work has also yielded a date of 52 ± 7 OSL kBP for a low-density MSA artifact scatter exposed in a modern quarry on the shores of Lake Ngami (Brook et al. 2008).

The transition to microlithic industries and the emergence of the LSA in Botswana appears to have taken place between 40 and 30 ka (Fig. 9.2), although the association of microlithic assemblages with fish fossils at White Paintings rock shelter in the Tsodilo Hills suggests this transition may have taken place earlier (Robbins et al. 2000).

While Walker (1998) notes that the strong association of Stone Age sites with the Kalahari's hydrological systems may partly be a function of archeological visibility, the role of water availability within this environment most likely was, and still is, extremely important to the survival of all species. The dynamics of the hydrological system during

Fig. 9.4 Stone tools made on local silcrete and found in the Ntwetwe Pan, Makgadikgadi found eroding out of the pan floor. The age and functionality of these tools remains unknown. Use as cores, bifacial tools, and/or symbolic artifacts have all been suggested (K. Kuman personal communication)

Table 9.1 Reported Stone Age sites in the Kalahari (see Fig. 9.3)

Location	Site description	References
1. Windhoek	MSA at spring	Fock (1954), Clark (1982)
2. Serowe	ESA on terrace of Metsemasweu River	Ebert et al. (1976)
3. Okwa Valley	ESA on river terraces	Aldiss (1987)
4. Kedia	MSA hunting camp on edge of pan	Cooke and Patterson (1960b)
5. Orapa	MSA factory site on beach ridge	Cohen (1974)
6. Lethakane Well	MSA/LSA site on pan edge	Cooke and Patterson (1960b)
7. Nata	LSA/MSA hunting camp on lake shoreline	Bond and Summers (1954)
8. Nata	MSA site on Nata river	Cooke (1967)
9. Ngxaishini pan	ESAR and MSA site with fossilized faunal remains	Robbins and Murphy (1998)
10. Makgadikgadi	ESA site	McFarlane and Segadika (2001)
11. Kudiakam pan	MSA at Kudiakam pan	Robbins (1987)
12. Toromoja/Gwi	LSA site on Boteti	Denbow and Campbell (1980), Helgren (1984)
13. Maklamabedi Drift	MSA in bed of Boteti river	Van Waarden (1988)
14. Samedupi drift	ESA quarry site on banks of Boteti	Wayland (1950), Cooke (1979)
15. Nhabe river	ESA to LSA sites	Campbell (1988), Robbins et al. (2008)
16. Lake Ngami	MSA and LSA at Toteng	Cooke (1979), Robbins (1984), Brook et al. (2008)
17. Lake Ngami	MSA factory site on southeast shore	Cooke and Patterson (1960a)
18. Savuti	MSA factory site along base of outcrops; ESA on raised beaches; LSA and rock paintings	Hitchcock (1982), Campbell (1970), Robbins (1987)
19. Dobe valley	ESA to LSA and dated MS/LSA site at #Gi pan edge	Yellen (1971), Brooks and Yellen (1977), Helgren and Brooks (1983), Brooks et al. (1990)
20. Xai Xai	LSA site at pan edge	Yellen (1971), Wilmsen (1978)
21. Tsodilo	MSA and LSA rock shelter sites	Junod (1963), Denbow and Campbell (1980), Robbins et al. (2000)
22. Victoria Falls	ESA to LSA at on terraces	Clark et al. (1950), Clark (1975)
23. Mumbwa	MSA cave site	Barham (1996, 2000)
24. Twin Rivers	MSA cave site	Clark and Brown (2001)
25. Upper Zambezi	MSA to LSA sites along river	Phillipson (1978)
26. Cunene	MSA site on banks of river	Nicoll (2010)

MIS 6-2 must therefore be considered critical to understanding the regional dynamics of hominin populations.

Subcontinental Records

With the exception of Drotsky's Cave, where speleothem growth has been dated by radiocarbon and U/Th techniques (Cooke and Verhagen 1977; Shaw and Cooke 1986; Brook et al. 1996) and the sporadic occurrence of datable freshwater shells and calcretes (Cooke and Verstappen 1984; Shaw 1985; Shaw and Thomas 1988; Shaw et al. 1992), there are almost no long records with which to tie in the environmental record of landscape and hydrological change in the Kalahari for the duration of MIS 6-2. Attempts to place the record of hydrological change within the context of concurrent continental-scale change are hampered by poor resolution. Though at the broadest scale, "wet" phases can be related throughout MIS 6-2 in both the northern and southern hemispheres of Africa (Burrough et al. 2009a, b). Recently published records from lake Malawi suggest a significant divergence in lake conditions between lake Malawi (which is subject to "megadrought" conditions during 115–95 ka) (Scholz et al. 2007, 2011) and the Middle Kalahari, where megalake Makgadikgadi is at least sporadically high during MIS 5d-5b (Fig. 9.2). On more recent timescales, however, better chronological resolution enables some analysis of regional leads and lags within paleoenvironmental records to be ascertained.

An interesting spatial divergence between patterns of environmental change in the equatorial tropics and the southern African tropical-subtropical zone (cf. Partridge et al. 2004; Gasse et al. 2008; Burrough et al. 2009a; Thomas et al. 2009) can be observed during the most recent post-glacial period. For example, Gasse et al. (2008) note that the major shift in P-E balance toward wetter conditions in the southern tropics occurred at 18–16 ka and continued until 13–12.5 ka. This contrasts with the later, and more stepwise, response of the equatorial tropics where major monsoonal reactivation occurred at 15–14 and 11.5–10.5 ka (Gasse et al. 2008), and seems to have been driven by strong teleconnections with the northern hemisphere high latitudes (Gasse 2000; Stager et al. 2011). Further, late glacial aridity, reported for the equatorial tropics (Stager et al. 2011) is not widespread through the southern hemisphere tropics and subtropics, where at the very least contemporaneous pockets of increasing wetness were experienced (Thomas et al. 2012; Singarayer and Burrough 2015).

Mechanistically, this may be related, via teleconnections, to the effect of major high latitude events, including meltwater discharge, on the extent of the seasonal migration of the ITCZ (and associated CAB) (e.g., Tierney et al. 2011). Coupled ocean-atmosphere-sea ice GCM simulations have been used to examine East African water balance responses to high latitude North Atlantic meltwater pulses. These experiments demonstrate the potential for a stronger southern ITCZ, leading to wetter surface conditions over southern Africa and southern areas of East Africa, and drier conditions in northern East Africa (Thomas et al. 2009). Causal mechanisms, however, remain notoriously difficult to test, and cannot be examined robustly beyond the last glacial/interglacial transition, without a substantial increase in the quantity and precision of Quaternary environmental data from the southern African summer rainfall zone (Burrough et al. 2009a). Nonetheless, the regional divergence of environmental conditions during periods of rapid transition holds strong potential as a driver of population movements during the late glacial, particularly during times of catastrophic equatorial drought as has been suggested by Stager et al. (2011).

Conclusion

The drivers of climatic change during MIS 6-2 in the semi-arid southern African interior remain a significant challenge to Quaternary science, especially to developing hypotheses regarding the climatic mechanisms that drove environmental changes in the region. Despite this, a much better and more sophisticated understanding of paleoenvironmental records, and their implications in terms of environmental dynamics relevant to human behavior, is emerging. This is due in no small part to approaches that seek to find the appropriate interpretation of the growing body of geoproxy records, which in central southern Africa is critical, as these records dominate data sources in this dryland region. In the context of past population dynamics, overarching climatic theories remain largely irrelevant, since it is regional environmental (rather than climatic) variability to which hominins are likely to have responded. In the Kalahari, this variability is driven largely by the availability of surface water and as Walker (1998) notes, it was (and is) reliability, rather than quantity, of rainfall that was the critical factor in the development of biological and cultural systems.

References

Aldiss, D. (1987). A record of stone axes near Tswaane borehole, Ghanzi District. *Botswana Notes and Records, 19*, 41–43.

Andersson, L., Gumbricht, T., Hughes, D., Kniveton, D., Ringrose, S., Savenije, H., et al. (2003). Water flow dynamics in the Okavango River Basin and Delta – a prerequisite for the ecosystems of the Delta. *Physics and Chemistry of the Earth, Parts A/B/C, 28*, 1165–1172.

Avery, D. M. (2003). Early and Middle Pleistocene environments and hominid biogeography; micromammalian evidence from Kabwe, Twin Rivers and Mumbwa Caves in central Zambia. *Palaeogeography, Palaeoclimatology, Palaeoecology, 189*(1–2), 55–69.

Bailey, R. M., Thomas, D. S. G. (2014). A quantitative approach to understanding dated dune stratigraphies. *Earth Surface Processes and Landforms, 39*, 614–631.

Barham, L. S. (1996). Recent research on the Middle Stone Age at Mumbwa Caves, central Zambia. In G. Pwiti & R. Soper (Eds.), *Aspects of African archaeology* (pp. 191–200). Harare: University of Zimbabwe Publications.

Barham, L. (2000). The palaeobiogeography of central Africa. In L. Barham (Ed.), *The Middle Stone Age of Zambia, South Central Africa* (pp. 223–236). Bristol: Western Academic and Specialist Press.

Barham, L. (2001). Central Africa and the emergence of regional identity in the Middle Pleistocene. In L. Barham & K. Robson Brown (Eds.), *Human roots: Africa and Asia in the Middle Pleistocene* (pp. 65–80). Bristol: Western Academic & Specialist Press.

Bartlam-Brooks, H. L. A., Bonyongo, M. C., & Harris, S. (2011). Will reconnecting ecosystems allow long-distance mammal migrations to resume? A case study of a zebra *Equus burchelli* migration in Botswana. *ORYX, 45*(2), 210–216.

Berger A. (1992). Orbital variations and insolation database. IGBP PAGES/World data. Center-A for paleoclimatology data contribution series # 92-007. NOAA/NGDC, Paleoclimatology Program, Boulder CO, USA.

Bond, G. (1975). The geology and formation of the Victoria Falls. In D. W. Phillipson (Ed.), *Mosi-oa-Tunya – a handbook of the Victoria Falls region* (pp. 19–48). London: Longman.

Bond, G., & Summers, R. (1954). A late Stillbay hunting camp on the Nata River, Bechuanaland Protectorate. *South African Archaeological Bulletin, 9*, 89–95.

Brook, G. A., Cowart, J. B., & Marais, E. (1996). Wet and dry periods in the southern African summer rainfall zone during the last 300 kyr from speleothem, tufa and sand dune age data. *Palaeoecology of Africa, 24*, 147–158.

Brook, G. A., Srivastava, P., Brook, F. Z., Robbins, L. H., Campbell, A. C., & Murphy, M. L. (2008). OSL chronology for sediments and MSA artefacts at the Toteng Quarry, Kalahari Desert, Botswana. *South African Archaeological Bulletin, 63*, 151–158.

Brooks, A. (1984). San land-use patterns past and present: Implications for southern African prehistory. In M. Hall, G. Avery, D. M. Avery, M. L. Wilson, & A. J. B. Humphreys (Eds.), *Frontiers: Southern African archaeology today* (pp. 40–52). Oxford: British Archaeological Reports.

Brooks, A., & Yellen, J. E. (1977). Archaeological excavations at ≠ Gi: A preliminary report on the first two field seasons. *Botswana Notes and Records, 9*, 21–30.

Brooks, A. S., Hare, P. E., Kokis, J. E., Miller, G. H., Ernst, R. D., & Wendorf, F. (1990). Dating Pleistocene archeological sites by protein diagenesis in ostrich eggshell. *Science, 248*(4951), 60–64.

Burrough, S. L., & Thomas, D. S. G. (2008). Late Quaternary lake-level fluctuations in the Mababe Depression: Middle Kalahari palaeolakes and the role of Zambezi inflows. *Quaternary Research, 69*(3), 388–403.

Burrough, S. L., Thomas, D. S. G., Shaw, P. A., & Bailey, R. M. (2007). Multiphase Quaternary highstands at Lake Ngami, Kalahari, northern Botswana. *Palaeogeography, Palaeoclimatology, Palaeoecology, 253*(3–4), 280–299.

Burrough, S. L., Thomas, D. S. G., & Bailey, R. M. (2009a). Mega-Lake in the Kalahari: A Late Pleistocene record of the palaeolake Makgadikgadi system. *Quaternary Science Reviews, 28*, 1392–1411.

Burrough, S. L., Thomas, D. S. G., & Singarayer, J. S. (2009b). Late Quaternary hydrological dynamics in the Middle Kalahari: Forcing and feedbacks. *Earth-Science Reviews, 96*(4), 313–326.

Burrough, S. L., Breman, E., & Dodd, C. (2012). Can phytoliths provide an insight into past vegetation of the Middle Kalahari palaeolakes during the late Quaternary? *Journal of Arid Environments, 82*, 156–164.

Campbell, A. C. (1970). Notes on some rock paintings at Savuti. *Botswana Notes and Records, 2*, 15–23.

Campbell, A. C. (1988). *Archaeological impact assessment – Maun reservoir*. Gaborone: Snowy Mountains Engineering Corporation/ Department of Water Affairs.

Chase, B. M., Meadows, M. E. (2007). Late Quaternary dynamics of southern Africa's winter rainfall zone. *Earth-Science Reviews 84*, 103–138.

Chase, B. M., & Thomas, D. S. G. (2006). Late Quaternary dune accumulation along the western margin of South Africa: Distinguishing forcing mechanisms through the analysis of migratory dune forms. *Earth and Planetary Science Letters, 251*, 318–333.

Clark, J. D. (1975). Stone Age man in the Victoria Falls. In D. W. Phillipson (Ed.), *Mosi-oa-Tunya – a handbook of the Victoria Falls region* (pp. 28–47). London: Longman.

Clark, J. D. (1982). The cultures of the Middle Paleolithic/Middle Stone Age. In J. D. Clark (Ed.), *Cambridge history of Africa* (Vol. I, pp. 284–341)., From the earliest times to ca. 500 B.C London: Cambridge University Press.

Clark, J. D., & Brown, K. S. (2001). The Twin Rivers Kopje, Zambia: Stratigraphy, fauna, and artefact assemblages from the 1954 and 1956 excavations. *Journal of Archaeological Science, 28*(3), 305–330.

Clark, J. D., Oakley, K. P., Wells, L. H., & McClelland, J. A. C. (1950). New studies on Rhodesian man. *Journal of the Royal Anthropological Institute, 77*, 7–32.

Cohen, G. (1974). Stone Age artefacts from Orapa diamond mine, central Botswana. *Botswana Notes and Records, 6*, 1–5.

Cooke, C. K. (1967). A preliminary report on the Stone Age of the Nata River. *Arnoldia Rhodesia, 2*(40), 1–10.

Cooke, C. K., & Patterson, M. L. (1960a). A Middle Stone Age open site: Ngamiland, Bechuanaland Protectorate. *South African Archaeological Bulletin, 15*, 36–39.

Cooke, C. K., & Patterson, M. L. (1960b). Stone Age sites: Lake Dow area, Bechuanaland. *South African Archaeological Bulletin, 15*, 119–122.

Cooke, H. J. (1979). The origin of the Makgadikgadi Pans. *Botswana Notes and Records, 11*, 37–42.

Cooke, H. J., & Verhagen, B. T. (1977). *The dating of cave development: An example from Botswana*. Sheffield, United Kingdom: Paper presented at the Proceedings of the Seventh International Speleological Congress.

Cooke, H. J., & Verstappen, T. H. (1984). The landforms of the western Makgadikgadi basin in northern Botswana, with a consideration of the chronology of the evolution of Lake Palaeo-Makgadikgadi. *Zeitschrift fur Geomorphologie, 28*(1), 1–19.

Denbow, J. R., & Campbell, A. C. (1980). National Museum of Botswana: Archaeological research programme. *Nyame Akuma, 17*, 3–9.

Drake, N., & Bristow, C. (2006). Shorelines in the Sahara: Geomorphological evidence for an enhanced monsoon from palaeolake Megachad. *Holocene, 16*(6), 901–911.

Ebert, J. I., Ebert, M. C., Hitchcock, R. K., & Thoma, A. (1976). An Acheulian locality at Serowe, Botswana. *Botswana Notes and Records, 8*, 29–37.

Ellery, W. N., & McCarthy, T. S. (1998). Environmental change over two decades since dredging and excavation of the lower Boro River, Okavango Delta. *Botswana. Journal of Biogeography, 25*(2), 361–378.

Feathers, J. K. (1997). Luminescence dating of sediment samples from White Paintings Rockshelter, Botswana. *Quaternary Science Reviews, 16*, 321–331.

Flint, L. (2006). Contradictions and challenges in representing the past: The Kuomboka festival of western Zambia. *Journal of Southern African Studies, 32*(4), 701–717.

Flint, R. F., & Bond, G. (1968). Pleistocene sand ridges and pans in western Rhodesia. *Bulletin of the Geological Society of America, 79*, 299–314.

Fock, G. J. (1954). Stone balls in the Windhoek Museum. *South African Archaeological Bulletin, 9*, 108–109.

Gasse, F. (2000). Hydrological changes in the African tropics since the Last Glacial Maximum. *Quaternary Science Reviews, 19*, 189–211.

Gasse, F., Chalié, F., Vincens, A., Williams, M. A. J., & Williamson, D. (2008). Climatic patterns in equatorial and southern Africa from 30,000 to 10,000 years ago reconstructed from terrestrial and near-shore proxy data. *Quaternary Science Reviews, 27*(25–26), 2316–2340.

Grove, A. T. (1969). Landforms and climatic change in the Kalahari and Ngamiland. *Geographical Journal, 135*, 191–212.

Gumbricht, T., McCarthy, T. S., & Merry, C. L. (2001). The topography of the Okavango Delta, Botswana, and its tectonic and sedimentological implications. *South African Journal of Geology, 104*(3), 243–264.

Haddon, I. G., & McCarthy, T. S. (2005). The Mesozoic-Cenozoic interior sag basins of Central Africa: The Late-Cretaceous-Cenozoic Kalahari and Okavango basins. *Journal of African Earth Sciences, 43*, 316–333.

Helgren, D. M. (1984). Historical geomorphology and geoarchaeology in the southwestern Makgadikgadi Basin, Botswana. *Annals – Association of American Geographers, 74*(2), 298–307.

Helgren, D. M., & Brooks, A. S. (1983). Geoarchaeology at Gi, a Middle Stone Age and Later Stone Age site in the northwest Kalahari. *Journal of Archaeological Science, 10*, 181–197.

Hitchcock, R. K. (1982). Prehistoric hunter-gatherer adaptations. In R. R. Hitchcock & M. Smith (Eds.), *Settlement in Botswana – the historical development of a human landscape* (pp. 47–65). Gaborone: Botswana Society.

Hitchcock, R. K. (2002). Coping with uncertainty: Adaptive responses to drought and livestock disease in the Northern Kalahari. In D. Sporton & D. S. G. Thomas 2002 (Eds.), *Sustainable livelihoods in Kalahari environments: Contributions to global debates* (pp. 161–192). Oxford: Oxford University Press.

Ivester, A. H., Brook, G. A., Robbins, L. H., Campbell, A. C., Murphy, M. L., Marais, E. (2010). A sedimentary record of environmental change at Tsodilo Hills White Paintings Rock Shelter, Northwest Kalahari Desert, Botswana. *Palaeoecology of Africa, 30*, 53–78.

Jacobs, Z., Roberts, R. G., Galbraith, R. F., Deacon, H. J., Grün, R., Mackay, A., et al. (2008). Ages for the Middle Stone Age of southern Africa: Implications for human behavior and dispersal. *Science, 322*(5902), 733–735.

Jacobs, Z., & Roberts, R. G. (2009). Catalysts for Stone Age innovations: What might have triggered two short-lived bursts of technological and behavioral innovation in southern Africa during the Middle Stone Age? *Communicative and Integrative Biology, 2*(2), 191–193.

Jouzel, J., Masson-Delmotte, V., Cattani, O., Dreyfus, G., Falourd, S., et al. (2007). Orbital and millennial Antarctic climate variability over the past 800,000 years. *Science, 317*(5839), 793–797.

Junod, S. M. (1963). A note on pottery from Tsodilo. *Bechuanaland. South African Archaeological Bulletin, 18*, 20.

Kocurek, G. (1998). Aeolian system response to external forcing factors – a sequence stratigraphic view of the Saharan region. In A. S. Alsharhan, K. W. Glennie, G. L. Whittle, & C. G. Kendall (Eds.), *Quaternary deserts and climatic change* (pp. 327–337). Rotterdam: Balkema.

Lancaster, N. (2011). Desert dune processes and dynamics. In D. S. G. Thomas (Ed.), *Arid zone geomorphology: Process, form and change in dryland* (pp. 487–515). Chichester: Wiley-Blackwell.

Lister, L. A. (1979). The geomorphic evolution of Zimbabwe-Rhodesia. *Transactions of the Geological Society of South Africa, 82*, 363–370.

Livingstone, D. (1858). *Missionary travels and researches in South Africa*. New York: Harper Bros.

Maslin, M. A., & Christensen, B. (2007). Tectonics, orbital forcing, global climate change, and human evolution in Africa: Introduction to the African paleoclimate special volume. *Journal of Human Evolution, 53*(5), 443–464.

Mazvimavi, D., & Wolski, P. (2006). Long-term variations of annual flows of the Okavango and Zambezi Rivers. *Physics and Chemistry of the Earth, 31*(15–16), 944–951.

McCarthy, T. S., & Hancox, P. J. (2000). Wetlands. In T. C. Partridge & R. R. Maud (Eds.), *The Cenozoic of southern Africa* (Vol. 40, pp. 218–235). New York: Oxford University Press.

McCarthy, T. S., Cooper, G. J. R., Tyson, P. D., & Ellery, W. N. (2000). Seasonal flooding in the Okavango Delta – recent history and future prospects. *South African Journal of Science, 96*, 25–33.

McFarlane, M. J., & Segadika, P. (2001). Archaeological evidence for the reassessment of the ages of the Makgadikgadi palaeolakes. *Botswana Notes and Records, 33*, 83–92.

Nash, D. J., & McLaren, S. J. (2003). Kalahari valley calcretes: Their nature, origins, and environmental significance. *Quaternary International, 111*(1), 3–22.

Nash, D. J., Meadows, M. E., & Gulliver, V. L. (2006). Holocene environmental change in the Okavango Panhandle, northwest Botswana. *Quaternary Science Reviews, 25*(11–12), 1302–1322.

Nicholson, S. E. (1996). A review of climate dynamics and climate variability in eastern Africa. In T. C. Johnson & E. Odada (Eds.), *The limnology, climatology and paleoclimatology of the East African lakes* (pp. 25–56). Amsterdam: Gordon & Breach.

Nicoll, K. (2010). Geomorphic development and Middle Stone Age archaeology of the Lower Cunene River. *Namibia-Angola Border. Quaternary Science Reviews, 29*(11–12), 1419–1431.

North Greenland Ice Core Project members (2004). North Greenland Ice Core Project oxygen isotope data. *IGBP PAGES/World Data Center for Paleoclimatology. Data Contribution Series # 2004-059.* NOAA/NGDC Paleoclimatology Program, Boulder CO, USA.

Partridge, T. C., Scott, L., & Schneider, R. (2004). Between Agulhas and Benguela: Responses of southern African climates of the Late Pleistocene to current fluxes, orbital precession and the extent of the circum-Antarctic vortex. In R. W. Battarbee, F. Gasse, & C. E. Stickley (Eds.), *Past climate variability through Europe and Africa* (Vol. 6, pp. 45–67). Dordrecht: Springer.

Passarge, S. (1904). *Die Kalahari*. Berlin: Dietrich Riemer.

Phillipson, L. (1978). *The stone-age archaeology of the Upper Zambezi*. Nairobi: British Institute in Eastern Africa.

Robbins, L. H. (1984). Toteng, a Late Stone Age site along the Nchabe River, Ngamiland. *Botswana Notes and Records, 16*, 1–6.

Robbins, L. H. (1987). Stone Age archaeology in the northern Kalahari, Botswana: Savuti and Kudiakam Pan. *Current Anthropology, 28*(4), 567–569.

Robbins, L. H., & Murphy, M. L. (1998). The Early and Middle Stone Age. In P. Lane, A. Reid, & A. Segobye (Eds.), *Ditswa Mmung: The archaeology of Botswana* (pp. 50–64). Gaborone: Pula Press/The Botswana Society.

Robbins, L. H., Murphy, M. L., Brook, G. A., Ivester, A. H., Campbell, A. C., Klein, R. G., et al. (2000). Archaeology, palaeoenvironment, and chronology of the Tsodilo Hills White Paintings Rock Shelter, northwest Kalahari Desert. *Botswana. Journal of Archaeological Science, 27*(11), 1085–1113.

Robbins, L. H., Campbell, A. C., Murphy, M. L., Brook, G. A., Liang, F., Skaggs, S. A., et al. (2008). Recent archaeological research at Toteng, Botswana: Early domesticated livestock in the Kalahari. *Journal of African Archaeology, 6*(1), 131–149.

Robbins, L. H., Brook, G. A., Murphy, M. L., Ivester, A. H., & Campbell, A. C. (2016). The Kalahari during MIS 6-2 (190–12 ka): Archaeology, paleoenvironment and population dynamics. In S. C. Jones & B. A. Stewart (Eds.), *Africa from MIS 6-2: Population dynamics and paleoenvironments* (pp. 175–193). Dordrecht: Springer.

Scholz, C. A., Johnson, T. C., Cohen, A. S., King, J. W., Peck, J. A., Overpeck, J. T., et al. (2007). East African megadroughts between 135 and 75 thousand years ago and bearing on early-modern human origins. *Proceedings of the National Academy of Sciences of the United States of America, 104*(42), 16416–16421.

Scholz, C. A., Cohen, A. S., Johnson, T. C., King, J., Talbot, M. R., & Brown, E. T. (2011). Scientific drilling in the Great Rift Valley: The 2005 Lake Malawi scientific drilling project – an overview of the past 145,000 years of climate variability in southern hemisphere East Africa. *Palaeogeography, Palaeoclimatology, Palaeoecology, 303*(1–4), 3–19.

Scholz, C. H., Koczynski, T. A., & Hutchins, D. G. (1976). Evidence of incipient rifting in southern Africa. *Geophysical Journal of the Royal Astronomical Society, 44*, 135–144.

Scott, L., Holmgren, K., & Partridge, T. C. (2008). Reconciliation of vegetation and climatic interpretations of pollen profiles and other regional records from the last 60 thousand years in the savanna biome of southern Africa. *Palaeogeography, Palaeoclimatology, Palaeoecology, 257*(1–2), 198–206.

Scott, L., Neumann, F. H., Brook, G. A., Bousman, C. B., Norström, E., & Metwally, A. A. (2012). Terrestrial fossil-pollen evidence of climate change during the last 26 thousand years in southern Africa. *Quaternary Science Reviews, 32*, 100–118.

Shaw, P. (1985). Late Quaternary landforms and environmental change in northwest Botswana: The evidence of Lake Ngami and the Mababe Depression. *Transactions of the Institute of British Geographers, 10*(3), 333–346.

Shaw, P. A., & Cooke, H. J. (1986). Geomorphic evidence for the late Quaternary paleoclimates of the Middle Kalahari of northern Botswana. *Catena, 13*(4), 349–359.

Shaw, P., & Thomas, D. S. G. (1988). Lake Caprivi: A late Quaternary link between the Zambezi and Middle Kalahari drainage systems. *Zeitschrift fur Geomorphologie, 32*(3), 329–337.

Shaw, P., & Thomas, D. S. G. (1993). Geomorphological processes, environmental change and landscape sensitivity in the Kalahari region of southern Africa. In D. S. G. Thomas & R. J. Allison (Eds.), *landscape sensitivity* (pp. 83–95). Chichester: John Wiley and Sons Ltd.

Shaw, P. A., Thomas, D. S. G., & Nash, D. J. (1992). Late Quaternary fluvial activity in the dry valleys (mekgacha) of the Middle and Southern Kalahari, southern Africa. *Journal of Quaternary Science, 7*(4), 273–281.

Shaw, P. A., Bateman, M. D., Thomas, D. S. G., & Davies, F. (2003). Holocene fluctuations of Lake Ngami, Middle Kalahari: Chronology and responses to climatic change. *Quaternary International, 111*(1), 23–35.

Singarayer, J. S., Burrough, S. L. (2015). Interhemispheric dynamics of the African rainbelt during the late Quaternary. *Quaternary Science Reviews, 124*, 48–67.

Sporton, D., & Twyman, C. (2002). Policy, politics and livelihoods. In D. Sporton & D. S. G. Thomas (Eds.), *Sustainable livelihoods in Kalahari environments: Contributions to global debate* (pp. 39–66). Oxford: Oxford University Press.

Stager, J. C., Ryves, D. B., Chase, B. M., & Pausata, F. S. R. (2011). Catastrophic drought in the Afro-Asian monsoon region during Heinrich event 1. *Science, 331*(6022), 1299–1302.

Stokes, S., Thomas, D. S. G., & Washington, R. (1997). Multiple episodes of aridity in southern Africa since the last interglacial period. *Nature, 388*(6638), 154–158.

Stute, M., & Talma, A. S. (1998). Glacial temperatures and moisture transport regimes reconstructed from noble gas and $d^{18}O$, Stampriet aquifer, Namibia. *Isotope techniques in the study of past and current environmental changes in the hydrosphere and the atmosphere* (pp. 307–328). Vienna: IAEA Vienna Symposium.

Taljaard, J. J. (1981). *Upper air circulation, temperature and humidity over southern Africa*, Technical Paper 10. Pretoria: Government Printers.

Telfer, M. W., Bailey, R. M., Burrough, S. L., Stone, A. E. S., Thomas, D. S. G., & Wiggs, G. S. F. (2010). Understanding linear dune chronologies: Insights from a simple accumulation model. *Geomorphology, 120*(3–4), 195–208.

Thomas, D. S. G. (1992). Desert dune activity: Concepts and significance. *Journal of Arid Environments, 22*(1), 31–38.

Thomas, D. S. G., & Burrough, S. L. (2012). Interpreting geoproxies of late Quaternary climate change in African drylands: Implications for understanding environmental change and early human behaviour. *Quaternary International, 253*, 5–17.

Thomas, D. S. G., Burrough, S. L. (2013). Luminescence-based dune chronologies in southern Africa: Analysis and interpretation of dune database records across the subcontinent. *Quaternary International*.

Thomas, D. S. G., & Shaw, P. A. (1991). *The Kalahari environment*. Cambridge: Cambridge University Press.

Thomas, D. S. G., & Shaw, P. A. (2002). Late Quaternary environmental change in central southern Africa: New data, synthesis, issues and prospects. *Quaternary Science Reviews, 21*(7), 783–797.

Thomas, D. S. G., Bailey, R., Shaw, P. A., Durcan, J. A., & Singarayer, J. S. (2009). Late Quaternary highstands at Lake Chilwa, Malawi: Frequency, timing and possible forcing mechanisms in the last 44 ka. *Quaternary Science Reviews, 28*(5–6), 526–539.

Thomas, D. S. G., Burrough, S. L., & Parker, A. G. (2012). Extreme events as drivers of early human behaviour in Africa? The case for variability, not catastrophic drought. *Journal of Quaternary Science, 27*(1), 7–12.

Tierney, J. E., Russell, J. M., Sinninghe Damste, J. S., Huang, Y., & Verschuren, D. (2011). Late Quaternary behavior of the East African monsoon and the importance of the Congo Air Boundary. *Quaternary Science Reviews, 30*(7–8), 798–807.

Tyson, P. D. (1986). *Climatic change and variability in southern Africa*. Cape Town: Oxford University Press.

Tyson, P. D. (1999). Atmospheric circulation changes and palaeoclimates of southern Africa. *South African Journal of Science, 95*(4), 194–201.

Van Waarden, C. (1988). Mkalamabedi Drift 03-B4-1, Mitigation report. Unpublished report by Joint Venture Consultants/Ove Arup & Partners/Stewart Scott & Partners, Gaborone.

Walker, N. (1998). The Late Stone Age. In P. Lane, A. Reid, & A. Segobye (Eds.), *Ditswammung: The archaeology of Botswana*. Gaborone: Pula Press/The Botswana Society.

Wayland, E. J. (1950). From an archaeological notebook. *South African Archaeological Bulletin, 5*, 4–14.

Wilmsen, E. N. (1978). Prehistoric and historic antecedents of a contemporary Ngamiland community. *Botswana Notes and Records, 10*, 5–8.

Yellen, J. E. (1971). Archaeological excavations in western Ngamiland. *Botswana Notes and Records, 3*, 276.

Chapter 10
The Kalahari During MIS 6-2 (190–12 ka): Archaeology, Paleoenvironment, and Population Dynamics

Lawrence H. Robbins, George A. Brook, Michael L. Murphy, Andrew H. Ivester, and Alec C. Campbell

Abstract We present a synthesis of archaeological and paleoenvironmental information for the period MIS 6-2 in the Kalahari. Discussion centers on the implications of nine new, internally consistent OSL ages obtained from White Paintings Rock Shelter. These dates provide a better understanding of the Middle Stone Age (MSA) and Late Stone Age (LSA) sequence. In addition, the revised chronology dates 11 buried soil A-horizons that were formed during wetter periods. The buried A-horizons, along with dated speleothems and high lake levels in the Kalahari correlate with Antarctic warming events (A) and North Atlantic Heinrich events (H). We also discuss the implications of the Kalahari megalake, paleolake Makgadikgadi, for human populations and compare dated changes in the archaeological sequence at WPS with dates established in Khoisan genetic evolutionary studies.

Keywords White Paintings Shelter • Middle Stone Age • Early–Later Stone Age • Paleolake Makgadikgadi • Kudiakam Pan • Antarctic Warming Events (A) • Heinrich (H) Events

L.H. Robbins (✉)
Department of Anthropology, Michigan State University, 354 Baker Hall, East Lansing, MI 48824, USA
e-mail: lrobbins@msu.edu

G.A. Brook
Department of Geography, University of Georgia, Athens, GA 30602, USA

M.L. Murphy
Kalamazoo Valley Community College, 6767 West 'O' Ave, P.O. Box 49003-4070 Kalamazoo, MI 49009, USA

A.H. Ivester
Department of Geosciences, University of West Georgia, Carrollton, GA 30118, USA

A.C. Campbell (*Deceased*)
P.O. Box 306 Crocodile Pools, Gaborone, Botswana

Introduction

The Kalahari is one of the major drylands in the Southern Hemisphere. It extends from the Orange River in the south, north into the southern part of the Democratic Republic of the Congo, and west into the Etosha Pan area of Namibia (Thomas and Shaw 1991). Kalahari sands also cover large areas of Angola, Zimbabwe and Zambia (Fig. 10.1). This chapter focuses on research carried out in Botswana in an area often considered the "heartland" of the Kalahari.

The Kalahari is also one of the most significant areas in the world for anthropological research because of the San (Bushmen) who live there (Lee 1979; Silberbauer 1981). Studies of the San have had a major theoretical impact in anthropology, especially within archaeology, including topics as varied as: the reconstruction of subsistence, band aggregation and dispersal, exchange networks, projectile point style, site formation processes, and the interpretation of rock art, including Upper Paleolithic cave art (e.g., Yellen 1977; Wiessner 1977, 1983; Brooks 1984; Clottes and Lewis-Williams 1998). The Kalahari has also been the focus of considerable debate in anthropology that has challenged the validity of using San ethnographic evidence to interpret the archaeological record (Wilmsen 1989). In addition, pioneering research in the development of the "out-of-Africa theory" of the origin of modern humans was initially supported by studies of San MtDNA (Cann et al. 1987).

Despite the importance of the Kalahari, the Stone Age archaeological record for the period MIS 6-2 is poorly known in comparison to other areas in southern Africa. Only three sites containing MSA deposits have been extensively excavated in a vast area and there are comparatively few dates available. Fauna that dates to the period MIS 6-2 is rare, and no fossil human or hominin remains have been recovered (cf. Grine 2016). The dearth of well-dated information as well as other factors has contributed to the view that the Kalahari was an archaeological backwater. However, new findings, especially from the Tsodilo Hills in

Sacha C. Jones and Brian A. Stewart (eds.), *Africa from MIS 6-2: Population Dynamics and Paleoenvironments*, Vertebrate Paleobiology and Paleoanthropology, DOI 10.1007/978-94-017-7520-5_10

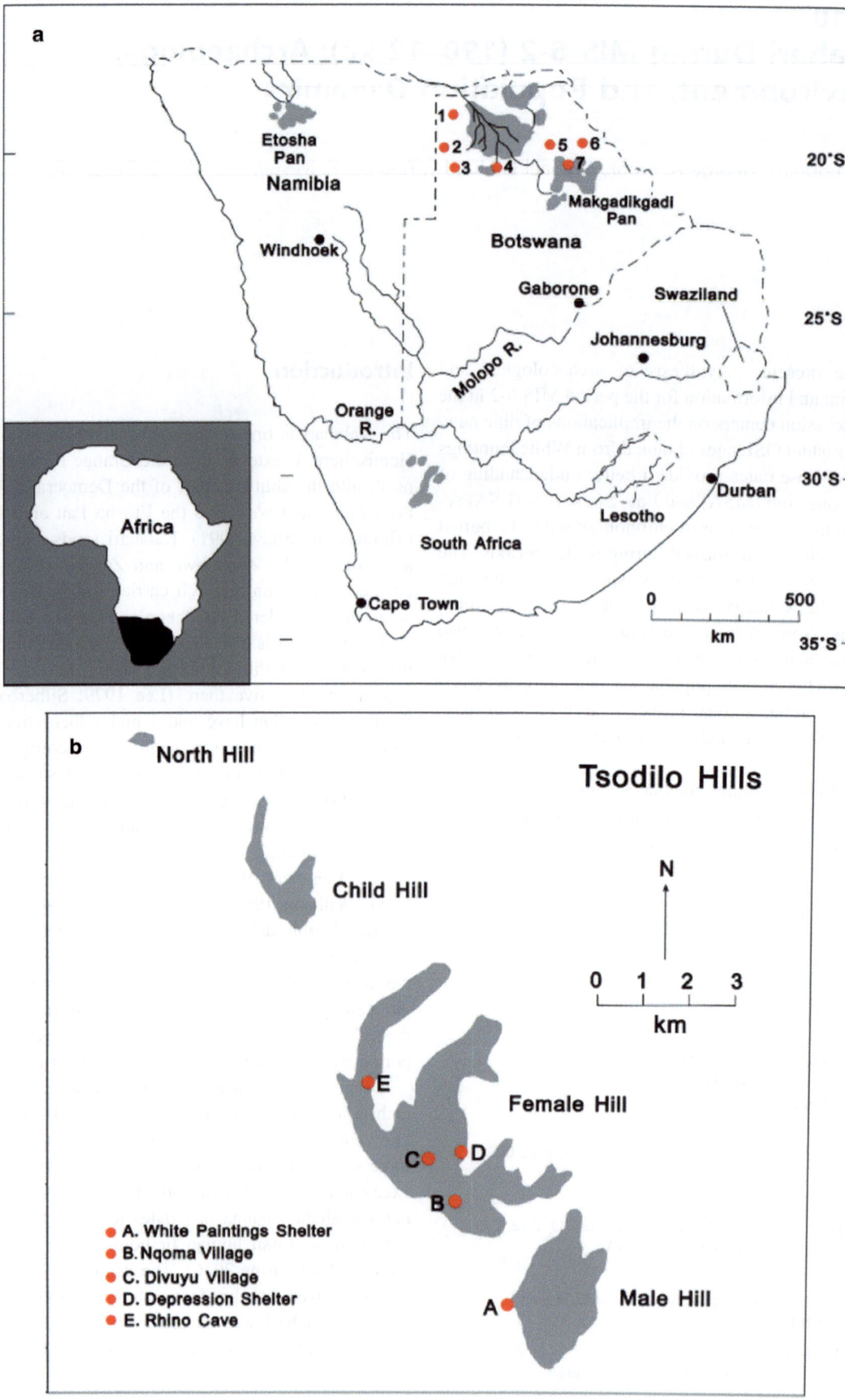

Fig. 10.1 Locality maps showing sites mentioned in the text. **a** *1* Tsodilo Hills, *2* #Gi Pan, *3* Gcwihaba Cave, *4* Toteng Quarry, *5* Kudiakam Pan, *6* Ngxaishini Pan, *7* Gutshaa Pan. **b** Sites at Tsodilo Hills

Table 10.1 Kalahari sites and their likely relationship to Marine Isotope Stages

MIS	Site	Archaeology	Comments
1–?	Ngxaishini Pan-	Acheulean, MSA, LSA, recent hunting blinds (Robbins and Murphy 1998)	Acheulean site with traces of fauna embedded in base of pan, not dated Surface collection
4–5?	Kudiakam Pan	MSA (Robbins 1988)	MSA points, scrapers, small bifaces, ostrich eggshell fragments Not dated Surface collection/sterile test pit
3–4/5	#Gi Pan	MSA, LSA (Brooks et al. 1990)	MSA, numerous points, extinct fauna MSA ostrich eggshell diagenesis = 65–85 ka, TL = 77 ± 11 ka Excavated site
3	Toteng Quarry wall	MSA (Brook et al. 2008)	Near Lake Ngami & Nchabe River Few artifacts in wall, OSL 51.5 ± 5.1 ka Overlying deposits lacking artifacts, OSL 26.5 ± 5.1 ka Surface collection on slope below quarry wall
1–5	WPS 7 m.	MSA, early LSA, LSA (Robbins et al. 2000; Ivester et al. 2010)	Radiocarbon, TL & OSL dates New internally consistent OSL chronology Excavated site
1–3	Depression Shelter	MSA?, early LSA, LSA (Brook et al. 2003)	Early LSA microliths and ground chromite Cupules & rock art Excavated site
1-4?	Rhino Cave	MSA, LSA (Robbins et al. 1996a, 2007; Brook et al. 2010; Coulson et al. 2011)	MSA points, specularite. Cupules & rock art Radiocarbon dates for LSA and cupules on wall Excavated site
1-3?	Gcwihaba (Drotsky's) Cave	LSA & Plio/Pleistocene fossil deposits (Yellen et al. 1987; Robbins et al. 1996b)	Dated terminal Pleistocene deposits with artifacts & fauna Base of test pit estimated age of ~30 ka Excavated site

Note WPS, Depression, and Rhino Cave also have deposits that contain Early Iron Age potsherds, while the Toteng Quarry contains Bambata pottery and livestock in the uppermost meter. The rock art mentioned here includes: red, herder paintings dating approximately to the first millennium AD; and white finger-drawn paintings dating to recent centuries (Campbell et al. 2010a)

northwest Botswana have helped to overturn this view, and are currently demonstrating the significance of the area (Brook et al. 2010; Campbell et al. 2010a; Ivester et al. 2010; Robbins and Murphy 2011).

Overview of Archaeological Sites

Early Stone Age (ESA) and MSA sites are widely distributed in the Kalahari as well as near its margins (Cook 1979; Thomas and Shaw 1991: Table 8.1; Robbins and Murphy 1998) (Fig. 10.1). Table 10.1 shows the general relationship of the best-known sites to Marine Isotope Stages. It can be seen that these sites are primarily rock shelters/caves and open-air pan margin sites (Robbins et al. 2010). Only five of the sites in Table 10.1 have been dated by radiocarbon and/or OSL. Most of the information comes from the Tsodilo Hills (Fig. 10.1). On the whole, the archaeological evidence from the Kalahari that fits within the parameters of this book is mainly known from the period MIS 5-2, with the bulk of the dated information coming from the Tsodilo Hills White Paintings Shelter (WPS).

Early Stone Age

ESA Acheulean handaxes along with flakes and cores have been found along the surface of river valleys such as the Nchabe, the Boteti and the Okwa, as well as at other areas both in and adjacent to the Kalahari (Fig. 10.1). These river valley surface sites are not dated and lack fauna as well as evidence of *in situ* contexts. However, the handaxes are likely to date to earlier than MIS 6 based on the discoveries of similar tools at dated sites in East and southern Africa. Thus, it is almost certain that early hominins inhabited parts of the Kalahari long before the onset of MIS 6 at 186 ka.

Along the Boteti River valley, nearby sources of silcrete from the riverbed were being used for removing large flakes from boulders. Acheulean raw material use along the river valleys may have been mainly "expedient" (using what was readily available), or was possibly constrained to locations close to the source area because long to moderate distance exchange systems had not yet developed.

No rock shelters or caves in Botswana have thus far yielded Acheulean artifacts, although they may exist in deep deposits underlying MSA levels, or at other unexcavated

Fig. 10.2 Embedded handaxe at Nxaishini Pan (Photograph L.H. Robbins)

Fig. 10.3 View of WPS showing white elephant on wall and surface squares (Photograph M.L. Murphy)

rock shelters or caves. Interestingly, Wonderwerk Cave, located just outside of the Kalahari in South Africa, has recently yielded the earliest archaeological evidence of fire in an Acheulean context dated to $\sim$1.0 Ma (Berna et al. 2012). Burned bone fragments and plant material were recovered from archaeological stratum 10 along with "a low density of lithic artifacts comprising only seven bifaces, 36 flakes, 15 cores, and 23 slabs of banded ironstone showing flake removals" (Berna et al. 2012: 2).

The only Acheulean site that we are aware of in Botswana that contains fauna is Ngxaishini Pan, located near Gweta. Figure 10.2 shows a possible butchering area where an embedded fossil bone can be seen near the pointed end of a large handaxe projecting from highly cemented calcrete at the base of the pan. Another area on the pan contains remnants of a fossil tusk and a scatter of flakes, also embedded in calcrete. As noted in Table 10.1, MSA and LSA artifacts also occur at this site, as well as recent hunting blinds. Elephants still come to this site as we saw them drinking water at night, directly in front of the stone hunting blinds. The availability of water in this pan during part of the dry season probably attracted animals such as elephants and other animals affording hunting and/or scavenging opportunities throughout much of the Stone Age.

Middle Stone Age

Fossil finds and genetic studies indicate that MSA peoples in southern Africa were anatomically modern humans. Only three excavated sites in the Kalahari of Botswana have yielded large samples of MSA artifacts: WPS, Rhino Cave and the pan margin site of #Gi. The latter site is the only MSA site that contained large numbers of extinct animals found in association with numerous MSA points. In contrast to the Acheulean river valley sites mentioned above, MSA peoples frequently used nonlocal raw materials for making tools, as documented at WPS and Rhino Cave.

We have also observed extensive MSA surface sites on pans, such as Kudiakam, where Acheulean sites do not occur, unless they are deeply buried. A shallow test pit excavated at Kudiakam Pan contained no artifacts (Robbins 1987). These pan sites contain scatters of MSA artifacts such as well-made unifacial and bifacial points, scrapers, denticulates, cores, flakes and blades. With the exception of occasional ostrich eggshell fragments (found at Kudiakam), faunal remains are generally absent. However, bone fragments have been observed at an MSA site in Nxai Pan National Park (Wadley personal communication). The abundance of MSA artifacts on the surfaces of previously uninhabited pans is suggestive of an expansion in population during the MSA or a change in adaptive strategies or both, most likely during MIS 5 and MIS 4 (see below).

In South Africa, the MSA includes bone points, decorated ostrich eggshell water containers, and tick shell (*Nassarius kraussianus*) beads in well-dated contexts (Henshilwood et al. 2004, 2011; Marean et al. 2007; Blackwell et al. 2008; Texier et al. 2010). These items, often viewed as signposts of modern human behavior, have not yet been found during the MSA in the Kalahari. However, ground specularite, burned stone tools and cupules have been reported for undated MSA deposits at Rhino Cave (Coulson et al. 2011). Cupules are small depressions, about the size of half of a tennis ball, or a little less, that have been intentionally ground/pecked into bedrock, or large boulders. They are widespread in Africa as well as in other parts of the world. There are numerous sites with cupules at Tsodilo Hills, but their specific ages are unknown (Campbell et al. 2010b). Rhino Cave contains over 300 cupules and Depression Shelter has over 1,000 cupules situated on a wall where nearby deposits contained 5 meters of archaeological deposits (Brook et al. 2003). Organic material was extracted from deposits on two of the cupules

Fig. 10.4 View of deep excavation at WPS (Photograph M.L. Murphy)

on the southwest wall of Rhino Cave. Brook et al. (2010: 1) have reported "minimum calibrated AMS ^{14}C ages of 6443–6204 to cal. yr B.P. (5580 ± 60 ^{14}C yr B.P) for one depression, and 3000–2761 cal. yr. B.P. (2830 ± 50 ^{14}C yr B.P) for the other." The cupules may well have been ground into the bedrock and large boulders at Tsodilo over a long period and their meaning to the people who created them remains enigmatic.

White Paintings Shelter

The Tsodilo Hills WPS is located about 45 km west of the Okavango River, which is the nearest permanently flowing body of water (Figs. 10.1, 10.2 and 10.3). Excavations have uncovered a 7 m-deep sequence (Fig. 10.4). As will be discussed, new OSL dates indicate that the site spans at least the last 90 ka, or MIS 5b-1. WPS is significant because it is the only extensively excavated site in the Kalahari that provides a detailed paleoenvironmental and archaeological sequence covering much of the LSA and MSA.

Eight divisions in the archaeological sequence were identified and 11 buried soil horizons were described in detail (Murphy 1999; Robbins et al. 2000, 2012; Ivester et al. 2010). A brief review of some of the archaeological findings that are of general importance will follow. The shelter has provided evidence of intensive fish procurement in the Kalahari along with the discovery of barbed bone spear or harpoon points used to spear fish. The barbed points are generally similar to those found at Lake Turkana in Kenya, and at other distant localities. Hundreds of fish bones of catfish (*Clarius*) and bream (tilapia) were found in LSA contexts in this currently dry area. Fish were either taken from a lake that formerly existed at Tsodilo near WPS or from the Xeidum/Ncamasere River located about 15 km to the north of the site (Brook 2010). The large size of some of the fish supports exploitation of spawning runs during seasonal migrations when rivers were in flood. The flooding of the Okavango River and fish spawning runs occur at present during the dry season in Botswana. The Okavango floods annually in response to increased rainfall to the north in Angola (see Burrough 2016).

While fish procurement was important, a diverse fauna was recovered. More than 40 kinds of animals were identified, including wetland species not found in the immediate area at present as well as two extinct species (Robbins et al. 2000: Table 3).

Other significant findings include a lengthy record of the use of bone arrow points found in deposits that extend from the historic period back to as early as about 35 ka in the early LSA (Robbins et al. 2012). One of the most continuous records of ostrich eggshell bead production was also uncovered (Robbins 1999; Robbins et al. 2000). Ostrich eggshell beads are still being made at Tsodilo and beads broken in manufacture were found in deposits extending from historic times back to ~31 ka. A bead that was broken in manufacture, and recovered from square 21 at a depth of 190–200 cm, was directly dated by AMS to 26,460 ± 300 ^{14}C BP (30,537–31,366 cal BP, AA 31729) (Robbins et al. 2000).

The discussion of WPS in this chapter primarily draws on information from the three lowermost parts of the sequence. Starting with the base: (1) MSA (c. 410/420–700 cm); (2) Overlying transitional early LSA/MSA large blade industry situated between ~300 and 410/420 cm; and (3) Lower Fish Deposits which can also be considered as a microlithic early LSA situated at a depth of ~210–280 cm. Stone artifacts are shown in Figs. 10.5, 10.6, and 10.7.

OSL Chronology

Ivester et al. (2010) have shown that the 7 m of sediment at WPS was deposited within about the last 100 ka and consists mostly of aeolian sand augmented by colluvial material from the adjacent hill and by detrital material spalled from the rock shelter wall and ceiling. Relative contributions from these sources varied through time as a result of climate and

Fig. 10.5 MSA points from WPS (Photograph M.L. Murphy)

environmental changes to produce 11 soil stratigraphic units (Fig. 10.8). Strata with finer and better sorted sands, lower gravel amounts, and more rapid depositional rates, were deposited during periods of greater aridity, while strata with coarser and more poorly sorted sands, more gravel, a stable to slowly accreting land surface, and geochemical evidence

Fig. 10.6 Artifacts from the WPS large blade industry (Photograph M.L. Murphy)

Fig. 10.7 Microliths from the Lower Fish Deposits at WPS (Photograph M.L. Murphy)

of enhanced pedogenesis, record periods of greater moisture availability. Chronologically mixed radiocarbon dates obtained previously at WPS (Robbins et al. 2000) had clearly demonstrated bioturbation and disturbance of the site. For this reason a profile in the undisturbed exposed section of the south wall of Square 23 was selected for collecting OSL and associated sediment samples. Nine OSL ages in correct stratigraphic order provided a chronology for the 11

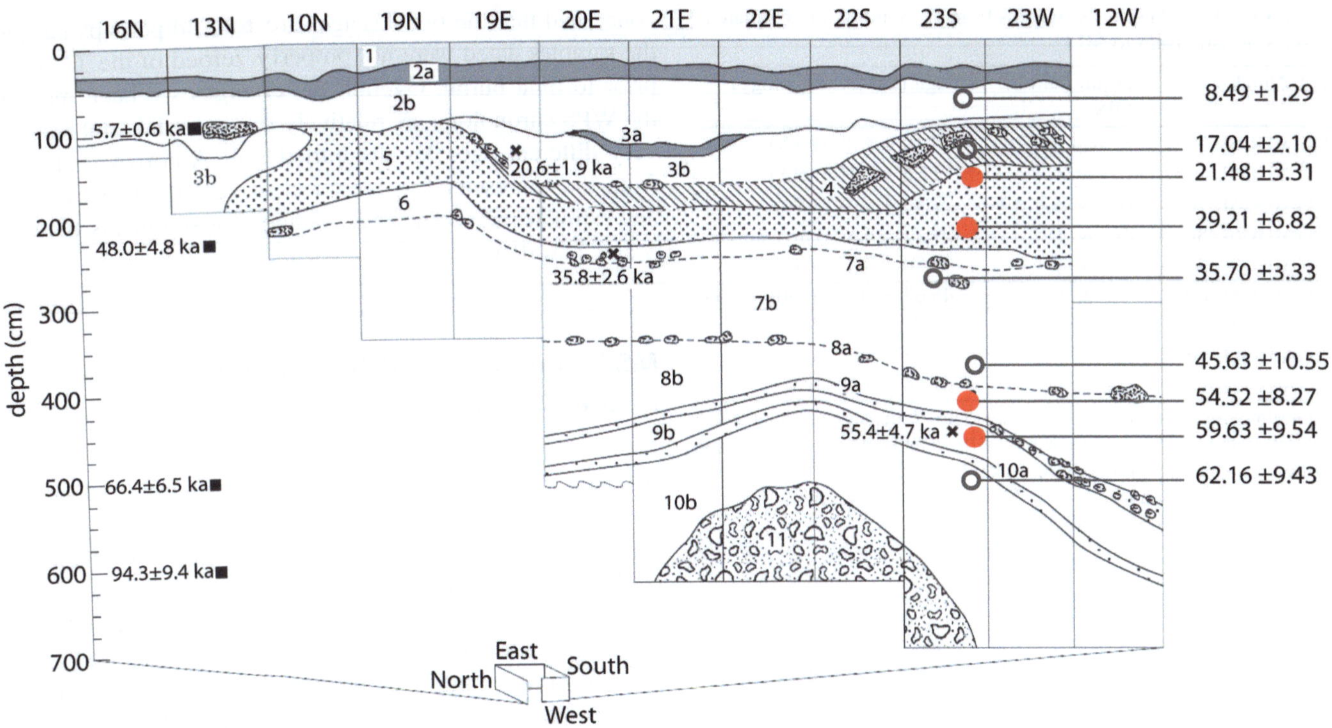

Fig. 10.8 WPS excavation squares 1–23 and wrap-around section showing the locations and ages of dated samples and the ages obtained. OSL samples dating close to the MIS 3-2 and MIS 4-3 boundaries are shaded. Square symbols show samples dated by TL at the University of Botswana and crosses show samples dated by OSL at the University of Washington. Soil stratigraphic units are numbered 1–11 from the surface downward and the subscript "a" is used to indicate a visible buried soil A-horizon or spall horizon. Unit 11 at the base of the section is an indurated, carbonate-cemented breccia (modified after Ivester et al. 2010)

soil horizons identified and revealed a trend toward more available moisture in the Pleistocene and Early to Middle Holocene, with a shift toward more arid conditions in the Late Holocene.

In this chapter, the OSL ages in Ivester et al. (2010) have been updated with improved corrections for sample depth, elevation, and the number of aliquots measured in the dating procedure. In addition, the age for the 75 cm depth is taken as the mean rather than the minimum age of the aliquots examined. Table 10.2 shows the new internally consistent OSL chronology obtained from sediments recovered from the profile of a single square. The procedures used for dating these samples are discussed elsewhere (Ivester et al. 2010).

It is important to mention here that Staurset and Coulson (2014) have demonstrated subsurface movement of lithic materials in MSA deposits at WPS through refitting of stone artifacts. The refitting work adds to the story of WPS and is consistent with previously reported radiocarbon and other data that help in assessing the extent of post-depositional movement of materials (Robbins et al. 2000; Ivester et al. 2010). We do not think that their analysis invalidates the OSL chronology, the associated results of the sedimentary analysis, or the correlations made in this chapter with other localities. Only 5 of their 34 refit groups indicate an interpreted minimum vertical displacement of more than 40 cm. The majority of their refits suggest a more typical vertical displacement of less than 30 cm. Some of the larger vertical displacements may in part reflect deposition along a slope, as some strata drop more than 1 m in elevation across the site. Nor does the refitting analysis refute the fact that almost all of the MSA points were found below the schist layer at 9a in Fig. 10.8. While the degree of post-depositional disturbance at WPS is spatially variable, the consistency of the OSL results suggests that the vicinity of the OSL samples is relatively intact and provides a reliable chronology for the site.

Figure 10.9 shows the depth-age relationship for this new suite of nine ages for samples from 75 to 510 cm depth. Linear regression produced the relationship: *Age (ka) = 1.067 + 12.5557*Depth (m)* (Eq. 1), which had an R^2 of 0.99. When the regression was forced through the origin (0, 0) the equation: *Age (ka) = 12.8513*Depth (m)* (Eq. 2) gave an R^2 of 0.95. This last equation was used to establish the chronology of the WPS sediment sequence. A linear age model that assumes a constant deposition rate was preferred over interpolation between individual OSL ages because the model has a very high R^2, and because the interpolation method does not take into account the large uncertainties associated with our, and indeed any, OSL ages.

Figure 10.9 also shows a series of four TL ages run at the University of Botswana and three OSL ages run at the

Table 10.2 OSL Ages for sediments from the South Wall of Square 23 at WPS: Relationship to MIS

Lab Number	Depth (cm) & MIS	Age (ka) (Archaeology)
UGA03OSL-102	75 & 1	8.49 ± 1.29 (LSA)
UGA03OSL-103	125 & 2	17.6 ± 2.10 (LSA)
UGA03OSL-101	*162 & 2/3*	*21.48 ± 3.31 (LSA)*
UGA03OSL-88	*215 & 3*	*29.21 ± 6.82 (early LSA)*
UGA03OSL-94	275 & 3	35.70 ± 3.33 (early LSA)
UGA03OSL-89	360 & 3	45.63 ± 10.55 (early LSA)
UGA03OSL-95	*415 & 3/4*	*54.52 ± 8.27 (MSA)*
UGA03OSL-96	*450 & 3/4*	*59.63 ± 9.54 (MSA)*
UGA03OSL-98	510 & 3/4	62.16 ± 9.43 (MSA)

Extending the depth-age model from 5 to 7 m suggests that MSA deposits between 5.8 and 7 m date to MIS 5a and MIS 5b with the base of the sediment column dating to 90 ka. MIS boundaries (1–2 = 12 ka, 2–3 = 24 ka, 3–4 = 59 ka, 4–5 = 74 ka, 5–6 = 130 ka, 6–7 = 190 ka) follow Martinson et al. (1987)

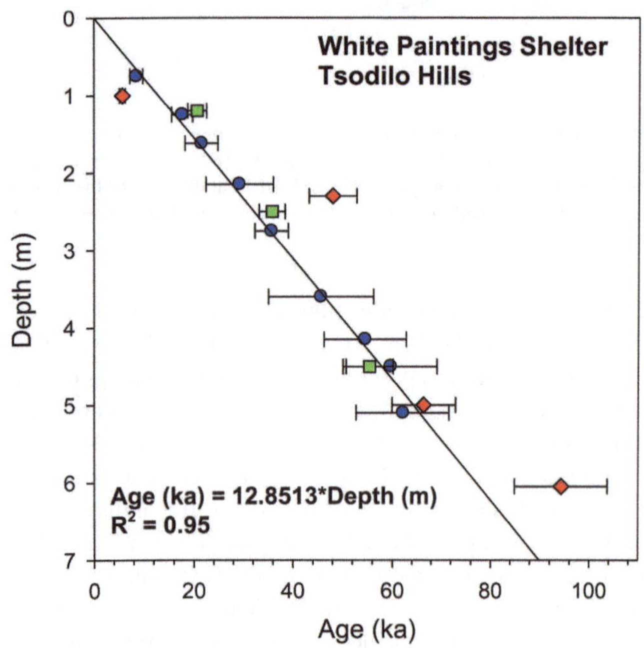

Fig. 10.9 OSL age-depth relationship for Square 23 at White Paintings Shelter. *Diamonds* show University of Botswana TL ages, *squares* show University of Washington OSL ages, *circles* show University of Georgia OSL ages

University of Washington, U.S.A. (Robbins et al. 2000). The Washington OSL ages match our OSL ages very well, as do two of the TL ages. However, two TL ages for 230 and 605 cm are 18 and 17 ka older than model predictions (48.0 ± 4.8 ka vs. 29.6 ka, and 94.3 ± 9.4 ka vs. 77.7 ka, respectively). Ivester et al. (2010) used the older of these two TL ages to estimate sediment age below c. 6 m. We have preferred to estimate these ages using the model developed from the new OSL age data presented here as we are concerned that the two TL ages are too old perhaps because the samples dated were not properly zeroed of the TL signal prior to final burial. Overall, the changes we have made in the WPS chronology are relatively minor and do not result in very different ages for the buried soil A-horizons at WPS from those presented in Ivester et al. (2010). However, our age estimates are somewhat younger for sediments between 6 and 7 m depth.

MSA and Early LSA Interpretations in the Kalahari

At WPS, MSA stone points with microwear evidence of impact damage were recovered from deposits dated between 55 ka (MIS 4) and older than 60 ka (MIS 5/4) (Donahue et al. 2002–2004). Damage on the point tips suggests that at least some functioned as spears. What the MSA and early LSA lithic assemblages from WPS suggest is that the people focused on the hunting of large migratory grazing and gregarious antelope during the late Pleistocene. Although fauna is poorly represented in the MSA levels at WPS, fauna from MSA levels at #Gi, which includes *Megalotragus, Pelorovis, Phacochoerus* and *Equus capensis*, supports this interpretation (Brooks et al. 1990). In addition, thinning of the bases of many MSA points from WPS and #Gi (Brooks et al. 1990) seems to show that at least some of these points were hafted.

Murphy (1999), following a model proposed by Deacon (1976), suggested that the seasonal movement of these large herds made it advantageous for MSA people in the Kalahari to live in larger groups that did not aggregate or disperse as modern San do. These groups may have occupied territories large enough to enable them to follow the game in its annual round. The use of spear hunting by large groups has been suggested and the use of prepared "tortoise" cores to produce MSA points is seen as a means to support this type of hunting strategy. Murphy (1999) also interpreted the increase in debitage and tools in these levels at WPS as the result of this demographic pattern and the use of the rock shelter as a temporary camp for a large number of individuals on a seasonal basis.

By contrast, the inception of what we labeled as the "Large Blade" transitional MSA/early LSA assemblages at WPS (300–410/420 cm) seems to be linked to a shift in mobility more closely resembling the aggregation and dispersal patterns observed among modern San groups. If our interpretation is correct, this may be closely related to a shift in hunting patterns away from large gregarious herd animals toward more solitary browsers. This was correlated with the use of blade cores to make large numbers of unretouched blades and large backed elements. Coupled with this technological shift is the demographic change discussed above, as well as changes in the incidence of nonlocal raw materials and the size and shape of unretouched flakes, which become more "blade-like" (Fig. 10.6). Continuity with the MSA is indicated as blades are also observed in the upper MSA assemblages at WPS indicating some continuity across the MSA/early LSA boundary. New formal tools were also introduced at this time including large backed forms anticipating microlithic forms that appear in the early LSA.

By the time that the early LSA "Lower Fish" (210–280 cm) assemblages were being deposited in MIS 3, the shift in hunting patterns appears complete, with the presence of backed microliths and bone points, which are interpreted as indicating essentially modern hunting techniques, possibly involving the use of bow and arrow (Robbins et al. 2012). The subsistence strategy shifted to include increased reliance on spawning fish and to the game animals that were historically available in the region, with the exception of wetland species and the extinct giant Cape equid (*Equus capensis*). By shifting to a wider array of animals, it is possible that people no longer had to rely exclusively on the seasonal movements of game. In addition, the early LSA diet included considerable collecting of ostrich eggs, tortoise and hunting springhare, all of which are important in historic and ethnographic contexts in the Kalahari.

During this period within MIS 3, c. 36–30 ka, (see Table 10.2) there is a marked increase in microlithic formal tools, unretouched bladelets and bladelet cores which coincided with a reduction in the length and relative thickness of unretouched non-quartz flakes, as well as a shift in raw material preferences to locally available quartz (ite) and imported jasper and chalcedony (Fig. 10.7). These changes may have been functionally related because of the smaller size of nodules and the exceptional flaking characteristics of the fine-grained raw materials, which may have necessitated a new means to reduce them.

Paleoenvironment

There are very few terrestrial records of environmental change from the Kalahari that cover the entire MIS 6-2 period (Burrough 2016). Figure 10.10 is a compilation of the longest and most reliably dated records, which are compared with glacial records from Greenland and Antarctica and with the WPS archaeological record. Evidence of wetter climatic conditions include OSL ages for high lake stands in the Makgadikgadi, Mababe, and Ngami basins of Botswana (Burrough et al. 2007, 2009; Burrough and Thomas 2008). The most comprehensive record is a synthesis of beach ridge data for paleolake Makgadikgadi including data from the Makgadikgadi, Ngami, and Mababe basins, and the Chobe enclave. This record suggests high lake levels at c. 8.5 ± 0.2, 17.1 ± 1.6, 26.8 ± 1.2, 38.7 ± 1.8, 64.2 ± 2.0, 92.2 ± 1.5, 104.6 ± 3.1 ka (Burrough et al. 2009).

A series of buried soil A-horizons at WPS document wetter intervals at 6.4–3.8, 11.6–10.9, 15.4–14.8, 23.8–22.8, 27.6–27.3, 35.6–33.4, 51.4–46.3, 60.7–58.7, 69.4–66.7, and 79.0–77.1 ka using the updated chronology for the site presented here (see also Ivester et al. 2010). The ages of the younger buried soils correlate with evidence from Tsodilo Hills of a large lake that existed immediately west of the WPS during the Late Pleistocene (Thomas et al. 2003). Luminescence ages for lakeshore sediments show that there was a sizeable body of water at 18 ka and again at 12–13 ka. Ages of speleothems in Lobatse, Gcwihaba (Drotsky's) and Bone caves in Botswana range from 1.5 ± 0.5 ka to 132.9 ± 26.6 ka and also indicate phases of increased moisture (Brook et al. 1998). Unfortunately, the U-series ages were determined by alpha counting so that uncertainties for the older ages are quite large, particularly those in MIS 5.

Evidence of drier climates includes ages of submerged speleothems in flooded caves and cenotes in northeast Namibia that were recovered by divers from depths varying from 3 to 40 m below present ground water level. U-series ages for these deposits indicate periods of lower ground water level at a variety of ages from 7.5 ± 0.3 to 129.9 ± 6.9 ka (Brook et al. 1998). Ground water levels were particularly low during MIS 5 with levels ranging from 16 to 40 m below present. One speleothem deposited during MIS 4 was 9 m underwater when recovered, while others dating to MIS 3 were from 5 to 13.5 m underwater suggesting less dramatic lowering of water tables during these periods. Two submerged speleothems dating to MIS 2 were both under 9 m of water when collected. MIS 1 speleothems indicate that water levels were 3–13 m lower in this period than today. Overall, the ages of the submerged speleothems generally confirm that interglacial MIS 1 (3–13 m) and MIS 5 (16–40 m) had more extreme drought conditions than glacial MIS 2 (9 m) and MIS 4 (9 m) or interstadial MIS 3 (5–13.5 m).

Dry conditions in the past are also indicated by ages of dune sand in linear dunes near Tsodilo Hills (Ivester et al. 2010). These ages suggest periods of drier climate and increased aeolian activity at 30–28, 35–34, 54–50, 63, 78, 88, and 98 ka. Importantly, in most cases the dunes were active in the time intervals separating periods of soil A-horizon development at WPS.

Extreme events in the late Quaternary, particularly major droughts induced by climate change, are being proposed as drivers of past human behavior in Africa (e.g., Mellars 2006; Scholz et al. 2007; Jacobs and Roberts 2009; Stager et al. 2011). However, there is considerable debate on how reliable information on "extreme events" really is. For example,

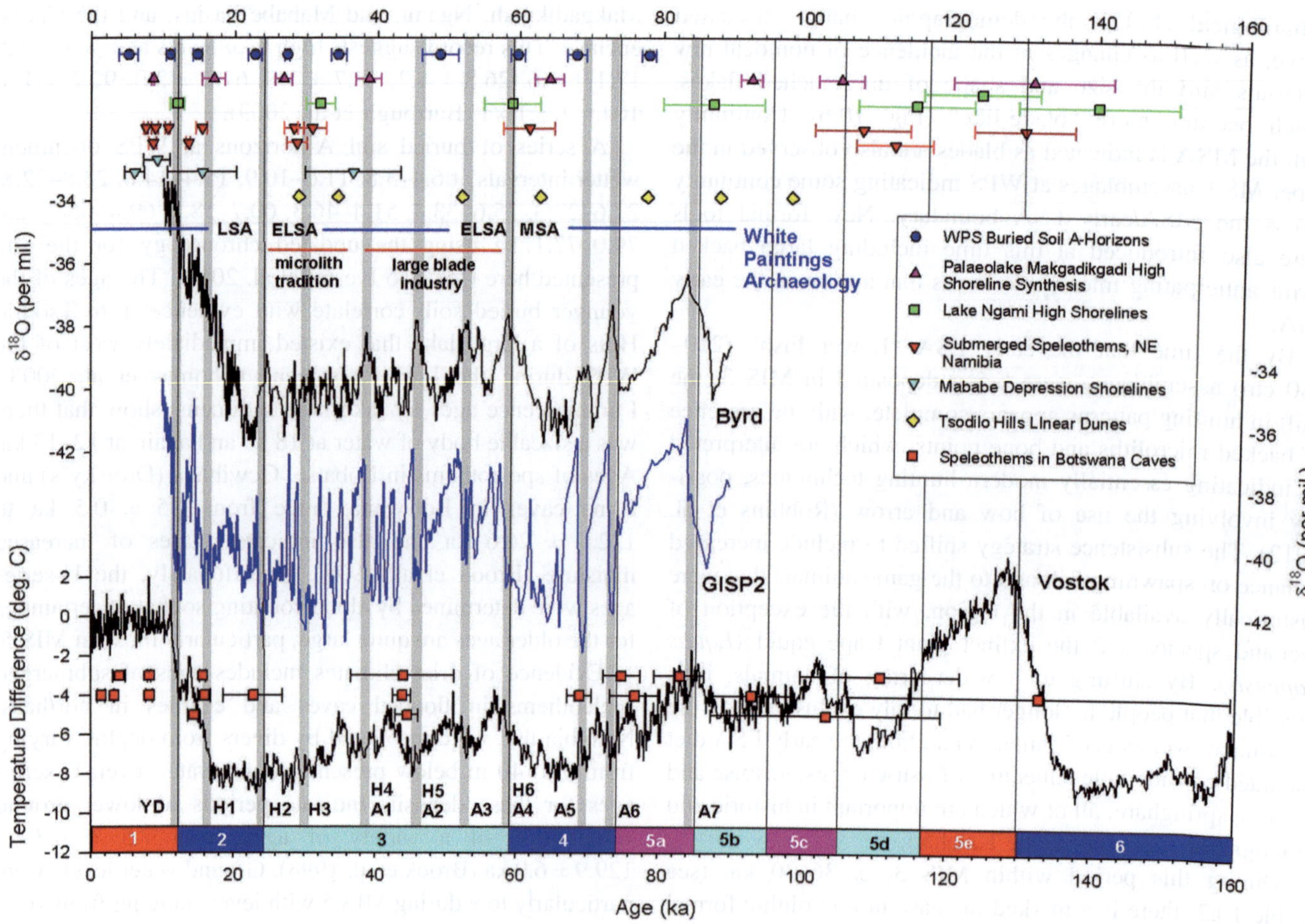

Fig. 10.10 The White Paintings Shelter archaeological and paleoenvironmental record compared with Greenland and Antarctic ice core data and a variety of paleoclimate proxies from the Kalahari. The Byrd and GISP2 ice core records are from Blunier and Brook (2001) who matched the methane records to create a common chronology. The Vostok temperature difference record is shown using the GT4 timescale (Petit et al. 1999, 2001) and so does not correlate exactly with the Byrd and GISP2 records. The Ngami, Mababe and paleolake Makgadikgadi synthesis data are from Burrough and Thomas (2008) and Burrough et al. (2007, 2009), the speleothem data from Brook et al. (1998), and the dune data from Ivester et al. (2010)

Thomas et al. (2012) examined a spatially extensive "catastrophic" drought at 17–16 ka, affecting much of the African and Asian monsoon region (Stager et al. 2011). After examining the paleoenvironmental records, they found that the scale and extent of the proposed drought was not supported by the data. Thomas and Burrough (2012) have emphasized the difficulty in interpreting geoproxy records of climate change. They note that paleoenvironmental/paleoclimate reconstructions from the southern African interior, including the Kalahari are often based on geoproxies that are difficult to interpret and have led to contradictions in conclusions. They suggest that contradictions between different geoproxy records may arise because of significant, short-term environmental variability within millennial timescales. These contradictions they suggest may point to the very variability in climate that drove innovation, survival and the capacity to move successfully (Maslin and Christensen 2007; Trauth et al. 2009).

Given this background, it is interesting to note that Marean (2010) contends that MIS 6 (190–130 ka) was very significant to human evolution as it may have been a time of increased aridity and cooling temperatures in Africa that led to a drastic decline in populations of early modern humans, which may have created an evolutionary bottleneck. In this scenario, a small population may have survived in a favorable area such as along the south coast of Africa (Marean 2010). Subsequently, when climatic conditions improved, the population of anatomically modern humans increased and spread. Does archaeological and paleoenvironmental evidence from the Kalahari provide support for this argument? Interestingly, no archaeological sites in the Kalahari of Botswana have been dated by OSL or other methods to MIS 6, although we have suggested that the Acheulean dates to before the beginning of MIS 6. The absence of dated sites may be due to insufficient research.

During MIS 6, paleolake Makgadikgadi beach ridge deposits indicate a lake phase at 131 ± 11 ka while deposits at Lake Ngami date to 140 ± 11 ka and lake deposits in the Dautsa Flats to 133 ± 12 ka (Burrough et al. 2007, 2009). Samples from the inner and outer lunettes at Aminuis Pan in east-central Namibia, from a depth of 6 m, provided OSL ages in the 140–160 ka range (Brook unpublished data), indicating deposition on the lunettes during the later part of MIS 6. The paleoenvironmental data for MIS 6 is sparse but what there is indicates both wet and dry conditions, not a prolonged mega-drought as suggested by Marean (2010). It appears that like many other claims for extreme events determining ancient human behavior, Marean's arguments must be challenged for lack of proof that the extreme event in question really did occur throughout the region in question.

There is good agreement between the dune and submerged stalagmite evidence for dry conditions (e.g., at 28 and 62 ka) in the Kalahari and this evidence suggests strongly that dry conditions correlate with colder intervals in the ice core records. For example, two speleothems deposited during MIS 5d are now 16 and 40 m underwater, while increased dune activity at 98, 88, 78, and 63 ka correlates with periods of cooler climate in MIS 5c, MIS 5b, the end of MIS 5a, and MIS 4 (Fig. 10.8). In contrast, the speleothem evidence for wet conditions shows significant groupings at 104–92, 84–68, 46–40, 17–13, and 9–1 ka, during warmer periods in the ice sheet records (e.g., MIS 5c, MIS 5a, and MIS 1). Overall, the speleothem, high lake and buried soil A-horizon data correlate strongly with Antarctic warm events (A) in the southern Hemisphere and synchronous North Atlantic Heinrich events (H) (cf. Blunier and Brook 2001). The correlation between the speleothem record and H/A events is particularly striking in the case of the H1 (three speleothem ages), H2 (one age), H5/A2 (three ages), A5 (one age), A6 (one age), and A7 (one age) events. Speleothem ages also correspond with warm peaks in MIS 5c, MIS 5b and with the increasing temperatures associated with Termination II at c. 130 ka (Fig. 10.10). In addition, five of the WPS buried soil ages correspond with the YD, H1, H2, H6/A4, and A5 events, while the Makgadikgadi, Ngami and Mababe high lake ages correspond with the YD (one age), H1 (two ages), H3 (one age), H4/A1 (two ages), H6/A4 (one age) and A7 (one age) events. Given that the buried soil, high lake, and dune records are based on OSL dating where age uncertainties are usually ±10%, it is perhaps significant that these ages correlate strongly with the more reliable U-series ages of the speleothems either submerged (for the dunes) or subaerial (for the buried soils and lake shorelines).

Thomas et al. (2009) document three prominent high stands of Lake Chilwa in Malawi at 43.7–33.3 or 38.4–35.5 ka, 26.2–21.0 or 24.3–22.3 ka, and 17.9–12.0 or 16.2–15.1 ka (depending on the age grouping method used), plus two other high lake events at 11.01 ± 0.76 ka and 8.52 ± 0.56 ka indicated by single ages. They note that these correlate very well with Heinrich events, particularly H0 (YD), H4, H2, and H1. That wet intervals in the Kalahari and at Lake Chilwa correlate with each other and seem to be related to Heinrich events is perhaps not surprising given that both areas are located in the southern African summer rainfall zone.

The association between Heinrich events in the North Atlantic and wet intervals of climate in southern Africa may be due to anti-phasing of climate changes between the Northern and Southern Hemisphere. Blunier and Brook (2001) point out that over the last 90 ka the onset of seven major millennial scale warmings in Antarctica preceded the onset of Greenland warmings by 1.5–3 ka. The thermal anti-phasing between the southern and northern surface Atlantic is thought to result from the freshwater-induced collapse of the Atlantic thermohaline circulation. North Atlantic Deep Water (NADW) is associated with a considerable inter-hemispheric northward heat transport in the Atlantic. Reducing this transport results in immediate cooling in the (high latitude) Northern Hemisphere that is accompanied by a warming at least in some parts of the Southern Hemisphere.

During Heinrich events, the North Atlantic Ocean cooled due to slowing or cessation of the NADW flow while the South Atlantic and Indian Ocean became warmer (e.g., Sicre et al. 2005). Colder conditions in the Northern Hemisphere may have caused a southward shift in the Inter Tropical Convergence Zone (ITCZ) while the warmer oceans in the Southern Hemisphere weakened the South Atlantic Anticyclone and possibly also the South African Anticyclone and associated southeast trade and westerly winds, and increased moisture in air masses moving onto southern Africa. Together these changes may have brought generally warmer and wetter conditions to the subcontinent explaining why H/A events increased lake levels and soil development, as well as drip waters entering caves leading to increased speleothem growth. In contrast, the ITCZ may have migrated northwards in the intervals between Heinrich events and the colder southern oceans led to a strengthening of anticyclones. As a result, air masses coming off the oceans were drier and the African subcontinent became generally more arid and windier activating dunes and lowering ground water levels. This last change allowed speleothems to be deposited in deeper parts of caves to be submerged later by rising water levels.

In fact, to explain the relationship between high lake stands of Lake Chilwa and Heinrich events, Thomas et al. (2009) used a coupled ocean–atmosphere–sea ice general circulation model, the Hadley Centre model (HadCM3) (Gordon et al. 2000), to determine how the surface water balance in East and southern Africa would respond to a shutdown of the Atlantic thermohaline circulation (THC) caused by a Heinrich iceberg-discharge event. Significantly, this simulation predicted wetter conditions over southern Africa, including the Kalahari. As expected, in the model the collapse of the Atlantic thermohaline circulation resulted in less heat being transported northward in the Atlantic, resulting in much higher sea surface temperatures off the southwest and east coasts of Africa, particularly during the austral summer. Land temperatures did not rise as much. Thomas et al. (2009) suggest that the greater temperature difference between land and ocean in summer and the increase in atmospheric moisture from the warmer Atlantic and Indian Oceans combined with a southerly shift in the ITCZ to produce wetter conditions over much of southern Africa.

Population Dynamics: Ideas for Consideration

As noted earlier, the abundance of MSA artifacts around many previously uninhabited pans may suggest an expansion in population possibly during MIS 5 and MIS 4. In fact, from 85 to 65 ka, including all of MIS 5a and the first part of MIS 4, there were three prominent Antarctic warm events: A7, A6 and A5. Each was associated with speleothem growth. In addition, A7 corresponds with high lake conditions at Lake Ngami, and A5 with a buried soil A-horizon at WPS. During this 20 ka period, the only evidence of dry conditions is increased dune activity at c. 78 ka (Fig. 10.8). This evidence of increased moisture may explain why the archaeology during this time interval indicates larger human populations. It might also explain why large migratory herds of antelope were the focus of hunting by MSA peoples in the Kalahari, as the animals may have been drawn to the area by the more abundant surface water resources.

From 65 to 59 ka, during the later part of MIS 4, the submerged speleothem and dune evidence indicates drier conditions, as does the lack of speleothem deposition in air-filled caves in Botswana. This dry period ended at the MIS 4-3 boundary (c. 59 ka), which marks the beginning of the early LSA, and may have signaled the beginning of the MSA/early LSA large blade industry. This industry persisted through four Antarctic warm events, three accompanied by recognized Heinrich events (A3/H6, A3, A2/H5, and A1/H4) eventually ending c. 38 ka. The lake, buried soil, and speleothem evidence during this interval indicates relatively high moisture levels with the only evidence of drier conditions being dune activity from 54 to 50 ka (Fig. 10.10). Despite increased moisture, the large blade industry is thought to record a shift in hunting patterns away from large gregarious herd animals toward more solitary browsers. This shift may well have begun during the dry period within MIS 4 and simply been carried over into MIS 3 despite an increase in rainfall. Adoption of a microlith tradition at WPS from c. 38 to 24 ka was accompanied by a trend toward colder and drier conditions, as indicated by dune activity from 35–34 to 30–28 ka, and by submerged speleothems in NE Namibia dating in the range 31–28 ka. Lack of evidence for subaerial speleothem deposition in Botswana caves also indicates drier conditions. As mentioned above, we believe that the use of backed microliths, and bone points, along with collecting, is evidence that by this time some elements of Kalahari San hunting and subsistence techniques were in place, possibly including the use of the bow and arrow.

The data in Fig. 10.10 make clear that there were rapid and substantial shifts in climate in southern Africa throughout the MIS 6-2 period in response to Bond cycles, Dansgaard–Oeschger (D-O) events, Antarctic warm events (A), and Heinrich events (H). This was particularly the case during MIS 3 when the changes were most profound and most frequent, occurring on average about every 6 ka. Thus, conditions were never stable. That is not to say that conditions were always difficult for the inhabitants of the area. At times, the Kalahari had huge, deep, permanent lakes, and numerous smaller lakes that would have facilitated the movement of animals and humans through the area and also provided the resources for larger populations.

Paleolake Makgadikgadi, along with other very large lakes in Etosha Pan, Lake Ngami and the Mababe Depression, as well as a multitude of small lakes one to several kilometer long in pans, transformed the Kalahari landscape at various times in the past, and must undoubtedly have had a significant impact on human populations during MIS 6-2 (Fig. 10.1). We will explore some of the possible implications of the ancient lakes for population dynamics.

The ancient Makgadikgadi lake is estimated to have covered $\sim$66,000 km^2 at its maximum (Burrough et al. 2009; Burrough 2016). By comparison, this lake was larger in surface area than Lake Michigan, the third largest of the Great Lakes (USA), covering $\sim$57,800 km^2.

What were the main effects of a lake that was about 250 km long and 150 km wide and possibly up to 90 m deep, on the mobility and seasonal migration/dispersal patterns of human populations? Did the lake form a barrier and isolate populations or did it concentrate them along the lake edge, particularly during the MSA? In addition, would such an extensive water body contribute to the movement of people/artifacts to the east or west of the lake? An isolation model might be supported by the discovery of contrasting artifact assemblages that were localized on different sides of the lake, whereas extensive concentrations of artifacts near the former shorelines might indicate aggregation near the lake edge. We have previously suggested seasonal aggregation for Kudiakam Pan (Fig. 10.1) where extensive MSA artifact surface scatters were found spread over at least 4.5 km on the eastern side of the Pan (Robbins 1988). The largest scatter extended "over an area as large as a football field" (Robbins 1988: 41). Yet, the artifact assemblage also contrasts in several ways with the MSA at WPS. For example, at Kudiakam Pan large denticulate tools were emphasized and some of the MSA points had been reworked on their tips with "scraper" retouch, most likely after their use as points. Unfortunately, the site has not been dated, nor did we find fossil bones associated with the MSA. MSA surface scatters also occur on the present-day northwest shore of Ntwetwe Pan and within the Makgadikgadi Pans National Park where points, blades and scrapers have been found similar to those at Kudiakam Pan. In addition, MSA and LSA artifacts, as well as hunting blinds have been found at Gutshaa Pan (Fig. 10.1).

Would the high lake have disrupted or encouraged exchange of raw materials? For example, Boteti River bed silcrete sources located about 400 km southeast of Tsodilo Hills would have been flooded at such times. Murphy (1999: Tables 74–77) has clearly shown that there was a substantial increase in the proportions of nonlocal silcrete and multi-colored chert debitage during the latter part of the MSA sequence at WPS. In contrast, the Large Blade industry that overlies the MSA and the early microlithic LSA deposits show substantial increases in the proportions of local quartz debitage. MSA peoples at Rhino Cave were also using nonlocal raw materials extensively, including chert, jasper and chalcedony (Robbins et al. 1996a, 2000). Murphy (1999) has emphasized the need to establish whether MSA peoples at Tsodilo were traveling to distant sources of raw materials or were involved in exchange networks.

Interestingly, two of the dated high lake stages (64.2 ± 2.0 ka and 92.2 ± 1.5 ka) fall within (or near) the MSA period at WPS and the youngest of these corresponds closely with the termination of the Howiesons Poort in South Africa (Jacobs et al. 2008; Burrough et al. 2009). Makgadikgadi lake levels were also high at 26.8 ± 1.2 ka and 38.7 ± 1.8 ka (Burrough et al. 2009) during the early LSA at WPS. As discussed previously, the buried soil A-horizons at WPS correlate well with the Makgadikgadi high lake levels. Howiesons Poort artifacts have not been found in Botswana and were not in WPS deposits spanning the Howiesons Poort time range 64.8–59.5 ka (Jacobs et al. 2008). The Still Bay (71.9–71.0 ka), which has a more restricted distribution in South Africa, also has not been found in the Kalahari, including WPS. Both of these traditions are very significant in the discussion of the origin of modern human behavior and are the major "cultural" components of the out-of-Africa theory (Texier et al. 2010). As noted by Jacobs et al. (2008: 733), "genetic studies of expansions, migrations, and isolations of modern populations within Africa and their initial exodus out-of-Africa" have been temporally associated with the Still Bay and Howiesons Poort. Did the Makgadikgadi megalake have an impact, as a barrier, on the assumed spread of the Howiesons Poort to the north, or is the absence of the Howiesons Poort, as well as the Still Bay, simply a product of the comparative lack of research in the Kalahari and the low number of excavated sites? Alternatively, cultural influence/population movements may have occurred to the east or west of the lake.

Future models should also consider seasonal fresh fish procurement in relation to the enlarged/flooded waterways that were connected to the expanded lakes of the Kalahari. WPS contains a long record of fish exploitation during seasonal spawning migrations (Robbins et al. 2000). Genetic evolutionary studies of fish have indicated that paleolake Makgadikgadi was the source of the serranochromine cichlid fish adaptive radiation that spread into the extensive river systems including the Okavango (Joyce et al. 2005). While no dates are available that indicate the timing of the fish radiation, data from WPS show that unidentifiable cichlids, as well as *Clarius sp.* (catfish) were present throughout most of the deposits and extend as far back as the MSA. Whereas only 25 fish bones were recovered from the MSA deposits, both cichlids and *Clarius sp.* were evident at 410–420 cm depth (K.M. Stewart, personal communication to LHR). We suggest an age of 53–64 ka for the earliest cichlid finds, hinting at an early age for the cichlid radiation, if the cichlids from the deposits were serranochromines. The deepest unidentifiable (that could not be identified to genus) fish bones were recovered from 620–630 to 630–640 cm, which is well into the MSA deposits. If Square 12 deposits are of comparable age to deposits in Square 23, these fish bones date in the range 77.8–80.4 ka based on Eq. 1 above, and 79.7–82.2 ka based on Eq. 2. The early LSA at WPS and the Holocene LSA contain barbed bone points that were almost certainly used to spear fish. It would be useful to determine what other fish procurement methods were used along Kalahari rivers and lakes and when they were developed, including use of dugouts and rafts.

A final comment is that natural selection was also operating as an evolutionary mechanism on early human populations in the Kalahari. Although this subject is one of the areas of low visibility in the archaeological record, especially in the absence of fossil human remains, we have learned that the paleolake at Tsodilo (dated to the time of the early LSA in MIS 3 and 2) contained mollusks that were hosts to parasites causing intestinal and urinal schistosomiasis (Robbins et al. 2000). These mollusks were also found in Holocene archaeological deposits at Lake Ngami, which was previously connected to the Kalahari Megalake during the Pleistocene (Robbins et al. 2009). Almost certainly these health risks, as well as others would have had negative impacts on MSA and LSA populations in the Kalahari.

Archaeological and MIS Boundaries

Maslin and Christensen (2007) have provided a comprehensive synthesis of climate change and human evolution in Africa. They note that a major issue for understanding the influence of global climate transitions on African climate is the lack of high-resolution continental records, particularly for southern Africa. Paleoenvironmental information has either come from paleoanthropological sites or from deep-sea sediment records containing regional and global climate signals but there has been insufficient effort to link the two (see also Carr et al. 2016). They argue that ocean records, including MIS boundaries do not necessarily reflect environmental variations in Africa.

Bearing in mind the reservations expressed by Maslin and Christensen (2007), we ask to what extent changes in the archaeological record in the Kalahari correspond to MIS boundaries? This question is difficult to evaluate because of the large standard deviations in the OSL dates in Table 10.2 and the fact that few dated sites are known in a vast area. It is also clear that significant changes in stone artifact "traditions" can occur independently of climatic and environmental changes and there also can be a significant lag time between changing "cultural" adaptations and climatic changes. Jacobs et al. (2008: 734), who worked with a large number of OSL dated sites, mainly in South Africa, state that they "cannot identify any particular climatic attribute that is consistently and uniquely associated with any MSA industry."

Nonetheless, there may be instances in the Kalahari where there is a correspondence between archaeological events and MIS boundaries (Table 10.2, shading). For example, OSL ages at or near the termination of the MSA at WPS may correspond to the MIS 4/3 boundary at about 59 ka. In addition the termination of the early LSA Lower Fish deposits could correspond to the MIS 3/2 boundary at about 24 ka. Elsewhere in the Kalahari, at Gcwihaba (Drotsky's) Cave, there is a thick charcoal layer situated in the northeast entry chamber between 50 and 80 cm. It contains LSA artifacts, an abundance of burned ostrich eggshell pieces and (*Pyxicephalus adspersus*) bullfrog bones. Very little archaeological evidence was found overlying this dense occupation. The top of the charcoal layer at 50 cm was radiocarbon dated to 13 kBP (11,240 ± 60 ^{14}C BP), while the base between 70 and 80 cm dated to 14.7–15.0 kBP (12,450 ± 80 ^{14}C BP) (Robbins et al. 1996b). These ages place the upper part of the charcoal layer close to the MIS 2/1 boundary at 12 ka ago. However, the correspondence could be fortuitous. In other instances (e.g., the MSA levels between 600 and 700 cm at WPS and the MSA at #Gi Pan) there does not appear to be any relationship between significant archaeological events that have been dated and MIS boundaries.

Kalahari Archaeology and Evolutionary Genetics

Evolutionary genetics is of major importance in documenting the African origin of modern humans and their spread both within and out-of-Africa (Soares et al. 2016). Genetic samples from the San have figured prominently in this research in a number of recent studies (Tishkoff et al. 2007; Behar et al. 2008; Schuster et al. 2010; Henn et al. 2011). As noted above, archaeological findings outside of the Kalahari in southern Africa (the Still Bay and Howiesons Poort) have been broadly linked to the genetic evidence regarding the origin and spread of modern humans (Jacobs et al. 2008).

We will walk down an uncertain path here and ask if there is any site-specific archaeological evidence in the Kalahari that would support these genetic studies? The answer is not yet, because of the lack of human fossils from the Kalahari and the unlikely chance that ancient DNA will be preserved. In addition, how does one interpret the meaning of similarities between OSL dates that mark substantial changes in the archaeological record and the estimated dates for population divergences that are based on genetic studies of Khoisan peoples? A case in point follows.

Henn et al. (2011: 6) state, "our Y-chromosome microsatellite data from the highly diverged, basal Khoe-San A3b1-M51 clade, found exclusively in southern Africa, has a most recent common ancestor estimated at 40,000 y ago (SI Appendix, Fig. S11). If the time to the most recent common ancestor (TMRCA) *within* the Khoe-San population is 40,000 y ago, then this is among the first genetic evidence that the Khoe-San have continuously occupied southern Africa for at least the entire span of the Upper Paleolithic". The 40 ka date agrees with the WPS archaeological record (we use the African Stone Age terminology, LSA, rather than the Upper Paleolithic). At WPS, soil A-horizon 7a has an estimated age of 33–36 ka that correlates reasonably well with H4 and A1. This OSL age marks the beginning of an early LSA microlithic tradition (Lower Fish deposits) that overlies the MSA/early LSA large blade assemblage. LSA microlithic tools continue throughout most of the overlying more recent deposits at WPS (Robbins et al. 2000). Moreover, most authorities have reasoned that ancestral Khoe-San populations were the LSA occupants of southern Africa (Deacon and Deacon 1999).

In this case, there is consistency between paleoclimatic changes, Khoe-San prehistory as seen in the archaeological record, and a genetic study. The difficulty in evaluating the meaning of the above comparative dating evidence will ultimately center on determining whether or not the potential agreement between genetic, OSL and MIS/H event ages and the associated changes in the archaeological sequence are entirely, or partially coincidental. Perhaps the evidence can be supported or refuted by future work resulting in a larger sample size from additional well-dated sites in the Kalahari.

Acknowledgments Our research was supported by grants from the National Science Foundation and the National Geographic Society. Permission to do the research was granted by the Office of the President in Botswana. We are grateful to staff members of the National Museum of Botswana as well to as students, local peoples and other volunteers for assisting with the fieldwork. We thank Doctors Alison Brooks and John Yellen for re-exposing the main excavation section at WPS and for collecting the nine OSL samples discussed in this paper. We also thank Guoqiang Li who helped with the figures and David Thomas and two other reviewers who provided extremely valuable suggestions on how to improve this chapter. Above all, we wish to acknowledge the leadership and contributions of the late Alec Campbell to all of our research at Tsodilo and other locations in Botswana.

References

Backwell, L., d'Errico, F., & Wadley, L. (2008). Middle Stone Age bone tools from the Howiesons Poort layers, Sibudu Cave, South Africa. *Journal of Archaeological Science, 35*, 1566–1580.

Behar, D. M., Villems, R., Soodyall, H., Blue-Smith, J., Pereira, L., Metspalu, E., et al. (2008). The dawn of human matrilineal diversity. *The American Journal of Human Genetics, 82*, 1–11.

Berna, F., Goldberg, P., Kolska Horwitz, L., Brink, J., Holt, S., Bamford, M., et al. (2012). Microstratigraphic evidence of *in situ* fire in the Acheulean strata of Wonderwerk Cave, Northern Cape province, South Africa. *PNAS, 109*, E1215–E1220.

Blunier, T., & Brook, E. J. (2001). Timing of millennial-scale climate change in Antarctica and Greenland during the Last Glacial Period. *Science, 291*, 109–112.

Brook, G. A. (2010). The paleoenvironment of Tsodilo. In A. C. Campbell, L. H. Robbins, & M. Taylor (Eds.), *Tsodilo Hills, copper bracelet of the Kalahari* (pp. 30–43). East Lansing & Gaborone: Michigan State University Press & The Botswana Society.

Brook, G. A., Cowart, J. B., & Brandt, S. A. (1998). Comparison of Quaternary environmental change in eastern and southern Africa using cave speleothem, tufa and rock shelter sediment data. In A. S. Alsharhan, K. W. Glennie, G. L. Whittle, & C. G. S. C. Kendall (Eds.), *Quaternary deserts and climatic change* (pp. 239–249). Rotterdam: Balkema.

Brook, G. A., Robbins, L. H., & Campbell, A. C. (2003). Forty thousand years of environmental change in the Kalahari as evidenced by sediments in the Depression Rock Shelter, Tsodilo Hills, Botswana. *Nyame Akuma, 59*, 2–10.

Brook, G. A., Srivastava, P., Brook, F. Z., Robbins, L. H., Campbell, A. C., & Murphy, M. L. (2008). OSL chronology for sediments and MSA artefacts at the Toteng Quarry, Kalahari Desert, Botswana. *South African Archaeological Bulletin, 63*, 151–158.

Brook, G. A., Railsback, L. B., Campbell, A. C., Robbins, L. H., Murphy, M. L., & Hodgins, G., et al. (2010). Radiocarbon ages for coatings on cupules ground in quartzite bedrock at Rhino Cave in the Kalahari Desert of Botswana, and their paleoclimatic significance. *Geoarchaeology, 26*, 1–22.

Brooks, A. S. (1984). San land-use patterns, past and present: Implications for southern African prehistory. In M. Hall, G. Avery, D. M. Avery, M. L. Wilson, & A. J. B. Humphreys (Eds.), *Frontiers: Southern African archaeology today* (BAR International Series No. 207) (pp. 40–52). Oxford: Archaeopress.

Brooks, A. S., Hare, P. E., Kokis, J. E., Miller, G. H., Ernst, R. D., & Wendorf, F. (1990). Dating archaeological sites by protein diagenesis in ostrich eggshell. *Science, 248*, 60–64.

Burrough, S. L. (2016). Late Quaternary environmental change and human occupation of the southern African interior. In S. C. Jones & B. A. Stewart (Eds.), *Africa from MIS 6-2: Population dynamics and paleoenvironments* (pp. 161–174). Dordrecht: Springer.

Burrough, S. L., & Thomas, D. S. G. (2008). Late Quaternary lake-level fluctuations in the Mababe Depression: Middle Kalahari palaeolakes and the role of Zambezi inflows. *Quaternary Research, 69*, 388–403.

Burrough, S. L., Thomas, D. S. G., Shaw, P. A., & Bailey, R. M. (2007). Multiple Quaternary highstands at Lake Ngami, Kalahari, northern Botswana. *Palaeogeography, Palaeoclimatology, Palaeoecology, 253*, 280–299.

Burrough, S. L., Thomas, D. S. G., & Bailey, R. M. (2009). Mega-lake in the Kalahari: A Late Pleistocene record of the Palaeolake Makgadikgadi system. *Quaternary Science Reviews, 28*, 1392–1411.

Campbell, A. C., Robbins, L. H., & Taylor, M. (2010a). *Tsodilo Hills, copper bracelet of the Kalahari.* East Lansing & Gaborone: Michigan State University Press & The Botswana Society.

Campbell, A. C., Robbins, L. H., & Murphy, M. L. (2010b). Cupules and grooves. In A. C. Campbell, L. H. Robbins, & M. Taylor (Eds.), *Tsodilo Hills, copper bracelet of the Kalahari* (pp. 64–71). East Lansing & Gaborone: Michigan State University Press & The Botswana Society.

Cann, R. L., Stoneking, M., & Wilson, A. (1987). Mitochondrial DNA and human evolution. *Nature, 325*, 31–36.

Carr, A. S., Chase, B. M., & Mackay, A. (2016). Mid to Late Quaternary landscape and environmental dynamics in the Middle Stone Age of southern South Africa. In S. C. Jones & B. A. Stewart (Eds.), *Africa from MIS 6-2: Population dynamics and paleoenvironments* (pp. 23–47). Dordrecht: Springer.

Clottes, J., & Lewis-Williams, D. (1998). *The shamans of prehistory, trance and magic in the painted caves*. New York: Harry N. Abrams.

Cooke, C. K. (1979). The Stone Age in Botswana: A preliminary survey. *Arnoldia, 8*, 1–32.

Coulson, S., Staurset, S., & Walker, N. (2011). Ritualized behavior in the Middle Stone Age: Evidence from Rhino Cave, Tsodilo Hills, Botswana. *Paleoanthropology, 2011*, 18–61.

Deacon, H. J. (1976). *Where hunters gathered: A study of Holocene Stone Age peoples in the eastern Cape* (South African Archaeological Society Monograph Series No. 1). Cape Town: South African Archaeological Society.

Deacon, H. J., & Deacon, J. (1999). *Human beginnings in South Africa.* Walnut Creek: Altamira Press.

Donahue, R. E., Murphy, M. L., & Robbins, L. H. (2002–2004). Lithic microwear analysis of Middle Stone Age artifacts from White Paintings Rock Shelter, Botswana. *Journal of Field Archaeology, 29*(1–2), 155–163.

Gordon, C., Cooper, C., Senior, C. A., Banks, H., Gregory, J. M., Johns, T. C., et al. (2000). The simulation of SST, sea ice extents and ocean heat transports in a version of the Hadley Centre coupled model without flux adjustments. *Climate Dynamics, 16*, 147–168.

Grine, F. E. (2016). The Late Quaternary hominins of Africa: The skeletal evidence from MIS 6-2. In S. C. Jones & B. A. Stewart (Eds.), *Africa from MIS 6-2: Population dynamics and paleoenvironments* (pp. 323–381). Dordrecht: Springer.

Henn, B. M., Gignoux, C. R., Jobin, M., Granka, J. M., Macpherson, J. M., Kidd, J. M., et al. (2011). Hunter-gatherer genomic diversity suggests a southern African origin for modern humans. *PNAS, 108*, 5154–5162.

Henshilwood, C., d'Errico, F, Vanhaeren, M., van Niekerk, K., & Jacobs, Z. (2004). Middle Stone Age beads from South Africa. *Science, 304*, 404.

Henshilwood, C., d'Errico, F., van Niekerk, K. L., Coquinot, Y., Jacobs, Z., Lauritzen, S.-Z., et al. (2011). A 100,000-year-old ochre-processing workshop at Blombos Cave, South Africa. *Science, 334*, 219–222.

Ivester, A. H., Brook, G. A., Robbins, L. H., Campbell, A. C., Murphy, M. L., & Marais, E. (2010). A sedimentary record of environmental change at Tsodilo Hills White Paintings Rock Shelter, Northwest Kalahari Desert, Botswana. In J. Runge (Ed.), *Palaeoecology of Africa* (pp. 53–78). Boca Raton: Balkema (CRC Press).

Jacobs, Z., & Roberts, R. G. (2009). Catalysts for Stone Age innovations: What might have triggered two short-lived bursts of technological and behavioral innovation in southern Africa during the Middle Stone Age? *Communicative and Integrative Biology, 2*, 191–193.

Jacobs, Z., Roberts, R. G., Galbraith, R. G., Deacon, H. J., Grün, R., Mackay, A., et al. (2008). Ages for the Middle Stone Age of southern Africa: Implications for human behavior and dispersal. *Science, 322*, 733–735.

Joyce, E. A., Lunt, D. L., Bills, R., Turner, G. G., Katongo, C., Duftner, N., et al. (2005). An extant cichlid fish radiation emerged in an extinct Pleistocene lake. *Nature, 435*, 90–95.

Lee, R. B. (1979). *The !Kung San*. Cambridge: Cambridge University Press.

Marean, C. W. (2010). When the sea saved humanity. *Scientific American, 303*(2), 54–61.

Marean, C. W., Bar-Matthews, M., Bernatchez, J., Fisher, E., Goldberg, P., Herries, A. I. R., et al. (2007). Early human use of marine resources and pigment in South Africa during the Middle Pleistocene. *Nature, 449*, 905–908.

Martinson, D. G., Pisias, N. G., Hays, J. D., Imbrie, J., Moore, T. C., & Shackleton, N. J. (1987). Age dating and the orbital theory of the ice ages: Development of a high resolution 0 to 300,000-year chronostratigraphy. *Quaternary Research, 27*, 1–29.

Maslin, M. A., & Christensen, B. (2007). Tectonics, orbital forcing, global climate change, and human evolution in Africa: Introduction to the African paleoclimate special volume. *Journal of Human Evolution, 53*, 443–464.

Mellars, P. (2006). Why did modern human populations disperse from Africa c. 60,000 years ago? A new model. *Proceedings of the National Academy of Sciences of the United States of America, 103*, 9381–9386.

Murphy, M. L. (1999). Changing human behavior: The contribution of the White Paintings Rock Shelter to an understanding of changing lithic reduction, raw material exchange and hunter-gatherer mobility in the interior regions of southern Africa during the Middle and Early Late Stone Age. Ph.D. Dissertation, Michigan State University.

Petit, J. R., Jouzel, J., Raynaud, D., Barkov, N. I., Barnola, J. M., Basile, I., et al. (1999). Climate and atmospheric history of the past 420,000 years from the Vostok ice core, Antarctica. *Nature, 399*, 429–436.

Petit, J. R., Jouzel, J., Raynaud, D., Barkov, N. I., Barnola, J. M., Basile, I., et al. (2001). Vostok ice core data for 420,000 years, IGBP pages/world data center for paleoclimatology data contribution series #2001-076. Boulder CO, USA: NOAA/NGDC Paleoclimatology Program.

Robbins, L. H. (1987). Stone Age archaeology in the Northern Kalahari, Botswana: Savuti and Kudiakam Pan. *Current Anthropology, 28*, 567–569.

Robbins, L. H. (1988). The Middle Stone Age of Kudiakam Pan. *Botswana Notes and Records, 20*, 47.

Robbins, L. H. (1999). Direct dating of worked ostrich eggshell in the Kalahari. *Nyame Akuma, Bulletin of African Archaeology, 52*, 11–16.

Robbins, L. H., & Murphy, M. L. (1998). The Early and Middle Stone Age. In P. Lande, A. Reid, & A. Segobye (Eds.), *Ditswa mmung: The archaeology of Botswana* (pp. 50–64). Gaborone: Pula Press & the Botswana Society.

Robbins, L. H., & Murphy, M. L. (2011). An overview of the Later and Middle Stone Age at Tsodilo Hills. *Botswana Notes and Records, 43*, 130–139.

Robbins, L. H., Murphy, M. L., Campbell, A. C., & Brook, G. A. (1996a). Excavations at the Tsodilo Hills Rhino Cave. *Botswana Notes and Records, 28*, 23–45.

Robbins, L. H., Murphy, M. L., Stevens, N. J., Brook, G. A., Ivester, A. H., Haberyan, K. A., et al. (1996b). Paleoenvironment and archaeology of Drotsky's Cave: Western Kalahari Desert, Botswana. *Journal of Archaeological Science, 23*, 7–22.

Robbins, L. H., Murphy, M. L., Brook, G. A., Ivester, A. H., Campbell, A. C., Klein, R. G., et al. (2000). Archaeology, palaeoenvironment, and chronology of the Tsodilo Hills White Paintings Rock Shelter, northwest Kalahari Desert, Botswana. *Journal of Archaeological Science, 27*, 1085–1113.

Robbins, L. H., Campbell, A. C., Brook, G. A., & Murphy, M. L. (2007). World's oldest ritual Site? The "Python Cave" at Tsodilo Hills World Heritage site, Botswana. *Nyame Akuma, 67*, 2–6.

Robbins, L. H., Campbell, A. C., Murphy, M. L., Brook, G. A., Mabuse, A. A., Hitchcock, R. K., et al. (2009). Mogapelwa: Archaeology, paleoenvironment and oral traditions at Lake Ngami, Botswana. *South African Archaeological Bulletin, 64*, 13–32.

Robbins, L. H., Murphy, M., & Campbell, A. (2010). Windows into the past: Excavating Stone Age shelters. In A. Campbell, L. Robbins, & M. Taylor (Eds.), *Tsodilo Hills, copper bracelet of the Kalahari* (pp. 50–63). East Lansing & Gaborone: Michigan State University Press & The Botswana Society.

Robbins, L. H., Campbell, A. C., Brook, G. A., Murphy, M. L., & Hitchcock, R. K. (2012). The antiquity of the in the Kalahari Desert: Bone point from White Paintings Rock Shelter, Botswana. *Journal of African Archaeology, 10*, 7–20.

Schuster, S. C., Miller, W., Ratan, A., Thomsho, R., Giardine, B., Kasson, L. R., et al. (2010). Complete Khoisan and Bantu genomes from southern Africa. *Nature, 463*, 943–947.

Scholz, C. A., Johnson, T. C., Cohen, A. S., King, J. W., Peck, J. A., Overpeck, J. T., et al. (2007). East African megadroughts between 135 and 75 thousand years ago and bearing on early-modern human origins. *Proceedings of the National Academy of Sciences, 104*, 16416–16421.

Sicre, M. A., Labeyrie, L., Ezat, U., Duprat, J., Turon, J. L., Schmidt, S., et al. (2005). Mid-latitude southern Indian Ocean response to Northern Hemisphere Heinrich events. *Earth and Planetary Science Letters, 240*, 724–731.

Silberbauer, G. B. (1981). *Hunter and habitat in the central Kalahari Desert*. Cambridge: Cambridge University Press.

Soares, P., Rito, T., Pereira, L., & Richards, M. B. (2016). A genetic perspective on African prehistory. In S. C. Jones & B. A. Stewart (Eds.), *Africa from MIS 6-2: Population dynamics and paleoenvironments* (pp. 383–405). Dordrecht: Springer.

Stager, J. C., Ryves, D. B., Chase, B. M., & Pausata, F. S. R. (2011). Catastrophic drought in the Afro-Asian monsoon region during Heinrich event 1. *Science, 331*, 1299–1302.

Staurset, S., & Coulson, S. (2014). Sub-surface movement of stone artefacts at White Paintings Shelter, Tsodilo Hills, Botswana: Implications for the Middle Stone Age Chronology of central southern Africa. *Journal of Human Evolution, 75*, 153–165.

Texier, P.-J., Porraz, G., Parkington, J., Rigaud, J.-P., Poggenpoel, C., Miller, C., et al. (2010). A Howiesons Poort tradition of engraving ostrich eggshell containers dated to 60,000 years ago at Diepkloof Rock Shelter, South Africa. *PNAS, 107*, 6180–6185.

Thomas, D. S. G., & Burrough, S. L. (2012). Interpreting geoproxies of late Quaternary climate change in African drylands: Implications for understanding environmental change and early human behaviour. *Quaternary International, 253*, 5–17.

Thomas, D. S. G., & Shaw, P. A. (1991). *The Kalahari environment*. Cambridge: Cambridge University Press.

Thomas, D. S. G., Brook, G., Shaw, P., Bateman, M., Haberyan, K., Appleton, C., et al. (2003). Late Pleistocene wetting and drying in the NW Kalahari: An integrated study from the Tsodilo Hills, Botswana. *Quaternary International, 104*, 53–67.

Thomas, D. S. G., Bailey, R., Shaw, P. A., Durcan, J. A., & Singarayer, J. S. (2009). Late Quaternary highstands at Lake Chilwa, Malawi: Frequency, timing and possible forcing mechanisms in the last 44 ka. *Quaternary Science Reviews, 28*, 526–539.

Thomas, D. S. G., Burrough, S. L., & Parker, A. G. (2012). Extreme events as drivers of early human behaviour in Africa? The case for variability, not catastrophic drought. *Journal of Quaternary Science, 27*, 7–12.

Tishkoff, S. A., Gonder, M. K., Henn, B. M., Mortensen, H., Knight, A., Gignoux, C., et al. (2007). History of click-speaking populations of Africa inferred from mtDNA and Y chromosome genetic variation. *Molecular Biology and Evolution, 24*, 2180–2195.

Trauth, M. H., Larrasoaña, J. C., & Mudelsee, M. (2009). Trends, rhythms and events in Plio-Pleistocene African climate. *Quaternary Science Reviews, 28*, 399–411.

Wiessner, P. W. (1977). Hxaro: A regional system of reciprocity for reducing risk among the Kung San. Vol. I. Ph.D. Dissertation, University Microfilms International, University of Michigan.

Wiessner, P. W. (1983). Style and social information in Kalahari San projectile points. *American Antiquity, 48*, 253–276.

Wilmsen, E. N. (1989). *Land filled with flies: A political economy of the Kalahari*. Chicago: University of Chicago Press.

Yellen, J. E. (1977). *Archaeological approaches to the present*. New York: Academic Press.

Yellen, J. E., Brooks, A. S., Stuckenrath, R., & Welbourne, R. (1987). A terminal Pleistocene assemblage from Drotsky's Cave, western Ngamiland, Botswana. *Botswana Notes and Records, 19*, 1–6.

Chapter 11
Paleoenvironments, Sea Levels, and Land Use in Namaqualand, South Africa, During MIS 6-2

Genevieve Dewar and Brian A. Stewart

Abstract In order to expand on the potential range of early human experiences and adaptive strategies, it is first necessary to determine the paleoenvironmental signatures for a given region of study. In this paper we report on proxy terrestrial, marine, and sea level data in order to reconstruct past environments of Namaqualand, South Africa, during MIS 6-2. Although this semiarid southern extension of the Namib Desert is a prime area to investigate early modern human adaptive innovations, environmental and human history of Namaqualand has been largely neglected. We present environmental, chronological, and subsistence data from recent excavations at Spitzkloof Rockshelter A, and review equivalent data from other sites in the Succulent Karoo Biome. The presence of handaxes on the landscape point to a pre-MIS 6 presence in the region, but current evidence suggests that a more dedicated human occupation of the region likely began during MIS 5. Subsequent human dispersals into Namaqualand are recurrent but heavily pulsed and typically linked to humid stadial phases when sea levels were lower. We propose that the westward movement of the coastline potentially increased the carrying capacity of the region by promoting the colonization of grasses onto the coastal plain, attracting larger game. The mechanism driving this change can be attributed to either an increase in inland precipitation as the Benguela-cooled coastline moved west or reduced evapotranspiration due to lowered temperatures. The strongest evidence for this pattern is during MIS 2 when faunal and floral data indicate a cold but humid environment. Faunal species from Last Glacial Maximum (LGM) layers at Spitzkloof A and Apollo 11 include large ungulates such as *Equus capensis*, a moisture-loving species that disappears toward the end of MIS 2 ($\sim$14 ka) when conditions become more xeric.

Keywords Namaqualand • Late Pleistocene • Middle Stone Age • Paleoenvironments • Deserts • Hunter-gatherers • Climate change • Southern Africa

Introduction

Over the past two decades fossils of early *Homo sapiens* and genetics studies of modern populations have identified an African origin for our species at $\sim$200 ka (cf. Cann et al. 1987; McDougall et al. 2008). Yet the sparse distribution of well-excavated sites across this vast continent means that we are still fleshing out the evolutionary processes within Africa that led ultimately to the complex, highly plastic forms of behavior typical of recent and living humans. Evidence from a handful of African Late Pleistocene sites provide glimpses of sociocultural, technological, and subsistence innovations that include geometric art forms (Henshilwood et al. 2002; Mackay and Welz 2008; Texier et al. 2010), personal ornamentation (Henshilwood et al. 2004; d'Errico et al. 2005; Bouzouggar et al. 2007), compound adhesive manufacture (Wadley et al. 2009), living and work space preparation (Wadley 2010; Wadley et al. 2011), fishing (Yellen et al. 1995; Henshilwood et al. 2001; Robbins et al. 2016), and shellfish exploitation (Klein et al. 2004; Avery et al. 2008; Marean 2010). These finds, although tantalizing, are material symptoms of underlying causes and processes that remain obscure. In order to begin understanding the deeper evolutionary currents responsible for behavioral complexity, however defined, the unevenness of datasets across the African continent must be corrected. This is especially true

G. Dewar (✉)
Department of Anthropology, University of Toronto, Scarborough, 1265 Military Trail, Toronto, ON M1C 1A4, Canada
e-mail: gdewar@utsc.utoronto.ca

B.A. Stewart
Museum of Anthropological Archaeology, University of Michigan, Ruthven Museums Building, 1109 Geddes Avenue, Ann Arbor, MI, USA

Sacha C. Jones and Brian A. Stewart (eds.), *Africa from MIS 6-2: Population Dynamics and Paleoenvironments*, Vertebrate Paleobiology and Paleoanthropology, DOI 10.1007/978-94-017-7520-5_11

in Africa's more marginal biomes, which would have challenged humans to expand their behavioral repertoire (Dewar and Stewart 2012; Stewart et al. 2012, 2016).

When assessing the evolution of behavior it is imperative to first understand the ecological conditions of a particular region for a given time period in order to establish a baseline. This is important for interpreting the causes of specific behaviors or innovations as rooted in either social or environmental adaptations, or both. Environments change through time and so too will their attractiveness as a niche for subsistence resources, raw materials, or some unknown sociocultural relevance. In order to contribute to the accumulation of data from under-researched African regions we present data from the periods MIS 6-2 with focus on Namaqualand, South Africa. Namaqualand is a semiarid desert that is currently unpredictable and patchy in floral resources, and while rainfall arrives during the austral winter months, it is quite low at 50–250 mm per year (Desmet 2007).

In this paper we synthesize what we currently know about past environments of Namaqualand from MIS 6-2. Because this region has received very little research, it is necessary to include data from geographically disparate sites in order to develop a tentative yet meaningful picture (see Appendix A). We present paleoenvironmental, chronological and subsistence data from recent excavations at Spitzkloof Rockshelter A (hereafter Spitzkloof A). Spitzkloof A (28° 51.790 S, 17° 04.65270 E) is located in southern Africa's Succulent Karoo Biome, and more specifically in the hinterland or "Richtersveld" of northern Namaqualand. The site is situated 30 km south of the Orange River and 30 km east of the Atlantic Ocean (Fig. 11.1). Results from Spitzkloof A are offered alongside previously published data

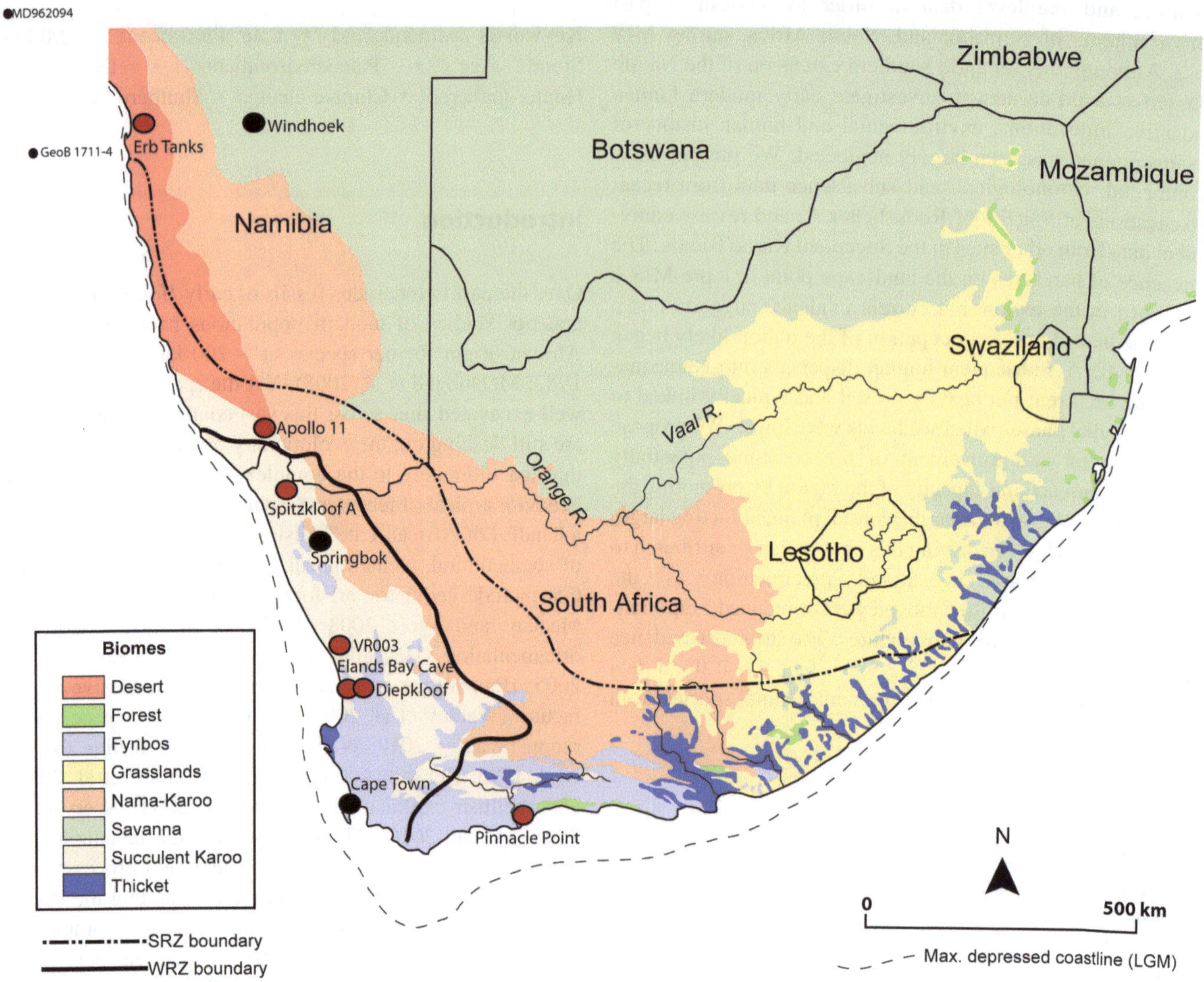

Fig. 11.1 Map of the study region identifying the biomes, archaological sites and marine cores referred to in the text. The winter rainfall zone (WRZ) and summer rainfall zone (SRZ) boundaries are indicated with the region between them receiving rainfall all year

from several other sites in the Succulent Karoo Biome with equivalent-aged deposits: Varsch River 003 (VR003) in southern Namaqualand, and the Namibian sites Apollo 11 Rockshelter and Erb Tanks.

In a study of the Holocene occupation of Namaqualand, Dewar (2008) notes that the region was inhabited during relatively cold humid phases including the Neoglacial (~4.2–1.4 ka) and the Little Ice Age (0.65–0.15 ka). Conversely, the region was occupied at a much lower density during warm and arid phases such as the Mid-Holocene Altithermal (7–4.2 ka) and Medieval Warm Epoch (~1.4–0.65 ka). Dewar concludes that water availability was the most likely factor driving settlement of the region. Here we test this hypothesis for the Pleistocene with the expectation that occupation will be linked to glacial and stadial phases, which were cold but more humid than current conditions (cf. Chase and Meadows 2007). The lowered sea levels that accompanied such phases may have played a particularly important role: with westward shifts of the arid coastline from its present position, precipitation could have reached larger areas of inland Namaqualand while the lowered temperatures could have increased the effectiveness of evapotranspiration.

Current Landscape

The Namaqualand coastal semidesert is the southern extension of the Namib Desert within the Succulent Karoo Biome. This region is demarcated by the Olifants River in the south (the boundary of the Fynbos Biome) while to the north it is defined by the Orange (Gariep) River (Fig. 11.1). The western edge is bounded by the Atlantic Ocean, while the eastern threshold borders the Bushmanland grasslands (Nama Karoo Biome) and is demarcated by the north–south-oriented Kamiesberg Mountains. These mountains range from 30 to 100 km inland and consist of granite gneiss that peak at 1,706 m above mean sea level (amsl). To the north, the Stinkfontein Mountain ranges consist of the Port Nolloth Group that unconformably overlies the Namaqua Metamorphic Group (Frimmel 2003) peaking at 1,377 m amsl. Numerous dry riverbeds crosscut the coastal plateau and flow very infrequently. The coastal plain or Sandveld consists of red Pleistocene sands that are overlain in areas with dynamic white aeolian sand of marine origin (Mucina et al. 2006). The Richtersveld includes the foothills and the Stinkfontein Mountains, with red Pleistocene fine-grained sand deposits and quartz gravel plains.

Namaqualand is within the Winter Rainfall Zone (WRZ), receiving >66% of its annual precipitation during the austral winter months. In the north, current mean annual rainfall is less than 50 mm with an increasing gradient to the south (up to 250 mm per annum) and east toward the mountains. The peaks of the Kamiesberg receive up to 400 mm per annum (Desmet 2007). The aridification of Namaqualand and the Namib Desert is caused by the interaction between cold sea surface temperatures and the westerly winds. Upwelling along the west coast in conjunction with the Benguela Current produces sea surface temperatures of between 11 and 17 °C (Eitel 2005). When humid southerly winds pass over the frigid sea, the air is cooled and cannot release precipitation until the winds have moved over the warm continent creating this west to east rain shadow gradient. Although rain is infrequent, when humidity ranges between 70 and 100% coastal fogs (known locally as "*Malmokkies*") form and can extend up to 90 km inland, providing an important source of hydration for flora and fauna. Although the official mean temperature for the entire Succulent Karoo Biome is 16.8 °C (Mucina et al. 2006), Namaqualand can exceed temperatures of 40 °C during the summer while winter minimums can fall to 7 °C and below. Very hot *Foehn* or "berg" winds can drive the daytime temperatures above 44 °C even in winter months.

The Succulent Karoo is one of eight terrestrial biomes in southern Africa, yet it has received relatively little archaeological attention compared to the productive Fynbos Biome of the Cape coastlands. Succulent Karoo vegetation is dominated by dwarf, succulent shrubs of which the Aizoaceae or "*Vygies*" are prominent as are Euphorbiaceae, Crassulaceae, and succulent members of Asteraceae, Iridaceae, and Hyacinthaceae (Mucina et al. 2006). Extravagant mass flowering of Asteraceae daisies occurs in spring. Grasses are rare and of C_3 type, while trees such as *Acacia karroo* are typically present along dry riverbanks. By necessity, flora and fauna are dry adapted species and occur in unpredictable and patchy distributions (Desmet 2007). The diversity of mammals and birds is very low compared to the rest of South Africa, but the region is a biodiversity hotspot for small reptiles and invertebrates.

Paleoenvironments, Sea Levels, and Settlement

According to Eitel (2005), the aridification of the Namib Desert and, by extension, Namaqualand, has its origin in the Miocene with the establishment of the Benguela Current. In 2007 Chase and Meadows synthesized the available paleoenvironmental proxy data to evaluate potential expansions of the WRZ using data from marine cores off the coast of Namibia, as well as terrestrial and other marine proxy data from South Africa's west coast. Their interpretation combined results from a wide range of datasets including pollen, aeolian, fluvial and lacustrine deposits, size and presence of mammals, and stable isotopes. They tested and seem to confirm an inverse relationship between temperature and humidity for the WRZ whereby glacial periods were humid and interglacials were dry. Scott et al. (2012) later revisited the well-dated terrestrial fossil pollen record and confirmed the inverse relationship between temperature and humidity.

Not only do glacial/interglacial periods reflect global temperatures they also affect the location of coastlines. The current South African shoreline was established ~9.0 ^{14}C kBP with small deviations during the Holocene. Throughout MIS 6-2, however, it fluctuated widely, ranging from +4 to −130 m amsl (Ramsey and Cooper 2002). Due to the gradual slope of the Southwest African Margin (continental shelf), even small changes in sea level will shift the location of the coastline. Cold glacial periods would have depressed sea levels exposing landmass that could be colonized by vegetation, while warm periods and concomitant sea level transgressions will have submerged land (cf. Compton 2011 for the southern coast). For example, a drop of 120 m would have extended the Namaqualand coastline westward by ~20 km. This could have impacted water availability in this semidesert by either: (a) exposing warm landmass shifting the rainfall shadow to the west and thereby increasing precipitation in regions that are today quite arid; or (b) evapotranspiration would have been less efficient in lowered temperatures. In either scenario increased water availability would have potentially supported a higher carrying capacity of flora and fauna.

Ramsey and Cooper (2002) evaluated the available sea level indicators for southern Africa in order to produce a well-constrained sea level curve from MIS 7-1. They rely largely on the dating of beachrock (aeolianite) using U-series and previously published dates from shoreline indicators primarily from the eastern Cape coast. In 2010, Fisher et al. developed a paleoscape model of changing sea level for the southern Cape coast at 1.5 ka intervals stretching back ~420 ka. They use integrated bathymetric datasets, GIS and a relative sea level curve (RSL) with ages extrapolated from oxygen isotope ratios from benthic foraminifera (the composite RSL constructed by Waelbroeck et al. 2002 correlates well with the localized geological data) to estimate the position of the coastline, and compared this predicted model with strontium isotope ratios from speleothems as an independent test of sea level. There are some incongruences between the two methods before 250 ka (MIS 7), but after this time they are more consistent. While this dataset was developed for the southern Cape, gross shifts in sea level will also affect the west coast and Namaqualand coastline. The Agulhas Bank of the southern Cape extends 300 km offshore, whereas the Southwest African Margin off Namaqualand reaches ~200 km offshore (Fig. 11.1). By contrast, the Eastern Cape continental shelf is very steep at only 3 km offshore (Fisher et al. 2010: Fig. 2). The Southwest African Margin is thus overall more similar in morphology to the Agulhas Bank than to the Eastern Cape, and we accordingly expect that offshore coastline movement in Namaqualand were more likely to be similar to Southern Cape values.

Pre-MIS 6 (>191 ka)

The Archaeology Contracts Office, a cultural resource management team based in Cape Town, has surveyed Namaqualand for over two decades (Dewar 2008; Dewar and Orton 2013; Halkett 2002, 2003, 2006a, b; Halkett and Dewar 2007; Halkett and Hart 1997, 1998; Halkett and Orton 2004, 2005a, b, 2007; Orton 2007, 2009; Orton and Halkett 2005, 2006; Webley 1992, 2002, 2009). Their extensive archaeological database (>1500 sites) makes clear that past populations used the landscape prior to MIS 6. For example, there are over 50 recorded Early Stone Age (ESA) open air sites identified by the presence of handaxes. The majority of these ESA sites reflect the quarrying of silcrete outcrops where thousands of artifacts were dropped. There is a ribbon of coarse silcrete outcrops along old marine terraces in northern Namaqualand (Roberts 2003) that extends across the Orange River into the Gemsbok region of Namibia (Corvinus 1983). The presence of large quartzite and quartz clasts at these silcrete quarry sites indicates that people must have transported these clasts to the marine terraces. There is no evidence for actual habitation sites although fossil bone has been found at one inland site near the town of Kleinsee (Orton personal communication), but few of these localities have been systematically sampled. All are situated within 5 km of the current coastline or along river valleys, the latter suggesting that early *Homo* followed corridors associated with fresh water or artifact deposits are covered with recent sands. Without proper chronological control, however, we cannot say whether artifacts that are currently near-coastal were deposited in a similar environmental setting, since sea levels have changed. No doubt many archaeological sites are currently submerged.

MIS 6 (191–130 ka)

Palaeoenvironmental datasets for the west coast of southern Africa that date to MIS 6 come from two marine cores (Fig. 11.1) off the southern and central coasts of Namibia: GeoB1711-4 and MD962094 (Shi et al. 2001; Stuut et al. 2002). The pollen from marine core GeoB177-4 identified high levels of Restionaceae, the evergreen family within the Fynbos Cape Floral Kingdom, in addition to a transitional desert/semidesert group Asteroideae, currently located near the Orange River (Shi et al. 2001). The term "desert/semidesert" is a poor descriptive though as it does not reflect the true water dependence of these families. The authors argue that the presence of these families is evidence for a northward expansion of the Cape floral elements during this stage and therefore reflects a humid signal. Stuut et al.'s (2002) study of grain size variations of terrigenous sediments from marine core MD962094 also suggests relatively humid conditions during MIS 6 based on a strong increase in the proportion of fluvial sediment deposits.

Sea levels during MIS 6 along the coast of Namaqualand are not yet well understood and thus for the time being we must rely on data from further afield. A U-series date from a sample of aeolianite located on the east coast of South Africa (182 ± 18 $^{230}Th/^{234}U$ ka: Pta-U430) was traced to a submerged beach rock facies at −3 m amsl (Ramsey and Cooper 2002). Strontium isotope ratios from Pinnacle Point on the southern Cape coast suggest two minor regression events at 189.7 and 173 ka. These events lead the paleoscape model that predicts low sea stands at 184.5 ka and 151 ka respectively (Fisher et al. 2010). While not yet resolved, both methods agree that there was a transgression at 167 ka to near modern coastal levels. Finally, a major drop in sea level is recorded until the end of MIS 6 with strontium isotope ratio data indicating a minimum sea level sometime between 155 and 150.5 ka, while the paleoscape model records two regressive peaks at 150.5 and 137 ka (Fisher et al. 2010).

There is currently no direct evidence for the occupation of Namaqualand during the penultimate glacial period. Surveys have identified over 90 Middle Stone Age (MSA) sites with large, heavily patinated blades, points and flakes as well as radial cores and Levallois reduction, but only a handful are diagnostic pieces and all relatively date to MIS 4 (see below). Site types include open air lithic scatters, quarries and food processing/habitation sites. There are two known rock shelters with MSA material on the surface with potentially very deep deposits: Spitzkloof (Dewar and Stewart 2012) and VR003 (Steele et al. 2012). There is also evidence for the reuse of the same silcrete outcrops that were exploited during the ESA, suggesting repeated use of the landscape (Dewar 2008; Dewar and Orton 2013).

MIS 5 (130–71 ka)

The transition to the Last Interglacial reflects a period of general aridification. Shi et al. (2001) record a sharp decline in Restionaceae and desert/semidesert pollen. These taxa were replaced by Kalahari dry forest, indicating an increased influence of the easterly trade winds and a reduced influence of the rain bearing westerlies. Only at the end of MIS 5 is there a slow return of the pollen spectra that signal humidity. Correspondingly, Stuut et al. (2002) note a drop in fluvial input during MIS 5e, with a return to moderate levels at MIS 5d followed by sharp increases in fluvial input during MIS 5c and 5a.

The paleoscape model and strontium isotope ratios from Pinnacle Point suggest that by 130 ka (MIS 5e) the coast had returned to near modern levels, with neither showing evidence for a major regression during MIS 5 (Fisher et al. 2010). Sea level data for the Last Interglacial in eastern South Africa is characterized by two sea level highstands at +4 m amsl separated by a −44 m lowstand. Based on two ionium dates of 110 ka and 98 ka at Klasies River Mouth, Hedley and Volman (1986) relate one of the +4 m amsl highstands to MIS 5c. At Sodwana Bay, U-series dated beachrock from a submerged shoreline at −44 m amsl yielded a date of 117 ± 7 $^{230}Th/^{234}U$ ka (Pta-U487). While Maud (1968), Hobday (1976), Cooper and Flores (1991) have identified two +4 m shoreline deposits in KwaZulu-Natal that are separated by the −44 m amsl unconformity, and suggest the other highstand was most likely deposited during MIS 5e (Ramsay and Cooper 2002). MIS 5c and 5a have been interpreted by Chappel and Shackleton (1986) to be close to the present sea level, which is supported by the paleoscape model with values ranging from +1.65 to +0.5 m amsl (Fisher et al. 2010: Table 1). The near modern sea level values remain until 72 ka where the strontium isotope data and paleoscape model agree that at there was a drop in sea level (Fisher et al. 2010).

The Last Interglacial has been proposed as the likely date for the small MSA assemblage from the near coastal rock shelter Boegoeberg 2 in Namaqualand. The corrected ^{14}C date on ostrich eggshell is 46,709–44,242 cal BP (44,200 ± 1200 ^{14}C BP, Pta-6956), but Klein et al. (1999) interpret the date as a minimum age and suggest that the occupation could be much older since the fauna implies true interglacial conditions; they propose substages 5e, 5b, or 5a. It is also possible that MSA processing sites located on the modern coastline date to this period, when sea level would have been similar. Processing sites are identified on the basis of the presence of large quantities of fossilized intertidal shellfish, tortoise and patinated lithics. This idea is perhaps reinforced by ethnographic research, which suggests that a typical daily foraging radius to the intertidal zones is likely to be no more than 8 and 10 km (Meehan 1982). Current evidence suggests that the sea level was further away ~45 ka (see below).

The oldest date in the greater Namib region associated with an archaeological deposit comes from Erb Tanks in central Namibia (McCall et al. 2011). An amino acid racemization date of 130 ka was recently obtained on ostrich eggshell from the very base of the shelter (McCall et al. 2011). Although this may be an indication that people were moving into the Namib Desert landscape when the sea levels were returning to near modern values, this date is a conspicuous outlier as noted by the authors. More dates are required to verify a human presence in early MIS 5. A more

Table 11.1 Number of specimens identified to species and their inferred diet (cf. Skinner and Chimimba 2005) for the Still Bay (SB), Howiesons Poort (HP) late MSA and Early LSA layers Apollo 11 (Vogelsang et al. 2010)

Species	Apollo 11 SB ~70 ka	Apollo 11 HP ~60 ka	Sptz A late MSA ~52–51 ka	Apollo 11 late MSA III ~43 ka	Apollo 11 Late MSA I ~30 ka	Apollo 11 early LSA ~25–14 ka	Diet
Bathyergus janetta	0	0	2	0	0	0	Browser
Raphicerus campestris	0	0	1	0	0	0	Browser
Oreotragus oreotragus	2	8	1	6	2	1	Browser
Silvicapra grimmia	0	0	1	0	0	0	Browser
Chersina angulata/Psammobates tentorius trimeni	2	1	41	1	1	0	Browser
Procavia sp.	3	2	0	4	11	10	Mixed
Lepus sp.	2	3	0	8	8	8	Mixed
Antidorcas marsupialis	2	1	1	3	1	0	Mixed
Phacocherus sp.	0	0	0	1	1	0	Grazer
Oryx gazella	0	0	2	0	0	0	Grazer
Equus zebra	2	5	0	4	5	3	Grazer
n	13	20	49	27	29	22	

These data are compared to the late MSA layers at Spitzkloof A (Dewar and Stewart 2012). *Note* In Vogelsang et al. (2010) the data are presented as relative abundance while this table presents the NISP

congruous date is 85 ka (also based on amino acid racemization) when paleoenvironmental indicators suggest that the region was returning to a more humid environment, although it is important to note that there is clearly evidence for vertical movement of ostrich eggshell at Erb Tanks (McCall et al. 2011). Unfortunately, there is no bone preserved from the MSA layers at Erb Tanks.

MIS 4 (71–57 ka)

Paleoenvironmental data for MIS 4 is restricted to marine proxies, a charcoal study, and faunal remains but nevertheless this period marks a distinct threshold reflecting a shift to a cooler and more humid environment. At 70 ka maximum sea surface temperatures (SST) dropped from interglacial temperatures of ~22 to ~20–19 °C, and maintained those lower values throughout both MIS 4 and 3 (Krist et al. 1999). There is an increase through time in Restionaceae and transitional desert/semidesert pollen taxa (cf. Asteroideae) at marine core GeoB1711-4 (Shi et al. 2001) while the coarse aeolian dust input increases dramatically (Stuut et al. 2002) suggesting increasing trade wind strength at the site of marine core MD962094. Charcoal analysis for the MIS 4 layers at Apollo 11 located 30 km north of the Orange River identify "a diverse array of woody vegetation reflecting an environment either very similar to or more favorable than today" (Vogelsang et al. 2010: 212). South of the study region, botanical remains from Diepkloof Rockshelter dating to ~65–50 ka reflect afromontane taxa (Chase and Meadows 2007) indicating a humid late MIS 4/early MIS 3, although this is within the Fynbos biome that currently receives more rainfall than Namaqualand.

After the regression at 72 ka, the sea level remained low along the south coast until 60 ka at which point the strontium isotopes and paleoscape model identify a transgression event (Fisher et al. 2010). The curve developed by Ramsey and Cooper (2002) lacks data for this period.

The presence of Still Bay (SB) points 6 km inland from the town of Koignass (Dewar and Orton 2013) and the discovery of a Howiesons Poort (HP) segment near the Tweepad farm suggest that MSA people inhabited the northern coast of Namaqualand sometime between 74 and 60 ka (cf. Jacobs et al. 2008). In southern Namaqualand, two sites near the Varsche River indicate occupation during MIS 4: the open site STF001 has bifacial points while VR003 has both bifacial points and HP segments (Mackay et al. 2010; Steele et al. 2012). The faunal data from VR003 is presented in Steele et al. (2012), but it has not yet been directly dated. The authors identify one confidently *in situ* layer representing the HP layer from test pit II-04. The sample of identified remains consists of 58 elements, with the arid adapted browsers *Chersina angulata* (Angulate tortoise) dominating at 76% (Steele et al. 2012). The small species list makes identifying the environment tentative but the two herbivores identified to species, *Chersina angulata* and *Cryptomys hottentotus*, at least suggest that browse was available within a potentially arid region.

A more robust sample that includes both SB and HP is found at Apollo 11 with Optically Stimulated Luminescence (OSL) dating these techno-complexes at 71 ± 3 ka (AP6) and

63 ± 2 ka (AP4) respectively, with an intermediate pulse of occupation at 67 ± 3 ka (AP5) (Vogelsang et al. 2010). The faunal data from the recent excavation (Vogelsang et al. 2010: Table 6), although limited, are discussed here because Thackeray's (1977) analysis of the bones from the original excavations did not separate the SB from the HP layers. In the SB layers, of the 32 identified fauna, small mammals (size 1 and 2; cf. Brain 1981: up to 80 kg) dominate the assemblage at 59% with large mammals (>30 kg) at 25%. All species in the sample live on the landscape today with the important exception of the large grazing equid. Using the dietary preferences of the identified species there is a high proportion of mixed feeders, followed by browsers and grazers respectively (Table 11.1). Similar to the charcoal signal from the same deposits, the fauna suggests an early MIS 4 environment that was similar to today, but with more water available to support woody vegetation and grasses.

The HP layers at Apollo 11 produced a sample of 35 identified elements, similarly dominated by small mammals at 57% of the assemblage followed by large ungulates at 23% (Vogelsang et al. 2010: Table 7). Mirroring the SB sample, the identified species are present on the landscape today with the addition of the equid, indicating a nearly modern environment but with increased water availability. Noting the dietary preferences of the identified species (Table 11.1) there is an increased proportion of grazers and browsers at the expense of the mixed feeders. As this dataset is small and the mixed feeders could consume graze or browse, the most powerful inference from this table is the increase in the presence of grazers that do not live on the landscape today.

Erb Tanks has also produced two dates at ~65 ka and one at 60 ka although the occupation lacks SB and HP diagnostic tools (McCall et al. 2011). There is no faunal sample from Erb Tanks for this stage.

MIS 3 (57–29 ka)

During MIS 3, multi-proxy data indicates that the environment was fluctuating from arid to humid within a climatic regime that was cool overall. The charcoal identified from a settlement hiatus bracketed by OSL dates (AP3: 57.9 ± 2.6 ka and AP2: 42.9 ± 2.7 ka) at Apollo 11 consists of a single family Chenopodiaceae (*Salsola* type), which indicate xeric conditions (Vogelsang et al. 2010). By contrast, Restionaceae and desert/semidesert taxa within marine core GeoN1711-4 signals an increase in humidity beginning at 50 ka peaking at 32 ka (Shi et al. 2001). In marine core MD962094, the coarse aeolian dust input initially drops but then fluctuates dramatically (Stuut et al. 2002).

The fossil assemblage from the hyena den Boegoeberg 1 on the Namaqualand coast includes large water-dependent grazers, such as *Connochaetes taurinus* (blue wildebeest) and *Redunca arundimun* (southern reedbuck), suggesting a moist climate, while the large size of the hyena bones suggests a cool environment (Klein et al. 1999). The calibrated ^{14}C dates on ostrich eggshell are 45,500–36,000 cal BP (37,000 ± 5000 ^{14}C BP, GX-22191), 42,000–36,000 cal BP (35,000 ± 3000 ^{14}C BP, GX-21190) and 40,000–34,700 cal BP (33,000 ± 2600 ^{14}C BP, GX-21189), and are interpreted by Klein et al. (1999) as a minimum age and likely representative of late MIS 4 or early MIS 3. It is possible though that the Boegoeberg 1 dates are not representing an infinite date but rather record relatively cool and moist conditions in mid-MIS 3 as shown by the offshore pollen record. Further evidence for high humidity is found at Kannikwa near Port Nolloth in northern Namaqualand where a peat bed is dated at 32,000–31,000 cal BP (27,900 ± 310 ^{14}C BP, Beaumont 1986). Additional support for mid to late MIS 3 humidity is seen in a composite distribution of ^{14}C dated evidence for increased humidity from within the Namib Desert as a whole (Lancaster 2002), with a humid peak ending at the MIS 3/2 boundary at 37,000–31,000 cal BP (32,000–26,000 ^{14}C BP).

The Ramsay and Cooper (2002) curve indicates a drop in sea level in the eastern Cape based on wetland peats with depths of −52 m and −46 m dating from 50,000 to 47,000 cal BP (45,200 ^{14}C BP, Pta-4140) and 44,000–42,000 cal BP (39,000 ± 1500 ^{14}C BP, Pta-4142) respectively. The paleoscape and strontium isotope data (Fisher et al. 2010) also suggests a mid-MIS 3 drop in sea level at ~52 ka while the paleoscape data identifies two shallow transgressions at 40 ka and 30 ka that the strontium isotope ratios do not register (Fisher et al. 2010: 1389, Fig. 4). An offshore marine shell at a depth of −78.4 m amsl records a rapid drop in sea level with a calibrated ^{14}C date of 31,500–30,800 cal BP (27,800 ± 440 ^{14}C BP, Pta-1104), from the Orange River Mouth. This regression event essentially continues through to the Last Glacial Maximum (LGM) in MIS 2.

OSL dates from Apollo 11 indicate that it was occupied in several pulses during MIS 3: 58 ± 3 ka (AP9) and 57 ± 3 ka (AP3) from the base of the late MSA complex; 43 ± 3 ka (AP2) from the middle of the complex; and 30 ± 1.4 ka (AP11) at the top of the complex (Vogelsang et al. 2010). Radiocarbon dates add mid to late occupational pulses occurring at ~37,000 cal BP, and ~32,000–29,000 cal BP (Vogelsang et al. 2010).

While there is no faunal data for the earliest MIS 3 occupation, Vogelsang et al. (2010: Table 6) present the identified remains for the ~43 ka pulse (Late MSA III) and the ~30 ka pulse (Late MSA I). The species list from these occupations mirrors the results from MIS 4 with the addition of a second grazing species (*Phacochoerus sp.*) supporting a more humid signal within an arid zone (Table 11.1).

Table 11.2 Radiocarbon ages of ostrich eggshell from Spitzkloof A, Namaqualand, South Africa

Lab no.	Context	Date in ^{14}C BP	Calibrated dates in cal BP
UBA-17609	Layer Nick	14,350 ± 10	17,274–17,093
UBA-17610	Layer Nick	14,400 ± 70	17,391–17,134
UBA-17611	Layer Nadja	15,200 ± 50	18,304–18,108
UBA-17612	Layer Jaird	16,250 ± 60	19,457–19,237
UBA-17613	Layer Dave	19,550 ± 60	23,415–23,132
UBA-17614	Layer Mark	19,750 ± 80	23,671–23,393
UBA-17615	Layer Julie	19,550 ± 60	23,415–23,132
UBA-17616	Layer Brian	>59,250	N/A
UBA-17617	Layer Brian	52,150 ± 800	N/A
UBA-17618	Layer Brian	51,150 ± 850	N/A

The ^{14}C dates were run at the ^{14}Chrono Centre at Queens University Belfast. Dates are calibrated using the software Calib 7.0 and the calibration curve Shcal13.14c for the southern hemisphere (Hogg et al. 2013). Note that the geological layers Dave, Mark, and Julie represent a single chronological layer. *Note* Experiments have shown that fossil ostrich eggshell is typically 180 ± 120 years too old (Vogel et al. 2001) and so 180 yr was subtracted before calibration

Radiocarbon dates measured on ostrich eggshell from layer Brian at Spitzkloof A returned ages of 52,150 ± 800 ^{14}C BP and 51,150 ± 850 ^{14}C BP (Table 11.2). A third date is likely infinite at >59,250 ^{14}C BP. The presence of gypsum nodules from this layer indicates climatic conditions ranging from arid to semiarid (Dregne 1976; Middleton 2003), but with enough moisture to have put the gypsum in solution (Dewar and Stewart 2012). The fauna from this occupation pulse consists of 810 identified specimens representing a minimum of fourteen different species that are all found on the landscape today (Dewar and Stewart 2012). Small mammals dominate the assemblage at 37% followed by tortoises at 35%. The identified species consists primarily of browsers (Table 11.1) but the presence of the *Oryx gazella* suggests that there was some grass available. Overall the species list suggests the environment at ~52–51 ka was very similar to MIS 4. Erb Tanks was occupied at 45 ka but fauna and other environmental indicators are absent (McCall et al. 2011).

MIS 2 (29–14 ka)

Increased or more effective precipitation during early MIS 2 (~28–20 ka) is recorded in Namibia from calcified reed beds and lacustrine deposits at Koichab Pan (Lancaster 1984), Narabeb (Teller and Lancaster 1986), Khommabes (Teller and Lancaster 1985) and Gobabeb (Vogel and Visser 1981). Charcoals from the Late Pleistocene layers at Apollo 11 are dominated by *Olea*, a woody species that lives in dry riverbeds of the central highlands (Vogelsang et al. 2010). This species is not found near Apollo 11 today. *Olea europaea* ssp *africana* is a frost and drought tolerant species that at first glance could signal a cool and arid landscape. Alternatively, *Olea* pollen from a hyrax midden in the Brandberg (Dâures Massif, Namibia) co-occurring with *Stoebe* type, dwarf shrubland taxa, *Artemeisia* and fern pollen dating to the LGM (~21 ka) has been interpreted by Scott et al. (2004) as indicating a cool moist signal. Although the authors do caution that this may not necessarily reflect increased precipitation but rather a drop in average temperature reducing evaporation, which would also render rainfall more effective (Scott et al. 2004).

The marine core data also support a wet early MIS 2. Shi et al. (2001) record the highest percentages of Restionaceae pollen in core GeoB1711-4 at ~24 ka, declining until ~19 ka and then finally dropping off to negligible values at ~14 ka (*contra* Scott et al. 2004 who did not find Restionaceae in the Brandberg). Fluvial sediments at MD962094 and trade wind proxy data from GeoN1711-4, MD962094, and GeoB1706 mimic the marine core pollen data with high values at the onset of MIS 2 that steadily decrease through time (Stuut et al. 2002).

Further south at Elands Bay Cave, the LGM layers (~25–21.5 ka) are marked by maximum values of pollen from woodland taxa and the lowest xeric karroid and Strandveld pollen values (Meadows and Baxter 1999). Charcoal studies from the same deposits substantiate this pattern of increased humidity with the presence of afromontane species such as *Celtis Africana* and *Grewia occidentalis*, which are intolerant of drought. Pollen from rock hyrax middens in the Cederberg Mountains suggest a shift at ~16 ka from a glacial vegetation consisting of *Stoebe/Elytropappus* shrubs and fynbos elements (Ericaceae and Proteaceae) to a Holocene vegetation signal with a mosaic of fynbos, thicket, and succulent vegetation (Scott and Woodborne 2007a, b). The authors interpret this shift as a result of increasing temperatures and reduced precipitation (although there is marked variability within the LGM).

Local data comes from the Eksteenfontein spring 18 km northeast of Spitzkloof A (Scott et al. 1995, 2012) where *Stoebe/Elytropappus* pollen samples indicate the region was still fairly cool from ~15.2 to 13.6 ka, but warming by ~12.5 ka. Scott et al. (2012) suggest that this period also reflects reduced moisture from a cold dry fynbos to a more modern arid environment.

In Durban Bay (Eastern Cape) a wetland peat located at −22 m amsl produced a calibrated date of 30,000–28,000 cal BP (24,950 ± 950 ^{14}C BP, GaK-1390) (King 1972 in Ramsey and Cooper 2002). By 20,000 cal BP (16,990 ± 160 ^{14}C BP, Pta-182) the sea had dropped to a maximum of −130 m amsl based on a dated *Pecten sp.* shell from Cape St. Francis (Vogel and Marais 1971), while submerged material ranging from −100 to −90 m amsl dating to ~13 ka indicates a slow post-LGM transgression.

The paleoscape model and strontium isotopes also identify a shallow transgression at the MIS 3/2 boundary while the paleoscape model confirms a rapid drop in sea level beginning in early MIS 2 with a peak ~20 ka. Unexpectedly, the LGM peak is not captured by the strontium isotope data (Fisher et al. 2010). A marine shell near the mouth of the Orange River mouth indicates that at 18,900–18,000 cal BP the Namaqualand coastline was located −87.2 m amsl (Vogel and Visser 1981).

There are two pulses of occupation at Spitzkloof A during MIS 2 (Table 11.2). The first pulse is identified from three ostrich eggshell ^{14}C dates at ~23,500–23,000 cal BP. A second pulse is registered by four ^{14}C dates ranging from ~19,000 to 17,000 cal BP, bracketing the period when the coastline would have been near-maximum distance away. Preliminary analysis of the fauna from the ~23,000 cal BP layers suggests an increase in the diversity of species with the addition of grazing equids and alcelaphines, a third species of tortoise *Homopus signatus signatus* and even a fish vertebra. While these few elements represent a small sample together they indicate a likely increase in fresh water availability.

The lithic scatter AK2006-001G along the coast of Namaqualand, though undated has artifacts typical of Late Pleistocene microlithic assemblages that occur between ~20 and ~9.5 ka in South Africa, Lesotho and Swaziland (Orton 2008). Orton (2008) argues that AK2006-001G was likely deposited between 17,000 and 11,000 BP (~20,500–13,000 cal BP). At Apollo 11, ^{14}C dates identify occupation at ~25,000 cal BP, ~22,000 cal BP, and 17,000–15,000 cal BP. Thackeray's (1977) "mean ungulate body mass index" analysis at Apollo 11 correlates positively with rainfall and was high during MIS 2, suggesting that both primary productivity and carrying capacity were higher than present day. The presence of *Equus capensis* at Apollo 11 until ~14 ka has been interpreted as evidence for humid conditions until latest MIS 2 when xeric conditions then dominated (Thackeray 1979). Erb Tanks also has punctuated dates at 25, 20, 15 and 12 ka (McCall et al. 2011) that are similar to Spitzkloof A and Apollo 11.

Synthesizing the Data: When Did People Occupy Namaqualand?

Pre-MIS 6 (>191 ka)

Current evidence for the occupation of Namaqualand before MIS 6 comes from the presence of handaxes along the marine terrace and inland river valleys. All we can say is that populations were using the landscape to some degree and provisioning themselves with quartzite, quartz and silcrete raw materials at quarry sites. When these individuals were present, what the environment was like and where the coastline lay are currently unknown and the foci of future research.

MIS 6 (191–130 ka)

Palaeoenvironmental proxy data from marine cores off the coast of Namibia indicate that the penultimate glacial period grew more humid through time based on the presence of flora that require higher water availability than are present in Namibia today and the high input of fluvial sediments. The Southern Cape and by proxy the Namaqualand coastline experienced flux during the first half of MIS 6 but the later half of this stage experienced a lowered sea level during the penultimate glacial maximum. This indicates that during much of MIS 6, the Namaqualand shoreline would have been much further west than it is today which has two implications: (1) the exposed coastal plain had the potential to increase carrying capacity, especially as precipitation would have been more effective through a shifting rain shadow or less efficient evapotranspiration; and (2) any coastal or near-coastal sites deposited during MIS 6 are now likely to be submerged. This hypothesis will be tested when more precise datasets are available for this stage. While it seems that MIS 6 would be a good time to occupy Namaqualand, to date there are no known sites but hopefully continued survey along inland river terraces will change this.

MIS 5 (130–71 ka)

Namaqualand experienced fluctuations in both temperature and humidity during MIS 5. The aridity that ushered in this stage became moderate during MIS 5d, but pollen and fluvial inputs signaling increasing humidity only occur later during MIS 5c and 5a.

Overall the sea level was near modern values for much of this stage. Two minor sea level highstands likely occurred during MIS 5e and 5c with a return to modern sea level by MIS 5a. While a major regression event is registered on the Eastern Cape at ~117 ka, there is no evidence for a substantial drop in sea level from the Pinnacle Point/Aghulas Bank data until the end of this stage at 72 ka. It thus remains unclear whether the Namaqualand coastline experienced the ~117 ka event. Direct measurement of the Southwestern African Margin can answer this question and a paleoscape model for this stretch of the continental shelf is currently being generated.

There is currently no directly dated evidence for human occupation of Namaqualand during MIS 5. However, the shellfish processing sites with fossilized material and heavily patinated lithics may date to this stage since they are unlikely to have been deposited beyond an 8–10 km foraging radius of the intertidal zone (cf. Meehan 1982). This is also the isotope stage during which Klein et al. (1999) infer human occupation at Boegoeberg 2, specifically during MIS 5e, 5b or 5a. Erb Tanks in Namibia has produced two dates at either end of MIS 5, but there is a strong possibility that the earlier date is erroneous. The later date of 85 ka, which is associated with evidence of increasing humidity, represents the earliest firmly dated evidence for occupation of the greater Namib region, at least for the time being. Of interest is the close correspondence between this age and that recently obtained date (~83 ka) for the earliest sustained human presence in the Maloti-Drakensberg, another challenging environment (Stewart et al. 2012, 2016).

MIS 4 (71–57 ka)

Paleoenvironmental data from the marine cores and charcoal from Apollo 11 reflect a shift to a cooler more humid environment starting at ~70 ka. Just before this, the paleoscape model and strontium isotope ratios predict a corresponding regression in sea level and thus a westerly expansion of the coastal plain, opening up the landscape to flora and fauna. The end of this stage is marked by a transgression event straddling the MIS 4/3 boundary that would have drowned this newly expanded coastal plain. Although, the fauna from Apollo 11 seem to suggest little change in species through this 14 ka and thus reflects a relatively muted environmental change, while the presence of woody vegetation and grasses suggests that it was more humid than today. Interestingly, the faunal remains from Varsch River 003 in southern Namaqualand suggest the end of MIS 4 was potentially semiarid and provided enough browse to attract tortoises. A Namaqualand-specific paleoscape model and larger datasets of faunal material are required to test these patterns. SB points and HP segments are found in a range of localities across the study region including open air sites and shelters, although compared to other southern Africa landscapes we know very little about these techno-complexes in Namaqualand. The increased number of sites and site types from this stage suggests a more consistent use of the landscape than was seen during MIS 5.

MIS 3 (57–29 ka)

Multi-proxy data from MIS 3 indicate that it was a stage in flux. Apollo 11 was occupied at ~58 ka when the shoreline was relatively close to modern values and the fauna indicates a modern-like environment but with the addition of equids. By 57 ka, Apollo 11 was abandoned and the charcoal sample reflects aridity until ~43 ka. Contradictorily the marine offshore pollen suggests increased and more effective precipitation by 50 ka peaking at 32 ka, while the fauna at Boegoeberg 1 potentially also indicate a wet and cool period from 45.5 to 35 ka. Data from both the eastern and southern coasts identify a low sea stand at ~52–50 ka, when Spitzkloof A was occupied and cemented gypsum deposits indicate an arid environment with enough moisture to put the gypsum into solution. The fauna from this layer is dominated by arid adapted browsing species yet the presence of gemsbok indicates the availability of some grass. The occupation of Erb Tanks (~45 ka) and Apollo 11 (43, 37 and 30 ka) also correlates with low or lowering sea levels and the presence of additional grazing species at Apollo 11 indicates a potential for more effective precipitation. Overall we see occupation of the greater region during both low and high sea stands and contradictory signals for humidity especially for early to mid-MIS 3.

For now, the most parsimonious answer is that this was a period of fluctuating environmental conditions and thus conflicting proxy signals and intermittent occupation of the region, particularly during early MIS 3. Only with the colder and potentially more humid conditions of late MIS 3 do occupational pulses become more frequent.

MIS 2 (29–14 ka)

Globally, MIS 2 exposed humans to one of the coldest and driest environments of the Pleistocene – the LGM. Yet most proxy data suggests that Namaqualand likely enjoyed some periods of increased water availability, with xeric conditions only recorded during latest MIS 2. LGM flora and fauna from

Apollo 11 suggest a cold but humid environment with increased primary productivity and carrying capacity, although the charcoal signal is complicated. The environment deteriorates by 14 ka with the introduction of more xeric conditions and the extinction of *Equus capensis*. Flora and fauna from the Spitzkloof region, including the Eksteenfontein spring produce a cool/humid signal until it changes at ~13 ka, a pattern supported by the marine core proxies. From 30 to 20 ka sea levels potentially dropped to a −130 m amsl, while a submerged marine shell near the Orange River Mouth records a lowstand of −87.2 m a.m.s.l at ~19–18.5 ka.

Occupations at Spitzkloof A, Apollo 11 and Erb Tanks are pulsed from ~22 to 13 ka but are more frequent than any previous stage. During this interval sea level would have been at its lowest with the coastline up to ~20 km further west than present day, opening up a vast tract of land available to be colonized by flora and fauna. Preliminary evidence that the flora and fauna did take advantage of the coastal plain is the presence of grazing ungulates at Apollo 11 and Spitzkloof A and a high "mean ungulate body-mass index" that correlates positively with precipitation.

Conclusion

The paleoenvironmental and settlement signals we possess for Namaqualand are currently patchy at best, primarily due to the dearth of large research projects in the region. This is currently being redressed through our ongoing excavations at Spitzkloof A and B in northern Namaqualand (Dewar and Stewart 2012) and the Varsche River project in southern Namaqualand (Mackay et al. 2010; Steele et al. 2012). Environments during MIS 6-2 have been preliminarily reconstructed based on proxy data from distant marine and terrestrial sources. The resulting picture tentatively supports the very broad correlation in the WRZ of glacials and stadials as phases of humidity, while interglacials were more arid (cf. Chase and Meadows 2007). One cautionary note is that the limited data for Namaqualand does not reflect particularly strong differences between glacial and interglacial periods and so continued fine-grained research in the region is required to verify the pattern. Currently the proxies used to identify past conditions remain imperfectly aligned, as best shown by the time lag between the sea level curves and the contradictory signals in MIS 3, but future work will hopefully address these issues.

As sea level is linked to global temperatures, glacial period shorelines are for the most part submerged. The majority of archaeological reconstructions of glacial periods will therefore have to reflect inland settlement and subsistence strategies. If human occupation of Namaqualand is linked to the availability of water (cf. Dewar 2008), as we suspect, then we would expect to see increased visibility of populations on the landscape during humid periods when the ocean is regressing and the potential for increased carrying capacity is highest. The current number of dated sites is too small to confidently verify this assertion but so far the majority of occupation pulses do seem to correlate with expanded coastlines. Although tentative, we certainly do have evidence for the occupation of Namaqualand during humid periods (from MIS 4-2). Particularly informative are sites dating to MIS 3, a period that is poorly known in South Africa, especially from the far better studied southern Cape coastline. So far the data indicates that Namaqualand was more often occupied when it was generally cooler and more humid than it is today, conditions that at times could have supported grasslands and thus large game. However of equal interest is that people were present when the region was not as "easy", during MIS 5 and parts of MIS 3 for example. Future research will expand our datasets to hopefully provide much more detailed understandings of when humans occupied different parts of Namaqualand, why humans were drawn or pushed into the region, and the environmental conditions that prevailed both when people were there and, importantly, absent. For example, the rare grasses in Namaqualand today are of the C_3 type, while those of neighboring Bushmanland are C_4. We can thus use stable isotope analysis to test whether environmental changes simply increased the availability of local species or were more complex and involved shifting rainfall zones with a concomitant westward expansion of Bushmanland grasses. Most crucially, improving our knowledge of the adaptive strategies involved in colonizing and mastering such shifting environments is essential for illuminating the processes underlying modern human behavioral evolution.

Acknowledgments The authors wish to thank the Social Sciences and Humanities Research Council of Canada for funding this project (grant #490993) and The South African Heritage Resources Agency for the excavation permit (permit #80/09/11/008/51b). We would also like to thank Dr. Paula Reimer at Queen's University Belfast and the Chrono Centre for Climate, The Environment, and Chronology laboratory for rapid turnaround on our radiocarbon dates. Finally, we would like to thank Dave Halkett and our other colleagues at the Archaeology Contracts Office without whose guidance none of this research would be possible.

Appendix A

The proxy data presented in this study, the associated dates, implications and sources of the data. The dates are presented as they were in their published form while the calibrated dates column reflects calculations for this study.

Marine isotope stage 6: Glacial 191–130 ka	Data	Signal	Date	Calibrated dates at 1σ (for this study) cal BP[b]	Implication	References
	Pollen[a]	High percentage of *Restionaceae* and desert/semi-desert taxa			Humid	Shi et al. (2001)
	Terrigenous sediments[a]	Grain size suggests increase in proportion of fluvial sediment deposits			Humid	Stuut et al. (2002)
Southern Cape sea level	Strontium isotopes	Slight increase in $^{87}Sr/^{86}Sr$	189.7 ka		Minor regression	Fisher et al. (2010)
Southern Cape sea level	Paleoscape model: Bathymetry and GIS[a]	>30 km from modern coastline	184.5 ka		Minor regression	Fisher et al. (2010)
Eastern Cape sea level	Uranium series dating of aeolianite	Beachrock at −3 m amsl	182,000 ± 18,000 (Pta-U430)		Minor transgression	Ramsey and Cooper (2002)
Southern Cape sea level	Strontium isotopes	Slight increase in $^{87}Sr/^{86}Sr$	173 ka		Minor regression	Fisher et al. (2010)
Southern Cape sea level	Paleoscape model: Bathymetry and GIS[a]	~4.81 km from modern coastline	167 ka		Minor transgression	Fisher et al. (2010)
Southern Cape sea level	Strontium isotopes	Slight decrease in $^{87}Sr/^{86}Sr$	167 ka		Minor transgression	Fisher et al. (2010)
Southern Cape sea level	Strontium isotopes	Peak high ratio of $^{87}Sr/^{86}Sr$ at ~152 ka	Between 155 and 150.5 ka		Major regression	Fisher et al. (2010)
Southern Cape sea level	Paleoscape model: Bathymetry and GIS[a]	Maxima peak/~91.11 km from modern coastline	150.5 ka		Major regression peak	Fisher et al. (2010)
Southern Cape sea level	Paleoscape model: Bathymetry and GIS[a]	Maxima peak/~96.51 km from modern coastline	137 ka		Major regression peak	Fisher et al. (2010)
Marine isotope stage 5: Last Interglacial 130–71 ka	**Data**	**Signal**	**Date**	**Calibrated dates at 1σ (for this study) cal BP[b]**	**Implication**	**References**
	Pollen[a]	Sharp decline of *Restionaceae* and desert/semi-desert taxa, replaced by Kalahari dry forest taxa, but they rebound at the end of this stage			Arid with an increase in humidity by MIS 5a	Shi et al. (2001)
	Terrigenous sediments[a]	Drop in fluvial input at stage MIS 5e, moderate at MIS 5d, sharp increases during MIS 5c and MIS 5a			Arid with slowly increasing humidity and humid peaks at MIS 5c and 5a	Stuut et al. (2002)
Southern Cape sea level	Paleoscape model: Bathymetry and GIS[a]	Minima/~1 km from modern coastline	~130 ka		Major transgression	Fisher et al. (2010)
Southern Cape sea level	Strontium isotopes	Decrease in $^{87}Sr/^{86}Sr$ to near modern values	~130 ka		Major transgression	Fisher et al. (2010)
	Shellfish	Boegoeberg 2: shellfish suggests coastline is near modern location	MIS5e, b or a?			Klein et al. (1999)
Eastern Cape sea level	Inferred date	+4 m amsl highstand	MIS5e?		Minor transgression	Ramsey and Cooper (2002)
Eastern Cape sea level	Uranium series date of aeolianite	−44 m amsl lowstand	117,00 ± 7,000 (Pta-U487)		Regression	Ramsey and Cooper (2002)
Eastern Cape sea level	Ionium dates from Klasies River Mouth	+4 m amsl highstand	110 ka and 98 ka		Minor transgression	Hendley and Volman (1986), Ramsey and Cooper (2002)

(continued)

Marine isotope stage 5: Last Interglacial 130–71 ka	Data	Signal	Date	Calibrated dates at 1σ (for this study) cal BP[b]	Implication	References
	Amino acid racemization date	Erb Tanks	85 ka		Presence of people in the landscape	McCall et al. (2011)
Southern Cape sea level	Paleoscape model: Bathymetry and GIS[a]	Peak/15.56 km from modern coastline	~72.5 ka		Regression	Fisher et al. (2010)
Southern Cape sea level	Strontium isotopes	Slight increase in $^{87}Sr/^{86}Sr$	~72 ka		Regression	Fisher et al. (2010)
Marine isotope stage 4: Glacial 71–57 ka	**Data**	**Signal**	**Date**	**Calibrated dates at 1σ (for this study) cal BP[b]**	**Implication**	**References**
	Pollen[a]	Steady increase in *Restionaceae* and desert/semi-desert taxa			Increasing humidity	Shi et al. (2001)
	Terrigenous sediments[a]	Peak input of aeolian dust and trade winds, but winds reduce before the end of the stage			Increasing humidity	Stuut et al. (2002)
	OSL date and fauna	Apollo 11: Still Bay points and arid adapted species + equids (grazers)	70.7 ± 2.6 ka (AP6)		Presence of people on the landscape in a modern-like environment with grass available: more humid?	Vogelsang et al. (2010)
	Relative dating	Namaqualand coast: Still Bay artefacts	~70 ka		Presence of people on the landscape	Dewar (2008)
	Relative dating	VR3: Still Bay artefacts	~70 ka		Presence of people on the landscape	Steele et al. (2012)
	Relative dating	STF001: Still Bay points	~70 ka		Presence of people on the landscape	Mackay et al. (2010)
	OSL date	Apollo 11	66.9 ± 2.6 (AP5)		Presence of people on the landscape	Vogelsang et al. (2010)
	Amino acid racemization date on eggshell	Erb Tanks	65 ka		Presence of people on the landscape	McCall et al. (2011)
	OSL date and fauna	Apollo 11: Howieson's Poort and arid adapted species + equids (grazers)	63.2 ± 2.3 ka (AP4)		Presence of people on a modern-like landscape but more humid with grass available?	Vogelsang et al. (2010)
	Amino acid racemization date on eggshell	Erb Tanks	60 ka		Presence of people on the landscape	McCall et al. (2011)
Southern Cape sea level	Strontium isotopes	Decreasing $^{87}Sr/^{86}Sr$ for a short period	~60 ka		Transgression	Fisher et al. (2010)

(continued)

Marine isotope stage 4: Glacial 71–57 ka	Data	Signal	Date	Calibrated dates at 1σ (for this study) cal BP[b]	Implication	References
Southern Cape sea level	Paleoscape model: Bathymetry and GIS[a]	Shoreline steadily returns to near modern values for a short period	~60 ka		Transgression	Fisher et al. (2010)
	Relative dating	Namaqualand coast: Howieson's Poort	~60 ka		Presence of people on the landscape	Dewar (2008)
	Relative dating and fauna	VR3: Howieson's Poort and browsing tortoises dominate	~60 ka		Presence of people on an arid landscape with browse available	Steele et al. (2012)
	Charcoal	Apollo 11: Diverse array of woody vegetation similar to modern environment or more favorable	58 ± 3 ka (AP9)		Semiarid and or slightly more humid	Vogelsang et al. (2010)
Marine isotope stage 3: Stadial 57–29 ka	**Data**	**Signal**	**Date**	**Calibrated dates at 1σ cal BP (for this study)[b]**	**Implication**	**References**
	OSL and Charcoal	Apollo 11: Single species of Xeric taxa *Chenopodiaceae* during occupation hiatus	Between 57.9 ± 2.6 ka (AP3) and 42.9 ± 2.7 ka (AP2)		Arid	Vogelsang et al. (2010)
	Pollen[a]	Decreasing but fluctuating levels of *Restionaceae* and desert/semi-desert species	~57–50 ka		Drying?	Shi et al. (2001)
Southern Cape sea level	Paleoscape model: Bathymetry and GIS[a]	Coastline shifts to ~18 km from the modern shore	~57–40 ka		Slight regression	Fisher et al. (2010)
Southern Cape sea level	Strontium isotopes	Slight decrease in $^{87}Sr/^{86}Sr$ ratios	~55 ka		Slight transgression	Fisher et al. (2010)
	Pollen[a]	Increasing proportion of *Restionaceae* and desert/semi-desert taxa with a peak at 32 ka	50–29 ka		Increasing humidity	Shi et al. (2001)
Southern Cape sea level		Steady increase in $^{87}Sr/^{86}Sr$	~55–27 ka		Regression	Fisher et al. (2010)
	Terrigenous sediments[a]	Rapidly fluctuating aeolian dust input and trade winds			Instability?	Stuut et al. (2002)
	^{14}C on Ostrich eggshell, fauna and gypsum	SptzA arid adapted species, primarily browsers but Gemsbok suggests some grass. Gypsum crystals present	~52–51 ka		People on the landscape in a (semiarid) modern-like environment but with increased humidity	Dewar and Stewart (2012, this volume): Table 11.1
Eastern Cape sea level	^{14}C dated Wetland Peats	−52 m amsl lowstand	45,200 ± 2,000 ^{14}C BP (Pta-4140)	49,968–47,070	Regression	Ramsey and Cooper (2002)
	^{14}C on ostrich eggshell and fauna	Boegoeberg1: Large hyenas and water-dependent grazing species	37,220 ± 5,010 ^{14}C BP (GX-22191)	45,433–36,161	Cool/humid with grass	Klein et al. (1999)
Eastern Cape sea level	^{14}C dated wetland Peats	−46 m amsl lowstand	39,100 ± 1,530 ^{14}C BP (Pta-4142)	44,286–41,875	Regression	Ramsey and Cooper (2002)
	Amino acid racemization dates on eggshell	Erb Tanks	45 ka		Presence of people on the landscape	McCall et al. (2011)
	OSL	Apollo 11	43 ± 3 ka (AP2)		Presence of people on the landscape	Vogelsang et al. (2010)
	^{14}C on ostrich eggshell and fauna	Boegoeberg1: Large hyenas and water-dependent grazing species	34,990 ± 3,110 ^{14}C BP (GX-21190)	42,061–36,101	Cool/humid with grass	Klein et al. (1999)
	^{14}C on ostrich eggshell and fauna	Boegoeberg1: Large hyenas and water-dependent grazing species	33,230 ± 2,630 ^{14}C BP (GX-21189)	40,118–34,755	Cool/humid with grass	Klein et al. (1999)

(continued)

Marine isotope stage 3: Stadial 57–29 ka	Data	Signal	Date	Calibrated dates at 1σ cal BP (for this study)[b]	Implication	References
Southern Cape sea level	Paleoscape model: Bathymetry and GIS[a]	Slight shift to ~10 km from modern shoreline	~40 ka		Shallow transgression	Fisher et al. (2010)
	Calibrated ^{14}C dates	Apollo 11 occupational pulses	~37 cal BP		Presence of people on the landscape	Vogelsang et al. (2010)
Southern Cape sea level	Paleoscape model: Bathymetry and GIS[a]	Coastline moves to ~25 km from modern shore	~32 ka		Slight regression	Fisher et al. (2010)
	Calibrated ^{14}C dates	Apollo 11 occupational pulses	~32–29 calBP		Presence of people on the landscape	Vogelsang et al. (2010)
	^{14}C dated peat bed	Peat bed at Kannikwa near Port Nolloth	27,900 ± 310 ^{14}C BP	31,998–31,269	High humidity	Beaumont (1986)
Orange River Mouth sea level	^{14}C marine shell	−78.4 m amsl lowstand	27,800 ± 440 ^{14}C BP (Pta-1104)	30,881–31,506[d]	Regression	Vogel and Visser (1981)
	OSL and fauna	Apollo 11: arid adapted species + equids and warthog (grazers)	30 ± 1.4 ka (AP11)		Landscape is slightly more humid than today. Some grass available?	Vogelsang et al. (2010)
Southern Cape sea level	Paleoscape model: Bathymetry and GIS[a]	Shore returns a few km to ~22 km from modern coast	~30 ka		Shallow transgression	Fisher et al. (2010)
Marine isotope stage 2: Last Glacial Maximum 29–14 ka	**Data**	**Signal**	**Date**	**Calibrated dates at 1σ cal BP (for this study)[b]**	**Implication**	**References**
Eastern Cape sea level	Durban Bay: ^{14}C dated Wetland Peats	−22 m amsl stand	24,950 ± 950 ^{14}C BP (GaK-1390)	29,949–27,997	Transgression	Ramsey and Cooper (2002)
	Pollen and charcoal	Elands Bay Cave: Woodland taxa peak and xeric taxa minimum + drought-intolerant species	20.5–17.8 ^{14}C kBP	~25,098–24,314 to 21,904–21,110[c]	High humidity	Meadows and Baxter (1999)
	Pollen[a]	Peak percentage of *Restionaceae*	~24 ka		High humidity	Shi et al. (2001)
	Terrigenous sediments and trade wind proxies[a]	Peak fluvial activity and trade winds	~24 ka		High humidity	Stuut et al. (2002)
	^{14}C Hyrax dung Pollen	*Olea*, *Stoebe* type, *Artemesia*, and fern pollen co-occurring with dwarf shrubs	17,000 ± 190 ^{14}C BP (Pta-8902)	20,739–20,248	Cool and moist or increased evapotranspiration	Scott et al. (2004)
Southern Cape sea level	^{14}C *Pecten sp.* shell	−130 m amsl maximum	16,990 ± 160 ^{14}C yrs BP (Pta-182)	20,044–19,605[d]	Last Glacial Maximum peak	Vogel and Marais (1971)
	Pollen[a]	*Restionaceae* percentages declining	~19–14 ka		Declining humidity	Shi et al. (2001)
	Terrigenous sediments and trade wind proxies[a]	Fluvial activity and trade wind curves declining	~19–14 ka		High humidity declining through time	Stuut et al. (2002)
Orange River Mouth sea level	^{14}C marine shell	−87.2 m amsl lowstand	16,100 ± 160 ^{14}C BP (Pta-1105)	18,982–18,611[d]	Regression	Vogel and Visser (1981)
	Pollen	Cederberg Mountains: increasing fynbos, thicket and succulent vegetation	13,000 ± 130 ^{14}C BP (Pta-5896) to 11,390 ± 100 ^{14}C BP (Pta-6041)	15,695–15,289 to 13,281–13,100	Increasing temperatures and reduced precipitation	Scott and Woodborne (2007a, b)
	Pollen	Eksteenfontein spring: *Stoebe/Elytopappus* indicate cool temperatures and increase of Karoo-like environment	15.2–13.6 calBP (extrapolated dates)		Cool and humid replaced by aridity	Scott et al. (1995)

[a]Dates were extrapolated using the oxygen isotope curve
[b]This study calibrated ^{14}C dates using the software Calib 7.0 and the Shcal13.14c calibration curve (Hogg et al. 2013; Reimer et al. 2013)
[c]No error provided, dates were calibrated with an approximate error of ±300 yrs
[d]Marine shell was calibrated using the calibration curve Marine13.14c with a ΔR value of 146 ± 85 (Dewar et al. 2012; Reimer et al. 2013)

References

Avery, G., Halkett, D., Orton, J., Steele, T., Tusenius, M., & Klein, R. (2008). The Ysterfontein 1 Middle Stone Age rock shelter and the evolution of coastal foraging. *South African Archaeological Society Goodwin Series, 10*, 66–89.

Beaumont, P. B. (1986). Where did all the young men go during ^{18}O stage 2? *Palaeoecology of Africa, 17*, 79–86.

Brain, C. K. (1981). *The hunters or the hunted?: An introduction to cave taphonomy*. Chicago: University of Chicago Press.

Bouzzouggar, A., Barton, N., Vanhaeren, M., d'Errico, F., Collcutt, S., Higham, T., et al. (2007). 82,000-year old shell beads from North Africa and implications for the origins of modern human behavior. *Proceedings of the National Academy of Sciences of the United States of America, 104*(24), 9964–9969.

Cann, R. L., Stoneking, M., & Wilson, A. C. (1987). Mitochondrial DNA and human evolution. *Nature, 325*(6099), 6–31.

Chappell, J., & Shackleton, N. (1986). Oxygen isotopes and sea level. *Nature, 324*, 137–140.

Chase, B., & Meadows, M. (2007). Late Quaternary dynamics of southern Africa's winter rainfall zone. *Earth-Science Reviews, 84*, 103–138.

Cooper, J., & Flores, R. (1991). Shoreline deposits and diagenesis from two Late Pleistocene highstands near +5 and +6 m, Durban, South Africa. *Marine Geology, 97*, 325–343.

Compton, J. S. (2011). Pleistocene sea-level fluctuations and human evolution on the southern coastal plain of South Africa. *Quaternary Science Reviews, 30*, 506–527.

Corvinus, G. (1983). *The raised beaches of the west coast of South West Africa/Namibia: An interpretation of their archaeological and paleontological data*. Munchen: Verlag C.H. Beck.

d'Errico, F., Henshilwood, C., Vanhaeren, M., & van Niekerk, K. (2005). *Nassarius kraussianus* shell beads from Blombos Cave: Evidence for symbolic behavior in the Middle Stone Age. *Journal of Human Evolution, 48*, 3–24.

Desmet, P. (2007). Namaqualand - a brief overview of the physical and floristic environment. *Journal of Arid Environments, 70*, 570–587.

Dewar, G. (2008). *The archaeology of the coastal desert of Namaqualand, South Africa: A regional synthesis* (BAR International Series No. 1761). Oxford: Archaeopress.

Dewar, G., & Orton, J. (2013). Subsistence, settlement, and material culture on the central Namaqualand coastline. In A. Jerardino, D. Braun, & A. Malan (Eds.), *The archaeology of the west coast of South Africa* (Cambridge Monographs in African Archaeology, BAR International Series No. 2526). Oxford: Archaeopress.

Dewar, G., & Stewart, B. (2012). Preliminary results of excavations at Spitzkloof Rockshelter, Richtersveld, South Africa. *Quaternary International, 270*, 30–39.

Dewar, G., Reimer, P. J., Sealy, J., & Woodborne, S. (2012). Holocene marine reservoir effect correction (ΔR) for the west coast of South Africa. *The Holocene, 22*(12), 1438–1446.

Dregne, H. (1976). *Soils of arid regions*. Amsterdam: Elsevier.

Eitel, B. (2005). Environmental history of the Namib Desert. In M. Smith & P. Hesse (Eds.), *23°S: Archaeology and environmental history of the southern deserts* (pp. 45–55). Canberra: National Museum of Australia.

Fischer, E., Bar-Matthews, M., Jerardino, A., & Marean, C. (2010). Middle and Late Pleistocene paleoscape modeling along the southern coast of South Africa. *Quaternary Science Reviews, 29*, 1382–1398.

Frimmel, H. (2003). Port Nolloth group including the Stinkfontein and Hilda subgroups. In M. R. Johnson (Ed.), *The catalogue of South African lithostratigraphic units* (pp. 35–37). Pretoria: South African Committee for Stratigraphy.

Halkett, D. (2002). *Phase 1 archaeological survey: Assessment of mining blocks in the BMC and KN areas, Namaqualand*. Unpublished report prepared for De Beers Consolidated Mines. Cape Town: Archaeology Contracts Office.

Halkett, D. (2003). *A report on the archaeological mitigation program at De Beers Namaqualand Mines: March 2002 to June 2003*. Unpublished report prepared for De Beers Consolidated Mines. Cape Town: Archaeology Contracts Office.

Halkett, D. (2006a). *Phase 1 archaeological assessment of mining targets in the BMC and KNC mining areas, Namaqualand, April/May 2006*. Unpublished report prepared for De Beers Consolidated Mines. Cape Town: Archaeology Contracts Office.

Halkett, D. (2006b). *Phase 1 archaeological assessment of mining targets in the BMC mining area, Namaqualand, July/August 2006*. Unpublished report prepared for De Beers Consolidated Mines. Cape Town: Archaeology Contracts Office.

Halkett, D., & Dewar, G. (2007). *Mitigation of archaeological sites within the Buffels Marine and Koingnaas Complexes, Namaqualand, July to October 2006*. Unpublished report prepared for De Beers Consolidated Mines. Cape Town: Archaeology Contracts Office.

Halkett, D., & Hart, T. (1997). *An archaeological assessment of the coastal strip, and a proposed heritage management plan for: De Beers Namaqualand Mines* (Vols. 1 and 2). Unpublished report prepared for De Beers Consolidated Mines. Cape Town: Archaeology Contracts Office.

Halkett, D., & Hart, T. (1998). *Phase 1 archaeological survey: Assessment of the proposed 1998 mining blocks on the Buffels Marine (BMC), Koingnaas (KN) and Buffels Inland (BIC) complexes*. Unpublished report prepared for De Beers Consolidated Mines. Cape Town: Archaeology Contracts Office.

Halkett, D., & Orton, J. (2004). *Phase 1 archaeological survey: Assessment of mining blocks and prospecting trenches in the BMC and KNC areas, Namaqualand*. Unpublished report prepared for De Beers Consolidated Mines. Cape Town: Archaeology Contracts Office.

Halkett, D., & Orton, J. (2005a). *Phase 1 archaeological assessment of mining targets in the BMC and KNC mining areas, Namaqualand, April and September 2005*. Unpublished report prepared for De Beers Consolidated Mines. University of Cape Town: Archaeology Contracts Office.

Halkett, D., & Orton, J. (2005b). *Phase 1 archaeological assessment of mining targets in the BMC and KNC mining areas, Namaqualand, November/December 2005*. Unpublished report prepared for De Beers Consolidated Mines. Cape Town: Archaeology Contracts Office.

Halkett, D., & Orton, J. (2007). *Phase 1 archaeological assessment of mining block extensions in the KNC mining area, Namaqualand, May 2007*. Unpublished report prepared for De Beers Consolidated Mines. Cape Town: Archaeology Contracts Office.

Hendey, Q., & Volman, T. (1986). Last interglacial sea levels and coastal caves in the Cape Province, South Africa. *Quaternary Research, 25*, 189–198.

Henshilwood, C. S., Sealy, J. C., Yates, R. J., Cruz-Uribe, K., Goldberg, P., Grine, F. E., et al. (2001). Blombos cave, southern Cape, South Africa: Preliminary report on the 1992–1999 excavations of the Middle Stone Age levels. *Journal of Archaeological Science, 28*(5), 421–444.

Henshilwood, C., d'Errico, F., Yates, R., Jacobs, Z., Tribolo, C., Duller, G. A. T., et al. (2002). Emergence of modern human behavior: Middle Stone Age engravings from South Africa. *Science, 295*, 1278–1280.

Henshilwood, C. S., d'Errico, F., Vanhaeren, M., van Niekerk, K., & Jacobs, Z. (2004). Middle Stone Age shell beads from South Africa. *Science, 384*, 404.

Hobday, D. (1976). Quaternary sedimentation and development of the lagoonal complex, Lake St Lucia, Zululand. *Annals of the South African Museum, 71*, 93–113.

Hogg, A. G., Hua, Q., Blackwell, P. G., Buck, C. E., Guilderson, T. P., Heaton, T. J., et al. (2013). SHCal13 southern hemisphere calibration, 0–50,000 years calBP. *Radiocarbon, 55*(4), 1889–1903.

Jacobs, Z., Roberts, R., Galbraith, R. F., Deacon, J., Grun, R., Mackay, A., et al. (2008). Ages for the Middle Stone Age of southern Africa: Implications for human behaviour and dispersal. *Science, 322*, 733–735.

Klein, R. G., Cruz-Uribe, K., Halkett, D., Hart, T., & Parkington, J. E. (1999). Paleoenvironmental and human behavioral implications of the Boegoeberg 1 Late Pleistocene hyena den, Northern Cape Province. *South Africa. Quaternary Research, 52*(3), 393–403.

Klein, R., Avery, G., Cruz-Uribe, K., Halkett, D., Parkington, J., Steele, T., et al. (2004). The Ysterfontein 1 Middle Stone Age site, South Africa, and early human exploitation of coastal resources. *Proceedings of the National Academy of Sciences of the United States of America, 10*(16), 5708–5715.

Kirst, G. J., Schneider, R. R., Muller, P. J., von Storch, I., & Wefer, G. (1999). Late Quaternary temperature variability in the Benguela Current System derived from alkenones. *Quaternary Research, 52* (1), 92–103.

Lancaster, N. (1984). Paleoenvironments in the Tsondab Valley, central Namib Desert. *Palaeoecology of Africa, 16*, 411–419.

Lancaster, N. (2002). How dry was dry?: Late Pleistocene palaeoclimates in the Namib Desert. *Quaternary Science Reviews, 21*(7), 769–782.

Mackay, A., & Welz, A. (2008). Engraved ochre from a Middle Stone Age context at Klein Kliphuis in the Western Cape of South Africa. *Journal of Archaeological Science, 35*, 1521–1532.

Mackay, A., Orton, J., Schwortz, S., & Steele, T. (2010). Soutfontein (SFT)-001: Preliminary report on an open-air site rich in bifacial points, southern Namaqualand, South Africa. *South African Archaeological Bulletin, 65*, 84–95.

Marean, C. (2010). Pinnacle Point Cave 13B (Western Cape Province, South Africa) in context: The Cape Floral kingdom, shellfish, and modern human origins. *Journal of Human Evolution*, *59*, 425–443.

Maud, R. (1968). Quaternary geomorphology and soil formation in coastal Natal. *Zeitschrift fur Geomorphologie, N.F Suppl, 7*, 155–199.

McCall, G., Marks, T., Thomas, J., Eller, M., Horn III, S., Horowitz, R., et al. (2011). Erb Tanks: Middle and Later Stone Age rockshelter in the central Namib Desert, western Namibia. *Paleoanthropology, 2011*, 398–421.

Meadows, M., & Baxter, A. (1999). Late Quaternary palaeoenvironments of the southwestern Cape, South Africa: A regional synthesis. *Quaternary International, 57*(58), 193–206.

Meehan, B. (1982). *Shell bed to shell midden*. Australian Institute of Aboriginal Studies, Atlantic Highlands, N.J. Canberra: Humanities Press.

Middleton, G. (2003). *Encyclopaedia of sediments and sedimentary rocks*. Dordrecht: Kluwer Academic Publishers.

McDougall, I., Brown, F., & Fleagle, J. (2008). Sapropels and the age of hominins Omo I and II, Kibish. *Ethiopia. Journal of Human Evolution, 55*(3), 409–420.

Mucina, L., Jürgens, N., Le Roux, A., Rutherford, M. C., Schmeidel, U., Esler, K., et al. (2006). Succulent Karoo biome. In L. Mucina & M. C. Rutherford (Eds.), *The vegetation of South Africa, Lesotho and Swaziland* (Strelitzia No. 19) (pp. 220–299). Pretoria: South African National Biodiversity Institute.

Orton, J. (2007). *Mitigation of archaeological sites within the Buffels Marine, Buffels Inland and Koingnaas Complexes, Namaqualand, August to September 2007*. Unpublished report prepared for De Beers Consolidated Mines. Cape Town: Archaeology Contracts Office.

Orton, J. (2008). A late Pleistocene microlithic Later Stone Age assemblage from coastal Namaqualand, South Africa. *Before Farming online version, 2008*(1), 1–9.

Orton, J. (2009). *Archaeological mitigation on erven 13 and 14, Hondeklipbaai, Namakwa Magisterial District, Northern Cape*. Unpublished report prepared for HKB Eiendomme BK. Cape Town: Archaeology Contracts Office.

Orton, J., & Halkett, D. (2005). *A report on the archaeological mitigation program at De Beers Namaqualand Mines, August to September 2004*. Unpublished report prepared for De Beers Consolidated Mines. Cape Town: Archaeology Contracts Office.

Orton, J., & Halkett, D. (2006). *Mitigation of archaeological sites within the Buffels Marine and Koingnaas Complexes, Namaqualand, September 2005 to May 2006*. Unpublished report prepared for De Beers Consolidated Mines. Cape Town: Archaeology Contracts Office.

Ramsay, P., & Cooper, J. (2002). Late Quaternary sea-level change in South Africa. *Quaternary Research, 57*, 82–90.

Reimer, P. J., Bard, E., Bayliss, A., Beck, J. W., Blackwell, P. G., Bronk Ramsey, C., et al. (2013). IntCal13 and Marine13 radiocarbon age calibration curves 0–50000 years calBP. *Radiocarbon, 55*, 1869–1887.

Robbins, L. H., Brook, G. A., Murphy, M. L., Ivester, A. H., & Campbell, A. C. (2016). The Kalahari during MIS 6-2 (190–12 ka): Archaeology, paleoenvironment and population dynamics. In S. C. Jones & B. A. Stewart (Eds.), *Africa from MIS 6-2: Population dynamics and paleoenvironments* (pp. 175–193). Dordrecht: Springer.

Roberts, D. (2003). *Age, genesis and significance of South African coastal belt silcretes*. Pretoria: Council for Geoscience.

Scott, L., Steenkamp, M., & Beaumont, P. (1995). Palaeoenvironmental conditions in South Africa at the Pleistocene-Holocene transition. *Quaternary Science Reviews, 14*(9), 937–947.

Scott, L., & Woodborne, S. (2007a). Pollen analysis and dating of late Quaternary faecal deposits (hyraceum) in the Cederberg, Western Cape, South Africa. *Review of Palaeobotany and Palynology, 144*, 123–134.

Scott, L., & Woodborne, S. (2007b). Vegetation history inferred from pollen in late Quaternary faecal deposits (hyraceum) in the Cape winter-rain region and its bearing on past climates in South Africa. *Quaternary Science Reviews, 26*, 941–953.

Scott, L., Marais, E., & Brook, G. A. (2004). Fossil hyrax dung and evidence of Late Pleistocene and Holocene vegetation types in the Namib Desert. *Journal of Quaternary Science, 19*, 829–832.

Scott, L., Neumann, F., Brook, G., Bousman, C. B., Norstrom, E., & Metwally, A. (2012). Terrestrial fossil-pollen evidence of climate change during the last 26 thousand years in southern Africa. *Quaternary Science Reviews, 32*, 100–118.

Shi, N., Schneider, R., Beug, H.-J., & Dupont, L. (2001). Southeast trade wind variations during the last 135 kyr: Evidence from pollen spectra in eastern South Atlantic sediments. *Earth and Planetary Science Letters, 187*, 311–321.

Skinner, J. D., & Chimimba, C. T. (2005). *The mammals of the southern African sub-region*. Cambridge: Cambridge University Press.

Steele, T., Mackay, A., Orton, J., & Schwortz, S. (2012). Varsche Rivier 003, a new Middle Stone Age site in southern Namaqualand, South Africa. *South African Archaeological Bulletin, 67*, 108–119.

Stewart, B. A., Dewar, G. I., Morley, M. W., Inglis, R., Wheeler, M., Jacobs, Z., et al. (2012). Afromontane foragers of the late Pleistocene: Site formation, chronology and occupational pulsing at Melikane Rockshelter, Lesotho. *Quaternary International, 270*, 40–60.

Stewart, B. A., Dewar, G., Morley, M. W., Parker, A., & Allott, L. (2016). Follow the Senqu: Maloti-Drakensberg paleoenvironments and implications for modern human dispersals into mountain systems. In S. C. Jones & B. A. Stewart (Eds.), *Africa from MIS*

6-2: Population dynamics and paleoenvironments (pp. 247–271). Dordrecht: Springer.

Stuut, J.-B., Maarten, A., Schneider, R., Weltje, G., Jansen, J. H., & Postma, G. (2002). A 300 kyr record of aridity and wind strength in southwestern Africa: Inferences from grain-size distributions of sediments on Walvis Ridge, SE Atlantic. *Marine Geology, 180*, 221–233.

Teller, J., & Lancaster, N. (1985). History of sediments at Khommabes, central Namib Desert. *Madoqua, 14*, 267–278.

Teller, J., & Lancaster, N. (1986). Lacustrine sediments at Narabeb in the central Namib Desert, Namibia. *Palaeogeography, Palaeoclimatology, Palaeoecology, 56*, 177–195.

Texier, P., Porraz, G., Parkington, J., Rigaud, J., Poggenpoel, C., Miller, C., et al. (2010). A Howiesons Poort tradition of engraving ostrich eggshell containers dated to 60,000 years ago at Diepkloof Rock Shelter, South Africa. *Proceedings of the National Academy of Sciences, 107*, 6180–6185.

Thackeray, J. (1977). *Towards a clarification of the extent to which environmental change contributed to terminal Pleistocene extinctions in southern Africa*. M.Sc Dissertation, University of Cape Town.

Thackeray, J. (1979). An analysis of faunal remains from archaeological sites in southern South West Africa (Namibia). *South African Archaeological Bulletin, 34*, 18–33.

Vogel, J., & Marais, M. (1971). Pretoria radiocarbon dates. *Radiocarbon, 13*, 378–394.

Vogel, J., & Visser, E. (1981). Pretoria radiocarbon dates II. *Radiocarbon, 23*, 43–80.

Vogel, J., Visser, E., & Fuls, A. (2001). Suitability of ostrich eggshell for radiocarbon dating. *Radiocarbon, 43*, 133–137.

Vogelsang, R., Richter, J., Jacobs, Z., Eichhorn, B., Linseele, V., & Roberts, R. (2010). New excavations of Middle Stone Age deposits at Apollo 11 rockshelter, Namibia: Stratigraphy, archaeology, chronology and past environments. *Journal of African Archaeology, 8*(2), 185–218.

Wadley, L. (2010). Were snares and traps used in the Middle Stone Age and does it matter? A review and a case study from Sibudu, South Africa. *Journal of Human Evolution, 58*, 179–192.

Wadley, L., Hodgkiss, T., & Grant, M. (2009). Implications for complex cognition from the hafting tools with compound adhesives in the Middle Stone Age, South Africa. *PNAS, 106*(24), 9590–9594.

Wadley, L., Sievers, C., Bamford, M., Goldberg, P., Berna, F., & Miller, C. (2011). Middle Stone Age bedding construction and settlement patterns at *Sibudu, South Africa. Science, 334*, 1388–1391.

Waelbroeck, C., Labeyrie, L., Michel, E., Duplessy, J. C., McManus, J. F., Lambeck, K., et al. (2002). Sea-level and deep water temperature changes derived from benthic foraminifera isotopic records. *Quaternary Science Reviews, 21*, 295–305.

Webley, L. (1992). The history and archaeology of pastoralism and hunter-gatherer settlement in the north-western Cape, South Africa. Ph.D. dissertation, University of Cape Town.

Webley, L. (2002). The re-excavation of Spoegrivier Cave on the west coast of South Africa. *Annals of the Eastern Cape Museums, 2*, 19–49.

Webley, L. (2009). *Archaeological impact assessment: Port Nolloth borrow pits, Richtersveld Municipality, Northern Cape*. Unpublished report prepared for Richtersveld Municipality. Cape Town: Archaeology Contracts Office.

Yellen, J., Brooks, A., Cornelissen, E., Mehlman, M., & Stewart, K. (1995). A Middle Stone Age worked bone industry from Katanda, Upper Semliki Valley, Zaire. *Science, 268*, 553–556.

Part III

Grasslands, Woodlands and Rainforests

Chapter 12
Human Evolution in Late Quaternary Eastern Africa

Marta Mirazón Lahr and Robert A. Foley

Abstract Eastern Africa (broadly Ethiopia, Somalia, South Sudan, Kenya, Uganda, and Tanzania) has yielded the earliest fossils of modern humans, the earliest evidence for Mode 3 technologies (Middle Stone Age), and is one of the areas in which modern humans may well have been endemic. This paper reviews the genetic, archaeological, and fossil evidence for the evolution of modern humans across MIS 6-2 in eastern Africa, and places this into the context of Middle Pleistocene human evolution, the development of the Middle Stone Age across the continent, and climatic change over the last two glacial cycles. We argue that while there is a paucity of well-dated sites that reduces the resolution of any interpretation, the available evidence indicates a major role for eastern Africa as an area of endemism, most probably related to the interaction of mosaic environments and refugia. We show that the evolution of modern humans has roots that extend well before MIS 6, and propose four overlapping stages, making this a much more prolonged process than has traditionally been described. There is a broad relationship between evolutionary history and major climatic oscillations; nevertheless, a closer examination reveals a more complex pattern. There are periods of synchrony and asynchrony in both contextual and evolutionary/behavioral changes, and these show variable links to both northern and southern Africa. Although eastern, northern and southern Africa (with central and western being largely unknown) show similarities and ultimately the same evolutionary and behavioral outcome, they also exhibit independent trajectories that require further research to throw light on the processes involved.

M. Mirazón Lahr (✉) · R.A. Foley
Leverhulme Centre for Human Evolutionary Studies, Department of Archaeology and Anthropology, University of Cambridge, The Henry Wellcome Building, Fitzwilliam Street, Cambridge, CB2 1QH, UK
e-mail: mbml1@cam.ac.uk

Keywords Anatomically modern humans • East African prehistory • Middle Stone Age • Paleoclimate • Marine Isotope Stages • African biogeography

Introduction

In Africa, the period MIS 6-2 (190–11 ka) coincides with the emergence of modern humans and the beginnings of diversity among populations of *Homo sapiens*. Evidence and theoretical considerations suggest eastern Africa played a central role in this process. This paper reviews the evidence for human evolution and prehistory in eastern Africa during MIS 6-2. That evidence comes from the hominin fossil record, from archaeological sites, and from inferences made from human genetics. Here we summarize that evidence, and link it to what the record suggests is happening in other parts of Africa at the time. While the chronological framework is MIS 6-2, we will also explore how events across this period may have their roots in earlier times. We outline the phases of evolutionary change seen in eastern Africa, and how these relate to major climatic changes. In the final section, we discuss how more detailed chronological and geographical analyses indicate that the relationships between climate change and hominin responses to that change across Africa are complex and varied.

Eastern Africa: Habitat Mosaics and Biogeographical Context

Africa is dominated by two major biomes – the tropical African rainforests and the Sahara desert. The rest of the continent can be described as a mosaic of woodlands, bushlands, grasslands, as well as special biomes such as coasts and mountains. If Africa is divided into a few general regions – north, south, central, etc., for the most part each of these regions is dominated by one habitat, or by a mosaic of two or

Sacha C. Jones and Brian A. Stewart (eds.), *Africa from MIS 6-2: Population Dynamics and Paleoenvironments*, Vertebrate Paleobiology and Paleoanthropology, DOI 10.1007/978-94-017-7520-5_12

three. Eastern Africa is very different. Within a relatively small area – approximately 7 million square km – virtually every habitat can be found. Tropical rainforests occur in the western part, and in small pockets in other areas; deserts occur in the Horn of Africa and northern Kenya. Much of the rest is made up of a mosaic of savannah types – woodlands, bush, grasslands, and so on. In addition, the rifting process has created a whole series of very localized habitats, ranging from montane regions with alpine vegetation to high escarpments with high-relief topography, to lake basins (Kingdon 1989). These are perhaps the most significant, as they not only provide a source of water, but also a focus for large mammal activity, high biomass, aquatic resources, and a large hinterland of fluvial habitats, with often rich gallery forests. The lakes are highly variable, from ones as small as Elementeita (~4 km long) to those as large as Turkana (290 km long).

A consideration of the role of East Africa in recent human evolution is therefore not about a dominant habitat, but about the mosaic of habitats that exists at a geographical scale that is integral to human adaptation. For example, a 100 km radius drawn around Lake Nakuru (a reasonable and possibly conservative estimate of the area a hunter-gatherer community would know, even if it is about twice or three times their estimated annual home range area [Kelly 1995]) would embrace two other lake basins (Baringo and Naivasha), the foothills of mount Kenya, and the highlands to the east and west of the Rift Valley. This habitat diversity has created one of the world's biodiversity hot spots (Jenkins et al. 2013).

Furthermore, it is not just a question of habitats, but also the relationship of these to the rest of Africa. Evolutionary patterns are strongly influenced by overall biogeography – barriers and isolation, corridors and contact zones. On the one hand, East Africa has points of contact with all the major regions of Africa: the Sahara and Sahel, the tropical forests to the west along the western Rift Valley, and southern Africa through the Rift Valley. On the other, its high relief and habitat heterogeneity also creates the conditions for local adaptation and isolation, which seems to have created unique localized responses to global climatic processes in the past (Mithen 1996; Lahr 2012b).

In this paper, therefore, we consider the role of East Africa as a biogeographical region characterized by mosaic habitats, rather than as a single biome.

Prolonged Origins: Current Perspectives on the Chronology and Geography of Modern Human Evolution

The supposed "point of origin" for modern humans has moved around considerably since the emergence of the "out of Africa" paradigm. The original mitochondrial DNA (mtDNA)-based model (Cann et al. 1987) placed the origins at ~150 ka. Subsequent developments that integrated the mtDNA model with anatomical and archaeological evidence compressed all significant events to the same time horizon. This collapsed event horizon – genetic origin, behavioral and anatomical modernity, and dispersals – became the dominant interpretation. With changing dates in all fields, the timing of the horizon could vary considerably – Klein (1992), for example, considered it to be late (~50 ka), while Mithen (1997) placed it much earlier (~100 ka).

With time, it became clear that there was no single horizon that encompassed all the significant evolutionary changes that occurred with the appearance of modern humans. Genetic coalescence dates have tended to lie between 200 and 150 ka for mtDNA (female lineages), and between 150 and 100 ka for the Y chromosome (Underhill and Kivisild 2007), although Poznik et al. (2013) have recently proposed a much more uniform coalescence, between ~100 and ~150 ka. Genetic estimates for the divergences of populations in Africa have settled to between MIS 6 and MIS 4 (150–70 ka) (Tishkoff et al. 2007), while an important population expansion may be inferred to have taken place 70–60 ka (Behar et al. 2008), of which the "out of Africa" dispersal was part. Anatomical evidence is equally spread over a long period of time. The earliest specimens that show morphological features only observed among *H. sapiens*, and which can therefore be considered within the modern human lineage, are between 200 and 150 ka (White et al. 2003; McDougall et al. 2005; Fleagle et al. 2008). Although these specimens are anatomically modern humans, they do not match any later regional morphologies. The fossil record of humans in the Late Pleistocene (130–11 ka) of Africa is too incomplete to draw definitive patterns, but the scant data available suggest regionalization of morphological patterns took place much later. In that context, the evolutionary history of many of the world's populations today is comparatively shallow and often Holocene (Lahr 1996).

While anatomical modernity and adaptation is chronologically dispersed, until recently the archaeological situation appeared more compressed, with a strong signal occurring between 50 and 40 ka (Mellars 1991, 2006; Klein 1992). This underpinned the idea of a "human revolution." However, subsequent research showed that this was largely driven by dating limitations and a European bias in the data. Discoveries in Africa have extended the time range of significant behavioral innovations to the period between at least 100 and 70 ka (Grotte des Pigeons [Bouzouggar et al. 2007; d'Errico et al. 2009], Blombos [Henshilwood et al. 2011]), with strong claims for significant changes by 160 ka (Pinnacle Point [Marean et al. 2007]) and even suggestions of behaviors as old as 300–250 ka which presage later human traditions (Barham 1998, 2002; Tryon and McBrearty 2002; Shea 2011).

The outcome is that the "origin" of modern humans occurs across a period of at least a quarter of a million years (Foley and Lahr 1997; McBrearty and Brooks 2000). This "extended chronology" provides an important context for this paper. Where, in the early days of the "out of Africa" paradigm, the problem was to find a suitable context and explanation for major events within a short period of time, the challenge now is to place the events into a cumulative sequence of changing ecological, demographic, and technological conditions.

African Regionality: Targeting the Location of Human Origins

Equally important is the issue of the location of human origins. While it is widely accepted that the origins of modern humans lie in Africa, this conclusion is perhaps not as significant as it at first appears. Africa comprises about half the habitable Old World, so in effect we know in which half of the world humans evolved. There has been considerable effort to pin down human origins further. On the basis of fossil evidence, East Africa has the strongest case for human phenotypic evolution (White et al. 2003; Fleagle et al. 2008). Genetic evidence originally also favored an East African origin (Ramachandran et al. 2005), or more broadly a tropical African one (Patin et al. 2009). However, more recent analyses, using much richer data sets with 10^5 SNPS, have either indicated a complex origin (Schuster et al. 2010) or a southern African one (Henn et al. 2011). However, the genetic evidence for a southern African origin of modern humans relies heavily on the current location of one of the few surviving hunter-gatherer populations of the continent, while presumably peoples with equally or similarly distinctive genetic make-up lived across Africa until recently and would have suggested a different spatial distribution.

The archaeological evidence is also equivocal. On the one hand, the earliest evidence for Mode 3 technologies, associated (though not uniquely) with modern human behavior, is found penecontemporaneously across much of Africa at ~270 ka (Morgan and Renne 2008), making a local origin difficult to establish. On the other hand, significant changes in both technology (bone points, hafting, microliths, fishing) and material culture (beads, use of pigments) appear as singular local traditions after 130 ka in southern and northern Africa (Bouzouggar et al. 2007; Marean et al. 2007; Brown et al. 2009; Henshilwood et al. 2011). The latter would suggest localized processes, particularly in southern Africa, but they also point to the possibility that such "advanced" markers did not persist and were short-term occurrences (Jacobs et al. 2008; Porraz et al. 2013), and therefore are not markers of species-level evolutionary processes. To complicate matters further, the relationship between the changes in the later Middle Stone Age (MSA) and the origins of the Later Stone Age (LSA) remain unclear. If the latter were taken as an important marker of fully modern behaviors and economies, then the earliest evidence now comes from Mumba Cave in East Africa (Gliganic et al. 2012).

Each line of evidence is subject to problems: the fossil evidence is very sparse, the archaeological evidence is multifaceted and open to a number of evolutionary interpretations, while the genetic evidence is subject to the overall critiques of phylogeography, namely that it assumes little geographic movement in populations over tens of thousands of years, as well as poor chronological parameters, and therefore makes inferences from the current distribution of peoples that ignore both major migrations and extinction.

Theoretical arguments have also been put forward for locating modern human origins in Africa in general (Foley 1989), East Africa (Lahr 2012b), South Africa (Marean 2011; Compton 2011), and in combination with other areas (Conard 2008). The case for East Africa is based on its hinge-like distribution in relation to other parts of the continent, its rich ecology, and its topographic complexity, which allows for refugia and refugial networks (Basell 2008; Lahr 2012b). The case for South Africa relates to the use of coastal resources and the high diversity and density of geophyte flora, and the role these played in providing a stable resource base during MIS 6. Current evidence would therefore suggest a model of human origins in which hominin populations throughout the continent were subject to very variable conditions across the MIS 6-2 period – and probably substantially before, and responded both demographically – dispersals and contractions – and adaptively to these, resulting in a complex series of changes. The sum of these describes the evolution of modern humans.

The precise location of human origins is an important issue for a number of reasons. Among these, the most important is toward the understanding of the factors that led to the evolution of certain traits instead of others. The location of the ancestral population will specify the conditions under which the human phenotype was initially adaptive, and so was favored, prior to its dispersal to a wider range of environments. Was this a particular habitat – coastal South Africa, for example – or a combination of habitats such as is implied by the mosaic hypothesis? Which element of the phenotype or associated behaviors was first linked to these conditions? In addition, it may be that the precise ecology is less important than biogeographical relationships – whether the origins lie in the network of refugia in East Africa (Lahr 2012b), or in a relatively homogenous but large region such as the Cape Province. This might have set the demographic conditions for isolation, drift, or selection that gave rise to humans. Finally, evolutionary theory would suggest that new species arise in relatively small parts of the parent species range, and so a full account of the evolutionary processes requires knowledge of the geographical range of the earliest modern humans and its immediate ancestors.

Here we focus specifically on the evidence for East Africa, and whether this region – or part of it – fulfills any of the necessary conditions.

The Evidence for Modern Human Evolution in East Africa

Genetic Evidence

There is overwhelming evidence that living humans are descended from an African population of hominins at some point toward the end of the Middle Pleistocene (Tishkoff et al. 2009). The chronology of our African genetic history is that a founder population lived 200–150 ka; that some African populations today (Khoisan, Hadza, Sandawe and possibly Central African pygmies) have broadly identifiable evolutionary histories as populations stretching back to >100 ka (Gonder et al. 2007; Tishkoff et al. 2007; Quintana-Murci et al. 2008); and that there was a demographic expansion of certain groups both across Africa and beyond after 70 ka.

Of greater interest here, however, is the evidence for geographical structure within the African populations of the last 150 ka, with this providing insights not just into when modern humans evolved, but also where.

The strongest case for a more specific place of origins has been made for South Africa. The basis for this is an analysis of 580,000 SNPs, which showed lower linkage disequilibrium, higher Fst, and a more distinct ancestral genetic structure among Khoisan than East African hunter-gatherers (Hadza and Sandawe) (Henn et al. 2011). On these grounds it has been argued that the Khoisan are closer to the earliest modern human populations, and their current geography thus represents a possible point of origin. However, this assumes that the Khoisan and their ancestral lineage have remained in the same place for more than 100 ka. In addition, the East African hunter-gatherer populations studied are two very small and isolated ones, both of which exhibit a recent and ongoing population bottleneck, while range contraction and extinction across much of Africa, especially East Africa, may have erased geographical signals elsewhere. What the South African evidence shows is that the hunter-gatherer populations there have admixed to a lesser extent than the few other groups that still live by hunting and gathering in Africa, and thus retain a more private and deeper genetic signature than any other.

The problem with using genetics to identify precise geographical locations derives from the general difficulties of phylogeography (Nielsen and Beaumont 2009). High mobility, low and fluctuating population sizes, and extinction of intermediate groups will reduce the geographical resolution of genetic data. It could be that there has been too much change – and that is certainly the case in eastern Africa – to locate reliably spatial origins. Furthermore, it is not necessarily the case that identifying the probable regions of gene coalescence is the same as locating human origins. For this, the fossil and archaeological records remain the primary source of information.

Fossil Evidence

We can consider the fossil hominin evidence under three headings – the East African evidence for modern humans, that for modern humans elsewhere in Africa, and the longer term context for these (Fig. 12.1).

East African Modern Humans

The earliest known modern humans have been found in East Africa. Before describing them, it should be stressed that "anatomically modern human" is an informal term, but largely conforms to being a member of *Homo sapiens*. Classically, this is defined in terms of a number of unique derived traits; these reflect the increasingly globular shape of the braincase (compared to the longer and lower shape seen in archaic hominins), and reduction of face size and its projection. The key traits are parietals that are more curved, which together with increased flexion of the cranial base increase the height of the cranium, a curved and vertically oriented frontal, non-continuous supraorbital torus that may be significantly reduced, a pronounced canine fossa, and a chin (mental eminence); there is also a general gracility to the skeleton (Lahr 1996).

Most of these features, however, are continuous, and taken in isolation they lack threshold parameters that define a modern shape from an archaic one. Thus, while supraorbital tori are usually small, in some individuals they may be large, and so on. However, two key traits are observed in all living and recent humans while being absent in all known archaic hominins. These are the shape of the cranial vault as determined by its relative height, and a chin. Thus the identification of these two features together for the first time in the fossil record is critical to recognizing the beginnings of the lineage. These traits are first observed in the Ethiopian fossils of Omo Kibish (Day and Stringer 1991b) and Herto (White et al. 2003), even if in other aspects of their

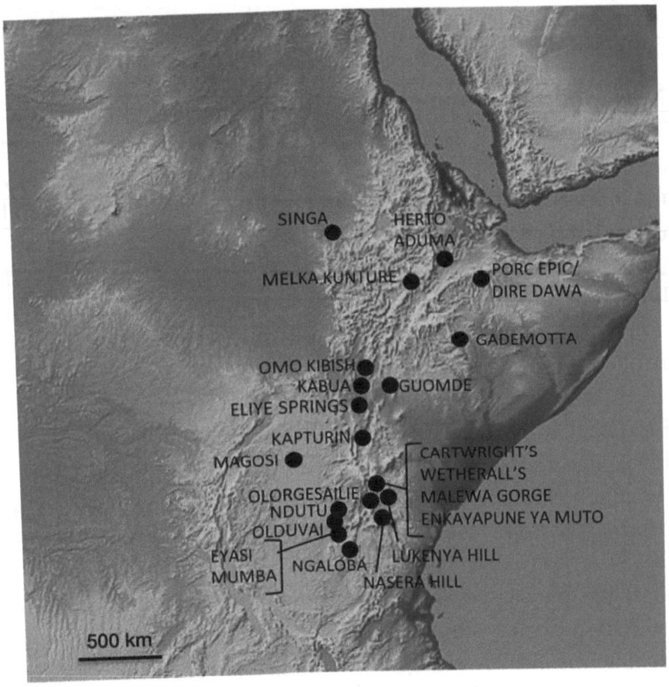

Fig. 12.1 Map of eastern Africa, showing distribution of primary paleoanthropological sites relevant to the evolution of humans during the later Quaternary

morphology they may appear outside the range observed in living populations.

Omo Kibish
The two fossils from Omo Kibish, Omo 1 and 2 are dated to 194 ka through $^{40}Ar/^{39}Ar$ and the geomorphological association of sediments with the Mediterranean sapropel record (McDougall et al. 2005; Fleagle et al. 2008). Found by Richard Leakey in 1967, their significance has not always been appreciated (Leakey 1969). In Omo 1, the occipital is strikingly modern (rounded), the parietals have their maximum breadth relatively high, there is only a weak supraorbital torus, and a chin is present. Omo 2, on the other hand, has a more angled occipital, the maximum breadth of the skull is lower, but the calvarium is high and arched (Day and Stringer 1991a). Most scholars agree that Omo 1 has a modern human morphology, while the affinity of Omo 2 remains unclear.

Herto
From the site of Herto there are three specimens, including a very complete adult cranium (BOU-VP-16/1). The latter has a number of derived features which place it within anatomically modern humans (height of the vault, supraorbital morphology, canine fossa, and chin), but it is also remarkably robust and large, with certain features of the occipital, palate breadth and mastoids that are outside recent human variation. Herto is dated to 160–154 ka ($^{40}Ar/^{39}Ar$) (White et al. 2003).

Singa
The Singa hominin comes from Sudan, and is dated by ESR and U/Th to at least 133 ka. It is relatively complete, and its morphology has been considered either modern or a mixture of archaic and modern traits (e.g., curved frontal bone, high cranial vault, broad parietals with marked bossing, and a very thick temporal bone). Some of these features are remarkable in relation to recent morphological patterns, such as the cranial bone thickness, and have been considered to be pathological. However, a reasonable interpretation is that Singa belongs to a population, similar to the slightly earlier Herto, that was part of the evolving lineage of *Homo sapiens* as shown by the presence of the key unique derived modern traits, but whose range of variation was different from that observed today (Grün and Stringer 1991; McDermott et al. 1996; Spoor et al. 1998).

Mumba
Mumba is a rock shelter in Tanzania, with a long stratigraphy that shows an early transition from the MSA to the LSA (Mehlman 1991). Only a few isolated hominin teeth were found, U-series dated to 130–109 ka, and their morphology is consistent with *Homo sapiens* (Bräuer and Mehlman 1988).

Aduma
Four incomplete crania were recovered from Aduma, Ethiopia. ADU-VP-1/3 is a vault with similar occipital morphology and proportions to Omo 1, LH18 and Skhūl 5, but with very distinctive sagittal keeling with parasagittal depressions not observed in the earlier crania (Haile-Selassie et al. 2003). Interestingly, this fossil shows a suprainiac fossa, which although not completely absent in modern human populations, is observed at high frequency only among Neanderthals. The site has been dated to late MIS 5 (105–79 ka) through various methods ($^{40}Ar/^{39}Ar$, U-series, OSL, TL).

Dire Dawa
A fragment of human mandible was excavated from the site of Porc-Epic (Dire Dawa) in 1934. Vallois (1951) described this as having "Neanderthaloid" characteristics. It has not been described subsequently, so it is difficult to establish its characteristics, but the Neanderthaloid attribution was framed in a now obsolete model, and it is more likely that this is a modern human with perhaps some archaic features. The mandible was found in association with MSA artifacts dated by obsidian hydration to 77–60 ka (Clark et al. 1984b).

Garba III, Melka Kunture
Three cranial fragments were recovered from contexts most probably dating to MIS 5e or earlier. Although they are well provenanced, they are not sufficiently complete to allow discrimination between a modern or more archaic form of *H. sapiens* (Chavaillon et al. 1987; Mussi et al. 2014).

The fossil evidence indicates that modern humans appear earliest in eastern Africa (200–160 ka). While that is significant, it should also be noted that the evidence is extremely scant and temporally widespread. We are thus left with the strong indication of a population that shares unique anatomical features with modern humans, but little evidence for the range of variation of which this might be part.

Modern Humans Across Africa

There is a considerable chronological break between the first East African appearance of modern humans and their presence in other parts of the continent; in fact, the only three other fossil remains dated to MIS 6 in Africa are from Morocco (Jebel Irhoud and Kébibat) (Hublin 2001), whose affinities to modern humans remain a subject of study, and the fossil from Ngaloba (LH18), Tanzania. The latter comes from Laetoli, Tanzania, and has been dated to MIS 6 on the basis of U-series and fauna and archaeological associations. It consists of a relatively complete cranium, with a combination of derived and archaic features (e.g., thick supraorbital tori, which are nevertheless clearly separated into medial and lateral sections, overall cranial breadth, and a curved occipital) (Day et al. 1980; Magori and Day 1983; Rightmire 1988), which may indicate either a morphologically very variable early human population in eastern Africa, or the existence of several distinctive groups within the region at the time.

However, what is striking is that when modern humans become more frequent in MIS 5, they are extremely widespread. Besides the fossils of Aduma, Dire Dawa and Mumba, *Homo sapiens* fossils dating to MIS 5-4 have been found in North Africa (Mugharet el'Aliya, Dar-es Soltan 2, El Harhoura 1, Témara, Haua Fteah, Taramsa Hill) and South Africa (Klasies River Mouth, Border Cave, Blombos Cave, Die Kelders Cave 1, Equus Cave), although the latter are mainly teeth (see Lahr [2012a] for a summary table and sources). To this should be added the remains from Skhūl and Qafzeh (Schwarcz et al. 1988; Stringer et al. 1989) from the Levant, which although technically not part of Africa, is nonetheless its biogeographical extension during MIS 5. The importance of this distribution is that while there may be reasons for seeing modern humans as regional in MIS 6, the same cannot be said for MIS 5, although the considerable range inherent in most of the dates of these fossils does not allow us to establish the tempo and mode of the demographic changes that may lie behind this.

At face value, the fossil record suggests that modern humans are not only ubiquitous after MIS 5, but also the only survivors on the continent. However, in the context of Africa representing half of the habitable world at the time, direct evidence remains extremely rare, and often altogether absent – for example, there are no East African specimens unambiguously dated to MIS 4 and MIS 3, only a handful in northern and southern Africa. A recent analysis of the 11 ka fossil from Iwo Eleru, Nigeria (Harvati et al. 2011), and the new late Pleistocene/early Holocene human remains from West Turkana (http://in-africa.org) suggest that at least some archaic traits persist in Africa until the recent past.

Non-Modern Humans in the Middle Pleistocene of Africa

The earliest modern humans in eastern Africa should be set against a background of other hominins, with whom they must share relationships, although not necessarily ancestral (Fig. 12.2). The evidence is remarkably sparse – 15 specimens of very variable completeness to represent 500 ka of human evolution in Africa. These hominins are known from both eastern and southern Africa from 700 to 600 ka (Bodo, Broken Hill [Kabwe], Elandsfontein), and persist in both areas until possibly as late as MIS 8-7 (Hoedjiespunt and Florisbad in southern Africa, Eliye Springs and Guomde in East Africa). These remains can broadly be grouped within the taxon *Homo heidelbergensis* (*H. rhodesiensis* to some), a robust hominin with an enlarged cranial capacity (Lahr 2013).

It has been argued that the Middle Pleistocene hominins of Africa display considerable chronological trends, in particular, the further enlargement of the brain (Bräuer 1992). To some this is simply the time-transgressive nature of *H. heidelbergensis* (Rightmire 1998), but others have recognized a further taxon, *H. helmei* (Lahr and Foley 2001). This disputed taxon (McBrearty and Brooks 2000) comprises fossils from South Africa from ~250 ka (Florisbad), eastern Africa from 300–200 ka (Eliye Springs, Guomde, Ngaloba), and, in our view, northern Africa ~200 ka (Jebel Irhoud) (see also Lahr [2012a] for summary tables and sources).

While there is general consensus about the broad pattern of human evolution in the Middle Pleistocene of Africa – an evolving lineage with a directional trend toward larger brains, and ultimately a modern human morphology, and with a broad Afro-Eurasian distribution in the earlier phases – there is less consensus about the phylogenetic relationships. To some, there is continuity across the spatial and temporal range of post-*Homo erectus* hominins until the evolution of modern humans and Neanderthals in Africa and Europe ~200 ka, and so only three recognized taxa (*H. heidelbergensis, H. sapiens*, and *H. neanderthalensis*) (Rightmire 2009; Endicott et al. 2010). Others would identify an early divergence of the archaic forms, with distinctive African and European lineages (i.e., dividing *H. heidelbergensis* [Europe] from *H. rhodesiensis* [Africa] [Hublin 2009]), and so recognize another taxon. We have argued for a third scenario, in which the relationship between African and Eurasian hominin lineages in the Middle Pleistocene is punctuated by a discrete number of dispersal events. This model argues for neither continuous contact nor isolation, but for distinct episodes of genetic and cultural exchange between hominins in Africa and Europe, so that the dispersal of *H. heidelbergensis* at ~700 ka would be separate from that of the last common ancestor (*H. helmei*) between Neanderthals and

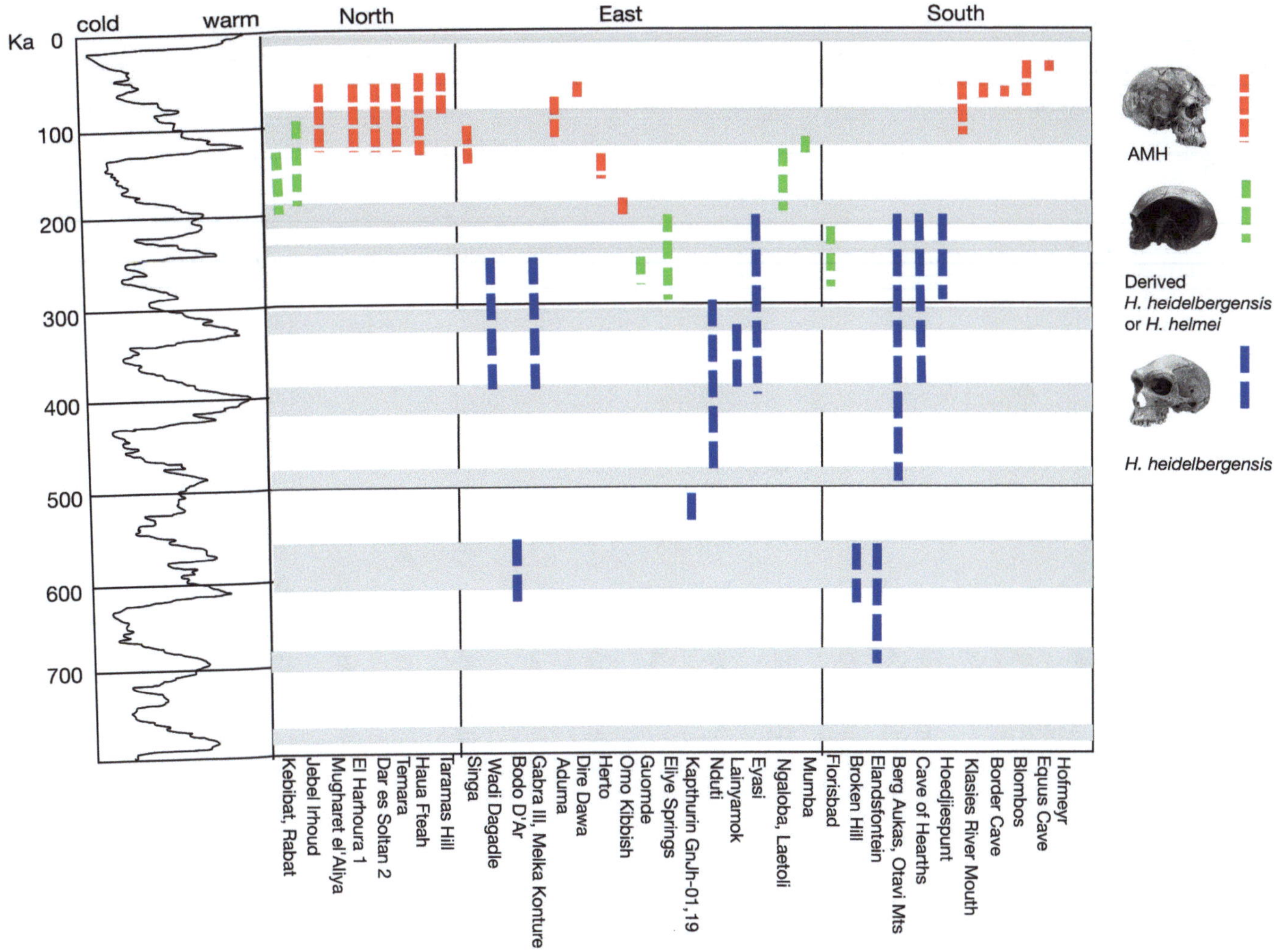

Fig. 12.2 Chronological and geographical distribution of fossil hominins in Africa during the Middle and Late Pleistocene. Three types of hominin are recognized – *H. heidelbergensis* (sometimes distinguished from the European form as *H. rhodesiensis*); an advanced form of *H. heidelbergensis* with larger brain sizes (sometimes known as *H. helmei*); and *H. sapiens* or anatomically modern humans. All lineages are widespread across Africa. The $\delta^{18}O$ curve is shown, with warmer periods to the *right*. *Shaded bands* indicate warmer marine isotope stages. See text for discussion and sources

modern humans 400–300 ka (Foley and Lahr 1997; Lahr and Foley 1998). Apart from differences in names, the key difference lies in the number of significant evolutionary divergences in the evolution of *Homo* in the Middle Pleistocene.

Archaeological Evidence

It is always the case that the durability of stone makes the archaeological record richer than the fossil record, and this is also the case with the evolution of modern humans in eastern Africa. However, it must still be stressed that there are very few well-dated sites (Fig. 12.3) (Herries 2011; Barham 2012; Morgan and Renne 2008). While the Middle Stone Age is well known for its extensive surface distributions, dating MIS 6-2 sites remains a problem.

Eastern Africa

MIS 6

Securely dated MIS 6 sites in East Africa are scarce. On the MIS 6/7 boundary (195 ± 5 ka) is Omo Kibish, already discussed on account of its evidence for the earliest modern humans. Although not directly associated with the fossils, the lithics from both excavated and surface collections are MSA, with Levallois and discoid reduction strategies, and a number of bifacial pieces that vary from foliates, to picks to handaxes. The assemblage suggests a typical MSA/Mode 3 industry, with the characteristic African signal of foliate points on a prepared core background (Shea et al. 2002). Further north in Ethiopia, well-dated sites at Gademotta and Kulkuletti also provide evidence for MSA in MIS 6. Again, bifacial and unifacial points, side scrapers, convergent scrapers, notched pieces and denticulates, on a largely

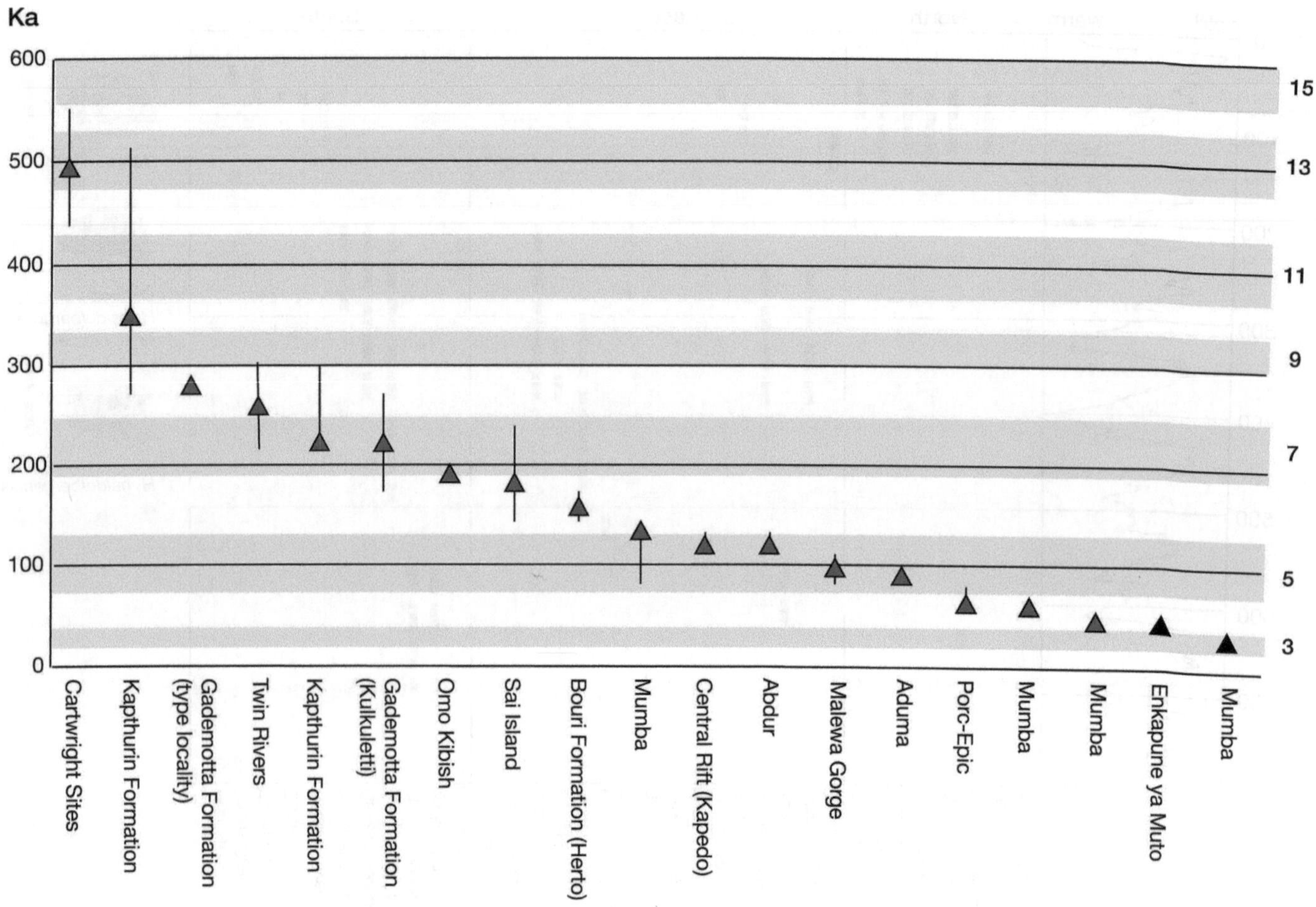

Fig. 12.3 Dated sites in eastern and eastern central Africa from the MSA and earliest MSA. Several sites were not included as the dates are likely to be misleading, especially older radiocarbon dates giving minimum ages (e.g., Lukenya Hill), or obsidian hydration dates (Prospect Farm), which are also likely to be underestimates. Rusinga has dates using modern techniques, but provides only an indication of MSA assemblages older than MIS 2. *Gray triangles* indicate MSA localities, and *black triangles* indicate LSA localities. See text for discussion and sources. *Shaded bands* indicate warmer marine isotope stages (MIS 1-13)

Levallois background, predominate among the tools, although there are other elements more associated with Mode 4 technologies (Morgan and Renne 2008). Also in Ethiopia, the site of Herto (again, with an anatomically modern human in the vicinity of the assemblages) has a MSA record (Clark et al. 1984a). Here the assemblages are surface finds, and include both MSA lithics and Acheulean handaxes. These dates indicate either the coexistence of Acheulean and MSA traditions as late as 160–154 ka, or the existence of "transitional" or multicomponent industries.

MIS 6 archaeological localities probably occur much more widely, from the Red Sea coast to Eyasi, but have seldom been securely dated. The existence of MSA much earlier (back to MIS 8) at sites such as Kapthurin (Tryon and McBrearty 2002) and Twin Rivers (Barham and Smart 1996) makes it probable that ~100 ka later, the techno-complex was firmly established, but questions remain about the timing of the disappearance of Mode 2 industries.

MIS 5

MIS 5 is a period lasting ~ 60 kyr, usually subdivided into a series of substages (a–e), which comprise three warm phases separated by two cooler intervals. Climatic conditions across this period were spatially, as well as temporally variable, although the quantity and quality of sites does not allow that to be taken into account at present.

In northern and southern Africa, MIS 5 is characterized by a noticeable increase in archaeological sites, and this is echoed in the earliest evidence for hominin complex behaviors across this period (Clark 1988; Brown et al. 2012; d'Errico et al. 2012). Surprisingly, however, there is an almost complete dearth of information in East Africa. This could be due to a difficulty in securely dating this period (virtually impossible until the development of OSL), and that the very large number of undated Late Pleistocene sites in East Africa include large numbers of MIS 5 records. Alternatively, it is possible that MIS 5 is largely absent or erased in the region's sedimentary history. However, we cannot exclude the possibility that the dearth of sites reflects a period of very low hominin population densities. Given the relevance of MIS 5 to models of early modern evolution and diversification, this would be a challenging observation.

Tryon et al. (2008) recently reported on a series of localities from Kapedo, located between the Baringo Basin and the Suguta Valley in northern Kenya. The site has been dated to 135–123 ka. The lithics, both surface and excavated, are MSA, and indicate a diverse set of reduction strategies, both Levallois and non-Levallois, and few formal tools. One possible interpretation is that the eastern African MSA at this time lacks the derived, diagnostic features of later assemblages from southern (microlithization, etc.) and northern (tanging, etc.) Africa.

Also of early MIS 5 age is the MSA industry from Bed VI at Mumba and the Eyasi lake sediments, Tanzania (Prendergast and Luque 2007). An outstanding aspect of these assemblages is the suggestion of very wide foraging/trade networks as revealed by the sourcing of obsidian to the Naivasha basin (Merrick et al. 1994), while its techno-typological idiosyncrasies (high frequency of high and notched/concave scrapers, bifacially retouched pieces, heavy-duty tools, and rare use of Levallois methods) have led to its identification as a regional MSA variant (the Sanzako industry) (Mehlman 1989; Mabulla 2007).

At the other end of MIS 5 is the site of Aduma, in the Middle Awash, Ethiopia. With a date between 105 and 79 ka, it is as distant from Kapedo in time as Kapedo is from Omo Kibish and Herto. Yellen et al. (2005) propose that the industry should be recognized as a distinctive regional type, within a broad MSA technology, which may be evidence that by this stage some of the more diagnostic regionalization of the MSA had already begun to occur in eastern Africa (Yellen et al. 2005).

MIS 4 and MIS 3

Lasting between ~74 and ~25 ka, MIS 4 and MIS 3 jointly cover another ~50 ka period, and, archaeologically, the establishment of LSA industries throughout Africa. However, prior to this, certain African industries show technical innovations that introduce what later become typical LSA artifacts into otherwise MSA assemblages. The best-known cases of this are the Howiesons Poort (HP) industries of southern Africa (Soriano et al. 2007; Henshilwood 2012). A similar very early occurrence of LSA elements within an MSA assemblage is observed at Porc-Epic, Ethiopia, dated to 77–61 ka by obsidian hydration (likely a minimum date), thus either contemporaneous to or earlier than the HP (Clark et al. 1984b). The assemblage is *in situ,* includes refits and no evidence of bioturbation, yet, it contains both typical MSA methods and products (Levallois and discoid reduction, Mode 3 retouched points, etc.) and backed bladelets and geometric artifacts (Pleurdeau 2006). However, this site remains the only eastern African site dated to between 80 and 70 ka.

The best record of the MSA-LSA transition is provided by Mumba Cave in Tanzania. In a recent reanalysis and dating of the site, Gliganic et al. (2012) showed that the nature of the transition is poorly understood, as it involves a number of different components – occurrence of backing and microliths, reduction in Levallois methods and products, changes in scraper types, and the appearance of new elements, such as ostrich eggshell beads. The evidence from Mumba indicates that this may occur between 60 and 49 ka. Diez-Martin et al. (2009) have suggested that there is considerable continuity between the MSA and LSA at the site, both in terms of technology and typology (microliths increase).

Enkapune Ya Muto, in Kenya, is probably the earliest fully LSA assemblage in Africa, dated to ~50 ka (Ambrose 1998). The dates have been questioned (Diez-Martin et al. 2009), but both the nature of the transition and the age are broadly in line with those from Mumba, and possibly also from the Naisiusiu Beds at Olduvai Gorge (Skinner et al. 2003). However, the dates of the earliest LSA should not be seen as the same as the end of the MSA, as it is clear from recent work by Tryon on Rusinga Island (Tryon et al. 2010; Faith et al. 2016), as well as older work at Lukenya Hill (Ambrose 1998). The transition from MSA to LSA may be diachronic.

MIS 2

MIS 2 starts at 24 ka, and covers the period of the Last Glacial Maximum (LGM) through to deglaciation. It is known that many parts of the world saw reduction and contraction of populations at this time, followed by post-LGM dispersals (Gamble 2004). Once again, the East African evidence is surprisingly poor, making it difficult to assess whether its poverty reflects a lack of known and dated sites or of prehistoric populations.

One of the few sites to provide extensive and dated evidence is Lukenya Hill, on the Athi Kipiti Plains in southern Kenya. This is an inselberg, with rock shelters dotted around the area, dating from 21 ka (Marean 1992) to the Holocene. The lithics indicate a complete LSA record, with a fully microlithic technology. The fauna from the site, dominated by alcelaphines as prey species, shows that during this period the region was covered by extensive grasslands, probably more arid than today.

Like the rest of the tropical world, East Africa experienced a major rise in precipitation, with almost universal lake level rises, by the end of MIS 2. The very end of the Pleistocene was a phase of major rainfall, expansion of lakes, woodland and forest that continued into the early Holocene, a phase known as the Holocene Climatic Optimum. For the high lake stands of the Rift Valley there is a rich archaeological record, showing high population densities, effective fishing communities, a sophisticated technology (bone harpoons), and most probably extensive cultural and social networks.

The Archaeology of East Africa 200–10 ka in the Context of the Rest of Africa

MIS 6 does not represent a simple threshold in the evolutionary processes that shape the evolution of *H. sapiens*. While MIS 6 is the stage for the earliest evidence of modern humans, it is clear from sites such as Jebel Irhoud and Ngaloba that the populations of Africa at that time were morphologically variable, and possibly multiple.

Equally, while MIS 6 has some significant MSA sites, Mode 3 technologies have a much longer history. At Kapthurin, Twin Rivers and Gademotta, well-dated horizons put the technology back to between 280 and 265 ka, and Cartwright's Farm and Kapthurin may indicate an even earlier date (Morgan and Renne 2008). The evolution of modern humans – biologically and behaviorally – has roots back into MIS 9-7, and the occurrence of sites with very early MSA as distant as Sai Island (Van Peer et al. 2003), Twin Rivers (Barham and Smart 1996) and Gademotta suggests a broad early distribution of populations sharing a technological template, even if East Africa was the center of it. As with the fossil hominins, the persistence of Acheulean technology alongside the MSA also indicates continuity of parallel traditions, although, again, whether this is of traits or populations is not known.

The most striking observation of the MIS 5-4 period is the marked contrast between East, South and North Africa. These two last regions witness a significant increase in archaeological sites in MIS 5, consistent with the amelioration of climatic conditions throughout the continent. This denser record is then the context for multiple technical and behavioral innovations. The period from 80 to 60 ka includes the flourishing of Still Bay and Howiesons Poort traditions in southern Africa and Aterian in the Maghreb, evidence for advanced composite technologies (Sibudu), and symbolic use of material culture (Blombos, Diepkloof, Grotte de Pigeons) (Vanhaeren et al. 2006; Bouzouggar et al. 2007; d'Errico and Henshilwood 2007; Jacobs et al. 2008; Wadley 2008; Wadley et al. 2009; Texier et al. 2010; Henshilwood et al. 2011). Despite the presence of parallel technological lithic innovations at Porc-Epic Cave in Ethiopia (Pleurdeau 2006; Rosso et al. 2014), nothing like the South African record of this time has been found in East Africa. Whether this is to do with preservation and the scale of research activity, or a real difference remains to be seen, but it emphasizes the regional nature of the African MSA – something we can see additionally in the variation in point shapes and sizes, one of the diagnostic traits of the African MSA. Lastly, this regionality is also expressed in the first records of the LSA. The early East African dates of Mumba and Enkapune Ya Muto so far predate all others in the continent (including eastern Africa) by a substantial period of time (Vermeersch et al. 1982; Barker et al. 2012; Villa et al. 2012).

The nature of the affinities of the stone tool assemblages informs us about biogeographical patterns. Little has been done by way of continental comparisons, but it is becoming clear that at least tropical (equatorial) Africa has significant links to the North during MIS 5, if not before. Traditionally, North African prehistory has been treated as part of the Mediterranean – and hence European – sphere of influence, but increased work in the Sahara has led to a shift of balance. Various traits can now be used to link East African and Saharan MSA technologies (and in particular to the Lupemban) (Van Peer and Vermeersch 2007; Foley et al. 2013; Van Peer 2016). This is critical for considering patterns of dispersal across Africa, and so the basis for the diasporas beyond Africa.

The African Evolutionary Pattern in MIS 6-2

While recognizing the limitations of the data – sparse samples (especially fossils), lack of chronological precision, and in some cases limited geographical resolution (genetics), we can explore the pattern of human evolution in eastern Africa over MIS 6-2. We identify four main phases to the evolution of modern humans.

MIS 9-7 and the Modern Human Evolutionary Context

The first phase lies beyond MIS 6. This conclusion derives from a convergence of evidence from several disciplines – archaeology because the MSA is the behavioral context, and that stretches back at least 300 ka; genetics because the nuclear genomic horizon is $\sim$400 ka or more, and the last common ancestor with Eurasian hominins is between 500 and 350 ka; and human paleontology because this shows that, although morphologically modern forms were present by 190 ka at the beginning of MIS 6, that there is continuity with older specimens (later *H. heidelbergensis*, or *H. helmei*) (Bräuer 1984).

In terms of the scale of change, perhaps the most important point to note is that prior to MIS 6, the most significant events are changes in lithic technology – the loss of the characteristic Mode 2 forms, and the development of Mode 3. Although the extent to which the latter is gradual or more discontinuous, local or geographically widespread, remains unclear, it has been argued that it represents a major cognitive and behavioral change in a hominin lineage (Foley and Lahr 1997). If this is the case, then the longer African record suggests that the early shifts toward modernity were behavioral in the first place, with other traits – such as anatomy – lagging behind.

The evidence also suggests both diversity and continuity. Sites such as Sai Island indicate that the Sangoan tradition was part of the early MSA (Van Peer et al. 2003), while evidence from Kapthurin shows a different signature, namely the production of large blades (Tryon 2006). The implication is that this prolonged period was one of diversification of traditions as the Mode 2 strategies fell into disuse and new methods were experimented upon, and this diversity lies at the heart of the development of the later MSA.

If we look at the context in which this occurs, approximately two thirds of the period between MIS 13-7 (500–200 ka) was "warm and wet," and one third was "cold and dry" (using the classic glacial cycle model). These wet/dry phases were packaged in average periods of 55 ka and 40 ka respectively. Even taking a more conservative perspective (MIS 8-7), wet and dry phases are of similar length (56 ka of cold conditions, followed by 54 ka of warm). The implication is that an understanding of the selective conditions of human origins in Africa has to incorporate extensive periods of both warmer/wetter and colder/drier environments, and that simple climatic deterministic interpretations are unlikely to offer significant insights. We will consider this point in more detail in relation to later periods.

MIS 6 – the Evolution of the Modern Human Lineage

MIS 6 lasts 70–60 ka, and is, like the last glacial cycle, variable in conditions, with the second half of the stage colder/drier than the first. The key finding related to MIS 6 is the presence of the first anatomically modern humans (Omo Kibish, Herto). The contrast – morphologically – between these two samples is striking and informative. It is clear that while there is an underlying affinity in the presence of unique derived modern human features, the two samples are sufficiently different to suggest either considerable evolutionary change through time, or, more likely, that hominin populations were significantly fragmented during this period, both hypotheses further supported by the inclusion of yet more different individuals (Jebel Irhoud, Ngaloba, Singa). The fact that there were high levels of variation both within the modern human lineage, and between it and other late hominin groups across Africa (Harvati et al. 2011; Lahr 2013) is consistent with periods of small localized populations under both strong selection and genetic drift. From an archaeological point of view, this observation of high variation may extend to the persistence of terminal Acheulean traditions into periods when sites like Pinnacle Point testify to the occurrence of novel behaviors, such as use of coastal resources (Marean et al. 2007).

MIS 5 – African Dispersals

MIS 5 is best discussed at a Pan-African level, as the East African data are so sparse. During this period (56 ka duration), modern humans occur very widely across Africa, and beyond in the Levant, which was at that time probably part of the African biome. The significantly greater density of archaeological sites, now consistently MSA, testifies to this as a critical early phase of modern human expansion and colonization of the whole, or most, of Africa (as so often, we have little information about West and Central Africa [but see Cornelissen 2016; Taylor 2016]). MSA lithic assemblages from this time point to an intensification of technological production, likely expression of individual/group identities, as well as increased regional (and possibly short-lived) trajectories (Jacobs et al. 2008).

The role played by East Africa in this phase is unclear, largely because the evidence is so scarce. Whether this reflects a lack of MIS 5 exposures across the region, or – rather less likely – an absence of the warm/wet conditions of the interglacial occurring elsewhere in Africa, or a genuine population reduction, is not clear. One possibility is that while equatorial and eastern Africa play a key role in the terminal stages of the Acheulean, emergence of the modern human lineage and appearance of the derived Mode 3 technologies, the dispersals across Africa exposed early human populations to new conditions which prompted localized demographic growth, innovations in behavior, and cultural diversification of populations at a regional level.

MIS 4-2

The fourth phase covers MIS 4-2, which is effectively the last glacial period, from its relatively cool start, through a warmer middle phase, to the extreme cold of the LGM and the rapid rise in temperatures and precipitation at the end of the Pleistocene. It has been suggested that the earlier part – MIS 4 (74–60 ka) was a period of contraction and fragmentation for human populations (Lahr and Foley 1998). This may have served to intensify the regionalization of African populations, and, from a biological point of view, lead to the basis for much of later human diversity. Sadly, the fossil evidence for this period is remarkably sparse, so that this is an inference based on human genetics, and requires, in particular, refinement of its chronological and geographical bases. The ephemeral nature of South African technological traditions would be consistent with this point of view, a pattern that may also characterize the rich Maghrebian record (Jacobs et al. 2012). The two well-dated East African sites, a cave (Porc-Epic) and rock shelter (Mumba) (see above), offer the only glimpses into the intensification and regional innovation of MSA traditions at this time.

The second part of this phase, MIS 3, is strongly associated with the first major dispersals beyond Africa, one almost certainly associated with Mode 3 technologies, and a predominantly southern dispersal to Asia – for instance, possibly across the Bab el Mandeb (Rose et al. 2011; Delagnes et al. 2012). Given the contiguous nature of eastern Africa with both the Nile and Red Sea exits from Africa, the nature of the populations in the region must have been significant contributors to the founding gene pools of Eurasians.

MIS 2 is a period of extreme aridity across much of Africa. There is evidence from the Central Sahara for major depopulation (di Lernia 2013), a process that is likely to have extended to most desertic and semidesertic regions, with demographic concentrations near permanent water sources. It could be that East Africa, with its range of environments and high relief, served as a refugium for populations during this period. Certainly, the rise in lake levels at the end of the Pleistocene is associated with an upsurge in archaeological sites and the widespread distribution of new cultural traditions.

Evolutionary Geography Models

The framework for exploring the patterns and processes of the evolution of modern humans over the last 200 ka must, at the broadest scale, be the global patterns of climatic variability, and their tropical expression. While there is an overall congruence between the climatic stages of the last 200 ka and events in hominin evolution in eastern Africa, it is unlikely to be sufficient when looking for explanations for the observed events. This is partly because of paucity of data, but another factor is that the use of marine isotope stages as the unit of analysis underplays local variation and the role of monsoonal changes. This is particularly the case in eastern Africa, where emerging data from lake cores indicate that different basins at different times responded to orbital-scale precessional insolation changes, or sea surface temperature variation, or shifts in the position of the Intertropical Convergence Zone and monsoon strength (Trauth et al. 2001; Bergner and Trauth 2004; Lynch-Stieglitz 2004), creating an even greater mosaic landscape.

In a recent paper, Blome et al. (2012) provide a meta-analysis of environmental and paleoanthropological patterns across Africa. Using marine and terrestrial proxies they show that across the period 150–30 ka, as expected, aridity and humidity varied considerably in space and time. Although their data for East Africa are relatively thin (only three terrestrial environmental localities, compared to 23 for North Africa), we can nonetheless use their reconstruction to examine briefly how a more finely graded analysis can provide new insights.

Figure 12.4 is derived from Blome et al. (2012). It firstly shows the distribution of wet and dry phases between 150 and 30 ka for East Africa. It can be seen that this does not match the marine isotope record very closely, or rather shows greater levels of oscillation. Second, it shows the times (in 5 ka units) when East Africa shared its predominant climatic condition with the three other regions of Africa the authors delineate (North, South, West/Central/Southeast). From this it can be seen that there is only one 5 ka unit across the whole period when all parts of the continent shared the same pattern of aridity (150–145 ka), and there is no period when it is uniformly wet. This implies that environmental linkages with human evolution and prehistory must be resolved to relatively narrow time frames and be regionally specific. When we look at the particular relationships, we can see that East Africa shares wet/dry conditions most frequently with North Africa (54%), and least with South Africa (37.5%). If we place this in terms of northern and southern similarities, the northern half of the continent was in step in 7 of the 5 ka bins (central, north and east), while this was the case in only 3 of the 5 ka bins for the southern hemisphere.

These observations show the complexity of environmental interactions across the African continent, and underscore the findings of Scholz et al. (2007), that African lakes do not rise and fall synchronously. How this plays out in evolutionary terms should be a major part of the research agenda. One possibility is that the asynchrony offers the opportunity for dispersal in response to adverse conditions, resulting in an inverse relationship with good conditions in refugia (Lahr 2012b). An alternative model would align dispersal patterns (and conversely contractions) across regions sharing the same broad climatic conditions. The current dataset, especially for tropical Africa, does not permit testing of these alternatives.

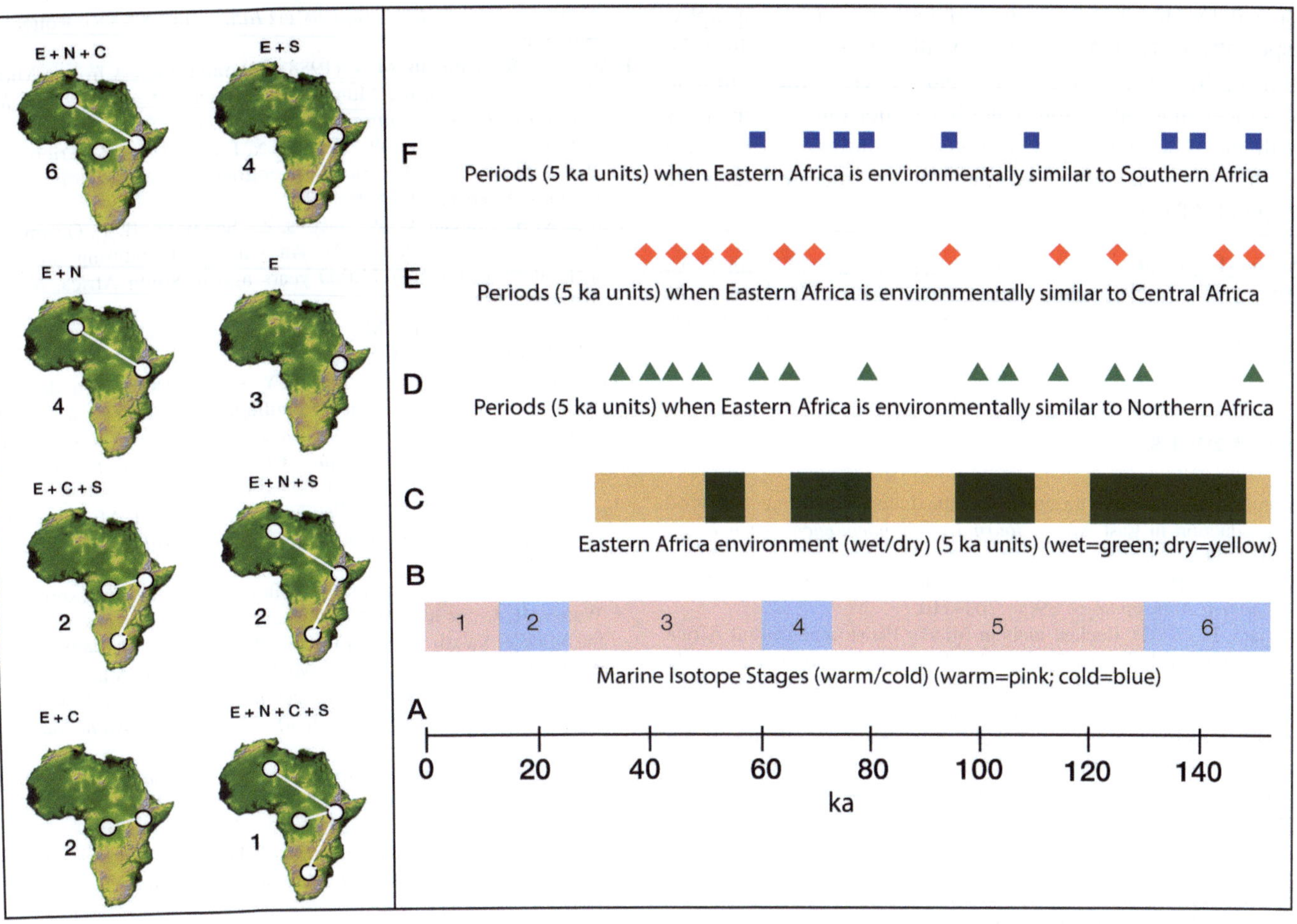

Fig. 12.4 Dynamics of environmental relationships between eastern Africa and other parts of the continent, 150–30 ka. The maps of Africa (*left*) show patterns of shared humidity/aridity between eastern Africa and other parts of Africa. *White circles* indicate shared humidity/aridity. E = eastern, N = northern, C = central, S = southern. Each map indicates a particular configuration of regional relationships. *Numbers* indicate the number of 5 ka bins in which that particular biogeographic configuration occurs. Maps are ordered in frequency of occurrence, from *top left* to *bottom right*. The most frequent configuration (six 5 ka periods) is E+N+C, the least frequent complete synchrony, which only occurs once (150–145 ka). To the right, the patterns of environmental similarity are shown chronologically: *A* Timescale, in kyr; *B* MIS 7-1; *C* Environmental reconstruction in eastern Africa, 150–30 ka (data from Blome et al. 2012, divided into 5 ka bins), with *green shading* indicating relatively humid phases, and *brown shading* indicating more arid phases. *D–F* Measures of similarity of humidity/aridity between eastern Africa and other parts of the continent. *Markers* indicate phases (in 5 ka bins) when eastern Africa shares relative humidity or aridity with northern ([D], *green triangles*), central ([E], *red diamonds*), and southern Africa ([F], *blue squares*). Absence of markers indicates regions are out of phase with eastern Africa (*G*)

Conclusions

In this paper we have attempted to set out the evidence for evolutionary and behavioral patterns among hominin populations in eastern Africa during the last 200 ka. Sadly, perhaps the strongest conclusion is that although there is a considerable number of sites known, relatively few of these are well dated, and so while general patterns can be seen, chronological precision is elusive. However, we can highlight a number of provisional, more positive, conclusions. The first is that the fossil record for eastern Africa shows both periods of synchronicity and asynchronicity with the glacial/aridity model, suggesting a complex relationship between African and global climate. Second, although we can see strong evidence for the development of modern humans and their behaviors in the context of the MSA during MIS 6-2, nonetheless, it is clear that the roots of this change lie much earlier. A third related point is that the evidence strongly points to what should be called an 'extended model of human origins and evolution'. The evolution of *Homo sapiens* is the cumulative product of a series of processes and events that occur from MIS 8 or 9 at least, and continue into MIS 2. Fourth, across the range of climatic and environmental changes that have occurred during the last 300 ka, there is both a signal of wet and dry oscillations that shape large regional patterns and a very localized response shaped by the mosaic nature of habitats and topography, especially in East Africa. And finally, in terms of the relationship between eastern Africa and the rest of the continent, it is clear that the region is neither in nor out of synchrony with the other regions, but linked at various times to the north, the south, and the west, and plays a unique role as the hinge of Africa.

Acknowledgments This research was supported by the ERC *In-Africa* Project (ERC 295907), and the Leverhulme Trust. We thank Sacha Jones and the three anonymous reviewers for helpful comments and suggestions.

References

Ambrose, S. (1998). Chronology of the Later Stone Age and food production in East Africa. *Journal of Archaeological Science, 25*, 377–392.

Barham, L. (1998). Possible early pigment use in south-central Africa. *Current Anthropology, 39*(5), 703–710.

Barham, L. (2002). Backed tools in Middle Pleistocene central Africa and their evolutionary significance. *Journal of Human Evolution, 43* (5), 585–603.

Barham, L. (2012). Clarifying some fundamental errors in Herries' "A chronological perspective on the Acheulian and its transition to the Middle Stone Age in Southern Africa: The question of the Fauresmith" (2011). *International Journal of Evolutionary Biology, 5*, 1–5.

Barham, L., & Smart, P. L. (1996). An early date for the Middle Stone Age of central Zambia. *Journal of Human Evolution, 30*(3), 287–290.

Barker, G., Bennett, P., Farr, L., Hill, E., Hunt, C., Lucarini, G., et al. (2012). The Cyrenaican Prehistory Project 2012: The fifth season of investigations of the Haua Fteah cave. *Libyan Studies, 43*, 115–136.

Basell, L. (2008). Middle Stone Age (MSA) site distributions in eastern Africa and their relationship to Quaternary environmental change, refugia and the evolution of *Homo sapiens*. *Quaternary Science Reviews, 27*, 2484–2498.

Behar, D. M., Villems, R., Soodyall, H., Blue-Smith, J., Pereira, L., Metspalu, E., et al. (2008). The dawn of human matrilineal diversity. *American Journal of Human Genetics, 82*(5), 1130–1140.

Bergner, A., & Trauth, M. H. (2004). Comparison of the hydrological and hydrochemical evolution of Lake Naivasha (Kenya) during three highstands between 175 and 60 kyr BP. *Palaeogeography, Palaeoclimatology, Palaeoecology, 215*, 17–36.

Blome, M. W., Cohen, A. S., Tryon, C. A., Brooks, A. S., & Russell, J. (2012). The environmental context for the origins of modern human diversity: A synthesis of regional variability in African climate 150,000-30,000 years ago. *Journal of Human Evolution, 62* (5), 563–592. doi:10.1016/j.jhevol.2012.01.011.

Bouzouggar, A., Barton, N., Vanhaeren, M., d'Errico, F., Collcutt, S., Higham, T. F., et al. (2007). 82,000-year-old shell beads from North Africa and implications for the origins of modern human behavior. *Proceedings of the National Academy of Sciences of the United States of America, 104*(24), 9964–9969.

Bräuer, G. (1984). A craniological approach to the origin of anatomically modern *Homo sapiens* in Africa and implications for the appearance of modern Europeans. In F. H. Smith & F. Spencer (Eds.), *The origin of modern humans: A world survey of the fossil evidence* (pp. 327–410). New York: Alan Liss.

Bräuer, G. (1992). Africa's place in the evolution of *Homo sapiens*. In G. Bräuer & F. H. Smith (Eds.), *Continuity or replacement? Controversies in Homo sapiens evolution* (pp. 83–98). Rotterdam: Balkema.

Bräuer, G., & Mehlman, M. J. (1988). Hominid molars from a Middle Stone Age level at the Mumba Rock Shelter. *Tanzania. American Journal of Physical Anthropology, 75*(1), 69–76.

Brown, K. S., Marean, C. W., Herries, A. I. R., Jacobs, Z., Tribolo, C., Braun, D., et al. (2009). Fire as an engineering tool of early modern humans. *Science, 325*, 859–862.

Brown, K. S., Marean, C. W., Jacobs, Z., Schoville, B. J., Oestmo, S., Fisher, E. C., et al. (2012). An early and enduring advanced technology originating 71,000 years ago in South Africa. *Nature, 491*, 590–593.

Cann, R. L., Stoneking, M., & Wilson, A. C. (1987). Mitochondrial DNA and human evolution. *Nature, 325*, 31–35.

Chavaillon, J., Hours, F., & Coppens, Y. (1987). Découverte de restes humains fossiles associés à un outillage Acheuléen final à Melka Kunturé (Ethiopie). *Comptes rendus de l'Académie des sciences. Série 2, Mécanique, Physique, Chimie, Sciences de l'Univers. Sciences de la Terre, 304*(10), 539–547.

Clark, J. D. (1988). The Middle Stone Age of East Africa and the beginnings of regional identity. *Journal of World Prehistory, 2*(3), 235–305.

Clark, J. D., Asfaw, B., Assefa, G., Harris, J. W. K., Kurashina, H., Walter, R. C., et al. (1984a). Palaeoanthropological discoveries in the Middle Awash Valley, Ethiopia. *Nature, 307*, 421–428.

Clark, J. D., Williamson, K. D., Michels, J. W., & Marean, C. A. (1984b). A Middle Stone Age occupation site at Porc Epic Cave, Dire Dawa (east-central Ethiopia). *The African Archaeological Review, 2*(1), 37–71.

Compton, J. S. (2011). Pleistocene sea-level fluctuations and human evolution on the southern coastal plain of South Africa. *Quaternary Science Reviews, 30*, 506–527.

Conard, N. J. (2008). A critical view of the evidence for a southern African origin of behavioural modernity. *Goodwin Series, 10*, 175–179.

Cornelissen, E. (2016). The later Pleistocene in the northeastern Central African rainforest. In S. C. Jones & B. A. Stewart (Eds.), *Africa from MIS 6-2: Population dynamics and paleoenvironments* (pp. 301–319). Dordrecht: Springer.

d'Errico, F., & Henshilwood, C. S. (2007). Additional evidence for bone technology in the southern African Middle Stone Age. *Journal of Human Evolution, 52*(2), 142–163.

d'Errico, F., Vanhaeren, M., Barton, N., Bouzouggar, A., Mienis, H., Richter, D., et al. (2009). Additional evidence on the use of personal ornaments in the Middle Paleolithic of North Africa. *Proceedings of the National Academy of Sciences of the United States of America, 106*(38), 16051–16056.

d'Errico, F., Backwell, L. R., & Wadley, L. (2012). Identifying regional variability in Middle Stone Age bone technology: The case of Sibudu Cave. *Journal of Archaeological Science, 39*(7), 2479–2495.

Day, M. H., & Stringer, C. B. (1991a). Les restes craniens d'Omo-Kibish et leur classification à l'interieur du genre *Homo*. *L'Anthropologie, 95*, 573–594.

Day, M. H., & Stringer, C. B. (1991b). The Omo Kibish cranial remains and classification within the Genus *Homo*. *Anthropologie, 95*, 573–594.

Day, M. H., Leakey, M. D., & Magori, C. (1980). A new hominid fossil skull (L.H. 18) from the Ngaloba Beds, Laetoli, northern Tanzania. *Nature, 284*, 55–56.

Delagnes, A., Tribolo, C., Bertran, P., Brenet, M., Crassard, R., Jaubert, J., et al. (2012). Inland human settlement in southern Arabia 55,000 years ago. New evidence from the Wadi Surdud Middle Paleolithic site complex, western Yemen. *Journal of Human Evolution, 63*(3), 452–474.

di Lernia, S. (2013). The cultural sequence. In S. di Lernia (Ed.), *The Uan Afuda Cave. Hunter-gatherer societies of Central Sahara* (pp. 39–50). Firenze: Edizioni All'Insegna del Giglio.

Diez-Martin, F., Domínguez-Rodrigo, M., Sánchez, P., Mabulla, A. Z. P., Tarriño, A., Barba, R., et al. (2009). The Middle to Later Stone Age technological transition in East Africa. New data from Mumba Rockshelter Bed V (Tanzania) and their implications for the origin of modern human behavior. *Journal of African. Archaeology, 7*(2), 147–173.

Endicott, P., Ho, S. Y. W., & Stringer, C. (2010). Using genetic evidence to evaluate four palaeoanthropological hypotheses for the timing of Neanderthal and modern human origins. *Journal of Human Evolution, 59*(1), 87–95.

Faith, J. T., Tryon, C., & Peppe, D. J. (2016). Environmental change, ungulate biogeography, and their implications for early human dispersals in equatorial East Africa. In S. C. Jones & B. A. Stewart (Eds.), *Africa from MIS 6-2: Population dynamics and paleoenvironments* (pp. 233–245). Dordrecht: Springer.

Fleagle, J. G., Assefa, Z., Brown, F. H., & Shea, J. J. (2008). Paleoanthropology of the Kibish Formation, southern Ethiopia: Introduction. *Journal of Human Evolution, 55*(3), 360–365.

Foley, R. A. (1989). The ecological conditions of speciation: A comparative approach to the origins of anatomically modern humans. In P. A. Mellars & C. B. Stringer (Eds.), *The human revolution: Behavioural and biological perspectives on the origins of modern humans* (pp. 298–318). Edinburgh: Edinburgh University Press.

Foley, R. A., & Mirazón Lahr, M. (1997). Mode 3 technologies and the evolution of modern humans. *Cambridge Archaeological Journal, 7*, 3–36.

Foley, R. A., Maílllo-Fernández, J. M. & Mirazón Lahr, M. (2013). The Middle Stone Age of the Central Sahara: Biogeographical opportunities and Technological strategies in later human evolution. *Quaternary International, 300*, 153–170.

Gamble, C. (2004). Climate change and evolving human diversity in Europe during the last glacial. *Philosophical Transactions of the Royal Society B: Biological Sciences, 359*(1442), 243–254.

Gliganic, L. A., Jacobs, Z., Roberts, R. G., Domínguez-Rodrigo, M., & Mabulla, A. Z. P. (2012). New ages for Middle and Later Stone Age deposits at Mumba rockshelter, Tanzania: Optically stimulated luminescence dating of quartz and feldspar grains. *Journal of Human Evolution, 62*(4), 533–547.

Gonder, M. K., Mortensen, H. M., Reed, F. A., de Sousa, A., & Tishkoff, S. A. (2007). Whole-mtDNA genome sequence analysis of ancient African lineages. *Molecular Biology and Evolution, 24* (3), 757–768.

Grün, R., & Stringer, C. B. (1991). Electron spin resonance dating and the evolution of modern humans. *Archaeometry, 33*, 153–199.

Haile-Selassie, Y., Asfaw, B., & White, T. D. (2003). Hominid cranial remains from Upper Pleistocene deposits at Aduma, Middle Awash. *Ethiopia. American Journal of Physical Anthropology, 123*(1), 1–10.

Harvati, K., Stringer, C., Grün, R., Aubert, M., Allsworth-Jones, P., & Folorunso, C. A. (2011). The Later Stone Age calvaria from Iwo Eleru, Nigeria: Morphology and chronology. *PLoS ONE, 6*(9), e24024. doi:10.1371/journal.pone.0024024.

Henn, B. M., Gignoux, C. R., Jobin, M., Granka, J. M., Macpherson, J. M., Kidd, J. M., et al. (2011). Hunter-gatherer genomic diversity suggests a southern African origin for modern humans. *PNAS, 108* (13), 5154–5162.

Henshilwood, C. S. (2012). Late Pleistocene techno-traditions in southern Africa: A review of the Still Bay and Howiesons Poort, c. 75–59 ka. *Journal of World Prehistory, 25*, 205–237.

Henshilwood, C. S., d'Errico, F., van Niekerk, K. L., Coquinot, Y., Jacobs, Z., Lauritzen, S. E., et al. (2011). A 100,000-year-old ochre-processing workshop at Blombos Cave, South Africa. *Science, 334*, 219–222.

Herries, A. I. R. (2011). A chronological perspective on the Acheulian and its transition to the Middle Stone Age in Southern Africa: The question of the Fauresmith. *International Journal of Evolutionary Biology, 2011*(2716), 1–25. http://doi.org/10.1560/IJPS_54_4_291.

Hublin, J.-J. (2001). Northwestern African Middle Pleistocene hominids and their bearing on the emergence of *Homo sapiens*. In L. Barham & K. Robson-Brown (Eds.), *Human roots. Africa and Asia in the Middle Pleistocene* (pp. 99–121) Bristol: CHERUB, Western Academic & Specialist Press.

Hublin, J.-J. (2009). The origin of Neandertals. *PNAS, 106*, 16022–16027.

Jacobs, Z., Roberts, R., Galbraith, R., & Deacon, H. (2008). Ages for the Middle Stone Age of southern Africa: Implications for human behavior and dispersal. *Science, 322*, 733–735.

Jacobs, Z. Z., Roberts, R. G. R., Nespoulet, R. R., Hajraoui, El, M. A. M., & Debénath, A. A. (2012). Single-grain OSL chronologies for Middle Palaeolithic deposits at El Mnasra and El Harhoura 2, Morocco: Implications for Late Pleistocene human-environment interactions along the Atlantic coast of northwest Africa. *Journal of Human Evolution, 62*(3), 18. http://doi.org/10.1016/j.jhevol.2011.12.001.

Jenkins, C. N., Pimm, S. L., & Joppa, L. N. (2013). Global patterns of terrestrial vertebrate diversity and conservation. *PNAS, 110*(28), e2602–e2610. doi:10.1073/pnas.1302251110.

Kelly, R. (1995). *The foraging spectrum*. Washington, DC: Smithsonian Institution.

Kingdon, J. (1989). *Island Africa*. London: Academic Press.

Klein, R. G. (1992). The archaeology of modern human origins. *Evolutionary Anthropology, 1*, 5–14.

Leakey, R. E. F. (1969). Early *Homo sapiens* remains from the Omo River region of South-west Ethiopia. *Nature, 222*, 1132–1133.

Lynch-Stieglitz, J. (2004). Hemispheric asynchrony of abrupt climate change. *Science, 304*, 1919–1920.

Mabulla, A. Z. P. (2007). Hunting and foraging in the Eyasi Basin, Northern Tanzania: Past, present and future prospects. *African Archaeological Review, 24*, 15–33.

Magori, C. C., & Day, M. H. (1983). Laetoli Hominid 18: An early *Homo sapiens* skull. *Journal of Human Evolution, 12*, 747–753.

Marean, C. W. (1992). Implications of Late Quaternary mammalian fauna from Lukenya Hill (south-central Kenya) for paleoenvironmental change and faunal extinctions. *Quaternary Research, 37*(2), 239–255.

Marean, C. W. (2011). Coastal South Africa and the coevolution of the modern human lineage and the coastal adaptation. In N. F. Bruno, J. A. Haws, & L. G. Davis (Eds.), *Trekking the Shore* (pp. 421–440). New York: Springer.

Marean, C. W., Bar-Matthews, M., Bernatchez, J., Fisher, E., Goldberg, P., Herries, A. I. R., et al. (2007). Early human use of marine resources and pigment in South Africa during the Middle Pleistocene. *Nature, 449*, 905–908.

McBrearty, S., & Brooks, A. S. (2000). The revolution that wasn't: A new interpretation of the origin of modern human behavior. *Journal of Human Evolution, 39*(5), 453–563.

McDermott, F., Stringer, C., Grün, R., & Williams, C. T. (1996). New Late-Pleistocene Uranium-Thorium and ESR dates for the Singa hominid (Sudan). *Journal of Human Evolution, 31*, 507–516.

McDougall, I., Brown, F. H., & Fleagle, J. G. (2005). Stratigraphic placement and age of modern humans from Kibish, Ethiopia. *Nature, 433*, 733–736.

Mehlman, M. (1989). *Later Quaternary archaeological sequences in northern Tanzania*. Urbana-Champaign: University of Illinois.

Mehlman, M. J. (1991). Context for the emergence of modern man in Eastern Africa: Some new Tanzanian evidence. In J. D. Clark (Ed.),

Cultural beginnings: Approaches to understanding early hominid lifeways in the African savanna (Vol. 19) (pp. 177–196). Bonn: Forschung- Institut fur Vor- und Fruhgeschichte, Romisch-Germanisches Zentralmuseum, Monographien.

Mellars, P. (1991). Cognitive changes and the emergence of modern humans in Europe. *Cambridge Archaeological Journal, 1*, 63–76.

Mellars, P. (2006). Archaeology and the dispersal of modern humans in Europe: Deconstructing the "Aurignacian." *Evolutionary Anthropology: Issues, News, and Reviews, 15*(5), 167.

Merrick, H. V., Brown, F. H., & Nash, W. P. (1994). Use and movement of obsidian in the Early and Middle Stone Ages of Kenya and northern Tanzania. In S. T. Childs (Ed.), *Society, culture and technology in Africa.* MASCA Research Papers in Science and Archaeology, vol. 11 (supplement), pp. 29–44. Philadelphia: University of Pennsylvania Museum of Archaeology and Anthropology.

Mirazón Lahr, M. (1996). *The evolution of human diversity*. Cambridge: Cambridge University Press.

Mirazón Lahr, M. (2012a). The evolution of modern human diversity in Africa. http://www.in-africa.org/origins-2/after-the-origin/Accessed October 27, 2013.

Mirazón Lahr, M. (2012b). Why East Africa could be driving human evolution? http://www.in-africa.org/origins-2/why-east-africa/ Accessed October 27, 2013.

Mirazón Lahr, M. (2013). Genetic and fossil evidence for modern human origins. In P. Lane & P. Mitchell (Eds.), *Oxford handbook of African archaeology* (pp. 325–340). Oxford: Oxford University Press.

Mirazón Lahr, M., & Foley, R. A. (1998). Towards a theory of modern human origins: geography, demography, and diversity in recent human evolution. *American Journal of Physical Anthropology, Suppl, 27*, 137–176.

Mirazón Lahr, M., & Foley, R. A. (2001). Mode 3, *Homo helmei*, and the pattern of human evolution in the Middle Pleistocene. In L. Barham & K. Robson Brown (Eds.), *Human roots: Africa and Asia in the Middle Pleistocene* (pp. 23–39). Bristol: CHERUB, Western Academic & Specialist Press.

Mithen, S. (1996). *The prehistory of the mind: A search for the origins of art, religion and science*. London: Thames and Hudson.

Mithen, S. (1997). The evolving mind. *Cambridge Archaeological Journal, 7*, 269–286.

Morgan, L. E., & Renne, P. R. (2008). Diachronous dawn of Africa's Middle Stone Age: New ^{40}Ar/^{39}Ar ages from the Ethiopian Rift. *Geology, 36*, 967–970.

Mussi, M., Altamura, F., Macchiarelli, R., Melis, R. T., & Spinapolice, E. E. (2014). Garba III (Melka Kunture, Ethiopia): A MSA site with archaic *Homo sapiens* remains revisited. *Quaternary International, 343*, 28–39.

Nielsen, R., & Beaumont, M. A. (2009). Statistical inferences in phylogeography. *Molecular Ecology, 18*, 1034–1047.

Patin, E., Laval, G., Barreiro, L.B., Salas, A., Semino, O., Santachiara-Benerecetti, S., et al. (2009). Inferring the demographic history of African farmers and Pygmy hunter–gatherers using a multilocus resequencing data set. *PLoS Genetics, 5*(4), e1000448. doi:10.1371/journal.pgen.1000448.

Pleurdeau, D. (2006). Human technical behavior in the African Middle Stone Age: The lithic assemblage of Porc-Epic Cave (Dire Dawa, Ethiopia). *African Archaeological Review, 22*(4), 177–197.

Porraz, G., Parkington, J. E., Rigaud, J.-P., Miller, C. E., Poggenpoel, C., Tribolo, C., et al. (2013). The MSA sequence of Diepkloof and the history of southern African Late Pleistocene populations. *Journal of Archaeological Science, 40*(9), 3542–3552.

Poznik, G. D., Henn, B. M., Yee, M.-C., Sliwerska, E., Euskirchen, G. M., Lin, A. A., et al. (2013). Sequencing Y chromosomes resolves discrepancy in time to common ancestor of males versus females. *Science, 341*, 562–565.

Prendergast, M. E., & Luque, L. (2007). New excavations at Mumba rockshelter. *Journal of African Archaeology, 5*, 217.

Quintana-Murci, L., Quach, H., Harmant, C., Luca, F., Massonnet, B., Patin, E., et al. (2008). Maternal traces of deep common ancestry and asymmetric gene flow between Pygmy hunter-gatherers and Bantu-speaking farmers. *Proceedings of the National Academy of Sciences of the United States of America, 105*(5), 1596–1601.

Ramachandran, S., Deshpande, O., Roseman, C. C., Rosenberg, N. A., Feldman, M. W., & Cavalli-Sforza, L. L. (2005). Support from the relationship of genetic and geographic distance in human populations for a serial founder effect originating in Africa. *Proceedings of the National Academy of Sciences of the United States of America, 102*, 15942–15947.

Rightmire, G. P. (1988). *Homo erectus* and later Middle Pleistocene humans. *Annual Review of Anthropology, 17*, 239–259.

Rightmire, G. P. (1998). Human evolution in the Middle Pleistocene: The role of *Homo heidelbergensis*. *Evolutionary Anthropology: Issues, News, and Reviews, 6*, 218–227.

Rightmire, G. P. (2009). Middle and later Pleistocene hominins in Africa and Southwest Asia. *Proceedings of the National Academy of Sciences of the United States of America, 106*, 16046–16050.

Rose, J., Usik, V., Marks, A., Hilbert, Y., & Galletti, C. (2011). The Nubian Complex of Dhofar, Oman: An African Middle Stone Age industry in southern Arabia. *PLoS ONE, 6*(11), e28239. doi:10.1371/journal.pone.0028239.

Rosso, D. E., d'Errico, F., & Zilhao, J. (2014). Stratigraphic and spatial distribution of ochre and ochre processing tools at Porc-Epic Cave, Dire Dawa, Ethiopia. *Quaternary International, 343*, 85–99.

Scholz, C. A., Johnson, T. C., Cohen, A. S., King, J. W., Peck, J. A., Overpeck, J. T., et al. (2007). East African megadroughts between 135 and 75 thousand years ago and bearing on early-modern human origins. *Proceedings of the National Academy of Sciences of the United States of America, 104*(42), 16416–16421.

Schuster, S. C., Miller, W., Ratan, A., Tomsho, L. P., Giardine, B., Kasson, L. R., et al. (2010). Complete Khoisan and Bantu genomes from southern Africa. *Nature, 463*, 943–947.

Schwarcz, H. P., Grün, R., Vandermeersch, B., Bar-Yosef, O., Vallois, H., & Tchernov, E. (1988). ESR dates for the hominid burial site of Qafzeh in Israel. *Journal of Human Evolution, 17*, 733–737.

Shea, J. J. (2011). *Homo sapiens* is as *Homo sapiens* was. *Current Anthropology, 52*, 1–35.

Shea, J. J., Fleagle, J. G., Brown, F., & Assefa, Z. (2002). Archaeological reconnaissance of the Omo Kibish Formation, Ethiopia. *Journal of Human Evolution, 42*, A33–A34.

Skinner, A. R., Hay, R. L., Masao, F., & Blackwell, B. A. B. (2003). Dating the Naisiuslu Beds, Olduvai Gorge, by electron spin resonance. *Quaternary Science Reviews, 22*, 1361–1366.

Soriano, S., Villa, P., & Wadley, L. (2007). Blade technology and tool forms in the Middle Stone Age of South Africa: The Howiesons Poort and post-Howiesons Poort at Rose Cottage Cave. *Journal of Archaeological Science, 34*, 681–703.

Spoor, F., Stringer, C., & Zonneveld, F. (1998). Rare temporal bone pathology of the Singa calvaria from Sudan. *American Journal of Physical Anthropology, 107*, 41–50.

Stringer, C. B., Grün, R., Schwarcz, H. P., & Goldberg, P. (1989). ESR dates for the hominid burial site of Es Skhul in Israel. *Nature, 338*, 756–758.

Taylor, N. (2016). Across rainforests and woodlands: A systematic re-appraisal of the Lupemban Middle Stone Age in Central Africa. In S. C. Jones & B. A. Stewart (Eds.), *Africa from MIS 6-2: Population dynamics and paleoenvironments* (pp. 273–299). Dordrecht: Springer.

Texier, P. J., Porraz, G., Parkington, J., Rigaud, J. P., Poggenpoel, C., Miller, C., et al. (2010). A Howiesons Poort tradition of engraving ostrich eggshell containers dated to 60,000 years ago at Diepkloof Rock Shelter, South Africa. *Proceedings of the National Academy of Sciences of the United States of America, 107*, 6180–6185.

Tishkoff, S. A., Gonder, M. K., Henn, B. M., Mortensen, H., Knight, A., Gignoux, C., et al. (2007). History of click-speaking populations of Africa inferred from mtDNA and Y chromosome genetic variation. *Molecular Biology and Evolution, 24*, 2180–2195.

Tishkoff, S. A., Reed, F. A., Friedlaender, F. R., Ehret, C., Ranciaro, A., Froment, A., et al. (2009). The genetic structure and history of Africans and African Americans. *Science, 324*, 1035–1044.

Trauth, M. H., Deino, A., & Strecker, M. R. (2001). Response of the East African climate to orbital forcing during the last interglacial (130–117 ka) and the early last glacial (117–60 ka). *Geology, 29*, 499–502.

Tryon, C. A. (2006). "Early" Middle Stone Age lithic technology of the Kapthurin Formation (Kenya). *Current Anthropology, 47*, 367–375.

Tryon, C. A., & McBrearty, S. (2002). The Middle Stone Age of the southern Kapthurin Formation, Baringo, Kenya. *Journal of Human Evolution, 42*, A37–A37.

Tryon, C., Roach, N., & Logan, M. (2008). The Middle Stone Age of the northern Kenyan Rift: Age and context of new archaeological sites from the Kapedo Tuffs. *Journal of Human Evolution, 55*, 652–664.

Tryon, C. A., Faith, J. T., Peppe, D. J., Fox, D. L., McNulty, K. P., Jenkins, K., et al. (2010). The Pleistocene archaeology and environments of the Wasiriya Beds, Rusinga Island. *Kenya. Journal of Human Evolution, 59*(6), 657–671.

Underhill, P. A., & Kivisild, T. (2007). Use of Y Chromosome and Mitochondrial DNA population structure in tracing human migrations. *Annual Review of Genetics, 41*, 539–564.

Vallois, H. V. (1951). La mandibule humaine fossile de la grotte du Porc Epic, près Dire Daoua (Abyssinie). *L'Anthropologie, 55*, 231–238.

Van Peer, P. (2016). Technological systems, population dynamics and historical process in the MSA of northern Africa. In S. C. Jones & B. A. Stewart (Eds.), *Africa from MIS 6-2: Population dynamics and paleoenvironments* (pp. 147–159). Dordrecht: Springer.

Van Peer, P., & Vermeersch, P. (2007). The place of northeast Africa in the early history of modern humans: New data and interpretations on the Middle Stone Age. In P. Mellars, K. Boyle, O. Bar-Yosef, & C. Stringer (Eds.), *Rethinking the human revolution* (pp. 187–198). Cambridge: McDonald Institute for Archaeological Research.

Van Peer, P., Fullagar, R., Stokes, S., Bailey, R., Moeyersons, J., Steenhoudt, F., et al. (2003). The Early to Middle Stone Age transition and the emergence of modern human behaviour at site 8-B-11, Sai Island, Sudan. *Journal of Human Evolution, 45*, 187–193.

Vanhaeren, M., d'Errico, F., Stringer, C., James, S. L., Todd, J. A., & Mienis, H. K. (2006). Middle Paleolithic shell beads in Israel and Algeria. *Science, 312*, 1785–1788.

Vermeersch, P., Otte, M., Gilot, E., Paulissen, E., Gijselings, G., & Drappier, D. (1982). Blade technology in the Egyptian Nile Valley: Some new evidence. *Science, 216*, 626–628.

Villa, P., Soriano, S., Tsanova, T., Degano, I., Higham, T. F. G., d'Errico, F., et al. (2012). Border Cave and the beginning of the Later Stone Age in South Africa. *PNAS, 109*, 13208–13213.

Wadley, L. (2008). The Howieson's Poort industry of Sibudu cave. *South African Archaeological Society Goodwin Series, 10*, 122–132.

Wadley, L., Hodgskiss, T., & Grant, M. (2009). Implications for complex cognition from the hafting of tools with compound adhesives in the Middle Stone Age, South Africa. *PNAS, 106*, 9590–9594.

White, T., Asfaw, B., DeGusta, D., Gilbert, H., Richards, G., Suwa, G., & Howell, F. (2003). Pleistocene *Homo sapiens* from Middle Awash, Ethiopia. *Nature, 423*, 742–747.

Yellen, J., Brooks, A., Helgren, D., Tappen, M., Ambrose, S., Bonnefille, R., et al. (2005). The archaeology of Aduma Middle Stone Age Sites in the Awash Valley, Ethiopia. *PaleoAnthropology, 2005*, 25–100.

Chapter 13
Environmental Change, Ungulate Biogeography, and Their Implications for Early Human Dispersals in Equatorial East Africa

J. Tyler Faith, Christian A. Tryon, and Daniel J. Peppe

Abstract To better understand the potential role of environmental change in mediating human dispersals across equatorial East Africa, this study examines the biogeographic histories of ungulates, including a summary of current knowledge and fossil evidence stemming from our fieldwork in the Kenyan portion of the Lake Victoria basin. Phylogeographic and paleontological evidence indicates that vegetation changes across Quaternary climate cycles mediated ungulate distributions and dispersals via the opening and closing of biogeographic barriers in equatorial East Africa. Dispersal capabilities would have been enhanced during phases of grassland expansion and diminished during phases of grassland contraction. We propose that the distribution and dispersal of diagnostic technological markers in the archaeological record may be similarly influenced by environmental changes. The Middle Stone Age record from the Lake Victoria region provides intriguing examples of possible environmentally mediated technological dispersals.

Keywords Grasslands • Lake Victoria • Late Pleistocene • Middle Stone Age • Paleoenvironments • Phylogeography • Range shift

Introduction

The prehistory of early modern humans is characterized by massive range expansions and population dispersals. Fossil and genetic evidence indicate an African origin of our species (*Homo sapiens*) ~200 ka (McDougall et al. 2005; Campbell and Tishkoff 2010; Brown et al. 2012; but see Weaver 2012 for an alternate interpretation). These early humans later diverged into multiple genetically (Campbell and Tishkoff 2010) and morphologically diverse (Gunz et al. 2009) populations during the Late Pleistocene (126–12 ka), likely reflecting intra-African population expansions and dispersals (Soares et al. 2012). At least one of these populations migrated out of Africa ~70 ka (Campbell and Tishkoff 2010), setting the stage for the expansion of humans across the globe and the replacement of all other hominin species, including Neanderthals (*Homo neanderthalensis*) in Eurasia, and the enigmatic "hobbit" (*Homo floresiensis*) in Indonesia.

East Africa features prominently in this history of early modern humans, as it provides the earliest fossil remains of *Homo sapiens* (Omo I and II from the Kibish Formation, Ethiopia, at ~195 ka; McDougall et al. 2005; Brown et al. 2012) and the probable source populations for their Late Pleistocene dispersals out of Africa (Soares et al. 2012, 2016). Understanding the factors underlying these dispersals is the subject of intense debate spanning the fields of archaeology, paleontology, genetics, and climatology, among others (Beyin 2011). One of the more common explanatory mechanisms includes climate-driven environmental change and its effects on human demography and the opening and closing of biogeographic barriers (e.g., Forster 2004; Finlayson 2005; Vaks et al. 2007; Carto et al. 2009; Trauth et al. 2010; Compton 2011; Blome et al. 2012; Eriksson et al. 2012; Soares et al. 2012, 2016). However, the precise role of environmental change in East African human dispersals is poorly understood, stemming from a lack of well-dated archaeological sites associated with detailed paleoenvironmental

J.T. Faith (✉)
School of Social Science, Archaeology Program, University of Queensland, Brisbane, QLD 4072, Australia
e-mail: j.faith@uq.edu.au

C.A. Tryon
Department of Anthropology, Harvard University, Peabody Museum of Archaeology and Ethnology, 11 Divinity Avenue, Cambridge, MA 02138, USA

D.J. Peppe
Terrestrial Paleoclimatology Research Group, Department of Geology, Baylor University, Waco, TX 76706, USA

Sacha C. Jones and Brian A. Stewart (eds.), *Africa from MIS 6-2: Population Dynamics and Paleoenvironments*, Vertebrate Paleobiology and Paleoanthropology, DOI 10.1007/978-94-017-7520-5_13

records. This is compounded by the high variability of the regional Middle Stone Age (MSA), which limits our ability to identify an archaeological signature of population movements (Tryon and Faith 2013).

While many fundamental questions concerning the role of environmental change in early human dispersals from East Africa remain unclear, a growing body of evidence indicates that Pleistocene environmental change played a central role in the biogeographic histories of associated East African faunas (e.g., Rodgers et al. 1982; Kingdon 1989; Marean 1992; Grubb et al. 1999; Wronski and Hausdorf 2008; Lorenzen et al. 2012; Faith et al. 2013). Not only does this imply that human population movements may be related to broader biogeographic patterns, but also that these insights into the relationships between changing climate, environments, and species ranges represent a useful starting point for formulating new hypotheses about human dispersals. Our aim here is to develop a framework for understanding the potential role of environmental change in Late Pleistocene human dispersals through an examination of ungulate biogeography. We focus on ungulates because they are well represented in the fossil record and because many ungulate species are clearly linked to specific habitats, making them strong candidates for exploring how environmental change mediates their biogeographic histories. This chapter provides a brief review of current knowledge together with an examination of paleontological and archaeological evidence stemming from our ongoing fieldwork in the Kenyan portion of the Lake Victoria basin.

Ungulate Biogeography

In common use, the ungulates represent a paraphyletic group of hoofed mammals, including in Africa the antelopes and buffalo (Bovidae), suids (Suidae), zebras (Equidae), giraffes (Giraffidae), rhinos (Rhinocerotidae), elephants (Elephantidae), and hippos (Hippopotamidae), among others. Ungulate diversity in Africa is exceptional (Turpie and Crowe 1994; du Toit and Cumming 1999), with the bovids (the most speciose family) alone represented by 82 extant species (IUCN 2012) and more than 100 fossil species (Gentry 2010).

The greatest diversity of African ungulates is found in equatorial East Africa (Turpie and Crowe 1994), which encompasses the boundaries between the Somalian, Sudanian, and Zambesian biogeographic regions for mammals, birds, reptiles, amphibians, and plants (Linder et al. 2012) (Fig. 13.1). These biogeographic regions, which are broadly similar to previously identified zones based on expert opinion (e.g., White 1983; Burgess et al. 2004), were statistically defined by Linder et al. (2012) using cluster analysis of species occurrences across 1° × 1° (latitude and longitude) cells for plants and vertebrate groups. We recognize that these zones are only a coarse measure of broad biogeographic patterns, and we use them here as a heuristic device to illustrate how biogeographic patterns changed in the past. Some of the ungulates characteristic of these regions include Grevy's zebra (*Equus grevyi*) in the xeric shrublands of the Somalian region, kob (*Kobus leche*) in the edaphic grasslands of the Sudanian region, and wildebeest (*Connochaetes taurinus*) in the secondary grasslands in the Zambesian region. All of these taxa are found in East Africa, but their ranges do not overlap, contributing to the high spatial turnover (beta diversity) observed more broadly for all East African mammals (Linder et al. 2012).

The main factors underlying these broad biogeographic regions include the complex interactions between climate and geomorphology (e.g., tectonics, topography, soil type) and their influence on the distribution of vegetation communities, which in turn mediate the distribution of faunas (e.g., Bell 1982; O'Brien and Peters 1999; Linder et al. 2012). For example, a drier climate characterized by bimodal rainfall distinguishes the Somalian region from both the Sudanian and Zambesian regions, which are wetter and receive primarily uni-modal rainfall (O'Brien and Peters 1999). This aridity in the Somalian region translates to the presence of dry scrub vegetation inhabited by arid-adapted ungulates, including Grevy's zebra and African wild ass (*Equus africanus*). At the same time, species ranges are also influenced by geomorphological factors. In addition to mediating vegetation cover and habitat suitability through altitudinal gradients, tectonically driven uplift and rifting can create dispersal barriers to biotic communities, as evidenced

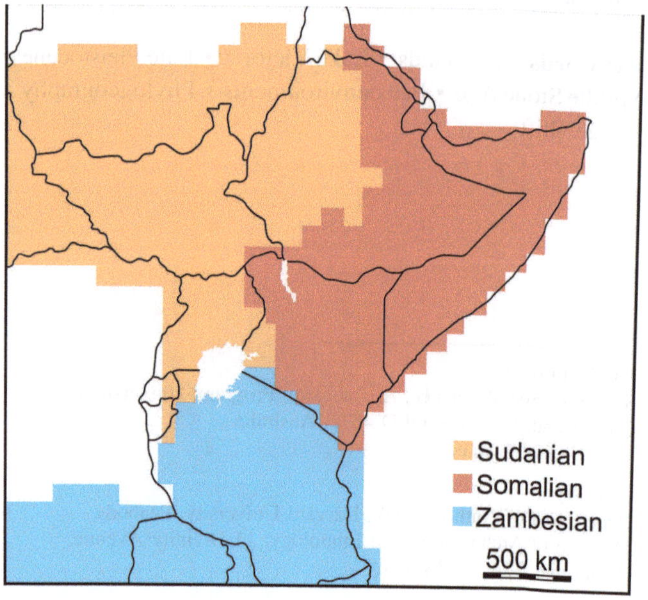

Fig. 13.1 The distribution of Sudanian, Somalian, and Zambesian biogeographic regions for mammals (after Linder et al. 2012)

by the close correspondence between the boundary of the Sudanian and Somalian regions in Ethiopia (Fig. 13.1) and the location of the Ethiopian Rift Valley, which bisects the Ethiopian highlands. Lastly, soil type can also determine broadscale biogeographic patterns. For example, the dominance of low-nutrient soils weathered from basement rock accounts for the Miombo (*Brachystegia*) woodlands and low-biomass ungulate communities characteristic of the Zambesian region, in stark contrast to the nutrient-rich volcanic soils in some portions of the Somalian region, which are associated with grasslands and high-biomass ungulate communities (Bell 1982; East 1984).

Not only does the combination of climatic and geomorphological variables in East Africa translate to biogeographic zonation of ungulate species, but the same is also true for populations within species. Lorenzen et al. (2012) show that many broadly distributed species, including warthog (*Phacochoerus africanus*), African buffalo (*Syncerus caffer*), hartebeest (*Alcelaphus buselaphus*), and giraffe (*Giraffa camelopardalis*), among others, display genetic substructuring that distinguishes between populations from the Sudanian savannas, and those in East Africa and southern Africa. Among some species there is also evidence for substructuring between East and southern African populations. The overall picture indicates that East Africa encompasses the boundaries of genetically distinct lineages from multiple regions.

A number of factors potentially contribute to this unique taxonomic and genetic diversity. High spatial variability in forage quantity and quality has been invoked as one possible driver of ungulate species richness in East Africa (du Toit and Cumming 1999), an argument supported by quantitative analyses showing extreme spatial heterogeneity in East African plant communities (Linder et al. 2012). However, this does not explain the convergence of genetically distinct lineages within species. In light of the phylogeographic evidence, Lorenzen et al. (2012) propose that environmental changes across late Quaternary glacial/interglacial cycles played an important role in the taxonomic and genetic diversity of East African ungulates. Equating glacial and interglacial conditions with dry versus humid climates, they suggest that the expansion of grassland vegetation during glacial phases would have facilitated interchange of ungulate populations throughout sub-Saharan Africa. During interglacials, the expansion of tropical forests across equatorial East Africa created a barrier that fragmented populations and restricted gene flow (see also Cowling et al. 2008). If so, then the distinct biogeographic regions and high spatial heterogeneity of biotic communities found in East Africa today (Linder et al. 2012) may be a Holocene phenomenon, with a more homogenous supercommunity characteristic of Pleistocene phases of grassland expansion.

The Fossil Record

Biogeographic evidence suggests that East Africa is a hub around which ungulate ranges expand, contract, and fragment across cycles of Quaternary environmental change. One of the key predictions generated from this is that during phases with expanded grassland cover, allopatric ungulate species from both north and south of the equator will converge in equatorial East Africa. The fossil record provides the requisite empirical evidence for testing this hypothesis. Compared to the record from southern Africa, which has been a focus of modern human origins research for decades, our understanding of the fossil history of East African ungulates over the last ~200 kyr is only beginning to come to light (Marean and Gifford-Gonzalez 1991; Marean 1992; Assefa 2006; Assefa et al. 2008; Faith et al. 2011, 2012, 2013, 2014). However, the emerging evidence provides compelling examples of climate-driven range shifts that are consistent with hypotheses derived from ungulate biogeography.

The Records from Rusinga and Mfangano Islands

Here we highlight some of the more prominent range shifts documented in the fossil record, with an emphasis on the Late Pleistocene ungulates from Rusinga and Mfangano Islands in Kenya's Lake Victoria basin (Fig. 13.2). We focus on these sites because: (1) all of the ungulate taxa involved in major Late Pleistocene range shifts are documented here (and in some cases elsewhere); and (2) the Lake Victoria region is strategically situated along the equator in East Africa, with geological evidence documenting massive expansions and contractions in lake area, including periodic desiccation, in response to late Quaternary climate change (Stager and Johnson 2008; Stager et al. 2002, 2011; Tryon et al. in press). Seismic profiles across Lake Victoria show at least four desiccation surfaces (Johnson et al. 1996; Stager and Johnson 2008), the most recent of which correspond to extreme aridity during Heinrich Event 1 from 17 to 16 ka and again during a subsequent dry phase from 15 to 14 ka (Stager et al. 2002, 2011); the other desiccation surfaces correspond to undated phases of previous aridity and the pre-lake surface. The vegetation of the Lake Victoria region was historically characterized by bushland, thicket, and forest (White 1983), although multiple lines of evidence indicate that phases of lake contraction were associated with expanded grasslands (Kendall 1969; Talbot and Laerdal 2000; Talbot et al. 2006; Tryon et al. in press).

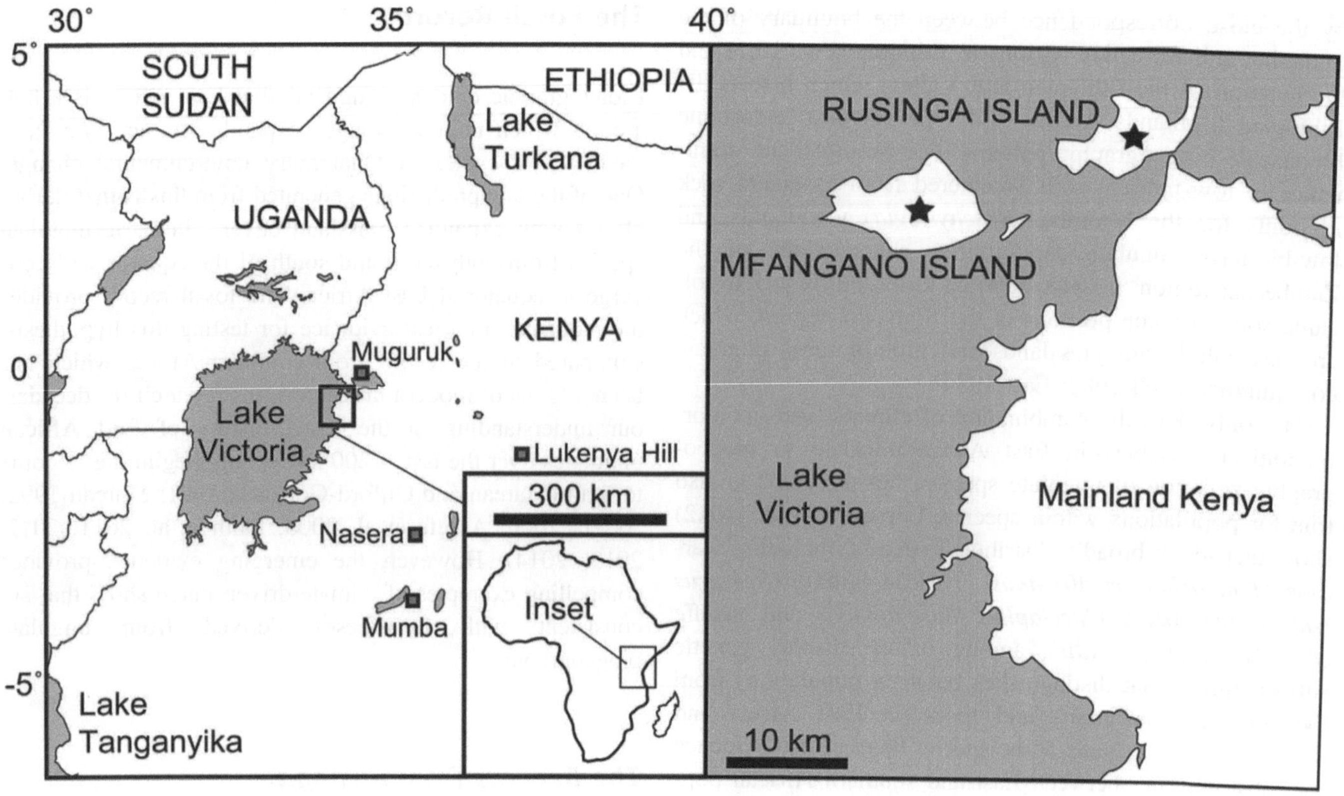

Fig. 13.2 The location of Rusinga and Mfangano islands in Kenyan Lake Victoria and nearby archaeological sites mentioned in the text

Details of our research on Rusinga Island and Mfangano Island have been published elsewhere (Faith et al. 2011, 2012, 2014; Tryon et al. 2010, 2012, 2014), and we provide only a brief summary here. Rusinga and Mfangano Islands are located within Lake Victoria (Fig. 13.2), with the former separated from the mainland by a narrow channel and the latter situated ~10 km from the mainland. On both islands, the poorly consolidated Pleistocene deposits, known as the Wasiriya Beds (Rusinga Island) and the Waware Beds (Mfangano Island), are characterized by weakly developed paleosols and tuffaceous fluvial sediments documenting a complex cut-and-fill system. The age of the Wasiriya Beds is constrained to between 100 and 33 ka, with the minimum provided by radiocarbon dates for intrusive gastropods and the maximum by geochemical analysis of tephra deposits, which suggest derivation from East African Rift System (EARS) volcanoes that began erupting ~100 ka (Blegen et al. 2015). The Waware Beds have a similar minimum age, also provided by radiocarbon age estimates for intrusive gastropods, whereas tephra correlations suggest a link to the Wasiriya Beds and a comparable maximum age (Blegen et al. 2015). The artifacts recovered from the Wasiriya Beds are typologically MSA (Fig. 13.3), and the same is likely the case for those from the Waware Beds, although a larger sample is needed to be certain.

Rusinga and Mfangano Islands have yielded some of the larger and better preserved fossil assemblages from this time period in East Africa (Table 13.1). Alcelaphine bovids and equids dominate the assemblages, indicating the presence of open grassland vegetation distinct from the historic vegetation (White 1983). Several extinct bovids are present, including *Rusingoryx atopocranion*, *Damaliscus hypsodon*, *Megalotragus* sp., *Syncerus antiquus*, and an unnamed impala, all of which are characterized by dental or postcranial adaptations to grazing in open habitats (Klein 1980, 1994; Faith et al. 2011, 2012, 2014). The presence of large gregarious grazers on Mfangano Island, some of which are migratory species, suggests a likely connection to the mainland when the deposits accumulated, requiring a ≥25 m reduction in lake level (Tryon et al. 2014). The precipitation decline needed to drive this would have set in motion a series of feedback mechanisms leading to a substantial reduction, if not complete desiccation, of Lake Victoria (Broecker et al. 1998; Milly 1999), leaving behind a topographically smooth grassy plain (Tryon et al. 2014). As detailed below, there is strong evidence that the expansion of grasslands and reduction in surface area of Lake Victoria facilitated interchange of ungulates from north and south of the equator.

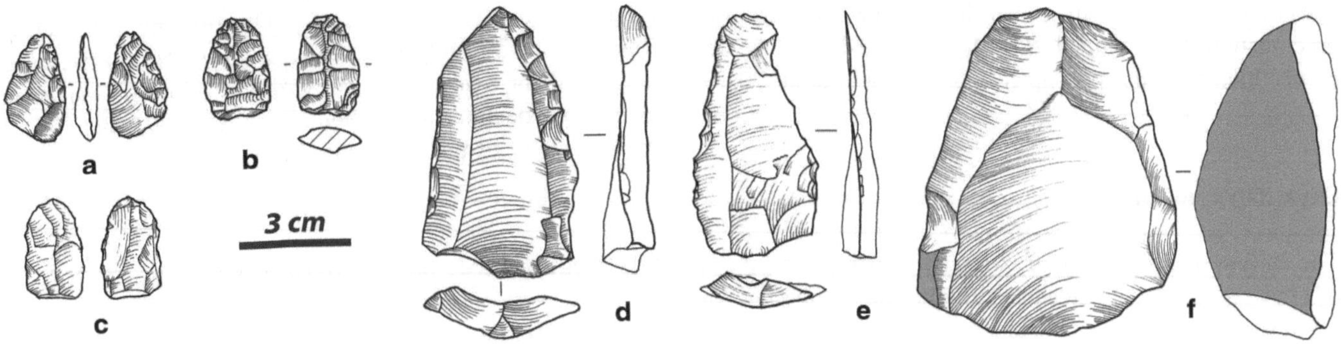

Fig. 13.3 Middle Stone Age artifacts from Rusinga and Mfangano Islands: **a–c** bifacial points, **d–e** Levallois blades variably retouched, **f** Levallois core

Table 13.1 The presence/absence of ungulate species from Rusinga and Mfangano islands as of our 2012 field season

Family	Taxon	Common name	Rusinga	Mfangano
Elephantidae	Elephantidae *cf. Loxodonta africana*	Elephant	X	–
Rhinocerotidae	*Ceratotherium simum*	White rhinoceros	X	–
Equidae	*Equus quagga*	Plains zebra	X	X
	Equus grevyi	Grevy's zebra	X	–
Suidae	*Kolpochoerus sp.†*	Extinct bushpig	X	–
	Phacochoerus sp.	Warthog	X	X
	Potamochoerus sp.	Bushpig	X	X
Hippopotamidae	*Hippopotamus sp.*	Hippopotamus	X	X
Bovidae	*Taurotragus oryx*	Eland	X	–
	Tragelaphus scriptus	Bushbuck	X	–
	Tragelaphus strepsiceros	Greater Kudu	X	–
	Tragelaphus cf. imberbis	?Lesser Kudu	X	–
	Oryx beisa	Oryx	X	–
	Redunca fulvorufula/redunca	Reedbuck	X	X
	Redunca arundinum/Kobus kob	Southern reedbuck/Kob	X	X
	Aepyceros sp. nov.*†*	Large extinct impala	X	–
	Connochaetes taurinus	Wildebeest	X	X
	Damaliscus hypsodon†	Small extinct alcelaphine	X	X
	Megalotragus sp.†	Giant wildebeest	X	–
	Rusingoryx atopocranion†	Extinct alcelaphine	X	X
	Alcelaphini cf. *Alcelaphus buselaphus*	Hartebeest	X	X
	Sylvicapra grimmia	Common duiker	X	–
	Gazella thomsoni	Thomson's Gazelle	X	X
	Syncerus antiquus†	Long-horn buffalo	X	X
	Syncerus caffer	African buffalo	X	X
	Ourebia ourebi	Oribi	X	X
	Oreotragus/Raphicerus	Klipspringer/Steenbok	–	X
	Madoqua sp.	Dik-dik	–	X

† = extinct

Grevy's Zebra (*Equus grevyi*)

Grevy's zebra historically ranged throughout arid to semi-arid grasslands and shrublands in the Horn of Africa (Fig. 13.4) (Williams 2002). Fossil remains are known from Rusinga Island (Table 13.1) and many other Middle-to-Late Pleistocene sites in southern Kenya and northern Tanzania, including Karungu, Lukenya Hill, Lainyamok, and Kisese II, documenting a south and westward range shift or expansion of at least 500 km (Faith et al. 2013). Although the species is primarily confined to the Somalian biogeographic region, its fossil occurrences are found in areas that are today part of the Somalian, Sudanian, and Zambesian biogeographic regions. At the onset of the Holocene, Grevy's zebra disappeared from southern Kenya and northern Tanzania and became rare in the fossil record, likely reflecting increased precipitation, the contraction of dry grassland habitats, and competition with mesic-adapted grazers (Faith et al. 2013).

Oryx (*Oryx beisa*)

The current distribution of oryx broadly corresponds to that of Grevy's zebra, but extends south into the Zambesian and northwest into the Sudanian biogeographic regions (Fig. 13.4). Oryx has a similar habitat preference as Grevy's zebra, and the two species are significantly associated in the fossil record (Faith et al. 2013). The presence of oryx on Rusinga Island (Table 13.1) indicates a ~250 km westward range shift or expansion. Like Grevy's zebra, its range and abundance declined throughout East African fossil sites at the beginning of the Holocene (Faith et al. 2013).

White Rhinoceros (*Ceratotherium simum*)

The white rhinoceros includes two subspecies, the northern white rhinoceros (*C. simum cottoni*) and southern white rhinoceros (*C. simum simum*), although some authorities

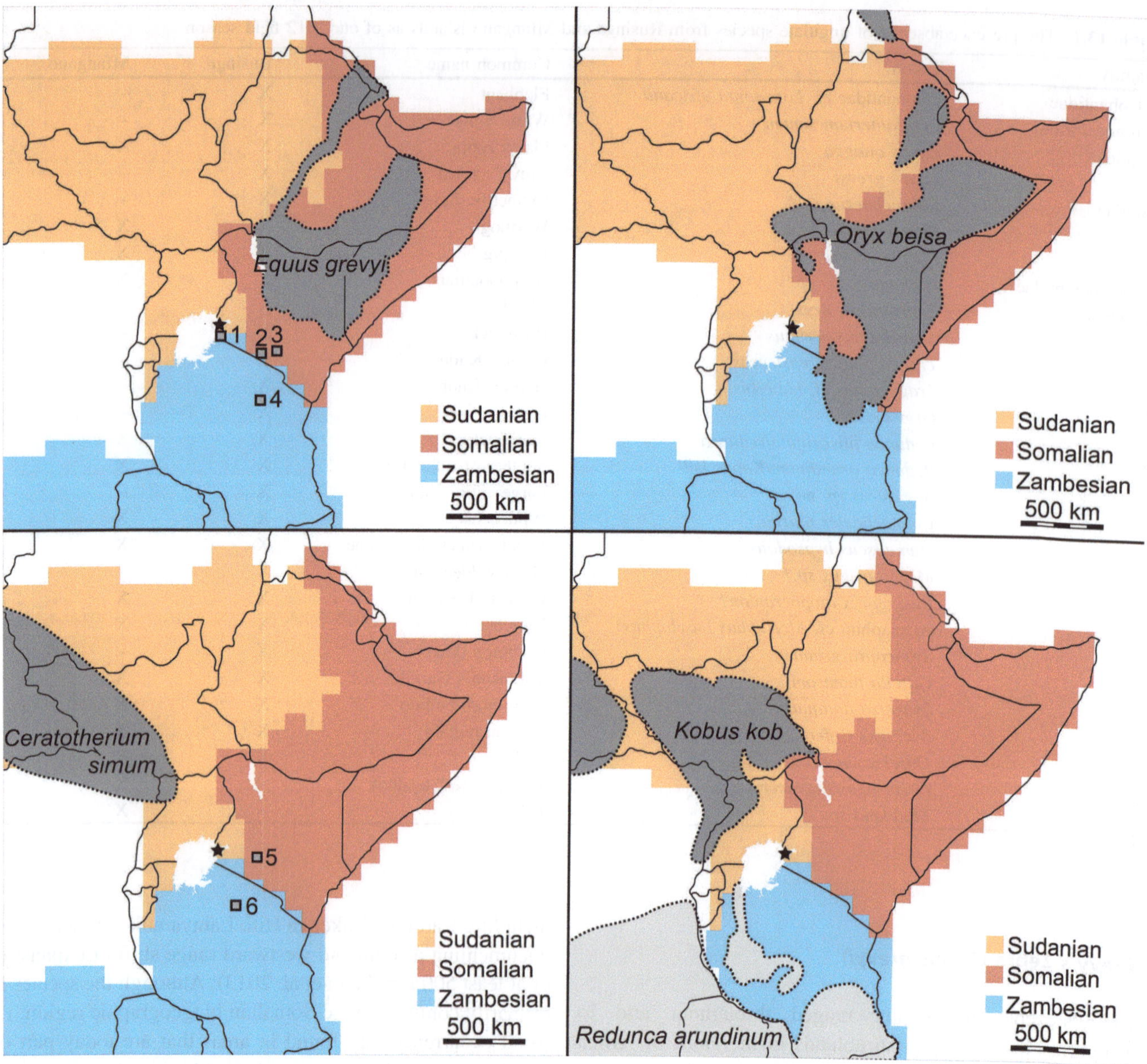

Fig. 13.4 The geographic ranges of Grevy's zebra (*Equus grevyi*), oryx (*Oryx beisa*), white rhinoceros (*Ceratotherium simum*), kob (*Kobus kob*), and southern reedbuck (*Redunca arundinum*) relative to their fossil occurrences on Rusinga and Mfangano islands (*star*). Range maps are from Williams (2002) for Grevy's zebra and the IUCN Red List (2012) for all other species. Boxes correspond to fossil occurrences mentioned in the text: (*1*) Karungu, (*2*) Lainyamok, (*3*) Lukenya Hill, (*4*) Kisese II, (*5*) Prolonged Drift, (*6*) Mumba shelter

place the former in its own species (*C. cottoni*) (Groves et al. 2010). Their geographic ranges are discontinuous, with the southern white rhinoceros historically known from southern Africa and the northern white rhinoceros known from central Africa northwest of Lake Albert (Fig. 13.4). Several white rhinoceros fossils have been recovered from Rusinga Island, ~430 km southeast of its historic range (assuming they belong to the northern taxon), with an even greater Late Pleistocene range extension (~775 km) documented at Mumba rockshelter in Tanzania (Mehlman 1989). As evidenced by an additional fossil occurrence from Prolonged Drift in central Kenya, a small and probably isolated population persisted in equatorial East Africa into the late Holocene (Gifford et al. 1980). The cause of its prehistoric disappearance from Kenya is not well understood, but could include long-term vegetation change, competition for forage and water with pastoralists and their livestock, or stochastic processes owing to small population size and isolation.

Kob (*Kobus kob*) or Southern Reedbuck (*Redunca arundinum*)

Both Rusinga and Mfangano Islands yield dental remains of a medium-sized reduncine, larger than Bohor reedbuck (*Redunca redunca*), but smaller than waterbuck (*Kobus ellipsiprymnus*), that are morphologically similar to and overlap in size with modern kob and southern reedbuck. Both taxa have a similar habitat requirement for fresh grasses and year-round access to water, which is consistent with the fluvial nature of the fossil deposits. The ranges of these two reduncines are allopatric (Fig. 13.4), with kob occurring north and west of Lake Victoria in the Sudanian savannas and southern reedbuck ranging from southern Africa to just south of the Lake Victoria basin. Although it is unclear whether the fossil specimens belong to kob, southern reedbuck, or both, the occurrence of either implies a ~350 km range shift to the southeast (kob) or northeast (southern reedbuck).

A Non-Analog Supercommunity

The convergence of various ungulate species with historically allopatric ranges on Rusinga and Mfangano Islands supports the notion of a homogenous East African supercommunity during Pleistocene phases of grassland expansion. It follows that the distinct biogeographic zonation found in East Africa today (Fig. 13.1; Linder et al. 2012) is ephemeral—probably limited to the Holocene—and closely linked to climate and its effects on vegetation structure. Other variables that contribute to the establishment of biogeographic zonation in East Africa (e.g., tectonics, topography, and soil type), unquestionably contributed to Late Pleistocene vegetation structure, but these are expected to be relatively stable over the timescale examined here, leaving climate as the most likely driver of first-order vegetation change.

While the non-analog supercommunity is consistent with enhanced ungulate dispersal capabilities related to expanded grasslands and a smaller Lake Victoria, it is also possible that environments at the time were highly productive and uniquely suited to supporting ungulate diversity. Across a range of East and southern African ecosystems, ungulate diversity and biomass peaks between 700 and 800 mm annual precipitation (Coe et al. 1976; East 1984; Faith 2013). A decline in annual precipitation from the ~1000 to 1200 mm/yr observed today, which is supported by the inferred decline of Lake Victoria (Tryon et al. 2014), could further underpin the diversity of the Late Pleistocene ungulate community. Reduced atmospheric CO_2 concentrations may have also contributed to the diversity, by enhancing foraging nutrient content, in turn supporting higher biomass large herbivore communities (Faith 2011).

The presence of extinct grazing bovids characterized by massive body mass (*Syncerus antiquus* and *Megalotragus* sp.) or extreme hypsodonty (*Damaliscus hypsodon*, *Rusingoryx atopocranion*, and the unnamed impala), of which *D. hypsodon* and *R. atopocranion* are dominant, also implies the presence of non-analog environments. We have previously interpreted their presence, together with that of Grevy's zebra and oryx, as evidence for aridity (Tryon et al. 2010, 2012; Faith et al. 2011, 2013), although this is at odds with the high diversity (see Faith 2013). A more detailed exploration of this conflict is beyond the scope of this study, but possible explanations could include the presence of a complex Late Pleistocene grazing succession (e.g., Brink and Lee-Thorp 1992) or an extinct migratory system (e.g., Marean 2010; Faith and Thompson 2013).

Climatic and Tectonic Drivers

Range shifts documented on Rusinga and Mfangano Islands illustrate the importance of vegetation change, particularly the expansion of grassland vegetation, in the biogeographic histories of ungulate populations. The mechanisms responsible for the expansion and contraction of grassland vegetation in equatorial East Africa during the Pleistocene include moisture availability, atmospheric CO_2 concentrations, and tectonically driven topographic shifts. The complex interaction of these variables would have played an important role in the opening and closing of dispersal corridors for ungulate populations, and we provide a summary of these mechanisms below.

Changes in moisture availability are sometimes equated with glacial versus interglacial climates, which are typically generalized as reflecting more arid (glacial) versus more humid (interglacial) conditions. However, although changes in global moisture availability across glacial/interglacial cycles may influence equatorial East African climate, we caution that this oversimplification ignores the complexities of regional climate dynamics. Moisture availability in equatorial East Africa is driven by the complex interplay between orbital forcing, the position of the Intertropical Convergence Zone (ITCZ), high-latitude climate events, tropical ocean temperature gradients, and regional tectonics (e.g., DeMenocal 1995, 2004, 2011; Trauth et al. 2003, 2005, 2007, 2009; Feakins et al. 2005, 2013; Sepulchre et al. 2006; Verschuren et al. 2009; Blome et al. 2012). In addition to these global and regional drivers, local factors may also play an important role in vegetation change. For example, today as much as 90% of water loss in Lake Victoria is due to evaporation and 80% of the input is from direct precipitation (rather than inflow from streams and rivers) (Piper et al. 1986; Crul 1995), such that precipitation change plays a key role in lake size and moisture availability (Broecker et al. 1998; Milly 1999). This sensitivity to precipitation means that small changes in local rainfall patterns could translate to more pronounced aridification and vegetation change in the Lake Victoria region compared to other areas of East Africa.

In addition to moisture availability, reduced atmospheric CO_2 concentrations during Pleistocene glacial phases may also facilitate the expansion of grasslands due to the competitive advantage of C_4 plants under such conditions (Prentice et al. 2011). For example, the paleoenvironmental record from Lake Challa on the Kenya/Tanzania border shows that both humid and arid intervals from 12 to 25 ka were dominated by C_4 vegetation, most of which are probably grasses, whereas C_3 vegetation only expands at the onset of the Holocene (Sinninghe Damsté et al. 2011). This may indicate that the past distribution of grassland habitats in equatorial East Africa is more sensitive to atmospheric CO_2 than to precipitation.

Over longer timescales (>100 kyr) through the Pleistocene, changes in ungulate ranges may also have been influenced by regional tectonics. Extension and uplift associated with the EARS has dramatically altered the East African landscape since the Oligocene by creating large rift lakes such as Lakes Malawi, Tanganyika, Kivu, Albert and Edwards, causing the opening and closure of connections between major lakes, changing flow directions of major rivers, and creating significant differences in topography (e.g., Rosendahl 1987; Chorowicz 2005). Lake Victoria, though not a true rift lake, was created when extension and uplift of the western and east arms of the EARS caused backponding of rivers into a topographic low between the two branches of the rift ~1.6–0.4 Ma (e.g., Kent 1944; Bishop and Trendall 1967; Ebinger 1989; Johnson et al. 1996; Talbot and Williams 2009). After its formation, continued rifting caused the lake basin to shift as much as 50 km eastward (Doornkamp and Temple 1966; Stager and Johnson 2008). Through the creation of rift-influenced lakes and significant topographic variability across the landscape, the rifting of the EARS may have influenced ungulate biogeographic histories by opening and closing potential dispersal corridors. Although precipitation and atmospheric CO_2 concentrations are the primary drivers of vegetation change, tectonically driven topographic changes probably also contributed to habitat fragmentation and variability across East Africa, which could have affected the ranges of ungulate taxa throughout the Pleistocene.

There is little doubt that climate and tectonics strongly influence East African paleoenvironments during the Pleistocene. However, as emphasized by Blome et al. (2012), determining the precise nature of environmental changes and their underlying mechanisms in the Lake Victoria basin and across East Africa will require improved chronological control and more refined paleoenvironmental records.

Implications for Early Human Dispersals

Biogeographic evidence suggests that climate-driven vegetation changes across equatorial East Africa played an important role in mediating the distribution of ungulate species, with the expansion of grasslands facilitating interchange across East Africa (Lorenzen et al. 2012). This is corroborated by the fossil records from Rusinga and Mfangano Islands, which show that ungulate species with historically allopatric ranges converged in equatorial East Africa at a time when Lake Victoria was much reduced, if not desiccated, and grasslands were widespread (Tryon et al. 2014). Presumably, the loss of grasslands and expansion of the equatorial forest belt at the onset of the Holocene and during previous humid intervals with high atmospheric CO_2 concentrations (Cowling et al. 2008) would have prompted range shifts and fragmentations in species distributions approaching those observed historically. What are the implications of these patterns for early modern human dispersals?

In the Paleolithic archaeological record, diagnostic lithic technology represents one of the few means of tracking human dispersals. On its own, the extent to which the movement of technological markers reflects the dispersal of human populations (and their genes) is uncertain, but it at least documents the dispersal of human behavioral repertoires. There is abundant ethnographic evidence showing that the subsistence behaviors and associated technologies of historic hunter-gatherers are very closely linked to the environment (Oswalt 1973; Kelly 1995; Binford 2001; Collard et al. 2005), in which case it is reasonable to suppose that the movement of technological markers linked to specific environments might parallel the broader patterns observed in the ungulates. Below we explore some possible links between the distributions of technologies in the Lake Victoria region in relation to paleoenvironmental change.

The Lake Victoria MSA

Previous discussions of distinct technological markers from the Lake Victoria region have focused on the Lupemban industry, which represents one of the more distinct MSA regional variants, characterized by the presence of large (>10 cm), thin, bifacially flaked lanceolate points. These lanceolates are well known from central Africa (Barham 2000; Mercader 2002), although their distribution extends to the eastern margins of Lake Victoria, including near-shore islands (Leakey and Owen 1945; Nenquin 1971; McBrearty 1988; Tryon et al. 2012; Taylor 2016). Where paleoenvironmental records are available, the Lupemban is typically associated with forested vegetation (Barham 2000; Mercader 2002; Taylor 2016).

The vegetation of the Lake Victoria region today represents a mosaic of endemic taxa from neighboring regions (White 1983), attesting to the past expansion and contraction of vegetation communities, a model consistent with the implications of the ungulates from Rusinga and Mfangano Islands (Table 13.1), the distribution of modern faunas (e.g., Kingdon 1989; Wronski and Hausdorf 2008), and paleo-vegetation models (e.g., Cowling et al. 2008). To the extent that the distribution of Lupemban artifacts is tied to forested paleoenvironments, we expect that its easternmost occurrences in the Lake Victoria basin to be associated with an expanded equatorial forest belt. As shown in Fig. 13.5, all of the Lupemban sites in the eastern portion of the Lake Victoria basin lie along now submerged portions of major rivers that traversed the basin prior to peak lake size sometime during the Middle or Late Pleistocene (Temple 1966; Scholz et al. 1998). These sites represent the easternmost limits of the Lupemban, and include Muguruk on the Winam Gulf in Kenya (McBrearty 1988), sites reported by Leakey and Owen (1945) in the Yala River valley of Kenya, and on Bugaia Island in Uganda (Nenquin 1971). The nearby rivers would have likely provided naturally wooded corridors facilitating the expansion of forest-adapted taxa and technologies, perhaps including the Lupemban, around the margins of Lake Victoria.

The MSA assemblages from the Wasiriya and Waware Beds of Rusinga and Mfangano Islands are typologically distinct from Lupemban assemblages in the Lake Victoria region or elsewhere, which is not surprising given the associated grassy paleoenvironment. In light of the evidence from ungulate biogeography, it is possible that the expansion of grasslands would facilitate the movement of people or technological traditions associated with these environments from broad East African regions north or south of the equator. In previous studies (Tryon et al. 2012; Tryon and Faith 2013), we identified one possible example of this. The lithic assemblages from Rusinga and Mfangano Islands (Fig. 13.3) include very small (2–4 cm in length) bifacially flaked points that overlap in size with those from MSA sites throughout the East African Rift System. These include the Late Pleistocene MSA or MSA/LSA assemblages from Nasera Rockshelter in Tanzania (~290 km southeast), site GvJm-16 at Lukenya Hill in Kenya (~340 km east southeast), and Aduma in Ethiopia (~1,390 km northeast). At all of these sites, the associated faunas or other paleoenvironmental records indicate grassland vegetation comparable to that inferred for Rusinga and Mfangano Islands, raising the possibility that these small points represent either a human population or (more likely) a technology associated with Late Pleistocene tropical grasslands. This possible connection to Rift Valley sites is further supported by the rare presence of obsidian from the Wasiriya Beds, the nearest outcrops of which are from Rift Valley sources ~250 km to the east.

The small MSA points found on Rusinga and Mfangano Islands and in the Rift Valley are rare to the west of Lake Victoria, reported only from Kibwera in Tanzania (Reid and Njau 1994), although unfortunately no illustrations or photographs are provided (Fig. 13.5). Assuming that points and other artifacts produced by hunter-gatherers are to some extent a reflection of the environment in which they were used, the geographic limits of different types of MSA point (small points versus Lupemban lanceolates) along the margins of present-day Lake Victoria (Fig. 13.5) is consistent with the hypothesis that this area is a nexus of environmental changes that facilitated the movement of people and their technological markers.

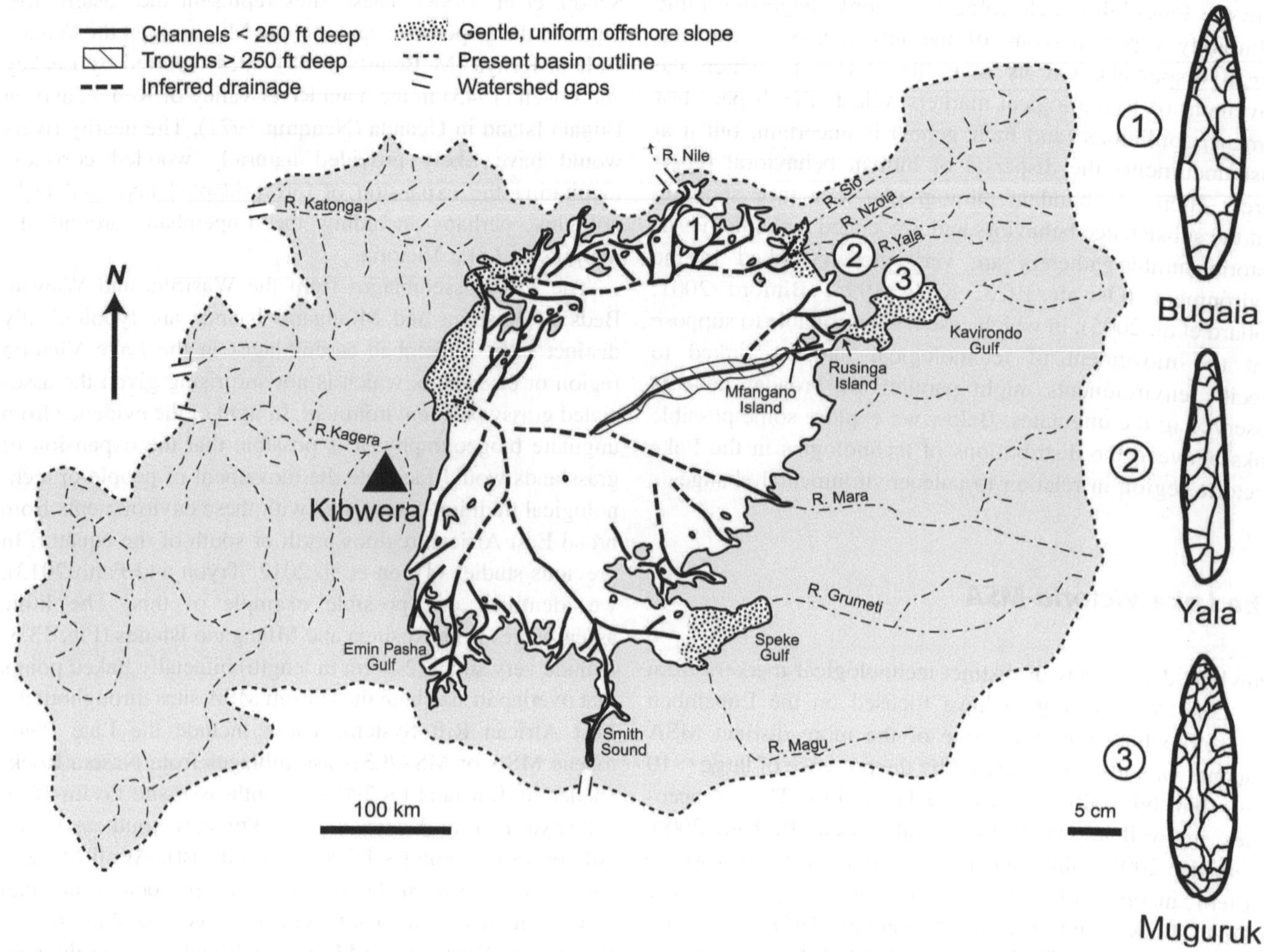

Fig. 13.5 Buried river channels across the Lake Victoria basin, simplified after Temple (1966). Note the proximity of the easternmost Lupemban lanceolate sites to major east-west flowing rivers that traverse the basin. Smaller MSA points are not reported in the western portion of the Lake Victoria basin except at Kibwera (illustrations unavailable). Artifact sketches after Leakey and Owen (1945), Nenquin (1971), and McBrearty (1988), and illustrated at same scale

Conclusions

Biogeographic evidence provides a compelling argument that environmental changes across late Quaternary climate cycles mediated the distributions and dispersals of ungulate species via the opening and closing of barriers and dispersal corridors in equatorial East Africa. For a broad range of savanna ungulates, dispersal capabilities would have been enhanced during phases of grassland expansion and diminished during phases of grassland contraction, leading to repeated range expansions and fragmentations.

These patterns are of importance to human origins research, as genetic evidence documents multiple East African dispersals of early modern humans during the Late Pleistocene, both within and out of Africa (Campbell and Tishkoff 2010; Soares et al. 2012, 2016). We propose that, as is clearly the case for ungulates, the distribution and dispersals of diagnostic technological markers in the archaeological record may also be mediated by environmental changes. The archaeological record from the Lake Victoria region provides some intriguing examples of possible environmentally mediated technological dispersals, but it is clear that a better sampled archaeological record and improved chronology is crucial to fleshing out these patterns. Although such limitations are an ever-present problem that can only be resolved through further fieldwork, the patterns explored here provide an initial framework for exploring the dispersal of early human populations migrations and expansions during a critical stage in human evolution.

Acknowledgments We thank Sacha Jones and Brian Stewart for inviting us to contribute to this volume and the many collaborators who have worked with us in the Lake Victoria basin, including Emily Beverly, Nick Blegen, Steve Driese, David Fox, Niki Garrett, Zenobia Jacobs, Kirsten Jenkins, Renaud Joannes-Boyau, Cara Roure Johnson, Kieran McNulty, Sheila Nightingale, David Patterson, and Alex Van

Plantinga. We also thank two anonymous reviewers and Sally Reynolds (reviewer) for helpful comments on a previous version of this chapter. Fieldwork was conducted under research permits NCST/RCD/12B/012/31 issued to JTF and NCST/5/002/R/576 issued to CAT and an exploration and excavation license issued by the National Museums of Kenya (NMK). Our fieldwork is made possible through the support of the NMK and the British Institute in East Africa and with financial support from the National Science Foundation (BCS-0841530, BCS-1013199, BCS-1013108), the Leakey Foundation, the National Geographic Society, the University of Queensland, New York University, and Baylor University. Lastly, none of this would have been possible without the support of Cornel Faith, Rhonda Kauffman, Violet Tryon, and Sholly Gunter.

References

Assefa, Z. (2006). Faunal remains from Porc-Epic: Paleoecological and zooarchaeological investigations from a Middle Stone Age site in Southeastern Ethiopia. *Journal of Human Evolution, 51*, 50–75.

Assefa, Z., Yirga, S., & Reed, K. E. (2008). The large-mammal fauna from the Kibish Formation. *Journal of Human Evolution, 55*, 501–512.

Barham, L. S. (2000). *The Middle Stone Age of Zambia, South-Central Africa*. Bristol: Western Academic & Specialist Press.

Bell, R. H. V. (1982). The effect of soil nutrient availability on community structure in African ecosystems. In B. J. Huntley & B. H. Walker (Eds.), *Ecology of tropical savannas* (pp. 193–216). New York: Springer-Verlag.

Beyin, A. (2011). Upper Pleistocene human dispersals out of Africa: A review of the current state of the debate. *International Journal of Evolutionary Biology, 2011* Article ID 615094, 17 doi: 10.406/2011/615094.

Binford, L. R. (2001). *Constructing frames of reference: An analytical method for archaeological theory building using ethnographic and environmental data sets*. Berkeley: University of California Press.

Bishop, W. W., & Trendall, A. F. (1967). Erosion-surfaces, tectonics, and volcanic activity in Uganda. *Quarterly Journal of the Geological Society of London, 122*, 385–420.

Blegen, N., Tryon, C. A., Faith, J. T., Peppe, D. J., Beverly, E. J., Li, B., et al. (2015). Distal tephras of the eastern Lake Victoria Basin, Equatorial East Africa: Correlations, chronology, and a context for early modern humans. *Quaternary Science Reviews, 122*, 89–111.

Blome, M. W., Cohen, A. S., Tryon, C. A., Brooks, A. S., & Russell, J. (2012). The environmental context for the origins of modern human diversity: A synthesis of regional variability in African climate 150,000–30,000 years ago. *Journal of Human Evolution, 62*, 563–592.

Brink, J. S., & Lee-Thorp, J. A. (1992). The feeding niche of an extinct springbok, *Antidorcas bondi* (Antelopini, Bovidae), and its Paleoenvironmental meaning. *South African Journal of Science, 88*, 227–229.

Broecker, W. C., Peteet, D., Hajdas, I., & Lin, J. (1998). Antiphasing between rainfall in Africa's Rift Valley and North America's Great Basin. *Quaternary Research, 50*, 12–20.

Brown, F. H., McDougall, I., & Fleagle, J. G. (2012). Correlation of the KHS Tuff of the Kibish formation to volcanic ash layers at other sites, and the age of early *Homo sapiens* (Omo I and Omo II). *Journal of Human Evolution, 63*, 577–585.

Burgess, N. D., Hales, J. D., Underwood, E., Dinerstein, E., Olson, D., Itoua, I., et al. (2004). *Terrestrial ecoregions of Africa and Madagascar: A continental assessment*. Washington, DC: Island Press.

Campbell, M. C., & Tishkoff, S. A. (2010). The evolution of human genetic and phenotypic variation in Africa. *Current Biology, 20*, R166–R173.

Carto, S. L., Weaver, A. J., Hetherington, R., Lam, Y., & Wiebe, E. C. (2009). Out of Africa and into an ice age: on the role of global climate change in the late Pleistocene migration of early modern humans out of Africa. *Journal of Human Evolution, 56*, 139–151.

Chorowicz, J. (2005). The East African rift system. *Journal of African Earth Sciences, 43*, 379–410.

Coe, M. J., Cumming, D. H., & Phillipson, L. (1976). Biomass and production of large African herbivores in relation to rainfall and primary production. *Oecologia, 22*, 341–354.

Collard, M., Kemery, M., & Banks, S. (2005). Causes of toolkit variation among hunter-gatherers: A test of four competing hypotheses. *Canadian Journal of Archaeology, 29*, 1–19.

Compton, J. S. (2011). Pleistocene sea-level fluctuations and human evolution on the southern coastal plain of South Africa. *Quaternary Science Reviews, 30*, 506–527.

Cowling, S. A., Cox, P. M., Jones, C. D., Maslin, M. A., Peros, M., & Spall, S. A. (2008). Simulated glacial and interglacial vegetation across Africa: Implications for species phylogenies and trans-African migration of plants and animals. *Global Change Biology, 14*, 827–840.

Crul, R. C. M. (1995). *Limnology and hydrology of Lake Victoria*. Paris: UNESCO.

DeMenocal, P. (1995). Plio-Pleistocene African climate. *Science, 270*, 53–59.

DeMenocal, P. (2004). African climate change and faunal evolution during the Pliocene-Pleistocene. *Earth and Planetary Science Letters, 220*, 3–24.

DeMenocal, P. (2011). Climate and human evolution. *Science, 331*, 540.

Doornkamp, J. C., & Temple, P. H. (1966). Surface, drainage and tectonic instability in part of Southern Uganda. *The Geographical Journal, 132*, 238–252.

du Toit, J. T., & Cumming, D. H. M. (1999). Functional significance of ungulate diversity in African savannas and the ecological implications of the spread of pastoralism. *Biodiversity and Conservation, 8*, 1643–1661.

East, R. (1984). Rainfall, soil nutrient status and biomass of large African savanna mammals. *African Journal of Ecology, 22*, 245–270.

Ebinger, C. J. (1989). Tectonic development of the western branch of the East-African Rift System. *Geological Society of America Bulletin, 101*, 885–903.

Eriksson, A., Betti, L., Friend, A. D., Lycett, S. J., Singarayer, J. S., von Cramon-Taubadel, N., et al. (2012). Late Pleistocene climate change and the global expansion of anatomically modern humans. *Proceedings of the National Academy of Sciences of the USA, 109*, 16089–16094.

Faith, J. T. (2011). Late Pleistocene climate change, nutrient cycling, and the megafaunal extinctions in North America. *Quaternary Science Reviews, 30*, 1675–1680.

Faith, J. T. (2013). Ungulate diversity and precipitation history since the Last Glacial Maximum in the Western Cape, South Africa. *Quaternary Science Reviews, 68*, 191–199.

Faith, J. T., & Thompson, J. C. (2013). Fossil evidence for seasonal calving and migration of extinct blue antelope (*Hippotragus leucophaeus*) in Southern Africa. *Journal of Biogeography, 40*, 2108–2118.

Faith, J. T., Choiniere, J. N., Tryon, C. A., Peppe, D. J., & Fox, D. L. (2011). Taxonomic status and paleoecology of *Rusingoryx atopocranion* (Mammalia, Artiodactyla), an extinct Pleistocene bovid from Rusinga Island, Kenya. *Quaternary Research, 75*, 697–707.

Faith, J. T., Potts, R., Plummer, T. W., Bishop, L. C., Marean, C. W., & Tryon, C. A. (2012). New perspectives on middle Pleistocene change in the large mammal faunas of East Africa: *Damaliscus hypsodon* sp. nov. (Mammalia, Artiodactyla) from Lainyamok, Kenya. *Palaeogeography, Palaeoclimatology, Palaeoecology, 361–362*, 84–93.

Faith, J. T., Tryon, C. A., Peppe, D. J., & Fox, D. L. (2013). The fossil history of Grévy's zebra (*Equus grevyi*) in Equatorial East Africa. *Journal of Biogeography, 40*, 359–369.

Faith, J. T., Tryon, C. A., Peppe, D. J., Beverly, E. J., & Blegen, N. (2014). Biogeographic and evolutionary implications of an extinct late Pleistocene impala from the Lake Victoria Basin. *Journal of Mammalian Evolution, 21*, 213–222.

Feakins, S. J., DeMenocal, P. B., & Eglinton, T. I. (2005). Biomarker records of late Neogene changes in Northeast African vegetation. *Geology, 33*, 977–980.

Feakins, S. J., Levin, N. E., Liddy, H. M., Sieracki, A., Eglinton, T. I., & Bonnefille, R. (2013). Northeast African vegetation change over 12 m.y. *Geology, 41*, 295–298.

Finlayson, C. (2005). Biogeography and the evolution of the genus *Homo*. *Trends in Ecology & Evolution, 20*, 457–463.

Forster, P. (2004). Ice Ages and the mitochondrial DNA chronology of human dispersals: A review. *Philosophical Transactions of the Royal Society B: Biological Sciences, 359*, 255–264.

Gentry, A. W. (2010). Bovidae. In L. Werdelin & W. J. Sanders (Eds.), *Cenozoic mammals of Africa* (pp. 741–796). Berkeley: University of California Press.

Gifford, D. P., Isaac, G. L., & Nelson, C. M. (1980). Evidence for predation and pastoralism at prolonged drift. *Azania, 15*, 57–108.

Groves, C. P., Fernando, P., & Robovský, J. (2010). The sixth rhino: A taxonomic re-assessment of the critically endangered northern white rhinoceros. *PLoS ONE, 5*, e9703.

Grubb, P., Sandrock, O., Kullmer, O., Kaiser, T. K., & Schrenk, F. (1999) Relationships between eastern and southern African mammal faunas. In: T. G. Bromage & F. Schrenk (Eds.), *African biogeography, climate change, & human evolution* (pp. 253–267). Oxford: Oxford University Press

Gunz, P., Bookstein, F. L., Mitteroecker, P., Stadlmayr, A., Seidler, H., & Weber, G. W. (2009). Early modern human diversity suggests subdivided population structure and a complex out-of-Africa scenario. *Proceedings of the National Academy of Sciences of the USA, 106*, 6094–6098.

IUCN (2012). *IUCN Red List of Threatened Species. Version 2012.2.* www.iucnredlist.org.

Johnson, T. C., Scholz, C. A., Talbot, M. R., Kelts, K., Ricketts, R. D., Ngobi, G., et al. (1996). Late Pleistocene desiccation of Lake Victoria and the rapid evolution of cichlid fishes. *Science, 273*, 1091–1093.

Kelly, R. L. (1995). *The foraging spectrum*. Washington DC: Smithsonian Institution Press.

Kendall, R. L. (1969). An ecological history of the Lake Victoria basin. *Ecological Monographs, 39*, 121–176.

Kent, P. E. (1944). The Miocene beds of Kavirondo, Kenya. *Quarterly Journal of the Geological Society of London, 100*, 85–118.

Kingdon, J. (1989). *Island Africa: The evolution of Africa's rare animals and plants*. Princeton: Princeton University Press.

Klein, R. G. (1980). Environmental and ecological implications of large mammals from Upper Pleistocene and Holocene sites in southern Africa. *Annals of the South African Museum, 81*, 223–283.

Klein, R. G. (1994). The long-horned African buffalo (*Pelorovis antiquus*) is an extinct species. *Journal of Archaeological Science, 21*, 725–733.

Leakey, L. S. B., & Owen, W. E. (1945). *A contribution to the study of the Tumbian culture in East Africa*. Nairobi: Coryndon Memorial Museum.

Linder, H. P., de Clerk, H. M., Born, J., Burgess, N., Fjeldså, J., & Rahbek, C. (2012). The partitioning of Africa: Statistically defined biogeographical regions in Sub-Saharan Africa. *Journal of Biogeography, 39*, 1189–1205.

Lorenzen, E. D., Heller, R., & Siegismund, H. R. (2012). Comparative phylogeography of African savannah ungulates. *Molecular Ecology, 21*, 3656–3670.

Marean, C. W. (1992). Implications of late Quaternary mammalian fauna from Lukenya Hill (South-Central Kenya) for paleoenvironmental change and faunal extinctions. *Quaternary Research, 37*, 239–255.

Marean, C. W. (2010). Pinnacle Point Cave 13B (Western Cape Province, South Africa) in context: The Cape Floral kingdom, shellfish, and modern human origins. *Journal of Human Evolution, 59*, 425–443.

Marean, C. W., & Gifford-Gonzalez, D. (1991). Late Quaternary extinct ungulates of East Africa and Palaeoenvironmental implications. *Nature, 350*, 418–420.

McBrearty, S. (1988). The Sangoan-Lupemban and Middle Stone Age sequence at the Muguruk site, Western Kenya. *World Archaeology, 19*, 379–420.

McDougall, I., Brown, F. H., & Fleagle, J. (2005). Stratigraphic placement and age of modern humans from Kibish, Ethiopia. *Nature, 433*, 733–736.

Mehlman, M. J. (1989). Later Quaternary archaeological sequences in northern Tanzania. (Ph.D. Dissertation, University of Illinois, 1989).

Mercader, J. (2002). Forest people: The role of African rainforests in human evolution and dispersal. *Evolutionary Anthropology, 11*, 117–124.

Milly, P. C. D. (1999). Comment on "Antiphasing between rainfall in Africa's Rift Valley and North America's Great Basin". *Quaternary Research, 51*, 104–107.

Nenquin, J. (1971). Archaeological prospections on the islands of Buvuma and Bugaia, Lake Victoria Nyanza (Uganda). *Proceedings of the Prehistoric Society, 37*, 381–418.

O'Brien, E. M., & Peters, C. R. (1999). Landforms, climate, ecogeographic mosaics, and the potential for hominid diversity in Pliocene Africa. In T. G. Bromage & F. Schrenk (Eds.), *African biogeography, climate change, and human evolution* (pp. 115–137). Oxford: Oxford University Press.

Oswalt, W. H. (1973). *Habitat and technology: The evolution of hunting*. New York: Holt, Rinehart and Winston Inc.

Piper, B. S., Plinston, D. T., & Sutcliffe, J. V. (1986). The water balance of Lake Victoria. *Hydrological Sciences Journal, 31*, 25–38.

Prentice, I. C., Harrison, S. P., & Bartlein, P. J. (2011). Global vegetation and terrestrial carbon cycle changes after the last ice age. *New Phytologist, 189*, 988–998.

Reid, D. A. M., & Njau, J. E. K. (1994). Archaeological research in the Karagwe District. *Nyame Akuma, 41*, 68–73.

Rodgers, W. A., Owen, C. F., & Homewood, K. M. (1982). Biogeography of East African forest mammals. *Journal of Biogeography, 9*, 41–54.

Rosendahl, B. R. (1987). Architecture of continental rifts with special reference to East Africa. *Annual Review of Earth and Planetary Sciences, 15*, 445–503.

Scholz, C. A., Johnson, T. C., Cattaneo, P., Malinga, H., & Shana, S. (1998). Initial results of 1995 IDEAL seismic reflection survey of Lake Victoria, Uganda and Tanzania. In J. T. Lehman (Ed.), *Environmental change and response in East African lakes* (pp. 47–58). Dordrecht: Kluwer Academic Publishers.

Sepulchre, P., Ramstein, G., Fluteau, F., Schuster, M., Tiercelin, J.-J., & Brunet, M. (2006). Tectonic uplift and Eastern Africa aridification. *Science, 313*, 1419–1423.

Sinninghe Damsté, J. S., Verschuren, D., Osssebaar, J., Blokker, J., van Houten, R., van der Meer, M. T. J., et al. (2011). A 25,000-year record of climate-induced changes in lowland vegetation of eastern equatorial Africa revealed by stable carbon-isotopic composition of fossil plant leaf waxes. *Earth and Planetary Science Letters, 302*, 236–246.

Soares, P., Alshamali, F., Pereira, J. B., Fernandes, V., Silva, N. M., Alfonso, C., et al. (2012). The expansion of mtDNA haplogroup L3 within and out of Africa. *Molecular Biology and Evolution, 29*, 915–927.

Soares, P., Rito, T., Pereira, L., & Richards, M. B. (2016). A genetic perspective on African prehistory. In S.C. Jones & B.A. Stewart (Eds.), *Africa from MIS 6-2: Population dynamics and paleoenvironments*. (pp. 383–405). Dordrecht: Springer.

Stager, J. C., & Johnson, T. C. (2008). The late Pleistocene desiccation of Lake Victoria and the origin of its endemic biota. *Hydrobiologia, 596*, 5–16.

Stager, J. C., Mayewski, P. A., & Meeker, L. D. (2002). Cooling cycles, Heinrich event 1, and the desiccation of Lake Victoria. *Palaeogeography, Palaeoclimatology, Palaeoecology, 183*, 169–178.

Stager, J. C., Ryves, D. B., Chase, B. M., & Pausata, F. S. R. (2011). Catastrophic drought in the Afro-Asian monsoon region during Heinrich event 1. *Science, 331*, 1299–1302.

Talbot, M. R., & Laerdal, T. (2000). The Late Pleistocene-Holocene paleolimnology of Lake Victoria, East Africa, based upon elemental and isotopic analyses of sedimentary organic matter. *Journal of Paleolimnology, 23*, 141–164.

Talbot, M. R., & Williams, M. A. (2009). Cenozoic evolution of the Nile Basin. In H. J. Dumont (Ed.), *The Nile: Origin, environments, limnology and human use* (pp. 37–60). Dordrecht: Springer.

Talbot, M. R., Jensen, N. B., Laerdal, T., & Filippi, M. L. (2006). Geochemical responses to a major transgression in giant African lakes. *Journal of Paleolimnology, 35*, 467–489.

Taylor, N. (2016). Across rainforests and woodlands: A systematic re-appraisal of the Lupemban Middle Stone Age in Central Africa. In S. C. Jones & B. A. Stewart (Eds.), *Africa from MIS 6-2: Population dynamics and paleoenvironments*. (pp. 273–299). Dordrecht: Springer.

Temple, P. H. (1966). Evidence of changes in the level of Lake Victoria and their significance. (Ph.D. Dissertation, University of London, 1966).

Trauth, M. H., Deino, A. L., Bergner, A. G. N., & Strecker, M. R. (2003). East African climate change and orbital forcing during the last 175 kyr BP. *Earth and Planetary Science Letters, 206*, 297–313.

Trauth, M. H., Maslin, M. A., Deino, A. L., & Strecker, M. R. (2005). Late Cenozoic moisture history of East Africa. *Science, 309*, 2051–2053.

Trauth, M. H., Maslin, M. A., Deino, A. L., Strecker, M. R., Bergner, A. G. N., & Dühnforth, M. (2007). High- and low-latitude forcing of Plio-Pleistocene East African climate and human evolution. *Journal of Human Evolution, 53*, 475–486.

Trauth, M. H., Larrasoaña, J. C., & Mudelsee, M. (2009). Trends, rhythms and events in Plio-Pleistocene African climate. *Quaternary Science Reviews, 28*, 399–411.

Trauth, M. H., Maslin, M. A., Deino, A. L., Junginger, A., Lesoloyia, M., Odada, E. O., et al. (2010). Human evolution in variable climate: The amplifier lakes of Eastern Africa. *Quaternary Science Reviews, 29*, 2981–2988.

Tryon, C. A., & Faith, J. T. (2013). Variability in the Middle Stone Age of Eastern Africa. *Current Anthropology, 54*, S234–S254.

Tryon, C. A., Faith, J. T., Peppe, D. J., Fox, D. L., McNulty, K. P., Jenkins, K., et al. (2010). The Pleistocene archaeology and environments of the Wasiriya Beds, Rusinga Island, Kenya. *Journal of Human Evolution, 59*, 657–671.

Tryon, C. A., Peppe, D. J., Faith, J. T., Van Plantinga, A., Nightengale, S., & Ogondo, J. (2012). Late Pleistocene artefacts and fauna from Rusinga and Mfangano islands, Lake Victoria, Kenya. *Azania: Archaeological Research in Africa, 47*, 14–38.

Tryon, C. A., Faith, J. T., Peppe, D. J., Keegan, W. F., Keegan, K. N., Jenkins, K. H., et al. (2014). Sites on the landscape: Paleoenvironmental context of late Pleistocene archaeological sites from the Lake Victoria basin, equatorial East Africa. *Quaternary International, 331*, 20–30.

Tryon, C. A., Faith, J. T., Peppe, D. J., Beverly, E. J., Blegen, N., Blumenthal, S., et al. (In Press). The Pleistocene history of the Lake Victoria basin. *Quaternary International*.

Turpie, J. K., & Crowe, T. M. (1994). Patterns of distribution, diversity and endemism of larger African mammals. *South African Journal of Zoology, 29*, 19–32.

Vaks, A., Bar-Matthews, M., Ayalon, A., Matthews, A., Halicz, L., & Frumkin, A. (2007). Desert speleothems reveal climatic window for African exodus of early modern humans. *Geology, 35*, 831–834.

Verschuren, D., Sinninghe Damsté, J. S., Moernaut, J., Kristen, I., Blaauw, M., Fagot, M., et al. (2009). Half-precessional dynamics of monsoon rainfall near the East African equator. *Nature, 462*, 637–641.

Weaver, T. D. (2012). Did a discrete event 200,000–100,000 years ago produce modern humans? *Journal of Human Evolution, 63*, 121–126.

White, F. (1983). *The vegetation of Africa*. Paris: UNESCO.

Williams, S. D. (2002). Status and action plan for Grevy's zebra (*Equus grevyi*). In P. D. Moehlman (Ed.), *Equids: Zebras, asses, and horses, status survey and conservation action plan* (pp. 11–27). Gland, Switzerland: IUCN.

Wronski, T., & Hausdorf, B. (2008). Distribution patterns of land snails in Ugandan rain forests support the existence of Pleistocene forest refugia. *Journal of Biogeography, 35*, 1759–1768.

Chapter 14
Follow the Senqu: Maloti-Drakensberg Paleoenvironments and Implications for Early Human Dispersals into Mountain Systems

Brian A. Stewart, Adrian G. Parker, Genevieve Dewar, Mike W. Morley, and Lucy F. Allott

Abstract The Maloti-Drakensberg Mountains are southern Africa's highest and give rise to South Africa's largest river, the Orange-Senqu. At Melikane Rockshelter in highland Lesotho (~1800 m a.s.l.), project AMEMSA (Adaptations to Marginal Environments in the Middle Stone Age) has documented a pulsed human presence since at least MIS 5. Melikane can be interrogated to understand when and why early modern humans chose to increase their altitudinal range. This paper presents the results of a multi-proxy paleoenvironmental analysis of this sequence. Vegetation shifts are registered against a background signal of C_3-dominated grasslands, suggesting fluctuations in temperature, humidity and atmospheric CO_2 within a generally cool highland environment with high moisture availability. Discussing Melikane in relation to other paleoenvironmental and archeological archives in the region, a model is developed linking highland population flux to prevailing climate. It is proposed that short-lived but acute episodes of rapid onset aridity saw interior groups disperse into the highlands to be nearer to the Orange-Senqu headwaters, perhaps via the river corridor itself.

Keywords Dispersals • Late Pleistocene • Later Stone Age • Lesotho • Middle Stone Age • Mountain foraging • Orange-Senqu River • Paleoenvironment

Introduction

Diverse lines of evidence suggest that Africa between MIS 6-2 experienced a complex demographic history (Cornelissen 2002; Batini et al. 2007, 2011; Tishkoff et al. 2007, 2009; Barham and Mitchell 2008; Behar et al. 2008; Atkinson et al. 2009; Castañeda et al. 2009; Pereira et al. 2010; Scheinfeldt et al. 2010; Drake et al. 2011; Henn et al. 2011; Blome et al. 2012; Pickrell et al. 2012; Schlebusch et al. 2012, 2013; Soares et al. 2012; Veeramah et al. 2012; Barbieri et al. 2013; Cancellieri and di Lernia 2013; Coulthard et al. 2013; Foley et al. 2013; Lombard et al. 2013; Mercarder et al. 2013; Rito et al. 2013). Frequent oscillations in climate, environment and natural resources contributed to variability in human adaptive strategies, including shifts in population sizes, structures and movements. Demographic flux appears to have been exacerbated in the continent's higher latitudes where environments are more water stressed than in equatorial East and tropical Africa (Blome et al. 2012).

In southernmost Africa, resource-rich coastal environments closely neighbor those of the less productive, more seasonal interior plateau. At the root of this ecological imbalance are the subcontinent's peripheral mountains that divide the coast and interior. Rising steeply from the narrow coastal forelands, the Great Escarpment and its associated ranges form a 5000 km-long arc from northern Angola to eastern Zimbabwe (Partridge and Maud 1987; Birkenhauer

B.A. Stewart (✉)
Museum of Anthropological Archaeology, University of Michigan, Ruthven Museums Building, 1109 Geddes Avenue, Ann Arbor, MI 48109, USA
e-mail: bastew@umich.edu

A.G. Parker
Human Origins and Palaeoenvironments Research Group, Department Of Social Sciences, Faculty of Humanities and Social Sciences, Oxford Brookes University, Headington, Oxford, OX3 0BP, UK

G. Dewar
Department of Anthropology, University of Toronto, Scarborough 1265 Military Trail, Toronto, ON M1C 1A4, Canada

M.W. Morley
Centre for Archaeological Science, University of Wollongong, Room 268, Building 41, Northfields Ave, Wollongong, NSW 2522, Australia

L.F. Allott
Institute of Archaeology, University College London, 31-34 Gordon Square, London WC1H 0PY UK

Sacha C. Jones and Brian A. Stewart (eds.), *Africa from MIS 6-2: Population Dynamics and Paleoenvironments*, Vertebrate Paleobiology and Paleoanthropology, DOI 10.1007/978-94-017-7520-5_14

1991; Moore and Blenkinsop 2006; Burke and Gunnell 2008; Clark et al. 2011). The mountains profoundly affect climates and environments on either side by confining most rainfall coastward and casting rain shadows far inland (Wellington 1955). With relatively high precipitation and low evaporation rates, their drainages are the subcontinent's major sources of fresh water, feeding all of southern Africa's perennial rivers (Clark et al. 2011). Within the mountain zone itself, high geological, topographical and biological diversity combine to produce microhabitats and high species endemism (Kingdon 1990; Cowling and Hilton-Taylor 1994, 1997; Davis et al. 1994; van Wyk and Smith 2001; Carbutt and Edwards 2003, 2006; Steenkamp et al. 2005; Mucina and Rutherford 2006).

The peripheral mountains' strong influence on resource distribution has major implications for prehistoric population dynamics, something archeologists have long recognized (e.g., Carter 1970, 1976; Parkington 1972, 1977, 1980; J. Deacon 1974; H. Deacon 1976, 1979). Areas of broken topography such as this are attractive to hunter-gatherers because they provide enhanced resource diversity per unit area of terrain and stable supplies of key resources (Harpending and Davis 1977; J. Deacon 1974; Mitchell 1990). One well established pattern for Holocene hunter-gatherer occupation of the mountains is to persist – and sometimes intensify – during times of resource stress. But their role in earlier prehistory – including MIS 6-2 – has gone largely unexplored. These long isotope stages were punctuated by numerous rapid climate change events that would have placed premiums on areas where key resources were comparatively stable. Surface water would have been particularly vital during short-lived but acute arid pulses with abrupt onsets (Partridge et al. 1993, 2004; Scott 1999; Holmgren et al. 2003). We expect that the well-watered, ecologically diverse peripheral mountains often featured prominently in human settlement decisions (cf. Mitchell 1990) through the instability of MIS 6-2.

To assess this, detailed paleoenvironmental frameworks are needed from local terrestrial archives that are scaffolded by robust chronologies. Previous efforts to reconstruct pre-Holocene population dynamics anywhere on the subcontinent have been impeded by a paucity of such archives (Deacon and Thackeray 1984; Beaumont 1986; Mitchell 1990). Here we present a well-dated multi-proxy paleoenvironmental record from the sedimentary sequence at Melikane, a rockshelter in the Maloti-Drakensberg Mountains of highland Lesotho (Fig. 14.1). The Maloti-Drakensberg is the highest and most widespread mountain system along the Great Escarpment, and is the source of the subcontinent's largest river – the Orange, known in Lesotho as the Senqu. A new radiometric chronology for Melikane shows that the sequence spans the last ~83 ka (Stewart et al. 2012). The site provides a record from MIS 5 onwards of foraging behavior and regional environmental conditions in southern Africa's highest mountains. We analyzed plant opal phytoliths, $\delta^{13}C$ of sediment organic matter (SOM) and archeological charcoals. Taken in unison, these proxies provide a comprehensive record of shifting environmental-vegetation dynamics with important implications for understanding early human dispersals into mountain systems.

Modeling Pleistocene Exploitation of the Highlands

Determining when, why and how Pleistocene foragers exploited highland Lesotho are key questions that have remained unanswered since Patrick Carter pioneered the area's archeology (Mitchell 2009; Stewart et al. 2012). For the Holocene, Carter (1970, 1976, 1978) envisaged a system of seasonal transhumance, with foragers spending summers in the highland Maloti-Drakensberg and winters at lower altitudes across the escarpment in the midlands of southern KwaZulu-Natal and northeastern Eastern Cape (Fig. 14.1). Some groups may have stayed entirely within the highlands year-round, moving seasonally along the Senqu corridor and its larger tributaries (Carter 1978). In contrast, Carter (1976) suspected that highland environments in the Late Pleistocene would have placed serious constraints on hunter-gatherers that demanded more seasonally constrained exploitation patterns, particularly during glacials and stadials when temperatures, snowlines and sour alpine grasses were all lower. During especially harsh climatic conditions, such as the Last Glacial Maximum (LGM), people likely abandoned the highlands altogether (Carter 1976).

Carter's model of seasonal mobility, like others proposed elsewhere along the Great Escarpment (Parkington 1972, 1977, 1980; Deacon 1976, 1979), met subsequent challenges (Cable 1984; Mazel 1984; Opperman 1987) and no consensus has yet emerged. In contrast, productive research has continued into the longer-term trends of Holocene population flux (see review in Mitchell 2009). Whereas issues of taphonomy and equifinality still impede the precision necessary for reconstructing seasonal movements, there is now a critical mass of well-dated Holocene archeological and paleoenvironmental sequences to map broader paleodemographic processes. Though much work remains before the same can be achieved for MIS 6-2, the pace of research into the cultures and environments of this period has recently accelerated. The southeastern part of the subcontinent now boasts a comparatively dense concentration of sites that together cover much of the Late Pleistocene (e.g., Deacon 1976; Carter 1978; Carter et al. 1988; Mitchell 1988, 1990, 1992, 1995, 1996a, b, 1998, 2008; Opperman 1988, 1992, 1996; Kaplan 1989, 1990; Opperman and Heydenrych 1990; Wadley 1997, 2006;

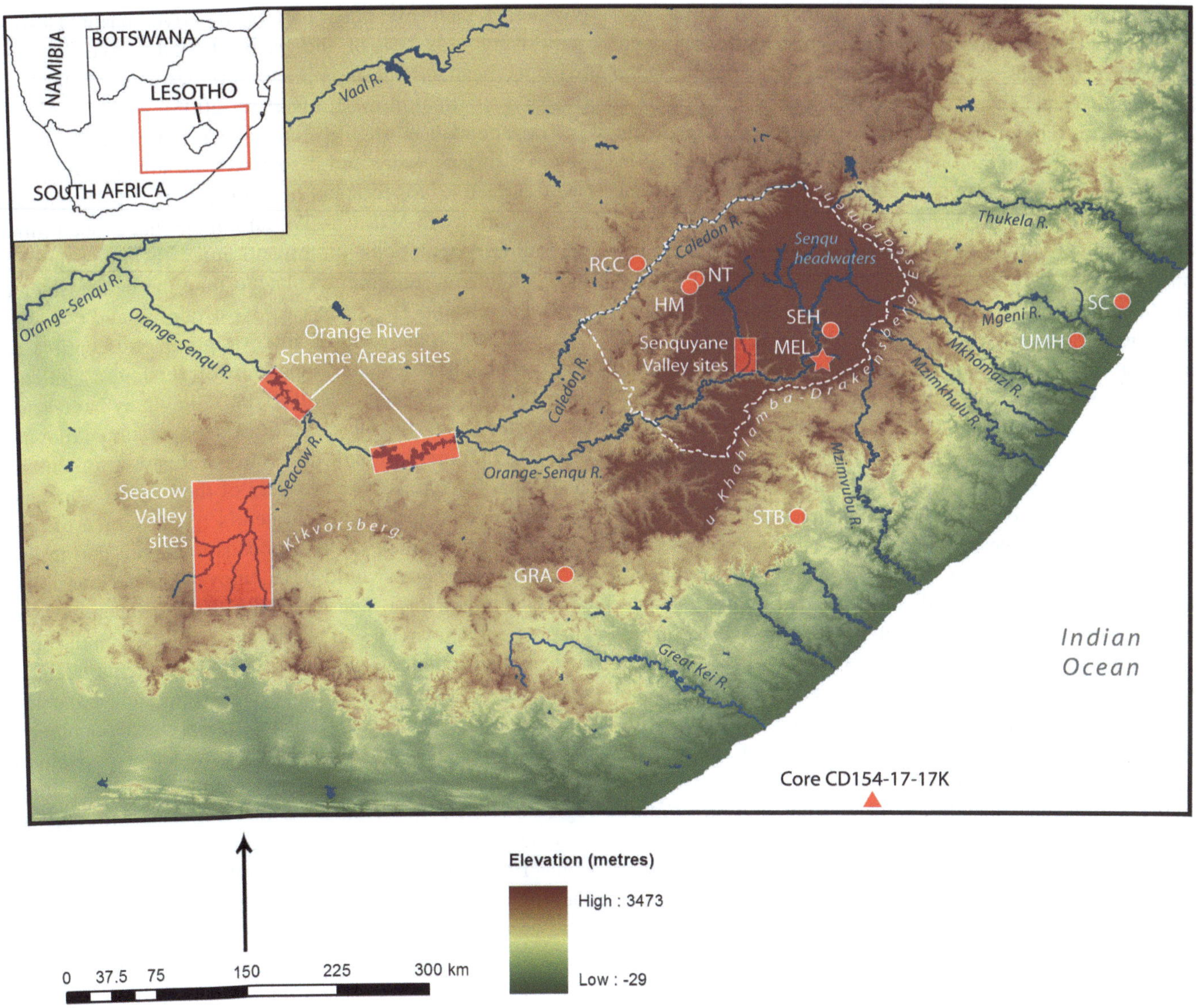

Fig. 14.1 Map of southeastern southern Africa with locales mentioned in the text. GRA: Grassridge; HM: Ha Makotoko; MEL: Melikane; NT: Ntloana Tsoana; RCC: Rose Cottage Cave; SC: Sibudu Cave; SEH: Sehonghong; STB: Strathalan B; UMH: Umhlatuzana

Mitchell and Arthur 2010; Stewart et al. 2012; Roberts et al. 2013) (Fig. 14.1). We are now in a position to ask why during specific periods of the Late Pleistocene lowland foragers chose to increase their altitudinal range.

Ethnographic and ethnohistorical accounts of hunter-gatherers engaging with mountain systems are rare (Aldenderfer 1998). But a growing global corpus of archeological data from such settings suggests that the motivations for, and modes of, upland exploitation were myriad (e.g., Wright et al. 1980; Bender and Wright 1988; Bettinger 1991; Black 1991; Benedict 1992; Madsen and Metcalf 2000; Walsh 2005; Aldenderfer 2006; Brantingham 2006; Walsh et al. 2006; Morgan 2009; Kornfeld et al. 2010; Morgan et al. 2012). The conventional view that medium and high mountains were strictly marginal habitats whose exploitation was a last resort option by lowland groups facing climatic or environmental 'deterioration' is ceding to models incorporating more realistic interplays of environmental and other (e.g., subsistence, demographic, social) variables. Similarly, the reasons for the Pleistocene exploitation of the Maloti-Drakensberg cannot be reduced to a single overarching driver, but rather involved diverse push and pull factors that varied through time and space. However, the region's paucity of Pleistocene research and deep antiquity of human settlement means that the number and resolution of datasets remain too low to move beyond climate-related inferences.

Elsewhere (Stewart et al. 2012) we have outlined the general environmental conditions in which we expect a Late Pleistocene human presence in the highlands to have intensified. In the broadest sense, we predicted that this would have coincided with phases when southeastern Africa was relatively (1) warm and/or (2) arid. While higher temperatures have obvious appeal to montane foragers, the latter expectation is related to southern Africa's uneven distribution of freshwater. The subcontinent has extremely steep gradients of rainfall and evaporation, with mean annual precipitation decreasing from east to west and mean annual potential evaporation increasing inversely (Lynch 2004; Schultz 2008). Roughly 60% of the subcontinent's precipitation falls on only 20% of its land surface and only ~9% of annual rainfall is converted into usable surface runoff or groundwater recharge (Midgley et al. 1994). Precipitation and runoff also vary markedly through time, both seasonally by rainfall zone and inter-annually within rainfall zones. Wet and dry years occur in cycles, with wet years heavily skewing the mean (Eamus et al. 2006). The inter-annual variation in runoff caused by this irregular rainfall is even more pronounced (Schultz et al. 2001). Evaporation rates fall under 1600 mm only in the highland Maloti-Drakensberg and the highest elevations of the southern Cape Fold Mountains (Lynch 2004). With upwards of 2000 mm of mean annual precipitation, the high uKhahlamba-Drakensberg escarpment is thus one of the subcontinent's only areas where evaporation does not exceed precipitation. Moreover, areas with the most precipitation and least evaporation are also those with the lowest inter-annual variability of rainfall and runoff (Schultz 2008). The freshwater resources generated by the high escarpment are therefore some of southern Africa's most abundant and consistent.

This surplus runoff constitutes the bulk of the base flow for a number of major rivers either side of the escarpment watershed (Eamus et al. 2006) (Fig. 14.1). Draining the steep eastern slopes are the Thukela, Mkhomazi, Mzimkhulu and Mzimvubu. These rivers fall rapidly until they reach the well-watered midlands of KwaZulu-Natal and Eastern Cape where they slow and broaden (Rivers-Moore et al. 2007). All runoff west of the escarpment contributes to the Orange–Senqu, which differs from these east-flowing rivers in important ways. The Senqu begins as high-altitude (~3200 m a.s.l.) bogs and small fast flowing streams in the steep basalts of northeastern Lesotho, but soon slows and widens as its bed cuts into the soft sandstones of southeastern Lesotho where Melikane is situated (Swanevelder 1981). The river then turns west to cross over 2000 km of increasingly arid interior plateau into the Atlantic. Annual rainfall falls rapidly from >1500 mm in the highland headwaters to ~300 mm in the central Karoo, then more gradually to <50 mm at its hyper-arid mouth (Earle et al. 2005). Before hydrological regulation began the late 19th century, the river's middle and lower reaches experienced regular droughts and some years ceased flowing completely (Cambray et al. 1986). Unlike the rivers flowing east from the escarpment, the Senqu's upper catchment is thus vastly more productive than any other part of the river basin. Despite comprising only 5% of the basin, Lesotho receives over twice its mean annual rainfall and supplies nearly half of the river's total streamflow (Earle et al. 2005).

The abundance, stability and accessibility of the upper Senqu's surface water and associated resources would have been attractive to foragers during pulses of heightened regional aridity in MIS 6-2. We anticipate that at such times the source populations for upland intensifications derived more often from the interior plateau than the better-watered coastal forelands. The steep east–west rainfall and evaporation gradients that rendered interior groups susceptible to aridity would have let them respond by dispersing east into the mountains. The Orange-Senqu River Valley may have acted as an important fluvial corridor for populations moving between the interior and the better-watered afromontane landscape. In contrast, the nearness of the Eastern Cape and Kwa-Zulu Natal to the subcontinent's major source of precipitation – the Indian Ocean – means that even during dry phases the forelands would have enjoyed greater humidity. Groups living in these areas may have had less incentive than inland populations to seek highland resources, and could reorient their settlement strategies locally. During particularly warm and/or humid phases, however, it is possible that some foreland groups were pushed upland in response to population growth. We thus posit a dual-source model of population influx to the Maloti-Drakensberg contingent on prevailing regional climate. We emphasize that this does not apply to seasonal exploitation of the uplands (cf. Carter 1970, 1978; Cable 1984; Opperman 1987), but rather to the longer-term patterns of intensification within which seasonal uses may have occurred.

The Site and Research Area

Melikane

Melikane is a large (44 × 21 m, 7.5 m high) rockshelter in the Maloti-Drakensberg highlands of southeastern Lesotho (Fig. 14.2). It is situated along the south side of the Melikane River, a tributary of the Senqu, at an altitude of 1860 m a.s.l. The Melikane River rises from the high Lesotho plateau buttressing the uKhahlamba–Drakensberg Escarpment. In these upper reaches the river meanders through broad alpine grassland valleys. But along its lower stretch, particularly between the rockshelter and the Senqu confluence, the river has cut a steep canyon-like ravine that offers a sheltered environment for dense riverine scrub vegetation (Fig. 14.3).

The site was originally excavated in 1974 by Patrick Carter (1978). In 2007, Carter's trench was reopened to obtain samples for a subcontinent-wide chronological study

Fig. 14.2 Views northeast from (**a**) and southwest into (**b**) Melikane Rockshelter

Fig. 14.3 A view west down the canyon-like, heavily wooded and well sheltered lower Melikane River Valley

using optically stimulated luminescence (OSL) (Jacobs and Roberts 2008; Jacobs et al. 2008). The following year we commenced new stratigraphic excavations at the site for an ongoing project investigating the evolution of modern human adaptive flexibility (Dewar and Stewart 2012; Stewart et al. 2012). Thirty layers comprising four major depositional modes were differentiated (Facies A–D) (Stewart et al. 2012) (Fig. 14.4). A vertical sediment column of bulk samples from surface to bedrock was taken at 10 cm intervals; these form the basis for the phytolith and SOM $\delta^{13}C$ analyses (Fig. 14.4). A new chronological framework based on single-grain OSL and AMS ^{14}C shows that the sequence was deposited in sporadic pulses ~83–80, ~61, ~50, ~46–38, ~24, ~9 and ~3 ka (Stewart et al. 2012) (Fig. 14.4). These pulses appear to have been interspersed with long periods of little or no sediment deposition or human occupation.

Landscape, Climate and Vegetation

The Maloti-Drakensberg Mountains extend for ~55,000 km^2 over most of Lesotho and adjacent parts of South Africa's KwaZulu-Natal and Eastern Cape Provinces (Fig. 14.1). The tallest peaks exceed 3000 m a.s.l. Melikane is one of numerous rockshelters in the Maloti-Drakensberg that formed from the differential erosion of a suite of sandstones (Clarens Formation) (Johnson et al. 1996; Schlüter 2006). Deep fluvial incision acting on these sandstones and overlying basalts created intricate mountain topography with steep valleys and dramatic scarps. The most spectacular and well-known escarpment is the uKhahlamba-Drakensberg, which forms the eastern flank of the system and the international border between Lesotho and South Africa (Fig. 14.1). However, the Maloti-Drakensberg is in fact bounded on all sides by scarps, forming a well-defined roughly rectangular mountain massif (Moore and Blenkinsop 2006). An important exception is the southwestern corner through which the Senqu River exits the mountains via a broad, deeply incised fluvial corridor (Fig. 14.1).

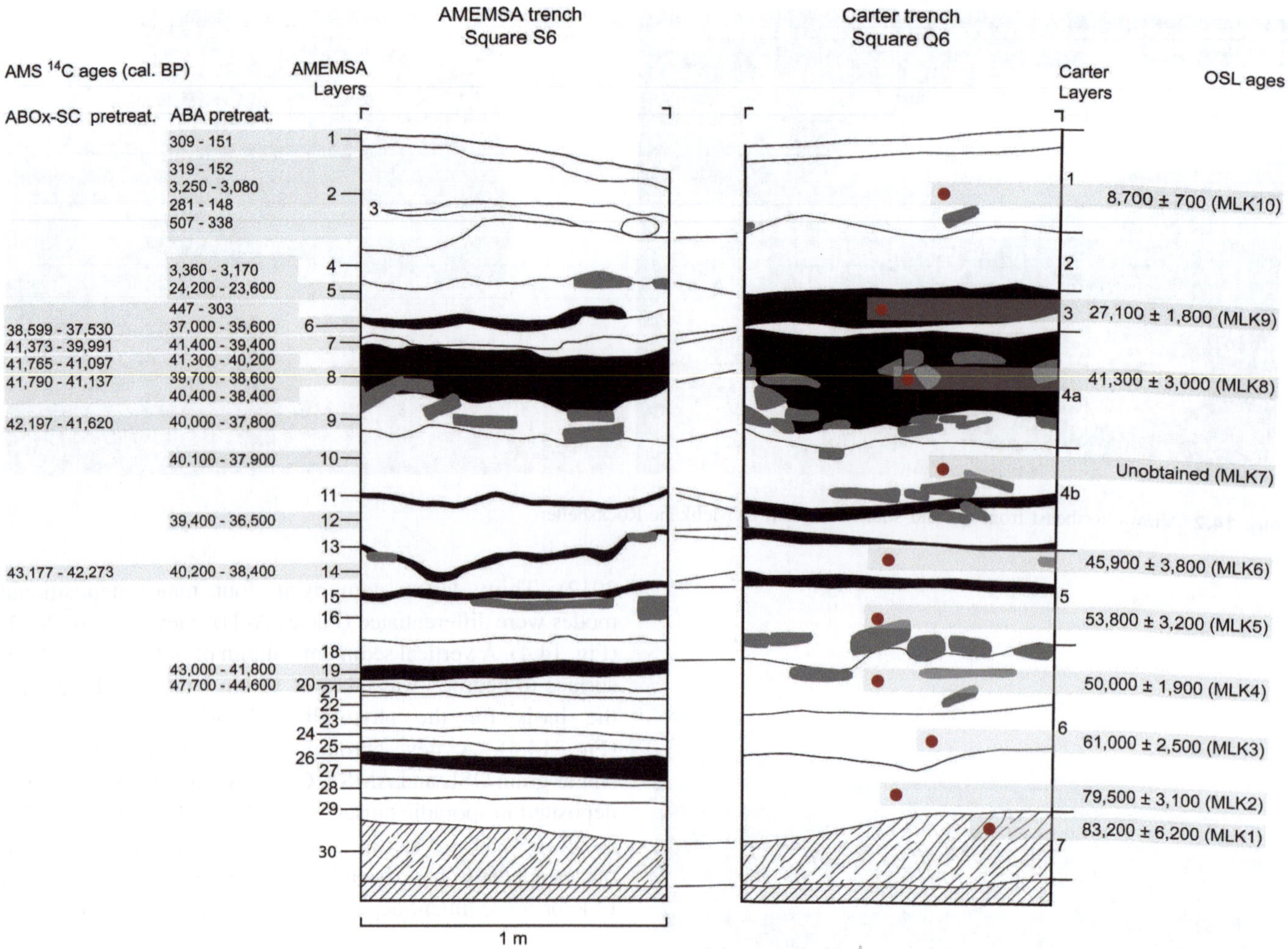

Fig. 14.4 The west profile of our excavation trench (*left*) and east profile of P. Carter's excavation trench (*right*) at Melikane showing stratigraphy and associated ^{14}C and OSL dates

The present-day climate and ecology of the Maloti-Drakensberg varies widely by season, locale and altitude. The overall climate is temperate, subhumid and continental. Summers (October-March) are generally warm and receive roughly three-quarters of the region's annual rainfall (Tyson et al. 1976; Killick 1978). Winters (May–August) are cool to cold and extremely dry, though the eastern highlands experience frequent snow from April to September. Rainfall is highest along the uKhahlamba-Drakensberg escarpment's eastern slopes and summit, where mean annual precipitation (MAP) can exceed 1600 mm (Killick 1963; Schulze 1979; Sene et al. 1998). Temperatures vary drastically on both a seasonal and diurnal basis, and by altitude. Mean annual temperatures range from ~15 °C in lowland Lesotho to ≤6 °C in the high Drakensberg (Grab 1994, 1997). The lowlands produce mean mid-summer maximum temperatures of 29 °C and mid-winter minimum temperatures 4.3 °C, with respective values for the highlands of 17 °C and –6.1 °C (Grab and Nash 2010). Frost is widespread and ranges from ~31 days per year in the lowlands to ~150 days per year in the highlands (Schulze 2008). Ground freezing in the high Maloti-Drakensberg is estimated to occur up to 200 days per year (Grab 1997).

The Maloti-Drakensberg are lower in altitude but higher in latitude than the mountain systems of tropical East and northeast Africa, with comparatively cooler temperatures reached at lower elevations. Consequently, vegetation distribution is mainly conditioned by altitude and aspect, variations in which produce an ecology with marked vertical differentiation. Particularly significant are altitudinal differences in the proportions of plant taxa following C_4 *versus* C_3 photosynthetic pathways. The dominance of C_4 or C_3 vegetation is primarily a function of temperature during the growing season (Vogel et al. 1978; Ehleringer et al. 1997), though precipitation and ambient CO_2/O_2 ratios are also important (Ficken et al. 2002). In general, plants following the C_4 pathway possess high tolerances to hot and dry climates with high irradiance and low atmospheric CO_2 (pCO_2) (Ehleringer et al. 1997; Retallack 2001). These include tropical grasses, sedges and xeric herbs (O'Leary 1981). C_3 plants are less CO_2 and water efficient, and thus prefer cool, moist climates (Alexandre et al. 1997). They include most woody taxa such as trees, herbs and shrubs, as well as temperate grasses and sedges. As elsewhere in the summer rainfall zone, the present-day Maloti-Drakensberg is dominated by C_4 grasses with C_3 trees and shrubs at lower altitudes and along the river corridors. However, the balance shifts at high elevations (>2100 m a.s.l.) where C_3 grasses have an adaptive advantage and outcompete C_4 taxa.

The research area's major vegetation zones, from lowest to highest altitude, are as follows. Along the boulder-strewn Senqu River Valley and its many tributaries at ~1600–1900 m a.s.l. is Senqu Montane Shrubland (Mucina and Rutherford 2006). This is a *Cymbopogon-Themeda-Eragrostis* grassland dominated by C_4 Panicoid grasses but with some C_4 Chloridoid taxa also present. Important grass taxa include *Cymbopogon phrinodis*, *Heteropogon contortis*, *Setaria flabellate*, *Themeda trianda*, *Tristachya hispida* (all C_4 Panicoids) and *Eragrostis spp.* (a C_4 Chloridoid). However, numerous species of tree and evergreen shrub (all C_3) also thrive here (Bawden and Carroll 1968; Mucina and Rutherford 2006). These C_3 taxa are dominated by *Rhus erosa*, *Olea europaea* and *Diospyros austro-africana*. Rarer thickets dominated by *Leucosidea sericea*, *Kiggelaria africana* and *Rhamnus prinoides* (all C_3) occur in more sheltered valleys (Mucina and Rutherford 2006). This zone is effectively an eastern intrusion of lowland Lesotho taxa into the highlands along the Senqu corridor (Jacot Guillarmod 1971). *Cymbopogon-Themeda-Eragrostis* grassland also extends up onto the gently rolling plateaux that overlook the deep river valleys, though here trees and woody shrubs are comparatively rare.

Lesotho Highland Basalt Grassland occurs above the river valleys and plateau shelves at altitudes of between ~1900–2900 m (Mucina and Rutherford 2006). This is a dense, short C_4 Panicoid and Chloridoid grassland with patchy shrublands dominated by *Passerina montana* (Mucina and Rutherford 2006). Although some ericaceous and composite taxa occur in the basalt grasslands, trees are mostly absent. Due to its large altitudinal range (~1000 m), this zone contains two altitude-specific vegetation belts. At lower elevations, between ~1900–2100 m on southern (cooler) slopes but reaching up to ~2700 m on northern (warmer) slopes, grasses are dominated by C_4 Panicoids, notably *Themeda triandra*. This C_4 species is renowned for its excellent pasturage quality. Other important C_4 taxa that occur in this lower *Themeda*-dominated belt include *Andropogon spp.* (a Panicoid), *Eragrostis spp.* and *Microchloa caffra* (both Chloridoids).

In the upper belt of the basalt grasslands – above ~2100 m on southern slopes and ~2700 m on northern slopes – C_3 grasses become dominant. These *Festuca-Merxmuellera* grasses are shorter and less palatable ('sour' or *letsiri* in SeSotho) than *Themeda. Festuca caprina* is dominant, but *Festuca rubra* and *Festuca costata* are also important. *Danthonia disticha*, a C_3 Arundinoid, occurs on thin, rocky soils. C_3 dominance continues at altitudes exceeding 2900 m, although at these highest elevations the basalt grasslands give way to Drakensberg Afroalpine Heathland (Mucina and Rutherford 2006; termed Alpine Heath by Killick 1990). This is a short heath and shrubland dominated by *Helichrysum trilineatum*, *Erica dominans* and *Eumorphia sericea* with dwarf bushes (all C_3) (Killick 1978; Mucina and Rutherford 2006). The heathlands are interspersed with *Merxmuellera*-dominated C_3 grasses (Killick 1990). Embedded within this zone are patches of aquatic and hygrophilous vegetation that include streambank communities and numerous alpine peat-forming bogs (Killick 1978, 1990). The latter occur either as basalt fissure seepage bogs on mountain slopes or as thufur-covered sponge bogs at riverheads that regulate headwater flow into Senqu River system (Grab 1997; Jacot Guillarmod 1971; Killick 1978, 1990; van Zinderen Bakker and Werger 1974).

The Maloti-Drakensberg's alpine grass- and heathlands are unique among southern Africa's summer rainfall zones as C_3 predominance is otherwise restricted to areas with winter rainfall (the Western Cape) (Vogel et al. 1978). Such close proximity of vegetation zones dominated by physiologically dissimilar taxa makes the region particularly well suited to paleoenvironmental reconstruction. Because the competitive balance between C_4 and C_3 is determined by prevailing climate, shifts in the proportions of C_4/C_3 taxa through time can be used to infer climatic changes acting to alter local vegetation dynamics. In general, colder phases encourage the down-slope expansion of C_3 taxa at the expense of C_4, whereas warmer phases provoke the reverse. The result is a vertically migrating 'front' of C_4/C_3 dominance, which registers in various proxies (e.g., phytoliths, $\delta^{13}C$ signatures) at terrestrial archives including rockshelters.

In fact, the first African rockshelter at which this principle was applied was Melikane. In a seminal study, Vogel (1983) investigated Late Pleistocene and Holocene paleoenvironments using dietary $\delta^{13}C$ composition of equid teeth from Carter's original excavation. His results were useful in demonstrating the dominance during the Holocene and Pleistocene of C_4 and C_3 taxa, respectively. However, opportunities for obtaining better resolution were limited by Carter's coarse excavation methods and the limits of radiocarbon dating. A series of more recent applications in the Maloti-Drakensberg (Smith et al. 2002; Grab et al. 2005; Parker et al. 2011; Roberts et al. 2013) and further afield in the highlands of East Africa (Ambrose and Sikes 1991; Street-Perrott et al. 1997; Huang et al. 1999; Olago et al. 1999; Wooller et al. 2003) have focused almost exclusively on the Holocene/terminal Pleistocene. Since Vogel's (1983) pioneering study, therefore, very little research has been conducted on deeper Pleistocene paleoenvironments in the research area, and none using C_4/C_3 proportions.

Materials and Methods

Phytoliths

Recent work on phytolith morphotypes from modern vegetation in southern Africa has yielded new information for the application of plant biogenic silica studies in paleoenvironmental reconstruction (Mercader et al. 2010; Cordova 2013). These studies suggest that a number of morphotypes traditionally used to separate C_3 and C_4 vegetation, especially grass silica short cell morphotypes, are found across a number of C_3 and C_4 grass tribes. There is thus greater redundancy in morphotypes than previously thought. Nevertheless, some morphotypes and morphotype groups can be attributed to particular grass subfamilies, sedges and woody taxa. In particular, short cell short-saddle forms are attributed to chloridoids, lobates to panicoids, papillae and achenes to sedges, and circular rugose/globular granulates to dicot trees and shrubs (Mercader et al. 2010, 2013). It should be noted, though, that Cordova (2013) designates some lobate forms to C_3 grass subfamilies.

Subsamples were prepared for phytolith analysis using the methods outlined in Parker et al. (2011). Ten grams of sediment per sample were sieved through a 2 mm sieve in order to remove the coarse fraction prior to phytolith extraction. Carbonates were removed using 5% HCl followed by 30% wv H_2O_2 to remove organics. Samples were deflocculated using 50 ml 2% Calgon and 250 ml distilled water, then shaken continuously for 30 min before being passed through a 212 μm sieve. Heavy liquid separation using sodium polytungstate (2.35 s.g.) was employed to separate the phytoliths from the heavier inorganic residue. The samples were diluted to a specific gravity of 1 and then passed through a 5 μm vacuum filtration system (sensu Theunissen 1994) to remove the clay and finest silt fractions. Samples were mounted onto microscope slides using Canada Balsam and identified at ×400 magnification using a Nikon Eclipse E400 light microscope. Overall phytolith preservation was good, but in some samples phytoliths were either entirely absent or had undergone dissolution rendering viable counting impossible. The number of phytoliths counted varied between 301 and 459 per sample.

Two phytolith indices were employed to help with data interpretation: the D/P ratio and the climatic index (Ic%). The D/P ratio is the ratio of ligneous dicotyledon morphotypes (D) to Poaceae morphotypes (P), and is used as a proxy of tree cover density (Alexandre et al. 1997). The value 1 indicates maximum tree cover and zero none. The climatic index (Ic%) indicates the influence of climate on the ratio of C_3 to C_4 morphotypes based on the proportion of pooids to the sum of all short cell grass forms (Ic % = pooid/pooid + chloridoid + panicoid) (Twiss 1987). Higher Ic% values indicate more C_3 pooid grasses and thus cooler conditions.

Sediment Organic Matter $\delta^{13}C$

Sediment samples were prepared for isotopic analysis at Oxford Brookes University's Human Origins and Paleoenvironments (HOPE) Research Group laboratory following procedures adapted from Ambrose and Norr (1993) and Smith (1997). After sieving each sample through 2 mm and 2 μm mesh, the sediment (~1 g) was pretreated with 2 M HCl to remove carbonates. The analyses were performed at the Godwin Laboratory at the University of Cambridge. The samples were rinsed with deionized water, freeze-dried, weighed into aluminum cups and measured using an automated elemental analyzer coupled in continuous-flow mode to an isotope-ratio-monitoring mass spectrometer (Costech elemental analyser coupled to a Finnigan Delta V mass spectrometer). Stable carbon isotope concentrations were measured relative to the VPDB international scale and are reported as permil δ values. The IAEA standard of caffeine and in-house laboratory standards of nylon, alanine and bovine liver were employed for calibration. Replicate analyses of international and laboratory standards suggest that measurement errors for $\delta^{13}C$ are less than ±0.2‰. However, sediments are likely to produce greater intra-sample uncertainty due to high variance in organic content (Parker et al. 2011; Roberts et al. 2013). Each sediment sample was run in triplicate.

It is well established that decreased temperatures during the growing season promote C_3 vegetation. However, Bousman (1991) demonstrated that a positive relationship also exists between C_3 grass percentage and rainfall. Working in the westernmost extension of the uKhahlamba-Drakensberg Escarpment (the Kikvorsberg; Fig. 14.1), he showed that as available moisture increases C_3 grasses outcompetes C_4 grasses and other C_3 non-grass species. He developed a formula for predicting the proportion of C_3 grasses in the landscape based on the bulk $\delta^{13}C$ values. We employ a version of Bousman's Index that is modified to account for anthropogenic influences on SOM $\delta^{13}C$ in the Melikane sequence. Bousman's formula allowed for alteration of $\delta^{13}C$ between plant and SOM of 3%, but we have used a more conservative 2% following Wedin et al. (1995): % of C_3 plants = $(\delta^{13}C - 2 + 12.5)/-0.14$. These values are checked against the phytolith Ic%, which provides an independent estimation of C_3/C_4 ratios as noted above.

Charcoals

The entire charcoal assemblage from Melikane was first examined and quantified to isolate deposits containing fragments suitable for analysis. The initial goal was to analyze 200 fragments from each stratigraphic unit to gain an overview of woody vegetation represented (cf. Allott 2004, 2005, 2006). However, preliminary inspection revealed high inter-layer variability in charcoal abundance and poor preservation of internal charcoal structure. This is likely a result of repeated throughput of water originating from fissures in the rear shelter wall at times both during and after the accumulation of occupation debris (Stewart et al. 2012). Although wood charcoal is comparatively inert to weathering processes, the effects of saturation and drying can be detrimental to preservation. Water flow or saturation introduces small particles of sediment into wood charcoal that can obscure and damage internal anatomical features used for identification. Most layers produced fewer than 200 fragments suitable for identification. Therefore, rather than examining changes in the relative abundance of specific taxa, we rely on habitat preferences and requirements in order to make inferences about past regional climate and vegetation conditions.

Five layers were selected that span the bulk of the Late Pleistocene sequence: Layers 24 (~61 ka), 13 (~42 ka), 11 (~42 ka), 8 (~41.3 ka) and 5 (24 ka). Where possible more than 100 charcoal fragments were identified from each layer for a total of 445 fragments. Nevertheless, the majority of layers produced very few charcoal fragments with only three layers (13, 11 and 8) of the five containing in excess of 100 fragments. Fragments were fractured along three planes to reveal transverse, tangential longitudinal and radial longitudinal sections (Leney and Casteel 1975). Characteristics of anatomical features visible in each were recorded with reference to the IAWA list of microscopic features for hardwood (Wheeler et al. 1989). Taxonomic identifications have been provided through comparison with modern reference material developed at the University of the Witwatersrand (Dowson 1988; Wadley et al. 1992; Esterhuysen 1996; Allott 2004, 2005) and reference atlases/resources (Kromhout 1975; Eichhorn 2002; Inside wood 2004 onwards). Where possible, fragments were identified to family, genus or species. Secure species identifications from wood anatomical features alone are rarely possible, though when a genus is represented by a single species within the region it has been named as the most likely origin of the charcoal. Family identifications are given when similarities between genera within a family are too great for differentiation or when anatomical features are not sufficiently clear to refine the identification.

Results

The results of the phytolith results are presented in Figs. 14.5 and 14.6. The former shows all phytolith morphotypes organized according to depth from surface, while the latter presents phytolith family summaries chronologically with ages of samples from undated layers interpolated using depth. The SOM $\delta^{13}C$ results are presented in Table 14.1 and Fig. 14.6, where they are presented graphically alongside the summary phytolith results. Table 14.2 presents the results of the charcoal analyses, while Table 14.3 gives habitat preferences of the woody taxa represented.

Several general observations can be made. First, both the phytolith and SOM $\delta^{13}C$ results clearly show that the entire Late Pleistocene portion of the Melikane sequence (Layers 3 and below) is heavily dominated by C_3 grassland taxa (Ic 60–95%) (Fig. 14.6). Only with the onset of the Holocene (Layer 2) do phytolith C_4 indicators outnumber C_3 (Ic 40%). This is consistent with Vogel's (1983) original findings at Melikane and recent SOM $\delta^{13}C$ data from the Lesotho lowlands (Roberts et al. 2013), and reaffirms that Late Pleistocene temperature depressions provoked the lowering of C_3 alpine grasses. Considering that today C_4 grasses reach up to 2700 m on north-facing slopes, and assuming a temperature drop of −0.6 °C/100 m (Smith et al. 2002), this suggests that temperatures were at least 5 °C cooler throughout the Late Pleistocene at Melikane, a north-facing shelter at an altitude of 1860 m a.s.l.. Second, with the exception of the uppermost part of the sequence there is good agreement between the phytolith and SOM $\delta^{13}C$ records, which show a close correspondence between the Ic % values and the Bousman Index values (Fig. 14.6). This gives us confidence that these records reflect changes in landscape vegetation dynamics rather than purely stochastic anthropogenic inputs (although the selective collection of vegetation will be reflected heavily in the charcoal record). Third, there are differences between the records in the magnitude of changes. The phytoliths and to a lesser extent the SOM $\delta^{13}C$ show variability through time, with changes apparent within the C_3 dominated signals. In contrast, the shrubs and trees identified in the wood charcoal assemblage show a degree of continuity (Table 14.2). This may indicate either overall stability of local vegetation or that humans preferentially and consistently selected these woods for fuel through time. Since shifts within the phytolith record make vegetation stability unlikely, we favor selection continuity as a more probable interpretation.

MIS 5a

One sediment sample – MLK 22 (~83 ka; Layer 29) – was taken from Melikane's deepest strata deposited during MIS 5a. The $\delta^{13}C$ value for this sample is −23.42‰ (Table 14.1). This is a strongly C_3 signal, with the resulting Bousman Index value suggesting >93% of the signal is derived from C_3 vegetation with minor but extant C_4 input. This is supported by the phytoliths, which show a dominance of pooid short cells and an Ic% value of >95% (Fig. 14.6). Only a trace of panicoid types and no short-saddle chloridoids were present. Also present in low abundances are papillae and achene phytoliths from sedges. The tree cover density index (D/P ratio), which indicates the proportion of woody taxa to grasses, is low at ~0.1 (Fig. 14.6). This suggests that this C_3 grass-dominated landscape hosted a low woodland element. A suitable charcoal sample for MIS 5a could not be obtained.

MIS 4

MIS 4 is represented by three samples: MLK21 (~73 ka extrapolated; Layer 26 lower), MLK20 (~67 ka extrapolated; Layer 26 upper) and MLK19 (~61 ka). The oldest MIS 4 sample (MLK21) registers a very slight positive shift to −23.19‰ (from −23.42‰ in late MIS 5a) (Table 14.1), with correspondingly small negative shifts in the Bousman Index (~90%) and phytolith Ic% (~90 %) (Fig. 14.6). There is then a 1.8‰ positive shift in $\delta^{13}C$ values to −21.41‰ in MLK20, followed by a slight negative shift in MLK19 back to −22.46‰. The corresponding Bousman Index value for MLK20 suggests a shift in the vegetation composition with an increase in the C_4 component to ~30%. The phytoliths corroborate this with Ic% values of ~70% (Fig. 14.6) The short cell grassland phytoliths in MLK20 account for 20% of the assemblage. Panicoid lobates increase slightly to ~5% when compared with MIS 5a, along with a trace of short body saddle chloridoids. Chloridoid phytoliths are absent from the youngest MIS 4 sample (MLK19). The presence of sedges was also noted but in low numbers. There is an increase in ligneous (woody) dicot phytoliths when compared with MIS 5a to ~7% in MLK20 and 5% in MLK19. The D/P ratio in MLK19 is also higher than for MIS 5a, rising to 0.2 (~20% woody vegetation) (Fig. 14.6). This suggests more trees were present in the landscape and/or that a greater proportion of woody taxa were being deliberately brought into the site. However, the overall proportion of woody phytoliths remains low.

A small but informative charcoal sample was obtained from the youngest MIS 4 sample MLK19 (Layer 24). Four taxa were recorded: *Buddleja* cf. *salviifolia*; *Rhamnus* sp.; *Leucosidea sericea*; and *Protea* sp. (Table 14.2). Of these,

Table 14.1 SOM $\delta^{13}C$ data for the Melikane samples showing depth from surface, mean, standard deviation and percentage of elemental carbon for each set of triplicate samples

Sample	Depth from surface (m)	Mean $\delta^{13}C$ (‰)	SD (‰)	C(%)
MLK1	0.1	–	–	–
MLK2	0.2	−24.01	0.05	0.71
MLK3	0.3	−23.19	0.07	0.27
MLK4	0.4	−22.48	0.04	2.84
MLK5	0.5	−23.12	0.13	5.78
MLK6	0.6	−22.91	0.09	2.73
MLK7	0.7	−22.71	0.06	3.34
MLK8	0.8	−22.58	0.10	5.69
MLK9	0.9	−23.00	0.11	0.55
MLK10	1	−22.77	0.08	0.47
MLK11	1.1	−23.89	0.01	3.08
MLK12	1.2	−23.41	0.06	1.63
MLK13	1.3	−23.68	0.05	0.96
MLK14	1.4	−23.43	0.04	1.97
MLK15	1.5	−22.88	0.08	0.97
MLK16	1.6	−23.42	0.12	0.48
MLK17	1.7	−22.92	0.05	0.83
MLK18	1.8	−23.41	0.10	1.46
MLK19	1.9	−22.46	0.04	1.38
MLK20	2	−21.41	0.04	4.90
MLK21	2.1	−23.19	0.03	1.93
MLK22	2.2	−23.42	0.01	0.71

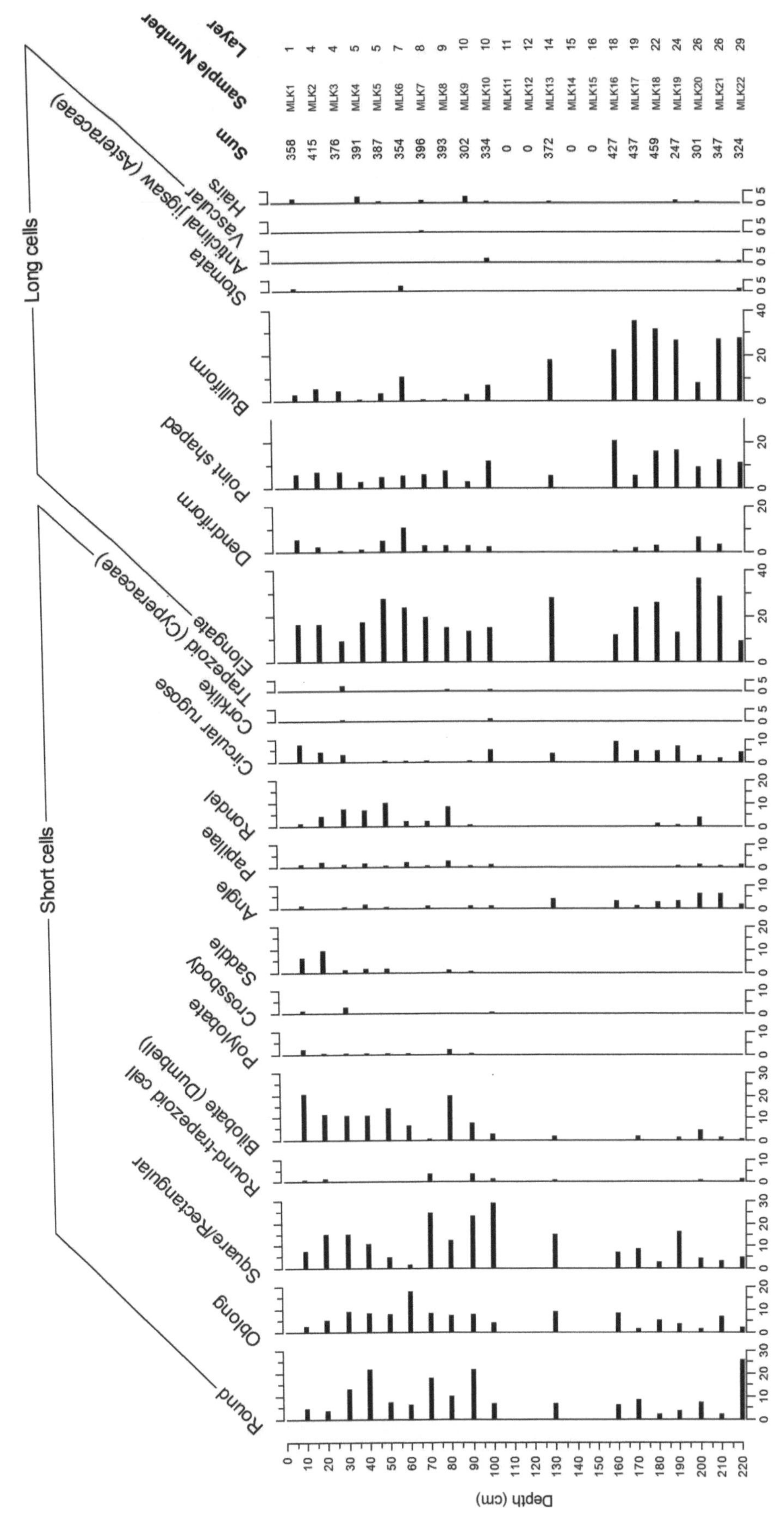

Fig. 14.5 All phytolith morphotypes (arranged by depth from surface)

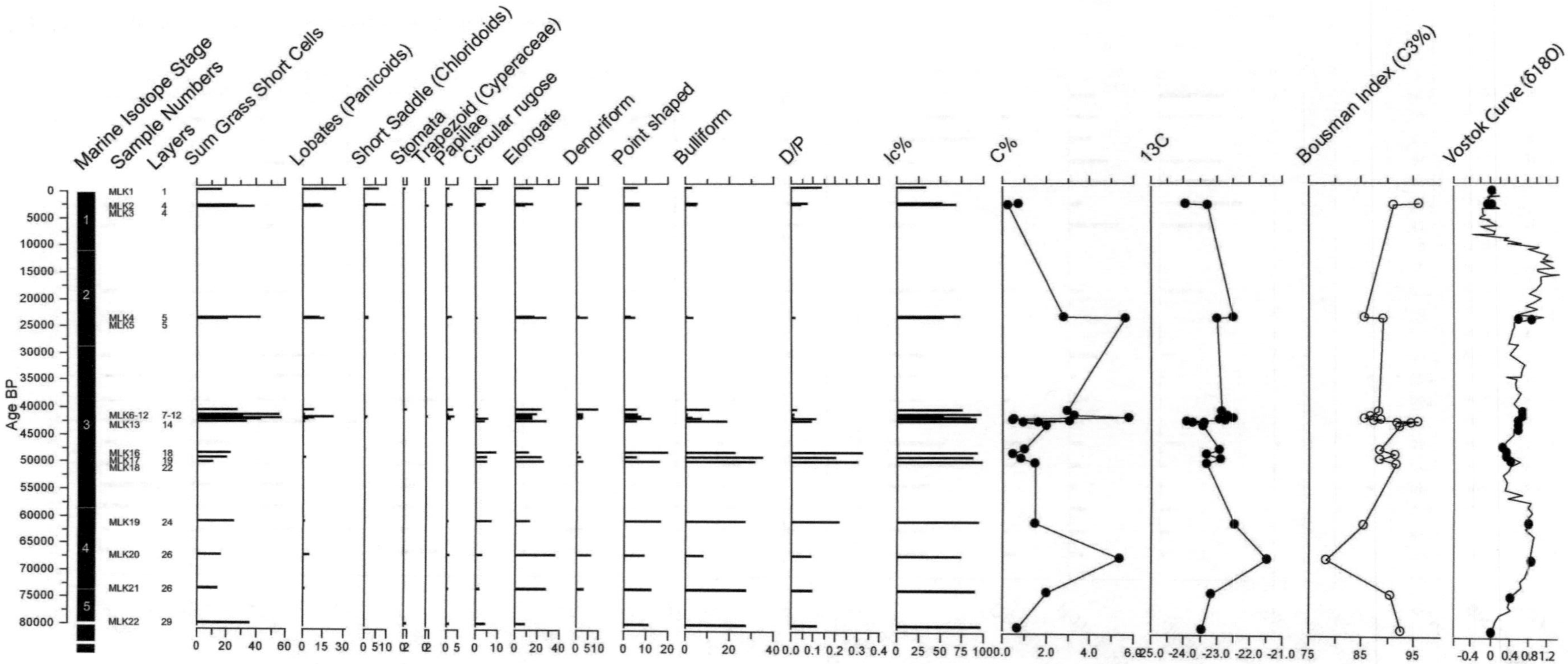

Fig. 14.6 Summaries of phytolith families and morphotypes, and the phytolith D/P and Ic indices juxtaposed with the SOM $\delta^{13}C$ results (C%, $\delta^{13}C$ and Bousman Index). These results are arranged chronologically with ages for undated layers interpolated by depth. Vostok data from Jouzel et al. (2007)

Table 14.2 Taxonomic identifications of the charcoal samples selected for analysis by layer. Details on relevant ^{14}C and OSL dates for these layers are also shown

Layer	C14 Date (cal BP)	OSL Date (BP)	*Buddleja salviifolia*	*Rhamnus* sp.	*Rapanea melanophloeos*	*Erica* sp.	*Erica drakensbergensis*	*Protea* sp.	*Leucosidea sericea*	*Heteromorpha* sp.	Rosaceae	*Olea europea*	cf. Fabaceae	*Grewia* sp.	*Maytenus* sp.	Indeterminate poor preservation
5	24,200–23,600	27,100 ± 1800						1	14			10				
5	24,200–23,600	27,100 ± 1800					1									
5	24,200–23,600	27,100 ± 1800					11									
8	41,765–41,097 41,790–41,137	41,300 ± 3000	32	49	19	3										
11	Bracketed between 42,197–41,620 (Layer 14) and 43,177–42,273 (Layer 9)	Bracketed between 41,300 ± 3,000 (Layer 8) and 45,900 ± 3,800 (Layer 14)	25	29				1	23	4	1					4
11	Bracketed between 42,197–41,620 (Layer 14) and 43,177–42,273 (Layer 9)	Bracketed between 41,300 ± 3,000 (Layer 8) and 45,900 ± 3,800 (Layer 14)	14	15											47	
13	Bracketed between 42,197–41,620 (Layer 14) and 43,177–42,273 (Layer 9)	Bracketed between 41,300 ± 3,000 (Layer 8) and 45,900 ± 3,800 (Layer 14)	52	19	10			13				1	1	2		3
24	No	61,000 ± 2500	23	1				3	14							

Table 14.3 Habitat preferences of the woody taxa identified in the charcoal assemblage

Identified Taxa	Habitat Preferences	Known firewood?
Medium to tall trees		
Rapanea melanophloeos	Occurs in evergreen forest, in riverine forest and sometimes in drier coastal and mountain forests	
Protea sp.	Drier, rocky ground or mountain grassland, sheltered locations with available moisture	Y
Tall Shrubs or small trees		
Buddleja cf. salviifolia	Occurs at the margins of or in evergreen forest, on rocky mountain slopes and along water courses at higher altitudes	Y
Rhamnus sp.	Widespread and locally common at med-high alt, along watercourses, in riverine forest and at margins of evergreen forest	
Heteromorpha sp.	Occurs over wide range of altitudes, often fringing evergreen forest, also found in wooded ravines, on hillsides and rocky outcrops	
Olea europea	Variety of habitats usually near water, stream banks, riverine fringes also open woodland, among rocks and in mountain ravines	Y
Grewia sp.	More typical of open grassland vegetation	
Small Shrubs		
Erica sp.	Mountain ravines, rocky grassy slopes, moist places, sometimes near streams	Y
Erica drakensbergensis	In moist places, forest margins, rocky grassy slopes, up to 1800 m	
Leucosidea sericea	Pioneer species of mountainsides, kloofs, valley bottoms and along streams at higher altitudes, often occurs in dense stands of multistemmed, somewhat straggling shrubs	Y
Gymnosporia sp.	Occurs in rocky places in grassland associated with forest	Y
Other taxa Identified to family		
Rosaceae	Includes a broad range of potential taxa	
cf. Fabaceae	Includes a broad range of potential taxa	

both *Buddleja* cf. *salviifolia* and *Rhamnus* sp. favor evergreen forest margins or grow along watercourses (Table 14.3). The *Rhamnus* specimens, although unidentifiable to species, are almost certainly *Rhamnus prinoides*, which is common in present-day afromontane forest settings (including Lesotho) at medium and high altitudes. *Buddleja* cf. *salviifolia* and *Rhamnus* sp. are found in similar habitats to *Leucosidea sericea*, which is a pioneer species of higher altitude (≥1000 m) mountainsides occurring along running streams (Table 14.3). All three taxa prefer relatively humid environments, consistent with the phytolith and $\delta^{13}C$ results for this later MIS 4 sample. The fourth taxon, *Protea* sp., tends to grow on slightly drier, rocky ground or mountain grassland, though it too prefers sheltered locations where moisture is available.

MIS 3

A ~10 kyr hiatus in deposition occurred between MIS 4 (MLK19) and MIS 3 (MLK18). The MIS 3 strata at Melikane (Layers 22-6) span from ~50 to 38 ka and appear to have been deposited two broad pulses: ~50 ka and ~46–38 ka. Thirteen samples were examined, four from the former pulse (MLK18–15) and nine from the latter (MLK14-6).

Isotopic values ~50 ka (MLK 18-15) range between −23.42 and −22.88‰ (Table 14.1). The Bousman Index and Ic% values show little variation in the proportion of C_3 (>90%) to C_4 (<10%). However, an increasing proportion of the isotope C_3 signal likely derives from tree and shrub elements of the vegetation. The phytolith record from the lower three samples (MLK18-16) shows an increase D/P from 0.2 to 0.3, indicating a greater proportion of woody phytolith morphotypes (30%) to grassland (70%) (Fig. 14.6). These values are the highest shown in the entire sequence and suggest an increase in the regional woodland component and/or an increase in the selection of woody material that was brought into the site. The phytolith and SOM $\delta^{13}C$ values suggest a greater cover of woodland than in MIS 4, with C_3 taxa dominating the grassland flora under cool, wet/humid conditions. The uppermost sample (MLK15) in the older MIS 3 sediments contained very poorly preserved phytoliths with major dissolution noted (Fig. 14.5). Its isotopic value (−22.88‰) gives some insight into a C_3 dominated landscape with a C_4 component. Panicoid levels are lower than those in MIS 4, while chloridoid phytoliths are present but in trace amounts (Fig. 14.6).

Between ~46–38 ka (MLK14-6) the SOM $\delta^{13}C$ values range between −23.89 and −22.56‰ (Table 14.1). The values become more depleted from MLK14 to MLK11 to reach the lowest values recorded in the Melikane sequence

(−23.89‰). Values fluctuate after this and become progressively more enriched, with a +1.31‰ shift by MLK8 (Layer 9). Phytolith preservation was extremely variable in these samples with a near complete absence of biogenic silica in MLK14, 12 and 11 (Fig. 14.5). This may result from the complete dissolution of phytoliths or the sediment being constituted of decomposed and weathered roof collapse material. Phytoliths show an increase in panicoid lobate morphotypes (up to 20%) with a small trace of chloridoids. The Bousman Index values show high C_3 values (90–95%). The Ic% shows close correlation up to MLK8 where it diverges (55%) from Bousman's value (90%) (Fig. 14.6). The most likely explanation for the difference in the two values is an enhanced C_3 component in the isotope record being derived from woody taxa collected for firewood. The D/P values fall considerably over this time period to levels <0.05 between ~42 and 38 ka, suggesting major reductions in regional woodland.

Suitable charcoal samples were obtained from the younger MIS 3 pulses (~46–38 ka). Charcoals from MLK13 (Layer 14) and MLK11 (Layer 11) – both ~43–42 ka – include a broader range of taxa (Table 14.2). In addition to those mentioned from MIS 4, there is evidence for further evergreen trees and shrubs that grow in riverine forests. These include species such as *Rapanea melanophloeos, Olea europaea* (Layer 13) and *Heteromorpha* sp. (Layer 11), as well as several that prefer more open grassland vegetation such as *Grewia* sp. and Fabaceae taxa (Layer 13) (Table 14.3). Later, at ~41 ka, there is continued evidence for evergreen, riverine forest taxa in Layer 8 (MLK7), but also *Erica* sp., a genus that includes species which are common components of high-altitude vegetation in the Maloti-Drakensberg (Mucina and Rutherford 2006) (Table 14.3). Shrubs within this large genus provide excellent fuel and were likely targeted for firewood when they became more accessible.

MIS 2

MIS 2 is represented by two samples – MLK5 and 4 – from Layer 5 (~24 ka). $\delta^{13}C$ values range from −23.12‰ (MLK5) to −22.48‰ (MLK4) (Table 14.1). Both are still strong C_3 pooid grassland signals, but with a slight increase in the C_4 component. The Bousman Index values indicate ~80–85% C_3 *versus* 20–15% C_4. The phytolith record likewise shows an increase in chloridoid (5%) and panicoid forms (~15%), along with sedges (~1%) (Fig. 14.6). Woody taxa phytolith values drop to the lowest levels in the sequence, accounting for <1% of the total sum. The D/P values are also the lowest from the sequence at <0.1 (Fig. 14.6). Given the low D/P values it is suggested that the C_3 component from woody taxa in the region is low. However, as in late MIS 3, the presence of wood charcoal may suggest that collected firewood has biased the C_3 signal slightly, enhancing it over a background grassland with an increased C_4 component.

The charcoal record from MIS 2 (Layer 5) differs slightly from MIS 4 and 3. *Buddleja cf. salviifolia* and *Rhamnus* sp., which have been recorded throughout the earlier deposits, are absent (Table 14.2). There is continued evidence for *Erica* sp., *Leucosidea sericea, Protea* sp. and *Olea europea*. With the exception of *Protea* sp., which is more indicative of open scrub, each of these are common components of riverine vegetation (Table 14.3). Although it is difficult to interpret the absence of the taxa noted in the earlier layers, it is interesting that some of those that do remain are frost tolerant and pioneer species. All of the taxa noted in this layer provide good sources of fuel.

Discussion

Regional Paleoenvironments and Population Flux

The Melikane sequence shows that humans were recurrently visiting the broken topography of the Maloti-Drakensberg highlands from at least MIS 5a. It is important to note, however, that these were not the first people to venture into these uplands. A handful of handaxes and cleavers from undated open-air contexts suggest a Middle Pleistocene hominin presence, although this was extremely ephemeral (Carter 1978). The phytolith and SOM $\delta^{13}C$ data suggest that highland environment ~83–80 ka was relatively stable, humid and cool. The landscape around the site was a C_3 dominated grassland environment with some C_4 elements most likely derived from tall, hydric, heliophyte panicoid grasses. Some woody elements from dicotyledonous trees and shrubs were also present, but in low abundance and most likely restricted to river corridors. The rockshelter itself also appears to have been relatively stable at this time. Previously published geoarchaeological data (Stewart et al. 2012) indicate a low-energy rockshelter environment evidenced by well-stratified sediments and sharp bounding surfaces recorded in this part of the profile. The fissures present at the rear of the rockshelter that today allow water ingress had not developed at this time, and the interior of the site was a dry, inactive environment well protected from mechanisms of erosion related to water influx.

Paleoenvironmental conditions in wider southeastern southern Africa for late MIS 5 remain poorly resolved. At present the nearest continuous paleoenvironmental archive to Melikane is the newly published Indian Ocean marine core CD154-17-17 K off the Eastern Cape's east coast (Ziegler et al. 2013) (Fig. 14.1). This record, which spans the last ~100 kyr, suggests that sea surface temperatures (SSTs) in the southwestern Indian Ocean during late MIS 5 were warmer than subsequent MIS 4-2, but cooler than today. This is consistent with two other marine cores (RC17-69 and MD 96-2077) extracted slightly further north off southern KwaZulu-Natal (Prell and Hutson 1979; Bard and Rickaby 2009). Core CD154-17-17 K also shows that discharge from rivers draining the southeastern escarpment and adjacent forelands during late MIS 5 were highly variable. Substages MIS 5c and b appear to have been relatively moist with several abrupt humid episodes registered by pulses of fluvial discharge. But at the outset of MIS 5a ~83 ka there is a shift to drier conditions that appear to persist for several millennia (Ziegler et al. 2013).

This peak aridity event ~83–80 ka corresponds to the earliest occupational pulse at Melikane. It is unclear what caused this dryness as warm southwestern Indian Ocean SSTs typically induce greater humidity across southeastern Africa (Reason and Mulenga 1999; Chase 2010; Dupont et al. 2011). However, this relationship is complex and at times warm anomalies in the western Indian Ocean correlate strongly with reduced (rather than increased) summer rainfall in southern Africa (Mason and Jury 1997; Landman and Mason 1999). Similarly, Thackeray (1987, 1988) has demonstrated that no linear relationship exists between temperature and rainfall in southern Africa. It is possible, for example, that conditions ~83–80 ka were akin to the mid-Holocene, when much of the subcontinent was simultaneously very warm and arid. A number of inland records situated further north in the Kalahari Basin also register lower precipitation ~80 ka, including various cave speleothems in the Kalahari that cease growing ~83–77 (Brook et al. 1996, 1997, 1998). If areas both inland and coastward of the Maloti-Drakensberg were arid, this contrasts with the highland zone around Melikane, which was humid C_3-dominated grassland with a minor woody component. This may suggest that some lowland populations were drawn into the highlands to exploit this better-watered zone as a temporary refugium. Clearly, though, more terrestrial archives and archeological records are needed to develop a sharper picture of regional ecology and population dynamics during late MIS 5.

Melikane appears to have been largely or entirely abandoned over the next ~20 kyr, an interval encompassing terminal MIS 5 and most of MIS 4. Core CD154-17-17 K suggests that after the dry phase ~83–80 ka the region became increasingly humid in terminal MIS 5, with pulses of heavy fluvial discharge into the Indian Ocean towards the transition to MIS 4 (Ziegler et al. 2013). Humidity remained high throughout MIS 4, though with fewer peak periods of increased discharge and sediment deposition than terminal MIS 5. High moisture availability is consistent with the relatively high SSTs estimated for MIS 4 in core MD79254 off Mozambique (van Campo et al. 1990; Chase 2010), though more southerly cores closer to the research area give colder SST estimates for this period (Prell and Hutson 1979; Bard and Rickaby 2009). Still, the signature of high MIS 4 humidity in core CD154-17-17 K agrees well with the terrestrial archive from Sibudu (Fig. 14.1), where multiple proxies suggest that during the Howiesons Poort occupations in late MIS 4 (~65–62 ka) the area hosted a woodland savanna crosscut by riparian evergreen forests (Allott 2006; Sievers 2006). Whereas winters appear to have been slightly drier and colder than today, summer temperatures and precipitation levels were comparable (Bruch et al. 2012).

Melikane's abandonment ~80–61 ka need not imply total depopulation of the highlands, but there is little to suggest a human presence in the area at this time. This interval saw two highly distinctive stone tool industries – the Still Bay and the Howiesons Poort – appear throughout the subcontinent's coastal forelands (Henshilwood 2012; Sealy 2016). Yet no characteristic Still Bay bifacial points have been found in the Maloti-Drakensberg despite several intensive field surveys (Carter 1978; Parkington and Poggenpoel 1980; Bousman 1988; Mitchell 1996a, b; Mitchell and Arthur 2010; Dewar and Stewart 2011), and the only Howiesons Poort backed artifacts discovered to date are those at Melikane and several lithic scatters noted by Carter below the Eastern Cape Drakensberg near Kenegha Poort (Mitchell 2009:130). According to Antarctic ice cores, the bulk of MIS 4 saw depressed temperatures with the coldest phase between ~70 and ~62 ka when mean temperature was 8.7 °C below present (Jouzel et al. 2007). Such severe cold earlier in MIS 4 probably discouraged highland exploitation, and indeed this upland hiatus contrasts with lower elevation sites either side of the Maloti-Drakensberg such as Sibudu and Rose Cottage Cave (Fig. 14.1) where humans *were* present ~70–62 ka (Jacobs et al. 2008; Pienaar et al. 2008).

When Melikane was reoccupied at ~61 ka, conditions may have warmed sufficiently for Howiesons Poort tool-makers to make higher altitude incursions. The phytoliths show that the highland landscape at this time was still dominated by grasses, but that temperatures and the proportion of woodland had increased. The charcoal record suggests that the latter included evergreen forest along the riverine corridors, with pioneer, open ground woody vegetation also present. The SOM $\delta^{13}C$ values and the Bousman Index values likewise suggest the local vegetation signal was still dominated by the C_3 component. This would have largely been sourced from pooid grass taxa, though an element will have also been derived from C_3 tree and shrub taxa. However, there was now more C_4 than was present in MIS 5a. The increase in the C_4 component (up to 30%) is corroborated by the increase in morphotypes attributed to panicoids along with the trace chloridoid signal (in MLK19). More C_4 taxa suggests slightly warmer temperatures, a shift in the vegetation belt with slightly drier conditions relative to MIS 5a, or lowered pCO_2 levels (Street-Perrott et al. 1997). We consider the latter more likely considering the higher D/P ratio, the presence of charcoal from water-loving taxa and evidence for widespread humidity across southern Africa during MIS 4 (Chase 2010). The Melikane River Valley directly below the site would have been a good source of wood for fuel and other uses (Fig. 14.3).

Melikane witnessed a further 10 kyr period of abandonment in terminal MIS 4 and early MIS 3 until ~50 ka. There is abundant evidence that the wider region experienced initial cold followed by increasingly warm and dry conditions during this hiatus. Fluvial discharge in core CD154-17-17 K diminishes markedly relative to MIS 4 and terminal MIS 5 (Ziegler et al. 2013). Though details vary, marine cores for which SSTs have been estimated register a drop in temperatures at some point between ~60–50 ka (van Campo et al. 1990; Bard and Rickaby 2009). At Sibudu, makers of post-Howiesons Poort ('Sibudan') technology intensively occupied the site across the MIS 4/3 transition around 58 ka. Diverse proxy datasets suggest a riverine forest setting that included frost tolerant afromontane vegetation signaling the coldest conditions in the entire Sibudu sequence. Also making a first appearance are several bushveld taxa, which today grow in more northerly, drier regions of the subcontinent. Gradual warming and further drying through this pulse at Sibudu is evidenced by the replacement of small, solitary browsing ungulates by medium and large grazers (Clark and Plug 2008), increasing frequencies of seeds from deciduous trees (Sievers 2006) and higher magnetic susceptibility values for the upper sediments (Herries 2006). Widespread and well-dated colluvial mantles (the Masotcheni Formation) deposited across inland KwaZulu-Natal during phases of hillslope instability ~56–52 ka (Botha and Partridge 2000) may indicate that this drying trend continued after – and might have played a role in – Sibudu's abandonment.

Though these are the conditions (relatively warm and dry) under which we expect highland settlement to intensify, Melikane was not reoccupied until ~50 ka. Humans were present, however, at the nearby highland site of Sehonghong at ~58 ka, and at ~56 ka at both Ntloana Tsoana and Rose Cottage Cave farther west in lowland Lesotho and the adjacent eastern Free State, respectively (Jacobs et al. 2008). Moreover, it is important to note that virtually no rockshelter sites anywhere in southern Africa have dated occupations between ~56 and ~50 ka. One possibility is that settlement preferences shifted away from rockshelters towards open-air locales, perhaps in response to warmer temperatures and reduced precipitation. Open-air sites with Late Pleistocene artifacts abound in the highlands (Carter 1978; Bousman 1988; Mitchell 1996a, b; Dewar and Stewart 2011), but until these can be dated the lack of highland occupation during this interval this does not fit model predictions.

Melikane was reoccupied in mid-MIS 3, first at ~50 ka and later ~46–38 ka. These were the site's most enduring human occupations, and the phytolith and SOM $\delta^{13}C$ records signal dramatic changes in the highland environment. The earlier pulse ~50 ka appears to have been warmer and wetter than either MIS 5a or 4, supporting the highest tree/shrub densities of any period registered at Melikane. Higher temperatures, increased summer rainfall and greater vegetation density are also inferred for the broadly coeval (~48 ka) late MSA levels at Sibudu as compared to earlier periods (Glenny 2006; Wadley 2006; Bruch et al. 2012). The interior likewise appears to have enjoyed higher precipitation at ~50 ka, with rainfall levels exceeding present day. Widespread warming and humidification may have encouraged exploitation of the highlands by forelands groups, who introduced a Sibudu late MSA-like lithic technology that included convergent Levallois points.

Very different conditions ensued at Melikane in the longer pulse between ~46 and 38 ka. The phytoliths indicate a reduction in woodland and an expansion of grassland. The increase in panicoids suggests more open conditions with tall, heliophyte grasses. These may have been more abundant due to slightly drier conditions and/or a lowering of pCO_2 levels during mid-MIS 3. The charcoal record shows evidence for evergreen trees and shrubs associated with riverine forests, along with taxa that prefer more open grassland conditions including *Grewia* sp. and Fabaceae (Tables 14.2 and 14.3). From ~43 ka woodland cover diminishes further, with the absence of some evergreen moist forest types and the appearance of taxa that may suggest a depression in the alpine belt due to colder conditions (*Erica* sp.). Reduced climatic stability is consistent with evidence for major changes in site formation processes at this time (Stewart et al. 2012). These include recurrent influxes of colluvial gravels alternating with episodes of roof collapse, and substantial physical and chemical transformations of the sediments (including phytolith dissolution) from increased water throughput via newly formed fissures.

This record of mid-MIS 3 instability accords well with core CD154-17-17 K, which shows a transition at ~50 ka to a period of wetter pulses punctuated by drier episodes for the remainder of MIS 3 (Ziegler et al. 2013). The timing of these wet events in southern Africa has been linked to bipolar responses to North Atlantic cold episodes (Heinrich Events), with corresponding increases of fluvial discharge into the Indian Ocean (Ziegler et a. 2013). Partridge et al. (2004) have shown that each Heinrich Event is preceded by ~3–4 kyr of rapid onset arid phases linked to Antarctic cooling. In KwaZulu-Natal's Masotcheni Formation, these changes in aridity and wetness are reflected in alternating phases of colluviation and pedogenesis resulting from phases of hillslope activity and stability, respectively (Botha and Partridge 2000). Colluvial episodes signaling heightened aridity have been OSL and ^{14}C dated to ~46–37 ka (Botha and Partridge 2000; Botha et al. 1992; Clarke et al. 2003; Temme et al. 2008). These are punctuated by numerous paleosols reflecting brief shifts to more humid conditions. The Masotcheni sediments also appear to register the drop in temperatures at ~42 ka that is apparent at Melikane (Temme et al. 2008). On the interior plateau there is likewise evidence that this period was unstable if predominantly arid. A recent multi-proxy analysis of a thoroughly dated (using OSL) overbank alluvial-paleosol succession at Erkroon near Bloemfontein indicate conditions between ~46 and ~32 ka were more arid (estimated MAP: ~200–400 mm) than present-day (MAP ~400–600 mm), with a brief phase of intervening humidity at ~42 ka (Lyons et al. 2014). This agrees well with ^{14}C dated portion of the inland record at Tswaing Crater further to the north, where a major reduction in rainfall is apparent beginning ~47 ka, followed by slight humidification ~42 ka and then another arid phase ~40–35 ka (Kristen et al. 2007).

This period of instability and recurrent aridity corresponds closely with Melikane's most intensive pulse of human occupation ~46–38 ka. During this 8 kyr interval over a third of the stratigraphic sequence was deposited. Recent work at nearby Sehonghong (Fig. 14.1) suggests that a substantial portion of that sequence also dates to this time (Jacobs et al. 2008). The concurrence between this intensification of highland exploitation and climatic instability in the wider region is striking. We suggest that this could reflect dispersals of lowland populations towards the upper reaches of the Senqu River in order to buffer against recurrent shortfalls of accessible surface water and dependent plants and animals. Preliminary support for this hypothesis comes from the paleoenvironmental proxy results presented here. While phytoliths and SOM $\delta^{13}C$ often reflect vegetation changes in the broader landscape, charcoals are more firmly tied to human selective processes and provide windows on locally available fuel woods. Unlike the phytoliths and SOM $\delta^{13}C$ which both show environmental changes, the charcoals exhibit a high degree of continuity through the sequence. As noted, most of the taxa represented are evergreen shrubs and small trees that favor wooded ravines often in association with watercourses. Ravines and crevices provide sheltered locations with increased available atmospheric moisture, and the presence of streams may have had an additional ameliorating effect enabling vegetation to remain at altitudes and in climatic conditions that were otherwise unsuitable. The Senqu and Melikane River Valleys could have provided such a refuge, offering people reliable sources of surface water, food, fuel and high-quality toolstone. Importantly, these riparian charcoals continue even after 42 ka when the phytoliths and SOM $\delta^{13}C$ suggest reduced temperatures, woody vegetation and rainfall. The upper Senqu and its tributaries like the Melikane thus appear to have sustained ample water flow through this unpredictable time.

Humans were again absent from Melikane in late MIS 3 and earliest MIS 2 between ~38 and 24 ka. After a pronounced peak of river discharge ~38 ka, core CD154-17-17 K suggests slightly diminished variability in this interval compared to ~50–38 ka (Ziegler et al. 2013). The peak at 38 ka coincides with the termination of the protracted mid-MIS 3 occupation at Melikane and a brief occupational pulse at Sibudu by makers of final MSA technology. A range of studies suggest that Sibudu's setting at this time was a complex mosaic of open savanna/woodland with patches of riparian forest similar to today, albeit slightly drier and cooler (Allott 2006; Glenny 2006; Bruch et al. 2012). At Erfkroon in the eastern interior, the aforementioned arid phase ~46–32 was followed by humid conditions ~32–28 ka, after which increasingly arid conditions again set in on the lead up to the LGM. However, currently very little additional paleoecological information is available for this interval, though our ongoing work at Sehonghong is augmenting this picture (Loftus et al. 2015).

Occupation during MIS 2 at Melikane occurred just before the LGM. The paleoenvironmental record from the site suggests colder conditions than in MIS 3. Grassland cover with a strong C_3 component most likely derived from alpine sour grasses is present along with some panicoids, which contribute to a low C_4 component. The presence of C_4 taxa may reflect lower pCO_2 levels during the onset of the LGM, as well as the presence of some tall, hydric, heliophyte grasses and sedges. The sediments from this level (Layer 5) comprise anthropogenic materials mixed with colluvial sediments and host bedrock attrition materials derived from roof fall debris (Stewart et al. 2012). This supports the notion of colder and drier conditions with material derived from landscape erosion as well as freeze-thaw and weathering processes. Tree cover is much reduced, with trees and shrubs likely tightly restricted to the river corridors or more sheltered areas of the landscape where, as with the ~46–38 ka climatic downturn, there was clearly sufficient surface water to support them. Cold and frost tolerant taxa are present, including *Leucosidea sericea* and *Protea* sp. At Rose Cottage Cave in the Caledon Valley, late MIS 3 and early MIS 2 is also marked by *Protea* sp. with *Leucosidea sericea* and other heathland species (Wadley et al. 1992). The composition of the youngest layer analyzed at Melikane is comparable with the addition of Ericaceous taxa. It must be noted, however, that inferring climate change from the presence of *Erica* sp. is limited by our inability to identify this to the species level; it is possible, for example, that these specimens derive from *Erica cafforum*, a streamside species that occurs in valleys together with *Buddleja* and *Leucosidea* (Mucina and Rutherford 2006).

At Melikane there is an absence of human activity from ~24 ka until the start of the Holocene. Cold and semi-arid to arid climatic conditions prevailed in the Drakensberg from ~24 ka, during what Lewis (1996) has named the Bottlenek Stadial, which broadly corresponds with the LGM. During this period there is evidence for niche glaciation and extensive periglacial conditions across the region. Small cirque glaciers were fed by snow-blow at altitudes as low as 2100 m at Mt. Enterprise in the Eastern Cape Drakensberg, South Africa (Lewis and Illgner 2001), and 3100 m in the Sekhokong range of eastern Lesotho. The onset of glaciation in the Eastern Cape Drakensberg has been dated to ~24 ka (Lewis and Hanvey 1993), while in the Sekhokong range glacial moraines have been dated between 19.4 and 14.7 ka (Mills et al. 2009). Based on glacial landform elevations in the Eastern Cape Drakensberg, Lewis and Illgner (2001) estimate mean annual temperatures during the LGM were at least ~10 °C lower than present-day, with permanent snowlines above 2100 m, and precipitation reduced by 70% (Lewis 2008). Pollen data from Strathalan Cave B in the Eastern Cape Drakensberg (Fig. 14.1) indicates alpine environments after 24 ka at 1800 m. The site was abandoned shortly thereafter and was not reoccupied until the early Holocene (Opperman and Heydenrych 1990). A similar occupational hiatus from early MIS 2 (28 ka) to the terminal Pleistocene (12.5 ka) is apparent at the recently reexcavated site of Ha Makotoko in lowland Lesotho to the west of our research area (P. Mitchell, personal communication). The cold, harsh conditions would have forced human populations out of high elevation regions and displaced the vegetation belts to much lower altitudes. It should be noted, however, that a relatively ephemeral human presence during the LGM is registered at nearby Sehonghong (Mitchell 1995). Between Heinrich Stadial 2 and 1 core CD154-17-17 K records very little discharge from rivers draining southeastern southern Africa (Ziegler et al. 2013).

Finally, Melikane's uppermost layers reveal mixed and reworked colluvial components that may have been formed during the late glacial with reworking and bioturbation during the Holocene. The higher chloridoid component is more indicative of early Holocene environments, which is supported by other $\delta^{13}C$ records in the region (Grab et al. 2005; Roberts et al. 2013).

Following the Senqu

Most authors have assumed that prehistoric groups moved between the highland Maloti-Drakensberg and the Eastern Cape or KwaZulu-Natal midlands or coasts (Carter 1970, 1978; Cable 1984; Opperman 1987). However, we suspect that the interior plateau also played an important role. Above we put forward a model linking variability of population movements into the highlands to prevailing climate. During phases of heightened warmth or humidity we suggest that incoming populations *did* usually derive from the forelands. Such movements may have resulted in Melikane's initial occupational pulse at ~80 ka, and the subsequent pulses at ~61 and ~50 ka. Greater ecological productivity at these times might have led to higher population densities in the southeastern forelands and simultaneously rendered the highlands more hospitable, creating a push/pull effect. It seems likely that such groups were more permanently based along the coastal belt (e.g., Stapleton and Hewitt 1927, 1928; Davies 1975; Kaplan 1989, 1990; Wadley 2006; Fisher et al. 2013) rather than in the midlands or Drakensberg foothills where MSA sites are rare and often ephemeral (Deacon 1976; Derricourt 1977; Mazel 1982; Opperman 1987; Opperman and Heydenrych 1990).

During drier periods, on the other hand, we anticipate that dispersals more often originated on the interior plateau of central South Africa. These may be responsible for Melikane's most enduring pulse ~46–38 ka and the pre-LGM pulse at ~24 ka. Steep east-west rainfall and evaporation gradients would have allowed interior populations evading desiccation to move eastwards into the more stable, productive and heterogeneous mountain country closer to the Orange-Senqu headwaters. The river itself may have served as a conduit into the Maloti-Drakensberg via the natural gap incised by its valley through the massif's southwest corner (Fig. 14.1). Along this axis of movement, unlike that into the forelands, there is abundant evidence for Late Pleistocene activity. In the highlands themselves, dense clusters of open-air and rockshelter sites – including Melikane, Sehonghong and others – center on the uppermost Senqu and its headwater tributaries (Carter 1978; Bousman 1988; Carter et al. 1988; Mitchell 1996a, b; Dewar and Stewart 2011; Stewart et al. 2012). Westward across the mountains similar clusters occur along the Caledon and one of its major tributaries, the Phuthiatsana, both of which ultimately feed the Orange-Senqu (Mitchell and Steinberg 1992; Mitchell 1994; Wadley 1997; Mitchell and Arthur 2010). Further downstream in the eastern and central Karoo, Sampson (1985) has documented thousands of open-air MSA occurrences in his comprehensive surveys of Orange River Scheme Area and Seacow River Valley, another large tributary (Fig. 14.1). These appear to reflect successive humid-phase recolonizations of the central interior by MSA groups who through time became increasingly tethered to the Orange-Senqu, its tributaries and their confluences (Sampson 1985). We think it plausible that in the drier abandonment phases between these desert occupations some of these groups dispersed east into the Maloti-Drakensberg. Finally, we note that during the Holocene when cultural connections and contacts are easier to trace, links between the central Karoo and the Lesotho highlands are numerous and well documented (Humphreys 1991; Mitchell 1996c, 1999).

Conclusion

This paper has presented the results of several paleoenvironmental proxy analyses from the Late Pleistocene sequence at Melikane Rockshelter in highland Lesotho. We discussed these results in relation to other terrestrial and marine archives to explore paleoenvironmental changes in southeastern southern Africa and their possible implications for regional population dynamics. Our expectations that a human presence in the highland Maloti-Drakensberg intensified when the wider region was relatively warm and/or arid appear to be largely supported. Warm (and variably humid) conditions conducive to upland exploitation seem characterize the wider region ~83–80 ka, ~50 ka and possibly also ~61 ka. In contrast, the occupational pulses ~46–38 and ~24 ka seem to correspond to drier (and colder) phases. However, deviations from our anticipated pattern also exist, namely the occupational hiatuses during the periods of acute aridity ~56–50 ka and across the LGM. While the former hiatus may simply echo the wider absence of rockshelter archeology across the subcontinent at this time, the latter is probably rooted in the LGM temperature plunge. We also posited a dual-source model of Pleistocene population movements into the Maloti-Drakensberg linked to prevailing climate. Testing this model is beyond the scope of this paper, and will require further fieldwork along with comparative studies of lithic technology between highland sites and others in possible lowland source areas. Integrating such analyses with paleoenvironmental records like those presented here as well as new, more continuous regional archives should yield further insights into the evolution of human engagements with mountain systems.

Acknowledgments Our work at Melikane was made possible because of an excavation permit generously granted (to BAS) by the Protection and Preservation Committee (PPC) of the Lesotho Department of Culture. We thank the PPC, and especially bo-Mme Moliehi 'Maneo Ntene, Puseletso Moremi and the late Ntsema Khitsane for their support. The project 'Adaptations to Marginal Environments in the Middle Stone Age' (AMEMSA) is supported by grants from the McDonald Institute for Archaeological Research, the University of Cambridge, the British Academy, the Wenner-Gren Foundation, the Prehistoric Society, and the Social Sciences and Humanities Research Council of Canada. We also gratefully acknowledge Catherine Kneale for assistance in processing the SOM δ^{13}C samples. Finally, we thank Francis Thackeray, Britt Bousman and a third anonymous reviewer, as well as Peter Mitchell, for helping us substantially improve this paper.

References

Aldenderfer, M. S. (1998). *Montane Foragers: Asana and the South-Central Andean Archaic*. Iowa City: University of Iowa Press.

Aldenderfer, M. (2006). Modelling plateau peoples: The early human use of the world's high plateaux. *World Archaeology, 38*(3), 357–370.

Alexandre, A., Meunier, J.-D., Lezine, A.-M., Vincens, A., & Schwartz, D. (1997). Phytoliths: Indicators of grassland dynamics during the late Holocene in intertropical Africa. *Palaeogeography, Palaeoclimatology, Palaeoecology, 136*(1), 213–229.

Allott, L. F. (2004). Changing environments in oxygen isotope stage 3: Reconstructions using archaeological charcoal from Sibudu Cave. *South African Journal of Science, 100*(3–4), 179–184.

Allott, L. F. (2005). Palaeoenvironments of the Middle Stone Age at Sibudu Cave, KwaZulu-Natal, South Africa: An analysis of archaeological charcoal. Ph.D. Dissertation, University of the Witwatersrand, 2005.

Allott, L. F. (2006). Archaeological charcoal as a window on palaeovegetation and wood use during the Middle Stone Age at Sibudu Cave. *Southern African Humanities, 18*(1), 173–201.

Ambrose, S. H., & Norr, L. (1993). Experimental evidence for the relationship of the carbon isotope ratios of whole diet and dietary protein to those of bone collagen and carbonate. In J. B. Lambert & G. Grupe (Eds.), *Prehistoric human bone* (pp. 1–37). Berlin: Springer-Verlag.

Ambrose, S. H., & Sikes, N. E. (1991). Soil carbon isotope evidence for Holocene habitat change in the Kenya Rift Valley. *Science, 253* (5026), 1402–1405.

Atkinson, Q. D., Russell, D. G., & Drummond, A. J. (2009). Bayesian coalescent inference of major human mitochondrial DNA haplogroup expansions in Africa. *Proceedings of the Royal Society B: Biological Sciences, 276*(1655), 367–373.

Barbieri, C., Vicente, M., Rocha, J., Mpoloka, S. W., Stoneking, M., & Pakendorf, B. (2013). Ancient substructure in early mtDNA lineages of southern Africa. *The American Journal of Human Genetics, 92*(2), 285–292.

Bard, E., & Rickaby, R. E. (2009). Migration of the subtropical front as a modulator of glacial climate. *Nature, 460*(7253), 380–383.

Barham, L., & Mitchell, P. (2008). *The first Africans: African archaeology from the earliest toolmakers to most recent foragers*. Cambridge: Cambridge University Press.

Batini, C., Coia, V., Battaggia, C., Rocha, J., Pilkington, M. M., Spedini, G., et al. (2007). Phylogeography of the human mitochondrial L1c haplogroup: Genetic signatures of the prehistory of Central Africa. *Molecular Phylogenetics and Evolution, 43*(2), 635–644.

Batini, C., Lopes, J., Behar, D. M., Calafell, F., Jorde, L. B., van der Veen, L., et al. (2011). Insights into the demographic history of African pygmies from complete mitochondrial genomes. *Molecular Biology and Evolution, 28*(2), 1099–1110.

Bawden, M. G., & Carroll, D. M. (1968). *The land resources of Lesotho*. London: United Kingdom Directorate of Overseas Surveys.

Beaumont, P. B. (1986). Where did all the young men go during 0–18 stage 2? *Palaeoecology of Africa, 17*, 79–86.

Behar, D. M., Villems, R., Soodyall, H., Blue-Smith, J., Pereira, L., Metspalu, E., et al. (2008). The dawn of human matrilineal diversity. *The American Journal of Human Genetics, 82*(5), 1130–1140.

Bender, S. J., & Wright, G. A. (1988). High-altitude occupations, cultural process, and high plains prehistory: Retrospect and prospect. *American Anthropologist, 90*(3), 619–639.

Benedict, J. B. (1992). Footprints in the snow: High-altitude cultural ecology of the Colorado front range. *USA Arctic and Alpine Research, 24*(1), 1–16.

Bettinger, R. L. (1991). Aboriginal occupation at high altitude: Alpine villages in the White Mountains of Eastern California. *American Anthropologist, 93*(3), 656–679.

Birkenhauer, J. (1991). *The Great Escarpment of southern Africa and its coastal forelands. A re-appraisal*. München: Institut für Geographie der Universität.

Black, K. D. (1991). Archaic continuity in the Colorado Rockies: The Mountain Tradition. *Plains Anthropologist, 36*(133), 1–29.

Blome, M. W., Cohen, A. S., Tryon, C. A., Brooks, A. S., & Russell, J. (2012). The environmental context for the origins of modern human diversity: A synthesis of regional variability in African climate 150,000-30,000 years ago. *Journal of Human Evolution, 62* (5), 563–592.

Botha, G. A., & Partridge, T. C. (2000). Colluvial deposits. *Oxford Monographs on Geology and Geophysics, 40*, 88–99.

Botha, G. A., Scott, L., Vogel, J. C., & Von Brunn, V. (1992). Palaeosols and palaeoenvironments during the Late Pleistocene hypothermal in Northern Natal. *South African Journal of Science, 88*(9), 508–512.

Bousman, C. B. (1988). Prehistoric settlement patterns in the Senqunyane Valley, Lesotho. *South African Archaeological Bulletin, 43* (147), 33–37.

Bousman, C. B. (1991). Holocene paleoecology and Later Stone Age hunter-gatherer adaptations in the South African interior plateau. Ph.D. Dissertation, Southern Methodist University, 1991.

Brantingham, S., & Xing, G. (2006). Peopling of the Northern Tibetan Plateau. *World Archaeology, 38*(3), 387–414.

Brook, G. A., Cowart, J. B., & Marais, E. (1996). Wet and dry periods in the Southern African summer rainfall zone during the last 300 kyr from speleothem, tufa and sand dune age data. *Palaeoecology of Africa, 24*, 147–158.

Brook, G. A., Cowart, J. B., Brandt, S. A., & Scott, L. (1997). Quaternary climatic change in southern and Eastern Africa during the last 300 ka: The evidence from caves in Somalia and the Transvaal region of South Africa. *Zeitschrift für Geomorphologie NF Supplementband, 108*, 15–48.

Brook, G. A., Cowart, J. B., & Brandt, S. A. (1998). Comparison of Quaternary environmental change in eastern and southern Africa using cave speleothem, tufa and rock shelter sediment data. In A. Alsharan, K. W. Glennie, G. L. Wintle, & C. G. St C. Kendall (Eds.), *Quaternary deserts and climate change* (pp. 239–250). Rotterdam: Balkema.

Bruch, A. A., Sievers, C., & Wadley, L. (2012). Quantification of climate and vegetation from Southern African Middle Stone Age sites: An application using Late Pleistocene plant material from Sibudu, South Africa. *Quaternary Science Reviews, 45*, 7–17.

Burke, K., & Gunnell, Y. (2008). The African erosion surface: A continental-scale synthesis of geomorphology, tectonics, and environmental change over the past 180 million years. *Geological Society of America Memoirs, 201*, 1–66.

Cable, J. H. C. (1984). *Economy and technology in Late Stone Age southern Natal*. BAR International Series, 201. Oxford: British Archaeological Reports.

Cambray, J. A., Davies, B. R., Ashton, P. J., Agnew, J. D., De Moor, F. C., & Skelton, P. H. (1986). The Orange-Vaal river system. In B. R. Davies & K. F. Walker (Eds.), *The ecology of river systems* (pp. 89–161). Monographiae Biologicae 60. Dordrecht: Junk.

Cancellieri, E., & di Lernia, S. (2013). Middle Stone Age human occupation and dispersals in the Messak Plateau (SW Libya, Central Sahara). *Quaternary International, 300*, 142–152.

Carbutt, C., & Edwards, T. J. (2003). The flora of the Drakensberg Alpine Centre. *Edinburgh Journal of Botany, 60*(03), 581–607.

Carbutt, C., & Edwards, T. J. (2006). The endemic and near-endemic angiosperms of the Drakensberg Alpine Centre. *South African Journal of Botany, 72*(1), 105–132.

Carter, P. L. (1970). Late Stone Age exploitation patterns in southern Natal. *South African Archaeological Bulletin, 25*, 55–58.

Carter, P. L. (1976). The effects of climate change on settlement in Eastern Lesotho during the Middle and Later Stone Age. *World Archaeology, 8*, 197–206.

Carter, P. L. (1978). The prehistory of eastern Lesotho. Ph.D. Dissertation, University of Cambridge, 1987.

Carter, P. L., Mitchell, P. J., & Vinnicombe, P. (1988). *Sehonghong: The Middle and Later Stone Age industrial sequence at a Lesotho rock-shelter*. BAR International Series 409. Oxford: British Archaeological Reports.

Castañeda, I. S., Mulitza, S., Schefuß, E., Lopes dos Santos, R. A., Sinninghe Damsté, J. S., & Schouten, S. (2009). Wet phases in the Sahara/Sahel region and human migration patterns in North Africa. *Proceedings of the National Academy of Sciences, 106*(48), 20159–20163.

Chase, B. M. (2010). South African palaeoenvironments during marine oxygen isotope Stage 4: A context for the Howiesons Poort and Still Bay industries. *Journal of Archaeological Science, 37*(6), 1359–1366.

Clark, J. L., & Plug, I. (2008). Animal exploitation strategies during the South African Middle Stone Age: Howiesons Poort and post-Howiesons Poort fauna from Sibudu Cave. *Journal of Human Evolution, 54*(6), 886–898.

Clark, V. R., Barker, N. P., & Mucina, L. (2011). The Great Escarpment of Southern Africa: A new frontier for biodiversity exploration. *Biodiversity and Conservation, 20*(12), 2543–2561.

Clarke, M. L., Vogel, J. C., Botha, G. A., & Wintle, A. G. (2003). Late Quaternary hillslope evolution recorded in Eastern South African colluvial badlands. *Palaeogeography, Palaeoclimatology, Palaeoecology, 197*(3–4), 199–212.

Cordova, C. E. (2013). C_3 poaceae and restionaceae phytoliths as potential proxies for reconstructing winter rainfall in South Africa. *Quaternary International, 287*, 121–140.

Cornelissen, E. (2002). Human responses to changing environments in Central Africa between 40,000 and 12,000 BP. *Journal of World Prehistory, 16*(3), 197–235.

Coulthard, T. J., Ramirez, J. A., Barton, N., Rogerson, M., & Brücher, T. (2013). Were rivers flowing across the Sahara during the Last Interglacial? Implications for human migration through Africa. *PLoS ONE, 8*(9), e74834.

Cowling, R. M. & Hilton-Taylor, C. (1994). Patterns of plant diversity and endemism in Southern Africa: An overview. In B. J. Huntley (Ed.), *Botanical diversity in southern Africa. Strelitzia, 1*, 31–52.

Cowling, R. M. & Hilton-Taylor, C. (1997). Phytogeography, flora and endemism. In R. M. Cowling, D. M. Richardson & S. M. Pierce (Eds.), *Vegetation of southern Africa* (pp. 43–61). Cambridge: Cambridge University Press.

Davies, O. (1975). Excavations at Shongweni South Cave: The oldest evidence to date for cultigens in southern africa. *Southern African Humanities, 22*(2), 627–662.

Davis, S. D., Heywood, V. H., & Hamilton, A. C. (Eds.) (1994). *Centres of plant diversity. A guide and strategy for their conservation*. Vol. 1. Europe, Africa, South West Asia and the Middle East. Gland, Switzerland: IUCN and WWF.

Deacon, H. J. (1976). *Where hunters gathered: A study of Holocene Stone Age people in the Eastern Cape*. Claremont: South African Archaeological Society.

Deacon, H. J. (1979). Excavations at Boomplaas Cave: A sequence through the Upper Pleistocene and Holocene in South Africa. *World Archaeology, 10*(3), 241–257.

Deacon, H. J., & Thackeray, J. F. (1984). Late Pleistocene environmental changes and implications for the archaeological record in Southern Africa. In J. C. Vogel (Ed.), *Late Cainozoic palaeoclimates of the southern hemisphere* (pp. 375–390). Rotterdam: Balkema.

Deacon, J. (1974). Patterning in the radiocarbon dates for the Wilton/Smithfield complex in Southern Africa. *The South African Archaeological Bulletin, 29*, 3–18.

Derricourt, R. M. (1977). *Prehistoric man in the Ciskei and Transkei*. Cape Town: Struik.

Dewar, G., & Stewart, B. A. (2011). *2009 archaeological field survey of the Sehonghong and Senqu Rivers, Thaba-Tseka District*. Lesotho: Unpublished report.

Dewar, G., & Stewart, B. A. (2012). Preliminary results of excavations at Spitzkloof Rockshelter, Richtersveld, South Africa. *Quaternary International, 270*, 30–39.

Dowson, T. A. (1988). Shifting vegetation zones in the Holocene and later Pleistocene: Preliminary charcoal evidence from Jubilee Shelter, Magaliesberg, Southern Transvaal. *Palaeoecology of Africa, 19*, 233–239.

Drake, N. A., Blench, R. M., Armitage, S. J., Bristow, C. S., & White, K. H. (2011). Ancient watercourses and biogeography of the Sahara explain the peopling of the desert. *Proceedings of the National Academy of Sciences, 108*(2), 458–462.

Dupont, L. M., Caley, T., Kim, J.-H., Castañeda, I., Malaizé, B., & Giraudeau, J. (2011). Glacial-interglacial vegetation dynamics in South Eastern Africa coupled to sea surface temperature variations in the Western Indian Ocean. *Climate of the Past, 7*, 1209–1224.

Eamus, D., Hatton, T., Cook, P., & Colvin, C. (2006). *Ecohydrology: Vegetation function, water and resource management*. Collingwood, Australia: CSIRO Publishing.

Earle, A., Malzbender, D., Turton, A., & Manzungu, E. (2005). *A preliminary basin profile of the Orange/Senqu River*. Pretoria: AWIRU, University of Pretoria.

Ehleringer, J. R., Cerling, T. E., & Helliker, B. R. (1997). C_4 photosynthesis, atmospheric CO_2, and climate. *Oecologia, 112*(3), 285–299.

Eichhorn, B. (2002). Anthrakologische untersuchungen zur vegetationgeshichte des Kaokolandes, nordwest Namibia. Ph.D. Dissertation, University of Köln, 2002.

Esterhuysen, A. B. (1996). Palaeoenvironmental reconstruction from Pleistocene to present: An analysis of charcoal from sites in the Eastern Free State and Lesotho. Ph.D. Dissertation, University of the Witwatersrand, 1996.

Ficken, K. J., Wooller, M. J., Swain, D. L., Street-Perrott, F. A., & Eglinton, G. (2002). Reconstruction of a subalpine grass-dominated ecosystem, Lake Rutundu, Mount Kenya: A novel multi-proxy approach. *Palaeogeography, Palaeoclimatology, Palaeoecology, 177*(1), 137–149.

Fisher, E. C., Albert, R.-M., Botha, G., Cawthra, H. C., Esteban, I., Harris, J., et al. (2013). Archaeological reconnaissance for Middle Stone Age sites along the Pondoland coast, South Africa. *PaleoAnthropology, 2013*, 104–137.

Foley, R. A., Maíllo-Fernández, J. M., & Lahr, M. M. (2013). The Middle Stone Age of the central Sahara: Biogeographical opportunities and technological strategies in later human evolution. *Quaternary International, 300*, 153–170.

Glenny, W. (2006). Report on the micromammal assemblage analysis from Sibudu Cave. *Southern African Humanities, 18*, 279–288.

Grab, S. W. (1994). Thufur in the Mohlesi Valley, Lesotho, Southern Africa. *Permafrost and Periglacial Processes, 5*(2), 111–118.

Grab, S. W. (1997). Thermal regime for a thufa apex and its adjoining depression, Mashai Valley, Lesotho. *Permafrost and Periglacial Processes, 8*(4), 437–445.

Grab, S. W., & Nash, D. J. (2010). Documentary evidence of climate variability during cold seasons in Lesotho, Southern Africa, 1833-1900. *Climate Dynamics, 34*(4), 473–499.

Grab, S., Scott, L., Rossouw, L., & Meyer, S. (2005). Holocene palaeoenvironments inferred from a sedimentary sequence in the Tsoaing River Basin, Western Lesotho. *Catena, 61*(1), 49–62.

Harpending, H., & Davis, H. (1977). Some implications for hunter-gatherer ecology derived from the spatial structure of resources. *World Archaeology, 8*(3), 275–286.

Henn, B. M., Gignoux, C. R., Jobin, M., Granka, J. M., Macpherson, J. M., Kidd, J. M., et al. (2011). Hunter-gatherer genomic diversity suggests a Southern African origin for modern humans. *Proceedings of the National Academy of Sciences, 108*(13), 5154–5162.

Henshilwood, C. S. (2012). Late Pleistocene techno-traditions in Southern Africa: A review of the Still Bay and Howiesons Poort, c. 75–59 ka. *Journal of World Prehistory, 25*(3–4), 205–237.

Herries, A. I. R. (2006). Archaeomagnetic evidence for climate change at Sibudu Cave. *Southern African Humanities, 18*, 131–147.

Holmgren, K., Lee-Thorp, J. A., Cooper, G. R. J., Lundblad, K., Partridge, T. C., Scott, L., et al. (2003). Persistent millennial-scale climatic variability over the past 25,000 years in Southern Africa. *Quaternary Science Reviews, 22*(21–22), 2311–2326.

Huang, Y., Street-Perrott, F. A., Perrott, R. A., Metzger, P., & Eglinton, G. (1999). Glacial-interglacial environmental changes inferred from molecular and compound-specific $\delta^{13}C$ analyses of sediments from Sacred Lake, Mt. Kenya. *Geochimica Et Cosmochimica Acta, 63* (9), 1383–1404.

Humphreys, A. J. B. (1991). On the distribution and dating of bifacial tanged and barbed arrowheads in the interior of South Africa. *The South African Archaeological Bulletin, 46*, 41–43.

InsideWood (2004-onwards). Retrieved May 14, 2012 from http://insidewood.lib.ncsu.edu/search.

Jacobs, Z., & Richard, R. G. (2008). Testing times: Old and new chronologies for the Howieson's Poort and Still Bay industries in environmental context. *South African Archaeological Bulletin Goodwin Series, 10*, 9–34.

Jacobs, Z., Roberts, R. G., Galbraith, R. F., Deacon, H. J., Grün, R., Mackay, A., et al. (2008). Ages for the Middle Stone Age of Southern Africa: Implications for human behavior and dispersal. *Science, 322*(5902), 733–735.

Jacot Guillarmod, A. (1971). *Flora of Lesotho (Basutoland)*. Lehre: Verlag Von Cramer.

Johnson, M. R., Van Vuuren, C. J., Hegenberger, W. F., Key, R., & Show, U. (1996). Stratigraphy of the Karoo Supergroup in Southern Africa: An overview. *Journal of African Earth Sciences, 23*(1), 3–15.

Jouzel, J., Masson-Delmotte, V., Cattani, O., Dreyfus, G., Falourd, S., Hoffmann, G., et al. (2007). Orbital and millennial Antarctic climate variability over the past 800,000 years. *Science, 317*(5839), 793–796.

Kaplan, J. M. (1989). 45000 years of hunter-gatherer history in Natal as seen from Umhlatuzana Rock Shelter. *South African Archaeological Society Goodwin Series, 6*, 7–16.

Kaplan, J. M. (1990). The Umhlatuzana Rock Shelter sequence: 100 000 years of Stone Age history. *Natal Museum Journal of Humanities, 2*, 1–94.

Killick, D. J. B. (1963). An account of the plant ecology of the Cathedral Peak area of the Natal Drakensberg. *Memoirs of the Botanical Survey of South Africa, 34*, 1–178.

Killick, D. J. B. (1978). The afro-alpine region. In M. J. A. Werger (Ed.), *Biogeography and ecology of Southern Africa* (pp. 515–560). The Hague: Junk.

Killick, D. J. B. (1990). *A field guide to the flora of the Natal Drakensberg*. Johannesburg: Jonathan Ball.

Kingdon, J. (1990). *Island Africa: The evolution of Africa's rare animals and plants*. London: Collins.

Kornfeld, M., Frison, G. C., Larson, M. L., & Bradley, B. A. (2010). *Prehistoric hunter-gatherers of the High Plains and Rockies*. Walnut Creek: Left Coast Press.

Kristen, I., Fuhrmann, A., Thorpe, J., Rohl, U., Wilkes, H., & Oberhansli, H. (2007). Hydrological changes in Southern Africa over the last

200 ka as recorded in lake sediments from the Tswaing impact crater. *South African Journal of Geology, 110*(2–3), 311–326.

Kromhout, C. P. (1975). 'n Sleutel vir die mikroskopiese uitkenning van die vernaamste inheemse houtesoorte van Suid-Afrika. *Department of Forestry Bulletin, 50*, 1–140.

Landman, W. A., & Mason, S. J. (1999). Change in the association between Indian Ocean sea-surface temperatures and summer rainfall over South Africa and Namibia. *International Journal of Climatology, 19*(13), 1477–1492.

Leney, L., & Casteel, R. W. (1975). Simplified procedure for examining charcoal specimens for identification. *Journal of Archaeological Science, 2*(2), 153–159.

Lewis, C. A. (1996). Periglacial features. In C. A. Lewis (Ed.), *The geomorphology of the Eastern Cape, South Africa* (pp. 103–119). Grahamstown: NISC.

Lewis, C. A. (2008). Late Quaternary climatic changes, and associated human responses, during the Last ~45 000 yr in the Eastern and adjoining Western Cape, South Africa. *Earth-Science Reviews, 88* (3–4), 167–187.

Lewis, C. A., & Hanvey, P. M. (1993). The remains of rock glaciers in Bottelnek, East Cape Drakensberg, South Africa. *Transactions of the Royal Society of South Africa, 48*, 265–289.

Lewis, C. A., & Illgner, P. M. (2001). Late Quaternary glaciation in Southern Africa: Moraine ridges and glacial deposits at mount enterprise in the Drakensberg of the Eastern Cape Province, South Africa. *Journal of Quaternary Science, 16*(4), 365–374.

Loftus, E., Stewart, B. A., Dewar, G., & Lee-Thorp, J. (2015). Stable isotope evidence of MIS 3 to middle Holocene palaeoenvironments from Sehonghong, eastern Lesotho. *Journal of Quaternary Science, 30*(8), 805–816.

Lombard, M., Schlebusch, C., & Soodyall, H. (2013). Bridging disciplines to better elucidate the evolution of early *Homo sapiens* in Southern Africa. *South African Journal of Science, 109*(11/12), 27–34.

Lynch, S. D. (2004). Development of a raster database of annual, monthly and daily rainfall for Southern Africa: Report to the water research commission. WRC Report No. 1156/1/03. Pretoria: Water Research Commission.

Lyons, R., Tooth, S., & Duller, G. A. T. (2014). Late Quaternary climatic changes revealed by luminescence dating, mineral magnetism and diffuse reflectance spectroscopy of river terrace palaeosols: A new form of geoproxy data for the Southern African interior. *Quaternary Science Reviews, 95*, 43–59.

Madsen, D. B., & Metcalf, M. D. (2000). *Intermountain archaeology*. Salt Lake City: University of Utah Press.

Mason, S. J., & Jury, M. R. (1997). Climatic variability and change over Southern Africa: A reflection on underlying processes. *Progress in Physical Geography, 21*(1), 23–50.

Mazel, A. D. (1982). Evidence for pre-Later Stone Age occupation of the Natal Drakensberg. *Southern African Humanities, 25*(1), 61–65.

Mazel, A.D. (1984). Through the keyhole: A preliminary peep at the lithic composition of Later Stone Age sites in the central and upper Tugela River Basin, Natal, South Africa. In M. Hall, G. Avery, D. M. Wilson, & A. J. B. Humphreys (Eds.), *Frontiers: Southern African archaeology today* (pp. 182–193). BAR International Series 207. Oxford: British Archaeological Reports.

Mercader, J., Astudillo, F., Barkworth, M., Bennett, T., Esselmont, C., Kinyanjui, R., et al. (2010). Poaceae phytoliths from the Niassa Rift, Mozambique. *Journal of Archaeological Science, 37*(8), 1953–1967.

Mercader, J., Bennett, T., Esselmont, C., Simpson, S., & Walde, D. (2013). Phytoliths from Middle Stone Age habitats in the Mozambican Rift (105–29 ka). *Journal of Human Evolution, 64*(5), 328–336.

Midgley, D. C., Pitman, W. V., & Middleton, B. J. (1994). The surface water resources of South Africa 1990, volumes 1 to 6. WRC Report Nos. 298/1.1/94 to 298/6.1/94 (text) and 298/1.2/94 to 298/6.2/94 (maps). Pretoria: Water Research Commission.

Mills, S. C., Grab, S. W., & Carr, S. J. (2009). Recognition and palaeoclimatic implications of late quaternary niche glaciation in Eastern Lesotho. *Journal of Quaternary Science, 24*(7), 647–663.

Mitchell, P. J. (1988). The Late Pleistocene early microlithic assemblages of Southern Africa. *World Archaeology, 20*(1), 27–39.

Mitchell, P. J. (1990). A palaeoecological model for archaeological site distribution in Southern Africa during the Upper Pleniglacial and Late Glacial. In C. Gamble & O. Soffer (Eds.), *The world at 18,000 BP: Low latitudes* (pp. 189–205). London: Unwin Hyman.

Mitchell, P. J. (1994). The archaeology of the Phuthiatsana-ea-Thaba Bosiu Basin, Lesotho, Southern Africa: Changes in Later Stone Age regional demography. *Antiquity, 68*, 83–83.

Mitchell, P. J. (1995). Revisiting the Robberg: New results and a revision of old ideas at Sehonghong Rock Shelter, Lesotho. *South African Archaeological Bulletin, 50*, 28–38.

Mitchell, P. J. (1996a). The Late Quaternary landscape at Sehonghong in the Lesotho Highlands, Southern Africa. *Antiquity, 70*, 623–638.

Mitchell, P. J. (1996b). The Late Quaternary of the Lesotho Highlands, Southern Africa: Preliminary results and future potential of ongoing research at Sehonghong Shelter. *Quaternary International, 33*, 35–43.

Mitchell, P. J. (1996c). Prehistoric exchange and interaction in southeastern Southern Africa: Marine shells and ostrich eggshell. *African Archaeological Review, 13*(1), 35–76.

Mitchell, P. J. (1998). The archaeology of the Alfred County Cave, KwaZulu-Natal. *Natal Museum Journal of Humanities, 10*, 1–17.

Mitchell, P. J. (1999). Pressure-flaked points in Lesotho: Dating, distribution and diversity. *South African Archaeological Bulletin, 54*, 90–96.

Mitchell, P. J. (2008). Developing the archaeology of marine isotope Stage 3. *South African Archaeological Society Goodwin Series, 10*, 52–65.

Mitchell, P. J. (2009). Gathering together a history of the People of the Eland: Towards an archaeology of Maloti-Drakensberg hunter-gatherers. In P. J. Mitchell & B. W. Smith (Eds.), *The eland's people: New perspectives in the rock art of the Maloti-Drakensberg Bushmen* (pp. 99–139). Johannesburg: Wits University Press.

Mitchell, P., & Arthur, C. (2010). Archaeological fieldwork in the Metolong Dam catchment, Lesotho, 2008–10. *Nyame Akuma, 74*, 51–62.

Mitchell, P. J., & Steinberg, J. M. (1992). Ntloana Tsoana: A Middle Stone Age sequence from Western Lesotho. *South African Archaeological Bulletin, 47*, 26–33.

Moore, A., & Blenkinsop, T. (2006). Scarp retreat versus pinned drainage divide in the formation of the Drakensberg Escarpment, Southern Africa. *South African Journal of Geology, 109*(4), 599–610.

Morgan, C. (2009). Climate change, uncertainty and prehistoric hunter-gatherer mobility. *Journal of Anthropological Archaeology, 28*(4), 382–396.

Morgan, C., Losey, A., & Adams, R. (2012). High-altitude hunter-gatherer residential occupations in Wyoming's Wind River Range. *North American Archaeologist, 33*(1), 35–79.

Mucina, L., & Rutherford, M. C. (2006). *The vegetation of South Africa, Lesotho and Swaziland*. Strelitzia 19. Pretoria: South African National Biodiversity Institute.

Olago, D. O., Street-Perrott, F. A., Perrott, R. A., Ivanovich, M., & Harkness, D. D. (1999). Late Quaternary glacial-interglacial cycle of climatic and environmental change on Mount Kenya. *Kenya. Journal of African Earth Sciences, 29*(3), 593–618.

O'Leary, M. H. (1981). Carbon isotope fractionation in plants. *Phytochemistry, 20*(4), 553–567.

Opperman, H. (1987). *The Later Stone Age of the Drakensberg Range and its foothills*. BAR International Series 339. Oxford: British Archaeological Reports.

Opperman, H. (1988). An excavation of a Middle Stone Age deposit in Grassridge Rockshelter, Sterkstroom District, Cape Province. *Fort Hare Papers, 9*, 51–61.

Opperman, H. (1992). A report on the results of a test pit in Strathalan Cave B, Maclear District, North-Eastern Cape. *Southern African Field Archaeology, 1*, 98–102.

Opperman, H. (1996). Strathalan Cave B, north-eastern Cape Province, South Africa: Evidence for human behaviour 29000–26000 years ago. *Quaternary International, 33*, 45–53.

Opperman, H., & Heydenrych, B. (1990). A 22 000 year-old Middle Stone Age camp site with plant food remains from the North-Eastern Cape. *South African Archaeological Bulletin, 45*, 93–99.

Parker, A. G., Lee-Thorp, J., & Mitchell, P. J. (2011). Late Holocene Neoglacial conditions from the Lesotho Highlands, Southern Africa: Phytolith and stable carbon isotope evidence from the archaeological site of Likoaeng. *Proceedings of the Geologists' Association, 122*(1), 201–211.

Parkington, J. E. (1972). Seasonal mobility in the Late Stone Age. *African Studies, 31*(4), 223–244.

Parkington, J. E. (1977). Follow the San: An analysis of seasonality in prehistory of the South Western Cape. Ph.D. Dissertation, University of Cambridge, 1977.

Parkington, J. E. (1980). Time and place: Some observations on spatial and temporal patterning in the Later Stone Age sequence in Southern Africa. *South African Archaeological Bulletin, 35*, 73–83.

Parkington, J. E., & Poggenpoel, C. (1980). *Senqu: An archaeological reconnaissance along the southern perimeter road*. Unpublished report.

Partridge, T. C., & Maud, R. R. (1987). Geomorphic evolution of Southern Africa since the Mesozoic. *South African Journal of Geology, 90*(2), 179–208.

Partridge, T. C., Kerr, S. J., Metcalfe, S. E., Scott, L., Talma, A. S., & Vogel, J. C. (1993). The Pretoria Saltpan: A 200,000 year Southern African lacustrine sequence. *Palaeogeography, Palaeoclimatology, Palaeoecology, 101*(3–4), 317–337.

Partridge, T. C., Scott, L., & Schneider, R. R. (2004). Between Agulhas and Benguela: Responses of Southern African climates of the Late Pleistocene to current fluxes, orbital precession and the extent of the circum-Antarctic vortex. In R. W. Battarbee, F. Gasse, & C. E. Stickley (Eds.), *Past climate variability through Europe and Africa* (pp. 45–68). Berlin: Springer Verlag.

Pereira, L., Silva, N. M., Franco-Duarte, F., Fernandes, V., Pereira, J. B., Costa, M. D., et al. (2010). Population expansion in the North African Late Pleistocene signalled by mitochondrial DNA haplogroup U6. *BMC Evolutionary Biology, 10*(1), 390.

Pickrell, J. K., Patterson, N., Barbieri, C., Berthold, F., Gerlach, L., Güldemann, T., et al. (2012). The genetic prehistory of Southern Africa. *Nature Communications, 3*, 1143.

Pienaar, M., Woodborne, S., & Wadley, L. (2008). Optically stimulated luminescence dating at Rose Cottage Cave. *South African Journal of Science, 104*(1–2), 65–70.

Prell, W. L., & Hutson, W. H. (1979). Zonal temperature-anomaly maps of Indian Ocean surface waters: Modern and ice-age patterns. *Science, 206*(4417), 454–456.

Reason, C. J. C., & Mulenga, H. (1999). Relationships between South African rainfall and SST anomalies in the southwest Indian Ocean. *International Journal of Climatology, 19*(15), 1651–1673.

Retallack, G. J. (2001). A 300-million-year record of atmospheric carbon dioxide from fossil plant cuticles. *Nature, 411*(6835), 287–290.

Rito, T., Richards, M. B., Fernandes, V., Alshamali, F., Cerny, V., Pereira, P., et al. (2013). The first modern human dispersals across Africa. *PLoS ONE, 8*(11), e80031.

Rivers-Moore, N. A., Goodman, P. S., & Nkosi, M. R. (2007). An assessment of the freshwater natural capital in Kwazulu-Natal for conservation planning. *Water SA, 33*(5), 665–674.

Roberts, P., Lee-Thorpe, J. A., Mitchell, P. J., & Arthur, C. (2013). Stable carbon isotopic evidence for climate change across the Late Pleistocene to early Holocene from Lesotho, Southern Africa. *Journal of Quaternary Science, 28*(4), 360–369.

Sampson, C.G., (1985). *Atlas of Stone Age settlement in the Central and Upper Seacow Valley.* Bloemfontein: Memoirs van die Nasionale Museum Bloemfontein 20: 1–112.

Scheinfeldt, L. B., Soi, S., & Tishkoff, S. A. (2010). Working toward a synthesis of archaeological, linguistic, and genetic data for inferring African population history. *Proceedings of the National Academy of Sciences, 107*(Supplement 2), 8931–8938.

Schlebusch, C. M., Skoglund, P., Sjödin, P., Gattepaille, L. M., Hernandez, D., Jay, F., et al. (2012). Genomic variation in seven Khoe-San groups reveals adaptation and complex African history. *Science, 338*(6105), 374–379.

Schlebusch, C. M., Lombard, M., & Soodyall, H. (2013). MtDNA control region variation affirms diversity and deep sub-structure in populations from Southern Africa. *BMC Evolutionary Biology, 13*(1), 56.

Schlüter, T. (2006). *Geological atlas of Africa, with notes on stratigraphy, tectonics, economic geology, geohazards and geosites of each country*. Berlin: Springer.

Schulze, R. E. (1979). *Hydrology and water resources of the Drakensberg*. Pietermaritzburg: Natal Town and Regional Planning Commission.

Schulze, R. E. (2008). South African atlas of climatology and agrohydrology. WRC Report No. 1489/1/06. Pretoria: Water Research Commission.

Schulze, R., Meigh, J., & Horan, M. (2001). Present and potential future vulnerability of eastern and Southern Africa's hydrology and water resources. *South African Journal of Science, 97*, 150–160.

Scott, L. (1999). Vegetation history and climate in the Savanna Biome, South Africa, since 190,000 ka: A comparison of pollen data from the Tswaing Crater (the Pretoria Saltpan) and Wonderkrater. *Quaternary International, 57*, 215–223.

Sealy, J. (2016). Cultural change, demography, and the archaeology of the last 100 kyr in Southern Africa. In S. C. Jones & B. A. Stewart (Eds.), *Africa from MIS 6-2: Population dynamics and paleoenvironments*. (pp. 65–75). Dordrecht: Springer.

Sene, K. J., Jones, D. A., Meigh, J. R., & Farquharson, F. A. K. (1998). Rainfall and flow variations in the Lesotho Highlands. *International Journal of Climatology, 18*(3), 329–345.

Sievers, C. (2006). Seeds from the Middle Stone Age layers at Sibudu Cave. *Southern African Humanities, 18*(1), 203–222.

Smith, J. M. (1997). Stable isotope analysis of fauna and soils from sites in the Eastern Free State and Western Lesotho, Southern Africa: A palaeoenvironmental interpretation. MSc Thesis, University of Cape Town, 1997.

Smith, J. M., Lee-Thorp, J. A., & Sealy, J. C. (2002). Stable carbon and oxygen isotopic evidence for late Pleistocene to middle Holocene climatic fluctuations in the interior of southern Africa. *Journal of Quaternary Science, 17*(7), 683–695.

Soares, P., Alshamali, F., Pereira, J. B., Fernandes, V., Silva, N. M., Afonso, C., et al. (2012). The expansion of MtDNA haplogroup L3 within and out of Africa. *Molecular Biology and Evolution, 29*(3), 915–927.

Stapleton, P., & Hewitt, J. (1927). Stone implements from a rock shelter at Howieson's Poort, near Grahamstown. *South African Journal of Science, 24*, 574–587.

Stapleton, P., & Hewitt, J. (1928). Stone implements from Howieson's Poort near Grahamstown. *South African Journal of Science, 25*, 399–409.

Steenkamp, Y., Van Wyk, A. E., Smith, G. F., & Steyn, H. (2005). Floristic endemism in southern Africa: A numerical classification at generic level. *Biologiske Skrifter, 55*, 253–271.

Stewart, B. A., Dewar, G. I., Morley, M. W., Inglis, R. H., Wheeler, M., Jacobs, Z., et al. (2012). Afromontane foragers of the Late Pleistocene: Site formation, chronology and occupational pulsing at Melikane Rockshelter, Lesotho. *Quaternary International, 270*, 40–60.

Street-Perrott, F. A., Huang, Y., Perrott, R. A., Eglinton, G., Barker, P., Khelifa, L. B., et al. (1997). Impact of lower atmospheric carbon dioxide on tropical mountain ecosystems. *Science, 278*(5342), 1422–1426.

Swanevelder, C. J. (1981). Utilising South Africa's largest river: The physiographic background to the Orange River Scheme. *GeoJournal, 2*(2), 29–40.

Temme, A. J. A. M., Baartman, J. E. M., Botha, G. A., Veldkamp, A., Jongmans, A. G., & Wallinga, J. (2008). Climate controls on late Pleistocene landscape evolution of the Okhombe Valley, KwaZulu-Natal, South Africa. *Geomorphology, 99*(1), 280–295.

Theunissen, J. D. (1994). A method for isolating and preparing silica bodies in grasses for scanning electron microscopy. *Biotechnic and Histochemistry, 69*(5), 291–294.

Thackeray, J. F. (1987). Late Quaternary environmental changes inferred from small mammalian fauna, Southern Africa. *Climatic Change, 10*, 285–305.

Thackeray, J. F. (1988). Quantification of climatic change during the Late Quaternary in Southern Africa. *Palaeoecology of Africa, 19*, 317–324.

Tishkoff, S. A., Gonder, M. K., Henn, B. M., Mortensen, H., Knight, A., Gignoux, C., et al. (2007). History of click-speaking populations of Africa inferred from mtDNA and Y chromosome genetic variation. *Molecular Biology and Evolution, 24*(10), 2180–2195.

Tishkoff, S. A., Reed, F. A., Friedlaender, F. R., Ehret, C., Ranciaro, A., Froment, A., et al. (2009). The genetic structure and history of Africans and African Americans. *Science, 324*(5930), 1035–1044.

Twiss, P. C. (1987). Grass-opal phytoliths as climatic indicators of the Great Plains Pleistocene. *Kansas Geological Survey Guidebook Series, 5*, 179–185.

Tyson, P. D., Preston-Whyte, R. A., & Schulze, R. E. (1976). *The Climate of the Drakensberg*. Pietermaritzburg: Town and Regional Planning Commission.

van Campo, E., Duplessy, J. C., Prell, W. L., Barratt, N., & Sabatier, R. (1990). Comparison of terrestrial and marine temperature estimates for the past 135 kyr off Southeast Africa: A test for GCM simulations of palaeoclimate. *Nature, 348*(6298), 209–212.

van Wyk, A. E., & Smith, G. F. (2001). *Regions of floristic endemism in Southern Africa: A review with emphasis on succulents*. Pretoria: Umdaus Press.

van Zinderen Bakker, E. M., & Werger, M. J. (1974). Environment, vegetation and phytogeography of the high-altitude bogs of Lesotho. *Plant Ecology, 29*(1), 37–49.

Vogel, J. C. (1983). Isotopic evidence for the past climates and vegetation of southern Africa. *Bothalia, 14*(3–4), 391–394.

Vogel, J. C., Fuls, A., & Ellis, R. P. (1978). Geographical distribution of kranz grasses in South Africa. *South African Journal of Science, 74*, 209–215.

Veeramah, K. R., Wegmann, D., Woerner, A., Mendez, F. L., Watkins, J. C., Destro-Bisol, G., et al. (2012). An early divergence of KhoeSan ancestors from those of other modern humans is supported by an ABC-based analysis of autosomal resequencing data. *Molecular Biology and Evolution, 29*(2), 617–630.

Wadley, L. (1997). Rose Cottage Cave: Archaeological work 1987 to 1997. *South African Journal of Science, 93*(10), 439–444.

Wadley, L. (2006). Partners in grime: Results of multi-disciplinary archaeology at Sibudu Cave. *Southern African Humanities, 18*(1), 315–341.

Wadley, L., Esterhuysen, A. B., & Jeannerat, C. (1992). Vegetation changes in the eastern Orange Free State: The Holocene and later Pleistocene evidence from charcoal studies at Rose Cottage Cave. *South African Journal of Science, 88*(11), 558–563.

Walsh, K. (2005). Risk and marginality at high altitudes: New interpretations from fieldwork on the Faravel Plateau, Hautes-Alpes. *Antiquity, 79*(304), 289–305.

Walsh, K., Richer, S., & De Beaulieu, J. L. (2006). Attitudes to altitude: Changing meanings and perceptions within a 'marginal' alpine landscape - the integration of palaeoecological and archaeological data in a high-altitude landscape in the French Alps. *World Archaeology, 38*(3), 436–454.

Wedin, D. A., Tieszen, L. L., Dewey, B., & Pastor, J. (1995). Carbon isotope dynamics during grass decomposition and soil organic matter formation. *Ecology, 76*(5), 1383–1392.

Wellington, J. H. (1955). *Southern Africa: A geographical study*. Cambridge: Cambridge University Press.

Wheeler, E., Baas, P., & Gasson, P. (1989). IAWA list of microscopy features for hardwood identification. *IAWA Bulletin, 10*(3), 219–332.

Wooller, M. J., Swain, D. L., Ficken, K. J., Agnew, A. D. Q., Street-Perrott, F. A., & Eglinton, G. (2003). Late Quaternary vegetation changes around Lake Rutundu, Mount Kenya, East Africa: Evidence from grass cuticles, pollen and stable carbon isotopes. *Journal of Quaternary Science, 18*(1), 3–15.

Wright, G. A., Bender, S., & Reeve, S. (1980). High country adaptations. *Plains Anthropologist, 25*, 181–197.

Ziegler, M., Simon, M. H., Hall, I. R., Barker, S., Stringer, C., & Zahn, R. (2013). Development of Middle Stone Age innovation linked to rapid climate change. *Nature Communications, 4*, 1905. doi:10.1038/ncomms2897.

Chapter 15
Across Rainforests and Woodlands: A Systematic Reappraisal of the Lupemban Middle Stone Age in Central Africa

Nicholas Taylor

Abstract The Central African Middle Stone Age (MSA) is very poorly understood in comparison to the higher-resolution records of East and southern Africa. Severe taphonomic barriers to the construction of reliable chrono-stratigraphic, techno-typological, and paleoenvironmental frameworks continue to inhibit any nuanced understanding of post-MIS 6 technological change and behavioral adaptations. This chapter reviews existing knowledge of the earlier part of MIS 6-2 in the rainforests and woodlands of Central Africa from the perspective of the MSA Lupemban industry. Archaeological sequences on the woodland fringes of the Congo Basin bear witness to a technological shift characterized by the replacement of hand-held (Mode 2) Acheulean implements by distinctive tools suitable for hafting (Mode 3). While Mode 2 technology is absent from the contemporary equatorial rainforest zone, Mode 3 tools, including bifacial lanceolate points, core axes, and backed blades, are found across the region as the MSA Lupemban industry. As the earliest sustained archaeological signature in Central Africa, U-series dates of ~260 ka for the industry at Twin Rivers (Zambia) suggest the initial dispersal of pre-*sapiens* hominins into the equatorial forest belt during MIS 7. The development of sophisticated composite technologies in this ecological context bears directly upon current debates about the origins of behavioral and cognitive complexity in archaic *Homo sapiens*. In this chapter, current knowledge of the Lupemban is explored systematically with special reference to the hypothesis that it represents a late Middle Pleistocene rainforest and woodland adapted technology. A new site database is drawn upon to critically reassess the industry's geographical distribution, stratigraphic integrity, chronological position, and paleoenvironmental associations, from which its potential evolutionary significance is reconsidered.

Keywords Congo Basin • Composite technology • Chronology • Evolutionary significance • Geographical distribution • Paleoecology • Rainforest and woodland foraging • Stratigraphic integrity

Introduction

Closed evergreen forest today covers c. 3.6 million km^2, or some 12%, of the African continent in a belt stretching from the Atlantic coast to the western Rift Valley, between 4° N and 4° S (Moeyersons and Roche 1982; Mayaux et al. 2003). Correlating principally with the Congo River Basin and its lowland margins in Central Africa (Fig. 15.1), climax rainforest is characterized by a 35–45 m high interlocking canopy of *Caesalpiniaceae, Gilbertiodendron,* and *Scorodophloeus* trees beneath which thrives a thin understory of heliophobic flora (e.g., mosses, sedges, ferns, lianas) (White 1983).[1] This ecosystem is among the world's most biodiverse (Bashige and Debonnet 2004) and supports a range of endemic large mammal species including the gorilla (*G. gorilla*, *G. beringei*), chimpanzee (*Pan troglodytes*), bonobo (*Pan paniscus*), bongo (*Tragelaphus eurycerus*), forest elephant (*Loxodonta cyclotis*), and okapi (*Okapia johnstoni*) (Sayer et al. 1992). The periphery of the rainforest belt is marked by ecological biomes distinctively different in floristic structure and faunal composition. The eastern fringe of the Congo Basin witnesses the altitudinal replacement of dense humid rainforest by montane forest, while the basin's

N. Taylor (✉)
Department of Archaeology, Classics and Egyptology, University of Liverpool, 12-14 Abercromby Square, Liverpool, L69 7WZ, UK
e-mail: nick828@gmail.com

[1]The term 'rainforest' is considered an unsatisfactory descriptor by some authors (e.g., White 1983: 46), but is retained here to refer to Central African vegetation zones previously termed Guineo-Congolian forest (White 1983) and closed evergreen forest (Mayaux et al. 2003).

Sacha C. Jones and Brian A. Stewart (eds.), *Africa from MIS 6-2: Population Dynamics and Paleoenvironments*, Vertebrate Paleobiology and Paleoanthropology, DOI 10.1007/978-94-017-7520-5_15

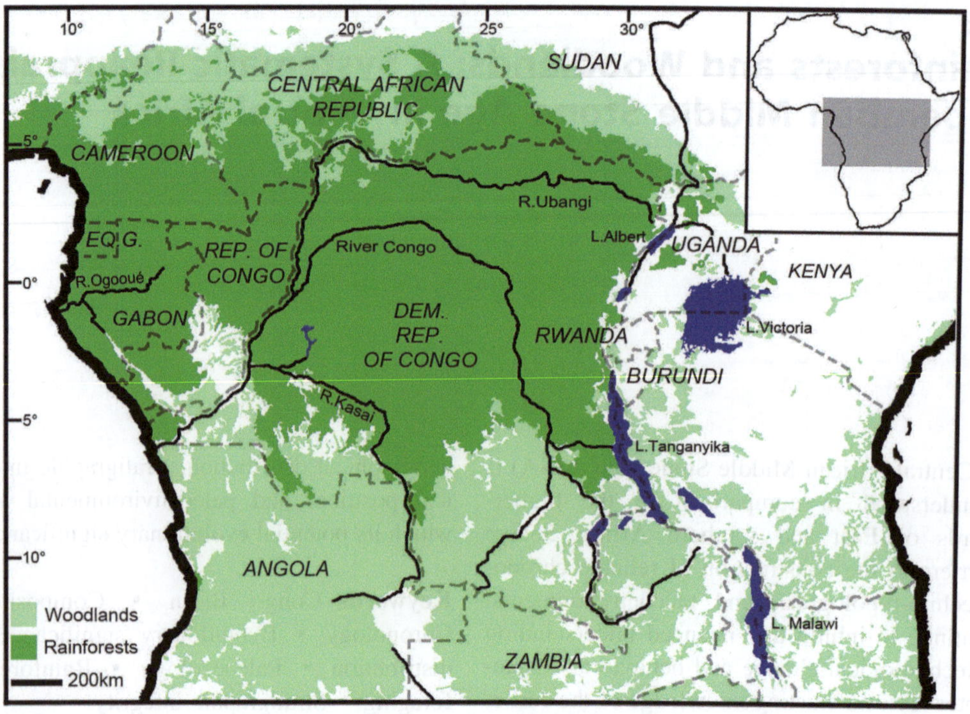

Fig. 15.1 Contemporary extent of equatorial rainforest and woodlands in Central Africa, based on Mayaux et al. (2003). "Rainforests" comprised by closed evergreen lowland forest; mangroves; mosaic forest/croplands; and mosaic forest/savanna. "Woodlands" incorporate closed deciduous forest and deciduous woodlands. Note that human activities have resulted in some spatial rearrangement and contraction of these biomes from their pre-industrial Holocene distributions

northern and southern plateaus are typified by a gradual transition from forest-savanna mosaic to deciduous woodlands of *Isoberlina* and *Brachystegia* (White 1983; Mayaux et al. 2003).

Across the current range of the rainforest belt and its woodland fringes, the Central African Stone Age suffers from an impoverished record that severely impedes the construction of any nuanced narrative for MIS 6-2. Serious archaeological interest in the deep prehistory of Central Africa was pioneered in the earlier part of the twentieth century (Colette 1929); however, a combination of human, ecological, and geomorphological obstacles has conspired against its continuous and systematic survey. Besides the region's dense vegetation, limited transport infrastructure, and periodic political upheaval present immediate logistical barriers to research-led fieldwork, while the rapid diagenesis of organic remains (Miller 1988; Kerbis et al. 1993; Tappen 1994), severe soil erosion (van Noten 1982), and drastic disruption of stratigraphies (e.g., Cahen 1976, 1978) have proven intractable problems (de Maret 1990; Cornelissen 2002; Mercader 2002). Stymied by deficiencies in basic datasets and geographic coverage, paleoanthropological enquiry in Central Africa necessarily remains focused on the problematic task of sequence building. Longstanding yet evolutionarily significant questions surrounding the technological and behavioral adaptations of Middle and Later Pleistocene hunter-gatherers have thus far been only tentatively explored (Barham 2001).

In spite of these significant problems, the Central African rainforest and woodland zone provides a strong signature of Stone Age activity (e.g., Clark 1959, 1964, 1982; Barham 2000, 2001; Cornelissen 2002). As well as quartz-based Later Stone Age (LSA) industries (see Cornelissen 2016), lithic assemblages assigned to the MSA Lupemban industry are represented widely across the region, leading to enduring speculation that the origins of rainforest hunting and gathering are embedded in its distinctive technologies (Clark 1959, 1963, 1988, 2001; McBrearty 1988; Barham 2001, 2002a; Mercader 2002; Mercader and Marti 2002; Taylor 2011). The appearance in the Lupemban of a range of novel tool forms, including prepared core technology (PCT) and lanceolate points, core axes, backed blades, and (rare) tranchets suitable for hafting as sophisticated composite (Mode 3) implements may reflect the development of a forest-adapted toolkit (Clark 1959; Barham 2001, 2002a). With precision U-series dates from the wooded southern margin of the Congo Basin now substantiating the initial appearance of the Lupemban by at least ~265 ka (Barham and Smart 1996; Barham 2000, 2001, 2002a, b), it has been posited the industry reflects an initial and sustained late Middle Pleistocene settlement of the equatorial African rainforest and woodland belt during MIS 7 (Barham 2001).

Ethnographic discourse underscores the extreme challenge tropical rainforest foraging poses for contemporary human communities. In spite of their floral and faunal richness, sources of carbohydrate and protein critical to human survival rarely feature in closed canopy rainforest ecosystems, and are unevenly distributed and difficult to access. This nutritional poverty previously led Bailey et al. (1989) to assert that dense jungles are impenetrable for hunter-gatherers in the absence of forager-farmer exchange networks (cf. Bailey and Headland 1991; Headland and Bailey 1991). This radical view is challenged by alternative (e.g., Bahuchet et al. 1991) and refined (e.g., Sato 2001; Yasuoka 2006, 2009) ethnographic data, as well as by emergent archaeological indications for the Late Pleistocene inhabitation of lowland rainforests (Mercader et al. 2000). Yet, the implication of Bailey et al.'s (1989) work remains compelling: specialized multicomponent tools and dedicated cooperative foraging strategies are prerequisites for the long-term exploitation of rainforests by mobile human foragers. The conceptualization and construction of hafted technologies formed from initially separate organic and inorganic components is now widely considered a key indicator of complex cognitive abilities (e.g., Wynn 2009; Ambrose 2010; Barham 2010; Lombard and Phillipson 2010; Wadley 2010a, b; Lombard 2012; Lombard and Haidle 2012). In this context, the possible early dispersal of Lupemban MSA hunter-gatherers into closed canopy rainforests bears directly on current debates about the evolution of behavioral and cognitive complexity in *Homo sapiens* before and during MIS 6-2 (Barham 2001). Starkly contrasting this perspective, Banks et al. (2006) and Marean and Assefa (2005) have argued rainforest habitats are simply too unproductive to have supported MSA foragers, and instead contend the geographical distribution of Lupemban technology reflects an MIS 6 adaptation to open grasslands.

A paucity of reliable primary data pertaining to the Central African MSA contrasts the potential significance of the higher-level evolutionary theories recently applied to it (e.g., Barham 2001), and continues to inhibit the development of broad scholarly consensus on the distribution, integrity, chronology, and paleoenvironmental adaptations of the Lupemban (cf. Barham 2001; Cornelissen 2002; Marean and Assefa 2005; Van Peer et al. 2004). This chapter addresses these issues through a fundamental and systematic top-down reanalysis of the data which underpins current knowledge of the industry, and structures hypotheses of its techno-ecological adaptations and evolutionary significance. A unique approach is taken in which discrete sources of information are synthesized to produce a new, comprehensive Lupemban site database, which is then subjected to sequential analyses that isolate robust patterns within the data. The appraisal of the resulting filtered data provides an effective platform for the holistic reevaluation of the industry. First, however, an overview of the Central African Stone Age record serves to contextualize the Lupemban and situate its technological innovations within the paleoanthropological framework of MIS 6-2.

The Early MSA in Central Africa, with Special Reference to the Lupemban

As observed across sub-Saharan Africa, the transition from the Early Stone Age (ESA) to the MSA reflects a fundamental technological shift marked by the widespread replacement of long-established ESA Acheulean hand-held (Mode 2) implements by regionally stylistic MSA tools of a size and shape suitable for hafting (Mode 3) (e.g., Clark 1988; McBrearty and Brooks 2000; Barham 2001; McBrearty 2001; Brooks et al. 2006; Rots and Van Peer 2006). The advent of hafted tool technology, in which separate organic and inorganic materials are combined to create more efficient composite implements (e.g., Barham 2010), is strongly suggested by the abundance of lithic artifacts made by prepared core technique (PCT) that characterizes the regional MSA (e.g., Clark 1988; McBrearty 2001; Brooks et al. 2006). Within this framework, the Lupemban industry is recognized as the distinctive regional expression of the MSA across Central Africa (e.g., Mortelmans 1957; Clark 1959, 1988; Cahen 1975; Barham 2000).

As defined by the geographical extent of the Congo Basin, smaller adjacent river basins (e.g., the Ogooué), and their respective bordering plateaus, Central Africa (Fig. 15.1) has been the subject of scientific archaeological enquiry since the earlier part of the twentieth century (e.g., Colette 1929). The early research undertaken by Jean Colette at the site of Kalina (now Gombe) Point, Democratic Republic of Congo (DRC) identified a sequence of Stone Age cultures that, from oldest to youngest, were named the Kalinian, Djokocian, and Ndolian (Bequaert 1938). Despite numerous subsequent changes in the observed composition and nomenclature of Central African Stone Age industries (Clark 1963: 45; 1971a; de Maret 1990; Table 15.1), these tool assemblages remain roughly equivalent to the technocomplexes known since the 1960s as the Sangoan, Lupemban, and as quartz-based LSA industries (incorporating subregional variants, e.g., the Tshitolian), which are widely acknowledged to characterize the regional archaeological sequence of Central Africa (Clark 1971a; de Maret 1990).

The Lupemban was first named after a lithic assemblage exposed by mining operations along the Lupemba stream in the Kasai Province of the DRC (Breuil 1944). Although initially described poorly (Clark 1971a: 71), the industry is now known to include lanceolate points, core axes, PCT points, and blade tools (see below). Where found in stratified

Table 15.1 Correlation of the various terms used for Central African lithic assemblages (after Cormack 1994: Fig. 1)

Relative chronology	Colette (1929)	O'Brien (1939)	Breuil (1944), Cabu (1953)	Leakey and Owen (1945)	Leakey (1949)	Mortelmans (1957), Clark (1959)
	–	–	–	–	–	Neolithic variants
	Leopoldian	–	–	–	–	–
Later	Ndolian	Late Tumbian	Tshitolian	Upper Tumbian	Lupemban	Tshitolian
↑	–	Upper Tumbian	–	–	Final Sangoan	Lupembo-Tshitolian
	Djokocian	Middle Tumbian	Lupemban	Middle Tumbian	Upper Sangoan	Upper Lupemban
Earlier	Kalinian	–	Djokocian	Lower Tumbian	Middle Sangoan	Lower Lupemban
	–	Proto Tumbian	Kalinian	Proto Tumbian	Lower Sangoan	Sangoan
	–	–	–	–	Proto Sangoan	–

sequences, the Lupemban has been observed to overlie assemblages attributed to the Sangoan industry (e.g., Clark 2001). The Sangoan itself lies above Mode 2 Acheulean industries at Nsongezi, Uganda (Wayland 1937; O'Brien 1939; Cole 1967), Asokrochona, Ghana (Nygaard and Talbot 1984) and Kalambo Falls, Zambia (Clark 1969, 2001). Characterized as an industry dominated by minimally reduced, thick, and heavy core tools including core axes, picks, choppers, and core-scrapers (e.g., Leakey and Owen 1945; McBrearty 1991), the Sangoan provides the earliest evidence for post-Acheulean technological change at some sites in Central Africa, but remains poorly understood and loosely defined (McBrearty 1988, 1991). Based on current data, the Sangoan lacks clear evidence of stone tools made by PCT (Clark 2001), and it may thus be either a Mode 2 or Mode 3 technology. While archaeological sequences on the margins of Central Africa attest to the presence of Acheulean and Sangoan assemblages, these technologies appear to be absent, and to have a limited presence, respectively, across the forested lowland interior of the region. The Lupemban is found widely across this area, and thus appears to constitute the earliest archaeological signature for the sustained prehistoric settlement of the Congo Basin (Barham 2001).

This general chrono-stratigraphic framework remains unrefined, however, on account of a raft of post-depositional processes that limit the material diversity of cultural assemblages and severely compromise the structural integrity of archaeological profiles. The degradation of organic materials dictates that attempts to define, correlate, and compare cultural sequences rely exclusively on lithic analyses (Clark 1971a). An equally widespread, yet more acute impediment results from the post-depositional reconfiguration of once discrete archaeological horizons into low-resolution subsurface "stone-lines" that confound the principle of stratigraphic superpositioning (e.g., Cahen 1976). Identifying the exact processes causing subsurface artifacts and other materials (e.g., charcoal) to disperse vertically and recompress as "stone-lines" has proven difficult (Brown et al. 2004). "Stone-lines" have been variously interpreted as reworked land surfaces (Lanfranchi and Schwartz 1991), erosional features (Schwartz 1996), or as the result of intense bioturbation by tree roots (Moeyersons 1978), and burrowing termites (McBrearty 1990) acting on weakly consolidated soils: mechanisms that may not be mutually exclusive (Mercader et al. 2002). Such severe and widespread disruption to time-depth relationships places drastic limits on the resolution of the Central African Stone Age record, constraining the definition, correlation, and comparison of the cultural content of prehistoric industries (Cahen 1976, 1978; Isaac 1982; McBrearty 1988).

Recent definitions of the Lupemban emphasize the co-occurrence of diverse lithic reduction strategies, including bifacial (Mode 2), PCT (Mode 3), blade (Mode 4), and backed blade (conceptual Mode 5) technologies (Clark 2001; Clark and Brown 2001; Barham 2002a). Large, bifacially flaked lanceolate points are considered the regionally distinctive *fossile directeur* of the industry (Mortelmans 1957; Clark 1969; Barham 2001; Mercader 2002). More common are the extremely heterogeneous tool classes of core axes and picks (Clark 2001: Fig. 2.6), while other tools suitable for hafting include small unifacial and bifacial PCT points, backed flakes and blades, trapezia, and rare tranchets (Barham 2000, 2001, 2002a). Previously forwarded hypotheses of Lupemban artifact function (Fig. 15.2) originate from the industry's conceptualization as a technological adaptation to rainforest foraging (Barham 2001: 70, 2002a; Clark 1959: 156–7, 1963; Miller 1988; Taylor 2009, 2011). The suitability of Lupemban lanceolate points, core axes, and backed blades as composite foraging tools is supported by ethnographic parallels (Hayden 1976; Miller 1988; Stout 2002), and by archaeological and experimental use-wear studies relating to morphometrically comparable MSA artifacts (e.g., Lombard 2004, 2005, 2008; Rots and Van Peer 2006; Shea et al. 2002; Shea 2006; Villa et al. 2009; Rots

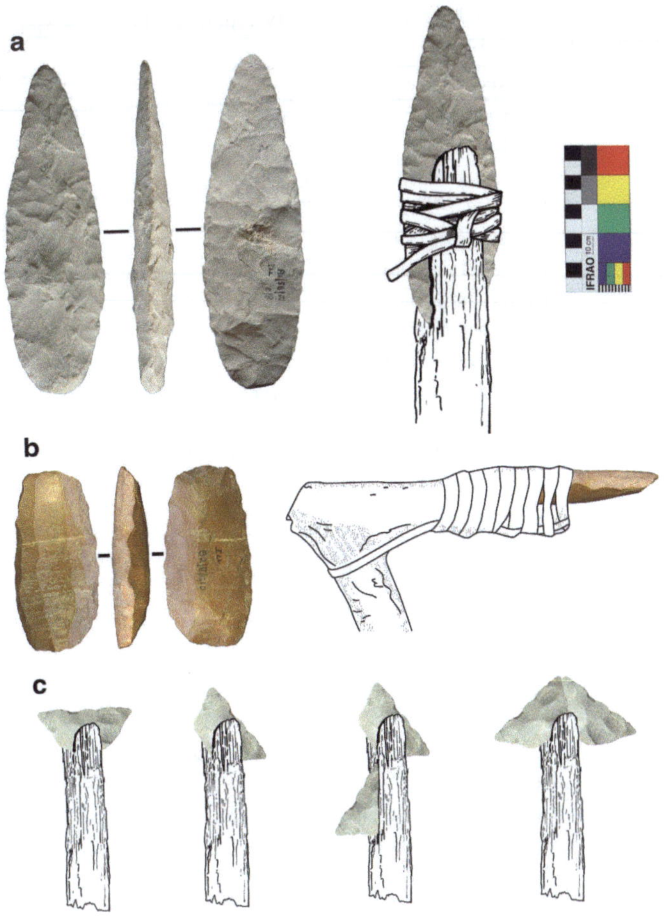

Fig. 15.2 Characteristic Lupemban artifacts and their hypothesized composite tool forms: **a** lanceolate point, slotted with bindings in axial arrangement, for use as a thrown or thrust spear; **b** parallel-sided core axe bound in a latero-distal arrangement as a heavy-duty adze; **c** geometric backed blades slotted and arranged variously as projectile armatures. All lithic artifacts excavated from Kalambo Falls, Zambia. Haft illustrations after Barham (2001, 2010). First published in *Before Farming [online version], 2011/1*, article 6

et al. 2011) in addition to detailed but numerically limited analysis of Lupemban tools (Taylor 2009).

Current evidence suggests the Lupemban MSA constitutes the earliest archaeological signature widespread across the modern Central African rainforest and woodland belt, and likely reflects the initial regional development of composite Mode 3 technological repertoires. New chronometric dating estimates establishing the industry's emergence during the late Middle Pleistocene (Barham 2000, 2001, 2012) fall within the ~400–200 ka range of estimates derived for the ESA-MSA transition from across sub-Saharan Africa (e.g., Howell et al. 1972; McHugh et al. 1988; Wendorf and Schild 1992; Schwarz and Grün 1993; Kuman et al. 1999; Liubin and Guede 2000; Grün and Beaumont 2001; Deino and McBrearty 2002; Tryon and McBrearty 2002; Van Peer et al. 2003; Beaumont and Vogel 2006; Porat et al. 2010). This indicates the Lupemban may represent a considerable portion of the paleoanthropological record before and during MIS 6-2, and is of central relevance to current debates about the evolution of complex cognition and behavior in archaic *Homo sapiens* (e.g., McBrearty and Brooks 2000; Deacon and Wurz 2001; d'Errico 2003; Henshilwood and Marean 2003; Mellars et al. 2007 and references therein; Klein 2008). In light of the importance afforded to these issues in current paleoanthropological research, the Lupemban merits holistic reconsideration.

A New Lupemban Site Database

The principle aim of this chapter is to review systematically and objectively available data bearing on the distribution, integrity, chronology, and paleoenvironment of the Lupemban, as a means through which to reassess current knowledge of the industry and recent hypotheses of its potential evolutionary significance. To undertake such a back-to-basics analysis first necessitates the isolation of all known occurrences of the industry across sub-Saharan Africa and their compilation into a comprehensive site database; a foundation from which localities and data relevant to each of the explored issues can be secondarily filtered out to identify robust patterns. The construction of an initial site database that is holistic, objective, and fully sourceable is integral to this process, with these characteristics being contingent upon the empirical and impartial discrimination of Lupemban sites from the wider background of African Stone Age localities and industries. This process, however, poses a considerable methodological challenge, in particular because as the expressed subjects of analysis, the use of geographical, stratigraphic, temporal, or paleoenvironmental criteria to define Lupemban sites would result only in a self-selected database, with subsequent analyses founded on circular reasoning.

The only pre-existing source of relevant information potentially serviceable as a Lupemban site database is Clark's *Atlas of African Prehistory* (1967), in which he presented a series of site distribution maps for various Stone Age industries – including the Acheulean, Sangoan, and Lupemban – accompanied by a site gazetteer. A number of limitations and contradictions, however, constrain the utility of this work and query its application to current Lupemban discourse. The absence of any source information corresponding to the Lupemban sites Clark mapped means it is impossible to identify individual localities,[2] verify the observational criteria upon which Lupemban affiliation was

[2] According to Clark (1965) the *Atlas* was never conceived as a functional database for the identification of individual sites.

Table 15.2 Alphabetized database of ninety-three sites incorporating Lupemban lithic assemblages, as previously reported. "Filters" attributed based on site, sequence or assemblage integrity, as detailed in text and presented in Table 15.3

No.	Site and country	Filter	Dated	Paleoecol	References
1	Aluia (ANG)	6	–	–	Clark (1963)
2	Ambriz (ANG)	1	–	–	Clark (1966)
3	Baboungue 2 (CAR)	4	–	–	Bayle des Hermens (1975, 1977)
4	Bacongo (RoC)	2	–	–	Droux and Bergaud (1937)
5	Basoko (DRC)	4	^{14}C	–	van Moorsel (1968)
6	Belgika (DRC)	4	^{14}C	–	van Moorsel (1968)
7	Boko Songho (RoC)	2	–	–	Droux and Kelley (1939)
8	Bomboli (CAR)	5	–	–	Bayle des Hermens (1975, 1977)
9	Boukoko (CAR)	5	–	–	Bayle des Hermens (1975, 1977)
10	Boundji (RoC)	3	^{14}C	–	Delibrias et al. (1983), Lanfranchi (1986)
11	Boungou (CAR)	3	–	–	Bayle des Hermens (1975, 1977)
12	Bout de Plateau (RoC)	3	^{14}C	–	Lanfranchi and Manima-Moubouha (1989)
13	Busenge (KEN)	3	–	–	Leakey and Owen (1945)
14	Caimbunji (ANG)	6	–	–	Clark (1963)
15	Camafufo (ANG)	6	–	–	Clark (1963)
16	Campion (RWA)	5	–	–	Nenquin (1967), van Noten (1982, 1983)
17	Capama (ANG)	6	–	–	Clark (1963)
18	Catete (ANG)	1	–	–	Clark (1966)
19	Catongula (ANG)	6	–	–	Clark (1963)
20	Cauma (ANG)	6	–	–	Clark (1963)
21	Chaminade (MAL)	4	–	–	Clark et al. (1972)
22	Dimba (DRC)	3	^{14}C	–	de Maret et al. (1977), Lavachery (1990)
23	Furi (ANG)	6	–	–	Clark (1963)
24	Gombe Point (DRC)	4	^{14}C	–	Colette (1933), Bequaert (1938), Cahen (1976)
25	Hinda Plateau (RoC)	3	^{14}C	–	Kouyoumontzakis et al. (1985)
26	Kabua (KEN)	2	–	–	Whitworth (1965)
27	Kadawa (KEN)	2	–	–	Leakey and Owen (1945)
28	Kalambo Falls (ZAM)	6	^{14}C/TT-OSL	Yes	Clark (2001), Barham et al. (2015), Duller et al. (2015)
29	Kamoa (DRC)	4	–	–	Cahen (1975)
30	Kango (GAB)	4	–	–	Bayle des Hermens et al. (1987), Clist (1989)
31	Khor Abu Anga (SUD)	2	–	–	Arkell (1949), Carlson (1967)
32	Kingabwa (DRC)	4	–	–	van Moorsel (1968)
33	Kinko (DRC)	5	–	–	van Moorsel (1970)
34	L'Angumu (DRC)	2	–	–	Bequaert (1943)
35	Leberu (RWA)	2	–	–	Nenquin (1967)
36	Lemba (DRC)	4	^{14}C	–	van Moorsel (1968)
37	Liba (RWA)	2	–	–	Nenquin (1967)
38	Lindili (GAB)	2	–	–	Assoko Ndong (1996)
39	Lodjo (DRC)	5	–	–	de Heinzelin de Braucourt (1948)
40	Lope (GAB)	2	–	–	Bayle des Hermens et al. (1987)
41	Loukoko II (RoC)	3	^{14}C	–	Pinçon (1991)
42	Lubudi (DRC)	4	–	–	van Moorsel (1968)
43	Lupemba (DRC)	3	–	–	Breuil (1944)
44	Lutunguru (DRC)	3	–	–	Passau (1946)
45	Mabalabala (ANG)	6	–	–	Clark (1963)
46	Maboue 5 (GAB)	2	^{14}C	–	Assoko Ndong (2002), Oslisly et al. (2006)
47	Makaw I (DRC)	5	–	–	van Moorsel (1970)
48	Malumba (ANG)	6	–	–	Clark (1963)
49	Marimba (ANG)	1	–	–	Clark (1966)
50	Masango (BUR)	5	–	–	Nenquin (1967), van Noten (1982, 1983)
51	Matafari (ANG)	6	–	–	Clark (1963)
52	Mavoio (ANG)	2	–	–	Clark (1966)
53	MbalaMbala (ANG)	6	–	–	Clark (1963)

(continued)

Table 15.2 (continued)

No.	Site and country	Filter	Dated	Paleoecol	References
54	Mbali (CAR)	3	–	–	Bayle des Hermens (1975, 1977)
55	Mbeji (KEN)	3	–	–	Leakey and Owen (1945)
56	Middle Kwango (DRC)	3	–	–	Cahen (1976)
57	Mimanga (ANG)	6	–	–	Clark (1963)
58	Moulandou Fouala (GAB)	1	–	–	Livingstone-Smith et al. (2007)
59	Mosumu (EG)	4	^{14}C	Yes	Mercader and Marti (1999, 2002)
60	Mpiaka (RoC)	5	–	–	Droux and Bergaud (1937)
61	M'Pila (RoC)	5	–	–	Droux and Bergaud (1937)
62	Mufo (ANG)	6	^{14}C	Yes	Clark (1963)
63	Mugera (BUR)	5	–	–	Nenquin (1967)
64	Muguruk (KEN)	6	–	–	Leakey and Owen (1945), McBrearty (1988)
65	Muhororo (RWA)	2	–	–	Nenquin (1967)
66	Mulehe (BUR)	2	–	–	Nenquin (1967)
67	Mur (KEN)	3	–	–	Leakey and Owen (1945)
68	Mushie (DRC)	5	–	–	van Moorsel (1970)
69	Musolexi (ANG)	6	–	–	Clark (1963)
70	Nemba (RWA)	2	–	–	Nenquin (1967)
71	Ndjole (GAB)	4	–	Yes	Farine (1965), Pommeret (1965b)
72	Ngoere Ngolo (CAR)	1	–	–	Bayle des Hermens (1975, 1977)
73	Ngouloukwa (CAR)	3	–	–	Bayle des Hermens (1975, 1977)
74	Nsongezi (UGA)	6	–	–	O'Brien (1939), van Riet Lowe (1952)
75	Nsongolo (ANG)	1	–	–	Clark (1966)
76	Nsontin (DRC)	5	–	–	van Moorsel (1970)
77	Nyankeru (BUR)	2	–	–	Nenquin (1967)
78	Nyarunazi (BUR)	2	–	–	Nenquin (1967), van Noten (1982, 1983)
79	Nzako (CAR)	5	–	–	Bayle des Hermens (1975, 1977)
80	Ober Awach (KEN)	3	–	–	Leakey and Owen (1945)
81	Okala (GAB)	3	^{14}C	–	Clist (1993, 1997)
82	Peperkorrel (NAM)	2	–	–	MacCalman and Viereck (1967)
83	Pointe Hollandaise (RoC)	5	–	–	Droux and Bergaud (1937)
84	Pointe Noire (RoC)	5	–	–	Droux and Kelley (1939)
85	Porte de l'Okanda (GAB)	4	–	–	Blankoff (1965), Pommeret (1965a)
86	Rumeza (BUR)	1	–	–	Nenquin (1967)
87	Rutongo (RWA)	2	–	–	Nenquin (1967)
88	Sai Island (SUD)	3	OSL	–	Van Peer et al. (2003, 2004)
89	Sango Bay (UGA)	3	–	–	Smith and Wayland (1923)
90	Seko (DRC)	5	–	–	van Moorsel (1970)
91	Tumba (DRC)	2	–	–	Menghin 1925
92	Twin Rivers (ZAM)	6	^{14}C/U-series	Yes	Clark and Brown (2001), Barham (2000, 2012)
93	Yala Alego (KEN)	3	–	–	Leakey and Owen (1945)

Country abbreviations: *ANG* Angola; *BUR* Burundi; *CAR* Central African Republic; *EG* Equatorial Guinea; *GAB* Gabon; *KEN* Kenya; *MAL* Malawi; *NAM* Namibia; *RoC* Republic of the Congo; *RWA* Rwanda; *SUD* Sudan; *UGA* Uganda; *ZAM* Zambia

resolved, or test and amend the data according to research developments since 1967. Careful scrutiny also highlights significant discrepancies between the Lupemban occurrences mapped by Clark (1967) and those listed in his gazetteer. This observation has ramifications for interpretations of the industry's geographical distribution and paleobiogeographical correlation, as detailed below.

In light of these difficulties, the approach taken to the construction of a site database here relies on the identification of sites that have yielded assemblages incorporating Stone Age materials previously published under "Lupemban" and correlating and derived terminologies, as previously applied across sub-Saharan Africa (Table 15.1). These include *Djokocian* (e.g., Colette 1929); *Lower Tumbian/Middle Tumbian* (Leakey and Owen 1945); *Middle Sangoan/Upper Sangoan/Final Sangoan* (Leakey 1949); *Lower Lupemban/Upper Lupemban*, (e.g., Mortelmans 1957); *Sangoan/Lower Lupemban* (e.g., Clark 1963); and *Lupembo-Tshitolian* (e.g., Clark 1959, 1963). No attempt is made at this stage to determine the veracity of Lupemban assemblages previously reported using these alternate terminologies, and while problems inherent in the utilization of Central African nomenclature for the purpose of database construction are acknowledged (Clark 1971a), these potential problems are inconsequential given the resulting database is subjected to further, detailed critical analyses (see below). Notably, West Africa has been omitted wholesale from this database and reexamination of the Lupemban on the basis of revisionist views (e.g., Wai-Ogosu 1973; Andah 1979; Allsworth-Jones 1987; Casey 2002) that agree previous claims for a West African expression of the industry (e.g., Davies 1954, 1961, 1967, 1976) are baseless.

According to these parameters, a comprehensive survey of African Stone Age literature generates a database of 93 Lupemban sites (Table 15.2) and a corresponding map showing their locations (Fig. 15.3). Aforementioned difficulties encountered in the identification of individual sites mapped by Clark (1967) mean it is not possible to determine accurately which, or how many, of the sites presented here (Table 15.2; Fig. 15.3) also feature in his *Atlas* map. The new database can nevertheless be considered exhaustive and representative of existing knowledge of the Lupemban, as dictated by research, fieldwork, and publication biases.

Given the analytical importance of the database to this chapter, brief analysis of its constituent published sources, as presented in Table 15.2, is warranted. Most notably, this reveals that a disproportionate number of sites (*n* = 42) were first reported in the 1960s, with a lesser but still considerable publication rate (*n* = 17) during the 1970s resulting in a cumulative 63% of all Lupemban sites initially described during a 20-year period. By contrast, far fewer new sites (*n* = 9), forming just 13% of known occurrences of the industry, have been described since 1990. This pattern likely result from several interlinked factors: Central African archaeology matured as a systematic and scientific venture in the mid-twentieth century (de Maret 1990), a period during which local institutions and infrastructure were conducive to primary fieldwork, and researchers were emboldened to undertake intra- and inter-regional comparative studies on vast scales unknown in more recent decades. Equally intriguing is the observation that six authors (J.D. Clark, J. Nenquin, R. de Bayle des Hermens, H. van Moorsel, L.S. B. Leakey, and G. Droux) are responsible for a majority (*n* = 67; 72%) of the Lupemban sites published to date, with J.D. Clark alone contributing to 24% of reported instances of the industry. This perhaps has implications for the consideration of whether the Lupemban is valid as a culturally and technologically distinct archaeological entity, as reviewed below. With these observations in mind, the new site database forms the essential foundation from which critical analyses of the distribution, integrity, chronology, and paleoenvironment of the Lupemban are now sequentially undertaken.

Geographical Distribution

The distribution of Lupemban sites is a key observation underpinning the longstanding theory of the industry as a prehistoric adaptation to rainforest and woodland environments. Clark's (1967) original Lupemban distribution map (Fig. 15.4a) shows 186 "certain" and "uncertain" localities that cluster in Central Africa (*n* = 158) and West Africa (*n* = 28), illustrating the Lupemban is associated with equatorial Africa, and closely correlated with current rainforest and woodland environments. This apparent spatial relationship has repeatedly been argued as a modern echo of a specialist Lupemban adaptation to rainforest habitats (Clark 1959, 1963, 1964, 1966, 1970, 1971a, b, 1982, 1988, 2001), a concept now of renewed interest given the industry's emergent Middle Pleistocene chronology (Barham and Smart 1996; Barham 2000, 2002a, 2012; Taylor 2009).

Clark's distribution map has been variously replicated, modified, and explored in subsequent research (e.g., Coles and Higgs 1969: Fig. 48; Barham 2000: Fig. 12.2; 2001: Fig. 6.6; Mercader 2002: Fig. 3; Taylor 2011: Fig. 6) and plays a fundamental role in structuring and sustaining the theory of the Lupemban as a rainforest and woodland industry. Until now, it has never been critically reappraised. Revision is particularly warranted since a significant discrepancy exists between the distributions of Clark's mapped Lupemban sites (as shown in Fig. 15.4a), and that resulting when sites attributed to the Lupemban in his accompanying site gazetteer are plotted (Fig. 15.4b).[3] The latter form an entirely different geographical signature centered not on the forest belt of equatorial Africa and the Congo Basin, but rather, the more open environments of the Zambezi River Basin in south Central Africa.[4] This inconsistency has already had a demonstrable research impact: in contrast to Coles and Higgs (1969), Mercader (2002), Barham (2000, 2001), and Taylor (2011), the gazetteer dataset, with its south Central African skew, has been employed by Banks et al. (2006: Fig. 6) as the foundation of their Lupemban spatial analysis. This unwitting data selection directly informs their correlation of the industry with Adams and Faure's (1997) modeled MIS 6 grassland biogeography for Central Africa, and divergent interpretation that the Lupemban represents a technological solution to savanna foraging (Marean and Assefa 2005: 110; Banks et al. 2006: 76), as discussed below.

[3]These differences likely stem from the assembly of data from disparate and informal sources (Clark 1965: 312).

[4]Lupemban sites listed in Clark's gazetteer frequently consist of informally named Zambian localities (e.g., Chambeshi Pontoon, Chilesha Railway Cutting) known not to have been the focus of methodical and scientific archaeological investigation, indicating this data may be unrepresentative of the industry.

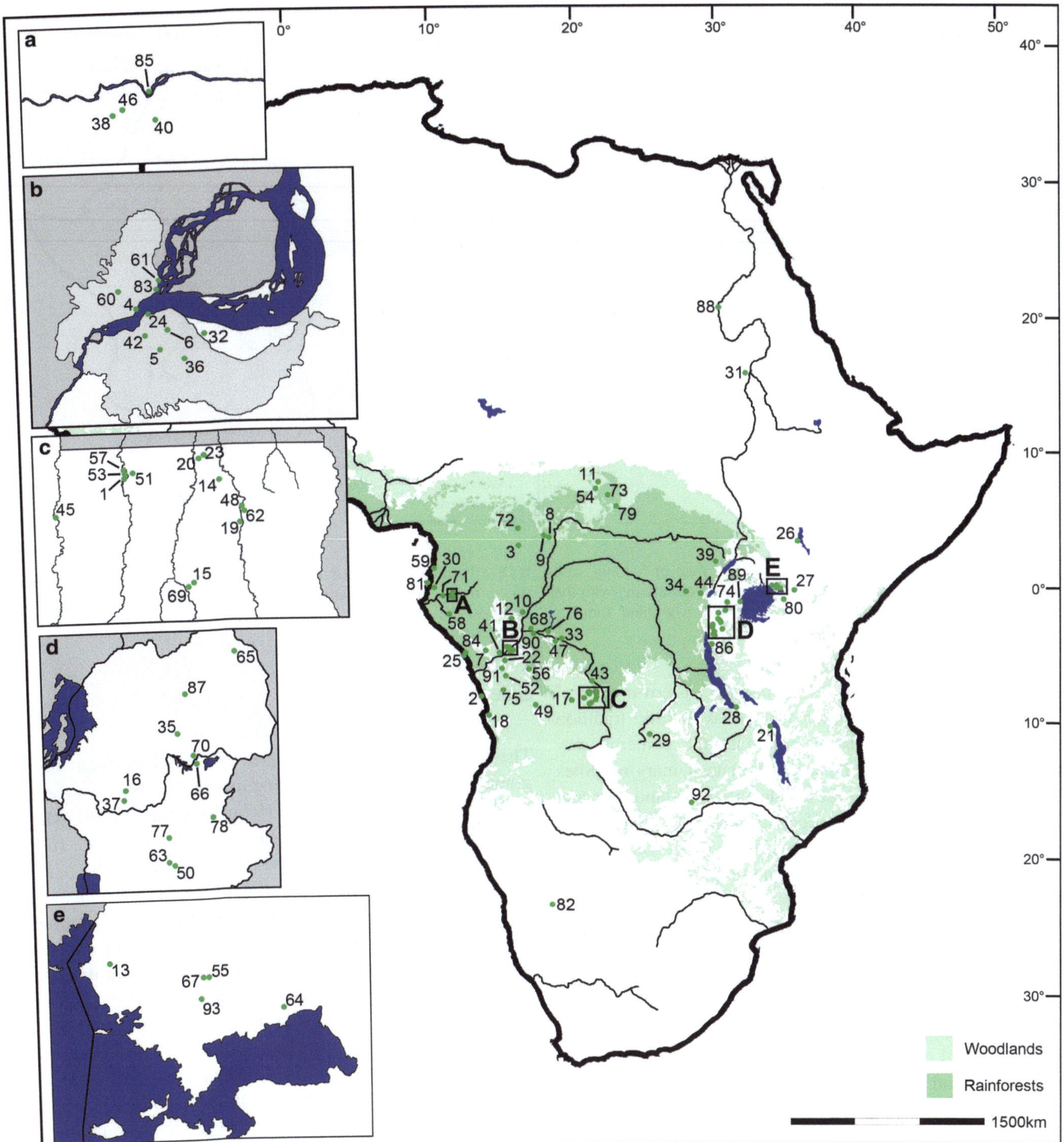

Fig. 15.3 Distribution of all published Lupemban sites. Numbered sites correlate with Table 15.2. Note general concordance with Clark's (1967) distribution (Fig. 15.4a) and correlation with contemporary Central African rainforest and woodland belt. *Inset* **a** Middle Ogooué River, Gabon; **b** Footprints of Brazzaville (RoC) and Kinshasa (DRC) bedside the Malebo Pool, Congo River; **c** Lunda Province, north east Angola; **d** Rwanda and Burundi; **e** western Kenya

Careful scrutiny of the new Lupemban site distribution (Fig. 15.3) and its comparison with Clark's (1967) data critically challenge the accuracy and reliability of the distribution derived from his Lupemban site gazetteer (Fig. 15.4b; Banks et al. 2006: Fig. 6), yet illustrates a degree of concordance confirming the essential robustness of his original (mapped) pattern (Fig. 15.4a). The industry as reported to date occurs in two broad pockets coinciding with

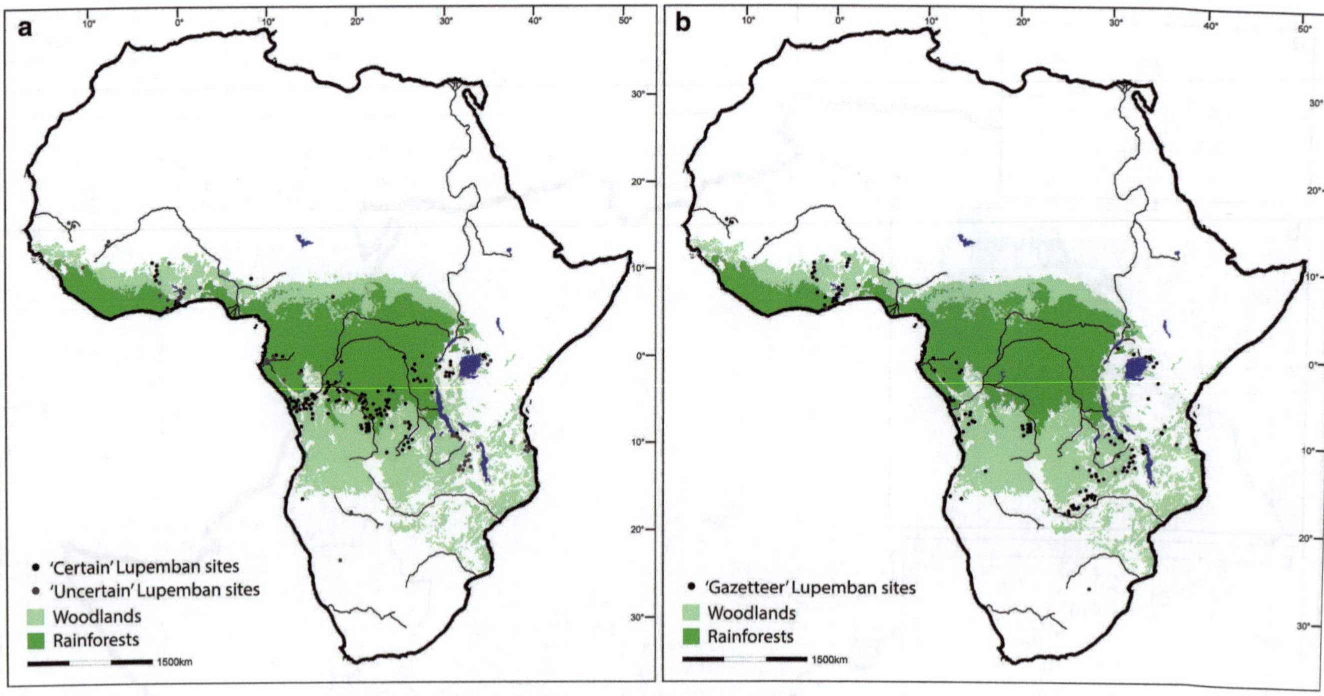

Fig. 15.4 **a** Clark's (1967) classic distribution map of 'certain' and 'uncertain' Lupemban sites with a principal cluster correlating with the Congo Basin and contemporary rainforest and woodland biomes. **b** Distribution of Lupemban sites listed in Clark's (1967) site gazetteer, with a main cluster associated with the more open environments of the Zambezi River Basin

the western half of Central Africa and the eastern margin of the Congo Basin, and is thus verified as a lithic tradition primarily observed in association with the contemporary lowland forest belt and its wooded plateau margins. Sites falling outside the rainforest belt to the south and east are, respectively, found in dense deciduous (*miombo*) woodland and montane forest environments (Fig. 15.3). Only a few outlying sites buck this general trend (e.g., Khor Abu Anga, Peperkorrel, Sai Island), while the apparent absence of Lupemban sites at the forested heart of the Congo Basin's cuvette likely reflects a fieldwork bias.

This data constitutes an important reaffirmation of the Lupemban's present-day techno-ecological correlation with the Central African rainforest and woodland belt. However, the substantial effect of Pleistocene climatic fluctuations on the extent and configuration of Central African biogeographical zones (e.g., Dupont et al. 2000) precludes the direct projection of modern vegetation patterns onto the prehistoric past (e.g., Barham 2001; Cornelissen 2002; Mercader 2002). Considered in isolation, the distribution of Lupemban archaeology then holds little interpretative value, yet is perhaps of some interest when utilized cautiously as a component of paleobiogeographic models (e.g., Barham 2001; Banks et al. 2006). Direct and indirect sources of paleoenvironmental evidence available for the reconstruction of Lupemban technological adaptations are explored later in this chapter.

Site Integrity

The post-depositional disturbance of Central African stratigraphic profiles is widely recognized to be both profound and widespread (e.g., Cahen 1976; McBrearty 1990; Mercader et al. 2002), placing such severe limits on the resolution of the record that Cahen (1978), Isaac (1982) and McBrearty (1988) have argued distinct post-Acheulean industries are effectively indivisible. Table 15.3 outlines six archaeological criteria relevant to the Central African Stone Age record and the Lupemban site database (Table 15.2). When applied sequentially to the database, these act as "filters" to exclude sites and generate cumulative refinements until a final residue of higher-resolution sites is isolated. The attribution of filters to individual sites is shown in Table 15.2,[5] and documented in Figs. 15.5, 15.6, 15.7, 15.8 and 15.9; as each filter is applied, sites gradually disappear from the mapped distributions (Fig. 15.3).

The filtration process begins with the full database of 93 Lupemban sites presented in Table 15.2 and Fig. 15.3.

[5]For most reported sites, a combination of criteria in Table 15.3 affects the relevance and quality of information that can be derived. While recognizing that the filters are not mutually exclusive, in each case site integrity is assessed according to the dominant criterion limiting knowledge.

Table 15.3 Six criteria applied sequentially (1-6) to filter Lupemban site data (Table 15.2) and the distribution map (Fig. 15.3). The application of each filter removes sites correlating with the distributions in Figs. 15.5, 15.6, 15.7, 15.8, 15.9

Filter No.	Criterion	Rationale
1	No stratigraphic information	Not possible to assess the archaeological context of reported assemblage
2	Surface archaeology	No chrono-stratigraphic resolution; impossible to assess unity of reported assemblage
3	Minimal available information	Stratigraphic and/or techno-typological description too poor to confirm cultural affiliations
4	'Stone-line' accumulations	Stratigraphy reflects severe vertical disturbance; assemblage has very low temporal and techno-typological integrity
5	Probable 'stone-line' accumulations	Stratigraphic description suggests vertical disturbance; assemblage likely retains very low temporal and techno-typological integrity
6	Geomorphological disturbance	Accumulative or post-depositional non-anthropogenic processes prevents derivation of high-resolution data

Filter 1 first isolates and removes all sites that have been only cursorily reported in publication, and for which absolutely no stratigraphic information is available ($n = 7$). Since these assemblages are effectively decontextualized on the basis of published information, it is impossible to certify their chrono-stratigraphic or techno-typological relevance to the Lupemban, or even the MSA. Forming just 8% of all sites, this filter has little effect on the overall distribution of the industry (Fig. 15.5), removing only geographically incongruous sites from northern Angola (Clark 1966), Burundi (Nenquin 1967), and the Central African Republic (CAR) (Bayle de Hermens 1975, 1977).

Filter 2 removes from the database all sites that are known to represent surface accumulations ($n = 20$). The techno-typological integrity of these assemblages is untestable in the absence of archaeological context but is likely to be extremely low. Surface assemblages comprise 22% of all sites and are widespread, such that their exclusion leaves the industry's geographical footprint effectively unaltered (Fig. 15.6). Geographical outliers in Kenya (Whitworth 1965), Namibia (MacCalman and Viereck 1967) and northern Sudan (Arkel 1949) are removed, but the principal clusters of affected sites are found along the lower reaches of the Congo River (a key example being the site of Tumba, DRC [e.g., Menghin 1925]), as well as beside the Middle Ogooué River, Gabon (e.g., Assoko Ndong 1996, 2002), and in Rwanda and Burundi (Nenquin 1967).

Filter 3 (Fig. 15.7) subtracts from the database sites whose stratigraphic and or techno-typological description lacks sufficient detail to confirm the Lupemban affinity of recovered assemblages (n = 19). This filter encompasses 20% of sites including the geographical outlier of Sai Island, Sudan (Van Peer et al. 2003) and clusters in western Kenya (Leakey and Owen 1945), the CAR (Bayle de Hermens 1975, 1977) and across western Central Africa, in the Republic of Congo (RoC) and DRC.

Passing the database through filter 4 (Fig. 15.8) removes all sites known to have experienced post-depositional disturbance leading to the formation of subsurface "stone-lines" ($n = 13$). "Stone-line" contexts typically exhibit extremely low spatiotemporal integrity and preclude the reliable isolation of Stone Age industries and the behavioral repertoires they represent (Cahen 1976; McBrearty 1990; Mercader and Marti 2002; Mercader et al. 2002). Definite examples of "stone-line" contexts take in 14% of all sites and markedly affect the overall distribution through the preferential removal of sites from lowland Central Africa. All remaining sites in Gabon are lost, as well as Mosumu, Equatorial Guinea (Mercader 2002), and all sites on the southern bank of the Malebo Pool in the DRC (van Moorsel 1968), including Gombe Point (Cahen 1976). Also affected are Kamoa (Cahen 1975) and the purported Lupemban site of Chaminade, Malawi (Clark et al. 1972), respectively located on and beyond the margins of the Congo Basin.

The drastic effect of "stone lines" on knowledge of the Lupemban and its distribution is compounded when sites highly likely to have undergone vertical disturbance (based on the examination of reported stratigraphies) are considered. Filter 5 (Fig. 15.9) affects 17% of all sites ($n = 16$), and again preferentially removes localities in the interior of the Congo Basin. Sites are affected on the northern edge of the Malebo Pool in the RoC (Droux and Bergaud 1937), and along the tributaries of the Congo River in the lower Kasai, DRC (van Moorsel 1970). A further four sites are lost on the northern margin of the Congo Basin, in an arc that follows the general course of the Ubangi River, while in Rwanda and Burundi the Lupemban seems to lie in "stone-lines" at Campion and Mugera (Nenquin 1967).

The application to the dataset of filters 1–5 leaves a residue of 17 published Lupemban sites (Fig. 15.9) that are, without exception, located beyond the margins of lowland Central Africa, on the eastern and southern rims of the Congo Basin. The final filter applicable to this remaining 18% of sites considers disruption caused by geomorphological processes during the accumulation of archaeological materials, or post-depositionally. However, since all Lupemban sites, sequences, and assemblages have been disturbed to some degree, the application of filter 6 results in the loss of *all*

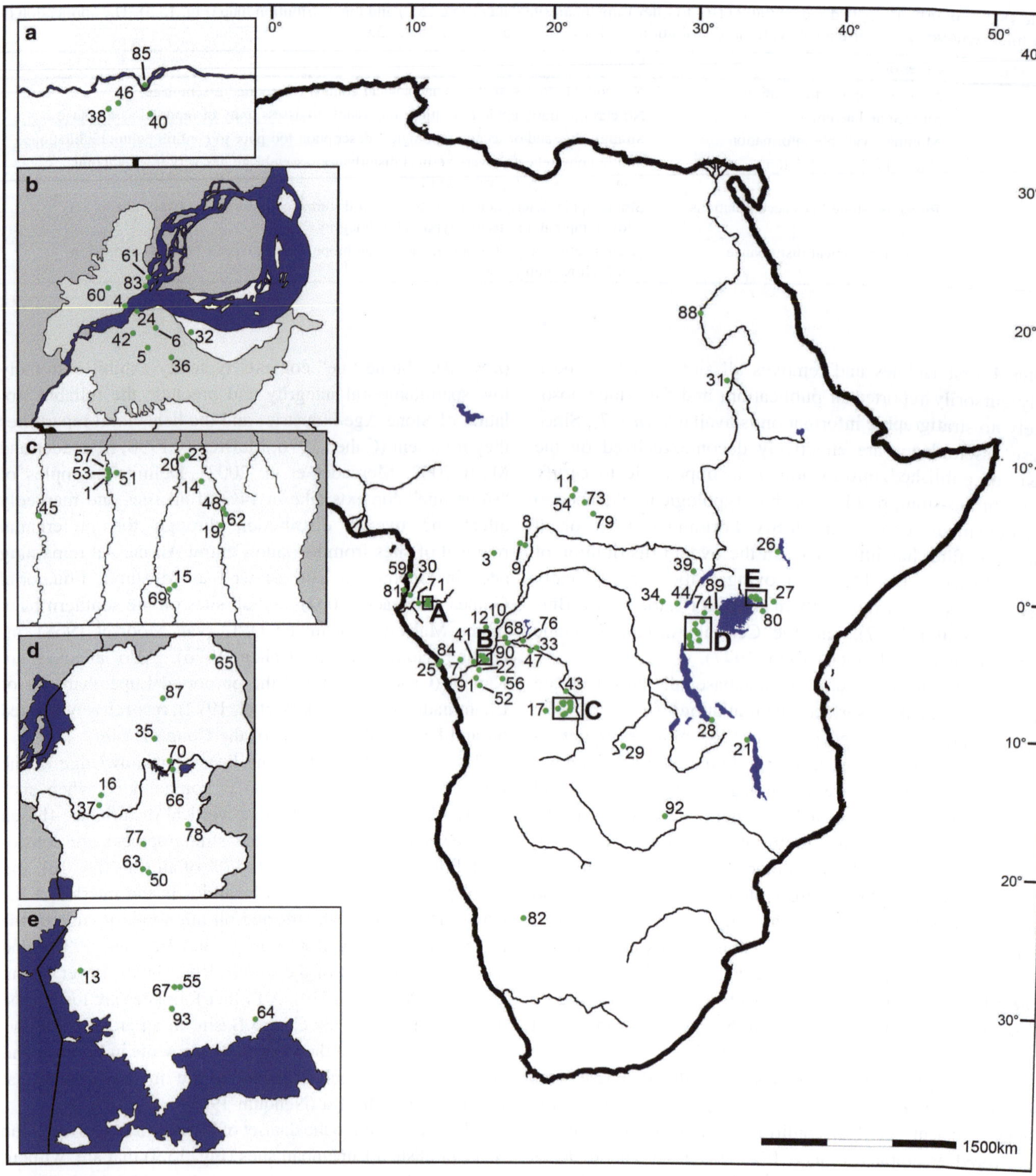

Fig. 15.5 Distribution of Lupemban sites after application of filter 1; sites lacking any stratigraphic information removed. Numbered sites correlate with Table 15.2. *Inset* **a** Middle Ogooué River, Gabon; **b** Brazzaville (RoC) and Kinshasa (DRC) on Malebo Pool, Congo River; **c** Lunda Province, northeast Angola; **d** Rwanda and Burundi; **e** Western Kenya

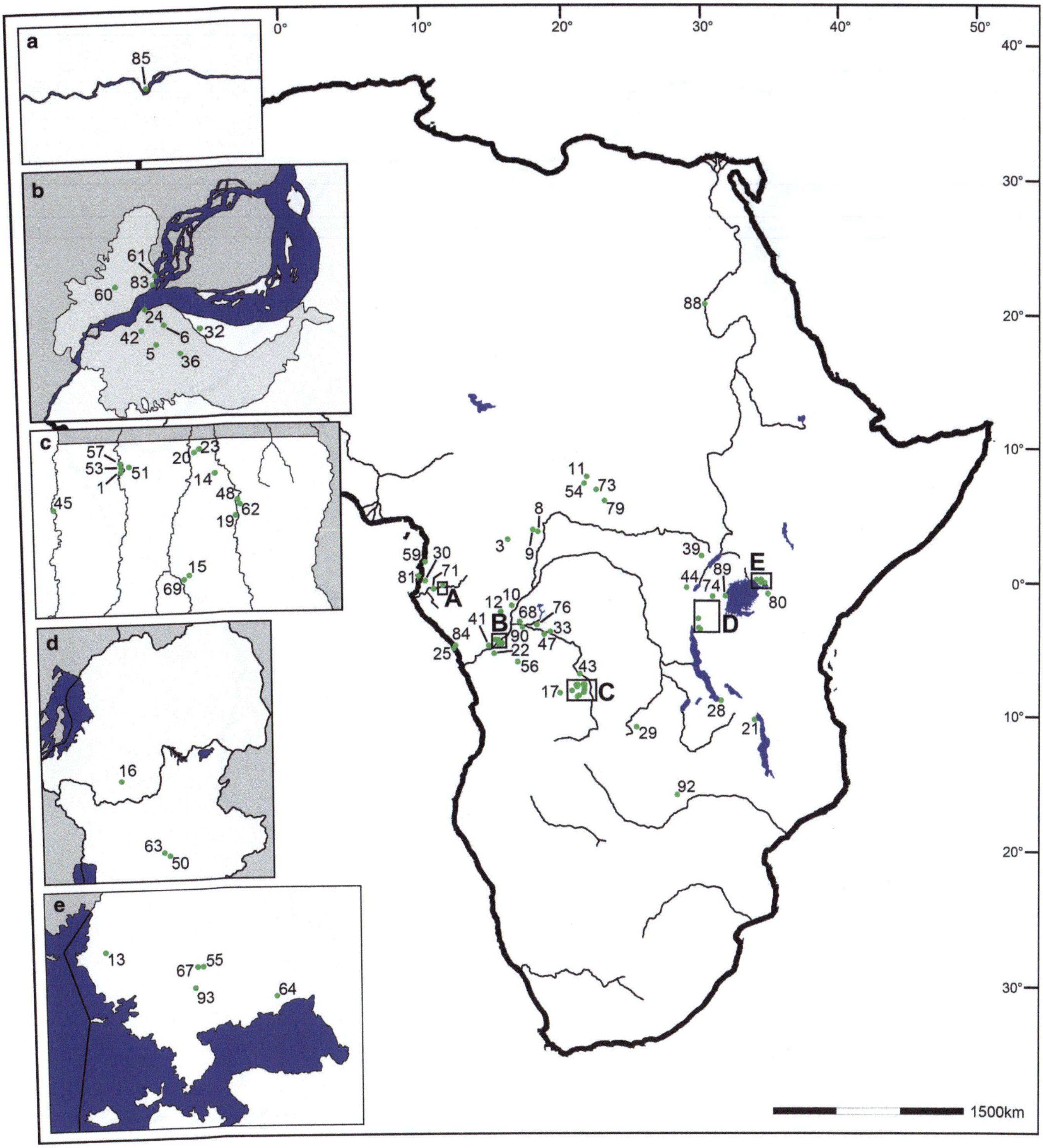

Fig. 15.6 Distribution of Lupemban sites after application of filter 2; known surface sites removed. Numbered sites correlate with Table 15.2

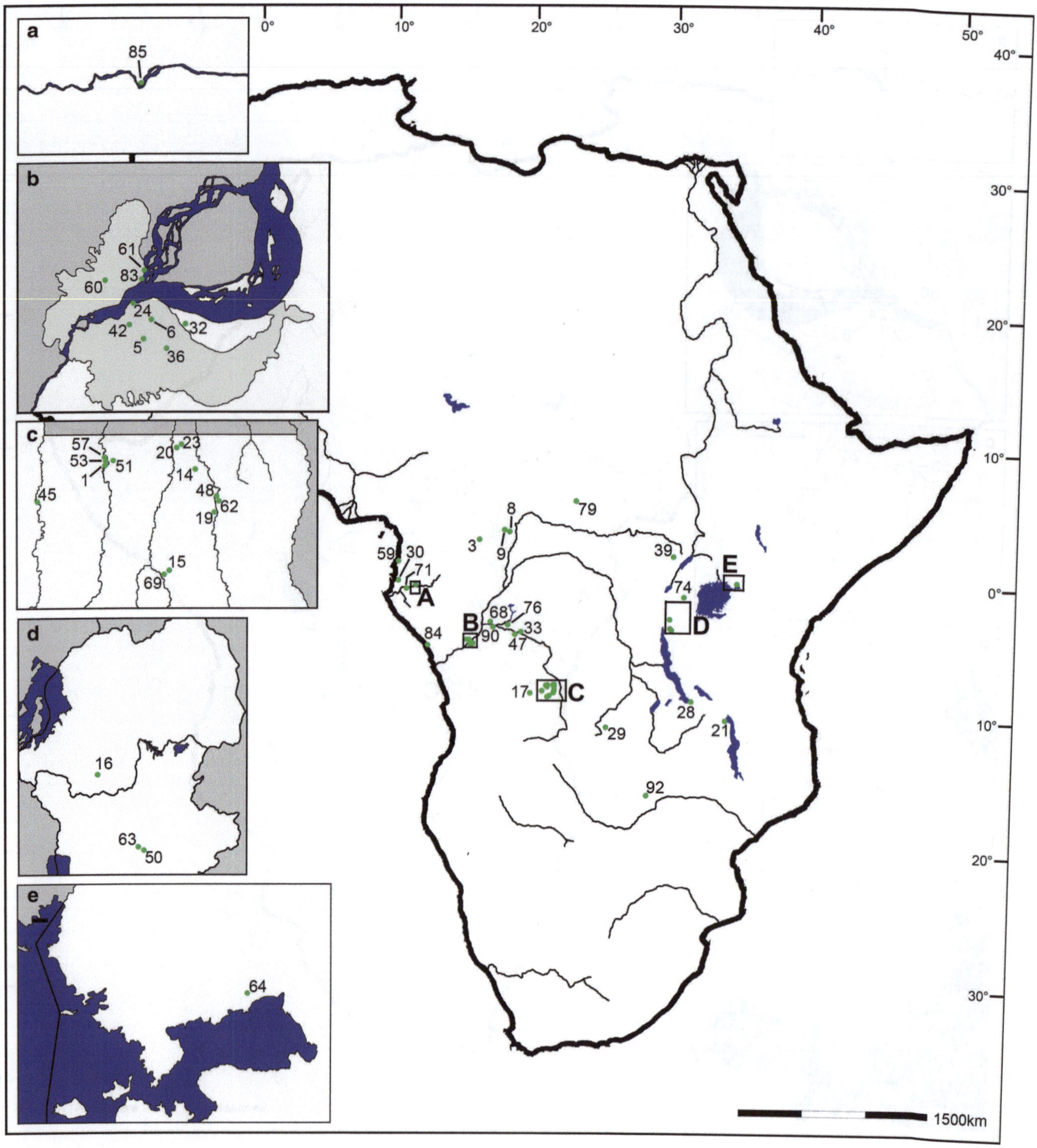

Fig. 15.7 Distribution of Lupemban sites after application of filter 3; minimally described sites/assemblages removed. Numbered sites correlate with Table 15.2

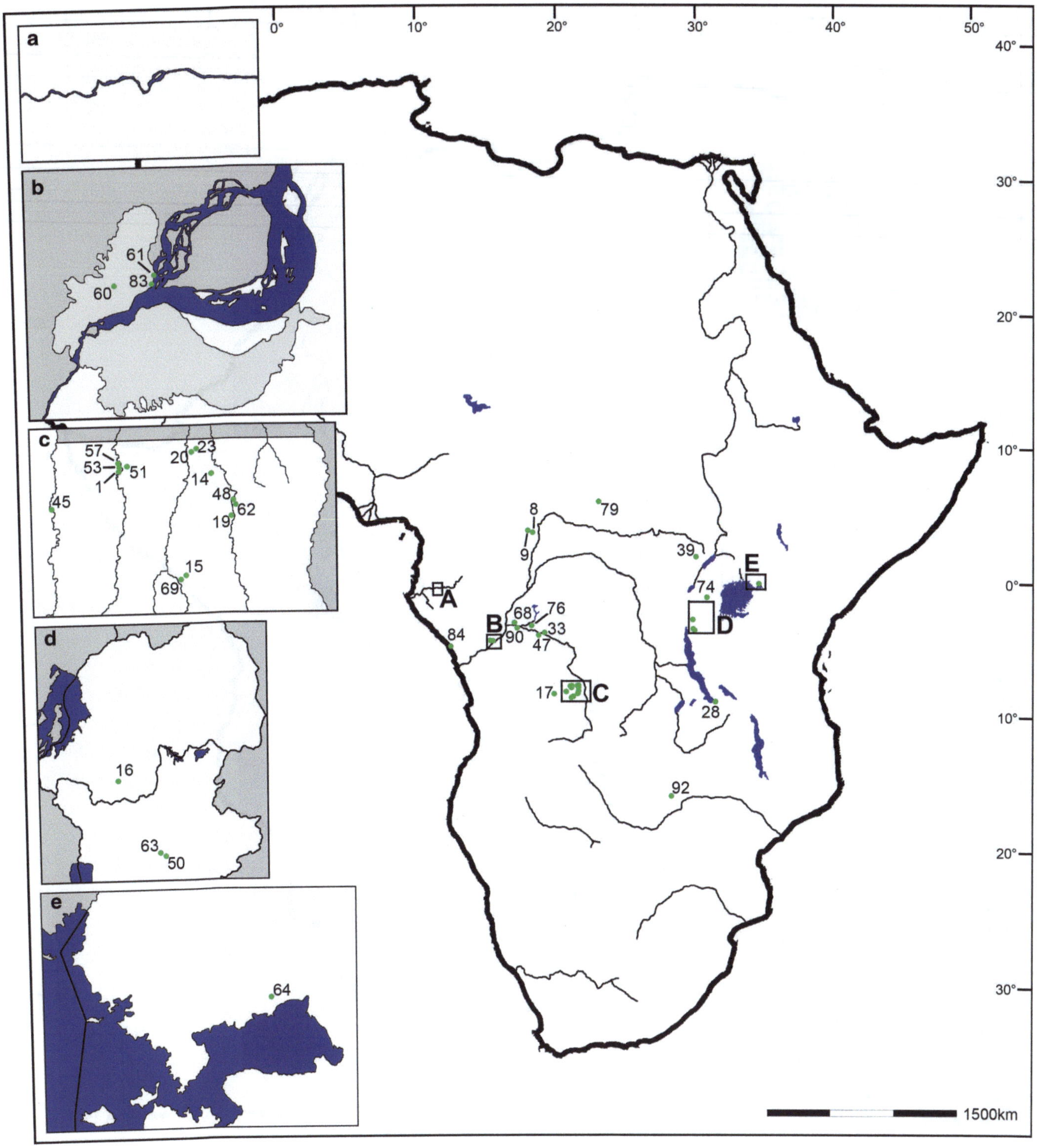

Fig. 15.8 Distribution of Lupemban sites after application of filter 4; "stone-line" sites removed. Numbered sites correlate with Table 15.2

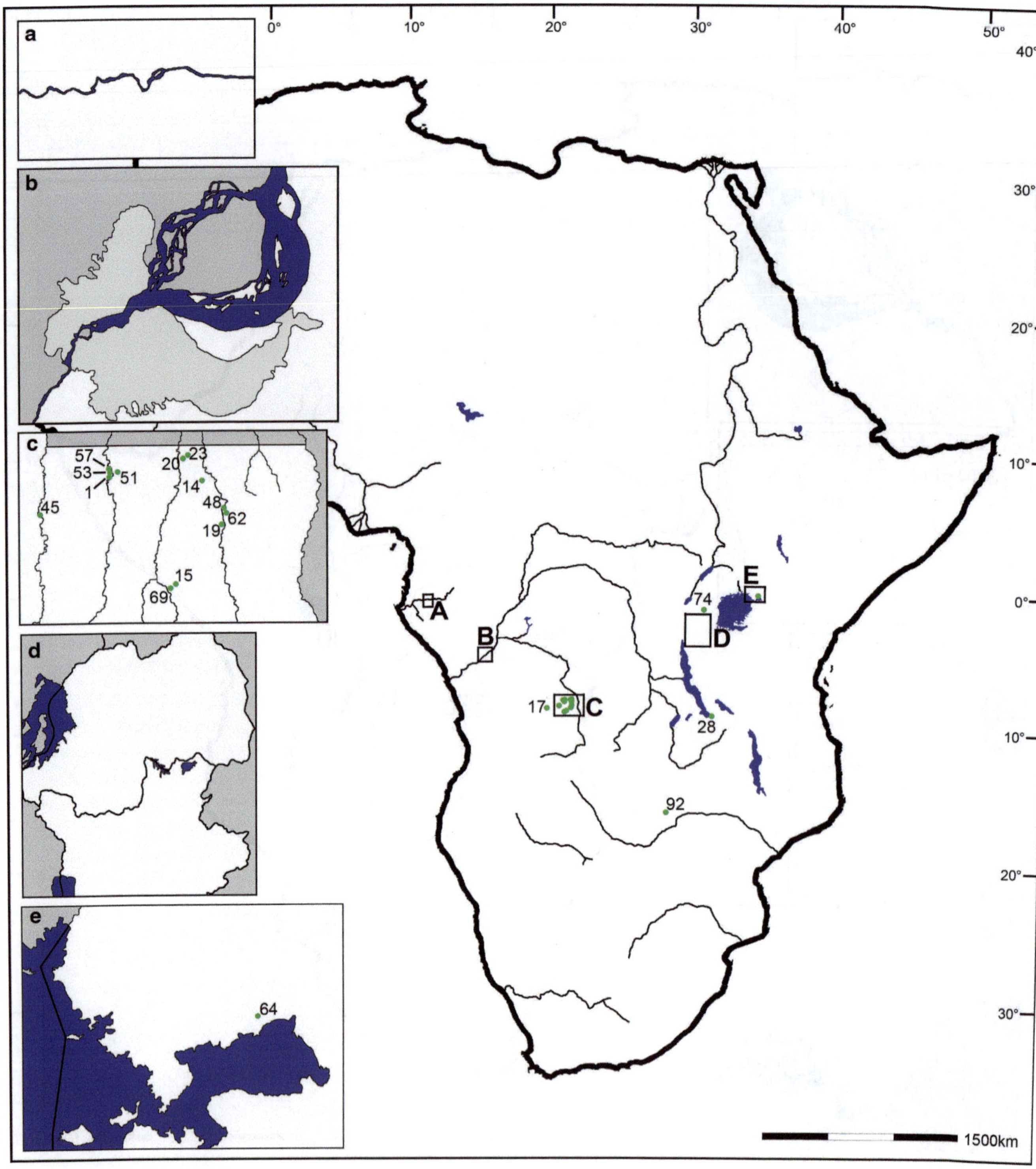

Fig. 15.9 Distribution of Lupemban sites after application of filter 5; probable "stone-line" sites removed. Numbered sites correlate with Table 15.2

remaining Lupemban sites from the database and distribution (not mapped). Despite this, variability in the nature and severity of the geomorphological processes affecting these residual sites dictates that within this subsample, some sites and assemblages exhibit greater integrity than others.

At a majority of higher integrity sites (Fig. 15.9) – including Nsongezi (Wayland 1937; O'Brien 1939; van Riet Lowe 1952; Posnansky 1962; Cole 1967), Muguruk (McBrearty 1988), and all sequences exposed by strip diamond mining in northern Angola (e.g., Musolexi, Camafufo, Catongula) (Clark 1963) – artifacts of Lupemban type are mixed variously with Sangoan or LSA implements in aggregate deposits with extremely coarse spatiotemporal resolution. Lacking a reliable stratigraphic framework, any attempt (e.g., Clark 1963) to extricate and define the Lupemban from these contexts is necessarily reliant on unproven progressivist assumptions of post-Acheulean techno-typological change. By contrast, at the collapsed cave site of Twin Rivers, Lupemban tools are held in a brecciated deposit capped by dated flowstones (see below) that ensure the industry's stratigraphic isolation (Barham 2000, 2001, 2002a, b, 2012; Clark and Brown 2001). However, the complete excavation of these deposits by dynamite in the 1950s (Clark 1971b; Clark and Brown 2001) confounds detailed study of their post-depositional disturbance and the detection of any synchronic or diachronic subtleties in Lupemban technological repertoires.

Kalambo Falls is arguably the key sequence for the Central African MSA. This discontinuous but stratified sequence captures a numerically large sample of Lupemban artifacts in alluvial and colluvial rubble horizons that are consistently separated by culturally sterile sands from underlying Sangoan and late Acheulean levels (Clark 1969, 1974, 2001). The variable lithostratigraphic context of these rubbles reflects a complex series of erosional and depositional processes that have introduced a few derived artifacts (Clark 2001: 83), but the overwhelming majority of material contained in these horizons is representative of the Lupemban industry (Clark 1969, 1974, 2001; Sheppard and Kleindienst 1996). Indeed, the technological distinctiveness of the Lupemban-bearing rubbles – which in contrast to Sangoan levels feature abundant evidence of PCT, blade-based tools, and lanceolate points – demonstrates beyond doubt that they are effectively isolated and relate to a chronologically later industrial stage (Sheppard and Kleindienst 1996; Clark 2001). Nevertheless, these secondary context deposits probably represent a significant chronological span and lack any internal spatial or temporal integrity. In the absence of fine stratigraphic and chronological control, Clark's (2001) distinction between earlier (Nakisasa) and later (Sisya) expressions of the Kalambo Falls Lupemban is best considered an artificial construct.

Assessment of Lupemban site integrity highlights that current knowledge of the industry rests on an extremely coarse-grained regional record that is plagued by the severe post-depositional reworking of lithic assemblages. Stratigraphic deficiencies are particularly severe across lowland Central Africa (e.g., Cahen 1978; Moeyersons 1978), where the industry is known only from surface accumulations and disrupted "stone-line" contexts of such low stratigraphic and cultural integrity that its isolation and definition is confounded. By contrast, a few imperfect yet better-stratified sequences found on the margins of the Congo Basin provide a basic observational framework. Yet, the lack of any intact, high-resolution profile sampling the industry precludes the derivation of precise data pertaining to its techno-typological composition and spatiotemporal variability, and hampers attempts to test its potential behavioral significance (Barham 2001).

Radiometric Dates

The chronological position of the Lupemban is central to the hypothesis that its technological innovations represent an early and evolutionarily significant adaptation to rainforest and woodland foraging (Barham 2001). A dearth of intact stratigraphic sequences sampling the Lupemban and general scarcity of datable organic materials, however, impedes the application of science-based dating techniques across Central Africa, to the extent that most sites remain undated. Indeed, of a total of 93 published sites with confirmed or purported Lupemban archaeology (Table 15.2), only 16 (17%) are associated with absolute dates: a record comprised by a total of 53 measured samples.

The age of most dated Lupemban assemblages ($n = 14$) and measured samples ($n = 32$) has been estimated based on the application of the radiocarbon (^{14}C) technique in the absence of longer-ranged methods. Derived from sequences across both forested lowland Central Africa (DRC, Equatorial Guinea, Gabon, RoC) and its wooded southern margin (Angola, Zambia), these estimates converge on the period ~20–40 ka and indicate a Late Pleistocene, MIS 2-3 timeframe for the industry (e.g., Barham 2001; Clark and Brown 2001: 4-6; Cornelissen 2002). This date range forms the historical paradigm for the chronology of the Lupemban; however, the ~45 ka technical limit of radiocarbon is now generally acknowledged to result only in infinite, minimum age estimates for the MSA (McBrearty and Brooks 2000). The post-depositional disturbance of sequences within the modern rainforest zone (see above) raises additional uncertainties about the reliability of lithic-charcoal stratigraphic associations (Cahen 1976, 1978; Cahen and Moeyersons 1977; McBrearty 1990), as highlighted by inverted ^{14}C dates at Gombe Point (Cahen et al. 1983), Mosumu (Mercader et al. 2002) and Dimba (de Maret et al. 1977). The radiocarbon record forms the only chronological framework for the Lupemban in Central Africa's lowland interior (e.g., Clist 1997; Cornelissen 2002), but is increasingly confronted by an emergent longer-range chronology for the industry on its periphery.

Dating techniques suited to the derivation of accurate age determinations beyond the limit of ^{14}C have until now been applied to just three Lupemban occurrences, located respectively beyond and on the margins of Central Africa: Sai Island (Sudan), and Kalambo Falls and Twin Rivers (both Zambia). This longer-ranged record comprises just 3% of all published sites and a total of 21 dated samples. At the geographically outlying site of Sai Island, an MSA industry initially ascribed by the excavators to the "Upper Sangoan" (Van Peer et al. 2003: 190), but latterly considered "clearly Lupemban" (Van Peer et al. 2004; Van Peer and Vermeersch 2007: 189), is bracketed by preliminary optically stimulated luminescence (OSL) dates of 182 ± 20 ka and 152 ± 10 ka (Van Peer et al. 2004). If the Lupemban affiliation of this material is confirmed, it may indicate the existence of characteristic Central African MSA technologies far outside the (much reduced) forest belt during MIS 6. Recent and ongoing efforts to resolve an accurate chronology for the ESA-MSA transition at Kalambo Falls through the application of single-grain OSL and thermally-transferred OSL (TT-OSL) techniques to previously unexcavated profiles in the sedimentary basin (Barham et al. 2009, 2015; Duller et al. 2015), have resulted in a provisional, very broad timeframe of 455±103 ka–39±2 ka for deposits encapsulating recognizably Mode 3 tool-bearing horizons. As cautiously reported by Barham et al. (2015), the numerically small artifact samples recovered lack the diagnostic technological attributes needed to distinguish the Lupemban from within the multiple Mode 3 horizons detected, and a more precise temporal placement for the industry at Kalambo Falls will hinge on the future refinement and extension of chronometric, stratigraphic and cultural sampling across the site. At the collapsed cave site of Twin Rivers, the application of thermal ionization mass spectroscopy (TIMS) U-series ($^{230}Th/^{234}U$) to speleothem (Barham and Smart 1996; Barham 2000: Table 10.1) resulted in dates of 200–170 ka for samples bracketing the Lupemban in F Block, and >400–170 ka in A Block, including a date of ~265 ka for a sample in direct association with the lithic assemblage (Barham 2000: 240). On this basis, Barham (2001) has contended that the age range of the Lupemban at Twin Rivers – and possibly across Central Africa – falls between ~270–170 ka, broadly coincident with the interglacial of MIS 7.

In his recent chronological review of the Acheulean – MSA transition in southern Africa, Herries (2011) argues the Twin Rivers chrono-stratigraphic record better reflects an age younger than 140 ka for the Lupemban but, as emphasized in Barham's (2012) comprehensive rejoinder, this assessment is based on a hasty and fundamental misreading of the published sequence and dates. Considered in accurate stratigraphic context, interpretation of the ten precision dates reported (Barham 2000, 2001, 2012) substantiates a solid pre-200 kyr age range for the Lupemban at Twin Rivers.

Considered broadly, the divergence in currently available age determinations for the Lupemban is thrown into stark relief

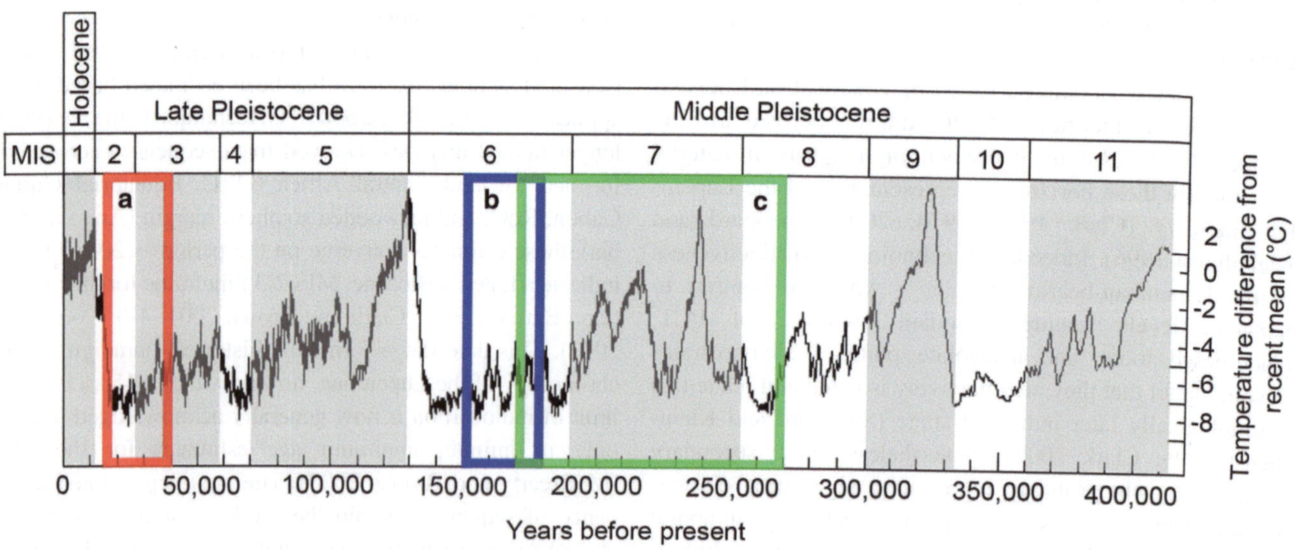

Fig. 15.10 Divergent age ranges reported for Lupemban assemblages superimposed on 400 kyr global temperature curve and correlated MIS chronology. **a** Late Pleistocene (MIS 2-3) age estimates based on radiocarbon technique; **b** MIS 6 age range for Lupemban, or 'Lupemban-like' material at Sai Island, based on preliminary OSL estimates (Van Peer et al. 2004); **c** principal span of Uranium-series dates for the Lupemban at Twin Rivers, Zambia, centering on the interglacial of MIS 7. Barham (2001) has interpreted the Twin Rivers data as reflecting the initial settlement of the Central African rainforest belt by pre-*sapiens* hominins during MIS 7 (~253–186 ka). Climate data after Petit et al. (1999). First published in *Before Farming [online version], 2011/1*, article 6

(Fig. 15.10), with longer-ranged dating techniques positioning the industry an order of magnitude older than infinite ^{14}C estimates. As one of only two Lupemban sites to which multiple dating techniques have been applied, Twin Rivers offers a glimpse into the direct influence of technical methods on this dichotomous record. When first dated by radiocarbon in the 1960s (Ferguson and Libby 1964: 335, 483; Clark and Brown 2001), travertine contained in Lupemban-bearing breccias yielded estimates of 22,800 ± 1000 ^{14}C BP and >33,200 uncal BP for the same sequence now robustly dated by U-series to 270–170 ka (Barham 2000, 2012). Such technically induced temporal variability, in combination with uneven regional coverage and a low number of dated samples means that full scholarly consensus on the age of the Lupemban is yet to emerge (cf. Barham 2001; Cornelissen 2002). Despite tacit agreement that late Middle Pleistocene dates provide a realistic age for the early Lupemban MSA on the margins of Central Africa (McBrearty and Brooks 2000; Mercader and Marti 2002; Rots and Van Peer 2006; Wadley 2007: 688), this low-resolution chronological record impedes attempts to model the paleobiogeographic associations of the Lupemban, as discussed below.

Paleoenvironmental Evidence

Longstanding speculation that the Lupemban represents a specialized adaptation to woodland and rainforest foraging is a central tenet underpinning its purported evolutionary significance (Barham 2001). As observed historically, the Lupemban correlates with Central Africa and contemporary rainforest and woodland biomes (Clark 1959, 1963, 1964, 1966, 1970, 1971a, b, 1982, 1988, 2001 and see above); however, oscillations in Pleistocene climate are known to have caused the significant spatial rearrangement of equatorial vegetation zones (e.g., Jahns et al. 1998) such that contemporary lithic-environment correlations (Fig. 15.3) cannot be projected onto the past (e.g., Barham 2001; Cornelissen 2002; Mercader 2002). Determining Lupemban paleoecological preferences is fraught with difficulties: poor organic preservation, a lack of faunal remains, and low stratigraphic integrity in combination hinder the acquisition of paleoenvironmental data directly from site sequences. In an effort to circumvent these issues, Mercader (2002) has inexplicitly advocated the use of archaeological deposit lithologies (e.g., for Muguruk) as local paleohabitat proxies; however, such information may correlate only generally and indirectly with past climate and vegetation cover, and is inadequately refined for this purpose. In fact, of 93 Lupemban sites currently known (Table 15.2), only five (5%) geographically disparate sites have to date yielded pollen or fauna enabling any degree of local paleohabitat reconstruction.

Located within the modern forested interior of lowland Central Africa, the sites of Mosumu (Mercader and Marti 1999) and Ndjole (Pommeret 1965a, b) have yielded palynological samples from "stone-lines" that allude to local habitats of humid rainforest during Lupemban occupation. This evidence must be treated cautiously, however, given these sequences' apparent post-depositional reworking. On the southern perimeter of the Congo Basin, sites now located in savanna woodland and woodland environments reflect more varied paleoecologies. At Twin Rivers, Lupemban-bearing breccias incorporate a small and fragmented faunal assemblage taking in zebra, wildebeest, and hartebeest, indicating a bushland or grassland habitat (Bishop and Reynolds 2000) during at least part of the c. 100 kyr timeframe encompassed by the deposits. The Upper Lupemban (or Lupembo-Tshitolian) at Mufo occurs in redistributed Kalahari sands alongside infrequent grass pollen and charcoal that hints at open savanna conditions (Clark 1963), but clearly relates to a mixed aggregate with poor techno-typological and temporal resolution. A more robust pollen sequence consistently featuring *Brachystegia* and other woodland taxa attests to the persistence of deciduous miombo woodlands throughout the better-stratified Acheulean, Sangoan and Lupemban sequence at Kalambo Falls (van Zinderen Bakker 1969; Taylor et al. 2001). In general, however, the variable and meager nature of direct paleoecological evidence derived from Lupemban sequences provides a wholly inadequate basis on which to determine the industry's association with particular environments, or gauge empirically its foraging adaptations.

Broader records derived from lacustrine and marine cores augment site-based data, and form the basis of attempts to reconstruct Central African paleobiogeography and model Lupemban paleoenvironments (Cornelissen 2002; Barham 2000, 2001), but have variably limited time depths and spatial fidelities. Available terrestrial records from lowland Central Africa, including Barombi-Mbo (Maley and Brenac 1998), and Bilanko and Ngamakala (Elenga et al. 1994) enable localized, landscape-scale habitat reconstructions only within the temporal confines of MIS 1-2. Encircling the modern rainforest belt, study of Lakes Chad (Leblanc et al. 2006), Bosumtwi (Shanahan et al. 2013), Tanganyika (Burnett et al. 2011), and Malawi (Scholz et al. 2011) has yielded progressively refined subregional paleoclimatic and paleoecological data for the Holocene, Last Glacial Maximum (LGM), and the last 90 kyr and 145 kyr, respectively, but are too spatially discontinuous for the high-resolution mapping of equatorial paleobiogeography prior to MIS 2. This shortcoming applies also to palynological sequences derived from deep marine sediment cores along the Atlantic coast of Central and West Africa, which offer more generalized, regional-scale paleoecological evidence for the last ~150 kyr (Dupont et al. 2000) and as far back as 400 ka (Jahns et al. 1998).

While limited in spatiotemporal resolution, indirect sources of paleoecological data provide critical evidence substantiating the expansion and contraction of Central African vegetation zones in synchrony with global shifts in rainfall and temperature through the Pleistocene (e.g., Dupont et al. 2000). As the only reconstructions of equatorial paleobiogeography spanning the entirety of MIS 6-2, Dupont et al.'s (2000) composite data and the Liberian marine pollen core reported by Jahns et al. (1998) are of critical interest, since they highlight the persistence of rainforest taxa and fern spores including *Alchornea*, *Bosqueia*, and *Cnestis* throughout the cool, dry glacial conditions of MIS 2, 3, 4, and 6. In response to reduced glacial precipitation, lowland rainforests fragment and shrink to form isolated refugia separated by enlarged savanna and woodland biomes (Hooghimistra and Agwu 1988; Fredoux 1994; Dupont et al. 2000; Cornelissen 2002; [but see Cowling et al. 2008]), before re-expanding to their modern limits, or beyond, under the warmer, wetter conditions of interglacial phases, including MIS 1, 5, and 7. These regional records nevertheless lack fine spatial resolution, such that the precise location, extent, and floral composition of rainforest refugia and adjacent biomes prior to MIS 2 remain somewhat speculative (Maley 1996).

Conditioned by this scarcity in available paleoecological data for mid-Pleistocene Central Africa, both Barham (2001: Fig. 6.6) and Banks et al. (2006: Fig. 6) have sought to model potential Lupemban paleoenvironmental associations by combining site distribution data with generalized, continental-scale paleobiogeographies mapped from MIS 1 and 2 onto previous climatic stages. Barham (2001) tentatively extended the MIS 7 temporal span of the Twin Rivers Lupemban across Clark's (1967) essentially accurate (see above) Central African distribution of the industry (Fig. 15.4a), and argued that the Lupemban correlates robustly with interglacial rainforest and woodland biomes, but only weakly with glacial environmental zones. The approach taken by Banks et al. (2006) forgoes at its interpretative core any dating evidence, and is based instead on the observation that the inaccurate (see above) south Central African distribution of Lupemban sites derived from Clark's (1967) gazetteer (Fig. 15.4b) fits better with Adams and Faure's (1997) savanna-dominated glacial paleobiogeographical model for MIS 6 (Banks et al. 2006: Fig. 6; Marean and Assefa 2005). While laudable, these efforts to model Lupemban paleobiogeographical associations rest on a series of assumptions that render their insights speculative at best. Beyond the variable representativeness of the harnessed Lupemban site distributions discussed above, the absence of a refined chronology for the industry necessitates paleobiogeographical models presuppose site contemporaneity (Barham 2001), while ~60-kyr-long periods of spatiotemporal variation in local habitats (e.g., Maley and Brenac 1998) are compressed into static approximations of vegetation cover.

Review of Lupemban paleoenvironmental data underscores that empirical information bearing on the habitats in which this technology was manufactured and deployed is scant, and emphasizes the scale of the challenge involved in determining unequivocally the nature of any specific – and spatiotemporally recurrent – lithic-environment relationship. The low number, geographical dispersion and generally poor stratigraphic integrity of sequences yielding paleoecological data inhibits the definitive correlation of discrete site occupation episodes with particular environments. The apparent persistence of dense miombo woodland throughout the Lupemban MSA at Kalambo Falls (van Zinderen Bakker 1969; Taylor et al. 2001) is exceptional, yet whether this single, geographically peripheral site mirrors wider Lupemban foraging preferences across lowland Central Africa is unclear. Constructing any detailed paleobiogeographical model for Central Africa during the likely mid-Pleistocene lifespan of the Lupemban is similarly problematic, being reliant on the attribution of Lupemban technologies to exceptionally broad temporal spans (MIS 6/7) and low-resolution paleoclimatic and paleoenvironmental datasets that ineffectively resolve habitat correlations at human scales.

Discussion and Conclusion

This chapter has focused on the Lupemban MSA of Central Africa, which likely represents an early and significant portion of the MIS 6-2 record across the contemporary rainforest and woodland zone. A clear contradiction exists between the potential evolutionary significance ascribed to the Lupemban as a marker for emergent cognitive and behavioral complexity in archaic *Homo sapiens* (e.g., Barham 2001, 2002a) and the extremely poor resolution of this regional record (e.g., Cahen 1976).

The Lupemban as a whole exhibits extraordinarily low chrono-stratigraphic integrity. Not only are intact, high-resolution profiles sampling the industry completely unknown (e.g., Barham 2002a), but also the industry almost vanishes when strict "quality control" criteria are applied to currently known sites and sequences (Figs. 15.5, 15.6, 15.7, 15.8 and 15.9). On this basis, it is reasonable to readdress (e.g., Cahen 1976, 1978; Isaac 1982; McBrearty 1988) whether the Lupemban in fact exists as a culturally and technologically distinct entity. Across the lowland interior of Central Africa, Stone Age sequences suffer disproportionately from profound post-depositional disturbance (Cahen 1976, 1978; Figs. 15.7 and 15.8), confounding attempts to isolate reliably and define precisely the industry, and to correlate and compare the cultural content of excavated assemblages. In contrast, on the periphery of the Congo Basin a few less disturbed sequences retain the essential stratigraphic resolution needed to tentatively isolate the Lupemban from preceding and succeeding industries (cf. Figs. 15.7, 15.8, 15.9). Uniting this highly variable regional record, Kalambo Falls (Clark 1969, 1974, 2001) captures the Central African ESA-MSA transition in a manner essentially mirroring historical observations and conceptions of post-Acheulean, Sangoan, and Lupemban industries drawn from more disturbed sites across the region (e.g., Colette 1929; Leakey and Owen 1945). That such a substantive chrono-stratigraphic and techno-typological match should occur across the equatorial zone by coincidence is exceedingly unlikely. Indeed, the fact that almost one-quarter of all published Lupemban assemblages from across Central Africa (including the key localities of Twin Rivers and Kalambo Falls) were originally excavated, handled, and reported by a single archaeologist – J.D. Clark – (Table 15.2) provides compelling evidence in support of a coherent Lupemban identity. By far the most parsimonious interpretation then considers the Lupemban as an archaeologically valid technocomplex, albeit the subject of an impoverished and structurally compromised regional record that has to date barred its refined definition, and knowledge of subtle variability in its spatial, synchronic, or diachronic composition (Clark 2001; Barham 2002a; Taylor 2011).

Myriad challenges remain before the techno-typological subtleties of the Lupemban are fully understood, and it must be acknowledged that the internal heterogeneity of the industry may reflect the reworking of once discrete technological sub-phases into a single cultural entity (Clark 2001). Despite this, where encountered in more reliable stratigraphic context (e.g., Twin Rivers, Kalambo Falls), analyses demonstrate the industry incorporates a range of artifact types suitable for the construction of a sophisticated Mode 3 foraging toolkit (Taylor 2011; Fig. 15.2): the first such lithic repertoire in the Central African record (Barham 2001). While empirical evidence linking Lupemban lanceolates, core axes, and blade-based tools to complex hunting and gathering subsistence strategies is currently limited (Taylor 2009; Rots et al. 2011), broad support is lent to the concept of their hafted use by a convergence of ethnographic and archaeological correlates, in combination with technological theory. Independently compiled geographical proxies for the distribution of Lupemban technology (Clark 1967; Figs. 15.3, 15.4a) confirm the regional nature of the industry (e.g., Clark 1988) and substantiate its correlation with the contemporary extent of the Central African woodland and rainforest belt (Fig. 15.3). The interpretation of this data warrants caution, yet these separate lines of evidence together indicate the development of composite technologies was integral to the first hominin settlement of Central Africa (Barham 2001), and validate the appropriateness of the longstanding hypothesis that the industry represents a prehistoric adaptation to closed canopy ecological biomes (Clark 1959, 1963, 1964, 1966, 1970, 1971a, b, 1982, 1988, 2001; Barham 2000, 2001; Taylor 2011).

Viewed objectively, the fragmentary and variable evidence currently available for the Lupemban renders ineffectivc attempts to determine the industry's evolutionary significance, and leaves the technocomplex open to the application of multiple and divergent interpretative frameworks. Most broadly, these are manifested as divisions between Middle / Late Pleistocene chronologies (cf. Cornelissen 2002; Barham 2012; Fig. 15.10) and interglacial rainforest / glacial savanna /diverse environments (cf. Barham 2001; Cornelissen 2002; Marean and Assefa 2005). As known from lowland Central Africa, the Lupemban may be a Late Pleistocene MIS 2-3 technology adapted to multifarious grassland, woodland, and rainforest habitats (e.g., Clist 1997; Cornelissen 2002). It remains possible that this geographical portion of the Lupemban record simply captures the latest phase or facies of the industry. Given the apparent similarity of reported Late Pleistocene (MIS 2-3) assemblages from the lowlands and Middle Pleistocene (MIS 7) Lupemban assemblages from the southern rim of the Congo Basin, this view however necessarily assumes ~150 kyr of essential technological stasis across Central Africa, even amidst a background of significant paleoenvironmental alterations. A more radical interpretation of the record of lowland Central Africa might consider that composite tool using mobile human foragers only dispersed into rainforests ~40 ka, or possibly very much later: an interpretation that implicitly casts the Lupemban as a late variant of the MSA of peripheral relevance to the evolution of *Homo sapiens*. Both interpretations are critically challenged, however, by the inexact archaeological association of site-based paleoecological samples, and by uncertainties over the stratigraphic and technical reliability of radiocarbon dating estimates that query the relevance of MIS 2-3 paleoenvironmental data and paleobiogeographical models.

Conceptions of the Lupemban as a MIS 6 savanna grassland-adapted technology (Marean and Assefa 2005; Banks et al. 2006) suggest the industry does not bear directly on issues of cognitive and behavioral complexity in archaic *Homo sapiens*. The MIS 6 dates from Sai Island, Sudan, might be taken in support of this view, but are preliminary and of uncertain geographical and industrial relevance to the Lupemban. Other lines of evidence backing this interpretation of the record are completely lacking: as presented by Banks et al. (2006: Fig. 6) the geographical correlation of Lupemban sites with MIS 6 paleobiogeography rests on spurious site distribution data and should be viewed with extreme caution. Barham's (2001, 2002a) interpretation of the Lupemban on the southern periphery of Central Africa as an MIS 7 composite technology adapted for rainforest foraging implies the industry is of central relevance to debates about the emergence of cognitive and behavioral complexity among pre-*sapiens* hominins. This view is based on a technically and stratigraphically robust late Middle Pleistocene chronology from Twin Rivers (Barham 2012), and lent circumstantial support by woodland paleoecological samples from Kalambo Falls and by the correlation of Lupemban sites with modeled interglacial stage rainforests (Barham 2001). While this account of the industry is best supported by current evidence, it remains to be demonstrated that data derived from geographically peripheral sites is representative of the industry across its entire range.

It is pertinent to consider whether polarized and uniform concepts of Lupemban foraging adaptations to closed (rainforest/woodland) (e.g., Clark 1959; Barham 2001), or open (savanna) environments (Marean and Assefa 2005; Banks et al. 2006) offer an appropriately nuanced paleoenvironmental basis for hypotheses of hominin behavioral adaptations. Indeed, available high-resolution environmental data pertaining to recent Holocene (Maley and Brenac 1998; Favier et al. 2004; Goetze et al. 2006; Ngomanda et al. 2009), and Pleistocene (Elenga et al. 1994; Cornelissen 2002; Shanahan et al. 2013) Central Africa highlight that equatorial ecosystems are rarely invariable at the small spatiotemporal scales experienced by humans. As Cornelissen (2002: 206–207) has forwarded, mosaic and ecotonal habitats bordering fragmented forest environments may provide a wealth of opportunities for hunter-gatherers, particularly in terms of a broadened nutritional and resource base obtainable from multiple, proximate habitats. It remains plausible, and perhaps even likely, that Lupemban composite technologies were a flexible, adaptive response to the penecontemporaneous exploitation of a variety of environmental settings, including both dense rainforest and rainforest-savanna mosaics. The latter are today increasingly considered dynamic environments that generate species diversity (Smith et al. 1997; Kark and van Rensburg 2006), and may in the past have been a draw point for Stone Age hunter-gatherers.

Systematic and holistic reviews such as that presented here can assist in the recognition of current research shortfalls and potentials. The immediate challenge is to both broaden and particularize knowledge of the Lupemban through the identification of less-disturbed archaeological profiles – especially in lowland Central Africa – and to combine this with a focused strategy of paleoecological and chronological sampling. Until now, Lupemban technology has in only two instances been reported within cave deposits (c. 2% of sites). While the time depth, stratigraphic integrity, and organic preservation potential of tropical cave and rock shelter contexts may be highly variable (e.g., Mercader et al. 2003), a targeted campaign to locate and assess the potential of new cave sites across the Congo Basin might yet result in the detection of intact Lupemban MSA sequences incorporating refined paleoecological records. The application of new and longer-range dating techniques, in particular OSL (e.g., Duller and Wintle 2012; Nicoll 2010) to the few existing sites retaining better resolution profiles (Fig. 15.9), would certainly serve as a cross check on ^{14}C-based dating estimates for the industry, and might substantiate the Lupemban as a late Middle Pleistocene technology across its wider geographical range. Moreover, the chronology of the Lupemban could be tightened at sites on the eastern periphery of Central Africa (e.g., Nsongezi, Muguruk) through their correlation with African or extra-continental volcanic events, should these deposits preserve cryptotephra (e.g., Lane et al. 2013).

Fundamental reappraisal of the Lupemban illustrates vividly the severe limits the Stone Age record imposes on attempts to develop any refined understanding of post-Acheulean technological and behavioral change across Central Africa before and during MIS 6-2. Contradicting its geographical position and potential evolutionary importance, the equatorial rainforest belt remains stubbornly peripheral to paleoanthropological debates centering on the behavioral emergence of *Homo sapiens*. Current evidence indicates the Lupemban marks the appearance of routine Mode 3 tool construction in Central Africa, an origin coinciding with the first sustained settlement of the region's lowland interior (Taylor 2011). Whether this technological development marks a precocious late Middle Pleistocene adaptation to rainforest foraging (e.g., Barham 2001, 2012) remains a valid, intriguing, and serious proposition deserving of persistent multidisciplinary enquiry.

Acknowledgments I would like to express my sincere thanks to Sacha Jones and Brian Stewart for their invitation to contribute to the Africa MIS 6-2 symposium, and this volume. I am especially grateful to Larry Barham and also to Els Cornelissen for discussions on the Central African MSA that have helped shape the content of this paper, and to Vicky Winton and three anonymous reviewers for their comments on an earlier draft. The work presented here was undertaken at the University of Liverpool as part of doctoral research funded by the Arts and Humanities Research Council (AHRC).

References

Adams, J. M., & Faure, H. (Eds.). (1997). *Review and atlas of palaeovegetation: Preliminary land ecosystem maps of the world since the Last Glacial Maximum*. Oak Ridge, Tennessee, USA: National Laboratory. http://www.esd.ornl.gov/projects/qen/adams1.html.

Allsworth-Jones, P. (1987). The earliest human settlement in West Africa and the Sahara. *West African Journal of Archaeology, 9*, 87–128.

Ambrose, S. H. (2010). Coevolution of composite-tool technology, constructive memory, and language: Implications for the evolution of modern human behaviour. *Current Anthropology, 51*(S1), S135–S147.

Andah, B. W. (1979). The early Palaeolithic in West Africa: The case of Asokrochona. *West African Journal of Archaeology, 9*, 87–110.

Arkell, A. J. (1949). *The Old Stone Age in the Anglo-Egyptian Sudan* (Sudan Antiquities Service Occasional Papers No. 1). Khartoum: Sudan Antiquities Service.

Assoko Ndong, A. (1996). Fouilles archéologiques dans les savanes de Lope (Ogooué-Ivindo) Gabon: Rapport de la Mission 1995. *Nyame Akuma, 45*, 24–29.

Assoko Ndong, A. (2002). Synthesis of recent archaeological data about the settlement of the fauna reserve of Lope during the Holocene. *L'anthropologie, 106*, 135–158.

Bailey, R. C., & Headland, T. N. (1991). The tropical rain forest: Is it a productive environment for human foragers? *Human Ecology, 19*, 261–285.

Bailey, R. C., Head, G., Jenike, M., Owen, B., Rechtman, R., & Zechenter, E. (1989). Hunting and gathering in tropical rain forests: Is it possible? *American Anthropologist, 9*, 59–82.

Bahuchet, S., McKey, D., & de Garine, I. (1991). Wild yams revisited: Is independence from agriculture possible for rainforest hunter-gatherers? *Human Ecology, 19*, 59–82.

Banks, W. E., d'Errico, F., Dibble, H., Krishtalka, L., West, D., & Olsezewski, D. I., et al. (2006). Eco-cultural niche modeling: New tools for reconstructing the geography and ecology of past human populations. *PaleoAnthropology, 2006*, 68–83.

Barham, L. S. (Ed.). (2000). *The Middle Stone Age of Zambia, South Central Africa*. Bristol: Western Academic and Specialist Press.

Barham, L. S. (2001). Central Africa and the emergence of regional identity in the Middle Pleistocene. In L. S. Barham & K. Robson-Brown (Eds.), *Human roots: Africa and Asia in the Middle Pleistocene* (pp. 65–80). Bristol: Western Academic and Specialist Press.

Barham, L. S. (2002a). Backed tools in the Middle Pleistocene Central Africa and their evolutionary significance. *Journal of Human Evolution, 43*, 585–603.

Barham, L. S. (2002b). Systematic pigment use in the Middle Pleistocene of south central Africa. *Current Anthropology, 43*(1), 181–190.

Barham, L. S. (2010). A technological fix for 'Dunbar's Dilemma'? In R. Dunbar, C. Gamble, & J. Gowlett (Eds.), *Social brain and distributed mind* (pp. 367–390). London: The British Academy.

Barham, L. S. (2012). Clarifying some fundamental errors in Herries' (2011) "A chronological perspective on the Acheulian and its transition to the Middle Stone Age in southern Africa: The question of the Fauresmith". *International Journal of Evolutionary Biology, 2012*, 5 pages.

Barham, L. S., & Smart, P. L. (1996). An early date for the Middle Stone Age of central Zambia. *Journal of Human Evolution, 30*, 287–290.

Barham, L., Tooth, S., Duller, G. A. T., Plater, A. J., & Turner, S. (2015). Excavations at site C North, Kalambo Falls, Zambia: New insights into the Mode 2/3 transition in south-central Africa. *Journal of African Archaeology, 13*(2). http://dx.doi.org/10.3213/2191-5784-10270.

Barham, L. S., Duller, G., Plater, A., Tooth, S., & Turner, S. (2009). Recent excavations at Kalambo Falls, Zambia. *Antiquity, 83*(322), project gallery.

Bashige, E., & Debonnet, G. (2004). Biodiversity in the DRC. In E. Bashige, G. Debonnet, A. Motingea Mangulu, M. Ntayingi, P. On'okoko, & S. A. Tshiluila (Eds.), *Nature and culture in the Democratic Republic of Congo* (pp. 35–64). Tervuren: Royal Museum for Central Africa.

Beaumont, P. B., & Vogel, J. C. (2006). On a timescale for the past million years of human history in central South Africa. *South African Journal of Science, 102*, 217–228.

Bequaert, M. (1938). *Les fouilles de Jean Colette à Kalina* (Annales du Musèe du Congo Belge, Tome I, Fasc. 2). Tervuren: Musèe Royal de l'Afrique Centrale.

Bequaert, M. (1943). Deux instruments en Pierre taillée de l'Angumu. *Institut Royal Colonial Belge, Bulletin des Séances, XIV-3*, 586–595.

Bishop, L. C., & Reynolds, S. C. (2000). Fauna from Twin Rivers. In L. S. Barham (Ed.), *The Middle Stone Age of Zambia* (pp. 217–222). Bristol: Western Academic and Specialist Press.

Blankoff, B. (1965). La préhistoire au Gabon. *Bulletin de la Société Préhistorique et Protohistorique Gabonaise, 1*, 4–5.

Breuil, H. (1944). Le Paléolithique au Congo Belge d'après les recherches de Docteur Cabu. *Transactions of the Royal Society of South Africa, 30*(2), 143–174.

Brooks, A. S., Yellen, J. E., Nevell, L., & Hartman, G. (2006). Projectile technologies of the African MSA: Implications for modern human origins. In E. Hovers & S. L. Kuhn (Eds.), *Transitions before the transition: Evolution and stability in the Middle Palaeolithic and Middle Stone Age* (pp. 233–256). New York: Springer.

Brown, D., McSweeney, K., & Helmke, P. A. (2004). Statistical, geochemical and morphological analysis of stone line formation in Uganda. *Geomorphology, 62*, 217–237.

Burnett, A. P., Soreghan, M. J., Scholz, C. A., & Brown, E. T. (2011). Tropical East African climate change and its relation to global climate: A record from Lake Tanganyika, Tropical East Africa, over the past 90+ kyr. *Palaeogeography, Palaeoclimatology, Palaeoecology, 303*, 155–167.

Cabu, F. (1953). Some aspects of the Stone Age in the Belgian Congo. In L. S. B. Leakey (Ed.), *Proceedings of the Pan African Congress on Prehistory, Nairobi 1947* (pp. 195–201). Oxford: Blackwell.

Cahen, D. (1975). *Le site archéologique de la Kamoa (Région du Shaba, République du Zaïre) de l'age de la Pierre Ancien a l'age du fer*. Tervuren: Musèe Royal de l'Afrique Centrale.

Cahen, D. (1976). Nouvelles fouilles à la Pointe de la Gombe (ex-pointe de Kalina), Kinshasa, Zaire. *L'anthropologie, 80*, 573–602.

Cahen, D. (1978). Vers une révision de la nomenclature des industries préhistoriques de l'Afrique Centrale. *L'anthropologie, 82*(1), 5–36.

Cahen, D., & Moeyersons, J. (1977). Subsurface movements of stone artefacts and their implications for the prehistory of Central Africa. *Nature, 266*, 812–815.

Cahen, D., Moeyersons, J., & Mook, W. G. (1983). Radiocarbon dates from Gombe Point (Kinshasa, Zaire) and their implications. *PACT, 85*, 441–452.

Carlson, R. L. (1967). Excavations at Khor Abu Anga and at sites in Nubia. *Current Anthropology, 8*(4), 352.

Casey, J. L. (2002). The archaeology of West Africa from the Pleistocene to the mid-Holocene. In J. Mercader (Ed.), *Under the canopy* (pp. 35–63). New York: Rutgers University Press.

Clark, J. D. (1959). *The prehistory of southern Africa*. New York: Plenum Press.

Clark, J. D. (1963). *Prehistoric cultures of northeast Angola and their significance in tropical Africa*. Lisbon: Companhis de Diamantes de Angola (DIAMANG).

Clark, J. D. (1964). The Sangoan culture of Equatoria: The implications of its stone equipment. In E. Ripoll-Perello (Ed.), *Miscellanea en homenaje al Abate Henri Breuil volume* (pp. 309–325). Barcelona: Instituto de Prehistoria y Arqueologia.

Clark, J. D. (1965). The atlas of African prehistory: A report on progress. In L. D. Cuscoy (Ed.), *Actas del V Congreso Panafricano de Prehistoria y de estudio del Quaternario (1963)* (pp. 311–328). Santa Cruz de Tenerife, Islas Canarias: Museo Arquólogico de Tenerife 5.

Clark, J. D. (1966). *The distribution of prehistoric culture in Angola*. Lisbon: Companhis de Diamantes de Angola (DIAMANG).
Clark, J. D. (1967). *The atlas of African prehistory*. Chicago: University of Chicago Press.
Clark, J. D. (1969). *Kalambo Falls prehistoric site. Volume I, the geology, palaeoecology and detailed stratigraphy of the excavations*. Cambridge: Cambridge University Press.
Clark, J. D. (1970). *The prehistory of Africa*. London: Thames and Hudson.
Clark, J. D. (1971a). Problems of archaeological nomenclature and definition in the Congo basin. *South African Archaeological Bulletin, 26*, 67–78.
Clark, J. D. (1971b). Human behavioral differences in southern Africa during the later Pleistocene. *American Anthropologist, 73*, 1211–1236.
Clark, J. D. (1974). *Kalambo Falls prehistoric site. Volume II, the later prehistoric cultures*. Cambridge: Cambridge University Press.
Clark, J. D. (1982). The cultures of the Middle Palaeolithic/ Middle Stone Age. In J. D. Clark (Ed.), *The Cambridge history of Africa* (Vol. 1, pp. 248–340). *From the earliest times to c.500 BC* Cambridge: Cambridge University Press.
Clark, J. D. (1988). The Middle Stone Age of East Africa and the beginnings of regional identity. *Journal of World Prehistory, 2*(3), 235–305.
Clark, J. D. (2001). *Kalambo Falls prehistoric site. Volume III, the earlier cultures: Middle and Earlier Stone Age*. Cambridge: Cambridge University Press.
Clark, J. D., & Brown, K. S. (2001). The Twin Rivers Kopje, Zambia: Stratigraphy, fauna, and artefact assemblages from the 1954 and 1956 excavations. *Journal of Archaeological Science, 28*, 305–330.
Clark, J. D., Haynes, C. V., & Mawby, J. E. (1972). Paleoanthropological investigations in the Lake Malawi Rift (1965–1966); an interim report. In H. J. Hugot (Ed.), *Actes du 6e Session du Congrès Panafricain de Préhistoire* (pp. 513–530). Les Imprimeries Réunies de Chambery.
Clist, B. (1989). Archaeology in Gabon 1886–1988. *African Archaeological Review, 7*, 59–95.
Clist, B. (1993). Archaeological fieldwork and labwork in Gabon during 1992. *Nyame Akuma, 39*, 32.
Clist, B. (1997). Le site d'Okala, Province de l'Estuaire, Gabon, et son importance pour la compréhension du passage à la sédentarisation en Afrique central. *Earth & Planetary Sciences, 325*, 151–156.
Cole, G. H. (1967). The later Acheulean and Sangoan of southern Uganda. In W. W. Bishop & J. D. Clark (Eds.), *Background to evolution in Africa* (pp. 481–528). Chicago: Chicago University Press.
Coles, J. M., & Higgs, E. S. (1969). *The archaeology of early man*. New York: Faber and Faber.
Colette, J. (1929). Le préhistorique dans le Bas-Congo. *Bulletin de la Societé Royale Belge d'Anthropologie et de Préhistoire, 44*, 41–47.
Colette, J. (1933). Essai biométrique sur la station préhistorique de Kalina (Congo Belge). *Xve Congrès International d'Anthropologie et d'Archéologie Préhistorique, Paris, 1931* (pp. 278–285). Paris: Nourry.
Cormack, J. L. (1994). Early Stone Age heavy duty implements of Africa. Ph.D. Dissertation, University of Liverpool.
Cornelissen, E. (2002). Human responses to changing environments in Central Africa between 40,000 and 12,000 BP. *Journal of World Prehistory, 16*(3), 197–235.
Cornelissen, E. (2016). The later Pleistocene in the northeastern Central African rainforest. In S. C. Jones & B. A. Stewart (Eds.), *Africa from MIS 6-2: Population dynamics and paleoenvironments* (pp. 301–319). Dordrecht: Springer.
Cowling, S. A., Cox, P. M., Jones, C. D., Maslin, M. A., Peros, M., & Spall, S. A. (2008). Simulated glacial and interglacial vegetation across Africa: Implications for species phylogenies and trans-African migration of plants and animals. *Global Change Biology, 14*, 827–840.
Davies, O. (1954). The Sangoan culture in Africa. *South African Journal of Science, May*, 273–277.
Davies, O. (1961). *Archaeology in Ghana*. Edinburgh: Thomas Nelson.
Davies, O. (1967). The Palaeolithic in West Africa. In O. Davies (Ed.), *West Africa before the Europeans: Archaeology and prehistory* (pp. 89–146). London: Methuen.
Davies, O. (1976). The Sangoan industries. *Annals of the Natal Museum, 22*, 885–911.
de Bayle des Hermens, R. (1975). *Recherches préhistoriques en République Centrafricaine*. Paris: Labethno.
de Bayle des Hermens, R. (1977). Premier aperçu du Paléolithique Inférieur en République Centrafricaine. *L'anthropologie, 71*(5–6), 435–466.
de Bayle des Hermens, R., Oslisly, R., & Peyrot, B. (1987). Premières séries de pierres taillées du Paléolithique Inférieur découvertes au Gabon. *Afrique Centrale. L'anthropologie, 91*(2), 693–698.
de Heinzelin de Braucourt, J. (1948). Industrie lithique des graviers aurifères de la Lodjo (Ituri, Congo belge), description de la collection V. Piret. *Bulletin du Musèe Royal d'Histoire naturelle de Belgique, XXIV-8*, 1–16.
Deacon, H. J., & Wurz, S. (2001). Middle Pleistocene populations of southern Africa and the emergence of modern behaviour. In L. S Barham & K. Robson-Brown (Eds.), *Human roots: Africa and Asia in the Middle Pleistocene* (pp. 55–65). Bristol: Western Academic and Specialist Press.
Deino, A. L., & McBrearty, S. (2002). $^{40}Ar/^{39}Ar$ dating of the Kapthurin Formation, Baringo, Kenya. *Journal of Human Evolution, 42*, 185–210.
Delibrias, G., Giresse, P., Lanfranchi, R., & Le Cocq, A. (1983). Datations des dépôts holorganiques quaternaires sur la bordure occidentale de la Cuvette congolaise (R.P. du Congo); corrélations avec les sédiments marins voisin. *Comptes Rendus de l'academie de sciences, Paris, 296*, 436–466.
de Maret, P. (1990). Phases and facies in the archaeology of Central Africa. In P. Robertshaw (Ed.), *The history of African Archaeology* (pp. 109–134). London: John Currey.
de Maret, P., van Noten, F., & Cahen, D. (1977). Radiocarbon dates from West Central Africa: A synthesis. *The Journal of African History, 18*(4), 481–505.
d'Errico, F. (2003). The invisible frontier: A multiple species model of the origin of behavioural modernity. *Evolutionary Anthropology, 12*, 18–202.
Droux, G., & Bergaud, G. (1937). Les gisements préhistoriques de la Pointe Hollandaise (Congo Brazzaville). *Bulletin de la Société des Recherches Congolaises, 28*, 137–142.
Droux, G., & Kelley, H. (1939). Recherches préhistoriques dans la région de Boko-Songho et à Pointe Noire (moyen-Congo). *Journal de société des africanistes, 4*, 71–84.
Duller, G. A. T., Tooth, S., Barham, L., & Tsukamoto, S. (2015). New investigations at Kalambo Falls, Zambia: Luminescence chronology, site formation, and archaeological significance. *Journal of Human Evolution, 85*, 111–125.
Duller, G. A. T., & Wintle, A. G. (2012). A review of the thermally transferred optically stimulated luminescence signal from quartz for dating sediments. *Quaternary Geochronology, 7*, 6–20.
Dupont, L. M., Jahns, S., Marret, F., & Ning, S. (2000). Vegetation change in equatorial West Africa: Time-slices for the last 150 ka. *Palaeogeography, Palaeoclimatology, Palaeoecology, 155*, 95–122.
Elenga, H., Schwartz, D., & Vincens, A. (1994). Pollen evidence of late Quaternary vegetation and inferred climate changes in Congo. *Palaeogeography, Palaeoclimatology, Palaeoecology, 109*, 345–356.

Farine, B. (1965). Recherches préhistoriques au Gabon. *Bulletin S.P.P. G., 3*, 68–84.

Favier, C., Chave, J., Fabing, A., Schwartz, D., & Dubois, M. A. (2004). Modelling forest-savanna mosaic dynamics in man-influenced environments: Effects of fire, climate and soil heterogeneity. *Ecological Modelling, 171*, 85–102.

Ferguson, G. J., & Libby, W. F. (1964). *UCLA radiocarbon dates III. Radiocarbon, 6*, 318–339.

Fredoux, A. (1994). Pollen analysis of a deep sea core in the Gulf of Guinea: Vegetation and climatic change during the last 225,000 years B.P. *Palaeogeography, Palaeoclimatology, Palaeoecology, 109*, 317–330.

Goetze, D., Hörsch, B., & Porembski, S. (2006). Dynamics of forest-savanna mosaics in north-eastern Ivory Coast from 1954 to 2002. *Journal of Biogeography, 33*, 653–664.

Grün, R., & Beaumont, P. B. (2001). Border Cave revisited: A revised ESR chronology. *Journal of Human Evolution, 40*, 457–482.

Hayden, B. D. (1976). Australian Western Desert lithic technology: An ethno-archaeological study of variability in material culture. Ph.D. Dissertation, University of Toronto.

Headland, T. N., & Bailey, R. C. (1991). Introduction: Have hunter gatherers ever lived in tropical rain forest independently of agriculture? *Human Ecology, 19*(2), 115–122.

Henshilwood, C. S., & Marean, C. W. (2003). The origin of modern human behaviour. *Current Anthropology, 44*(5), 627–651.

Herries, A. I. R. (2011). A chronological perspective on the Acheulian and its transition to the Middle Stone Age in southern Africa: The question of the Fauresmith. *International Journal of Evolutionary Biology, 2011*, 25 p.

Hooghiemistra, A., & Agwu, C. O. C. (1988). Changes in the vegetation and trade winds in equatorial northwest Africa 140,000–70,000 yrs B.P. as deduced from two marine pollen records. *Palaeogeography, Palaeoclimatology, Palaeoecology, 66*, 173–213.

Howell, F. C., Cole, G. H., Kleindienst, M. R., Szabo, B. J., & Oakley, K. P. (1972). Uranium series dating of bone from the Isimila prehistoric site. *Nature, 237*, 51–52.

Isaac, G. L. (1982). The earliest archaeological traces. In J. D. Clark (Ed.), *The Cambridge history of Africa* (Vol. 1, pp. 157–247). *From the earliest times to 500 BC* Cambridge: Cambridge University Press.

Jahns, S., Hills, M., & Sarnthein, M. (1998). Vegetation and climate history of west equatorial Africa based on a marine pollen record off Liberia (site GIK 16776) covering the last 400,000 years. *Review of Palaeobotany and Palynology, 102*, 277–288.

Kark, S., & van Rensburg, B. J. (2006). Ecotones: Marginal or central areas of transition? *Israel Journal of Ecological and Evolution, 52*, 29–53.

Kerbis, J., Wrangham, R., Carter, M., & Hauser, M. (1993). A contribution to tropical rain forest taphonomy: Retrieval and documentation of chimpanzee remains from Kibale forest, Uganda. *Journal of Human Evolution, 25*, 484–514.

Klein, R. G. (2008). Out of Africa and the evolution of human behaviour. *Evolutionary Anthropology, 17*, 267–281.

Kouyoumontzakis, G., Lanfranchi, R., & Giresse, P. (1985). Les datations radiométriques du quaternaire de la République Populaire du Congo. *Cahiers Congolais d'Anthropologie et Histoire, 10*, 11–31.

Kuman, K., Inbar, M., & Clarke, R. J. (1999). Paleoenvironments and cultural sequence of the Florisbad Middle Stone Age hominid site, South Africa. *Journal of Archaeological Science, 26*, 1409–1425.

Lane, C. S., Chorn, B. T., & Johnson, T. C. (2013). Ash from the Toba supereruption in Lake Malawi shows no volcanic winter in East Africa at 75 ka. *Proceedings of the National Academy of Sciences, 10*(20), 8025–8029.

Lanfranchi, R. (1986). Les industries préhistoriques congolaises dans le contexte du Quaternaire récent. In H. Faure, L. Faure, & E. S. Diop (Eds.), *Changement globaux en Afrique, passé, présent, futur. International Symposium, Dakar, 21–28 April, 1986* (pp. 247–249). Paris: ORSTON.

Lanfranchi, R., & Manima-Moubouha, A. (1989). Bilan de la recherche archéologique en R.P. du Congo 1975–1989. *Nsi, 6*, 67–68.

Lanfranchi, R., & Schwartz, D. (1991). Les remaniements de sols pendant le Quarternaire supérieur au Congo: Évolution des paysages dans la région de la Sangha. *Cahier ORSTOM, series Pedologie, XXV, I*(1), 11–24.

Lavachery, P. (1990). *L'Age de la Pierre Récent au Bas-Zaire: Étude du matériel lithique des missions Bequaert 1950–1952 et de Maret 1973* (Mémoire de licence, Université Libre de Bruxelles, Brussels). Brussels: Université Libre de Bruxelles.

Leakey, L. S. B. (1949). *Tentative study of the Pleistocene climatic changes and Stone Age cultural sequences in north eastern Angola.* Lisbon: Museo do Dundo Publicações Culturais 4.

Leakey, L. S. B., & Owen, W. E. (1945). *A contribution to the study of the Tumbian culture in East Africa (Coryndon Memorial Museum Occasional Paper 1).* Nairobi: Coryndon Memorial Museum.

Leblanc, M. J., Leduc, C., Stagnitti, F., van Oevelen, P. J., Jones, C., & Mofor, L. A., et al. (2006). Evidence for Megalake Chad, north-Central Africa, during the late Quaternary from satellite data. *Palaeogeography, Palaeoclimatology, Palaeoecology, 230*, 230–242.

Liubin, V. P., & Guédé, F. Y. (2000). *Paleolit Respubliki Kot d'Ivvuar (Zapadnaya Afrika) (The Palaeolithic of the Republic of Cote d'Ivoire (West Africa)* (Vol. 3). St. Petersburg: Russian Academy of Sciences, Institute of the History of Material Culture Proceedings.

Lombard, M. (2004). Distribution patterns of organic residues on Middle Stone Age points from Sibudu Cave, Kwazulu-Natal. *South Africa. South African Archaeological Bulletin, 59*(180), 37–44.

Lombard, M. (2005). Evidence of Hunting and hafting during the Middle Stone Age at Sibidu Cave, KwaZulu-Natal, South Africa: A multianalytical approach. *Journal of Human Evolution, 48*(3), 279–300.

Lombard, M. (2008). Finding resolution for the Howiesons Poort through the microscope: Micro-residue analysis of segments from Sibudu Cave, South Africa. *Journal of Archaeological Science, 35*, 26–41.

Lombard, M. (2012). Thinking through the Middle Stone Age of sub-Saharan Africa. *Quaternary International, 270*, 140–155.

Lombard, M., & Haidle, M. N. (2012). Thinking a bow-and-arrow set: Cognitive implications of Middle Stone Age bow and stone-tipped arrow technology. *Cambridge Archaeological Journal, 22*(2), 237–264.

Lombard, M., & Phillipson, L. (2010). Indications of bow and stone-tipped arrow use 64,000 years ago in KwaZulu-Natal, South Africa. *Antiquity, 84*, 1–14.

Livingstone-Smith, A., Assoko Ndong, A., & Cornelissen, E. (2007). Prospection archéologique dans le sud du Gabon. *Nyame Akuma, 67*, 26–35.

MacCalman, H. R., & Viereck, A. (1967). Peperkorrel, a factory site of Lupemban affinities from central south west Africa. *South African Archaeological Bulletin, 22*, 41–50.

Maley, J. (1996). The African rain forest – main characteristics of changes in vegetation and climate from the Upper Cretaceous to the Quaternary. *Proceedings of the Royal Society of Edinburgh Biological Sciences, 104B*, 31–73.

Maley, J., & Brenac, P. (1998). Vegetation dynamics, palaeoenvironments and climatic changes in the forests of western Cameroon during the last 28,000 years BP. *Review of Palaeobotany and Palynology, 99*, 157–187.

Marean, C. W., & Assefa, Z. (2005). The Middle and Upper Pleistocene African record for the biological and behavioural origins of modern humans. In A. B. Stahl (Ed.), *African Archaeology: A critical introduction* (pp. 93–129). Oxford: Blackwell.

Mayaux, P., Bartholomé, E., Massart, M., van Cutsem, C., Cabral, A., & Nonguierma, A., et al. (2003). *A land cover map of Africa*. Luxembourg: Office for Official Publications of the European Communities.

McBrearty, S. (1988). The Sangoan-Lupemban and Middle Stone Age sequence at the Muguruk site. *Western Kenya. World Archaeology, 19*(3), 388–420.

McBrearty, S. (1990). Consider the humble termite: Termites as agents of post-depositional disturbance at African archaeological sites. *Journal of Archaeological Science, 17*, 111–143.

McBrearty, S. (1991). Recent research in Kenya and its implications for the status of the Sangoan industry. In J. D. Clark (Ed.), *Cultural beginnings: Approaches to understanding early hominid lifeways in the African savanna* (pp. 159–176). Bonn: Germanisches Zentralmuseum.

McBreaty, S. (2001). The Middle Pleistocene of east Africa. In L. S Barham & K. Robson-Brown (Eds.), *Human roots: Africa and Asia in the Middle Pleistocene* (pp. 81–98). Bristol: Western Academic and Specialist Press.

McBrearty, S., & Brooks, A. S. (2000). The revolution that wasn't: A new interpretation of the origin of modern human behaviour. *Journal of Human Evolution, 39*, 453–563.

McHugh, W. P., Breed, C. S., Schaber, G. G., McCauley, J. F., & Szabo, B. J. (1988). Acheulean sites along the Radar Rivers, southern Egyptian Sahara. *Journal of Field Archaeology, 15*, 361–379.

Mellars, P., Boyle, K., Bar-Yosef, O., & Stringer, C. (Eds.). (2007). *Rethinking the human revolution, new behavioural and biological perspectives on the origin and dispersal of modern humans*. Cambridge: University of Cambridge.

Menghin, O. (1925). Die Tumbakultur am unteren Kongo und der westafrikanische Kultureis. *Anthropos, 20*, 516–557.

Mercader, J. (2002). Forest people: The role of African rainforests in human evolution and dispersal. *Evolutionary Anthropology, 11*, 117–124.

Mercader, J., & Marti, R. (1999). Middle Stone Age sites in the tropical forests of Equatorial Guinea. *Nyame Akuma, 51*, 14–24.

Mercader, J., & Marti, R. (2002). The Middle Stone Age occupation of Atlantic Central Africa. In J. Mercader (Ed.), *Under the canopy* (pp. 65–92). New York: Rutgers University Press.

Mercader, J., Runge, F., Vrydaghs, L., Doutrelpont, H., Ewango, C. E. N., & Juan-Tresseras, J. (2000). Phytoliths from archaeological sites in the tropical forest of Ituri, Democratic Republic of Congo. *Quaternary Research, 54*, 102–112.

Mercader, J., Marti, R., Martinez, J. L., & Brooks, A. (2002). The nature of 'stone lines' in the African quaternary record: Archaeological resolution at the rainforest site of Mosumu, Equatorial Guinea. *Quaternary International, 89*, 71–96.

Mercader, J., Marti, R., González, I. J., Sánchez, A., & Garcia, P. (2003). Archaeological site formation in rain forests: Insights from the Ituri Rock Shelters, Congo. *Journal of Archaeological Science, 30*, 45–65.

Miller, S. F. (1988). Patterns of environmental utilization by late prehistoric cultures in the southern Congo basin. In J. R. F. Bower & D. Lubell (Eds.), *Prehistoric cultures and environments in the late Quaternary of Africa* (BAR International Series No. 405) (pp. 127–143). Oxford: Archaeopress.

Moeyersons, J. (1978). The behaviour of stones and stone implements buried in consolidating and creeping Kalahari Sands. *Earth Surface Processes, 3*, 115–128.

Moeyersons, J., & Roche, E. (1982). Past and present environments. In F. van Noten (Ed.), *The archaeology of Central Africa* (pp. 15–25). Graz: Akademische Druck und Verlagsanstalt.

Mortelmans, G. (1957). La préhistoire du Congo Belge. *Revue de l'Université de Bruxelles 9e année, 2–3*, 119–71.

Nenquin, J. (1967). *Contributions to the study of the prehistoric cultures of Rwanda and Burundi (Musèe Royal de l'Afrique Centrale, Annales 80)*. Tervuren: Musèe Royal de l'Afrique Centrale.

Ngomanda, A., Chepstow-Lusty, A., Makaya, M., Favier, C., Schevin, P., & Maley, J., et al. (2009). Western equatorial African forest-savanna mosaics: A legacy of Holocene climate change? *Climate of the Past, 5*, 647–659.

Nicoll, K. (2010). Geomorphic development and Middle Stone Age archaeology of the Lower Cunene River, Namibia-Angola Border. *Quaternary Science Reviews, 29*, 1419–1431.

Nygaard, S., & Talbot, M. (1984). Stone Age archaeology and environment on the southern Accra plains, Ghana. *Norwegian Archaeological Review, 17*, 19–38.

O'Brien, T. P. (1939). *The prehistory of the Uganda Protectorate*. Cambridge: Cambridge University Press.

Oslisly, R., Doutrelepont, H., Fontugne, M., Forestier, H., Giresse, P., & Hatte, C., et al. (2006). Premiers résultats pluridisciplinaires d'une stratigraphie vieille de plus de 40,000 ans du site de Maboué 5 dans la réserve de la Lopé au Gabon. *Actes du XIVe Congrès de l'UISPP, Liege 2–8 Sept 2001, Prehistoire en Afrique* (BAR International Series No. 1522) (pp. 189–198). Oxford: Archaeopress.

Passau, G. (1946). Decouvertes préhistoriques dans le Haut-Ituri et dans la Haut-Lindi (Région nord est du Congo). *Institut Royal Colonial Belge. Bulletin des Séances, XVII-2*, 637–649.

Petit, J. R., Jouzel, J., Raynaud, D., Barnola, J-M., Basile, I., & Bender, M., et al. (1999). Climate and atmospheric history of the past 420,000 years from the Vostok ice core, Antarctica. *Nature, 399*, 429–436.

Pinçon, B. (1991). Archeologie des plateaux et collines Teke (République Populaire du Congo): De nouvelles données. *Nsi, 8* (9), 24–32.

Pommeret, Y. (1965a). *Civilisations préhistoriques au Gabon, Tome 1, Valée du Moyen Ogooué: Présentation de l'industrie lithique de traditions Sangoenne, Lupembienne et Néolithique* (Memoires de la Société Préhistorique et Protohistorique Gabonaise). Libreville: Centre Culturel Française Saint Exupury.

Pommeret, Y. (1965b). *Civilisations préhistoriques au Gabon, Tome 2, vallée du Moyen Ogooué: Notes préliminaires à propos du gisement Lupembien et Néolithique de Ndjolé* (Memoires de la Société Préhistorique et Protohistorique Gabonaise). Libreville: Centre Culturel Française Saint Exupery.

Porat, N., Chazan, M., Grün, R., Aubert, M., Eisenmann, V., & Kolska, L. (2010). New radiometric ages for the Fauresmith industry from Kathu Pan, southern Africa: Implications for the Earlier to Middle Stone Age transition. *Journal of Archaeological Science, 37*, 269–283.

Posnansky, M. (1962). Recent Palaeolithic discoveries in Uganda. In G. Mortelmans & J. Nenquin (Eds.), *Actes du IVe Congrès Panafricain de Préhistoire et de l'Étude du Quaternaire* (Séries in-8 Sciences Humaines No. 40) (pp. 207–214). Tervuren: Musée Royal de l'Afrique Centrale.

Rots, V., & Van Peer, P. (2006). Early evidence of complexity in lithic economy: Production and maintenance of hafted core axes at late Middle Pleistocene site 8-B-11, Sai Island (Sudan). *Journal of Archaeological Science, 33*, 360–371.

Rots, V., Van Peer, P., & Vermeersch, P. M. (2011). Aspects of tool production, use and hafting in Palaeolithic assemblages from northeast Africa. *Journal of Human Evolution, 60*, 637–664.

Sato, H. (2001). The potential of edible wild yams and yam-like plants as a staple food resource in the African tropical rain forest. *African Study Monographs Suppl., 26*, 123–134.

Sayer, J. A., Harcourt, C. S., & Collins, N. M. (1992). *The conservation atlas of tropical forests: Africa*. New York: Simon & Schuster.

Scholz, C. A., Cohen, A. S., Johnson, T. C., King, J., Talbot, M. R., & Brown, E. T. (2011). Scientific drilling in the Great Rift Valley: The 2005 Lake Malawi Scientific Drilling Project – an overview of the past 145,000 years of climate variability in southern hemisphere East Africa. *Palaeogeography, Palaeoclimatology, Palaeoecology, 303*, 3–19.

Schwarcz, H. P., & Grün, R. (1993). Electron spin resonance dating of tooth enamel from Bir Tarfawi. In F. Wendorf, R. Schild, & A. E. Close (Eds.), *Egypt during the last interglacial* (pp. 234–237). New York: Plenum.

Schwartz, D. (1996). Archeologie préhistorique et processus de formation des stone-lines en Afrique Centrale (Congo-Brazzaville et zones périphériques). *Geo-Eco-Trop, 10*(1–4), 15–38.

Shanahan, T. M., Peck, J. A., McKay, N., Heil Jnr, C. W., King, J., & Forman, S. L., et al. (2013). Age models for long lacustrine sediment records using multiple dating approaches – an example from Lake Bosumtwi, Ghana. *Quaternary Geochronology, 15*, 47–60.

Shea, J. J. (2006). The origins of lithic projectile point technology: Evidence from Africa, the Levant, and Europe. *Journal of Archaeological Science, 33*, 823–846.

Shea, J. J., Brown, K. S., & Davis, Z. J. (2002). Controlled experiments with Middle Palaeolithic spear points: Levallois points. In R. Mathieu (Ed.), *Experimental archaeology: Replicating past objects, behaviours and processes* (BAR International Series No. 1035) (pp. 55–72). Oxford: Archaeopress.

Sheppard, P. J., & Kleindeinst, M. R. (1996). Technological change in the Early and Middle Stone Age at Kalambo Falls. *African Archaeological Review, 13*(3), 171–196.

Smith, R. A., & Wayland, E. J. (1923). *Some primitive stone implements from Uganda (Geological Survey of Uganda Occasional Paper No 1)*. Kampala: Geological survey of Uganda.

Smith, T. B., Wayne, R. K., Girman, D. J., & Bruford, M. W. (1997). A role for ecotones in generating forest biodiversity. *Science, 276*, 1855–1857.

Stout, D. (2002). Skill and cognition in stone tool production: An ethnographic case study from Irian Jaya. *Current Anthropology, 43* (5), 693–722.

Tappen, M. (1994). Bone weathering in the tropical rain forest. *Journal of Human Evolution, 21*, 667–673.

Taylor, D., Marchant, R., & Hamilton, A. C. (2001). A reanalysis and interpretation of palynological data from the Kalambo Falls prehistoric site. In J. D. Clark (Ed.), *Kalambo Falls prehistoric site* (Vol. III, pp. 66–81). Cambridge: Cambridge University Press.

Taylor, N. J. (2009). The role of Lupemban core-axes in the hominin dispersal into the Congo basin, Central Africa: A functional assessment by use-wear analysis. Ph.D. Dissertation, University of Liverpool.

Taylor, N. J. (2011). The origins of hunting and gathering in the Congo basin: A perspective on the Middle Stone Age Lupemban industry. *Before Farming, 2011/1*, article 6.

Tryon, C. A., & McBrearty, S. (2002). Tephrostratigraphy and the Acheulean to Middle Stone Age transition in the Kapthurin Formation, Kenya. *Journal of Human Evolution, 42*, 211–235.

van Moorsel, H. (1968). *Atlas de préhistoire de la Plaine de Kinshasa*. Kinshasa: Université Lovanium.

van Moorsel, H. (1970). Recherches préhistoriques au pays de l'entre-fleuves Lukenie-Kasaï. *Études d'Histoire Africaine, I*, 7–36.

van Noten, F. (1982). *The archaeology of Central Africa*. Graz: Akademische Druck und Verlagsanstalt.

van Noten, F. (1983). *Histoire archéologique du Rwanda* (Musèe Royal de L'Afrique Centrale, Annales No. 112). Tervuren: Musèe Royal de L'Afrique Centrale.

Van Peer, P., Fullagar, R., Stokes, S., Bailey, R. M., Moeyersons, J., & Steenhoudt, F., et al. (2003). The Early to Middle Stone Age transition and the emergence of modern human behaviour at the site 8-B-11, Sai Island, Sudan. *Journal of Human Evolution, 45*, 187–193.

Van Peer, P., Rots, V., & Vroomans, J.-M. (2004). A story of colourful diggers and grinders: The Sangoan and Lupemban at site 8-B-11. *Sai Island, Northern Sudan, Before Farming, 2004*(3), 1–28.

Van Peer, P., & Vermeersch, P. M. (2007). The place of northeast Africa in the early history of modern humans: New data and interpretations on the Middle Stone Age. In P. Mellars, K. Boyle, O. Bar-Yosef, & C. Stringer (Eds.), *Rethinking the human revolution* (pp. 187–198). Cambridge: McDonald Institute for Archaeological Research.

van Riet Lowe, C. (1952). *The Pleistocene geology and prehistory of Uganda: Part II*. England: Geological Survey of Uganda.

van Zinderen Bakker, E. M. (1969). The Pleistocene vegetation and climate of the basin. In J. D. Clark (Ed.), *Kalambo Falls prehistoric site* (Vol. I, pp. 57–84). the geology, palaeoecology and detailed stratigraphy of the excavations Cambridge: Cambridge University Press.

Villa, P., Soressi, M., Henshilwood, C. S., & Mourre, V. (2009). The Still Bay points of Blombos Cave (South Africa). *Journal of Archaeological Science, 36*, 441–460.

Wadley, L. (2007). Announcing a Still Bay industry at Sibudu Cave, South Africa. *Journal of Human Evolution, 52*, 681–689.

Wadley, L. (2010a). Compound-adhesive manufacture as a behavioral proxy for complex cognition in the Middle Stone Age. *Current Anthropology, 51*(S1), S111–S119.

Wadley, L. (2010b). Were snares and traps used in the Middle Stone Age and does is matter? A review and a case study from Sibudu, South Africa. *Journal of Human Evolution, 58*, 179–192.

Wai-Ogosu, B. (1973). Was there a Sangoan industry in West Africa? *West African Journal of Archaeology, 3*, 191–196.

Wayland, E. J. (1937). The Stone Age cultures of Uganda. *Man, 37* (67), 55–56.

Wendorf, F., & Schild, R. (1992). The Middle Palaeolithic of North Africa: A status report. In F. Klees & R. Kuper (Eds.), *New light on the northeast African past* (pp. 40–78). Koln: Heinrich-Barth Institut.

White, F. (1983). *The vegetation of Africa*. Paris: UNESCO.

Whitworth, T. (1965). Artifacts from Turkana, northern Kenya. *South African Archaeological Bulletin, 20*, 75–78.

Wynn, T. (2009). Hafted spears and the archaeology of mind. *Proceedings of the National Academy of Sciences, 106*, 9544–9545.

Yasuoka, H. (2006). Long-term foraging expeditions (molongo) among the Baka hunter-gatherers in the northwestern Congo basin, with special reference to the "Wild Yam Question". *Human Ecology, 34*, 275–296.

Yasuoka, H. (2009). The variety of forest vegetation in south eastern Cameroon, with special reference to the availability of wild yams for the forest hunter-gatherers. *African Study Monographs, 30*(2), 89–119.

[illegible], J. A., Johnson, [illegible] (20[illegible]). [illegible] *[illegible] African [illegible]* [illegible] Stone [illegible].

Schild, [illegible] [illegible] [illegible] [illegible] [illegible] J. [illegible] [illegible] [illegible]

[illegible] sites [illegible] and cultural [illegible] overview of the past [illegible] 000 years. Climate variability [illegible] in East Africa. *Palaeogeography, Palaeoclimatology, Palaeoecology*, 213, [illegible].

[illegible] R. [illegible] [illegible] from the Fayum. In [illegible] (Eds.), [illegible] Egypt during the last interglacial (pp. [illegible]). New York: Plenum.

Schmidt, [illegible] (2009). [illegible] [illegible] et [illegible] de [illegible] [illegible] [illegible] [illegible] [illegible] [illegible].

[illegible] [illegible], [illegible] J. [illegible], McKay, [illegible], Heilbron, [illegible] [illegible] [illegible] models for [illegible] sediment records using multiple dating approaches: An example from Lake Bosumtwi, Ghana. *Quaternary Geochronology*, [illegible].

[illegible] J. [illegible] (2000). The origins of [illegible] technology [illegible] and Europe. *Journal of [illegible] Science*, [illegible].

[illegible] J. [illegible] [illegible] (20[illegible]) [illegible] [illegible]

[illegible] [illegible] [illegible] [illegible] [illegible] [illegible] *Journal of [illegible]* [illegible]

[illegible] Van [illegible] [illegible] *[illegible] Africa* [illegible] [illegible] Royale de l'Afrique Centrale, Annales, No. [illegible]. Tervuren: Musée royal de l'Afrique Centrale.

Van Peer, P., Fullagar, R., Stokes, S., Bailey, R. M., Moeyersons, J., & [illegible] [illegible] and the emergence of modern human behaviour at site 8-B-11, Sai Island, Sudan. *Journal of Human Evolution*, 45, 187–193.

Van Peer, P., Rots, V., & Vroomans, J.-M. (2004). A story of colourful diggers and grinders: The Sangoan and Lupemban at site 8-B-11, Sai Island, Northern Sudan. *Before Farming*, 2004(3), 1–28.

Van Peer, P., & Vermeersch, P. M. (2007). The place of northeast Africa in the early history of modern humans: New data and interpretations on the Middle Stone Age. In P. Mellars, K. Boyle, O. Bar-Yosef, & C. Stringer (Eds.), *Rethinking the human revolution* (pp. 187–198). Cambridge: McDonald Institute for Archaeological Research.

Van Riet Lowe, C. (1952). *The Pleistocene geology and prehistory of Uganda. Part II.* England: Geological Survey of Uganda.

van Zinderen Bakker, E. M. (1969). The Pleistocene vegetation and climate of the basin. In J. D. Clark (Ed.), *Kalambo Falls prehistoric site* (Vol. I, pp. [illegible]). [illegible] [illegible] [illegible]

[illegible]

[illegible]

Chapter 16
The Later Pleistocene in the Northeastern Central African Rainforest

Els Cornelissen

Abstract The reconstruction of Late Pleistocene population dynamics in the northeastern Central African rainforest is hampered by the scanty though intriguing environmental, archaeological, and human fossil records. The few well documented and dated sites combined with undated surface finds in museum collections are examined for patterning in their spatial, temporal, and technological distribution. The results are ambiguous and point to both continuity and discontinuity in occupation of forested environments prior to, during, and after MIS 2. Particularly striking is the absence of quartz microlithic industries or any Later Stone Age (LSA) assemblages in the western part of the region. This may be due to lack of suitable raw materials or low visibility of quartz scatters encountered during informal surveys. At the same time the analysis suggests potentially interesting avenues for future research. These include, for example, the role that riverine systems might have played in the patterning of both prehistoric and extant genetic relationships. Utilization of archaeologically perishable bone in lithic poor regions may also account for apparently disjunctive archaeological distributions.

Keywords Democratic Republic of Congo (DRC) • Ituri • Lupemban • Quartz microlithic industries • Later Stone Age • Fishing

E. Cornelissen (✉)
Heritage Studies, Royal Museum for Central Africa, Leuvensesteenweg 13, 3080 Tervuren, Belgium
e-mail: els.cornelissen@africamuseum.be

Introduction: Does Genetic Reconstruction Provide an Alternative to the Archaeological Record in Equatorial Forests?

Reconstruction of population dynamics over the last 70–200 kyr in the area covered today with equatorial forests has increasingly been incorporated into the field of genetic studies. African hunter-gatherer populations are sampled and examined as "important reservoirs of human genetic diversity" (Hammer et al. 2011: 1). Also, because of the sparse paleontological and archaeological record, an understanding of the genetic diversity would have the potential to provide a more valid and reliable source for reconstructing human demographic history on the African continent as well as their expansion into other regions of the world (for an overview, see Campbell and Tishkoff 2010). This chapter will look in more detail at the archaeological, paleontological, and environmental record of the northeastern part of the equatorial forest in the Democratic Republic of Congo (DRC) (Fig. 16.1). The area includes the Ituri forest where present-day hunter-gatherer inhabitants are considered to be direct descendants from ancestral populations. The focus is on MIS 2 including the Last Glacial Maximum (LGM), which might represent a crucial moment in their genetic history. New perspectives that genetic studies have generated may provide an enriched context within which to reconsider archaeological data. After a brief survey of available environmental and historical information from the Ituri and surrounding rainforest, an overview of the dated archaeological sites as well as additional material from undated sites available from museum collections in the Royal Museum for Central Africa (Belgium) is presented.

The equatorial forest poses a particular challenge for interpreting genetic data, which is relevant to any attempt to understand the pattern of genetic diversity observed between present-day eastern and western Central African hunter-gatherers commonly referred to as Pygmies.

Sacha C. Jones and Brian A. Stewart (eds.), *Africa from MIS 6-2: Population Dynamics and Paleoenvironments*, Vertebrate Paleobiology and Paleoanthropology, DOI 10.1007/978-94-017-7520-5_16

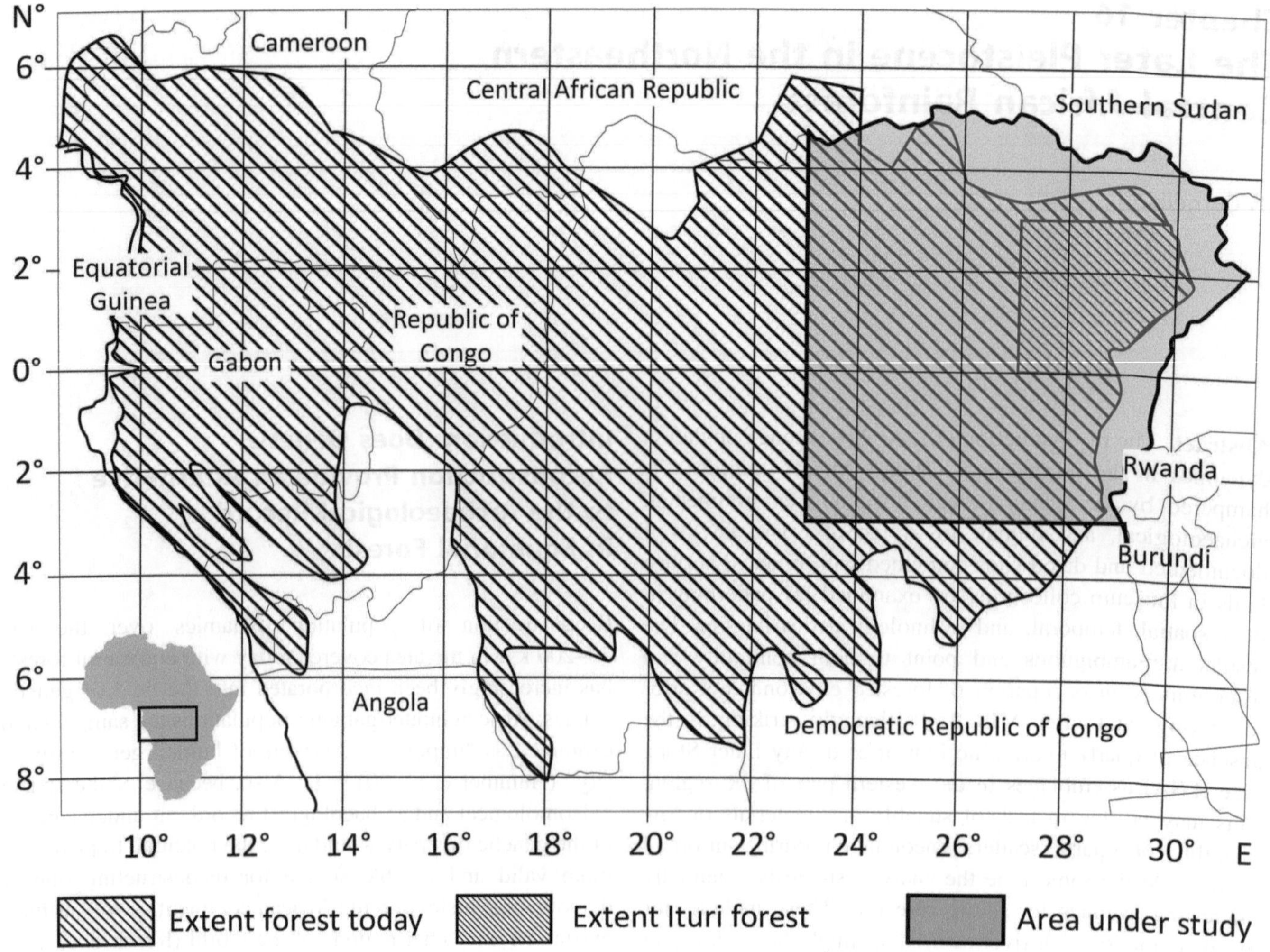

Fig. 16.1 Extent of Central African rainforest and location of study area

Among those living in the forests of western Atlantic Central Africa, various Pygmy groups are spread across Cameroon, Gabon, the northern Republic of Congo and the southern Central African Republic. Included in the eastern Pygmy group are the Mbuti who inhabit the Ituri forest in northeastern DRC. A comprehensive overview of the various groups of Central African hunter-gatherers is provided in Bahuchet (2012).

The current geographical distribution of Central African hunter-gatherers has been taken to represent continuity from prehistoric hunter-gatherers. This belief has been prevalent since the beginning of prehistoric research in Central Africa (de Maret 1990) and also occurs in more recent genetic studies (Cavalli-Sforza 1986). Consideration of time effects has become a prominent feature in genetic analyses (Destro-Bisol et al. 2004; Batini et al. 2007; Quintana-Murci et al. 2008; Patin et al. 2009; Verdu et al. 2009; Campbell and Tishkoff 2010; Batini et al. 2011a, b). These analyses suggest that the most recent common ancestor of Pygmy and farming populations would have lived from ~70–60 ka. Western and eastern Pygmies would subsequently have separated ~20 ka, and further genetic differentiation among the western Pygmies would have occurred within the last 2.8 kyr. The genetic variation suggestive of two branches among Central African hunter-gatherers, their short stature and subsistence strategy in equatorial forests have been explained in various ways (see, for instance, Batini et al. 2011a: 1100, 1107). Of interest here is the hypothesis that they are a fragmentation of an ancient population formerly spread over the entire rainforest. They became isolated into multiple groups when the forest itself fragmented into refugia during MIS 2 or the Last Glacial Maximum (LGM). The separated population continued to evolve into two related branches and, in the later Holocene, interacted with newly arriving farmer populations. Isolation may not have been total since after this, there seems to have been gene flow between western and eastern Pygmy groups (Patin et al. 2009). This may have been male mediated, and may explain the lack of shared recent mitochondrial ancestry among populations (Batini et al. 2011b).

The Mbuti population seems genetically relatively distant from other Pygmy and neighboring non-Pygmy populations. This feature may suggest that in Central Africa a now-extinct archaic form hybridized with modern humans (Hammer et al. 2011). Perhaps one of the reasons for observing more admixture among western Pygmies and neighboring farmer populations than among eastern Pygmies and their neighboring farmer populations, is the fact that a sufficient sample of comparative genetic material from northeastern DRC farmers is lacking (Batini et al. 2011a: 1108).

In sum, multiple genetic analyses indicate a complex set of evolutionary relationships among Central African hunter-gatherers. In certain scenarios, environmental history is invoked as one of the driving parameters to explain extant genetic patterns.

Evidence for Climatic-Induced Environmental Changes in the Northeastern Central African Forest at the End of the Pleistocene

The equatorial forest as we know it today has been shaped by both a long history of climatically induced environmental change and the impact of human presence and population pressure during more recent periods. The hypothesis that under intensified dry and cold conditions the forest retreated into refugia from which it expanded to recolonize areas under more clement (moist and warm) periods might, nevertheless, be an oversimplification of reality.

The Ituri Forest covers the upper watershed of the Ituri River between the equator and 3° N, and 27° E and 30° E longitude (Fig. 16.1, based on Hart et al. 1996: 546; Cerling et al. 2004: 6). Its current composition consists of two principal types of Guineo-Congolian rainforest: mixed moist semi-evergreen forest, and single-dominant evergreen forest dominated by *Gilbertiodendron dewevrei* (De Wild.) Léonard (Hart et al. 1996: 547). Swamp forest is mainly found along the Ituri River and its tributaries, which are part of the Congo basin river system. Human activity in the 20th century has resulted in the development of scattered patches of secondary forest of variable extent in the high forest (Terashima and Ichikawa 2003).

Sedimentological, pedological and geochemical studies in the Walikale and Osokari regions (Runge 2001) (Fig. 16.2) offer a deeper insight into late Quaternary landscape history and the dynamic history of forests in northeastern DRC. The results point to climate-induced changes such as retreat and fragmentation of tropical forests before (36–28 ka), during the (21–18 ka) interval and after (13–12 ka), the LGM (dating of LGM after Runge 2001: 29–30). During these cooler and drier periods the vegetation cover was most likely a savanna woodland mosaic with extended gallery forest along the rivers. This lends support to the concept of a fluvial refuge in the eastern DRC (Runge 2001). Other environmental data derive from archaeological sites. The Matupi cave yielded fauna throughout 40 kyr of occupation. For the Late Pleistocene, part of the stratigraphic sequence the biotope of hunted game was an open environment with access to more wooded and forested areas (Van Neer 1989); at some time between 14 and 3 ka, a shift to closed environments occurred. Pollen of *Gramineae* (Van Noten 1982: 36) were recovered at a depth of between 25 and 50 cm in Holocene deposits, and between 500 and 525 cm below the layers that were dated and beyond the range of radiocarbon dating.[1] In the Holocene section, grass pollen is accompanied by ferns found in moist forests or in gallery forests. Brook et al. (1990) also analyzed pollen embedded in cave speleothems at Matupi. They concluded that these pollens reflected the existence of a grass savanna environment in this currently densely forested area between 22 and 12 ka. Further evidence for the evolution of the Ituri Forest is provided by the analyses of phytoliths from three archaeological rock shelter sites north of Epulu (Mercader et al. 2000) (Fig. 16.2). These show that the late glacial forests may have had a more open canopy than today's counterpart, and may have been reduced – though not enough to have converted tropical forest to grassland. The phytoliths also indicate that the area was densely forested throughout the Holocene. Yet during the Holocene, changes in floristic composition are also observed. Although charcoal records documenting vegetation history over the last 4 kyr from Epulu (Hart et al. 1996; Hart 2001) (Fig. 16.2) reveal no major change in vegetation type – dominant species are all typical of moist tropical forest – the floristic composition changed radically. *Gilbertiodendron dewevrei*, which now occur in extensive monodominant stands, is absent in charcoal samples; whereas *Lebruniodendron leptanthum*, which is most frequent in charcoal, is completely absent from today's forest. This might be linked to a "process of opportunistic replacements of species that are ecological equivalents" (Hart et al. 1996: 558).

The available data, although scarce, show that during the Late Pleistocene and Holocene the rainforest in northeastern

[1]File nr 1421 of the *Archives Préhistoire* at the Royal Museum for Central Africa contains an exchange of letters dating from 1987 between F. Van Noten of the Museum and J. Huxtable from the Research Laboratory for Archaeology and the History of Art, Oxford University. A preliminary TL result in excess of 80 ka on burnt stone from a depth of −525 cm below surface is mentioned, but there is no formal communication or publication of the result.

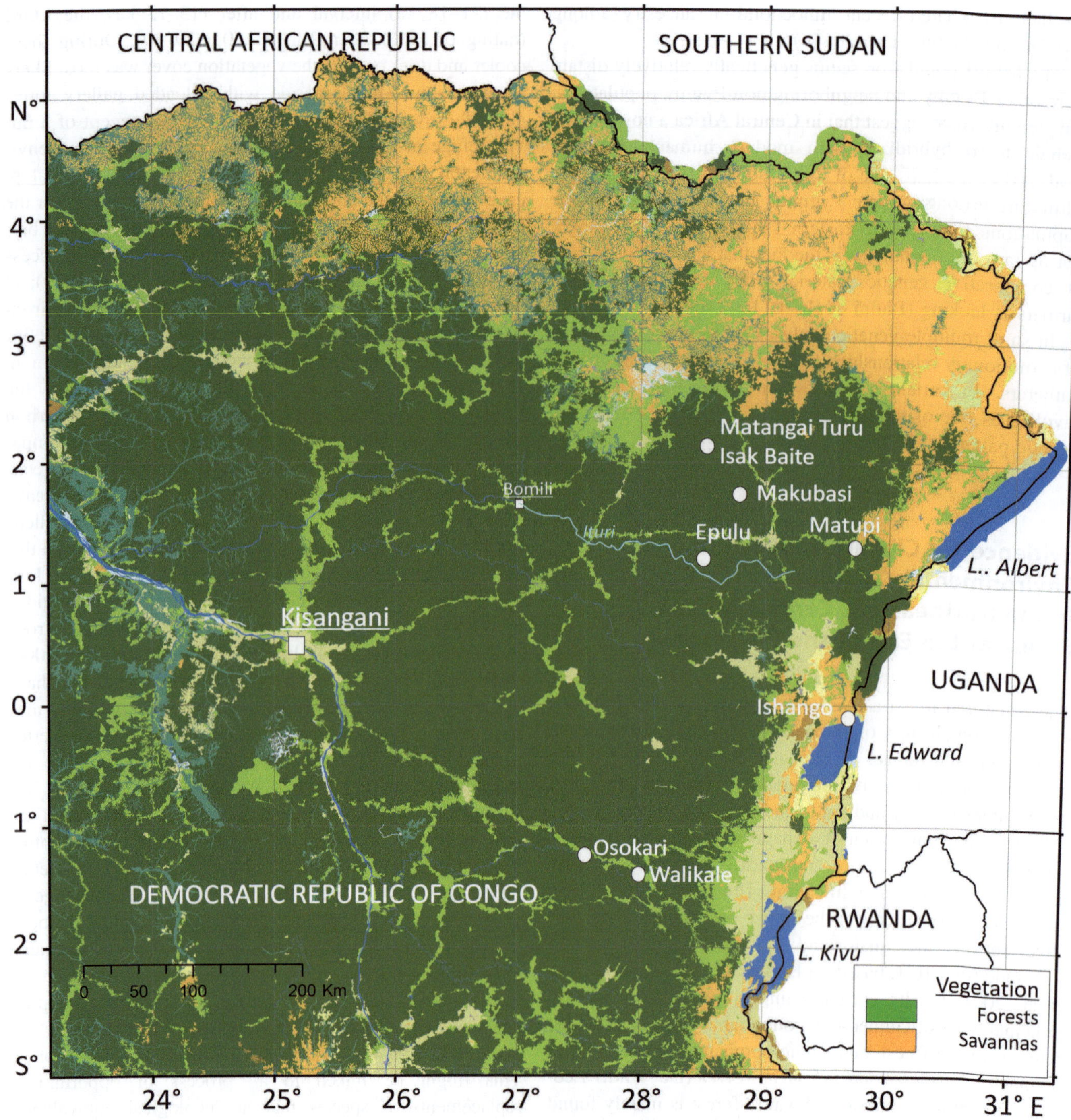

Fig. 16.2 Locations of sites with paleoenvironmental evidence and places cited in the text

DRC underwent multiple environmental changes. It is reasonable to suppose that subsistence strategies during the Late Pleistocene and early Holocene likewise incorporated a degree of flexibility in order to respond to changes in resource distribution and accessibility. The following analysis of the archaeological record examines the questions of if, and to what extent, material culture reflects adaptability to environmental changes through time.

Late Pleistocene Archaeological and Human Fossil Record of Northeastern DRC

As a first step toward determining the applicability of site distribution maps for reconstructing past hunter-gatherer behavior in the northeastern Congoese forest during and after

MIS 2, data drawn from dated and undated published records and unpublished information available from the collections stored at the Royal Museum for Central Africa (RMCA) was synthesized. Except for sites published by J. Mercader (e.g., Mercader 2002), and part of those surveyed by J. de Heinzelin in the Semliki Valley, all sites from northeastern DRC are represented in the RMCA collections. The older materials in the collection result from fortuitous and random sampling and were not driven by any research paradigm. Information in the accompanying archives indicates that items were collected during construction work, along road cuts, incidental to plantation work, geological survey, and mining activities. Some material was purchased from local inhabitants (on the history of the RMCA collections, see also de Maret 1990). Exact locations are often impossible to determine because of the vague descriptions of the circumstances in which they were found. These find spots are mapped in the center of the town or lake nearest to where the artifacts were collected. Sometimes, however, collection areas are indicated very precisely on hand drawn maps. Many localities have several spellings of their name. Checking and cross-checking of various sources of information is time-consuming but necessary; however, a certain degree of uncertainty in the precise location of these sites will always remain.

The resulting archaeological maps are a "work in progress." Yet, despite all caveats, the spatial distributions have the potential to reveal regional and more general patterns. They may also provide insight into the effect of environment on site visibility and can be used to situate dated sites into a broader regional context. Ongoing human conflict, health and logistical issues make this a high-risk area for research, which largely explains the lack of recent field data. For these reasons, museum collections, although far from perfect, are valuable as they may provide a basis for a broad scale assessment of site and artifact distribution patterns; however, they will never produce the fine-grained coverage that exists for other regions of the continent.

Currently the museum collections contain records of 309 find spots, located east of longitude 23° E, and between latitudes 3° S and 5° N (Fig. 16.3). Of these, 280 (91%) yielded stone artifacts, twice together with pottery and twice iron objects, which were reported from the same locality. More than half (160) of the entries consist of a single artifact find (Fig. 16.4). 111 sites contain between one and 100 stone artifacts. At nine sites, over 100 artifacts were collected. Unless otherwise specified in the following descriptions, none of these artifacts were found during either controlled surface collection or excavation.

For the period under consideration, it is possible to define a number of series of sites on the basis of both dated records (see below) from the Semliki Valley, and from the eastern edge of the forest. The first series comprise those with lithic industries with no bifacial component, and with MSA and LSA affiliation, the latter utilizing quartz as the preferred raw material. A further 15 undated sites can be added to this core group of dated sites. The second grouping of undated sites, combined with dated material from outside the region (Taylor 2016), point to the presence of material with a Lupemban affinity at a minimum of 14 localities. Bifacial technology is also present at an additional ten sites. The remainder of the localities that cannot be encompassed in either of these traditions will be discussed separately. Their spatial distribution may provide insights into both site visibility and feasibility of occupation in the northeastern part of today's Central African rainforest during this period.

Quartz Industries of MSA and LSA Affinity

Open-Air Sites in the Semliki Valley

Securely dated archaeological sites are rare in Central Africa, for both the Late Pleistocene (Cornelissen 2002; Mercader 2002) as well as the Holocene in the rainforest area today (Figs. 16.3, 16.4 and 16.6).

Among Pleistocene sites, those from Katanda (Brooks et al. 1995; Yellen et al. 1995; Yellen 1996, 1998), dated to 80–70 ka (Feathers and Migliorini 2001), are unique because of the presence of carefully worked bone in a MSA context. Archaeological material was found incorporated at the base of the Katanda sands overlying a truncated paleosol (de Heinzelin and Verniers 1996). The barbed bone points belong to a specific African tradition: the barbs are uniserial and enclosed, and relatively large, wide, and thick (Yellen 1998). The vertically compact "pavement" at Katanda 9 contained over 8000 lithics and 7000 mammal and fish remains in an area of c. 35 m^2. The lithic industry, primarily on quartz and quartzite, lacks formal stone tools that could be considered characteristic for either MSA or LSA. The presence of grindstones may hint at plant processing. Faunal evidence suggests a subsistence adaptation, almost indistinguishable from the later Ishango sites (Peters 1990; Mercader and Brooks 2001), jointly focusing on savanna and riverine fauna. Fish are abundant and the most common species, *Clarias* and *Synodontis,* are large – some exceed 30 kg in weight – adult, slow moving, bottom dwelling catfish, which enter shallow water to spawn. They were probably captured at this time when they are most vulnerable (Stewart 1989; Brooks et al. 1995; Yellen 1996). This understanding of fish behavior and the associated capture technology, similar to that of the Late Pleistocene Ishango sites, is suggestive of fish procurement rather than fishing. The behavior incorporates the use of close-range thrusting weapons, which are effective in both forest and in aquatic settings (Stewart 1989: 232–

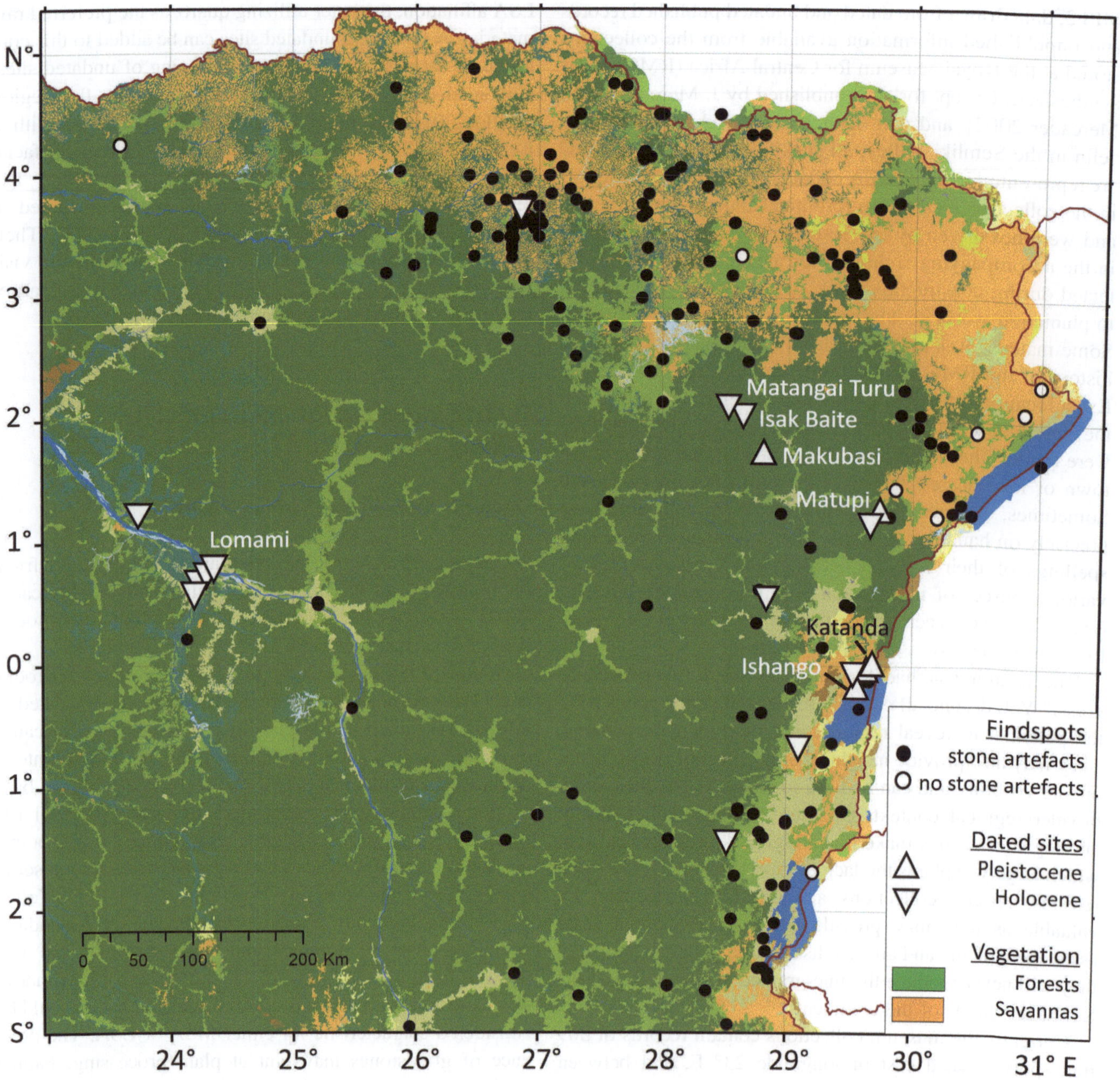

Fig. 16.3 Lithic component reported from dated sites and undated find spots in published and unpublished records from the collections of the Royal Museum of Central Africa

233). The fish present in the early Ishango assemblages (e.g., *Protopterus*, *Barbus* and *Lates*) are, again, most easily caught in low receding water conditions.

The late Pleistocene quartz microlithic assemblages at site Ishango 11 are dated to c. 25–20 ka. (de Heinzelin 1957; Brooks and Smith 1987; Boaz et al. 1990; Brooks and Robertshaw 1990; Mercader and Brooks 2001). Here, the three successive Ishangan levels contain material associated with every stage of the manufacture of various types of tools, such as uniserial and biserial barbed harpoons (Brooks and Smith 1987; Yellen 1996). This is followed by a gap in the archaeological record for the period between the end of the LGM and the next quartz microlithic industries, which post-date 7 ka. The bone component, so well developed in the Pleistocene levels, is sporadic in these later levels. For the entire region, only one polished implement "found in obscure stratigraphic circumstances" (de Heinzelin 1957: 78) is mentioned. Flat-based undecorated pottery, identified at the time as "non-Bantu" (de Heinzelin 1957: 79), was found in association with the later, post 7 ka, quartz microlithic industry.

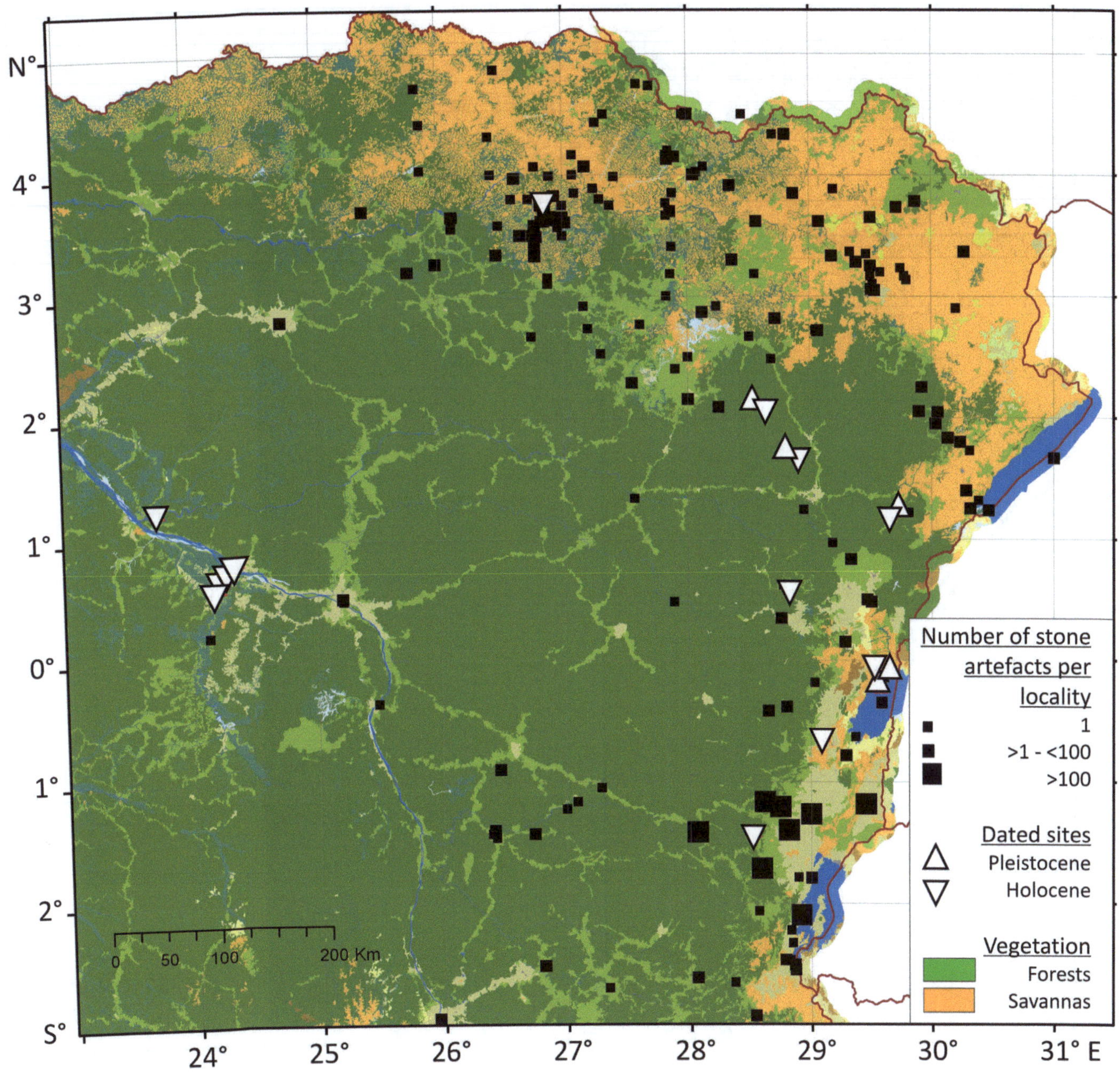

Fig. 16.4 Number of artifacts at undated find spots

At Katanda and Ishango, the techniques used to acquire fish were limited and might not have required an in-depth understanding of the fish themselves or their aquatic surroundings. Stewart (1989) considers these techniques as evidence of fish procurement rather than fishing. The importance of fish utilization and the significance of riverine contexts are, nevertheless, clearly attested. The sites show that by at least 70 ka hunter-gatherers on the eastern edge of the rainforest recognized bone as an adequate alternative for stone, and were successfully experimenting with aquatic food resources. Subsistence strategies relating to an aquatic environment may be important for exploring and exploiting an environment such as the rainforest with swamps and an abundant network of waterways (Fig. 16.5). Waterways provide an important means of access and communication. They thus increase the potential geographic scope and the ease with which food and raw materials may be obtained. The Katanda and Ishango sites indicate that their inhabitants were capable and had the necessary background to exploit water related resources.

The range of artifacts related to fishing and fish procurement in tropical freshwater Africa is well documented in ethnographic

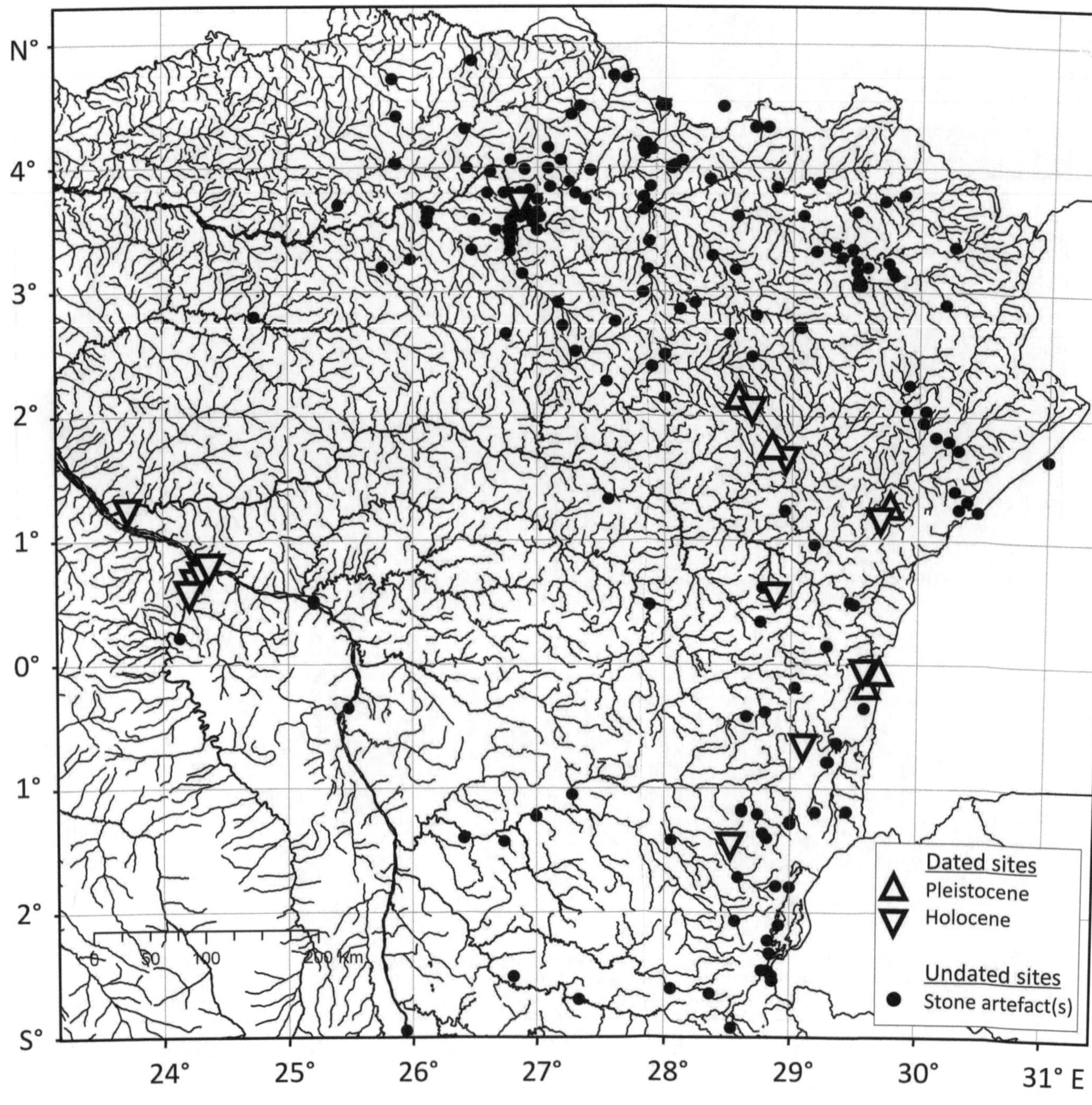

Fig. 16.5 Archaeological sites and present-day river network

records. These include: thrust baskets, stationary basket traps, weirs, small dip and scoop nets, counts nets, weights, floats, spears or harpoons, hooks or gorges and poison (Stewart 1989: 66–68). Unfortunately, "the most notable aspect of traditional technology is its potential lack of visibility in the archaeological record. Virtually all implements are constructed of grass, fibers, reeds or wood, none of which preserve except in exceptional circumstances; only lithic or bone spear/harpoon heads or hooks have good survival potential" (Stewart 1989: 77).

Quartz Microlithic Industries at the Rock Shelters of Matupi and the Ituri

Matupi Cave, located on Mount Hoyo (Van Noten 1977, 1982), yielded a sequence of lithics manufactured from milky veined quartz. From the deepest levels of a 5 m test pit, a nondescript lithic assemblage made from quartz was discovered. The industry has been dated to between 40 and 20 ka and may not be fully microlithic (Van Noten 1982;

Mercader and Brooks 2001). However, younger material (from 20 ka onwards) is microlithic. The macrolithic component includes: grindstones, hammerstones, anvils, utilized flakes, hollow scrapers and rare chisels. Raw materials here are mainly quartzite, sandstones or schist. Matupi is the only site in the area to have yielded a fragment of a decorated bored stone, or "kwé", from a dated context of ~20 ka. There is no mention of polished implements in the cave. Although not as abundant as in the Pleistocene levels, LSA geometric microliths, as well as multiple core types, are also present in the upper 30 cm, which post dates 3 ka. A few fragmentary bone tools also come from these levels (Van Noten 1982). As discussed in relation to the history of the forest, the occupants of the Matupi cave hunted in an open savanna with access to gallery forests. The environment became more forested at some time between 12 and 3 ka, but unfortunately there is a hiatus in sedimentation for this period, which prevents more precise dating of this event (Van Neer 1989: 82). Fish remains are very rare; a pectoral spine of *Clarias* sp. and a dorsal fin of *Tilapia* in the upper levels may indicate either infrequent fishing or poor bone preservation (Van Neer 1989: 78).

Phytolith analysis firmly situates the occupation of the Ituri sites in a rainforest environment throughout the Late Pleistocene and Holocene (Mercader et al. 2001). The upper levels at all Ituri sites contain unstandardized quartz shatter, probably attributable to the LSA, macrobotanical remains of *Canarium* and some *Elaïs*, faunal remains, later Iron Age pottery and sporadic iron tools (Mercader and Brooks 2001). Seven AMS radiocarbon dates indicate ages of between 1080 ± 41 ^{14}C BP and 715 ± 45 ^{14}C BP for these upper layers at various sites. The underlying units contain abundant LSA stone artifacts, small amounts of charcoal, phytoliths and *Canarium* endocarps, but no fauna. Five AMS radiocarbon dates from five sites set the oldest occurrence of these assemblages to c. 18 ka (Mercader and Brooks 2001: Table 1), 10, 6 and 3 ka.

A detailed analysis of lithic assemblages was conducted for three dated sites: Matangai Turu Northwest, Isak Baite Southwest, and Makubasi Southeast (Mercader and Brooks 2001). Except for the absence of centripetal debitage and pebble tools in the Pleistocene assemblages, reduction strategies yielded primarily simple debitage and no evidence of bipolar percussion. Formal tools, such as core- and side-scrapers, very rare geometrics, points and perforators, are similar throughout the entire Pleistocene–Holocene sequence at the Ituri sites.

Quartz Microlithic Industries from Undated Sites

Additional information on quartz microlithic industries comes from assemblages collected from the surface, road cuts and an excavation on the Kimaneza Hill in the region of the Mokoto lakes (Van Noten 1968b, 1982) (Fig. 16.6a). Here, segments, triangles, trapezoids and core scrapers were found together with flake blades, bladelets, and disc cores below 10 to 70 cm of black humus soil in a 10–20 cm thick reddish layer, which rests on a 5 cm thick undated volcanic ash. The sites at the southern edge of the area under study (Fig. 16.6a) were sampled in 1935–1936 by D. Gillès de Pelichy, a missionary of the congregation of White Fathers who had been charged to establish a mission post to the north of Lake Kivu in 1912. His surveys document localities with more than 100 artifacts (Fig. 16.4), as well as dense quartz occurrences to the west and south of the Mokoto lakes and the two sites near Butembo. He also collected artifacts in quartz at Mutibu, which lies close to the dated sites in the Ituri Forest. In fact, Gillès de Pelichy alone is responsible for the discovery of 13 of 18 sites under consideration here. wa Bitanko Kamuanga (1985–1986),[2] who analyzed the museum collections, has classified all of them as belonging to the LSA based on the presence of backed pieces, microliths and bipolar cores (cf. Bequaert 1945). Finally, an assemblage of 21 quartz artifacts that are possibly LSA comes from Tshampi in the Rwindi Valley.

The general distribution of dated and undated sites yielding quartz microlithic industries shows a concentration in the eastern half of the study area; in the western half (west of 28° E) no sites with quartz microlithic industries were reported.

Sites of Lupemban Affinity

Using *fossils directeurs* is neither the most efficient nor most reliable way of approaching the archaeological record; however, for a first assessment, it may suffice. "Lupemban" is often associated with finely retouched, long, dagger-like lanceolates up to 40 cm long, and with leaf-shaped, elongated, bifacially trimmed points (Taylor 2011, 2016). In this analysis, sites containing such lanceolates are termed "Lupemban" and, perhaps artificially, are analyzed separately from sites that lack lanceolates but that contain bifacially trimmed implements such as core-axes or handaxes (Fig. 16.6b).

In his survey of the Stone Age in the East of Central Africa, Van Noten (1982: 30–31) briefly describes the finds from the confluence of the Lodjo and Djitofu (Jitofu) rivers and from the Lutunguru and Angumu river valleys in dense equatorial forest. These sites deserve some special attention.

[2]The author has published as Muya, D. and as wa Bitanko Kamwanga, M. due to political changes in name giving policy in the Democratic Republic of Congo.

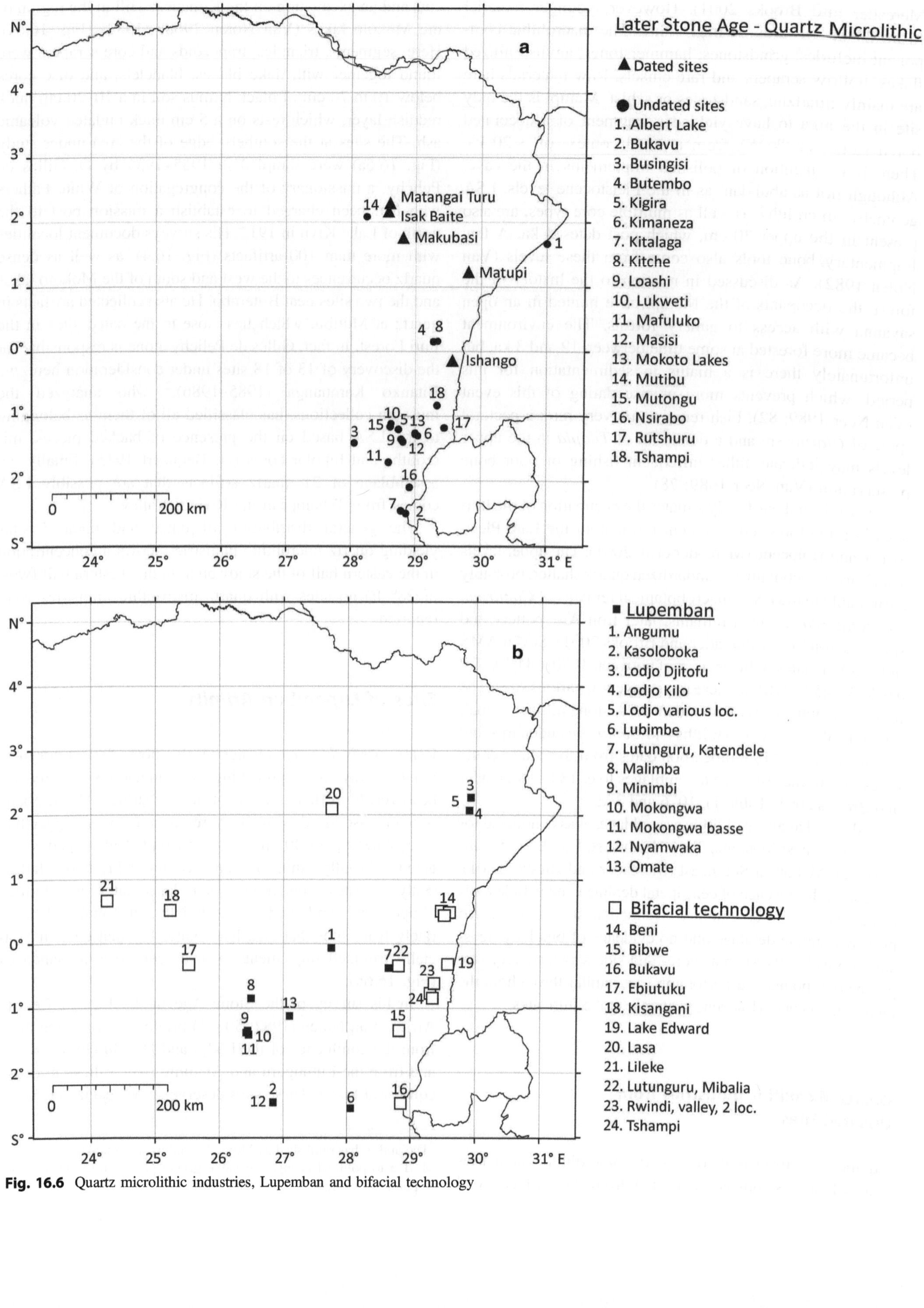

Fig. 16.6 Quartz microlithic industries, Lupemban and bifacial technology

The most northern occurence is in the mining area between Kilo and Moto along the Lodjo and Lindi rivers and their tributaries. Implements of Lupemban affinity were found here during mining activities and geological surveys. M. Piret, a technical engineer at the *Société des Mines d'Or de Kilo Moto*, carefully collected prehistoric artifacts in the auriferous fluvial deposits at the confluence of the Djitofu and Lodjo rivers (the latter, a tributary of the Ituri River). The artifacts are a mixture of cobbles that have polished cutting edges, perforated stones, hammerstones, various unpolished axes, laurel and oval shaped handaxes, lanceolates, bifacially flaked points and blades (De Haene 1949; de Heinzelin 1948). When Van Noten (1982) visited the site of Djitofu in 1973, he observed that the stratigraphy of the valley floors was totally obscured where the gravel had been dug during gold mining. He collected previously undocumented Levallois cores and flakes at the dumpsites. There are a few smaller assemblages from other localities in the Kilo Moto mining area. According to the description of the stratigraphy at Djitofu by its discoverer, M. Piret (cited in De Haene 1949), all artifacts were found in an approximately 10 cm thick layer of coarse auriferous gravel on top of the riverbed bedrock. This was overlain by 20–30 cm of less coarse gravel, covered by a 1.5 m thick layer of river sediment, upon which rested another 1.5–3 m of soil. Raw materials include: quartz, red veined quartz, diorite and schist, all of which are locally available (De Haene 1949). Amphibolite (de Heinzelin 1948), later identified as dolerite, was used for lanceolates and points.

The Lutunguru Valley is located west of Lake Edward. "G5", a tributary of the Katendele near Lutunguru, has yielded a finely finished lanceolate and a point made on a black volcanic rock that is also described as *silex* or flint in the literature. They were found during mining exploitation of river terraces, below a 3–4 m thick layer of sterile sediment (Passau 1946: 646–647).

A leaf-shaped and an elongated bifacial point, manufactured from local sandstone and quartzite respectively, were found when washing gravel from a small affluent of the Angumu (Bequaert 1943).

J. Palgen (1949), a mining engineer, collected a number of prehistoric stone tools during mining activities in the tin-ore rich and auriferous deposits of Maniema at various locations in today's equatorial forest. The museum collections contain a sample of 64 implements found in a river terrace of the Lubimbe near Kananga and from six other localities: two near Mokongwa, as well as others at Kasoloboka, Malimba, Minimbi and Omate (Fig. 16.6b).

Another J. Palgen (1950: 553) collection comes from Nyamwaka, where artifacts were found at a depth of 2–4 m in gravel, near the base at the interface with the bedrock. This is almost identical to the stratigraphic position of sites further north in the Kilo-Moto area. The condition of the artifacts varies from mint to abraded. Raw materials consist of regional quartzite varieties, some quartz, metamorphic schist and diorite. Of the more than 200 tools that were collected, 128 are described as points; 24 of these points measure between 40 and 44 cm. Palgen's (1950) drawings of these tools show a clear resemblance with Lupemban elongated lanceolates, and the author sees an affiliation with the Djokocian (Palgen 1950: 555), which is another name for Lupemban (Cahen 1978).

All of these Lupemban artifacts were found in a secondary context at a considerable depth below the present-day surface, and were exposed as a result of mining activities. The published descriptions of sites in northeastern DRC are not conclusive regarding the exact nature of their stratigraphic contexts. This is not at all surprising since even in controlled field situations, it is difficult to identify whether fluvial processes or bioturbation are responsible for stone line formation (see, for example, Mercader et al. 2002; Brown et al. 2004). Also, identifying the natural process that caused the material to accumulate in or above so-called stone lines might not be relevant because the typological composition of the archaeological assemblage makes it clear that in many cases several periods are indeed mixed together. Compared to the quartz microlithic sites, the Lupemban is distributed further west into the present-day forest (Fig. 16.6b). However, except for possible Holocene bifacial occurrences (see below), they do not extend beyond 26° E.

Sites Presenting Bifacial Technology, Pleistocene or Holocene

Further to the west in the area under consideration (Fig. 16.6b), on the banks of the Lomami River, stone artifacts at Lileke were tentatively attributed to the Lupemban or Tshitolian (Van Noten 1982: 30). Yet, during fieldwork in 2010, similar artifacts in the same fine-grained sandstone and chert were firmly associated with pottery (Smith et al. 2010). Another comparable small Tshitolian-like bifacial implement in chert was found to the south of Kisangani where the road to Ubundu crosses the Ebiutuku River. A rough-out of a possible bifacial implement or the edge of a radial core in fine-grained sandstone was recovered in the area around Kisangani. These sites are, in fact, the only instances of undoubtedly flaked stone from the western half of the area under consideration.

Continuing on west-east and north-south axes, a small handaxe-like quartz implement, measuring 6.4 cm, was recovered together with quartz debris from the surface at Lasa, in the Ituri Forest, by P. Schebesta (Lebzelter 1933).

Other bifacially trimmed implements were found around Beni. In the town center, a large ovate handaxe and core-axe were collected at the surface together with flaking debris. Various bifacial implements, including a core-axe and oval handaxe, and flaking debris in volcanic rock were collected by the geologist J. Lepersonne in the silts at the confluence of the Mibalia and the Lutunguru rivers.

A rough-out of a biface, a pick-core and two massive chunks in vein quartz were collected from an exposure on the southern edge of Lake Albert. The northern and southern shores of Lake Edward, the lower and upper Semliki valley as well as the southern shore of Lake Albert have been surveyed intensively. They yielded several indications for Early to Middle Stone Age artifacts of a probable Middle Pleistocene age (de Heinzelin 1961), as well as the even older Senga 5 site (Ludwig and Harris 1998), which may be in a reworked secondary context (de Heinzelin and Verniers 1996). Finds of a probable Late Pleistocene age were also reported from the valley of the Rwindi River and its outlet on Lake Edward. At Tshampi, they include a slender bifacially trimmed elongated point in fine-grained sandstone. Another bifacially trimmed point in quartz and a rough-out of a large biface in a heterogeneous rock were also recovered from the Rwindi area. Of the two bifacially trimmed implements with a square section from Bibwe, one is made from quartz and the other from quartzite. In the vicinity of Bukavu, a 5.5 cm long bifacial point in quartz was found together with a fragment of quartz at 20 cm below the surface.

Sites of Uncertain Affiliation

Flaked stone artifacts such as picks, chunks and flakes, utilized artifacts such as grindstones and hammerstones, polished implements and kwés manufactured from various raw materials were also collected. Although not indicative of a specific technique or period or industry, they attest to the presence of stone and hence potential for lithic production.

The spatial distribution of flaked material (Fig. 16.7a) indicates clusters along the eastern edge of the study area. West of 28° E and north of 1° S a total of four sites are located in the vicinity of Kisangani. Each of these yielded a single artifact, yet the heavily rolled artifacts may in fact be abraded natural stone.

A common feature in the spatial distribution of grindstones and hammerstones (Fig. 16.7b), kwés (Fig. 16.7c) and polished implements (Fig. 16.7d) is their absence west of 28° E and north of 2° N. The cluster of kwés northeast of Matupi lies in the area of the Kilo-Moto mining company; some of them were found in the same stratigraphic position as the Lupemban artifacts.

Most of the isolated finds consist of polished axes (104 out of 160 occurrences) (compare Figs. 16.4 and 16.7d). In 22 cases, two axes come from the same locality; while in one, 11 axes were reported together. In some instances hammerstones, grindstones, kwés and iron objects are spatially associated with polished implements. Polished axes from the northeastern area have been described as Uelian, an industry that has been analyzed in detail (Van Noten 1968a). The distribution of the Uelian (including various types of axes, adzes, chisels, polished blocks, natural forms with a polished cutting edge, polished blocks and a range of miscellaneous partially and completely polished tools) is concentrated outside the northern edge of the present-day forest (Fig. 16.7d). The southern outlier in this spatial distribution is a grounded axe-like implement in schist. Other items of stone, bone or pottery are totally absent.

Human Remains

Skeletal data from past human populations in the Ituri Forest and surrounding environments is indeed scant. At the Ishango sites, preservation in the strata that date from 20 to 25 ka is comparatively excellent. These materials are currently being restudied. The Ishango population has been interpreted as not showing direct links with any modern population in the area including hunter-gatherers. They stand out in their robustness in comparison to modern humans (Twiesselmann 1958; Boaz et al. 1990; Crevecoeur et al. 2010a, b). As noted elsewhere (Bouchneba and Crevecoeur 2009), Late Pleistocene modern human variation may have been greater than that among recent populations (see also Hammer et al. 2011). Other skeletal material of possible interest to more recent population dynamics, and hence relevant to discussions of discontinuity versus continuity in hunter-gatherer communities after MIS 2, are an Iron Age skeleton excavated at Ishango (Kanimba 1990) and a partially excavated burial in the rock shelter of Matangai Turu Northwest (Mercader et al. 2001). The remains from Matangai Turu Northwest date to the thirteenth century AD. Though anthropometric data are not conclusive in this case, when combined with the associated material culture they point to a hunter-gatherer affiliation. Some human bone has also been recovered from the upper layers of Matupi (Van Neer 1989: 25).

In sum, the scarce human skeletal remains do not argue in favor of direct continuity between the hunter-gatherers of today and in the inhabitants of the fringes of the forest between 20 and 25 ka.

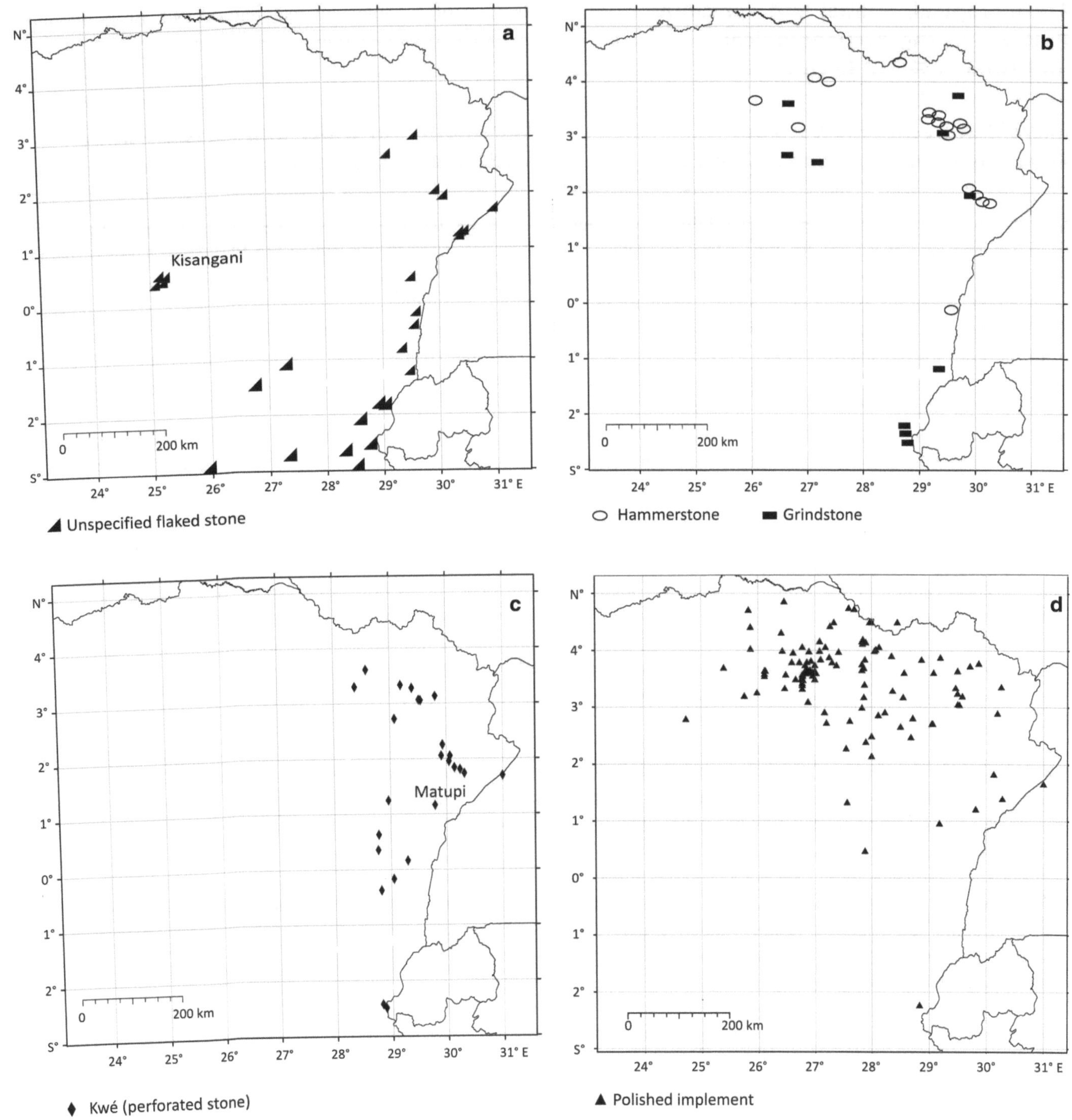

Fig. 16.7 Sites of unspecified affiliation

Technological, Spatial and Temporal Patterns in the Archaeological and Human Fossil Record

Microlithic Technology: A Poor Environmental and Temporal Marker

In a comparative analysis of quartz microlithic assemblages from Ishango and three of the Ituri sites, Mercader and Brooks (2001: 213–214) conclude that technological choices reflected in the lithic assemblages of the savannas in the Semliki Valley and those of the Ituri Forest are quite similar from the Late Pleistocene onwards. As has been observed elsewhere (Cornelissen 2003), LSA hunter-gatherers selected quartz as the preferential raw material. Thus, stone technology appears to be a very poor environmental indicator. As J. Mercader (2002: 123) observes, "neither glacial nor interglacial forests" seem to have been "a cultural, economic, or physical barrier to African foraging groups of the late Palaeolithic period." Quartz microlithic industries not only became an intentional technological choice in the Late Pleistocene, but they also exhibit a consistent longevity. Despite the chronological gap in radiocarbon dates between 15 and 7 ka at Ishango, and between 12 and 3 ka Matupi with a concomitant hiatus in sedimentation, the later Holocene lithic assemblages are also quartz based and microlithic. All Ituri rock shelters investigated yielded an unstandardized quartz shatter of probable LSA affiliation postdating 3 ka (Mercader and Brooks 2001). Due to the recurrent occupation in the restricted area of rock shelters, the possibility of admixture of later and earlier material at Matupi and Ituri rock shelters cannot be excluded. However, open-air sites like Ishango are less subject to the impact of reoccupation of the same space and they show the same pattern of continued quartz based microlithic technology.

From the quartz industries documented in the dated open-air sites in the Semliki Valley and in the cave sites of Matupi and the Ituri Forest, a dense pattern of quartz assemblages would be expected on surface collection maps; however, this is not the case. The explanation for this apparent anomaly may be due to collector's bias since the visibility of archaeological quartz to the inexperienced eye is low compared to for instance lanceolates or polished implements. Both de Pelichy and J. Mercader were aware of the archaeological relevance of quartz and located numerous sites. J. Mercader and coworkers (Mercader et al. 2001) report on 50 archaeological sites in just 8.5 km^2 in the Ituri Forest.

The absence of flaked quartz and of microlithic assemblages in the area between 23° E and 28° E, and 3° S and 2° N may thus be result of either surveying methods, absence of suitable raw material or lack of technological skills for its exploitation.

Lupemban: An Older and Possibly More Widespread Substratum

Given the fact that the sites of Katanda, Ishango, Matupi and Ituri are dated to the end of the Pleistocene and yielded no Lupemban artifacts, the Lupemban occurrences at Kilo Moto, Lutunguru and northern Maniema may represent a more ancient and more widely spread hunter-gatherers habitation of the region. There are, however, serious difficulties in associating Lupemban technology with dates (Taylor 2011, 2016).

In southeastern Central Africa, Lupemban artifacts represent an ancient industry that did not persist until the end of the Pleistocene. Yet, in southwestern and Atlantic Central Africa, the bifacial component of the Pleistocene Lupemban continued well into the more recent Lupembo-Tshitolian and Tshitolian industries, which are of Late Pleistocene and of Holocene age (Cornelissen 2002, 2013). Holocene bifacial technology, reminiscent of these western Central African sites, is also attested along the Lomami River in northeastern DRC. This requires further investigation.

The stratigraphic position of Lupemban artifacts near the top of the bedrock in fluvial sediments below more than 3 m of deposit is not of relevance, since iron objects as well as polished implements and kwés have been found in stratigraphic association. Noteworthy is the absence, in this context, of a microlithic component in quartz or any other material. How the Lupemban and other artifacts got into their current stratigraphic position is unclear. This may be explained by torrential and massive erosion of previously separated sites, which is unlikely to occur now as the area is under dense forest cover.

Site Invisibility Due to Absence of Suitable Raw Materials for Stone Artifact Production

Though there is no direct correlation between lithic technology, chronology, environment or the identity of the makers, site distribution may be relevant for an assessment of raw material availability. A consistent phenomenon observed in many of the spatial distribution patterns of various technological and typological features is the absence of sites in the southwestern part of the study area. In fact, west of the arbitrary dotted line in Fig. 16.8, localities with

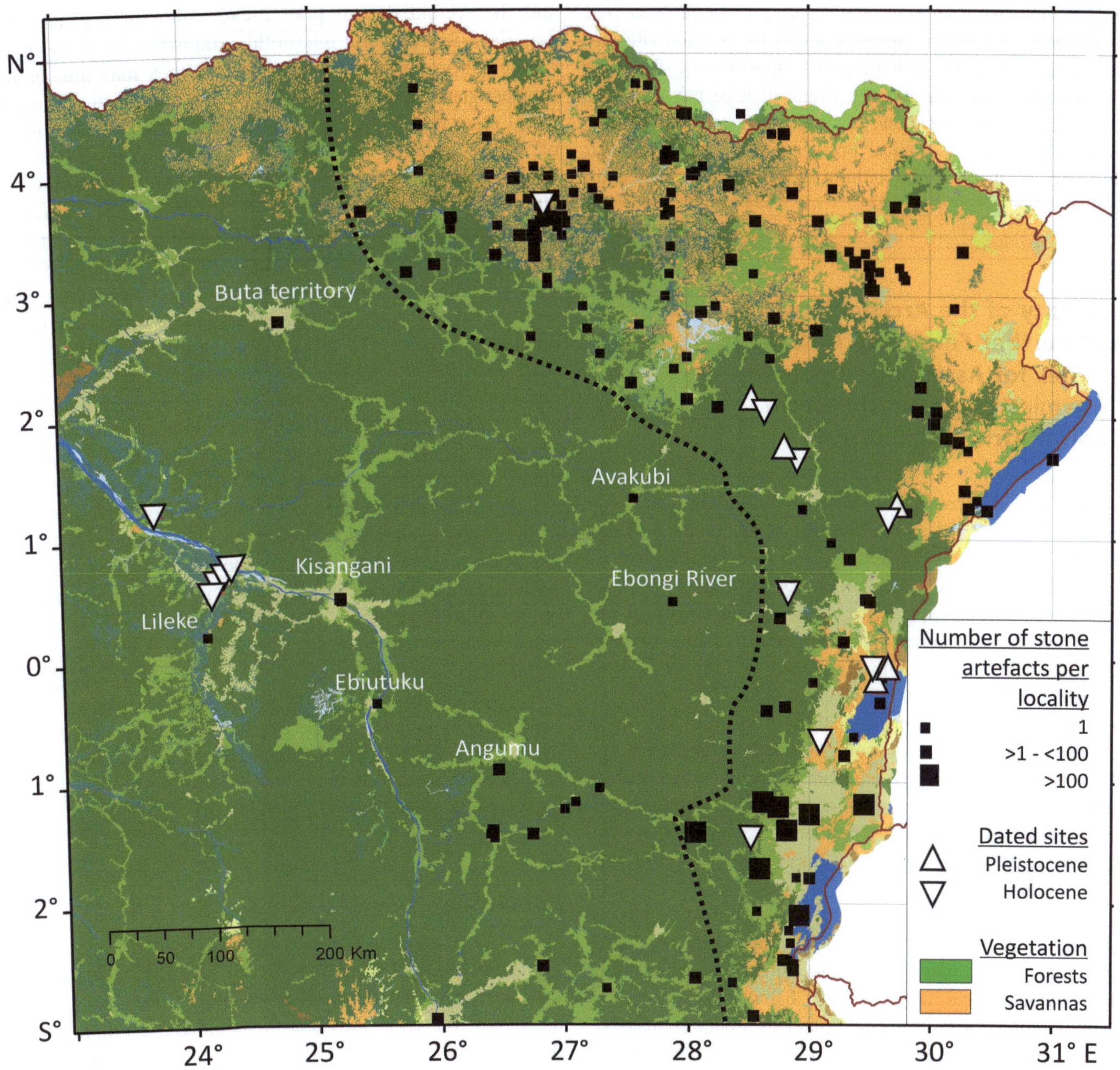

Fig. 16.8 General distribution pattern of the archaeological record in northeastern DRC

stone artifacts are extremely rare. There are a total of eight find spots in the vicinity of the city of Kisangani, of which three yielded a single clear artifact of unspecified affiliation. Lileke, and probably Ebiutuku as well, are of Holocene age. Sites south of Angumu yielded either Lupemban artifacts or flaked stone, but neither polished or utilized fragments nor microlithic industries on either quartz or any other raw material. North of Kisangani no flaked stone has been recorded; from various localities in the Buta territory, from Avakubi and from the Ebongi River only polished implements have been recovered.

The interpretation of these observations remains highly speculative. Taking present-day vegetation distribution and the river network into account, no particular barrier can be identified to explain this patterning. Low site visibility, especially for quartz microlithic scatters, may be a determining factor. However, lack of exploitable quartz in the area, or lack of the technological expertise necessary for producing microliths, either in quartz or any other raw material, cannot be excluded. It is also possible to hypothesize that there were no post-Lupemban stone-producing hunter-gatherers in the region prior to the arrival of populations using ground implements. If stone were not available to the hunter-gatherers, and if the Semliki sites are at all indicative of technological capacities of the Late Pleistocene communities, they would at least have had the capacity of exploiting bone for implements. But given the poor state of archaeological preservation in the rainforest, this part of the material culture inventory would remain undetectable.

Visibility of the Archaeological Record and Further Feasibility of Genetic Studies for Assessing Population Dynamics in the Northeastern Central African Rainforest

Environmental reconstruction of the Central African rainforest during MIS 2 and earlier points to climate induced changes that were responsible for variation in both floristic composition and forest distribution. During colder and drier periods, open landscapes with extended gallery forests are suggested. It is assumed that the present system of river networks mirrors that of the past, and this argues against isolated blocks of forests in which populations would have been separated and the circulation of people and ideas impeded. Access to fresh water would be guaranteed and food resources were likely abundant, especially when fishing or fish procurement became an essential part of the subsistence strategy. Riverine exploitation is attested in the archaeological record east of the Ituri Forest, in the Semliki Valley, from at least 70 ka.

The archaeological record in the region is generally limited to the lithic component of Lupemban and LSA quartz microlithic industries. Assessing the precise nature of their chronological correlation is hampered by lack of absolute dates for the Lupemban and its stratigraphic position in secondary contexts. The earliest LSA quartz microlithic industries date to the period prior to MIS 2. Hunter-gatherers operating in forested, wooded, riverine and open landscapes used them from the end of the Pleistocene onwards, throughout the Holocene. Though, exactly why and when their use was abandoned is unclear. While one might suppose that technological patterns can provide the basis for human demographic reconstruction, the technological continuity in quartz industries from MIS 2 onwards is at odds with the paleontological evidence from a single site, Ishango, which does not support the argument of direct continuity between the Late Pleistocene and present hunter-gatherer-fisher communities in the area.

Quartz implements and fishing may go unnoticed in the archaeological record; the first because of low field visibility and the second due to poor preservation of organic raw materials. Since quartz microlithic industries have no environmental or chronological attribution, they may thus be of little assistance for reconstructing population dynamics in the forest other than indicating the regional presence or absence of hunter-gatherer communities. In the case of quartz, intensive and intentional surveys may add to and eventually alter the distribution maps and the chronological record. Potentially, genetic studies may offer interesting perspectives on prehistoric fishing. Currently, genetic anthropological and linguistic studies primarily focus on the dichotomy between hunter-gatherers who consume agricultural products and their relation to farmers who in exchange procure forest products such as game and honey. Fish are not included from the equation. Today, fishing seems to be a specialized subsistence strategy among farming communities, and is only practiced to a lesser extent by hunter-gatherers. Mbuti hunter-gatherers have been observed (Ichikawa 1987: 98–99) to appreciate fish, though they seldom fish themselves. When it does occur, primarily women and children who practice fish poisoning and fish bailing in the dry season conduct it. This contemporary specialization may not necessarily mirror pre-agricultural patterns. If the Katanda and Ishango hunter-gatherer-fisher communities are at all indicative of the ability of ancestral populations near the rainforest to exploit fish, such groups were likely equipped for exploring and exploiting an aquatic riverine environment. Though difficult to assess with the current archaeological record, genetic studies may help in the reconstruction of subsistence practices at prior points in time.

Conclusion

The Late Pleistocene archaeological record of the northeastern Central African rainforest suffers from both a low number of recorded sites and their uneven distribution in time and space. Technological features and spatial distribution of the few dated and well-documented sites combined with undated sites from museum collections reveal continued use of quartz microlithic industries from 40 ka onwards, throughout MIS 2 and the Holocene, in forested as well as a variety of other environments. Quartz microlithic industries cannot be used to support the argument, as suggested by genetic reconstructions, of continuity of a hunter-gatherer population adapted to a forested environment in a refuge area. Nor does the admittedly scanty paleontological evidence support such a model. Lithic assemblages are the most visible part of the archaeological record under consideration, but their presence or characteristics may not be relevant for assessing past population dynamics in the region, even less so in areas where almost no stone tools have been found. In reconstructing past human habitation of lowland forests rich in waterways and perhaps poor in stone, the potential role of fishing in subsistence strategies and the use of bone as a raw material should be taken into account. These capacities are clearly attested at the eastern fringes of the forest from 70 ka onwards. Fishing and bone in the northeastern Central African lowland forest are, at present, undetectable by archaeology due to conservation issues. Therefore, extending the contemporaneous sampling framework for genetic studies to include fisher populations might offer interesting perspectives. Incorporating fishermen might identify new relevant genetic markers which may then be traced back into time in order to assess the impact of riverine resources exploitation on past population dynamics.

Acknowledgments I thank Brian Stewart and Sacha Jones for their invitation to the conference, and especially Brian for his editorial guidance. At the Royal Museum for Central Africa I would like to thank several colleagues; at the Prehistory and Archaeology Department Lien Speleers and Alexander Vral for their efforts in locating a first series of sites and verifying sources in our joint attempt to draw archaeological maps of the Democratic Republic of Congo. Alexandre Livingstone-Smith has been very patient in explaining the use of ArcGIS and graciously shared his data on the Lomami River. The Cartography Department has provided the necessary assistance in GIS and F. Mees of the Mineralogy unit ensured identifications of raw materials.

I am indebted to J. Mercader and two anonymous reviewers for their most useful comments on an earlier draft and to J. Yellen for editing the English.

The final responsibility for all flaws in interpretation and points of view remains, of course, mine alone.

References

Bahuchet, S. (2012). Changing language, remaining Pygmy. *Human Biology, 84*(1), 11–43.

Batini, C., Coia, V., Battaggia, C., Rocha, J., Pilkington, M. M., Spedini, G., et al. (2007). Phylogeography of the human mitochondrial L1c haplogroup: Genetic signatures of the prehistory of Central Africa. *Molecular Phylogenetics and Evolution, 43*, 635–644.

Batini, C., Ferri, G., Destro-Bisol, G., Brisighelli, F., Luiselli, D., Sa′nchez-Diz, P., et al. (2011a). Signatures of the preagricultural peopling processes in sub-Saharan Africa as revealed by the phylogeography of early Y chromosome lineages. *Molecular Biology and Evolution, 28*, 2603–2613.

Batini, C., Lopes, J., Behar, D. M., Calafell, F., Jorde, L. B., van der Veen, L., et al. (2011b). Insights into the demographic history of African pygmies from complete mitochondrial genomes. *Molecular Biology and Evolution, 28*, 1099–1110.

Bequaert, M. (1943). Deux instruments en pierre taillée de l'Angumu. *Bulletin des séances de l'Institut Royal Colonial Belge, 14*, 586–595.

Bequaert, M. (1945). Contribution à la connaissance des industries de la pierre taillée dans le nord-est du Bassin du Congo. *Bulletin de la Société Royale Belge d'Anthropologie et de Préhistoire, 56*, 154–172.

Boaz, N. T., Pavlakis, P. P., & Brooks, A. S. (1990). Late Pleistocene-Holocene human remains from the Upper Semliki Area, Zaire. In N. T. Boaz (Ed.), *Evolution of environments and Hominidae in the African western Rift Valley* (Virginia Museum of Natural History Memoir No. 1) (pp. 273–299). Springfield: Philipps Press.

Bouchneba, L., & Crevecoeur, I. (2009). The inner ear of Nazlet Khater 2 (Upper Paleolithic, Egypt). *Journal of Human Evolution, 56*, 257–262.

Brook, G. D., Bruney, D. A., & Cowart, J. B. (1990). Paleoenvironmental data for Ituri, Zaire, from sediments in Matupi Cave, Mt. Hoyo. In N. T. Boaz (Ed.), *Evolution of environments and Hominidae in the African western Rift Valley* (Virginia Museum of Natural History Memoir No. 1) (pp. 49–70). Springfield: Philipps Press.

Brooks, A. S., & Robertshaw, P. (1990). The Glacial Maximum in tropical Africa: 22,000–12,000 BP. In C. Gamble & O. Soper (Eds.), *The world at 18,000 BP* (Vol. 2, pp. 121–167)., *low latitudes* London: Unwin Hyman.

Brooks, A. S., & Smith, C. C. (1987). Ishango revisited: New age determinations and cultural interpretations. *African Archaeological Review, 6*, 65–78.

Brooks, A. S., Helgren, D. M., Cramer, J. S., Franklin, A., Hornyak, W., Keating, J. M. et al. (1995). Dating and context of three Middle Stone Age sites with bone points in the Upper Semliki Valley, Zaire. *Science, 268*, 548–553.

Brown, D. J., McSweeney, K., & Helmkeb, Ph A. (2004). Statistical, geochemical, and morphological analyses of stone line formation in Uganda. *Geomorphology, 62*, 217–237. doi:10.1016/j.geomorph.2004.03.002.

Cahen, D. (1978). Vers une révision de la nomenclature des industries préhistoriques de l'Afrique Centrale. *L'Anthropologie, 82*, 5–36.

Campbell, M. C., & Tishkoff, S. A. (2010). The evolution of human genetic and phenotypic variation in Africa. *Current Biology, 20*, 166–173.

Cavalli-Sforza, L. L. (1986). *African pygmies: An evaluation of the state of research*. Orlando: Academic Press.

Cerling, T. E., Hart, J. A., & Hart, T. B. (2004). Stable isotope ecology in the Ituri Forest. *Oecologia, 138*, 5–12.

Cornelissen, E. (2002). Human responses to changing environments in Central Africa between 40,000 and 12,000 BP. *Journal of World Prehistory, 16*, 197–235.

Cornelissen, E. (2003). On microlithic quartz industries at the end of the Pleistocene in Central Africa: The evidence from Shum Laka (NW Cameroon). *African Archaeological Review, 20*, 1–24.

Cornelissen, E. (2013). Hunting and gathering in Africa's tropical forests at the end of the Pleistocene and Early Holocene. In P. Mitchell & P. Lane (Eds.), *The Oxford handbook of African archaeology* (pp. 403–417). Oxford: Oxford University Press.

Crevecoeur, I., Semal, P., Cornelissen, E., & Brooks, A. (2010a). The Late Stone Age human remains from (Democratic Republic of): Contribution to the study of the African Late Pleistocene modern human diversity. *American Journal of Physical Anthropology* (Program of the 79th Annual Meeting of the American Association of Physical Anthropologists), *Meeting Abstracts*, p. 60 (http://physanth.org/annual-meeting/2010/79th-annual-meeting-2010/2010%20AAPA%20Abstracts.pdf).

Crevecoeur, I., Brooks, A. S., Cornelissen, E., Ribot, I., & Semal, P. (2010b). Les restes humains du site Late Stone Age d'Ishango (République Démocratique du Congo). Aperçus de la diversité passée des hommes modernes à la fin du Pléistocène supérieur/The human remains from the Late Stone Age site of Ishango (Democratic Republic of Congo). Insights on Late Pleistocene modern human diversity. *13th PanAfrican Archaeological Association Congress Abstracts* (Dakar).

De Haene, J. (1949). Découvertes préhistoriques en Haut-Ituri. *Zaïre, III*(9), 1003–1010.

de Heinzelin de Braucourt, J. (1948). Industrie lithique des graviers aurifères de la Lodjo, (Ituri, Congo Belge), description de la collection V. Piret. *Bulletin du Musée Royal d'Histoire Naturelle de Belgique, 24*(8), 1–16.

de Heinzelin de Braucourt, J. (1957). *Les fouilles d'Ishango* (Exploration du Parc National Albert, Fasc. 2). Brussels: Institut des Parcs Nationaux du Congo belge.

de Heinzelin de Braucourt, J. (1961). *Le paléolithique aux abords d'Ishango* (Exploration du parc naturel Albert, Fasc. 6). Brussels: Institut des Parcs Nationaux du Congo Belge et du Ruanda-Urundi.

de Heinzelin, J., & Verniers, J. (1996). *Realm of the Upper Semliki (Eastern Zaire): An essay on historical geology (Annales du Musée Royal de l'Afrique Centrale, Sciences Géologiques, 102)*. Tervuren: Musée Royal de l'Afrique Centrale.

de Maret, P. (1990). Phases and facies in the archaeology of Central Africa. In P. Robertshaw (Ed.), *A history of African archaeology* (pp. 109–134). London: James Currey.

Destro-Bisol, G., Coia, V., Boschi, I., Verginelli, F., Caglia, A., Pascali, V., et al. (2004). The analysis of variation of mtDNA hypervariable region 1 suggests that eastern and western pygmies diverged before the Bantu expansion. *The American Naturalist, 163*, 212–226.

Feathers, J. K., & Migliorini, E. (2001). Luminescence dating at Katanda – a reassessment. *Quaternary Science Reviews, 20*, 961–966.

Hammer, M. F., Woerner, A. E., Mendez, F. L., Watkins, J. C., & Wall, J. D. (2011). Genetic evidence for archaic admixture in Africa. *Proceedings of the National Academy of Sciences, 108*, 15123–15128.

Hart, T. B. (2001). Forest dynamics in the Ituri basin (DR Congo). Dominance, diversity and conservation. In W. Weber, L. White, A. Vedder & L. Naughton-Treves (Eds.), *African rain forest ecology and conservation. An interdisciplinary perspective* (pp. 154–164). Yale: Yale University Press.

Hart, T. B., Hart, J. A., Dechamps, R., Fournier, M., & Ataholo, M. (1996). Changes in forest composition over the last 4000 years in the Ituri basin, Zaire. In L. J. G. Van der Maesen & X. M. Van der Burgt (Eds.), *The biodiversity of African plants* (pp. 545–563). Dordrecht: Kluwer.

Ichikawa, M. (1987). Food restrictions of the Mbuti pygmies, eastern Zaire. *African Study Monographs, Supplementary Issue, 6*, 97–121.

Kanimba, M. (1990). Archaeological research on the age of metals in the Semliki area, Zaire. In N. T. Boaz (Ed.) *Evolution of environments and Hominidae in the African western Rift Valley* (Virginia Museum of Natural History Memoir No. 1) (pp. 301–316). Springfield: Philipps Press.

Lebzelter, V. (1933). Steinzeitliche funde aus Rwanda und vom Ituri. *Anthropos, 28*, 87–98.

Ludwig, B. V., & Harris, J. W. K. (1998). Towards a technological reassessment of East African Plio-Pleistocene lithic assemblages. In M. D. Petraglia & R. Korisettar (Eds.), *Early human behaviour in global context* (pp. 84–107). New York: Routledge.

Mercader, J. (2002). Forest people: The role of African rainforests in human evolution and dispersal. *Evolutionary Anthropology, 11*, 117–124.

Mercader, J., & Brooks, A. S. (2001). Across forests and savannas: Later Stone Age assemblages from Ituri and Semliki, Democratic Republic of Congo. *Journal of Anthropological Research, 5*, 197–217.

Mercader, J., Runge, F., Vrydaghs, L., Doutrelepont, H., Ewango, C. E. N., & Juan-Tresseras, J. (2000). Phytoliths from archaeological sites in the tropical forest of Ituri, Democratic Republic of Congo. *Quaternary Research, 54*, 102–112.

Mercader, J., Garralda, M. D., Pearson, O. M., & Bailey, R. C. (2001). Eight hundred-year-old human remains from the Ituri. *American Journal of Physical Anthropology, 115*, 24–37.

Mercader, J., Marti, R., Martinez, J. L., & Brooks, A. (2002). The nature of 'stone-lines' in the African Quaternary record: Archaeological resolution at the rainforest site of Mosumu, Equatorial Guinea. *Quaternary International, 89*, 71–96.

Palgen, J. (1949). Quelques pierres taillées du Maniema. *Bulletin de la Société de Préhistoire Française, XLVI (1/2)*, 47–50.

Palgen, J. (1950). Les gisements préhistoriques de Niamwaka (Congo Belge). *Bulletin de la Société de Préhistoire Française, XLVII* (11/12), 553–561.

Passau, G. (1946). *Découvertes préhistoriques dans le Haut-Ituri et dans la Haute Lindi* (pp. 637–649). XVII: Institut Royal Colonial Belge – Bulletin des Séances.

Patin, E., Laval, G., Barreiro, L. B., Salas, A, Semino, O., Santachiara-Benerecetti, S., et al. (2009). Inferring the demographic history of African farmers and pygmy hunter–gatherers using a multilocus resequencing data set. *PLoS Genet 5*(4). doi:10.1371/journal.pgen.1000448.

Peters, J. (1990). Late Pleistocene hunter-gatherers at Ishango (eastern Zaire): The faunal evidence. *Revue de Paléobiologie, 9*, 73–112.

Quintana-Murci, L., Quach, H., Harmant, C., Luca, F., Massonnet, B., Patin, E., et al. (2008). Maternal traces of deep common ancestry and asymmetric gene flow between Pygmy hunter – gatherers and Bantu-speaking farmers. *PNAS, 105*, 1596–1601.

Runge, J. (2001). Central African palaeoclimates and palaeoenvironments since 40 ka – an overview (with two maps). *Palaeoecology of Africa, 27*, 1–10.

Smith, A. L., Hubau, W., & Cornelissen E. (2010). *Boyekoli Ebale Congo 2010 (RDC) Archaeological survey along the Congo River.* Paper presented at the 13th Panafrican Association of Prehistory and Assimilated Disciplines, 20th conference of the Society of Africanist Archaeologists, Dakar.

Stewart, K. M. (1989). *Fishing sites of North and East Africa in the Late Pleistocene and Holocene* (Cambridge Monographs in African Archaeology No. 34, BAR International Series No. 521). Oxford: Archaeopress.

Taylor, N. (2011). The origins of hunting and gathering in the Congo basin: A perspective on the Middle Stone Age Lupemban industry. *Before Farming, 1*, article 6.

Taylor, N. (2016). Across rainforests and woodlands: A systematic re-appraisal of the Lupemban Middle Stone Age in Central Africa. In S. C. Jones & B. A. Stewart (Eds.), *Africa from MIS 6-2: Population dynamics and paleoenvironments* (pp. 273–299). Dordrecht: Springer.

Terashima, H., & Ichikawa, M. (2003). A comparative ethnobotany of the Mbuti and Efe hunter-gatherers in the Ituri Forest, Democratic Republic of Congo. *African Study Monographs, 24*(1, 2), 1–168.

Twiesselman, F. (1958). *Les ossements humains du gîte mésolithique d'Ishango (exploration du Parc National Albert, Mission J. de Heinzelin de Braucourt 1950) Institut des parcs nationaux du Congo Belge, Fascicule 5*. Institut des parcs nationaux du Congo Belge: Brussels.

Van Neer, W. (1989). *Contribution to the archaeozoology of Central Africa* (Zoologische wetenschappen No. 259). Tervuren: Annales Musée Royal de l'Afrique Centrale.

Van Noten, F. (1968a). *The Uelian: Culture with a Neolithic aspect, Uele-Basin (Congo Republic) an archaeological study* (Annales Musée Royal de l'Afrique Centrale, Sciences humaines No. 64). Tervuren: Annales Musée Royal de l'Afrique Centrale.

Van Noten, F. (1968b). Note sur l'âge de la pierre récente dans la région des lacs Mokoto (Kivu, Congo). *Bulletin de la Société Royale Belge Anthropologie et Préhistoire, 79*, 91–101.

Van Noten, F. (1977). Excavations at Matupi cave. *Antiquity, 51*, 35–40.

Van Noten, F. (1982). *The archaeology of Central Africa.* Graz: Akademische Druck- und Verlagsanstalt.

Verdu, P., Austerlitz, F., Estoup, A., Vitalis, R., Georges, M., Théry, S., et al. (2009). Origins and genetic diversity of pygmy hunter-gatherers from western Central Africa. *Current Biology, 19*, 312–318.

wa Bitanko Kamuanga, M. (1985–1986). *Préhistoire du Zaire oriental. Essai de synthèse des âges de la pierre taillée.* Ph.D. Dissertation, Université Catholique de Louvain, Louvain-la-Neuve.

Yellen, J. E. (1996). Behavioural and taphonomic patterning at Katanda 9: A Middle Stone Age site, Kivu Province, Zaire. *Journal of Archaeological Science, 23*, 915–932.

Yellen, J. E. (1998). Barbed bone points: Tradition and continuity in Saharan and Sub-Saharan Africa. *African Archaeological Review, 15*, 173–198.

Yellen, J. E., Brooks, A. S., Cornelissen, E., Mehlman, M. J., & Stewart, K. (1995). A Middle Stone Age worked bone industry from Katanda, Upper Semliki Valley, Zaire. *Science, 268*, 553–556.

[illegible] 19

[illegible] Kallio, [illegible]

[illegible]

Part IV

Broader Perspectives

Chapter 17
The Late Quaternary Hominins of Africa: The Skeletal Evidence from MIS 6-2

Frederick E. Grine

Abstract The late Quaternary African hominin fossil record provides a tantalizing glimpse into considerable temporal and geographic morphological diversity within the genus *Homo*. A total of 50 sites that can be constrained from MIS 6-2 have yielded specimens ranging from isolated teeth to nearly complete skeletons. However, only a dozen or so provide particularly informative or interesting evidence spanning this period of nearly 200 kyr. In addition to the rather paltry nature of the record, one of the seemingly more intractable problems that bedevil its interpretation is the nature of the chronometric record for many of the sites. The Late Pleistocene terrestrial climatic record for Africa is also rather patchy, making continent-wide generalizations difficult. Attempts to link large-scale environmental perturbations in Africa to patterns of human evolution and behavior are even more problematic. Although the African fossil (and archaeological) record is most often viewed from the perspective of a single lineage culminating in the appearance of *Homo sapiens* and thence modern humans, the degree of morphological diversity evident even in this meager assemblage can be rather striking. Some of this diversity may be related to geographic and/or temporal differences, but in other instances, there are noticeable differences among remains that are contemporaneous, or at least penecontemporaneous. The Late Pleistocene African hominin fossil record, despite its manifestly incomplete nature, finds consistency with an impressive array of genetic evidence that points to an African origin for our species, and it also has consilience with genetic data that indicate a coalescence of lineages to the common ancestor of *Homo sapiens* at around the beginning of MIS 6. Although multiple lines of genetic evidence indicate a deep separation of lineages, with the ancestors of the southern African Khoesan diverging early on from that which gave rise to all other groups, there is a notable paucity of human remains that predate MIS 2 that exhibit strong phenetic resemblance to recent African populations. A number of the human dental samples from Late Pleistocene South African sites possess morphological variants that characterize the teeth of the recent inhabitants of sub-Saharan Africa, but these similarities do not necessarily signify a close evolutionary relationship with any of these populations because they appear to be plesiomorphic.

Keywords Cranium • Dentition • Postcranial skeleton • Middle Stone Age • Later Stone Age • Middle Paleolithic • Upper Paleolithic

Introduction

Stable isotope studies have provided the basis for constructing global and regional climatic scales against which the human paleontological and archaeological records can be assessed (Behrensmeyer 2006). These paleoclimatic interpretations may be based on stable carbon isotopes extracted from paleosols (Cerling and Quade 1993; Wynn 2006), or the oxygen isotope records obtained from the calcitic shells produced by marine organisms such as foraminifera (McCrea 1950; Epstein et al. 1951, 1953; Emiliani 1966). In either case, these isotope studies are based on the same basic principles of chemistry. All atoms of an element have the same number of protons (i.e., they have the same atomic number), but they may differ in their numbers of neutrons (i.e., they differ in atomic mass). Elements such as carbon have two principal, or common isotopic forms (i.e., ^{12}C or ^{13}C), whereas oxygen has three (i.e., ^{16}O, ^{17}O and ^{18}O). In each instance the isotope is identified by a superscript number, which is the sum of the number of protons and neutrons in the nucleus. The higher mass isotopes are considerably more rare in nature, such that 98.9% of stable

F.E. Grine (✉)
Departments of Anthropology and Anatomical Sciences, Stony Brook University, Stony Brook, NY 11794-4364, USA
e-mail: frederick.grine@stonybrook.edu

Sacha C. Jones and Brian A. Stewart (eds.), *Africa from MIS 6-2: Population Dynamics and Paleoenvironments*, Vertebrate Paleobiology and Paleoanthropology, DOI 10.1007/978-94-017-7520-5_17

carbon in the atmosphere (in the form of CO_2) consists of ^{12}C and 99.7% of oxygen consists of ^{16}O.

The mass, or weight, differences of these elemental isotopes translate to different thermodynamic and kinetic properties (Hoefs 2009). Molecules that contain the more rare (higher-mass) isotope are thermodynamically more stable and slower to react in chemical reactions as well as in physical and biological processes such as the precipitation of calcium carbonate from water, photosynthesis and digestion. In such instances, the weaker bond formed by the lighter isotope affects molecular reaction kinetics; molecules with the lighter isotope react faster (using less energy), which results in its enrichment relative to the parent isotope ratio. Thus, for example, during the process of photosynthesis ^{12}C is enriched relative to ^{13}C as a kinetic effect of the lighter isotope proceeding more rapidly through the process. The enrichment, or fractionation of one isotope relative to another is expressed as a ratio represented as delta (δ) values that represent the difference between the observed isotope ratio and that within an internationally recognized standard in parts in parts *per mille* (or *permil = per thousand*).

In other instances, however, the heavier isotope can become enriched, such as in the precipitation of calcium carbonate from water. In this case, because ^{18}O is two neutrons heavier than ^{16}O, the latter takes less energy to vaporize and diffuse from water, resulting in the remaining water being enriched in ^{18}O. Because the physical process of evaporation results in an enrichment of ^{16}O relative to ^{18}O in atmospheric water vapor (hence rain and snow), and a concomitant increase of ^{18}O water, any precipitate from the water will be relatively enriched in the heavier ^{18}O. Since this process is temperature dependent, the changes in the light isotope ratios of oxygen can be used to determine the temperature of the water in which the precipitation of calcium carbonate occurred (Urey 1947). Thus, increased temperature results in progressively higher $^{18}O/^{16}O$ ratios in marine carbonate precipitates (Dansgaard 1964). This is the basis for the reconstruction of oceanic paleotemperatures and the recognition of oxygen isotope stages (OIS) or marine isotope stages (MIS) from the analyses of shells secreted by marine organisms (McCrea 1950; Epstein et al. 1951, 1953; Emiliani 1955, 1966).

Oxygen isotope ($\delta^{18}O_{calcite}$) records obtained from planktonic foraminifera from deep-sea cores in the Caribbean Sea and the Atlantic and Pacific Oceans revealed a cyclic temporal pattern that enabled Emiliani (1955) to recognize seven most recent climatic (paleotemperature) cycles or stages. He numbered these from the most recent (Holocene) downwards, with the odd-numbered stages representing warm periods (interglacial intervals) and the even-numbered designating cold episodes (glacial intervals). A decade later, Shackleton (1967) provided evidence that this $\delta^{18}O$ record is dominated by oceanic changes related to ice-sheet volume. Hence, Emiliani's oxygen isotope stages (OIS) are now widely referred to as Marine Isotope Stages (MIS). Importantly, Shackleton (1967) also observed that this record could be employed to construct a global stratigraphic framework for marine sediments. Since these pioneering studies, considerable advances have been made in refining the $\delta^{18}O$-based temporal/climatic record, providing an increasingly accurate geochronological calibration extending back several Ma, and the identification of short-term (i.e., c. 1 kyr), high-resolution and high-frequency Dansgaard-Oeschger events in the Pleistocene.[1] The isotope record has been extended to astronomically based models of orbital insolation and precession, resulting in the radiometrically calibrated spectral mapping project (SPECMAP) $\delta^{18}O$ record (Hays et al. 1976; Imbrie et al. 1984; Bond et al. 1993; Cramp and O'Sullivan 1999; Thompson and Goldstein 2006; Imbrie and McIntyre 2006).

Over one hundred marine isotope stages have been identified extending into the Miocene, but the most thoroughly documented and well-dated cover the latter half of the Quaternary (Imbrie et al. 1984, 1993; Martinson et al. 1987; Williams et al. 1988; Bassinot et al. 1994; Shackleton et al. 1990; Aitken and Stokes 1997; Andrews 2000; Wright 2000; Tzedakis 2003; Landais et al. 2004; Lisiecki and Raymo 2005; Lang and Wolff 2011). A number of studies have compiled stacked records from various sources and have used these data to refine the SPECMAP dates initially identified by Imbrie et al. (1984) for the isotope stages extending over the past 500 ka or beyond (Table 17.1). For example, the duration of the peak of the last Interglaciation (MIS 5e) has been determined between 124 and 111 ka by Martinson et al. (1987), 122 and 106 ka by Bassinot et al. (1994), and 130 and 119 ka by Hearty et al. (2007). The latter range comfortably accommodates the MIS 5e peak dates of 122 and 123 ka proposed by Wright (2000) and Lisiecki and Raymo (2005). For the most parts, such modifications have resulted in differences among various age determinations that are on the order of about 2%.

The marine paleotemperature signatures of the MISs and their dates of their onset over the past 0.3 Ma are provided in Fig. 17.1. The dates employed here to define stage boundaries are those of Lisiecki and Raymo (2005).

Marine records also have been interpreted in relation to Quaternary environmental changes in Africa (Hilgren 1991; deMenocal 1995, 2004; Camp and O'Sullivan 1999;

[1] Since the mid-1980s, the lower boundary of the Pleistocene Epoch (i.e., the beginning of the Quaternary Period) has been regarded as corresponding with the base of the Calabrian stratotype at 1.81 Ma. Recently, however, the International Union of Geological Sciences has recognized the base of the Gelasian stratotype, which corresponds to the Matuyama (C2r) chronozone, or the Gauss-Matuyama boundary, as defining the Pliocene-Pleistocene boundary at 2.588 Ma (Gibbard and Head 2009). This change is significant for discussions of hominin paleontology. Pending the outcome of appeals to this ruling, I continue to regard the base of the Pleistocene at 1.81 Ma.

Table 17.1 Dates (ka) determined for the onset of Mid-Late Pleistocene MIS stages by different workers

MIS	Stage					
	A	B	C	D	E	F
MIS 1	12	12	11	11	12	14
MIS 2	24	24	24	24	24	29
MIS 3	59	59	57	60	59	57
MIS 4	71	74	71	71	71	71
MIS 5a	80	79	79	79	82*	82*
MIS 5b	87	91	86	–	–	87*
MIS 5c	99	99	97	–	105*	96*
MIS 5d	107	111	106	–	–	109*
MIS 5e	122	124	122	–	122*	123*
MIS 5	128	139	127	130	128	130
MIS 6	(186)	190	186	190	186	191
MIS 7	(245)	244	242	244	245	243
MIS 8	(303)	–	301	301	303	300
MIS 9	(339)	–	334	334	339	337
MIS 10	362	–	364	364	362	374
MIS 11	(423)	–	427	427	423	424
MIS 12	(478)	–	474	474	478	478

Sources: *A* Imbrie et al. (1984: Table 6) SPECMAP with ages in parentheses interpolated between adjacent ages with reference to the stratigraphic level of the corresponding event in core V28-238; *B* Martinson et al. (1987: Table 2); *C* Bassinot et al. (1994: Table 4); *D* Aitken and Stokes (1997: Table 1.2); *E* Wright (2000: Fig. 4); *F* Lisieck and Raymo (2005)

*Dates are for the *peaks* of the MIS 5 substages; the other dates for MIS substages represent their onset

deMenocal et al. 2000; Schefuß et al. 2003, 2005; Weldeab et al. 2005; Adkins et al. 2006; Kröpelin et al. 2008). Undoubtedly, oceanic temperatures in the Northern Hemisphere affect the Southern Hemisphere as well (Crowley 1992), where reduced North Atlantic temperatures weaken African monsoonal circulation and tend to reduce rainfall. However, changes in Southern Hemisphere temperatures may precede changes in Northern Hemisphere ice volume, and although oceanic patterns are generally confirmed by terrestrial records, there are differences between them with regard to aspects of timing and intensity (Partridge et al. 1997; deMenocal et al. 2000; Drake and Bristow 2006; Armitage et al. 2007; Trauth et al. 2003, 2005, 2007).

While considerable attention has been paid to developing detailed terrestrial records in Europe (e.g., van Andel and Tzedakis 1996; Tzedakis et al. 1997; Watts et al. 2000; Voelker et al. 2002), there are comparatively few detailed terrestrial records for Africa (e.g., Pokras and Mix 1985; Stokes et al. 1997; Gasse 2000; Salzmann et al. 2002; Thomas and Shaw 2002; Zhao et al. 2003; Tierney et al. 2008, 2011; Bruch et al. 2012; Scholz et al. 2007, 2011). As noted by Tryon et al. (2010), we have, at best, an incomplete understanding of the spatial and temporal patterns of climatic change in Pleistocene Africa. Indeed, noticeable differences have been recorded for terrestrial responses within Africa even for very recent periods of time.

Thus, for example, the terminal Pleistocene to early Holocene "African Humid Period" (AHP) saw abundant North and East African lakes that were much higher than present; whereas lakes in South Africa stood at considerably lower levels (Hoelzmann et al. 2002; Schefuß et al. 2005; Weldeab et al. 2005; Kuper and Kröpelin 2006; Tierney et al. 2011). Additional evidence for time-transgressive terrestrial changes in the AHP has been documented even *within* the Sahara (e.g., deMenocal et al. 2000; Kuper and Kröpelin 2006; Cole et al. 2009), as well as between northern and southern Africa (Gasse 2000).

Moreover, while Trauth et al. (2003) claim that the MIS 6/5 boundary ("Termination II") is roughly synchronous across Africa, their own data (Trauth et al. 2003: Fig. 4) suggest otherwise. Furthermore, the data presented by Scholz et al. (2007) and Cohen et al. (2007) demonstrate multiple episodes of extreme hyperaridity in the Late Pleistocene, but that Lakes Malawi and Tanganyika experienced "mega-drought" conditions out of phase with the arid intervals elsewhere in the eastern Rift Valley system let alone sub-Saharan Africa. These differences suggest that climate changes within different areas of continental sub-regions (e.g., East Africa) were out of phase. Such discrepancies would have resulted in different parts of the same sub-region experiencing very different climate regimes.

This renders continent-wide generalization difficult if not perilous (Mitchell 2008; Chase 2010; Marean 2010). Scenarios that attempt to link large-scale environmental perturbations in Africa to patterns of human evolution and behavior (e.g., Scholz et al. 2007; Blome et al. 2012; Thomas and Burrough 2012; Ziegler et al. 2013) are even more problematic.

The hominin fossil record for the late Quaternary – especially the latter part of the Pleistocene – provides a tantalizing glimpse into considerable morphological diversity, temporally and geographically, within the genus *Homo*. Some of this almost certainly translates to potential lineage splitting and attendant taxonomic diversity. Considerable heed has been paid to the morphological and/or molecular evidence for later Pleistocene lineages in Eurasia, such as our relationships with *H. neanderthalensis* (e.g., Ponce de León and Zollikofer 2001; Harvati et al. 2004; Green et al. 2008, 2010; Noonan et al. 2006; Gunz et al. 2010; Hodgson et al. 2010; Currat and Excoffier 2011; Lalueza-Fox and Gilbert 2011) and the "Denisovans" (e.g., Krause et al. 2010; Reich et al. 2010, 2011; Meyer et al. 2012; Sawyer et al. 2015). Attention has been directed also to Southeast Asia in relation to the existence of late surviving *H. erectus* (Yokoyama et al. 2008; Indriati et al. 2011) and *H. floresiensis* (Brown et al. 2004; Aiello 2010).

By comparison, however, relatively little attention has been directed toward the morphological diversity that is evident within Africa in the later Pleistocene and Holocene (Manica et al. 2007; Crevecoeur et al. 2009; Gunz et al.

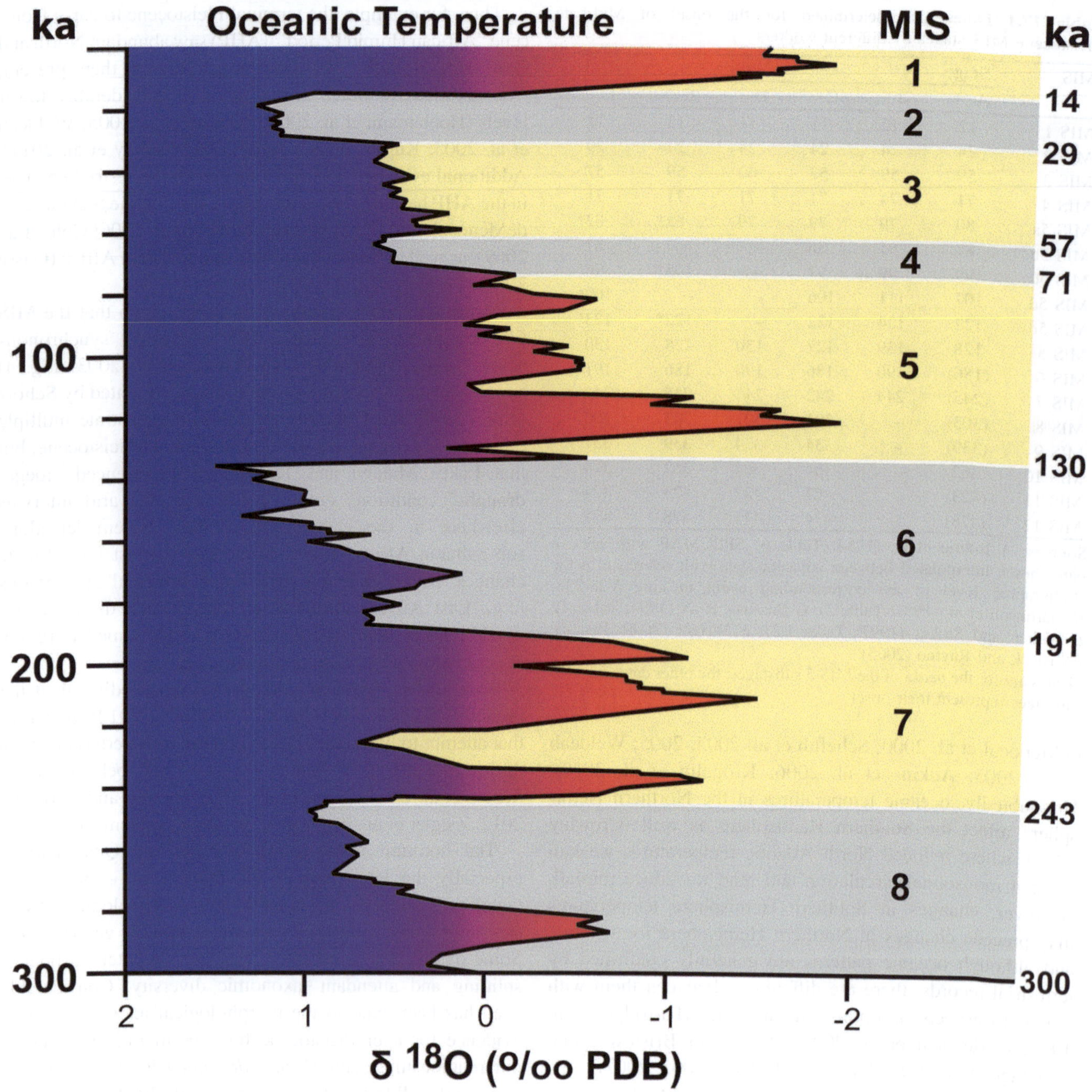

Fig. 17.1 Marine paleotemperature signatures reflecting global ice volume compiled from stacked oxygen isotope records of benthic foraminifera from deep-sea cores over the past 300 kyr. The $\delta^{18}O$ record reflects variation in parts *per mille* from the global reference standard (Vienna PeeDee Belemnite [VPDB]), which was obtained initially from the calcitic internal rostrums of the Cretaceous cephalopod, *Belemnitella*, preserved in the PeeDee Formation marine limestone of South Carolina (PDB). This has since been recalibrated by a laboratory in Vienna to the original sample (Coplen 1988, 1996). The standard is defined as zero (0‰). The other standard for oxygen ($^{18}O/^{16}O$) is Vienna Standard Mean Ocean Water (VSMOW), although this is somewhat of a misnomer since it now refers to pure water rather than the original seawater standard (Coplen 1988, 1996). PDB rather than VSMOW is used as the standard in this instance because the oxygen isotope data for benthic foramina derive from the calcium carbonate of their shells. The isotope plot and its relationship to the marine isotope stages is adapted from Martinson et al. (1987: Fig. 18), Porter (1989: Fig. 1), Kennett et al. (2000: Fig. 4) and Crosta and Shemesh (2002: Fig. 5). It is based upon data from an equatorial Pacific Core (V19-30) reported by Shackleton and Pisias (1985). The dates that correspond to the MIS stage boundaries are those of Lisiecki and Raymo (2005)

2009; Harvati et al. 2011). Rather, the African fossil (and archaeological) record is most commonly viewed from the perspective of a single lineage culminating in the appearance of *Homo sapiens* and thence modern humans (e.g., Bräuer 1984b, 2008; Smith 2002; Stringer 2002; Rightmire 2008, 2009; Pearson 2011). Thus, hominin fossils that date to perhaps 700–400 ka from sites such as Kabwe (Broken Hill), Elandsfontein, Bodo, Ndutu, Olduvai Gorge (OH 11), and Garba III (Melka Kunturé) have been viewed as representing a form that is related to us through a morphologically more derived intermediate. The latter is held to be represented by specimens such as the cranium from Florisbad.

This scheme remains the dominant one by which human evolution in Africa is interpreted, even if the names and hypodigms accorded these "grades" or "stages" differ somewhat among workers (e.g., Bräuer 1989a, 1992, 2008; Foley and Lahr 1997; Rightmire 1998, 2001, 2008, 2009; Kuman et al. 1999; McBrearty and Brooks 2000; Haile-Selassie et al. 2004). Some view these more "archaic" forms as representing separate species, with *Homo rhodesiensis* (or, arguably, *H. heidelbergensis*) being the appellation for the earlier, and *Homo helmei* the preferred name for the latter. Others view them as "grades" of *Homo sapiens* (i.e., "early archaic" *H. sapiens*, and "late archaic" or "pre-modern" *H. sapiens*). Regardless of the names that are applied, they are still viewed as representing "stages" of morphological evolution in a lineage leading inextricably to anatomically modern humans.

Fossils that are universally (or at least almost universally) recognized as being attributable to *Homo sapiens* – as defined by the possession of a number of morphological apomorphies that we see in ourselves – appear for the first time in Africa in the temporal span between MIS 7 and MIS 6, on the order of some 200 ka (Weaver 2012). However, at this time, as subsequently in the Pleistocene, the degree of morphological variation can be rather striking, and its significance continues to be a subject of discussion (e.g., Stringer 2007; Hammer et al. 2011).

One of the seemingly more intractable problems that bedevils the rigorous analysis and interpretation of the fossil evidence relating to the origin of *Homo sapiens* and the emergence of fully anatomically modern human morphology in Africa in the Late Pleistocene is the nature of the chronometric record for many of the potentially more interesting specimens (Millard 2008).

Despite various attempts to provide absolute ages for many African fossils that extend from MIS 6-2, a number of these have been dated solely by notoriously fallible techniques, such as the application of electron spin resonance (ESR) to mammalian tooth enamel (e.g., Grün 2006). ESR is problematic because variables such as background dose rate and sediment moisture history must be known or assumed; open systems (such as tooth enamel) only compound these troubles. Depending upon the uranium uptake model that is preferred (often preferred for less than compelling scientific reasons), just about any "date" is feasible. Indeed, Zhao et al. (2001) found significant differences in U-series dates obtained from speleothems and tooth enamel samples, indicating that uranium uptake history is "far more complex than existing models can handle." They cautioned that great care must be taken in the interpretation of ESR "dates" for fossil teeth. Similarly, Grine et al. (2012) have observed that dating studies based on ESR of tooth enamel from the South African Early Pleistocene karst cave site of Swartkrans have produced a bizarre range of age estimates: 4.38 Ma–36 ka. Another example of the malleability of ESR dating of more proximate interest relates to the age determinations that were made on two parts of a single bovid tooth excavated from adjacent stratigraphic levels (4–5 and 6) at the Late Pleistocene site of Die Kelders, South Africa. The initial assessment found these two conjoining pieces to be separated by $\sim$10 kyr in a site that likely dates to only $\sim$70 ka (Avery et al. 1997), but subsequent reanalysis found them to have identical Early Uptake and Linear Uptake ages (Schwarcz and Rink 2000).

Unfortunately, the validity of dates obtained by alternative methods, such as U-series analyses of open systems (e.g., vertebrate bone, dentine, and enamel), may be problematic (Hinz and Kohn 2010). Finally, the dates obtained by optically stimulated luminescence (OSL) of sediment particles of unknown moisture history are clearly open to question (Guérin et al. 2013; Tribolo et al. 2013). It is lamentable that many (if not most) of the Late Pleistocene sites and fossils of Africa owe their geochronological placement to such questionable and sometimes seemingly idiosyncratic methods. As a result, the ages of many of these fossils are loosely constrained, and this may render a number of evolutionary scenarios suspect. This imprecision necessarily complicates attempts to deduce possible relationships among demographic, cultural, morphological, and climatic variables.

The Late Pleistocene African human fossils that are reasonably constrained between MIS 6-2 are reviewed here (Table 17.2). A total of 50 sites that can be attributed to this interval have yielded hominin remains, ranging from isolated teeth to nearly complete skeletons. While this number might seem rather impressive, only a dozen or so have provided particularly informative or interesting specimens.

In this review, the sites and/or specimens are grouped geochronologically, insofar as possible, in relation to the marine isotope stages to which they pertain. For a number of

Table 17.2 African hominin-bearing sites dating from MIS 6 through MIS 2

Within MIS	Site/specimen	Site sge (ka)	Human remains
7–6	Omo (Kibish Fm.)	195	**Cranium; fragmentary skull and partial postcranial skeleton**
7–6	Kébibat	200–130	Skull fragments
6	Twin Rivers	178–139	Humerus fragment
6	Mumbwa Caves	172	Two teeth; two radius fragments; possible femoral diaphysis
6	Jebel Irhoud	160	**Two crania, juvenile mandible; fragmentary postcrania**
6	Herto	160–150	**Cranium; fragmentary cranial remains of five individuals**
6	Singa	145–133	**Calvaria**
6–3	Border Cave	170–56	Postcranial fragments
5e	Ngaloba Beds	129	**Cranium**
5e	Blind River	124–112	Femur
5e–4	Klasies River	115–58	**Multiple cranial, mandibular and postcranial fragments**
5e–5a	Sea Harvest	110–71	Manual distal phalanx; tooth
5d–5c	Grotte des Contrebandiers	110–92	Cranial fragments
5e–5c	Dar es-Soltan II	125–92*	**Incomplete skull; cranial fragments**
5c	Eyasi	104–92	**Partial cranium; mandibles; cranial fragments; teeth**
5c–3	Equus Cave	103–30	Eight teeth
5c–5a	Aduma	105–80	**Cranium; cranial fragments**
6–5c	Pinnacle Point	162–90	Parietal; tooth
5c–5a	Blombos	102–70	Nine teeth
5c–3	Ysterfontein 1	130–50	Three teeth
5c–3	Witkrans	100–50	Three teeth
5a–4	Plovers Lake	89–62	Postcranial fragments
5a	Haua Fteah	80–68	Two mandibular fragments
5a	Mumba Shelter	78–60	Teeth
4–3	Porc-Épic	78–36	Mandibular fragment
4	Die Kelders	74–59	24 teeth; mandibular fragment; 2 phalanges
4	Klipdrift Shelter	72–52	Isolated tooth
3	Sibudu	77–38	Phalanx; distal fibula
3	Diepkloof	61–48	Two toe bones; tooth
3	Mugharet el 'Aliya	57–27	Juvenile maxilla; isolated teeth
3	Nyamita	55–45	Partial humerus
3	Magubike Rock Shelter	42	Six isolated teeth
3	Nazlet Khater	38	**Skulls and postcranial skeletons**
3	Hofmeyr	36	**Cranium**
3	El Harhoura I	41–26	Mandible; isolated tooth
2	Ishango 11	26–20	Fragmentary crania and postcrania
2	Taramsa 1	70–24	Child skeleton
2	Leopard's Hill Cave	24–21	Isolated parietal
2	Lukenya Hill	24–22	**Partial calotte**
2	Tuinplaas	<20–11	Skull and partial postcranial skeleton
2	Deir El-Fakhuri (E71K1)	18	Two partial skeletons
2	Taza Cave I	16–14	Skull
2	Afalou-bou-Rhummel	15–11	**63 partial crania and skeletons**
2	Gebel Silsila 2A	14–13	Isolated frontal bone
2	Jebel Sahaba (117)	14–12	**58 partial skeletons**
2	Wadi Halfa (6B28 & 6B36)	14–10	Mandible (6B28); 37 partial skeletons (6B36)
2	Ifri n'Baroud	17–11	Single postcranial skeleton
2	Bushman Rock Shelter	13–12	Single infant mandible
2	Mlambalasi Rock Shelter	13–12	Partial postcranial skeleton (? *in situ*)
2	Grotte des Pigeons (Taforalt)	13–11	**200 skeletons**
2	Iwo Eleru	13–11	Incomplete skeleton and calvaria

The age ranges provided for a site are the maximum and minimum estimates for the site and do not include the confidence intervals for individual sample averages. The age ranges for the human specimens in a site when they do not extend throughout the thickness of the deposit, but are rather restricted to specific dated horizons within the site, are recorded in Tables 17.3, 17.4, 17.5 and 17.6. See text for details. The more significant specimens are designated in boldface type

*The Aterian- and hominin-bearing strata at Dar es-Soltan II have not been directly dated; these estimates are those that have been reported for the Aterian at proximate localities. The older date is likely too old; the range is more likely 107–96 ka

sites, the hominin fossil-bearing strata may overlap two (or more) isotope stages. In other instances, the stratigraphic derivation of the fossils is unclear, such that it is not possible to assign a given specimen to a particular MIS with certainty. In these cases, I have referred the sites and/or fossils to the geochronologically oldest possible stage simply for purposes of organization. Thus, it should be stressed that the order of appearance of the sites on the pages that follow does not necessarily imply their strict geochronological seriation.

Human Specimens Not Considered: The Undated and the Redated

A number of human specimens that have been regarded at one time or another as being of Late Pleistocene age are not considered here. These specimens fall into one of two categories that serve to exclude them from consideration. The first comprises those that are undated, but have been thought to date to the Late Pleistocene on less than adequate morphological criteria or on highly questionable geochronological evidence. The second category consists of specimens that were once thought to be related to the Middle Stone Age (MSA), but are demonstrably Holocene in age. It is reasonable to briefly discuss these specimens insofar as all have been regarded at one time or another as relevant to discussions of Late Pleistocene human evolution in Africa.

Specimens that May or May Not Be Relevant: The Undated

Specimens that comprise this category are the calvaria from Oranjemund, Namibia; Boskop, South Africa; Eliye Springs, Kenya, and the KNM-ER 999 femur and KNM-ER 3884 cranium from Ileret, Kenya. In addition, the sites of Mumbwa Caves, Zambia and Lukenya Hill, Kenya contain human bones for which radiocarbon and/or amino acid racemization dates pertaining to MIS 2 have been reported (Protsch 1975, 1977). Unfortunately, the veracity of these age determinations is extremely doubtful; the human remains could well relate to the Holocene. However, the Mumbwa Caves also preserve human specimens associated with the MSA that likely relate to MIS 6.

Oranjemund, Namibia

The Oranjemund calotte was found in 1988 on the beach between the Orange River and the Atlantic Ocean (28°36′ S, 16°26′ E) following an exceptional flood (Senut et al. 2000). Although Senut et al. (2000) acknowledge that the specimen is "difficult to date, having been found out of its geological context," they opine that it likely originated from an undated black clay horizon that outcrops along the banks of the river approximately 100 m away. The "main evidence" presented concerning the calotte's age is its morphology, which was said to "compare well with other African archaic sapients [sic] aged between 100,000 and 50,000 years" (Senut et al. 2000: 813). However, this assessment is based on comparisons of measurements of overall cranial size for three recent human population samples and of several frontal bone dimensions with fossils from Ngaloba (Tanzania), Florisbad (South Africa), and Zuttiyeh (Israel). In no instance do the cranial metrics align the Oranjemund specimen with fossils dated to c. 100–50 ka in preference to recent humans. Moreover, the Oranjemund calotte evinces no morphology than can be regarded as archaic. Its noticeably separate medial superciliary eminence and lateral supraorbital trigone do not, contrary to Senut et al. (2000), constitute a "well-marked supraorbital torus," but rather define a fully modern configuration.

Boskop, South Africa

The Boskop calotte was recovered in 1913 by workers digging a trench on a farm along the banks of the Mooi River in Gauteng Province, South Africa. The specimen was found at a shallow depth in disturbed riverine sediments (Haughton 1917). Subsequent finds included a mandibular fragment, a temporal bone, and parts of a poorly preserved skeleton. The specimen was initially thought to be of considerable antiquity, and was believed to be associated with the MSA on the basis of a single artifact reported to have come from the same site (van Riet Lowe 1954). Wells (1952) intimated this when he wrote that although the precise antiquity of the cranium had not been established, "there is now good evidence that the skull was derived from a Middle Stone Age horizon." Broom (1918) proposed that Boskop represented a distinct species, *Homo capensis*, which he diagnosed primarily on the basis of a poorly preserved fragment of mandibular corpus. However, with the possible exception of its comparatively large endocranial cavity, the specimen preserves no morphology that can be considered unusual for modern humans; indeed, a number of studies have concluded that Boskop (though perhaps suggestive of a distinct "physical type" to some) has discernible features in common with the recent Khoe-San (e.g., Broom 1923; Pycraft 1925; Wells 1959). The lack of any solid geological or archaeological context for these remains, and the fact that other animal bones from the same deposits were judged to have been secondarily redeposited by fluvial action (Haughton 1917) suggest that Boskop is best regarded as undated.

Eliye Springs, Kenya

The heavily mineralized human calvaria from Eliye Springs has featured prominently in discussions of the Middle to Late Pleistocene hominin fossil record, although it is wholly undated. The cranium lacks any primary geological context, having been discovered in 1985 by two German tourists who were "beach combing" along the western shore of Lake Turkana near the Eliye Springs resort (Bräuer and Leakey 1986a, b). Immediate investigation of the site by R.E. Leakey revealed that the specimen was unlikely to have been found in primary context, but rather to have been reworked by wave action from sediments that outcrop nearby. A few isolated faunal remains which exhibit the same heavily mineralization as the hominin cranium were collected in the immediate vicinity, but they provide no evidence of antiquity.

Although the calvaria from Eliye Springs lacks an absolute geochronological date, its morphological resemblance (or lack thereof) to other African specimens has led some workers to associate it with a date. Thus, in their initial assessment, Bräuer and Leakey (1986a: 251) concluded that because the cranium "exhibits clear relations to archaic *Homo sapiens* – in particular to the later representatives, i.e., to late archaic *Homo sapiens* – a late Middle to early Upper [sic] Pleistocene age for this hominid is a very probable assumption." This age was taken to encompass a span between 200 and 100 ka (Bräuer and Leakey 1986a, b), as illustrated by Bräuer (1989b: Fig. 8.2). Subsequently, Bräuer et al. (2003: 200) revised its presumed age. They concluded that "based on the morphological similarities to absolutely dated archaic *Homo sapiens* specimens, an age of about 200,000–300,000 years can be assumed for the Eliye Springs hominid." This range was expanded when Bräuer et al. (2004: 114) wrote, "comparisons of the cranium have shown close affinities to later Middle Pleistocene archaic *H. sapiens*, which, according to recent dating, might have existed between 300,000 and 150,000 years BP." Four years later, however, it was cited as having an age between Omo II at c. 200 ka and Florisbad at c. 260 ka (Bräuer 2008). In an interesting turn of phrase relating to its age, Bräuer (2008: 25) stated, "although it is not as well dated [as KNM-ER 3884, LH 18 and Florisbad], the cranium from Eliye Springs…might also belong to this group based on its combination of archaic and derived conditions." Thus, in the space of just over two decades, this cranium transmogrified from an unprovenanced and undated fossil to one with a rather secure placement between 260 and 200 ka. This was achieved wholly on the basis of morphological comparisons. As such, the use of this "date," or of the earlier date of 200–100 ka, as by Senut et al. (2000) in their assessment of the Oranjemund calvaria, in any discussion of human evolution in Africa is wholly circular.

Ileret, Kenya

Two hominin specimens of potential relevance from Ileret, Kenya also fall into the category of being poorly constrained geochronologically. These are the KNM-ER 999 femur and KNM-ER 3884 cranium. Although they are bracketed by age constraints, the brackets are so broad as to render the estimates of little meaning.

The KNM-ER 999 femur is essentially modern in appearance, with a high neck-shaft angle and a pronounced linea aspera (Day and Leakey 1974; Trinkaus 1993). Indeed, Trinkaus (1993) drew favorable comparisons between it and early *Homo sapiens* femora from Skhul and Qafzeh, noting that this pilastric configuration is unknown in the femora of archaic members of the genus. KNM-ER 999 was recovered in 1971 on the surface in locality 6A at Ileret, and was considered to have eroded from a horizon originally described as comprising part of the Guomde Formation (Bowen and Vondra 1973). These strata are now assigned to the Chari Member of the Koobi Fora Formation (Brown and Feibel 1986).

The KNM-ER 3884 cranium seems to evince a mixture of archaic morphology (e.g., a relatively receding frontal with a supraorbital region described as "torus-like") with more derived features such as a rounded occipital bone (Bräuer et al. 1992b; Bräuer 2001). KNM-ER 3884 was discovered in 1976 in Area 5 at Ileret. Most of the pieces were recovered from the surface, although some parts were excavated *in situ* from sediments also initially described as comprising part of the Guomde Formation. These too are now assigned to the Chari Member.

The strata from which these two fossils derive are some 8–10 m (or less) above the Chari Tuff, which defines the base of the Chari Member of the Koobi Fora Formation. The Chari Tuff is equivalent to Tuff L of the Shungura Formation, dated by $^{40}Ar/^{39}Ar$ to 1.38 Ma (McDougall and Brown 2006). The Chari Member extends to the base of the Holocene Galana Boi Beds. Unfortunately, the Chari stratigraphy remains one of the least well understood of any member of the Koobi Fora Formation because its lithological facies are extremely localized (Gathogo and Brown 2006). The fossils do not derive from the Holocene Galana Boi Formation because the sediments do not contain equivalent mollusk shells, and it is not certain where the sediments from which KNM-ER 999 and KNM-ER 3884 derived should be placed within that member – that is, between 1.38 Ma and 10 ka (Feibel personal communication). An iron manganese oxide cemented shell bed that lies above the *probable* hominin-bearing level is probably equivalent to the "orange and black stained bivalve packed bed" described by Gathogo and Brown (2006); if this is so, then the hominin fossils lie above the Silbo and Kale Tuffs, which would suggest an age of less than 0.75 Ma

(McDougall and Brown 2006). The strata from which these specimens came were tilted to the west and eroded before deposition of the overlying sediments of the Galana Boi Formation (Brown personal communication).

In the Omo-Turkana Basin, there are few exposures of strata that can be assigned to the time interval between the Galana Boi Formation and the upper part of the Chari Member of the Koobi Fora Formation. The best known of these is the Kibish Formation in southern Ethiopia, which ranges in age from 200 to ~6.5 ka (Brown and Fuller 2008). Although the Kibish Formation strata are separated from the Ileret region by some 125 km, it is perhaps noteworthy that they have not been structurally deformed and are still flat lying. This might be taken as evidence that the deformation that affects the Chari Member predates the Kibish Formation, and the fossils from the upper part of the Chari Member are greater than 200 ka (Brown personal communication).

Bräuer et al. (1997) reported direct U-series for KNM-ER 3884 and KNM-ER 999; the two cranial samples provided U-Th dates of 279 and 272 ka with minimum ages of 160 ka, and the single femoral sample yielded a U-Th age of 301 ka with a minimum of 205 ka. The U-Pa determinations for all three samples provided ages in excess of 180 ka. However, the older confidence limit for the U-Th estimates includes infinity, thus extending the method beyond its actual limits. The U-series estimates provided by Bräuer et al. (1997) accord with their stratigraphic derivation above the Silbo Tuff, and with their derivation from tilted strata. It is likely that they are between 750 and 200 ka. At present, It is not possible to place these specimens more precisely than this and, in any case, they are very likely beyond the geochronological scope of this contribution.

Specimens Once Thought to Be Relevant: The Redated

The second category of human remains excluded from consideration here includes those from South African sites such as Peers Cave, Cape Flats, and Canteen Kopje. The South African sites of Tuinplaas and Bushman Rock Shelter once considered to be of considerable antiquity have also been redated, and while both may ultimately prove to date to the Holocene, they are here (tentatively) attributed to MIS 2.

The Peers Cave rock shelter, also known by the name Skildergat, is located near the town of Fish Hoek on the Cape Peninsula of South Africa. It was the subject of a series of excavations by V.P. and B. Peers in the late 1920s. They uncovered a number of human skeletons of clear LSA origin, and one specimen (Fish Hoek 1, or Peers Cave 4) described as coming from the late MSA levels (Peers and Goodwin 1953). An uncalibrated conventional radiocarbon date of some 36.0 ^{14}C kBP was obtained for charcoal that lay beneath the burial (Anthony 1967; Vogel and Beaumont 1972), and Protsch (1974) reported a date of 35.0 ^{14}C kBP for bone from the level from which the human skeleton was believed to have derived. As a result, the specimen has featured in several studies as an example of Late Pleistocene human morphology (e.g., Rightmire 1978; van Vark 1984; Bräuer and Rosing 1989). However, a recent AMS assessment of the Peers Cave 4 cranium itself found that it actually dates to only 7,457–7,145 ^{14}C BP (Stynder et al. 2009).

In 1929, a human cranium and femoral fragment were found eroding from a sand dune in Philippi (the Cape Flats), near Cape Town. Drennan (1929) described this specimen as having "Australoid" affinities, and argued that the stone artifacts found eroding out with the bones attested to their MSA derivation. At the same time, however, Goodwin (1929) noted that these artifacts appeared to be a mixture of MSA (i.e., Still Bay) and LSA (i.e., Wilton) lithics and that "no valid proof" of the association of the MSA with the cranium was available. Nevertheless, Drennan (1929) persisted in reasoning that "in the light of the evidence afforded by the bones themselves it is difficult to escape the conclusion that this individual represents one of the human links between Mousterian and modern man." In particular, Drennan was struck by the relatively low, receding frontal – "it is not an exaggeration to describe the forehead as 'Neanderthaloid'" (1929: 422) – as attesting to its antiquity and its physical difference from the crania of the recent inhabitants of southern Africa. Drennan's assessment of the specimen's morphology and antiquity worked its way into the literature, with some workers accepting its morphological distinctiveness (e.g., Wells 1952, 1959). Although other workers expressed doubts about Drennan's assessment of the cranium's morphology (e.g., Brothwell 1963) or its MSA association (Klein 1970), some (e.g., Lahr 1996) have persisted in referring to its "late Pleistocene" age, despite Singer's (1993) observation that the specimen has been radiocarbon dated to 150 ± 60 ^{14}C BP. It is particularly interesting to note the emphasis that Drennan (1929) had placed on the low, receding frontal of the Cape Flats specimen in view of the recent conclusions that have been proffered by Harvati et al. (2011) about the affinities of the Iwo Eleru cranium from the terminal Pleistocene of Nigeria.

Several other isolated and variably mineralized specimens, such as the fragmentary cranium from Canteen Kopje, South Africa at one time regarded as being of MSA antiquity (Broom 1929b), are likely to be of recent derivation, although their absolute ages cannot be determined at present (Smith et al. 2012). These specimens have rightly fallen into obscurity.

Human Fossils Associated with MIS 6 (191–130 ka)

The earliest fossils that have been securely attributed to *Homo sapiens* appear during the transition from MIS 7-6 at ~195 ka. Marine isotope stage 6 was one of the longest and coldest glacial periods of the Northern Hemisphere (Petit et al. 1999) and, as such, is generally considered to have witnessed increased aridification in Africa, with desert environments – especially the Sahara, Namib, and Kalahari – expanding across the continent (Hetherington et al. 2008; Carto et al. 2009; Castañeda et al. 2009; Marean 2010). Marean (2010) has suggested that this may have hampered the movement of humans and other mammals between and even within subcontinental regions, but notes that this downturn in global temperature was likely to have had variable effects on the ecosystems of sub-Saharan Africa, with pulses of increased rainfall (see also Blome et al. 2012).

Some of the most complete and therefore informative crania and postcranial skeletons derive from sites that are dated to the transition to and within the early part of MIS 6. Eight hominin-bearing sites are grouped here as being associated with MIS 6 (Fig. 17.2; Table 17.3). The Omo Kibish fossils derive from the transition between MIS 7-6, and the Kébibat cranial fragments are possibly of this antiquity. The Border Cave postcranial bones, which are here judged likely to be the only MSA human remains from that site, may date anywhere from MIS 6-3. The eight sites are discussed below in the order in which they appear in Fig. 17.2 and Table 17.3.

Omo (Kibish Formation), Ethiopia (5°23′ N, 35°56′ E)

In 1967, a team led by R.E. Leakey discovered three fossil hominin specimens in the Member I sediments of the Kibish Formation along the Omo River in southern Ethiopia (Leakey 1969; Day 1969; Butzer 1969; Butzer et al. 1969). These fossils, which were described as contemporaneous, comprise an incomplete skull and associated partial postcranial skeleton (Omo I), a well-preserved calvaria (Omo II), and fragments of a neurocranium and face (Omo III). Some of the Omo I remains were reported to have been recovered *in situ* during an excavation of the site (Shea et al. 2007).

Although both Omo I and Omo II were found in the upper part of Member 1 of the Kibish Formation and below the KHS Tuff at the base of Member 2, the Omo II calvaria presents an overall more archaic morphology than the cranial remains of Omo I. Some have held this to signify a population with considerable individual variation (Day 1969; Rightmire 1976; Trinkaus 2005), or perhaps two different contemporaneous populations (Day and Stringer 1982, 1991). Others have suggested that Omo I is younger, and perhaps intrusive (e.g., Chavaillon 1982; Bräuer et al. 1997; Bräuer 2001; Klein 2009). Subsequent fieldwork between 1999 and 2003 led by J. G. Fleagle resulted in the discovery of a partial tibia and fibula of a fourth individual at another site in Member I (AHS-Kib-158-1; Pearson et al. 2008b). Importantly, this also led to the discovery of additional parts of the Omo I skeleton that had weathered out of the level excavated in 1967 at KHS. Some of these actually refitted bones discovered in 1967 (Pearson et al. 2008a). These discoveries, together with more comprehensive stratigraphic and geochronological analyses of the site (McDougall et al. 2005, 2008; Brown and Fuller 2008; F. Brown et al. 2012) have provided ample evidence for the contemporaneity of the Member 1 hominin specimens.

Table 17.3 African hominin-bearing sites and hominin fossils associated with MIS 6

Site/specimen	Country	Coordinates	Age (ka)
Omo (Kibish Fm.)	Ethiopia	05°23′ N, 35°56′ E	195
Kébibat	Morocco	34°02′ N, 06°51′ W	200–130
Twin Rivers Kopje	Zambia	15°31′ S, 28°11′ E	178–139
Mumbwa Caves	Zambia	15°01′ S, 26°59′ E	172
Jebel Irhoud	Morocco	31°56′ N, 08°52′ W	160–150
Herto	Ethiopia	10°15′ N, 40°33′ E	160–150
Singa	Sudan	13°00′ N, 33°55′ E	145–133
Border Cave	South Africa	27°01′ S, 31°59′ E	170–56

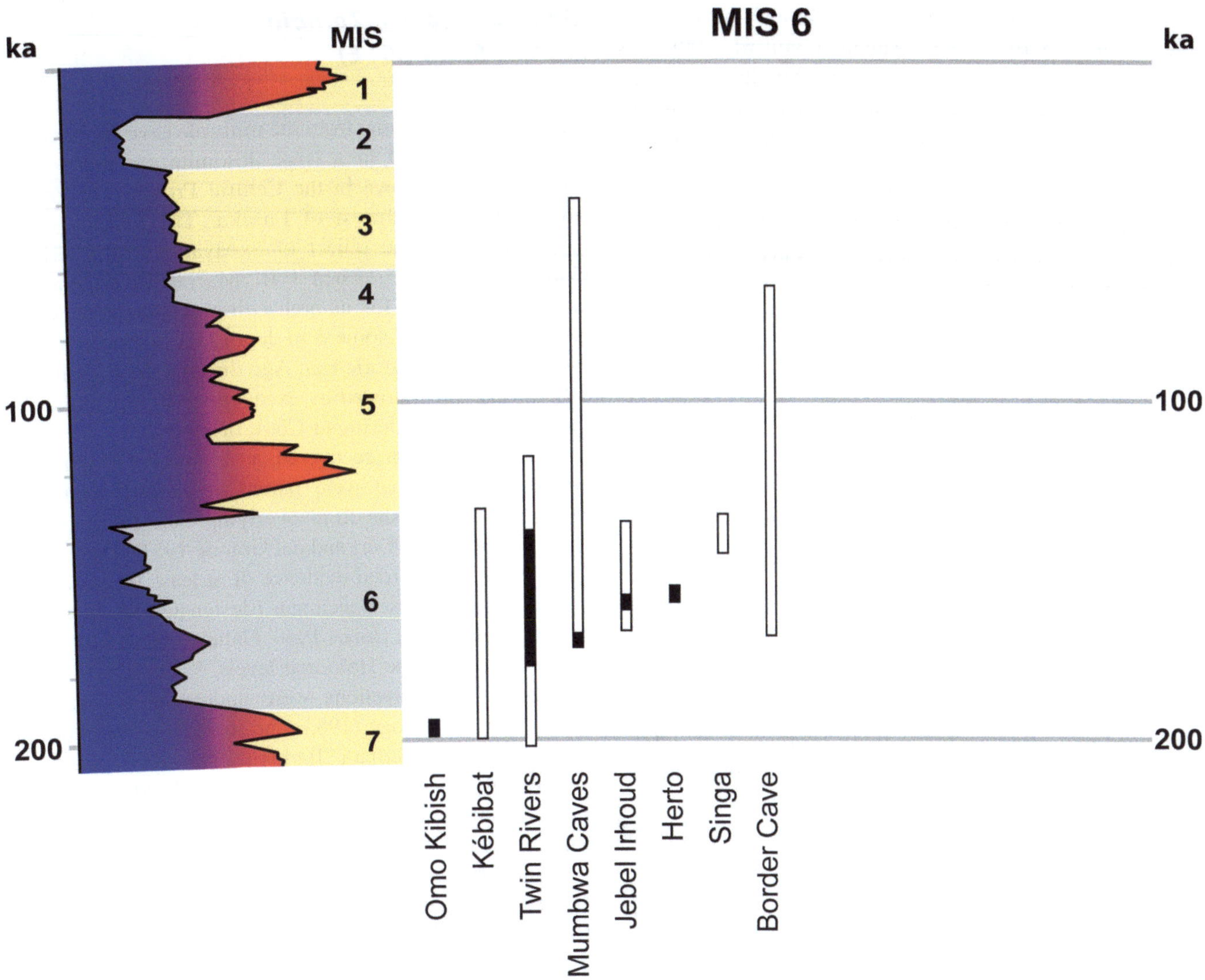

Fig. 17.2 Distribution of age determinations and/or estimates for hominin fossils and hominin-bearing sites that are most reasonably related to Marine Isotope Stage 6. *Open rectangles* represent possible age ranges deduced for a site or specimen; *solid rectangles* represent well-established dates for the hominin fossils

Member I of the Kibish Formation formed, probably in a short period of time, during a period of high water level in Lake Turkana correlated with Mediterranean sapropel 7 at c. 197 ka (McDougall et al. 2008). Ar–Ar dating of the Nakaákire Tuff just below the level of the fossils and the Aliyo Tuff well above to 196 and 104 ka, respectively, serves to bracket the fossils (McDougall et al. 2005, 2008; Millard 2008). More recently, F. Brown et al. (2012) have presented additional evidence that the KHS Tuff at the base of Member 2, which overlies the recovery sites of Omo I and Omo II, is best dated by correlation with other tuffs in East Africa to 172 ka. This more tightly brackets the Kibish Formation hominins between 197 and 172 ka, and their stratigraphic position within Member 1 places them much closer to 195 ka. In addition, Aubert et al. (2012) have conducted U-series dating on a fragment of the Omo I cranium, which yielded a minimum age of some 187–155 ka.

Kébibat, Morocco (34°02′ N, 6°51′ W)

Dynamite blasting in 1933 in the Mifsud-Guidice Quarry in a suburb of Rabat, Morocco resulted in the discovery of a fragmentary mandible, maxilla, and cranial vault of a juvenile individual (Rabat 1). Uranium-series (Th^{230}/U^{234}) dates on overlying seashells suggest an age in excess of 130 ka (Stearns and Thurber 1965). Although Millard (2008) has argued that nothing more precise than a Middle Pleistocene age can be assigned to the hominin remains, the biostratigraphic data are consistent with an age of some 200–130 ka (Hublin 2001).

The dentition, mandible, and cranial remains, that comprise Rabat I, have been described by several workers (e.g., Vallois 1945; Thoma and Vallois 1977; Saban 1975, 1977). This fossil was initially regarded as representing "evolved *Homo erectus*" by Saban (1975, 1977), but a more recent interpretation (Hublin 2001) attributes it to *Homo sapiens*.

Twin Rivers Kopje, Zambia (15°31′ S, 28°11′ E)

The site of Twin Rivers in eastern Zambia consists of several fissures filled with speleothem-cemented sediments that contain animal bones as well as MSA and LSA lithics. The fissures formed in a hill located between the Chikupi and Nangombi rivers, some 24 km southwest of Lusaka. The site was excavated by J.D. Clark in the 1950s (Clark and Brown 2001) and by L. Barham in the 1990s (Barham et al. 2000). Uranium-series and thermoluminescence (TL) dates from the speleothems have placed the MSA breccias in the range of c. 226–117 ka; the LSA material dates to between 60 ka and the Holocene (Barham et al. 2000).

The only human bone from the site is a fragment of humeral diaphysis described by Pearson (2000). Its sole noteworthy feature is its cortical thickness, which has been described as "strikingly thick relative to recent humans but similar to the thick-walled bones of other Pleistocene hominins" (Pearson 2000: 281).

The humeral fragment, which was recovered in the 1999 excavation, comes from sediments "within and beneath" an 8–10 cm thick ferro-manganese "crust" covering the top of Block F (Barham 2002). Uranium-series dates for two speleothem samples from Block F range between c. 178 and c. 139 ka (Barham et al. 2000: Table 10.1), but the presence of small lithic segments, which are a distinctive feature of the LSA in Zambia, may imply the possibility of intrusive contamination. Alternatively, as suggested by Barham (2002), they may simply attest to technological variability in the Lupemban. Given the striking thickness of the humeral cortex, it would seem unlikely to be intrusive from the LSA, so a date of c. 178–139 ka for it is accepted here as not being unreasonable.

Mumbwa Caves, Zambia (15°01′ S, 26°59′ E)

This site takes its name from the multiple cave entrances to a deep cavity formed in a large dolomite outcrop near the township of Mumbwa in the Central Province of Zambia, some 180 km northwest of Lusaka. The infilling of this cavity has been the object of excavation by a series of workers since 1925, when F.B. Macrae investigated the entrance of Cave I for its archaeological potential. Macrae dug a test trench of some 3 m depth that produced artifacts ranging from the historic/Iron Age through the LSA into the MSA. This was followed by excavations by Raymond Dart in 1930 and by J. Desmond Clark in 1939.

Dart's expedition recovered a number of fragmentary human skeletons that were buried at various depths into MSA-bearing deposits from overlying Late Pleistocene or Holocene horizons (Dart and del Grande 1931). Dart and del Grande (1931) recorded evidence of at least 16 individuals, but only four of these specimens (designated Mumbwa 1-4) have been preserved (Jones 1940; Gabel 1963). Clark (1942) found 17 teeth in the Holocene levels.

Subsequent excavations were undertaken in the early 1970s by D.K. Savage (1983) and in 1993–1996 under the direction of L.S. Barham (Barham 1995, 2000). Barham's expeditions yielded a small number of isolated human teeth and postcranial fragments, some of which come from the Holocene LSA strata (Pearson et al. 2000).

Although Protsch (1975, 1977) published a series of radiocarbon dates of between 20,450 ± 340 ^{14}C BP and 18,000 ± 370 ^{14}C BP for animal bones in apparent association with the four human specimens recovered by Dart, and a date of 19,780 ± 130 ^{14}C BP for bone collagen from Mumbwa 1, these dates are of very questionable validity. The LSA specimens could derive from MIS 2, as suggested by Protsch (1975, 1977), but they could equally derive from the Holocene. As such, Mumbwa Caves is not discussed below with reference to human skeletal remains associated with MIS 2 in sub-Saharan Africa.

However, the latest and most detailed excavations led by Barham recovered at least two teeth and two fragmentary radii that certainly derive from the MSA deposits – a third tooth is of possible MSA derivation and a partial femoral diaphysis is of probable MSA origin (Pearson et al. 2000). The MSA human remains are generally unremarkable. The teeth are somewhat large by comparison with modern southern African homologues, while the postcranial bones suggest individuals of comparatively small body size.

Thermoluminescence determinations on sand grains, burnt quartz, and calcite as well as ESR measurements on mammalian tooth enamel fragments suggest an age for the intermittent MSA occupations of the site of between 40 ka and somewhat greater than 172 ka (Barham and Debenham 2000). The MSA human remains derive from the lower units (X and XII), which are believed to date to $\sim$172 ka.

Jebel Irhoud, Morocco (31°56′ N, 8°52′ W)

Mining operations in 1961 and subsequent excavations from 1963 to 1969 of this cave site some 60 km southeast of Safi, Morocco resulted in the recovery of the skeletal remains of several individuals in association with MSA (Middle Paleolithic) artifacts. The first specimen (Irhoud 1) is represented by a nearly complete cranium, the second (Irhoud 2) by a calvaria, and the third (Irhoud 3) is a juvenile mandible. Excavations in 1969 uncovered a juvenile humeral diaphysis (Irhoud 4), and Hublin (2001) has more recently identified a fragmentary pelvis of another adolescent individual in a sample of breccia collected in 1969 by J. Texier.

Irhoud 1 and 2 were initially thought to have Neandertal affinities (Ennouchi 1962, 1963, 1968; Howell 1978; Bräuer 1984b), but other analyses have suggested instead a mixture of plesiomorphic traits and features shared with modern humans (Stringer 1974, 1978; Hublin 1992, 2001). When compared with the Skhul-Qafzeh group of fossils, the Irhoud specimens are only slightly more primitive (Hublin 2001). The teeth of the juvenile individual (Irhoud 3), which was also initially interpreted as belonging to a Neandertal (Ennouchi 1969), have been argued to represent the earliest evidence for slow, modern human-like development (Smith et al. 2007).

Electron spin resonance (ESR) spectroscopy on three artiodactyl teeth from a level immediately overlying one of the human specimens provided age ranges of 125–90 ka assuming an early uptake model, or 190–105 ka assuming a linear-uptake model (Grün and Stringer 1991). As noted by Millard (2008), however, these would appear to be ranges for the midpoint estimates rather than confidence ranges. More recent U-series and ESR dates on a tooth from the Irhoud 3 juvenile suggest an age of c. 160 ka (Smith et al. 2007). This is close to the maximum age of c. 150 ka suggested by Amani and Geraads (1993) on the basis of faunal comparisons.

Herto, Ethiopia (10°15′ N, 40°33′ E)

Six hominin specimens were initially recovered from the Upper Herto Member of the Bouri Formation in the Middle Awash of Ethiopia by a team led by J.D. Clark, T.D. White, B. Asfaw, and G. Wolde-Gabriel (White et al. 2003; Clark et al. 2003). Two of the specimens (BOU-VP-16/1 and, BOU-VP-16/2) comprise complete and partial adult crania, and a third (BOU-VP-16/5) is the cranium of a child. The other three specimens are isolated, fragmentary cranial bones, or teeth. The more complete fossils represent some of the most significant discoveries of early *Homo sapiens* made thus far, owing not only to the morphological information they possess, but also to their well-constrained geochronology and archaeological context. All of the specimens display evidence of human modification, such as cut marks, and the discoverers have interpreted this as resulting from mortuary practices that may or may not have included cannibalism.

The cranium is very large and shares with more ancient African fossils a wide interorbital breadth, a prognathic maxilla, and a short and angulated occipital. On the other hand, it displays a moderately domed forehead, a low and flat mid-face, and a globular calvaria, features that are more widely shared with more recent humans. White et al. (2003) attributed the Herto specimens to the "paleosubspecies" *Homo sapiens idaltu*.

These fossils have been dated between 160 and 154 ka on the basis of volcanic clasts from the fossil bearing unit, and by geochemical correlation of the overlying Waidedo Vitric Tuff (WAVT) with an unnamed tuff underlying the Konso Silver Tuff. Millard (2008) has recomputed the ages from the supplementary data provided in Clark et al. (2003) with stratigraphic ordering using OxCal, and has obtained an age constraint of 161–150 ka for the Herto fossils. Clark et al. (2003) and Morgan et al. (2009) have obtained $^{40}Ar/^{39}Ar$ ages ranging from 110 to 253 ka for obsidian clasts within the Herto Member. Although Brown et al. (2012a, b) have argued that the stratigraphic relation between the WAVT, which they observe as being probably near 172 ka, and the sedimentary sequence containing the Herto fossils has not been determined satisfactorily, a correlative age of c. 160 ka for the Herto hominin fossils seems likely pending further investigation.

Singa, Sudan (13°00′ N, 33°55′ E)

Singa 1 is a heavily mineralized partial human cranium. It was discovered in 1924 together with MSA artifacts eroding from a calcified hardpan (caliche) deposit on the west bank of the Blue Nile River some 320 km south of Khartoum.

The calcrete matrix that encrusted the cranium was dated by U-Th to c. 133 ka, which is within the ranges of ESR estimates from mammalian tooth enamel, being 159 and 89 ka on the basis of a late uptake and early uptake models, respectively (McDermott et al. 1996). Because the calcrete formed after the deposition of the calvaria, its age must constitute a minimum for the fossil (Millard 2008). However, the degree of preservation of the specimen would seem to suggest that it was not deposited too far in advance of the calcrete formation; thus, an age range of c. 145–133 ka would seem reasonable for Singa 1 (Millard 2008).

The cranium presents a number of modern features together with more archaic traits, such as a well-developed supraorbital torus. Its accentuated parietal bossing, which some early workers identified as being suggestive of Khoe-San affinities (Woodward 1938; Wells 1951), may be unduly developed as a result of cranial asymmetry and possibly some unidentified underlying pathology (Stringer et al. 1985; Spoor and Stringer 1998).

Border Cave, South Africa (27°01′ S, 31°59′ E)

This large cave is situated high on the steep western side of the Lebombo Mountains in KwaZulu-Natal, some 400 m from the border with Swaziland. It was discovered and first excavated in 1934 by R. Dart, who dug a thin, shallow trench into the deposit (Cooke et al. 1945). This was followed in 1940 by extensive digging for bat-guano by miners led by W.E. Horton, who discarded stone tools and bones in their dumps. B.D. Malan and colleagues renewed archaeological excavations in 1941–1942, and P. Beaumont undertook more extensive work in the 1970s. A deposit of nearly 5 m depth records MSA, LSA, and Iron Age occupations. In the upper part of the sequence, two layers dated to 44.0–42.0 ^{14}C kBP by AMS and conventional ^{14}C appear to mark the beginning of the LSA in South Africa (Bird et al. 2003; Villa et al. 2012), although this is some 20 ka earlier than evidenced at other dated sites (Deacon 1995; Vogel 2001; Bousman 2005). These early LSA layers have not yielded any human remains.

The cave contains thick deposits that preserve MSA 1, Howiesons Poort and MSA 3 artifacts (Butzer et al. 1978a). Estimates of the ages of the MSA-bearing horizons have been based on ESR of tooth enamel (Grün et al. 1990a, 2003; Grün and Beaumont 2001; Millard 2008) and/or amino acid racemization of eggshell (Miller et al. 1999). A series of unpublished luminescence dates cited by Grün and Beaumont (2001) also exist. These various estimates place the MSA sequence from perhaps 171–152 ka at the base to ~56 ka for MSA 3.

There are eight human specimens from Border Cave that have been said to derive from the MSA levels. A partial cranium (BC1) and an edentulous partial mandible (BC2) were found by Malan and colleagues in 1941 in the dumps created by Horton's guano diggings (Cooke et al. 1945). Although Beaumont (Beaumont et al. 1978; Beaumont 1980) has argued that these specimens derive from the MSA deposit – either layer 4BS or layer 5BS (BS = Brown Sands) – below the Howiesons Poort on the basis of tiny bits of adhering sediment, other workers (e.g., Klein 1983; Parkington 1990) have cited factors pertaining to their state of preservation which indicate that they may not derive from an MSA context. In particular, they are much better preserved than the faunal remains that have been excavated *in situ* from the MSA. Sillen and Morris (1996) obtained conflicting results in their assessment of the mineral crystallinity of BC2 bone, but their own sample provided a "young" signal in relation to the known MSA animal bones. While they obtained elevated crystallinity values for powdered samples of the BC 1 cranium taken by Kenneth Oakley, they had cause to doubt their validity in light of the mixed results they obtained from a similarly powdered sample from BC 2.

A nearly complete skeleton of a very young (c. 3-month-old) infant (BC 3) buried in a shallow grave was recovered in 1941 from the base of the Howiesons Poort sequence by Malan and colleagues (Cooke et al. 1945). It was posited to derive from a MSA layer immediately above. However, the state of preservation of the very fragile bones together with the fact that Sillen and Morris (1996) found the "splitting factor" of samples from BC 3 to differ noticeably from that of the *in situ* MSA fauna certainly indicate it to be intrusive from much younger levels. Sillen and Morris (1996) suggested that the BC3 infant is "associated with the Holocene" and given its clear association with grave goods in the form of a perforated *Conus* shell, they noted that it "would be archaeologically consistent with other [Holocene] sites in the region." Protsch (1975) reported one of his infamous isoleucine racemization dates for the BC 3 infant, but this estimate of c. 60 ka can be wholly disregarded.

An adult mandible (BC 5) was found by C. Powell, working with Beaumont in 1974. The jaw, which is nearly complete, is said to have come from a depression in the base of Layer 3WA (de Villiers 1975), and to be associated with fauna dated to c. 66 ka on the basis of ESR (Beaumont et al. 1978; Beaumont 1980; Grün and Beaumont 2001). It is stated to have been "recovered" while "collecting sediment samples… from the south face of Excavation 3A" (Grün et al. 2003: 155). More precisely, however, it was actually dislodged from the south face of the previous excavation wall during the process of cleaning it prior to the removal of sediment samples. It was *not* recovered using controlled excavation techniques. Sillen and Morris (1996) noted that BC 5 is very well preserved, with "both poor crystallinity and elevated nitrogen when compared with faunal bones from the 3WA" layer. In view of the discrepancies between BC 3 and BC 5, on the one hand, and the *in situ* MSA fauna on the other, Sillen and Morris (1996) concluded that these two human specimens cannot be connected to the MSA with any confidence.

Subsequent analysis by Grün et al. (2003) from a piece of tooth enamel from BC 5 yielded an ESR date of 74 ± 5 ka, which was held to corroborate their earlier assertions regarding its antiquity. However, this argument rests substantially (if not entirely) on the fact that their ESR date for BC 5 fits "exactly into the ESR age sequence of the faunal material" from the stratigraphic sequence that was determined by them (Grün et al. 2003: 165). In actual fact, examination of their data (see Grün et al. 2003: Fig. 9) reveals that the BC 5 value does not differ significantly from specimen 537c from Layer 2 BS, and that several of the faunal tooth enamel values from Layer 3WS cannot be differentiated from those for fragments from Layer 2 BS or even layer 1WA (said to be between 47 and 36 ka).

Almost all studies to date have concluded that the BC 1 cranium and the BC 2 and BC 5 mandibles are fully modern in morphology (de Villiers 1973, 1975; Rightmire 1979, 1984; de Villiers and Fatti 1982; Bräuer 1984a; Smith 1985; Fatti 1986; Smith et al. 1989). The only exceptions relate to the multivariate morphometric analyses undertaken by Cambell (1984), Ambergen and Schaafsma (1984) and van Vark et al. (1989). Cambell (1984) and Ambergen and Schaafsma (1984), who employed Rightmire's (1979) data in their analyses, indicated that BC-1 is statistically atypical for the modern southern African samples that were employed by Rightmire. van Vark et al. (1989), who included larger samples of modern humans from different geographical regions, argued that although BC-1 may look modern, it is "quite different from recent *Homo sapiens*." This same study also concluded that BC-1 is perhaps "only slightly more distinct from Asiatic *Homo erectus* than it is from recent *Homo sapiens*" (van Vark et al. 1989: 54)! As noted by Grine et al. (1991), such a patently absurd conclusion cannot but detract from the credibility of this kind of analysis.

Indeed, Rightmire (1979) has demonstrated distinct morphometric similarities of the Border Cave cranial and mandibular remains to the recent indigenous peoples of South Africa. This, in itself, is most unusual for cranial remains elsewhere in Africa that are in indisputable Late Pleistocene MSA contexts.

On the other hand, Morris (1992) briefly described several postcranial fragments – a humeral shaft, proximal ulna, and two metatarsals (from rays IV and V) – that had been discovered by P. Beaumont in 1987 from sediments that had slumped into the 1941–42 excavation pit. Because they derive from a disturbed context, they cannot be attributed to any particular stratum within the cave. Nevertheless, these bones, which have been referred to as BC 6, 7, and 8 (Grün and Beaumont 2001; Villa et al. 2012: SI page 2), are the most likely of all of the human remains from Border Cave to actually derive from the MSA (although where in the MSA sequence is impossible to determine at present). The reason for this is threefold.

In the first instance, they are generally more fragmentary that the other human remains and have the same macroscopic preservation as indisputable MSA faunal elements (Sillen and Morris 1996). In the second instance, samples from the humerus and ulna have the same degree of crystallinity and low nitrogen as the MSA fauna (Sillen and Morris 1996). Finally, unlike the other human remains, both the humerus and ulna display archaic features in common with other undeniable MSA hominins from South Africa (Pearson and Grine 1996; Pfeiffer and Zehr 1996). In particular, the proximal ulna is comparable to one from the MSA at Klasies River and to more ancient African homologues in the archaic pattern of the relative heights of the coronoid and olecranon processes (Pearson and Grine 1996), and the humeral diaphysis has thick cortical bone (Pearson and Grine 1996; Pfeiffer and Zehr 1996). Although morphology cannot be used to "date" a fossil, it is nonetheless noteworthy that these fragmentary postcranial bones are the only human specimens from Border Cave to differ in any notable manner from those of the Holocene (including the Iron Age and current) inhabitants of South Africa.

Human Fossils Associated with MIS 5 (130–71 ka)

Marine Isotope Stage 5 represents the penultimate interglacial period. Shackleton (1969) proposed its informal division into five substages. The first of these (5e) represents the interglacial maximum, when sea levels stood c. 6 m above the present and global temperatures were some 2 °C higher than today (Shackleton 1987; Jouzel et al. 1987). In sub-Saharan Africa, MIS 5 is thought to have been characterized by generally warm and wet conditions with a general expansion of forest over savannah regions (Pokras and Mix 1985). However, this stage is characterized by rather wide swings in oceanic temperature (Fig. 17.1), which must have been accompanied by significant climatic fluctuations across parts of Africa. Thus, while MIS 5 represents a humid interval in northern Africa (Castañeda et al. 2009), Lake Malawi experienced a lowstand for much of this time (Lyons et al. 2011; Scholz et al. 2011). Moreover, as noted above, regional differences in response to these global oceanic patterns of temperature fluctuation must have varied quite widely across different parts of Africa, making sweeping, continent-wide generalizations very difficult if not meaningless.

Fifteen hominin-bearing sites are grouped here as being associated or likely associated with MIS 5 (Fig. 17.3; Table 17.4). Some, such as Pinnacle Point appear to extend from MIS 6-5 (in this instance through to MIS 5c), but their human fossils derive from the latter. In other cases, such as Klasies River Main Site, the sequence and human remains

Table 17.4 African hominin-bearing sites and hominin fossils associated with MIS 5

Site/specimen	Country	Coordinates	Age (ka)
Ngaloba Beds (Laetoli)	Tanzania	03°13′ S, 35°11′ E	129
Blind River	South Africa	27°01′ S, 33°00′ E	118
Klasies River Main Site	South Africa	34°06′ S, 24°24′ E	115–58
Sea Harvest	South Africa	33°01′ S, 17°57′ E	110–71
Grotte de Contrebandiers	Morocco	33°55′ N, 07°00′ W	107–96
Dar es-Soltan II	Morocco	33°57′ N, 06°35′ W	?125–96
Eyasi	Tanzania	03°32′ S, 35°16′ E	104–92
Equus Cave	South Africa	27°37′ S, 24°38′ E	?103–30
Aduma	Ethiopia	10°25′ N, 40°31′ E	100–80
Pinnacle Point	South Africa	34°12′ S, 22°05′ E	90
Blombos	South Africa	34°25′ S, 21°13′ E	100–73
Ysterfontein 1	South Africa	33°20′ S, 18°09′ E	105–71
Witkrans	South Africa	27°37′ S, 24°38′ E	100–50
Plovers Lake	South Africa	25°58′ S, 27°46′ E	89–63
Haua Fteah	Libya	32°55′ N, 22°05′ E	80–68
Mumba Shelter	Tanzania	03°32′ S, 35°19′ E	78–60

extend upwards beyond MIS 5 (in this instance, into MIS 4). Other sites, such as Equus Cave, Ysterfontein 1, and Witkrans entail more loosely constrained dates that may extend from MIS 5-3. The 15 sites are discussed below in the order in which they appear in Fig. 17.3 and Table 17.4.

Ngaloba, Tanzania (3°13′ S, 35°11′ E)

In 1976, fossilized human cranial fragments were discovered *in situ* eroding out of the Ngaloba Beds at Laetoli, northern Tanzania (Day et al. 1980). The reassembled fragments formed a nearly complete adult cranium (LH 18) that displays a mixture of archaic and modern features (Day et al. 1980; Magori and Day 1983; Cohen 1996). In particular, the expanded parietals, rounded occiput, and low position of inion are seen as modern traits, while the receding frontal, large supraorbital tori, and thick vault bones are seen as archaic. It has been likened to specimens such as Omo I and Eyasi I (Magori and Day 1983).

The cranium derives from a horizon that includes a trachytic tuff that has been correlated with the "marker tuff" in the lower unit of the Ndutu Beds at Olduvai Gorge. Magori and Day (1983) stated that this has provided an estimate of some 120 ka for the cranium, but gave no basis for this assertion. Subsequently, Hay (1987) provided an age estimate of the "marker tuff" of 129 ± 4 ka on the basis of a U-Th date of bone from it. Manega (1993, 1995) revised the date to ~200 ka on the basis of amino acid racemization of ostrich eggshell. However, the dependency of this technique on temperature, as well as the moisture and pH histories of the matrix renders amino acid dating of material with unknown environmental histories a very perilous undertaking.

Blind River, South Africa (27°1′ S, 33°0′ E)

A nearly complete human femur lacking its proximal end was found by P. Laidler in 1933 just upstream from the mouth of the Blind River canyon, which opens to the Indian Ocean in East London, South Africa. The bone is fundamentally modern in morphology, and most of its features are compatible with those of indigenous South African populations (Wells 1935; Wang et al. 2008).

The specimen is an isolated find that is said to have come from an estuarine calcarenite of an old raised beach platform. Optically stimulated luminescence (OSL) assessments of two samples of sedimentary quartz grains extracted from just below the level from which the femur is said to have derived provided a weighted mean age of 118 ± 6 ka (Wang et al. 2008). This date gains some apparent support from the presence below the femoral horizon of marine/estuarine deposits at c. 10 m above current sea level. According to Wang et al. (2008), these deposits would correspond to the maximum marine transgression of MIS 5e.

Klasies River Main Site, South Africa (34°06′ S, 24°24′ E)

The "site" of Klasies River, also referred to as Klasies River Mouth (Singer and Wymer 1982), comprises five caves or rock shelters located between 0.5 and 2 km east of the mouth of the Klasies River on the Tsitsikamma Coast of South Africa. Its significance derives from the abundant archaeological debris and the 40 or so fragmentary human fossils from the thick MSA deposits of what has become known as

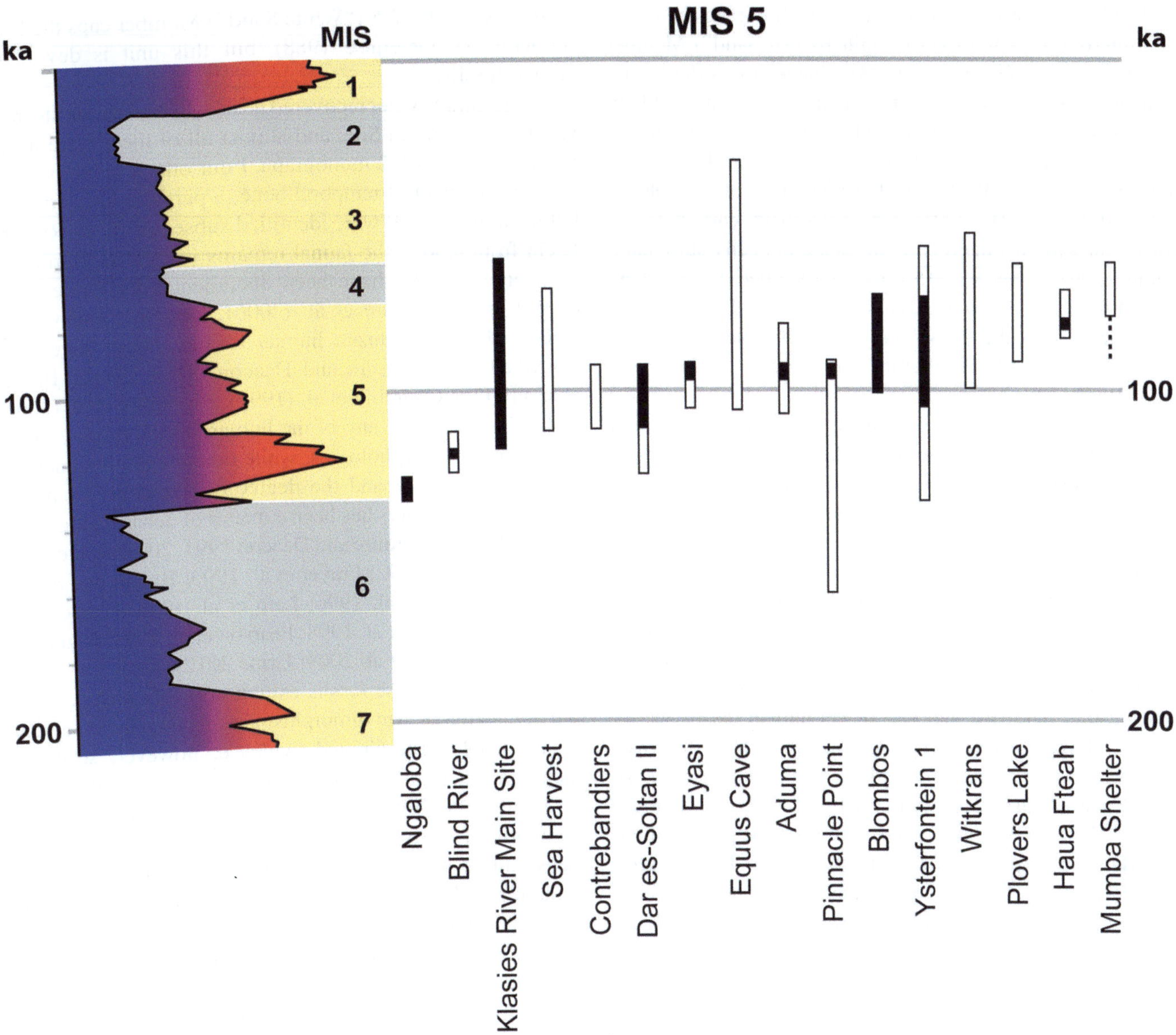

Fig. 17.3 Distribution of age determinations and/or estimates for hominin fossils and hominin-bearing sites that are most reasonably related to Marine Isotope Stage 5. *Open rectangles* represent possible age ranges deduced for a site or specimen; *solid rectangles* represent well-established dates for the hominin fossils; *dotted lines* indicate possible but not probable dates

the "Main Site." These artifacts and faunal remains have featured prominently in discussions over the emergence of modern human behaviors (e.g., Klein 1976, 1989; Deacon 1992; Klein and Cruz-Uribe 1996; Milo 1998; Wurz 1999, 2008; McCall 2006; Dusseldorp 2010; d'Errico et al. 2012). The hominin fossils, despite their generally fragmentary nature, have played a significant role in interpretations of human evolution (e.g., Singer and Wymer 1982; Smith 1992; Bräuer et al. 1992a; Frayer et al. 1993; Stringer and Bräuer 1994; Bräuer and Singer 1996; Churchill et al. 1996; Lam et al. 1996; Ahern and Smith 2004; Royer et al. 2009).

The Main Site consists of a c. 21 m deep deposit of sediments spread across a series of interrelated recesses referred to as Caves 1, 1A, 1B, and 2. Most of the work has focused on the Cave 1 and Cave 1A deposits. The first systematic excavation of the Main Site was undertaken by J. Wymer in 1967–1968. This resulted in the recovery of abundant stone tools and animal bones and the bulk of the human fossils found thus far (Singer and Wymer 1982). Subsequent excavations by H. Deacon in 1984–1995 provided an improved stratigraphic framework and a reasonable geochronology of the MSA deposits, together with some additional human fossils. The stratigraphic scheme provided by Deacon and Geleijnse (1988) recognized a number of members, and they managed to place most of the units described by Singer and Wymer (1982) into this context.

The lowermost sedimentary unit, which overlies bedrock, was referred to as the LBS ("Light Brown Sand") Member by Deacon and Geleijnse (1988). Shackleton (1982) and Deacon et al. (1988) have argued that it was most likely deposited c. 115–105 ka during MIS 5d, when sea level was very close to that at present, following the Last Interglacial transgression of MIS 5e. A U-series date of c. 108 ka obtained by Vogel (2001) for speleothem carbonates is consistent with this interpretation, as are optically stimulated luminescence dates of 115–110 ka recorded by Feathers (2002) for sand from Cave 1A. The two maxillae recovered by Deacon from the LBS Member are the oldest hominin fossils from the site (Rightmire and Deacon 1991; Bräuer et al. 1992a).

The LBS Member is overlain by the SAS ("Shell and Sand") Member, which comprises the bulk of the thickness of the deposit in Caves 1 and 1A (Deacon and Geleijnse 1988). It appears to have begun accumulation during MIS 5c at 105–92 ka (Shackleton 1982; Deacon et al. 1988),[2] and several dating methods have produced results consistent with an age of c.100–80 ka (Grün et al. 1990b; Vogel 2001; Feathers 2002). Although Millard (2008) has recalibrated the ESR ages from data in Grün et al. (1990a, b) to produce a wider range of dates, his recalibration is at odds with all other results. The vast majority of the human fossils recovered by Singer, Wymer, and Deacon derive from the SAS Member, and most are from its lower units (Deacon 2008).

In Cave 1A, the relatively thin RF ("Rock Fall") Member overlies the SAS Member. The oxygen isotope profile of the RF Member correlates with MIS 5a (84–74 ka) (Deacon et al. 1988). Vogel (2001) and Feathers (2002) have reported concordant U-series and luminescence dates of some 77 and 80–70 ka, respectively, for what seems to be the lower part of this unit. The stratigraphic sequence in Cave 1A is capped by the Upper Member. The majority of its thickness contains Howiesons Poort lithics, while the artifacts from its uppermost portion have been categorized as conforming to a distinctive post-Howiesons Poort MSA III by Singer and Wymer (1982; see also Wurz 2002; Villa et al. 2010). The oxygen isotope profile of the Upper Member is consistent with its deposition during MIS 4 at some 71 ka (Deacon et al. 1988). Several methods have produced dates between 65 and 40 ka (Grün et al. 1990b; Vogel 2001; Feathers 2002), but Jacobs et al. (2008a) obtained somewhat earlier dates of c. 65.5–63.4 ka for the Howiesons Poort, and a date of 57.9 ka for the post-Howiesons Poort. The RF and Upper Member deposits have yielded two human parietal fragments (Singer and Wymer 1982) and three isolated teeth (Rightmire and Deacon 1991).

In Cave 1, the WS ("White Sand") Member caps the SAS (Deacon and Geleijnse 1988), but this unit is devoid of human fossils.

Singer and Wymer recovered nearly 40 human fossils from Klasies River Main Site, and almost all of these were documented in their 1982 monograph. Four others (a lumbar vertebra, a fragment of temporal bone, a partial atlas, and a left hallucial metatarsal) were identified subsequently by Richard Klein from among the faunal remains recovered from those excavations. These have been documented by Grine et al. (1998) and Rightmire et al. (2006). Deacon's excavations recovered another dozen human fossils. These have been described by Rightmire and Deacon (1991, 2001), Bräuer et al. (1992a), Churchill et al. (1996), Rightmire et al. (2006) and Grine (2012). Many of the human fossils display manifestly modern morphologies, while others exhibit somewhat more archaic traits, and the degree of modernity or primitiveness of still others has been a matter of debate (Singer and Wymer 1982; Rightmire and Deacon 1991, 2001; Bräuer et al. 1992a, b; Smith 1992; Frayer et al. 1993; Bräuer and Singer 1996; Churchill et al. 1996; Lam et al. 1996; Pearson and Grine 1997; Grine et al. 1998; Pearson et al. 1998; Rightmire et al. 2006; Royer et al. 2009; Grine 2012).

A number of these fossils exhibit arguably archaic traits, and the degree of size dimorphism seems unmatched among recent people. The limited evidence, however, seems to suggest a pattern of general, albeit incomplete morphological modernity. As observed by Smith (1992: 148), the "somewhat primitive aspects of certain features in some specimens" might be expected in an otherwise morphologically modern population of this geological antiquity. The mosaic nature of human evolution is perhaps manifest by the Late Pleistocene representatives of the human lineage from Klasies. Alternatively, it is possible that these specimens do not represent individuals from a single population, but attest to different populations who utilized the site at different times.

Sea Harvest, South Africa (33°01′ S, 17°57′ E)

This locality comprises a dozen or so "pockets" of fossiliferous deposits overlain by MSA shell middens on the Atlantic coast of South Africa just north of Cape Town. These pockets, which were first investigated in 1969 by Q.B. Hendey (1984), most likely represent shallow solution cavities that were utilized as dens by hyenas. From them, G. Avery and R.G. Klein recovered a premolar and manual distal phalanx, which are morphometrically compatible with homologues of recent indigenous South Africans (Grine and Klein 1993).

The faunal (biochronological) and geological evidence are consistent in suggesting deposition during the relatively

[2]The ages for MIS 5c employed by Shackleton (1982) and Deacon et al. (1988) correspond to the SPECMAP dates recorded in Imbrie et al. (1984).

cool and moist conditions of MIS 5, corresponding perhaps to MIS 5c or MIS 5a (Hendey 1984; Grine and Klein 1993; Butzer 2004). Butzer (2004), however, has suggested that the Sea Harvest sequence is “an essentially complete record of MIS 5.”

La Grotte des Contrebandiers à Témara, Morocco (33°55′ N, 7°00′ W)

This site, also known Smugglers’ Cave, or simply as Témara (after the closest town), is located on the Atlantic coast of Morocco, some 17 km southwest of Rabat. It was discovered in 1955, and excavated in 1955–1957, 1967–1975, 1994, and 2007–2009 by different workers. The 1956 excavations yielded a mandible that was initially mistakenly thought to be associated with the Acheulean (Vallois and Roche 1958). The 1975 field season produced the back of a cranium in association with Aterian artifacts (Roche and Texier 1976; Debénath et al. 1986). A skull and partial skeleton of a child were discovered in these same strata in 2010 (Balter 2011). All of the human remains recovered thus far appear to derive from the Aterian layers, and they are, overall, morphologically modern, although the teeth are comparatively large (Ferembach 1976a, 1998; Debénath 2000; Hublin et al. 2012).

The antiquity of the human remains from the Aterian levels, as established by Jacobs et al. (2011) on the basis of single grain OSL determinations that average to 95.9 ± 4.1 and 107 ± 3.5 ka, are generally concordant with the bulk (multigrain) OSL estimates reported by Schwenninger et al. (2010). Scerri (2013: Table 2) has tabulated a very wide range of U-series, TL, and OSL dates reported for Aterian assemblages elsewhere in North Africa: from c. 145 ka at Ifri n’Ammar, Morocco (Richter et al. 2010) to c. 61 ka at Uan Tabu, Libya (Cremaschi et al. 1998). The OSL determinations for the Aterian at La Grotte des Contrebandiers à Témara fall very nearly in the middle of this range.

Dar es-Soltan II, Morocco (33°57′ N, 6°35′ W)

This cave site, situated on the Atlantic coast some 6 km southwest of Rabat, was first excavated in 1969 by A. Debénath. It has a deep (c. 8 m) archaeological deposit that comprises three Aterian horizons below Neolithic/Epipaleolithic occupations.

The remains of five hominin individuals reportedly come from the middle and lower Aterian strata. Unfortunately, most of these are not described. The most complete comprises a partial adult cranium and associated mandible (H 5) recovered from a sterile marine sand deposit (level 7) immediately below a layer containing Aterian artifacts (Debénath 1976, 1980). Ferembach (1976b) assigned the specimen to *Homo sapiens sapiens*, although noted that it has a relatively wide face and robust supraorbital relief. Bräuer and Rimbach (1990) have pointed out multivariate morphometric similarities with earlier specimens from North Africa and with later (Upper Paleolithic) specimens from Eurasia. Harvati and Hublin (2012) have argued on the basis of a 3D geometric morphometric analysis of the face that the H 5 specimen is similar to penecontemporaneous specimens from the Levantine site of Qafzeh. The teeth from this and the other Aterian specimens at the site (H4, NN, H6, H9, and H10) are comparatively large in relation to some modern human samples (Hublin et al. 2012).

Although there are no published dates for Dar es-Soltan II, the Aterian layers in the nearby (c. 200 m distant) Dar es-Soltan I cave have been dated by OSL to ~125 ka (Barton et al. 2009). This date is close to the oldest recorded for this industry (c. 145 ka at Ifri n’Ammar, Morocco [Richter et al. (2010)]), and is substantially older than the OLS dates of 107–96 ka (Schwenninger et al. 2010; Jacobs et al. 2011) for the Aterian at the nearby site of La Grotte des Contrebandiers à Témara. It is also significantly older than the ESR dates (Wrinn and Rink [2003], revised by Millard [2008]) of 57–27 ka for the Aterian at the proximate cave site of Mugharet el ‘Aliya. Because the Aterian levels in Dar es-Soltan II are themselves undated, it might be reasonable to place them within the limits of other Aterian deposits nearby at between 145 and 96 ka, with a somewhat greater probability of being in the range of 125–96 ka.

Eyasi, Tanzania (3°32′ S, 35°16′ E)

This “site,” from which the remains of a number of hominin individuals have been recovered, comprises several fossil-bearing localities on the shore of Lake Eyasi. The first human cranial remains (Eyasi 1-3) were recovered in the 1930s by Margit Kohl-Larsen, and subsequent discoveries, culminating in that by Dominquez-Rodrigo and colleagues, have brought the number of specimens to seven. All are fragmentary, with Eyasi 1 being the most complete; all appear to derive from strata (Member C of the Eyasi Beds) associated with the MSA (Mehlman 1987). Although earlier analyses of these fragments (e.g., Bräuer 1984b, Bräuer and Mabulla 1996) tended to emphasize their “archaic” qualities, Domínguez-Rodrigo et al. (2008) have stressed their more modern features, suggesting similarities to other early *Homo sapiens* specimens from sites such as Omo Kibish and Herto. It is perhaps noteworthy, however, that even Domínguez-Rodrigo et al. (2008) regard Eyasi 1 and 6 to show primitive features in a period where “other areas have yielded…substantially more modern-looking specimens.”

Initial dates for the Eyasi hominin fossils by Protsch (1981) using amino-acid racemization have been discredited (Mehlman 1984). Domínguez-Rodrigo et al. (2008) obtained ESR and U-series ($^{230}Th/^{234}U$) estimates of some 104 and 92 ka, respectively, for a bovid tooth found within reasonable proximity of the Eyasi 6 hominin frontal. Domínguez-Rodrigo et al. (2008) consider that the earlier finds could be similar or even older in age depending on whether they were over, or belonged to the top of, the red soils unit from which Eyasi 6 derives.

However, as noted by Millard (2008), ESR dates on tooth enamel cannot be relied upon without investigation of its uranium uptake history. As such, he has argued that these results most likely underestimate the true age; but by how much is uncertain.

Equus Cave, South Africa (27°37′ S, 24°38′ E)

This karst cave formed in the face of the Oxland tufa fan that built out from the Gaap Escarpment near Taung, some 193 km north of Kimberly, in the Northern Cape Province of South Africa. The site contains a rich record of vertebrate remains found in association with MSA artifacts. The bones appear to have been accumulated by hyenas (Klein et al. 1991). A number of isolated human teeth were recovered from these deposits during excavations led by P. Beaumont in 1978 and 1982, and a mandibular fragment found by C.K. Brain in 1971 on a scree slope below the cave may come from the same MSA horizons. The specimens are morphometrically unremarkable by comparison with recent homologues (Grine and Klein 1985), and the molars display enamel thickness and formation rates consistent with those of recent humans (Smith et al. 2006).

The Equus Cave fauna is dominated by grazing species, which suggests an open grassland, while pollens from coprolites indicate the environment to have been cooler and moister; this would perhaps correspond to the last glaciations (MIS 4-2) or to the colder intervals of MIS 5 (i.e., 5d and 5b) (Butzer et al. 1978b; Klein et al. 1991; Scott 1987). Millard (2008) reviewed the evidence pertaining to the geochronological age of the deposit, and concluded that it most likely belongs to the period between 103 and 30 ka.

Aduma, Ethiopia (10°25′ N, 40°31′ E)

Five human specimens were discovered on the surface of the MSA-bearing Ardu Beds at Aduma and Bouri along the Awash River of Ethiopia (Haile-Selassie et al. 2004). One partial cranium (ADU-VP-1/3) found partially *in situ* has the globular appearance of modern human homologues. The other four pieces from Aduma and the one from Bouri (BOU-VP-5/1) confirm this overall pattern of morphological modernity, although there is some resemblance as well to "premodern" specimens such as LH 18 (Haile-Selassie et al. 2004). Although Haile-Selassie et al. (2004) describe the presence of a suprainiac fossa in the Aduma specimen, this feature is not unique to (although it occurs in high frequencies in) Neandertals; similarly located depressions are observable on the occipitals of Middle Paleolithic *Homo sapiens* from Skhul and Qafzeh (Nara 1994), and in recent Khoe-San crania from the LSA of South Africa (personal observation).

A variety of dating methods (e.g., U-series, TL, and OSL) have been applied in attempts to determine the age of the Aduma deposits, but there are some inconsistencies in the results (Haile-Selassie et al. 2004). Argon-argon determinations on underlying pumice demonstrate that the deposits are younger than 180 ka (Yellen et al. 2005). Optically stimulated luminescence determinations of 91–93 ka are considered to be the most reliable, and are concordant with U-series estimates on associated bone of 105–79 ka (Yellen et al. 2005). An age range of c.100–80 ka is probable for these specimens, although Millard (2008) has placed little confidence on this estimate.

Pinnacle Point, South Africa (34°12′ S, 22°05′ E)

The locality of Pinnacle Point consists of a series of caves and rock shelters cut into cliffs around a small promontory that extends into the Indian Ocean near the town of Mossel Bay on the southern coast of South Africa. Survey work and excavations initiated by C. Marean and P. Nilssen explored three sites (PP 5-6, PP 9, and PP 13B), with the focus of the work being PP 13B. Pinnacle Point is perhaps best known for providing the oldest evidence (from PP 5-6) for heat treatment of rock to produce stone tools (Brown et al. 2009) and microlithic technology dating between 71 and 60 ka (Brown et al. 2012a, b), and (from PP 13B) the use of scraped and ground ochre and the early systematic exploitation of marine resources at c. 164 ka (Marean et al. 2007; Jerardino and Marean 2010; Watts 2010). It is one of the few archaeological sites in Africa with a depth of deposits to sample MIS 6.

The artifact- and faunal-bearing sediments at PP 13B do not comprise a deep, continuously stacked series, but rather a series of horizontally disconnected set of occupations that seem to span the period of time from c. 162 to 90 ka (Jacobs 2010; Bar-Matthews et al. 2010). PP 5-6 has a ~14 m vertical sequence of sediments dating from about 90 to 50 ka (K. Brown et al. 2009, 2012).

The only human remains recovered so far – a parietal fragment and a mandibular central incisor – likely derive from the upper part of the sequence in PP 13B. They were found in 2000 in deposits that had been disturbed, probably by fishermen or campers digging to produce a flat surface in the cave (Marean et al. 2004). The disturbed sediment was mostly from the MSA Brown Sand facies, which is almost certainly the source of the fossils (Marean et al. 2004). There is no LSA deposit above the areas where the fossils were found, and the cave appears to have been sealed to human occupation from 90 to 40 ka by a dune, as were most of the caves on the western side of the point (Bar-Matthews et al. 2010). Thus, there is reason to be confident that the human remains date to "at least 90 ka, but probably not much older than that" (Marean personal communication). The parietal fragment is thin, whereas the incisor is relatively large in comparison to recent African homologues (Marean et al. 2004).

Blombos Cave, South Africa (34°25′ S, 21°13′ E)

Blombos Cave is situated adjacent to the Indian Ocean along the southern Cape coast of South Africa, some 300 km east of Cape Town. This wave-cut shelter contains LSA deposits overlying some 2 m of stratified MSA deposits. The site has been excavated since 1992 under the leadership of C. Henshilwood (Henshilwood et al. 2001). The MSA-bearing layers have been designated M1, M2, and M3 from top to bottom.

The site is best known for the artifacts from the M1 and M3 phases. In the uppermost M1, these include pressure flaked bifacial Still Bay foliate points fashioned on heat treated stone, pierced gastropod shell beads, pieces of engraved ochre, and formal bone tools (Henshilwood et al. 2002, 2004; Mourre et al. 2010; Henshilwood 2012). The M3 phase has produced evidence of engraved ochre and abalone shells that appear to have been used to store and/or produce a pigment-rich mixture (Henshilwood et al. 2009, 2011).

Nine isolated human teeth have been recovered from the MSA strata: three from level M1, one from level M2, and five from the lowermost M3 (Grine et al. 2000; Grine and Henshilwood 2002). About half of the tooth crown diameters exceed those of recent Africans, while the rest fall among modern African sample means. The specimens are morphologically similar to those of recent Africans and to penecontemporaneous homologues from sites such as Die Kelders (Grine et al. 2000; Grine and Henshilwood 2002).

The upper part of the M1 Still Bay phase has been dated by OSL to 72.7 ± 3.1 ka (Jacobs et al. 2003a, b, 2013), with TL dates of 78 ± 6 and 74 ± 5 ka for this phase of occupation (Tribolo et al. 2006). The OSL dates for the M2 phase fall between 84.6 ± 5.8 and 76.8 ± 3.1 ka (Jacobs et al. 2006). The M3 phase dates to c. 100–94 ka on the basis of OSL and U-Th determinations (Henshilwood et al. 2009, 2011). Thus, the human remains from the MSA strata at Blombos would seem to date to between about 100 and 73 ka; that is from MIS 5c to the end of MIS 5a.

Ysterfontein 1, South Africa (33°20′ S, 18°09′ E)

This rock shelter on the Atlantic Coast of the Western Cape Province of South Africa contains nearly 4 m of stratified sands with classic MSA artifacts. The site was first visited in the early 1980s by G. Avery, and a series of excavations the late 1990s and early 2000s led by R.G. Klein, R. Yates and D. Halkett recovered evidence for the early exploitation of coastal resources (Halkett et al. 2003; Klein et al. 2004; Avery et al. 2008). This site has yielded the largest lithic assemblage attributable to MIS 5 from any site on the southwestern coast of South Africa, and this technocomplex is dominated by quadrilateral flakes and blades (Wurz 2012). Three unpublished human teeth have been recovered from these MSA deposits (Klein personal communication).

Although four OSL dates of 132–120 ka have been obtained from sand samples (Jacobs personal communication, cited in Avery et al. 2008), Avery et al. (2008) have argued that the site must have formed after the high sea level stand of MIS 5e (i.e., after 123 ka), and suggest that it was occupied sometime during MIS 5c to 5a (i.e., between 105 and 71 ka). Another perhaps somewhat less likely possibility discussed by Avery et al. (2008) is that it was occupied in MIS 3 between c. 59 and 50 ka.

Witkrans Cave, South Africa (27°37′ S, 24°38′ E)

This site, like Equus Cave, formed as a karst feature in a travertine apron that built out from the Gaap Escarpment near Taung, some 193 km north of Kimberly, in the Northern Cape Province of South Africa. Excavations by F. Peabody in 1947–1948 resulted in the recovery of a large number of MSA artifacts and faunal remains and, in association, three isolated human mandibular molars from Layer C. The teeth are similar in size and morphology to homologues from other South African MSA sites such as Die Kelders and Klasies River Main Site (McCrossin 1992).

Clark (1971) likened the fauna to the Florisian (Land Mammal Age), and the artifacts to the MSA 2 (pre-Howiesons Poort) of Klasies River Main Site. This would suggest an age of c. 100–80 ka (MIS 5c-4). Volman (1984), however, has argued that the Witkrans artifacts are attributable to the post-Howiesons Poort, MSA 3, and this, in turn, would suggest an age of c. 52 ka or less (i.e., attribution to MIS 3).

Plovers Lake, South Africa (25°58′ S, 27°46′ E)

The "internal deposit" of the Plovers Lake karst feature consists of loosely calcified sediments with faunal remains and MSA artifacts that are constrained by underlying and capping flowstones (de Ruiter et al. 2008). The human remains from this deposit comprise several isolated teeth and seven postcranial pieces from a single individual, of which only the tibia has been afforded either description or illustration, and this only in relation to its state of preservation (Backwell et al. 2012).

The age of the "internal" deposit is constrained by U-series dates of 88.7 ± 1.6 and 62.9 ± 1.3 ka for the underlying and capping flowstones, respectively. A single isochron ESR date of 75.6 ± 5.6 ka is consistent with this constraint (Backwell et al. 2012).

Haua Fteah, Libya (32°55′ N, 22°05′ E)

This large cave is situated on the Mediterranean coast of northeastern Libya (Cyrenaica). It was recognized by C.T. Houlder and C. McBurney in 1948 and excavated by the latter from 1951 to 1955. These excavations, which extended to a depth of some 13 m, uncovered levels that span occupations from recent centuries through the Holocene into the Middle Paleolithic (McBurney 1967; Barker et al. 2010; Hunt et al. 2010). McBurney (1967) recognized 35 archaeologically rich levels overlying what he referred to as the "Phase A" basal "Pre-Aurignacian." He divided the "Phase B" or "Levalloiso-Mousterian" occupation layers, which constitute the bulk of the Pleistocene deposit, into four units. This is overlain by "Phase C" Upper Paleolithic "Dabban" deposits and three subsequent occupation phases extending into the Graeco-Roman period. Current excavations at the site were initiated in 2007 under the direction of G. Barker, C. Hunt, and T. Reynolds (Barker et al. 2007, 2010, 2012; Hunt et al. 2010; Reynolds 2013; Jones et al. 2016). This work has provided an improved appreciation of the stratigraphy, archaeology, and age of the site.

Two human mandibular fragments – one young adult (Haua Fteah I) and one juvenile (Haua Fteah II) – were recovered in 1952 and 1955 from Layer XXXIII, one of the lower "Levalloiso-Mousterian" levels (McBurney et al. 1953a, b; McBurney 1958, 1967; Tobias 1967). Initial assessment of these jaws suggested similarities to the "Tabun group" of "Neandertaloids" (McBurney et al. 1953a, b; McBurney 1958). Tobias (1967) agreed with this assessment, viewing them as "part of an advanced Neandertaloid population" where the "Neandertaloids" represented a phase in the evolution of humans pertaining not only to Eurasia but also to sub-Saharan Africa. Tobias (1967) also saw some similarities to "remains described as a primitive variant of Neanderthal man (*Homo sapiens rhodesiensis*)." Rak (1998), however, astutely observed that both rami are symmetric in appearance with equal-sized condylar and coronoid processes separated by a deep and symmetrical notch, and that this "anatomy is diagnostic enough to confidently rule out the possibility that they are Neanderthals and strongly suggests that they are *Homo sapiens.*" Hublin (2000) subsequently pointed out some additional features (e.g., the lack of a retromolar space and the shape of the mandibular foramen) that argue against their Neandertal affinity.

McBurney (1967) estimated the "Phase A" basal "Pre-Aurignacian" layers to be 80–65 ka on the basis of climatic reconstructions. Some workers (e.g., Meignen 2007; Foley et al. 2013) have seemingly accepted McBurney's (1967) description of this as a blade-artifact assemblage, which would most closely resemble the "Amudian" at sites such as Qesem and Tabun in Israel (Garrod 1970; Jelinek 1990; Copeland 2000; Meignen 2000; Barkai et al. 2005). The relevant level at Tabun (Layer Ea of Garrod and Unit XI of Jelinek) has been dated by TL to some 264 ka (Mercier and Valladas 2003), and the Acheuleo-Yabrudian, of which the Amudian is a facies at Qesem has been dated by $^{230}Th/^{234}U$ TIMS to between 382 and 152 ka (Barkai et al. 2003). However, Reynolds' (2013) analysis of the "Pre-Aurignacian" lithics from Haua Fteah indicates that all of its "typological elements" can be found with variable degrees of expression in the overlying Levalloiso-Mousterian, and that it has closer affinities to the overlying Middle Paleolithic assemblage at Haua Fteah than to anything else. As such, the "Pre-Aurignacian" provides no information relating to the age of the human jaws that overlie it.

The lowermost "Levalloiso-Mousterian" layers (XXX–XXXVI) were interpreted as possibly representing the early cooler phase of the last glaciation on the basis of paleotemperature reconstructions from marine mollusk shells (Emiliani et al. 1964; McBurney 1967). Klein and Scott (1986), however, have argued that layers XXII–XXXV may equally well belong to MIS 4 and/or MIS 3 on the basis of faunal composition.

Vogel and Waterbolk (1963) obtained a conventional radiocarbon date for burnt bone from Layer XXXIII of 45,050 ± 3200 ^{14}C BP. Employing this date, McBurney (1967) considered this BII phase of occupation to have begun ~55 ka, and to have been followed by a stage (Phase

BIII) of "probably Aterian character." However, the validity of this radiocarbon date has been discounted by some workers (e.g.; Klein and Scott 1986; Hublin 2000; Klein 2009). Indeed, Lowe et al. (2012) have recently identified Campanian Ignimbrite deposits that date to 39.28 ± 0.11 ka (DeVivo et al. 2001; Fitzsimmons et al. 2013) in Haua Fteah. These ashes are within a continuous sequence of Upper Paleolithic ("Dabban") deposits that clearly overlie the "Levalloiso-Mousterian" layers from which the human jaws derive (Lowe et al. 2012). Thus, the mandibles must predate 40 ka; Klein (2009) has suggested that the jaws may be anywhere between 50 and 130 ka.

Although Hublin (2000) states that the hominin-bearing level underlies the Aterian, which might establish the upper age limit for it, Haua Fteah does not preserve an Aterian assemblage (Scerri 2013; Reynolds 2013). Indeed, even if the Aterian was evident in Haua Fteah (as Phase BIII overlying the jaws in Phase BII), dates for the Aterian in North Africa (synthesized in Scerri 2013) range from c. 145 ka at Ifri n'Ammar, Morocco (Richter et al. 2010), to c. 61 ka at Uan Tabu, Libya (Cremaschi et al. 1998).

More recent work at Haua Fteah, however, indicates that the lower "Levalloiso-Mousterian" levels at the site date to the latter part of MIS 5 (MIS 5a) or the earlier part of MIS 4 (Lane et al. 2011; Reynolds 2013; Douka et al. 2014). Lane et al. (2011) obtained an age of c. 67.5 ka for a sample of hidden, diminutive volcanic ash layers (cryptotephra) just above the level from which the human jaws were extracted; this tephra layer now has a more secure varve date of 68,620 ± 2060 BP (Douka et al. 2014). Additional work has suggested that these early Levalloiso-Mousterian levels at Haua Fteah date to ~80 ka (Douka et al. 2014). In particular, the human mandibles are likely to date to this interval (Barker personal communication).

Mumba Shelter, Tanzania (3°32′ S, 35°19′ E)

This rock shelter is located in close proximity to Lake Eyasi. It has yielded three isolated, comparatively small human molars in an MSA context (Bräuer and Mehlman 1988). The molars derive from the basal deposits (Bed VI-B) for which uranium-thorium and uranium-protactinium measurements on bone have yielded average estimates of some 132 and 109 ka, respectively (Bräuer and Mehlman 1988). However, as noted by Millard (2008), U-series dates on bone cannot be relied upon without investigation of their uranium uptake history. More recently, Gliganic et al. (2012) have obtained both optically stimulated and infrared stimulated luminescence (OSL and IRSL) measurements on single grains of quartz and multigrain aliquots of potassium feldspar from the MSA horizons. These have provided dates of 63.4 ± 5.7 and 73.6 ± 3.8 ka respectively for the Bed VI-A deposits that overlay the human teeth. The MSA human remains are unlikely to be much older given their degree of stratigraphic separation from the dated horizons.

Human Fossils Associated with MIS 4 (71–57 ka)

A major shift in global climatic conditions occurred from the warmer, interglacial conditions of MIS 5 to the significantly cooler glacial conditions of MIS 4. Some have speculated that this shift, at ~71 ka, coincides with the Toba super-eruption, which would have caused a long-lasting "volcanic winter" triggering the MIS 4 Ice Age due to the huge volume of gas and ash that it injected into the atmosphere (e.g., Rampino and Self 1992, 1993). However, this specific trigger event has been questioned by several studies, which found no evidence for persistent climatic perturbation associated with it (e.g., Zielinski et al. 1996; Schulz et al. 2002). In fact, the Mt. Toba eruption appears to precede the onset of MIS 4 by ~3 kyr or more (Mark et al. 2014). Moreover, ash from the Toba eruption (the Youngest Toba Tuff) has been identified in the sediment bed of Lake Malawi, and this is not accompanied by any change in sediment composition or evidence for substantial temperature change. This eruption does not seem to have significantly impacted the climate of East Africa (Lane et al. 2013).

In Africa, MIS 4 is generally asserted to have witnessed relatively dry conditions (e.g., Avery 1982; Dupont 2011), although there was almost certainly significant variation at the local scale (e.g., Chase 2010). Indeed, Adams (1997) has provided a model that predicts much of sub-Saharan Africa to have been covered in grassland at this time, and water levels in Lake Malawi varied throughout MIS 4 (Lyons et al. 2011; Scholz et al. 2011). Chase (2010) has argued that in southern Africa, there is a relationship between the climatic conditions of MIS 4 and the occurrence of archaeological traditions such as the Still Bay and Howiesons Poort.

Three sites in South Africa (Die Kelders, Klipdrift Shelter, and Sibudu) and one site in Ethiopia (Porc-Épic) have deposits that began to accumulate towards the onset of MIS 4. Deposition at Sibudu and Porc-Épic appears to have continued throughout much of the succeeding MIS 3. The human fossils from Die Kelders and Klipdrift Shelter most likely date to MIS 4, and the human jaw from Porc-Épic probably also dates to this interval (Table 17.5; Fig. 17.4). The human bones from Sibudu date to MIS 3.

Table 17.5 African hominin-bearing sites and hominin fossils associated with MIS 4

Site/specimen	Country	Coordinates	Age (ka)
Porc-Épic	Ethiopia	09°34′ N, 41°53′ E	78–36
Die Kelders	South Africa	34°32′ S, 19°22′ E	74–59
Klipdrift Shelter	South Africa	34°27′ S, 20°43′ E	64

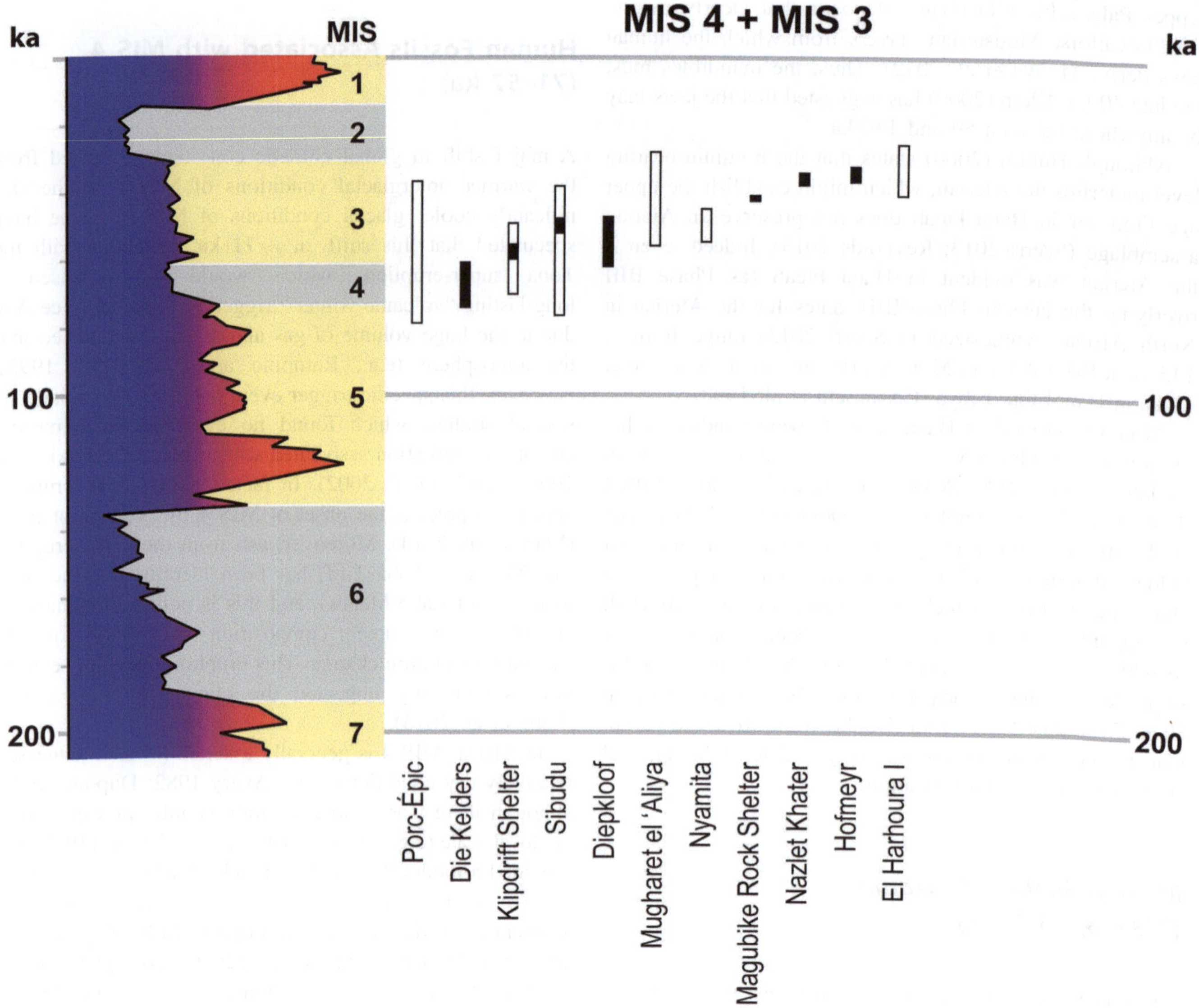

Fig. 17.4 Distribution of age determinations and/or estimates for hominin fossils and hominin-bearing sites that are most reasonably related to Marine Isotope Stages 4 and 3. *Open rectangles* represent possible age ranges deduced for a site or specimen; *solid rectangles* represent well-established dates for the hominin fossils

Thus, there are only three sites whose hominin remains are considered here as being associated (or likely associated) with MIS 4 (Fig. 17.4; Table 17.5). These three are discussed below in the order in which they appear in Fig. 17.4 and Table 17.5.

Porc-Épic, Ethiopia (9°34′ N, 41°53′ E)

This cave site is located in a steep cliff face some three km south of the town of Diré Dawa in southeastern Ethiopia. It was first excavated in 1929 by P. Teilhard de Chardin

(Teilhard de Chardin 1930). In 1933 a fragment of human mandible was recovered during excavations led by Teilhard de Chardin and P. Wernert (Breuil et al. 1951) The specimen was initially assessed by Vallois (1951) as an Ethiopian "neanderthaloid." Bräuer (1984a) described it as lacking specific Neandertal features (e.g., a retromolar gap), but he noted that it exhibits some archaic traits (e.g., a vertical symphysis without a chin).

Unfortunately, details of stratigraphy and specimen location from these early excavations were not well documented, and it is possible that some of the materials described as coming from the MSA strata are intrusive from the overlying LSA (Clark and Williamson 1984). In 1974, excavations were undertaken under the leadership of J. Desmond Clark with the object of reconstructing the stratigraphic sequence in the cave, and providing additional MSA artifacts (Pleurdeau 2005). Obsidian hydration dating provided ages for the MSA occupation of between 61.2 and 77.5 ka (Michels and Marean 1984). Unfortunately, the dated artifacts are from the 1933 excavation and lack adequate stratigraphic provenance, and AMS dating of gastropod shell yielded an age range of between >43.2 and c. 35.6 ^{14}C kBP (Assefa 2006). Although Millard (2008) concluded that a date no more precise that "later Pleistocene" can be attributed to the human mandible, it would seem likely to date to somewhere between 78 and 36 ka, and probably closer to the former if it does (as seems likely) come from the MSA levels.

Die Kelders, South Africa (34°32′ S, 19°22′ E)

This site, also known as Klipgat, consists of a pair of contiguous caves situated at sea level on the southeastern Atlantic coast of South Africa. Initial excavations by F. Schweizer between 1969 and 1973 revealed a rich LSA occupation and a thick series of MSA layers below. The bulk of the MSA is situated in Cave 1 (referred to as DK1). Schweitzer's excavations produced nine isolated human teeth from the MSA levels (levels 4–17), with most deriving from level 6 (Grine et al. 1991). Subsequent excavations of DK1 between 1992 and 1995 led by C.W. Marean, R.G. Klein, G. Avery and F.E. Grine resulted in the recovery of an additional 18 human specimens comprising isolated teeth, a small mandibular fragment and two manual phalanges (Grine 1998, 2000). These are distributed from level 4 through level 15, but most come from level 6.

As might be expected for teeth of this antiquity, most of the crowns tend to be large in comparison to recent African homologues. However, the majority of morphological variants displayed by the DK1 teeth characterize those of recent sub-Saharan Africans, and they resemble recent African homologues in a number of traits that have been used to define a sub-Saharan African regional complex. These morphological similarities, however, do not necessarily signify a close evolutionary relationship between them, because the crowns variants appear to be plesiomorphic (Grine 2000). The teeth also evince a modern human pattern of crown development (Smith et al. 2006).

Luminescence and early uptake model ESR dates place the deposits in the range of ~70–60 ka (Feathers and Bush 2000; Schwarcz and Rink 2000), and sedimentological considerations suggest an age for the entirety of the MSA in DK1 between some 74–59 ka, that is, within MIS 4 (Grine 2000). Millard (2008) has employed a Bayesian stratigraphic model incorporating the luminescence and early uptake ESR dates published by Feathers and Bush (2000) and Schwarcz and Rink (2000) to deduce a considerably wider range of possible ages of 85–48 ka. This is certainly within the realm of possibility, although the close proximity of the cave mouth to the present shoreline would seem to argue against deposition during MIS 5a, as the rise in sea level at this time would have scoured the cave (Grine et al. 1991; Grine 2000; Goldberg 2000).

Klipdrift Shelter, South Africa (34°27′ S, 20°43′ E)

This site, together with Klipdrift Cave and Klipdrift Cave Lower, comprises the Klipdrift Complex located in the De Hoop Nature Reserve, southern Cape Province, South Africa. The complex is situated some 12–15 m from the Indian Ocean coast. Test excavations under the leadership of C. Henshilwood and K. van Niekerk in 2011 in the Klipdrift Shelter revealed a c. 1.6 m deep deposit of horizontally stratified MSA deposits immediately below a steeply sloping, eroded surface. Subsequent excavations in 2012 and 2013 showed these layers to contain marine shells, terrestrial faunal remains (dominated by tortoise and small mammals), ash lenses and hearths, ochre, engraved ostrich eggshells, and lithic artifacts (Henshilwood et al. 2014).

The MSA levels at Klipdrift Shelter have been dated by single-grain optically stimulated luminescence (OSL). The artifacts in the uppermost level appear to relate to a post-Howiesons Poort industry, and these are associated with an OSL date of 51.7 ± 3.3 ka. The lowermost stratigraphic level at the site, which contains no anthropogenic material, has been dated to 71.6 ± 5.1 ka. The middle layers contain abundant artifacts attributed to the Howiesons Poort, and are dated to between 65 and 60 ka.

An isolated human mandibular second deciduous molar crown was recovered from one of the lowermost HP layers (designated PBE), and would appear to be most closely

Table 17.6 African hominin-bearing sites and hominin fossils associated with MIS 3

Site/specimen	Country	Coordinates	Age (ka)
Sibudu	South Africa	29°31′ S, 31°05′ E	48
Diepkloof Rock Shelter	South Africa	32°25′ S, 18°30′ E	61–48
Mugharet el 'Aliya	Morocco	35°45′ N, 05°56′ W	57–27
Nyamita	Kenya	00°24′ S, 34°10′ E	55–45
Magubike Rock Shelter	Tanzania	07°45′ S, 35°28′ E	42
Nazlet Khater	Egypt	26°47′ N, 31°21′ E	38
Hofmeyr	South Africa	25°58′ E, 31°34′ S	36
El Harhoura I	Morocco	33°57′ N, 06°55′ W	41–26

associated with a date of c. 64 ka (Henshilwood et al. 2014). Unlike penecontemporaneous homologues from Die Kelders, where most of the crowns tend to be large in comparison to recent African homologues, the Klipdrift molar is comparable in size to recent South African dm_2s. Like the other deciduous teeth that have been recovered from the Howiesons Poort levels at sites such as Klasies River Main Site (Grine 2012) and Diepkloof Rock Shelter (Verna et al. 2013), the Klipdrift Shelter crown is comparatively small (Harvati et al. 2015). The majority of morphological variants displayed by the Klipdrift molar also characterize those of recent sub-Saharan Africans, and it resembles recent African homologues in a number of traits that have been used to define a sub-Saharan African regional complex. These morphological similarities, however, appear to be plesiomorphic (Grine 2000). The sole exception pertains to the presence of a "mid-trigonid" crest on the Klipdrift crown; this variant is not unknown among modern San and South African populations, but it is comparatively rare (at some 3–8%).

Human Fossils Associated with MIS 3 (57–29 ka)

During the last glacial period of MIS 3, global climate underwent strong millennial-scale fluctuations. The Northern Hemisphere experienced repeated abrupt warming phases (Dansgaard-Oeschger events), characterized by shifts of up to 16 °C between stadial (cold phase) and interstadial (warm phase) climates that occurred sometimes on the order of a few decades (Dansgaard et al. 1993). These abrupt climatic shifts, although best expressed in the North Atlantic, were almost certainly important on a global scale, with effects seen in the South Atlantic as well (Crowley 1992; Stocker 1998; Compton 2011; Hessler et al. 2011). However, in the context of the abrupt climatic perturbations that characterized MIS 3, only sparse terrestrial paleoecological information is available, and much of this comes from sites located in Mediterranean Europe (Voelker et al. 2002; Ziegler et al. 2013).

North Africa experienced a generally humid period (Castañeda et al. 2009) and Lakes Malawi and Tanganyika experienced highstands throughout most of MIS 3 (Trauth et al. 2003; Lyons et al. 2011; Scholz et al. 2011), but evidence from Lake Tanganyika also reveals changes in lake water temperature and precipitation, indicating pulses of aridity that are consistent with Northern Hemisphere Heinrich events (Tierney et al. 2008). Vegetation and climatic changes for at least some of MIS 3 in South Africa have been recorded from fossil plant material in Sibudu Cave (Bruch et al. 2012). This work has shown that climatic changes were reflected most pronouncedly by winter temperatures (especially minimum winter temperatures), and to a lesser extent by changes in summer precipitation. Ecological trends seem to be affected mainly by the former.

Mitchell (2008) has reviewed the available paleoecological information for South Africa during MIS 3, and while he has identified a good number of archaeological sites that he argues date to MIS 3, the human paleontological record for this period in southern Africa – indeed, throughout Africa – remains woefully small. There are eight sites throughout the entire continent whose human remains date to MIS 3 (Table 17.6; Fig. 17.4). These are discussed below in the order in which they appear in Fig. 17.4 and Table 17.6.

Sibudu Cave, South Africa (29°31′ S, 31°05′ E)

This rock shelter, located some 15 km inland from the Indian Ocean in KwaZulu-Natal, was discovered and first excavated by A. Mazel in 1983. Current excavations, under the direction of L. Wadley, began in 1998 and these have demonstrated a stratified sequence of intermittent occupation with Still Bay, Howiesons Poort and final MSA phases (Wadley and Jacobs 2004, 2006; Wadley 2007). Indications are that the site was occupied throughout MIS 3, but during its wetter rather than its drier phases (Jacobs et al. 2008b). Wadley's efforts have provided evidence for the earliest recorded bow and arrow technology, as well as for bone needles (Backwell et al. 2008; Lombard and Phillipson 2010), the use of a heat-treated glue compound (Wadley et al. 2009), bedding (Wadley et al. 2011) and possible shell beads (d'Errico et al. 2008) in MSA context.

Despite its spectacular archaeological record, Sibudu Cave is virtually lacking in human remains. Two specimens thought to probably derive from a single individual have been recorded as possibly coming from the final MSA (Plug 2004). These bones comprise a distal fibula and a manual phalanx. Although some Iron Age pits had been dug into the final MSA deposits in the cave, extreme care was taken to clear these before commencing work in the MSA strata, and no human bone was ever recorded as coming from the Iron Age pits. Thus, although Plug (2004) considered that these two human bones were intrusive owing to their "relatively fresh" appearance, this possibility would seem to be reasonably remote (Wadley personal communication).

A series of OSL dates from various levels in Sibudu place occupation of the site from ~77.2 to 38.0 ka (Wadley and Jacobs 2006; Jacobs et al. 2008a, b). The levels of the horizons from which the human remains derive within the final MSA strata would indicate an age of c. 48 ka for these two bones (Wadley personal communication).

Diepkloof Rock Shelter, South Africa (32°25′ S, 18°30′ E)

This rock shelter is located approximately 180 km north of Cape Town and some 17 km from the Atlantic Ocean at Elands Bay in the Western Cape Province of South Africa. The site was first excavated by J. Parkington and C. Poggenpoel in 1973, and subsequent work has uncovered a deep record of MSA occupation – including Still Bay and Howiesons Poort horizons – that dates from MIS 5-3 (Parkington and Poggenpoel 1987). The site is perhaps best known for fragments of engraved ostrich eggshell from the Howiesons Poort levels (Texier et al. 2010).

Excavations in 2002 recovered a mandibular deciduous molar in the latest Howiesons Poort levels and two pedal phalanges from post-Howiesons Poort strata (Verna et al. 2013). The toe bones are indistinguishable from those of recent humans; the molar is morphologically unremarkable but comparatively large in relation to recent homologues.

Thermoluminescence dates of the Howiesons Poort levels yielded a weighted average of some 52 ± 5 ka, but with high environmental dose rate variability (Tribolo et al. 2009). In the sector where the human toe bones were found, OSL dates give consistent ages of between 61.3 ± 1.9 and 58.1 ± 1.9 ka and for this Howiesons Poort occupation, and the layer from which the tooth was extracted is dated by OSL to 60.5 ± 1.9 ka. The post-Howiesons Poort layers located above the layer that yielded the toe bones has been dated by OSL to between 55.4 ± 2.0 and 47.7 ± 1.7 ka (Tribolo et al. 2009, 2013).

Mugharet el 'Aliya, Morocco (35°45′ N, 5°56′ W)

This cave site, situated some 11 southwest of Tangier on Cap Ashakar, was first excavated in the 1930s by C. Coon, who recovered a juvenile human maxilla (Tangier 1) *ex situ*. Two isolated molars discovered subsequently likely belong to the same individual, and another isolated upper molar derives from a second individual (Minugh-Purvis 1993). The site, also referred to simply as Tangier or as Dar el 'Aliya, contains a thick accumulation of fauna and Aterian artifacts. The human remains are believed to derive from one of the uppermost Aterian levels (Layer 5). Although initial assessment of the remains suggested their Neandertal affinities (Şenyürek 1940), the maxilla shares features with those from other North African Aterian contexts (Hublin 1993), and these are of modern appearance (Minugh-Purvis 1993).

Although none of the human specimens was recovered *in situ*, they are all believed to have come from layer 5. Electron spin resonance dates from one mammal tooth each from layers 9, 6, and 5 have produced estimates of between 56 and 39 ka (Wrinn and Rink 2003). However, Millard (2008) has argued that the gamma dose rate from the sediment seems to have been miscalculated by Wrinn and Rink (2003), which would revise the dates for the human remains to 46–21 ka assuming early uranium uptake, and 57–27 ka assuming linear uptake. These age estimates are significantly younger than the OSL dates of 107–96 ka obtained by Schwenninger et al. (2010) and Jacobs et al. (2011) for the Aterian levels at the nearby Smugglers' Cave, and they are also substantially younger than the youngest TL and OSL dates of c. 61 ka recorded for *in situ* Aterian artifacts at Uan Tabu, Libya (Cremaschi et al. 1998).

Nyamita (Wasiriya Beds), Kenya (0°24′ S, 34°10′ E)

The artifact and bone-bearing Late Pleistocene Wasiriya Beds, which are discontinuously exposed around the perimeter of Rusinga Island on the northeastern edge of Lake Victoria, have been known since the 1940s (van Couvering 1972; Pickford 1986). However, the first systematic archaeological surveys and excavations of these exposures were initiated only in 2009 (Tryon et al. 2010). This work has focused on the recovery of artifacts and fossils from several localities along the southern margin of the island, including Wakondo and Nyamita. The MSA artifacts recovered from the Wasiriya Beds are typologically distinct from Lupemban assemblages in the region around Lake Victoria, and include Levallois flakes and very small bifacially flaked points that are similar to those from other tropical grassland MSA localities in the East African Rift (Tryon et al. 2010, 2012). Alcelaphines (e.g., wildebeest) and equids dominate the faunal assemblages, indicating the presence of an open grassland that is distinct from the historic vegetation. Several extinct bovids are represented, and these are characterized by dental or postcranial adaptations to grazing in open habitats (Faith et al. 2016).

A series of AMS radiocarbon dates of between c. 45 and 33 ka obtained from the carbonate fractions of intrusive gastropod shells at the Nyamita 2 and Nyamita 3 localities indicate that the Wasiriya beds predate MIS 2 (Tryon et al. 2010). The maximum age of the Wasiriya Beds would appear to be constrained by the geochemistry of the basal phonolitic tephra deposits at Nyamita and Wakondo, which suggest derivation from the Logonot and/or Suswa volcanoes that began erupting at $\sim$100 ka (Tyron et al. 2010).

The only human fossil that has been recovered from the Wasiriya Beds to date is a partial right humeral diaphysis. It was found during surface collecting at Nyamita in 2010 and identified as human in 2013 by J.T. Faith. The proximal end is missing, probably as a result of carnivore damage, and there is an old break approximately mid-shaft, just distal to the deltoid tuberosity (Faith personal communication). It is currently under study by O.M. Pearson. A combination of OSL, ESR, and U-Series age estimates of 75–55 ka for the stratigraphically lower Wakondo locality (Blegen et al. 2015; Faith personal communication) constrains the lower age limit for the specimen, and its upper limit is established by the radiocarbon dates for the intrusive snails at Nyamita. Thus, the specimen almost certainly dates to MIS 3, and an age between 55 and 45 ka would seem to be a reasonable estimate.

Magubike Rock Shelter, Tanzania (7°45′ S, 35°28′ E)

This rock shelter is located close to the village of the same name in the Iringa Region of southern Tanzania. Test excavations were undertaken by Pamela Willoughby in 2005, and these revealed a sequence containing historic/Iron Age and LSA artifacts separated by some 30 cm of mixed MSA and LSA artifacts from underlying MSA strata (Biittner et al. 2007; Biittner 2011; Collins and Willoughby 2010; Willoughby 2012). A half dozen isolated human maxillary teeth were recovered from the MSA deposits; four of the teeth derive from a 10 cm deep level just below the disturbed horizon, and the other two were separated by a further 10 cm of deposit (Willoughby 2012). Some of the teeth exhibit traits found in moderate to high frequencies in recent sub-Saharan African human samples (Willoughby personal communication), but this is to be expected since other Late Pleistocene human teeth from southern Africa do as well (Grine and Klein 1985, 1993; Grine 2000; Grine et al. 2000; Grine and Henshilwod 2002).

A radiocarbon date of 41,790 ± 690 ^{14}C BP, calibrated to 43,020 ± 830 cal BP (using CalPal), has been obtained from snail shells from the same level as the human teeth (Willoughby 2012), although which level was not specified. More recently, ESR determinations on snail shells from this layer suggest an age in excess of 200 ka (Willoughby personal communication). However, the extremely porous nature of gastropod shell makes it of highly questionable utility in uranium-based age assessments (Schellmann and Radtke 1999; Feathers 2002). This, together with other factors such as the incredible discrepancy between the ESR and ^{14}C determinations suggests that little reliance can be placed on these supposedly more ancient dates.

Nazlet Khater, Egypt (26°47′ N, 31°21′ E)

Nazlet Khater comprises eight geographically proximate archaeological sites associated with intermittent Middle and Upper Paleolithic chert-mining activities close to the Nile River in Egypt. The locality takes its name from a nearby village. Excavations in 1980–1982 recovered two burials (Nazlet Khater 1 and Nazlet Khater 2) in Upper Paleolithic context at the locality known as Boulder Hill. The burials are associated with the Upper Paleolithic chert-mining site of Nazlet Khater IV (Vermeersch 2002).

Nazlet Khater 1 (NK 1) is a very poorly preserved skeleton of an adult female associated with fetal bones (Vermeersch 2002). Owing to its state of preservation, the bones of NK 1 were never subjected to any morphometric analysis before they were submitted for ^{14}C dating. Nazlet Khater 2 is a nearly complete adult male skeleton (Vermeersch et al. 1984). Considerable attention has been paid to it (Thoma 1984; Pinhasi and Semal 2000; Bruner and Manzi 2002; Pinhasi 2002; Crevecoeur and Trinkaus 2004; Crevecoeur 2008a, 2009, 2012; Bouchneb and Crevecoeur 2009). Despite its overall modern appearance (the cranium is described as "robust," but it lacks a supraorbital torus and displays a projecting chin, rounded cranial form and modest dental dimensions), several archaic features are noteworthy. These include thick cranial vault bone, the broad ramus and robustness of the mandibular corpus, and the inner ear, which presents features that are unusual among recent humans, but occur with some frequency among Middle and Upper Paleolithic specimens from Eurasia (Crevecoeur and Trinkaus 2004; Bouchneb and Crevecoeur 2009; Crevecoeur 2012). The postcranial skeleton exhibits a few traits that may be plesiomorphic, such as the marked angulation of the radial neck and shaft (Crevecoeur 2012). Comparative morphometric analyses of Nazlet Khater 2 and other Late Pleistocene specimens paint a picture of variation that is distinctly elevated relative to that among recent humans (Crevecoeur et al. 2009).

The two burials are dated to the latter part of MIS 3 by AMS ^{14}C, OSL, and ESR (Stokes and Bailey 2002). In particular, the NK 1 skeleton is AMS-dated to 37,600 ± 350 ^{14}C BP, and NK 2 is dated to c. 38 ± 6 ka on the basis of ESR on tooth enamel fragments (Crevecoeur 2008a; Crevecoeur et al. 2009).

Hofmeyr, South Africa (25°58′ E, 31°34′ S)

The Hofmeyr skull was discovered in 1954 by a farmer digging for sand in a dry channel bed of the Vlekpoort River near the town of the same name, in the Eastern Cape Province of South Africa. This isolated partial human skull is without any associated archaeological or faunal evidence, and it lacks clear geological context. When it was discovered, the cranium was nearly complete, but subsequent mishandling has resulted in the loss of much of the lower face and occipital. The specimen has been partially reconstructed using mirror imaging, clay modeling, and reference-based methods from photographs and measurements recorded prior to the damage (Grine et al. 2010).

The cranium is overall morphologically modern, but it possesses a moderately strong supraorbital torus and projecting glabella as well as marked alveolar prognathism (Grine et al. 2007, 2010). Morphometric analyses showed it to have close affinities to penecontemporaneous Upper Paleolithic specimens from Eurasia and to reside at the edge of the range of variation for recent sub-Saharan Africans (Grine et al. 2007). Comparisons of it with penecontemporaneous remains from Egypt (Nazlet Khater) and Europe (Peştera cu Oase) point to a greater range of variability in Late Pleistocene human crania than may be evident today (Crevecoeur et al. 2009), although aboriginal Australians were not included in that analysis. The reconstructed cranium (Grine et al. 2010), which has yet to be studied in a comparative phenetic context, may reveal additional or novel insights into its morphometric affinities.

No excavation was undertaken at the time of the discovery, and within a decade the channel from which it came had become filled by silt following the construction of an anti-erosion dam downstream. This precludes any possibility of locating the original position of the skull or of directly dating the surrounding sediments. The bone lacks sufficient collagen to be amenable to direct AMS ^{14}C dating. As a result, the heavily indurated carbonate sand matrix that largely filled the endocranial cavity was analyzed using a combination of OSL and U-series dating methods to estimate the burial time of the skull at 36.2 ± 3.3 ka (Grine et al. 2007).

El Harhoura I, Morocco (33°57′ N, 6°55′ W)

The cave of Zouhrah at El Harhoura 1 is situated between Dar es-Soltan and Grotte des Contrebandiers. It was excavated in the 1970s by A. Debénath, who recovered a canine from the Aterian level. In 1977, a mandible and an isolated canine were found during a salvage operation; while their provenance is uncertain, it is believed that they also derive from the Aterian strata (Debénath 1979, 1980).

The lower Aterian level (Niveau 2), which is the source for the excavated human canine, has been dated to between 41.2 and 25.5 ka by thermoluminescence on the basis of material that derives from the superjacent level (Niveau 3) (Aouraghie and Abbassi 2002). These dates are in keeping with Millard's (2008) revised estimates for the Aterian at Mugharet el 'Aliya (i.e., 57–21 ka). However, they are younger than the dates for the aforementioned Aterian levels at Smugglers' Cave and Uan Tabu.

The teeth from the Aterian at El Harhoura I have been studied by Hublin et al. (2012), who commented upon their very large size.

Human Fossils Associated with MIS 2 (29–14 ka)

During the last part of MIS 3, global climate underwent strong millennial-scale fluctuations. The Northern Hemisphere experienced repeated and abrupt warming and cooling phases, which also affected the African climate. The end of MIS 3 witnessed the growth of ice sheets to their maximum positions between 33.0 and 26.5, and almost all were position at their maximum (the Last Glacial Maximum, or LGM) between 26.5 and 20 ka (Clark et al. 2009).

The growth of these ice sheets had a profound impact, causing desertification in many regions of Africa – especially an expansion of the Sahara and Kalahari/Namib deserts – and a drop in sea level that exposed variably wide swaths of land. Paleovegetation (pollen and plant macro-remains) and lake-level data suggest overall drought conditions in East

Table 17.7 African hominin-bearing sites and hominin fossils associated with MIS 2

Site/specimen	Country	Coordinates	Age (ka)
Ishango 11	Democratic Republic Congo	00°08′ S, 29°36′ E	26–20
Taramsa 1	Egypt	26°06′ N, 32°42′ E	24
Leopard's Hill Cave	Zambia	15°36′ S, 28°44′ E	24–22
Lukenya Hill	Kenya	01°29′ S, 37°04′ E	24–22
Tuinplaas	South Africa	25°00′ S, 28°36′ E	<20–11
Deir El-Fakhuri (E71K1) (Esna)	Egypt	25°22′ N, 32°28′ E	18
Taza Cave I	Algeria	36°42′ N, 05°32′ E	16–14
Afalou-bou-Rhummel	Algeria	36°29′ N, 03° 00′ E	15–11
Gebel Silsila 2A (Kom Ombo)	Egypt	24°28′ N, 32°57′ E	14–13
Jebel Sahaba (Site 117)	Sudan	21°59′ N, 31°20′ E	14–12
Wadi Halfa (Sites 6B28 & 6B36)	Sudan	21°57′ N, 31°20′ E	14–10
Ifri n'Baroud	Morocco	34°53′ N, 03°20′ E	13
Bushman Rock Shelter	South Africa	24°35′ S, 30°38′ E	13
Mlambalasi Rock Shelter	Tanzania	07°35′ S, 35°30′ E	13–12 (? *in situ*)
Grotte des Pigeons (Taforalt)	Morocco	34°48′ N, 02°24′ W	13–11
Iwo Eleru	Nigeria	07°26′ N, 05°08′ E	11

Africa (Farrera et al. 1999; Barker and Gasse 2003; Filippi and Talbot 2005). In South Africa, cooler and potentially wetter conditions prevailed (Partridge et al. 1999; Mills et al. 2012), and there is evidence for an expanded winter rainfall zone (Chase and Meadows 2007). In North Africa, the Maghreb appears to have been relatively cool and arid with expanded steppe and lower water levels in Lake Chad during the LGM; this was followed at ~15 ka by an increase of pollens indicative of higher humidity – the so-called "African Humid Period," or AHP (Close and Wendorf 1990; deMenocal et al. 2000). Recent evidence from the Gulf of Aden indicates that the AHP was very abrupt both in its initiation and termination (i.e., occurring within centuries) in the Horn of Africa (Tierney and deMenocal 2013).

However, while the terminal Pleistocene-early Holocene AHP witnessed abundant North African lakes and a variable rise in Central and East African lake levels, those in South Africa stood at considerably lower levels (Hoelzmann et al. 2002; Johnson et al. 2002; Thomas and Shaw 2002; Schefuß et al. 2005; Weldeab et al. 2005; Kuper and Kröpelin 2006; Garcin et al. 2009; Thomas et al. 2009; Tierney et al. 2011; Foerster et al. 2012; Lebamba et al. 2012). Moreover, time-transgressive terrestrial changes during the AHP have been documented even *within* the Sahara (e.g., deMenocal et al. 2000; Kuper and Kröpelin 2006; Cole et al. 2009; Larrasoaña 2012), as well as between northern and southern Africa (Gasse 2000; Gasse et al. 2008).

During the Last Glacial Maximum (LGM) the Maghreb is believed by some to have been largely depopulated c. 20–18 ka (Lubell 1984, 2001; Close and Wendorf 1990; Larrasoaña 2012). Similarly, Wadley (1993) has speculated, "the harsh conditions of the LGM over much of southern Africa may have resulted in widespread resource shortages."

A number of sites in North Africa (especially in Algeria) that date to the terminal Pleistocene-early Holocene have yielded fragmentary human remains associated with the microliths and backed bladelets that characterize the Epipaleolithic (Iberomaurusian) tradition there. These sites include Alain, Cap Tenes, Champlain, Ifri n'Ammar, and La Mouillah. Other sites, such as Columnata, contain large assemblages of human skeletons (c. 60 individuals; Chamla 1970) or very poorly preserved human remains (e.g., Tushka Site 8905, Locality A) that clearly date to the beginning of MIS 1. None of these sites is included in the present review.

Fifteen sites spread across the length and breadth of Africa contain human remains in reasonably well-dated contexts that help to document aspects of skeletal and dental variability during the period represented by MIS 2 (Table 17.7; Fig. 17.5). These 15 are discussed below in the order in which they appear in Fig. 17.5 and Table 17.7.

Ishango 11, Democratic Republic of the Congo (0°8′ S, 29°36′ E)

This site is located on the bank of the Semliki River at its exit from Lake Rutanzige (formerly Lake Edward) on the western side of the Democratic Republic of the Congo. It was discovered by H. Damas in 1935. He found a human jaw and several bone harpoons in a partially cemented sand-gravel matrix (Damas 1940). The site was excavated in the mid-1950s under the direction of J. de Heinzelin, and the mid-1980s by A. Brooks, J. Yellen and K. Misago. Ishango 11 preserves horizons that relate to early LSA and Holocene (Neolithic) occupations, and both preserve human skeletal

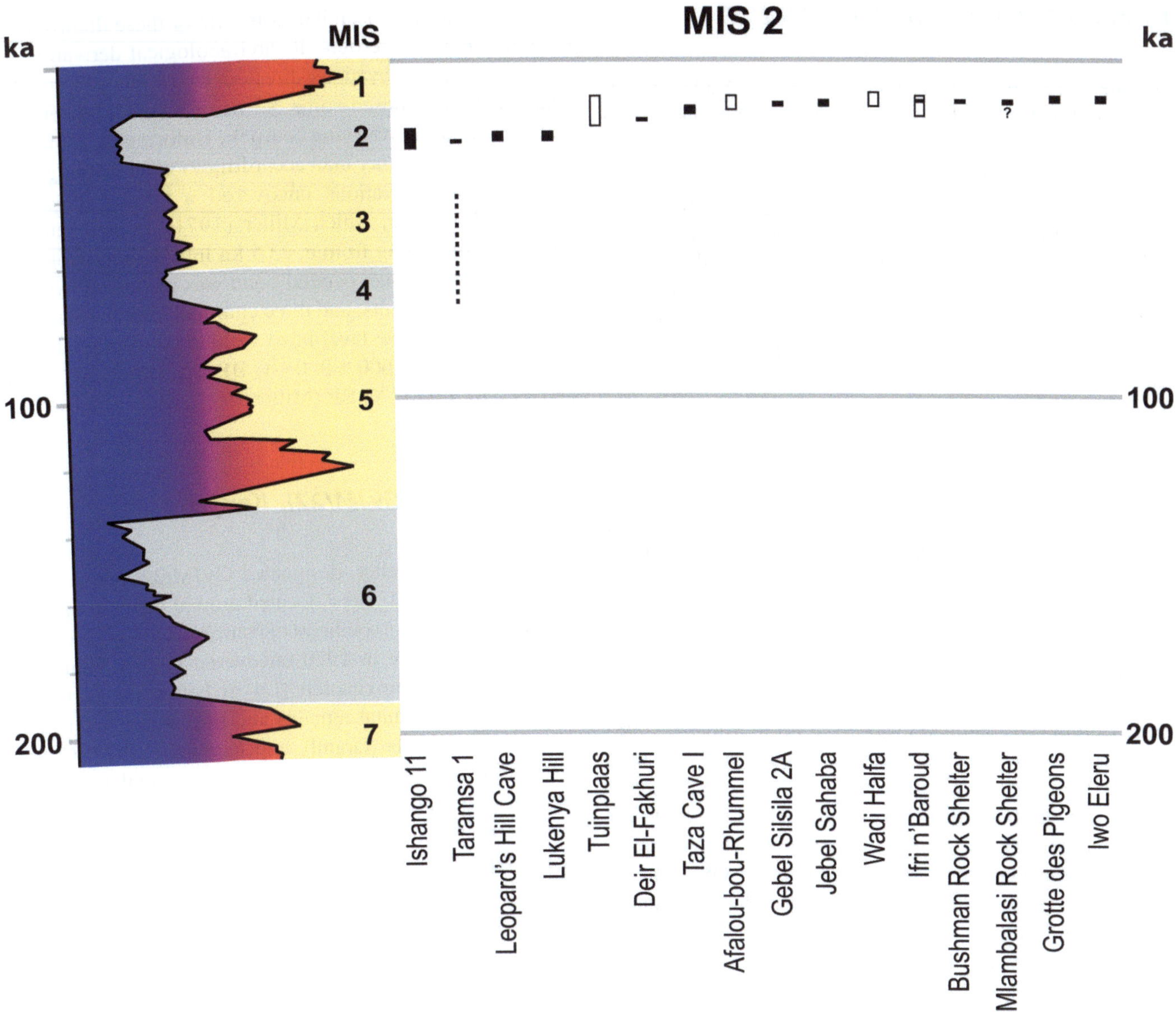

Fig. 17.5 Distribution of age determinations and/or estimates for human remains and human-bearing sites that are most reasonably related to Marine Isotope Stage 2. *Open rectangles* represent possible age ranges deduced for a site or specimen; *solid rectangles* represent well-established dates for the hominin fossils; *dotted lines* indicate possible but not probable dates; ? indicates likely disturbed context

remains. The LSA horizon, referred to as the Principal Fossiliferous Layer (Niveau Fossilifére Principal, or NFPr), contains a rich faunal assemblage and abundant archaeological debris, including quartz microliths, numerous barbed bone harpoon points, and engraved bone.

The human remains recovered by de Heinzelin from the NFPr comprise skull fragments, isolated teeth, and variably complete postcranial bones; 27 additional human fragments from this layer were uncovered by Brooks and collaborators (Twiesselmann 1958; de Heinzelin 1962; Boaz et al. 1990; Brooks et al. 1991). The entire collection represents five adults, one adolescent, three children, and two perinatal individuals (Crevecoeur personal communication). Various studies of these remains (e.g., Orban et al. 2001; Crevecoeur 2008b; Crevecoeur et al. 2010) indicate the crania to be comparatively robust, the teeth to be large, and the inner ear morphology to be unusual. The limb bone diaphyses are straight, and possess comparatively thick cortices. The appendicular remains exhibit considerable dimorphism in size and robustness, and considerable variation is also indicated by the crania and teeth (Crevecoeur et al. 2010).

A series of AMS dates on ostrich eggshell and mollusk shell from the NFPr layer range between 25,570 ± 350 ^{14}C BP and 19,920 ± 450 ^{14}C BP; these dates appear to be supported by amino acid racemization assessments on the same materials (Brooks and Smith 1987; Brooks et al. 1995; Brooks and Robertshaw 1990; Mercader and Brooks 2001; Brooks 2008).

Taramsa 1, Egypt (26 6′ N, 32°42′ E)

The site of Taramsa Hill is situated in the Nile Valley. It was discovered in 1985 and reported as site E85/2 in the earliest publications; it was subsequently referred to as Taramsa Hill (Van Peer et al. 2010). The hill is an isolated landform covered with chert cobbles, and the mining or extraction of chert for lithic tool production occurred here in what seems to have been five separate phases of activity in the Middle and Upper Paleolithic (Vermeersch et al. 1997). While surveying sector 91/01 of the hill for chert extraction sites in March 1994, a heavily weathered skull and very brittle partial skeleton of a child was discovered. Uniquely, it was found in a seated position in a collapsing ancient trench of phase 3 extraction debris.

The descriptions of the human child by Vermeersch et al. (1998) emphasized its anatomically modern affinities. This is particularly evident from the frontal bone, which exhibits a vertical squama and no supraorbital development, and the rounded occipital. At the same time, however, the face appears rather prognathic, and the teeth are comparatively large. The postcranial bones are slender rather than robust.

When the burial was first reported (Vermeersch et al. 1998), a weighted OSL average of 55.5 ± 3.7 ka from a wide range of imprecise estimates was given for it. Millard (2008) has remodeled these data to suggest a range of between some 73 and 41 ka. Most recently, Van Peer et al. (2010) have argued that the burial was covered by extraction waste for which an OSL date of 68.6 ± 8.0 ka is given. Interestingly, however, Van Peer et al. (2010) discount an OSL date of 24.3 ± 2.1 ka for sand grains that filled the endocranial cavity of the skull on the grounds that there are no other archaeological traces of this age on the hill. However, this sediment – as opposed to the disturbed sediments around the skeleton – would seem to be the most reasonable source for an age of the specimen itself inasmuch as the endocranial cavity would have filled within a reasonably short period of time after death. Thus, while an age of c. 70–40 ka is possible for this burial, it would seem that the discounted age of 24 ka is the most likely.

Leopard's Hill Cave, Zambia (15°36′ S, 28°44′ E)

This dolomitic cave is located approximately 55 km southeast of Lusaka, in south-central Zambia. It was initially explored in 1946 by J.D. Clark and R.P. Odendaal, who observed animal bones and stone tools in the dumps left behind by commercial guano miners. They recovered a fragment of human occipital from one of these dumps, but it is impossible to determine its archaeological derivation.

Subsequent excavations directed by L. Hodges and J.D. Clark in 1958 exposed some 3.2 m of Late Pleistocene and Holocene deposit overlying bedrock. Unfortunately, they dug in arbitrary spits rather than according to natural stratigraphic units. These excavations uncovered a number of animal bones and artifacts, which Miller (1971) divided into four LSA units extending from c. 23.5 ka into the mid-Holocene. A fragmentary human parietal bone was recovered from one of the lowermost units, and three charcoal samples that likely come from the same level have yielded conventional radiocarbon dates of 23,600 ± 360 ^{14}C BP, 22,600 ± 510 ^{14}C BP, and 21,550 ± 950 ^{14}C BP (Miller 1971; Klein 1984).

Lukenya Hill (GvJM/22), Kenya

This small rock shelter, designated GvJM/22, is one of five LSA archaeological "sites" located around the base of Lukenya Hill, some 37 km southeast of Nairobi. A test pit excavated by Richard Gramly in 1970 uncovered human frontal and parietal bones approximately 1.4 m below the surface in association with faunal remains and microlithic Later Stone Age (LSA) artifacts (Gramly and Rightmire 1973; Gramly 1976; Marean 1992; Kusimba 2001). The frontal is clearly of modern aspect, although the supraorbital torus is rather robust and the frontal "squame" (forehead) recedes somewhat more than is typical for modern sub-Saharan Africans (Gramly and Rightmire 1973). This is reminiscent of the condition described by Brothwell and Shaw (1971) for the terminal Pleistocene cranium from Iwo Eleru, Nigeria. Recent morphometric analysis of this partial calotte has shown it differ somewhat from recent African crania, but to have close affinity to some Neolithic specimens from North Africa (Tryon et al. 2015). This morphometric study demonstrates its relative distinctiveness from Iwo Eleru and Late Pleistocene African specimens such as those from Hofmeyr and Nazlet Khater.

Conventional dates of 17,700 ± 760 ^{14}C BP and 17,650 ± 800 ^{14}C BP were reported for collagen from associated animal bones, and an amino acid (aspartic acid) racemization date of c. 17.8 ka was reported for the parietal itself (Gramly and Rightmire 1973; Protsch 1975). However, the ^{14}C dates are considered questionable because of their reliance on collagen pretreatment techniques that have been shown to produce erroneous readings (Higham et al. 2006). The calotte derives from "occurrence E" and is associated with AMS ^{14}C dates determined from ostrich eggshell that indicated a modeled range of 23,576–22,887 years BP (Tryon et al. 2015).

Tuinplaas, South Africa (25°0′ S, 28°36′ E)

The fragmentary Tuinplaas skull and postcranial skeleton (TP 1) was discovered in 1929 at the Springbok Flats gravel quarry some 130 km north of Pretoria, in the Limpopo Province of South Africa. It was initially thought to be associated with MSA artifacts (Broom 1929a), and therefore a Late Pleistocene age seemed reasonable. However, van Riet Lowe (1929) observed that "linking of any of the implements…with the skeletal material is, in the absence of contributory evidence, wholly impossible." As noted by Pike et al. (2004), it was primarily the absence of LSA artifacts in the vicinity that led to the accepted consensus that the skeleton was associated with the MSA. Assessments of the human remains have stressed their similarities with the recent indigenous inhabitants of South Africa (Toerien and Hughes 1955; Hughes 1990; Pfeiffer et al. 1996; Houghton and Thackeray 2011), and U-series assessment of the TP 1 radius indicates it to be less than 20 ± 3 ka, and perhaps as young as 11.0 ± 0.7 ka (Pike et al. 2004).

Deir El-Fakhuri (Esna), Egypt (25°22′ N, 32°28′ E)

Five discrete surface concentrations of Late Paleolithic artifacts and faunal remains surround the abandoned Coptic monastery at Deir El-Fakhuri, some 11 km northwest of Esna, Egypt (Wendorf and Schild 1976). These sites were formed during the deposition of the sands and silts that comprise the Dibeira-Jer Formation, which represents an aggradation episode of the Nile River between c. 22–16 ka (Butzer and Hansen 1968; Wendorf et al. 1970, 1979). The microlithic flakes and blades at these sites have been attributed to the Fakhurian (Lubell 1974). Radiocarbon dates on shells of the aquatic bivalve *Unio* have been determined for some of the sites, and these range between 18,030 ± 330 ^{14}C BP for site E71K1, and 17,640 ± 300 ^{14}C BC for site E71K3 (Buckley and Willis 1969).

Two partial human skeletons were found weathering out of the western side of site E71K1 in the mid-1960s (Wendorf et al. 1970). One comprises the fragmentary pelvis and proximal femora of an adult male, and the other is represented by the fragmentary skull and partial postcranial skeleton of a late juvenile/young adult (Butler 1974).

Taza Cave I (Grotte de la Madeleine), Algeria (36°42′ N, 5°32′ E)

This small cave, situated on the Mediterranean coast in the Bédjaîa Bay of northern Algeria, was known originally as the Grotte de la Madeleine. It was first excavated by C. Arambourg in the 1920s (Arambourg et al. 1934). Its lowermost strata contain Middle Paleolithic artifacts, and the upper levels contain archaeological debris attributed to the Iberomaurusian. Excavations in 1990 under the direction of M. Medig and R. Meier uncovered a human skull in the lower part of the Iberomaurusian sequence (Medig et al. 1996). The skull, which is considered to be from an adult female, is rather diminutive in comparison with other Iberomaurusian specimens (Meier et al. 2003). A radiocarbon date of 16,100 ± 1400 ^{14}C BP is apparently associated with the human skull, although Barton et al. (2008) have questioned whether the specimen is in primary context. Even if it was buried from a higher level, radiocarbon dates for overlying Iberomaurusian strata constrain the specimen to at least 13,800 ± 130 ^{14}C BP.

Afalou-bou-Rhummel, Algeria (36°29′ N, 3°00′ E)

This large cave, located some 30 km east of Bédjaîa on the Mediterranean coast of Algeria, was first excavated by C. Arambourg in the late 1920s and subsequently by S. Hachi in the mid-1980s. A large number of human skeletons – some 53 in total with the majority represented by isolated crania – were uncovered by Arambourg (Arambourg et al. 1934; Camps 1969) in what he described as an "ossuary" in Level I. One more complete adult skeleton in association with the skull of an infant was recovered by him from the lowest level in the cave (Level III). Hachi's excavations uncovered an additional group of skeletons representing a minimum of eight individuals. The human remains are robust, and have been attributed by some workers to a distinctive cranial phenotype referred to as the "Mechta-Afalou" (Briggs 1955; Wendorf 1968; Vallois 1969; Ferembach 1985). These specimens have been the subject of a number of comparative studies (e.g., Irish 2000; Bruzek et al. 2004; Balzeau and Badawi-Fayad 2005), which concluded (unsurprisingly) that they share a distinct phenetic relationship with other North African Iberomaurusian people. Hachi (1996) obtained four dates between 15,000 and 11,000 ^{14}C BP that pertain to the human remains from the site.

Gebel Silsila 2A (Kom Ombo), Egypt (24°28′ N, 32°57′ E)

This open-air site is located on the eastern bank of the Nile River just north of the town of Kom Ombo. Later Pleistocene sands and gravels are exposed in this region of the Kom Ombo Plain in a series of channel complexes known as the Gebel Silsila Formation (Butzer and Hansen 1968). These complexes range from channel beds and point-bar sequences to levee and overbank deposits, and many contain Late Pleistocene lithic artifacts and animal bones (Vignard 1928; Sandford 1934; Churcher 1972; Butzer 1997). One such channel produced a human calvaria in 1926 in apparent association with Upper Paleolithic ("Sebilian") artifacts (Sandford 1934). Unfortunately, the specimen was never adequately described and has since been lost.

Subsequent exploration of the Kom Ombo deposits resulted in the discovery nearby of an isolated human frontal bone in 1963 in sediments related to an ancient branch of the Nile, the Fatira channel, which runs immediately to the east of the Gebel Silsila station (Reed 1965). The frontal bone is clearly of modern aspect, with fully separated medial and lateral superciliary eminences (Reed 1965). The specimen was found in direct association with "Middle Sebilian" artifacts, and two standard radiocarbon dates of 13,070 ± 160 ^{14}C BP and 13,560 ± 120 ^{14}C BP were obtained on charcoal and clam shell from the same layer (Reed 1965).

Soon after the discovery of the frontal bone, additional cranial fragments were found "immediately adjacent" to it but in a disturbed context. As such, it is not possible to relate these fragments to the frontal or the radiocarbon dates.

Jebel Sahaba (Site 117), Sudan (21°59′ N, 31°20′ E)

The "site" of Jebel Sahaba, which takes its name from a nearby hill, comprises two cemeteries on either side of the Nile River in northern Sudan, some 3 km north of the town of Wadi Halfa. A third penecontemporaneous cemetery situated close by is Tushka (Site 8905, Locality A), from which the very fragmentary remains of 12 adults and 1 child have been recovered.

Site 117 is the larger of the two cemeteries at Jebel Sahaba. It was discovered in 1962 by R. Paepe and J. Guichard during an archaeological survey of the region under the leadership of F. Wendorf for the UNESCO Aswan High Dam Salvage project. They discovered four human skeletons, and Wendorf returned 3 years later to begin extensive excavations that resulted in the recovery of 46 burials. In 1966, A. Marks excavated an additional eight burials, bringing the total from Site 117 to 58 skeletons (Wendorf 1968). Comprising this total are at least 24 adult females, 19 adult males, 3 adults of indeterminate sex, and a dozen juveniles ranging in age from infants to adolescents. Individual as well as multiple burials in shallow grave pits are present, and most were covered with sandstone slabs.

The skeletal assemblage from Site 117 is best known for its testament to violent death. About 40% of individuals exhibit some form of evidence for this in the form of "parry" fractures of the forearm bones, unhealed cut marks, or the not uncommon presence of lithic projectile points that would have been embedded in the bodies at time of death (Anderson 1968; Wendorf 1968; Wendorf and Schild 2004). Indeed, some bodies would have had between 15 and 30 projectile points embedded in them.

The skeletal assemblage has been widely studied (e.g., Anderson 1968; Irish and Turner 1990; Irish 2000, 2005; Franciscus 2003; Shackelford 2007; Starling and Stock 2007; Stock et al. 2011; Holliday 2015). Anderson (1968) concluded that the Jebel Sahaba skeletons were, as a group, quite homogeneous and that they were (as might be expected) most similar to presumably contemporaneous specimens from Wadi Halfa, and the Epipaleolithic and Mesolithic remains from the Maghreb sites of Afalou-bou-Rhummel and Grotte des Pigeons (Taforalt). He also argued that the crania exhibited features indicative of continuity with more recent (Holocene) skeletons from northern Africa (Anderson 1968). On the other hand, dental studies (Irish and Turner 1990; Irish 2000, 2005) have suggested that the Site 117 individuals were distinct not only from the Late Pleistocene Maghreb people but also from Holocene Nubians. These studies concluded that the skeletons from Jebel Sahaba showed phenetic affinities with recent West African populations. Franciscus (2003) also found similarities in maxillary morphology between the Jebel Sahaba sample and recent sub-Saharan Africans, and Holliday (2015) has suggested that the Jebel Sahaba sample was most similar in body form to recent sub-Saharan Africans, being distinct from penecontemporaries in other parts of North Africa or their successors in Nubia.

Although absolute (i.e., radiocarbon) dates for Site 117 were unavailable to Wendorf, he nevertheless was able to employ the artifacts (especially the microliths suggestive of the Qadan) and the geological setting of the site to suggest an age of 14–12 ka for the burials (Wendorf 1968). A single radiocarbon date of 13,740 ± 600 ^{14}C BP was subsequently obtained from bone of one of the skeletons (Close 1988), but poor collagen preservation at the site suggests that this determination may be somewhat unreliable. Most recently,

Antoine et al. (2013) employed apatite radiocarbon dating on nine samples of bone, tooth enamel and dentine from four skeletons to suggest that the site dates to at least 11.6 ka.

Wadi Halfa (Sites 6B28 and 6B36), Sudan (21°57′ N, 31°20′ E)

The "site" of Wadi Halfa, which takes its name from the nearby town in northern Sudan, was excavated in the mid-1960s during archaeological work in the region under the leadership of Gordon Hewes of the University of Colorado for the UNESCO Aswan High Dam Salvage project. It is located only approximately 3 km from the penecontemporaneous (i.e., terminal Pleistocene-early Holocene) cemeteries at Wadi Tushka (Site 8905, Locality A), and Jebel Sahaba (Site 117).

Wadi Halfa Site 6B28 (21°57′ N, 31°20′ E) is a wind-eroded surface occupation site on the eastern bank of the Nile River. A single edentulous human mandible was found exposed on the surface by Eugene McCluney in 1963; it was associated with Upper Paleolithic artifacts and several fossilized bovid bones. The specimen was afforded a very brief description by Armelagos (1964). No direct radiocarbon value has been published for Site 6B28, but it is thought to date to c. 15 ka on the basis of a ^{14}C determination for the adjacent site of 6B29.

Wadi Halfa Site 6B36 (21°57′ N, 31°19′ E), which is located on the western bank of the Nile almost opposite the town, is a graveyard containing the cranial and postcranial remains of some 37 individuals (13 adult males, 15 adult females, and there juveniles). These specimens have been studied extensively (e.g., (Hewes et al. 1964; Greene and Armelagos 1972; Carlson 1976; Armelagos et al. 1984: Smith and Shegev 1988; Irish and Turner 1990; Turner and Markowitz 1990; Irish 2005; Shackelford 2007). Most studies have concluded that this sample has strong phenetic affinities to the one from the physically and (supposedly) chronologically proximate site of Jebel Sahaba Site 117. No direct radiocarbon dates have been published for Wadi Halfa Site 6B36, but it is thought to be penecontemporaneous with Jebel Sahaba Site 117 on the basis of shared microlithic (Qadan) assemblages. A terminal Pleistocene age of c. 14–10 ka is probably reasonable.

Ifri n'Baroud (Gunpowder Cave), Morocco (34°53′ N, 3°20′ E)

This cave site is located some 40 km south of the Mediterranean coast in eastern Morocco. It was discovered and excavated in the mid-1990s under the direction of A. Mikdad and J. Eiwanger, who documented a deep sequence (c. 3 m) of Iberomaurusian deposits (Mikdad and Eiwanger 2000; Nami 2007). A single human burial, comprising the postcranial skeleton of a tall adult female, was uncovered in the upper portion of the sequence (Ben-Ncer 2004). Radiocarbon dates for the Iberomaurusian deposits at Ifri n'Baroud range from 16,777 ± 83 ^{14}C BP to 11,639 ± 58 ^{14}C BP (Görsdorf and Eiwanger 1998; Nami 2007), and the skeleton is seemingly associated with a date of c. 12,500 ^{14}C BP (Ben-Ncer 2004).

Bushman Rock Shelter, South Africa (24°35′ S, 30°38′ E)

Bushman Rock Shelter is located near Origstad, Mpumalanga Province, South Africa. Excavations in the 1960s produced faunal remains together with LSA and MSA artifacts from a c. 2 m thick sequence of strata (Louw et al. 1969). In 1969, a tourist guide found the partial mandible of a 6–8-month-old human infant protruding from a witness section. The jaw was believed to have derived from one of the lower MSA levels (levels 16 or 17). A morphometric analysis by H. de Villiers revealed a cluster of features aligning it with those of the modern Bantu-speaking population of South Africa (Protsch and de Villiers 1974). A series of radiocarbon samples analyzed by Vogel dated the MSA at the site to between 53 and 12 ka (Louw et al. 1969), although the more recent age limit was thought to result from LSA disturbance of the underling MSA (Mason 1969). Subsequently, Protsch claimed to have undertaken a nitrogen analysis on the infant mandible, placing it in level 16 or 17, and to have produced radiocarbon dates of 29.5 ^{14}C kBP on bone from these levels (Protsch and de Villiers 1974). However, these chemical and dating analyses are highly suspect, if they were ever even conducted. Moreover, radiocarbon evaluations of charcoal taken from levels 16 through 18 have yielded dates of 12,950 ± 70 ^{14}C BP to 12,500 ± 120 ^{14}C BP (Vogel et al. 1986).

Mlambalasi Rock Shelter, Tanzania (7°35′ S, 35°30′ E)

This site is located midway up a large escarpment some 50 km northwest of Iringa in southern Tanzania. It was discovered and first excavated by Paul Msemwa in 2001–2002. Subsequent test excavations under the direction of P. Willoughby in 2006 (Biittner et al. 2007) and 2010 (Willoughby 2012) uncovered an archaeological sequence consisting of historic/Iron Age deposits overlying terminal Pleistocene and Holocene LSA levels.

The skeletal remains of two adults and one juvenile have been recovered, but only one of these (the most complete adult, numbered B-1) "can be confidently attributed to the LSA" (Sawchuk and Willoughby 2015). This specimen was excavated over the course of two field seasons; the elements of the lower body were recovered in 2006 and the skull and upper body in 2010. The specimen is of indeterminate sex and appears to be of interest owing to its comparatively small body size (Sawchuk and Willoughby 2015). The B-1 skeleton, which is reportedly *in situ*, is associated with an AMS date of 12,765 ± 55 ^{14}C BP from a sample of charcoal found next to its right shoulder. Additional dates on charcoal and land snail shells from levels above and below the skeleton have provided ages of 12,940 ± 90 ^{14}C BP and 11,710 ± 90 ^{14}C BP, respectively. Although these would seem to serve as brackets for the date most closely associated with the skeleton itself, it is disconcerting that the "brackets" are inverted. This would seem to indicate disturbance and reworking of Late Pleistocene sediments through interment of the individual in later LSA or even Iron Age times.

Grotte des Pigeons (Taforalt), Morocco (34°48′ N, 2°24′ W)

The large cave site of Grotte des Pigeons is located near the village of Taforalt in the Beni Snassen Mountains of northeastern Morocco. It is approximately 40 km inland from the Mediterranean coast. A series of excavations undertaken in the mid-1940s by Ruhlmann, the early 1950s and 1969–1977 by Abbé Roche (1953, 1963, 1969), and since 2003 under the direction of A. Bouzouggar and N. Barton (Barton et al. 2005; Bouzouggar et al. 2007, 2008) have revealed a deep sequence of archaeologically rich Middle Paleolithic and Epipaleolithic (Iberomaurusian) deposits. The site is perhaps best known archaeologically for the presence of perforated mud snail (*Nassarius*) shells in the c. 82 ka Aterian deposits (Bouzouggar et al. 2007). Although Roche (1953) initially reported some fragmentary human bones from the Aterian levels, these are almost certainly an intrusive burial.

Numerous human skeletons have been recovered from the final Iberomaurusian deposits in the cave (Ferembach 1962; Roche 1963; Mariotti et al. 2009; Humphrey et al. 2012). The remains of close to 200 adults and juvenile individuals constitute the burials in the so-called Iberomaurusian Necropolis I and II (Ferembach 1962, 1965; Mariotti et al. 2009; Humphrey et al. 2012). The burials represent both primary and secondary depositions, and some elements show evidence of intentional postmortem modification such as ochre staining and cut marks (Mariotti et al. 2009; Humphrey et al. 2012). In a number of instances it is clear that earlier burials were disturbed or truncated by subsequent internments. Antemortem evulsion of the maxillary central incisors is evidenced by nearly all of the adult cranial remains (Humphrey and Bocaege 2008). While the antemortem evulsion of the incisor teeth appears to have been a common practice by the Iberomaurusian people of northwestern Africa, the deliberate postmortem manipulation of human skeletal remains has not been documented at any other Iberomaurusian site (Humphrey and Bocaege 2008; Humphrey et al. 2012).

Charcoal from a level immediately above the burials excavated by Roche yielded a radiocarbon date of 11,900 ± 240 ^{14}C BP, and a series of charcoals from a continuous sequence in the burial-containing Grey Series deposits are constrained between 12,675 ± 50 ^{14}C BP and 10,935 ± 40 ^{14}C BP (Barton et al. 2007, 2013; Bouzouggar et al. 2008).

Iwo Eleru, Nigeria (7°26′ N, 5°8′ E)

This rock shelter is situated some 180 km from the Gulf of Guinea in southwestern Nigeria. It was excavated in 1965 by Thurstan Shaw, who uncovered a rich LSA archaeological assemblage, which was divided on the basis of artifact analyses into two periods and four "time vector planes" (Shaw and Daniels 1984). The earliest of these is associated with radiocarbon dates between 11,200 ± 200 ^{14}C BP and 8,685 ± 120 ^{14}C BP (Shaw and Daniels 1984). Shaw's excavations uncovered a human burial, which is almost certainly not intrusive, and is associated with a conventional date of 11.2 ^{14}C kBP (Shaw and Daniels 1984). This date could possibly calibrate to c. 13 ka (Allsworth-Jones et al. 2010; Harvati et al. 2011). A recent U-series analysis of a long bone cortical fragment provided a range of dates between c. 16.3 and 11.7 ka (Harvati et al. 2011). However, in view of the questionable reliability of highly alterable bone in U-series dating and its absolute reliance upon the uptake model (Millard and Hedges 1995; Zhao et al. 2001; Hedges 2002; Pike et al. 2002), these results should be regarded with some skepticism. At best, they can be regarded as simply corroborating the published radiocarbon date.

The human remains (Iwo Eleru 1) comprise a very fragmentary and poorly preserved postcranial skeleton and a partial skull of what was judged to be an adult male (Brothwell and Shaw 1971). The specimen exhibits a fairly weakly developed chin. Brothwell described the cranial vault as being relatively long and low, and observed that the frontal "shows a moderate degree of recession." Importantly, he noted, "this degree of frontal sloping is atypical of the fairly large West African series available in London, but does occur" (Brothwell and Shaw 1971: 222). Indeed, as noted above, this is reminiscent of the condition described for the LSA cranium from Lukenya Hill, Kenya (Gramly and Rightmire 1973). It is also interesting to note the emphasis that Drennan (1929) placed on the low, receding frontal of the Cape Flats specimen.

Unsurprisingly, Brothwell's initial observations have been borne out by subsequent multivariate morphometric analyses. In these studies, the receding frontal serves in some manner to align the cranium somewhat with more archaic human remains (e.g., Omo II), but the cranium also finds alignment with specimens from the Upper Paleolithic of Eurasia (Stringer 1974; Harvati et al. 2011). However, the Iwo Eleru I calvaria suffers from some distortion and considerable reconstruction (Brothwell and Shaw 1971), and 3D geometric morphometric (3DGM) analyses are very sensitive to such distortion. The failure to correct for (or recognize) this sort of distortion in 3DGM studies can lead to highly questionable results (Spoor et al. 2008). The conclusions of Harvati et al. (2011: e24024), who attempted to relate their results to the study by Hammer et al. (2011) in finding "suggestions of a deep population substructure in Africa and a complex evolutionary process for the origin of modern humans," are questionable at best. Indeed, it is unclear why the analysis by Harvati et al. (2011) failed to include either the Lukenya Hill or Cape Flats crania in light of the receding frontals that have been observed for both (Drennan 1929; Gramly and Rightmire 1973).

Any suggestion of a deep population substructure in Africa must clearly come from genomic rather than cranial morphometric data, but, as discussed below, even the genetic evidence that has been proffered for archaic admixture such as this is very much open to question.

Discussion

From the foregoing review, it should be evident that deterministic interpretations which attempt to correlate morphological features portending either the emergence of our species or the subsequent emergence of full-blown anatomical modernity with any of the climatic vicissitudes that characterize the various MIS over the past 200 kyr will be difficult, if not downright impossible, to achieve. The dominant scheme by which the later Pleistocene African hominin fossils have been interpreted envisions a single lineage culminating in the emergence of *Homo sapiens* and thence modern humans (e.g., Stringer and Andrews 1988; Hublin 2001; Smith 2002; Stringer 2002; Bräuer 2008; Rightmire 2008, 2009; Pearson 2011). However, increasing attention is being directed to the morphological diversity that is manifest even within this admittedly rather paltry assemblage (e.g., Manica et al. 2007; Stringer 2007; Rightmire 2009; Crevecoeur et al. 2009; Gunz et al. 2009).

This diversity may be manifest among penecontemporaneous (or potentially penecontemporaneous) sites that are located variously in southern, eastern, and northern Africa (e.g., among Klasies River Main Site, Dar es-Soltan II, Eyasi and Aduma, or between Hofmeyr and Nazlet Khater). In these instances, the diversity may be related to geographic and/or temporal differences between the specimens. In other instances, however, there is considerable diversity among contemporaneous (or, at least seemingly contemporaneous) specimens within a single site (e.g., between the Omo I and Omo II crania from the Kibish Formation, and among the human fossils from the lower part of the SAS Member at Klasies River Main Site).

Thus, for example, the four adult mandibles from the SAS Member at Klasies evince considerable differences in both size (Royer et al. 2009) and symphyseal morphology (Lam et al. 1996) (Fig. 17.6). While they derive from separate horizons[3] that may represent discrete occupation events, they are generally regarded as penecontemporaneous. These jaws, like some other elements from Klasies River Main Site, have been argued to demonstrate higher levels of sexual dimorphism than living humans (Rightmire and Deacon 1991; Smith 1992, 1994; Lam et al. 1996; Royer et al. 2009). As noted by Royer et al. (2009), high levels of mandibular variation – especially in corpus height – are also characteristic of some other Pleistocene samples (e.g., the Middle Paleolithic sample from Skhūl, the Upper Paleolithic sample from Dolní Věstonice, and the Middle Pleistocene sample from Sima de los Huesos). To the extent that size variation can be used as a proxy for sexual dimorphism, this suggests that the degree of dimorphism exhibited at Klasies was not unique among Middle and Late Pleistocene hominins. Although Wymer (1982: 139) opined that this sample comprises the mandibles of *Homo sapiens* and a "Neanderthal type," Lam et al. (1996) concluded that it simply attests to a high degree of intrapopulation variability. However, it should be noted that the analysis by Lam et al. (1996) entailed only two dimensions of what is manifestly a three-dimensional structure. The jury may still be out over this issue.

Similarly, as discussed above, the differences in calvarial morphology between Omo I and Omo II (Fig. 17.7) have been interpreted by different workers to mean different things. Some have argued that it attests to a population with

[3]The best-represented element from the lower part of the SAS member is the mandibular corpus (n = 4). Of these four, the KRM 41815 jaw derives from the deposits in cave 1B (Layer 10 of Singer and Wymer 1982), while the others are from separate horizons in cave 1 (KRM 16424 from Layer 14+, KRM 13400 from Layer 14, and KRM 21776 from Layer 17 of Singer and Wymer 1982).

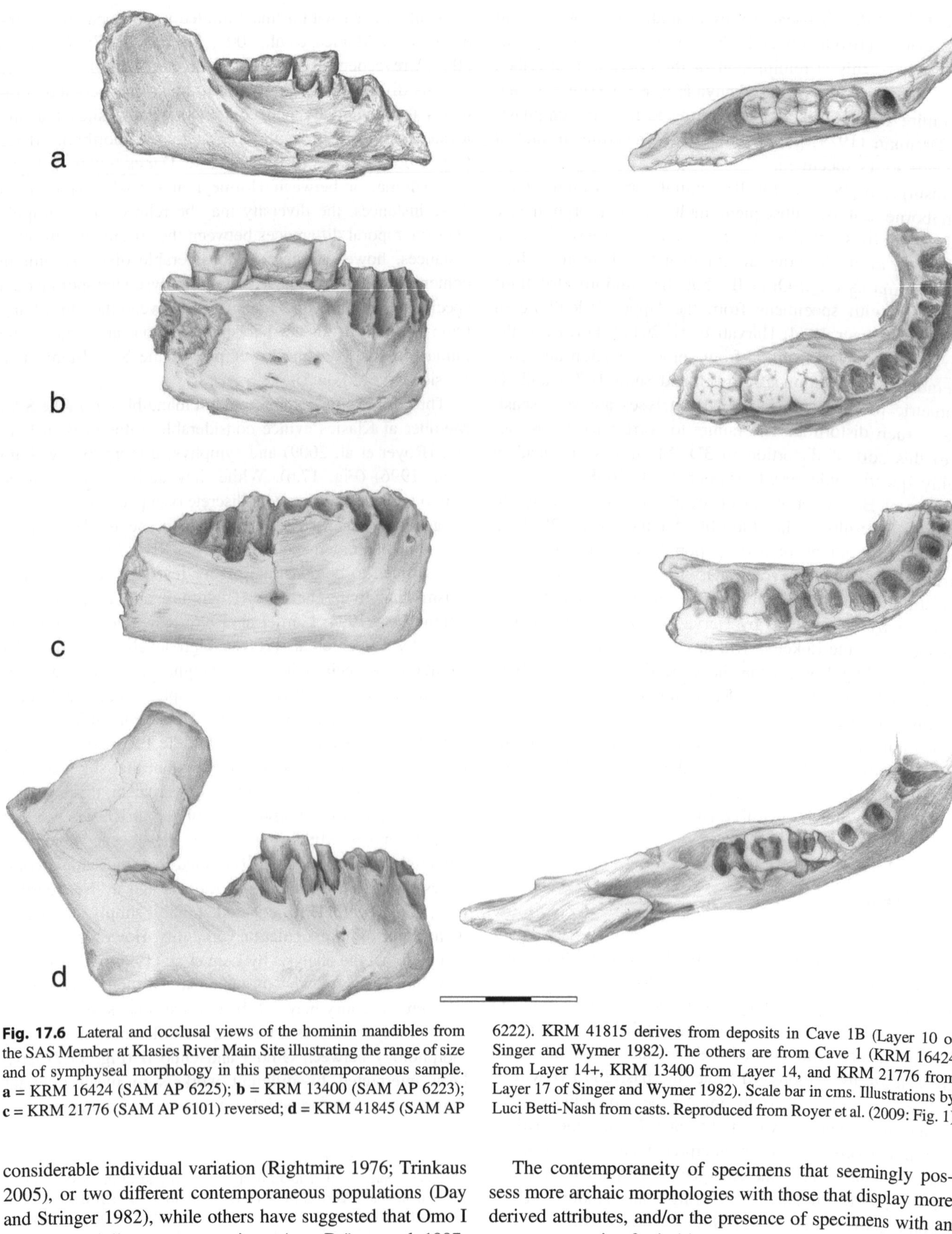

Fig. 17.6 Lateral and occlusal views of the hominin mandibles from the SAS Member at Klasies River Main Site illustrating the range of size and of symphyseal morphology in this penecontemporaneous sample. **a** = KRM 16424 (SAM AP 6225); **b** = KRM 13400 (SAM AP 6223); **c** = KRM 21776 (SAM AP 6101) reversed; **d** = KRM 41845 (SAM AP 6222). KRM 41815 derives from deposits in Cave 1B (Layer 10 of Singer and Wymer 1982). The others are from Cave 1 (KRM 16424 from Layer 14+, KRM 13400 from Layer 14, and KRM 21776 from Layer 17 of Singer and Wymer 1982). Scale bar in cms. Illustrations by Luci Betti-Nash from casts. Reproduced from Royer et al. (2009: Fig. 1)

considerable individual variation (Rightmire 1976; Trinkaus 2005), or two different contemporaneous populations (Day and Stringer 1982), while others have suggested that Omo I is a substantially younger specimen (e.g., Bräuer et al. 1997; Bräuer 2001; Klein 2009).

The contemporaneity of specimens that seemingly possess more archaic morphologies with those that display more derived attributes, and/or the presence of specimens with an apparent mosaic of primitive and derived features have been interpreted as being consistent with at least three different

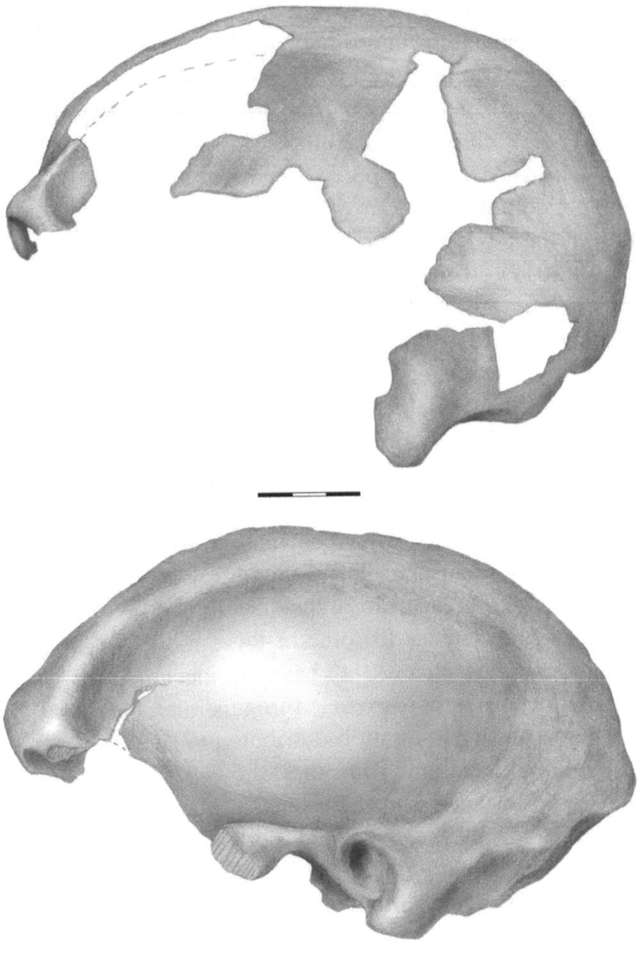

Fig. 17.7 Lateral views of the Omo I (*above*) and Omo II (*below*) calvaria from Member 1 of the Kibish Formation illustrating the differences in the frontal and occipital contours between these penecontemporaneous specimens. The image of Omo I is the right lateral view reversed. *Scale bar* in cms. Illustrations by Luci Betti-Nash from casts

scenarios. One sees this as reflecting the presence of (and possible admixture between) contemporaneous archaic and modern hominin populations, such as envisioned by Bräuer's (1992) "Hybridization and Replacement" hypothesis. A second interpretation, such as Smith's "Assimilation" hypothesis (Smith et al. 1989), holds that mosaic mixtures are to be expected in populations that represent part of an evolving lineage with roots in earlier, more archaic populations and descendants in later, more derived populations. Although the "Assimilation" hypothesis, as proposed, related to the assimilation of indigenous archaic Neandertal genes into the modern human populations expanding from Africa into Europe, this concept of genetic admixture could apply equally to archaic and modern populations in Africa. A third scenario might envision the presence of possibly contemporary archaic and more modern hominin populations with replacement of the former by the latter. This would be consistent with the strict "African Replacement" hypothesis of Stringer and Andrews (1988).

How does the late Quaternary African human fossil record accord with the results of studies of the modern human genome? An impressive array of evidence from mitochondrial DNA (mtDNA), the Y chromosome, non-coding autosomal microsatellites (short tandem repeats or STRs), and autosomal single nucleotide polymorphisms (SNPs) has been brought to bear on five germane questions. These relate to: (1) the geographic origin of the species *Homo sapiens*; (2) the age of the divergence of *H. sapiens* from the most recent common ancestor that we shared with our now-extinct sister species; (3) the possible assimilation of genes through introgression from contemporaneous archaic lineages into the *H. sapiens* genome, whether within or outside of Africa; (4) the phylogeographic structure of Africa populations and lineage divergence times; and (5) the timing of the most recent human migration out of Africa.

In the first instance, the late Quaternary African hominin fossil record, despite its manifestly incomplete nature, finds at least some consistency with an impressive array of genetic evidence that points to an African origin for our species (Cann et al. 1987; Ingman et al. 2000; Underhill et al. 2001; Salas et al. 2002; Frazer et al. 2007; Garrigan et al. 2007; Gonder et al. 2007; Behar et al. 2008; Li et al. 2008; Tishkoff et al. 2009). There is currently no question that the geochronologically oldest specimens attributable to *Homo sapiens* derive from sub-Saharan Africa (White et al. 2003; McDougall et al. 2008; Aubert et al. 2012; F. Brown et al. 2012).

In the second instance, most haploid mitochondrial DNA (mtDNA) and Y-chromosome genetic data indicate a coalescence of lineages to the most recent common ancestor of *Homo sapiens* at between 200 and 100 ka (Cann et al. 1987; Scozzari et al. 1999; Ingman et al. 2000; Underhill et al. 2000; Tang et al. 2002; Tishkoff and Verrelli 2003; Garrigan and Hammer 2006; Kivisild et al. 2006; Gonder et al. 2007; Behar et al. 2008). Previous estimates of the most recent female common ancestor have tended to fall closer to 200 ka, with a range of perhaps 240–150 ka (e.g., Cann et al. 1987; Ingman et al. 2000; Tang et al. 2002), while estimates of the most recent common male ancestor have been closer to 100 ka, with a range of some 50–142 ka (Pritchard et al. 1999; Thomson et al. 2000; Tang et al. 2002; Cruciani et al. 2011). Most recently, however, Poznik et al. (2013) analyzed new mtDNA and Y-chromosome genome data using comparable techniques, and found no statistically significant difference between the two loci. Both produce estimates for the most recent common ancestor (T_{MRCA}) dating to between 150 and 120 ka. However, as observed by Weaver (2012), these models need not be interpreted as being indicative of a severe population bottleneck at the time of coalescence. The same coalescence time could be achieved in a population of constant size with an effective population size equal to the harmonic mean effective population size of a severely bottlenecked population. For this reason, Weaver (2012) has noted that the coalescence time for human mtDNA and Y-chromosome need not be expected to correspond to the timing of the emergence of the species *Homo sapiens*.

However, these T_{MRCA} dates of 150–120 ka do place an upper bound on the time of population divergence among modern humans, such that prior to 120 ka there was a common ancestral population (early population divergence in Africa is discussed below). Inference of a bottleneck during MIS 4, MIS 5 or MIS 6 is eminently testable with genetic data, but current limited analyses have produced conflicting results (Henn et al. 2012). For example, employing the complete diploid genome sequences from an admittedly tiny sample of present day humans (N = 6), Li and Durbin (2011) found no evidence whatsoever for a bottleneck in human population size between 200 and 100 ka, although they did find strong evidence for both an African and non-African bottleneck occurring roughly 60–30 ka (see also Kidd et al. 2012).

In the third instance, the fossil record reveals the penecontemporaneous presence in the Late Pleistocene of individuals possessing more archaic morphologies and those with more derived attributes. While this is perhaps most reasonably (i.e., most parsimoniously) interpreted as simply attesting to the retention of some plesiomorphic traits (i.e., attesting to polymorphic allelic loci) in some populations and/or individuals, and a consequent enhanced degree of interindividual phenotypic variability, it also might be seen as being consistent with indirect evidence that suggests archaic admixture in the genomes of current Africans (e.g., Wall et al. 2009; Hammer et al. 2011; Lachance et al. 2012).

It should be noted, however, that the conclusions relating to "archaic" admixture in Africa by Hammer et al. (2011) derive from the analysis of sequence data from 61 regions of the genome in a sample of only three sub-Saharan populations. They suggested that these populations received a c. 2% contribution of genetic material from some extinct "archaic" population that would have split from the *Homo sapiens* lineage some 700 ka, and then interbred with a Central African population some 35 ka. Why it was deemed to have been Central African is unclear, and why it took nearly 665 kyr to accomplish this remains a mystery. Indeed, with reference to morphological studies that purport to show evidence in the terminal Pleistocene of Africa that is consistent with admixture of archaic lineages (e.g., Harvati et al. 2011), it is unclear how 2% archaic admixture in a few Central African hunter-gatherer populations manifest itself in the fossil record. While it is conceivable that discrete archaic features might be recorded in a fossil representative of the first filial (F_1) generation that resulted from such an introgessive cross, the presence of 2% archaic admixture inherited through tens of thousands of years would surely be morphologically indistinguishable from normal population variation.[4]

The scenario proposed by Hammer et al. (2011) might be seen to gain some support from the study by Mendez et al. (2013) of Y-chromosome composition in an African American who carries the apparent ancestral states for the single nucleotide polymorphisms (SNPs) that define the basal portion of the phylogenomic tree for this chromosome. Mendez et al. (2013) estimated the time to the most recent common ancestor for this Y-chromosome tree at a remarkably old age of between 581 and 237 ka. However, the ancient age estimated by Mendez et al. (2013) has been resoundingly criticized by Elhaik et al. (2014), who pointed out a number of statistical and methodological flaws with the study. Elhaik et al. (2014), in turn, estimated the Y-chromosome T_{MRCA} at c. 208 ka. Moreover, the studies by Hammer et al. (2011) and Mendez et al. (2013) – which emanate from the same lab – are heavily dependent upon the modeling of demographic population structure. Apart from the questions raised by Weaver (2012) with reference to some of these model assumptions, Eriksson and Manica (2011) have demonstrated that spatial population structure is expected to generate genetic patterns similar to those that might be attributed to ancient admixture through hybridization. Thus, they showed, with reference to the supposed admixture between Neandertals and modern Eurasians, their excess shared polymorphism is strongly linked to the structure of ancient populations, and is, in fact, compatible with scenarios in which no hybridization occurred. Currat and Excoffier (2011) also have presented strong statistical arguments in relation to paleodemography that question the degree of hybridization that has been inferred between archaic and modern populations.

[4]There are some theoretical expectations and empirical observations to the effect that admixture may result in the expression of increased morphological anomalies, or of morphologies that may be intermediate in nature but with heightened levels or unusual modes of variation (Ackermann 2010). Still, we have very little expectation of how to recognize a hybrid individual (or sample) in the paleontological record. A number of specimens from sites in Europe that date to MIS 3 have been posited to represent hybrids, with most being viewed as evincing an overall modern aspect with some Neandertal features. These include the remains from Abrigo do Lagar Velho, Portugal (Duarte et al. 1999; Bayle et al. 2010), Peştera cu Oase, Romaina (Trinkaus et al. 2003; Rougier et al. 2007), Peştera Muierii, Romania (Soficaru et al. 2006), Mladeč, Czech Republic (Frayer et al. 2006; Wolpoff et al. 2006) and Cioclovina, Romania (Soficaru et al. 2007). These claims have not gone unchallenged (e.g., Tattersall and Schwartz 1999; Harvati et al. 2007). Perhaps one of the more interesting specimens in this regard is the 40–30 ka mandibular corpus from Riparo di Mezzena, Italy. This fragment has been interepreted as a Neandertal on the basis of its mtDNA, which shows "a classic Neandertal motif with the diagnostic transversion 16256 C/A" (Condemi et al. 2013: 6). However, it also displays an incipient mental trigone (chin), a feature that long has been held to be a distinction of *Homo sapiens* (Schwartz and Tattersall 2000).

Finally, Gokcumen et al. (2013) have shown that significant levels of (balancing) selection can mimic signatures of archaic admixture, bringing into question the notion that the ancient Denisovans from the Altai Mountains in southern Siberia somehow contributed some 5% of the genomes of the modern inhabitants of Papua New Guinea (Meyer et al. 2012; Wall et al. 2013). Thus, archaic admixture, while a distinct possibility, should only be regarded as a plausible scenario for modern genomic structure after more likely explanations relating to population/demographic history and natural selection have been discounted. Although beguiling, it is at best overambitious to interpret one or another feature of a single human specimen as suggesting anything at all about the depth or nature of population substructure in Africa.

With regard to the fourth set of questions enumerated above, there is considerable genetic evidence to indicate a relatively deep separation of human populations in Africa, with the ancestors of the southern African Khoe-San diverging early on from the population that gave rise to all other groups (Hammer et al. 2001; Ingman and Gyllensten 2001; Salas et al. 2002; Semino et al. 2002; Torroni et al. 2006; Gonder et al. 2007; Tishkoff et al. 2007, 2009; Behar et al. 2008; Schuster et al. 2010; Gronau et al. 2011; Henn et al. 2011; Pickerell et al. 2012; Schlebush et al. 2012, 2013; Veeramah et al. 2012). Within the mtDNA L0 lineage, the deeply divergent branches (L0d and L0k) are carried almost exclusively by the Khoe-San or their current near neighbors. Unfortunately, an estimate of population divergence has not been computed, although the coalescence time will be older than the population divergence.

Nevertheless, because L0 and the deepest clades within it have a southern African distribution, it seems very reasonable to conclude that it had a southern African origin. Three estimates of population divergence between the Khoe-San and other Africans have been performed recently on varying amounts of autosomal/genome-wide data (Gronau et al. 2011; Veeramah et al. 2012; Schlebusch et al. 2012). Schlebusch et al. (2012) estimated that six of seven sampled Khoe-San language/ethnic groups shared a common ancestor that had diverged from all other sub-Saharan African populations probably by 100 ka. Models based on sequence data support a divergence of Khoe-San from other African populations between 130 and 110 ka. A southern African origin is consistent not only with other human genomic data, but also with variation in strains of the gastric bacterium *Helicobacter pylori*. The latter indicate origin with an African human host (Salama et al. 2000; Linz et al. 2007), and coalescence of the two strains associated with southern African Khoe-San populations (hpAfrica 1 and hpAfrica 2) at some 116–88 ka (Moodley et al. 2012).

Some Late Pleistocene fossils have been interpreted at one time or another as sharing distinctive morphological similarities with, and therefore being in the ancestry of specific recent African populations – for example, the accentuated parietal bosses of the Singa calvaria suggested Khoe-San affinities to Woodward (1938) and Wells (1951) – but there is a rather noticeable lack of phenetic resemblance between the craniodental remains (and postcranial fragments) from the Late Pleistocene with those of these modern Holocene groups. Thus, for example, the Hofmeyr cranium from South Africa only dates to some 36 ka, and genetic evidence suggests the existence of a Khoe-San lineage in southern Africa dating back well beyond that (Henn et al. 2011; Schlebusch et al. 2013), but the analyses of the skull undertaken thus far reveal no specific morphometric affinity with Holocene Khoe-San crania (Grine et al. 2007). It is, of course, possible that more detailed analysis of the reconstructed specimen (Grine et al. 2010) may emphasize the tantalizing linear morphometric similarities hinted at earlier. Similarly, a number of the human dental samples from Late Pleistocene South African sites (e.g., Die Kelders) display morphological variants that characterize the teeth of the recent inhabitants of sub-Saharan Africa, but these similarities do not necessarily signify a close evolutionary relationship because these features appear to be plesiomorphic (Grine 2000).

The lack of morphologies in any of the Late Pleistocene fossils that can serve as ties to specific recent population groups in Africa is perhaps not unexpected. There is no reason to suppose that genetic variants associated with aspects of hard tissue development and growth (e.g., RUNX2, which codes for a key transcription factor for osteoblast and chondrocyte differentiation [Fujita et al. 2004; Komori 2010]), whose phenotypic effects might manifest as distinctive craniodental features, should be temporally associated with the changes implicated with the splitting of other genetic lineages. To date, only those crania from Holocene LSA contexts in South Africa display morphological patterns that can be identified as distinctly Khoe-San (Stynder et al. 2007a, b).

The absence of fossils from the Late Pleistocene with identifiable Khoe-San affinities led Morris (2002, 2003) to suggest that the Khoe-San morphotype arose relatively late in South Africa. He has hypothesized that the ancestors of recent Khoe-San populations underwent a population (and hence phenotypic) bottleneck associated with the Last Glacial Maximum of MIS 2 at c. 29–14 ka. Although this is an interesting suggestion, there is, unfortunately, no genetic evidence for it.

Given the sheer size of Africa (over 30,000,000 km^2) and its extensive modern climatic variation, it is perhaps understandable that the continent remains very much a black box with regard to environmental influences on Late Pleistocene human evolution. We think we know that our species emerged somewhere in sub-Saharan Africa close to 200 ka, and we know something about the people and their archaeological traditions in the Holocene, but the mechanisms and processes that were involved over the past two hundred millennia are frustratingly opaque. The period MIS 6-2 holds the key to understanding our evolution in the world's second largest continent. Despite three quarters of a century of research and exploration, we have only begun to glimpse our recent heritage, but what has been detected thus far is a clear enticement for further exploration and research.

Acknowledgements I thank Brian Stewart for the invitation to provide this review, and Brian, Sacha Jones and Eric Delson for their persistence and patience. Thanks to Zelalem Assefa, Graeme Barker, Frank Brown, Isabelle Crevecoeur, Tyler Faith, Craig Feibel, John Fleagle, Chris Henshilwood, Jean-Jacques Hublin, Chris Hunt, Richard Klein, Christine Lane, Curtis Marean, Justin Pargeter, John Parkington, Tim Reynolds, Eleanor Scerri, John Shea, Lyn Wadley, Tim White, Pamela Willoughby and Sarah Wurz for information pertaining to specific Late Pleistocene sites and fossils. I am grateful to Jenna Cole for information and discussions concerning paleoclimate and late Quaternary African paleoenvironments, and to Brenna Henn for providing invaluable insights into the human genetic landscape of Africa. I am especially grateful to Eric Delson, John Fleagle, Tim Compton, Chris Stringer and Richard Klein for their comments and suggestions on the entire manuscript. The illustrations were skillfully executed by Luci Betti-Nash.

References

Ackermann, R. R. (2010). Phenotypic traits of primate hybrids: Recognizing admixture in the fossil record. *Evolutionary Anthropology, 19*, 258–270.

Adams, J. M. (1997). Global land environments since the Last Interglacial. Oak Ridge National Laboratory (TN, USA). http://www.esd.ornl.gov/ern/qen/nerc.html.

Adkins, J., deMenocal, P., & Eshel, G. (2006). The "African Humid Period" and the record of marine upwelling from excess ^{230}Th in ODP Hole 658C. *Paleoceanography, 21*, PA4203.

Ahern, J. C. M., & Smith, F. H. (2004). Adolescent archaic or adult moderns? Le Moustier 1 as a model for estimating age at death of fragmentary supraorbital fossils in the modern human origins debate. *Homo, 55*, 1–19.

Aiello, L. C. (2010). Five years of *Homo floresiensis*. *American Journal of Physical Anthropology, 142*, 167–179.

Aitken, M. J., & Stokes, S. (1997). Climatostratigraphy. In R. E. Taylor & M. J. Aitken (Eds.), *Chronometric dating in archaeology* (pp. 1–30). New York: Plenum.

Allsworth-Jones, P., Harvati, K., & Stringer, C. B. (2010). The archaeological context of the Iwo Eleru cranium from Nigeria and preliminary results of new morphometric studies. In P. Allsworth-Jones (Ed.), *West African archaeology: New developments, new perspectives* (BAR International Series No. 2164) (pp. 29–42). Oxford: Archaeopress.

Amani, F., & Geraads, D. (1993). The Mousterian site of Jebel-Irhoud, Morocco—new data on the fauna and biochronology, and description of a new human fossil. *Comptes rendus de l'Académie des sciences Série II, 316*, 847–852.

Ambergen, A. W., & Schaafsma, W. (1984). Interval estimates for posterior probabilities: Applications to Border Cave. In G. N. van Vark & W. W. Howells (Eds.), *Multivariate statistical methods in physical anthropology* (pp. 115–134). Dordrecht: Reidel.

Anderson, J. E. (1968). Late Paleolithic skeletal remains from Nubia. In F. Wendorf (Ed.), *The prehistory of Nubia* (2nd ed., pp. 996–1040). Dallas: Southern Methodist University Press.

Andrews, J. T. (2000). Dating glacial events and correlation to global climate change. In J. S. Noller, J. M. Sowers, & W. R. Lettis (Eds.), *Quaternary geology. Methods and applications* (pp. 447–455). Washington, DC: American Geophysical Union.

Anthony, B. W. (1967). Excavations at Peers Cave, Fish Hoek, South Africa. *Paleoecologia Africana, 2*, 58–59.

Antoine, D. M., Zazzo, A., & Friedman, R. (2013). Revisiting Jebel Sahaba: New apatite radiocarbon dates for one of the Nile valley's earliest cemeteries. *American Journal of Physical Anthropology, 150*(S56), 68.

Aouraghie, H., & Abbassi, M. (2002). Les rongeurs du site Atérien d'El Harhoura 1 (Témara, Maroc). *Quaternaire, 13*, 125–136.

Arambourg, C., Boule, M., Vallois, H., & Verneau, R. (1934). *Les Grottes Paléolithiques des Béni-Ségoual (Algérie)* (Archives de l'IPH No. 13). Paris: Masson.

Armelagos, G. (1964). A fossilized mandible from near Wadi Halfa, Sudan. *Man, 64*, 12–13.

Armelagos, G., Van Gerven, D., Martin, D., & Huss-Ashmore, R. (1984). Effects of nutritional change on the skeletal biology of northeast African (Sudanese Nubian) populations. In J. D. Clark & S. Brandt (Eds.), *From hunters to farmers: The causes and consequences of food production in Africa* (pp. 132–146). Berkeley: University of California Press.

Armitage, S. J., Drake, N. A., Stokes, S., El-Hawat, A., Salem, M., White, K., et al. (2007). Multiple phases of North African humidity recorded in lacustrine sediments from the Fazzan Basin, Libyan Sahara. *Quaternary Geochronology, 2*, 181–186.

Assefa, Z. (2006). Faunal remains from Porc-Épic: Paleoecological and zooarchaeological investigations from a Middle Stone Age site in southeastern Ethiopia. *Journal of Human Evolution, 51*, 50–75.

Aubert, M., Pike, A. W. G., Stringer, C., Bartsiokas, A., Kinsley, L., Eggins, S., et al. (2012). Confirmation of a late Middle Pleistocene age for the Omo Kibish 1 cranium by direct uranium-series dating. *Journal of Human Evolution, 63*, 704–710.

Avery, D. M. (1982). Micromammals as paleoenvironmental indicators and an interpretation of the Late Quaternary in the southern Cape Province, South Africa. *Annals of the South African Museum, 85*, 183–377.

Avery, G., Cruz-Uribe, K., Goldberg, P., Grine, F. E., Klein, R., Lenardi, M. J., et al. (1997). The 1992–93 excavations at the Die Kelders Middle and Later Stone Age cave site, South Africa. *Journal of Field Archaeology, 24*, 263–291.

Avery, G., Halkett, D., Orton, J., Steele, T., Tusenius, M., & Klein, R. (2008). The Ysterfontein 1 Middle Stone Age rock shelter and the evolution of coastal foraging. *South African Archaeological Society Goodwin Series, 10*, 66–89.

Backwell, L., d'Errico, F., & Wadley, L. (2008). Middle Stone Age bone tools from the Howiesons Poort layers, Sibudu Cave, South Africa. *Journal of Archaeological Science, 35*, 1566–1580.

Backwell, L. R., Parkinson, A. H., Roberts, E. M., d'Errico, F., & Huchet, J. B. (2012). Criteria for identifying bone modification by termites in the fossil record. *Paleogeography, Paleoclimatology, Paleoecology, 337*, 72–87.

Balter, M. (2011). Was North Africa the launch pad for modern human migration? *Science, 331*, 20–23.

Balzeau, A., & Badawi-Fayad, J. (2005). La morphologie externe et interne de la région supra-orbitaire est-elle corrélée à des contraintes biomécaniques? Analyses structurales des populations d'*Homo sapiens* d'Afalou Bou Rhummel (Algérie) et de Taforalt (Maroc). *Bulletins et Mémoires de la Société d'Anthropologie de Paris, 17*, 185–197.

Barham, L. (1995). The Mumbwa Caves Project, Zambia, 1993-94. *Nyame Akuma, 43*, 66–72.

Barham, L. (2000). The Mumbwa Caves 1993–1996. In: L. Barham (Ed.), *The Middle Stone Age of Zambia, South Central Africa* (pp. 11–49). Bristol: Western Academic & Specialist Press.

Barham, L. (2002). Backed tools in Middle Pleistocene central Africa and their evolutionary significance. *Journal of Human Evolution, 43*, 585–603.

Barham, L., & Debenham, N. (2000). Mumbwa Caves chronology. In L. Barham (Ed.), *The Middle Stone Age of Zambia, south central Africa* (pp. 43–49). Bristol, UK: Western Academic & Specialist Press.

Barham, L., Simms, M., Gilmour, M., & Debenham, N. (2000). Twin Rivers, excavation and behavioural record. In L. Barham (Ed.), *The Middle Stone Age of Zambia, south central Africa* (pp. 165–216). Bristol, UK: Western Academic & Specialist Press.

Barkai, R., Gopher, A., Lauritzen, S. E., & Frumkin, A. (2003). Uranium series dates from Qesem Cave, Israel, and the end of the Lower Paleolithic. *Nature, 423*, 977–979.

Barkai, R., Gopher, A., & Shimelmitz, R. (2005). Middle Pleistocene blade production in the Levant: An Amudian assemblage from Qesem Cave, Israel. *Eurasian Prehistory, 3*, 39–74.

Barker, G., Hunt, C., & Reynolds, T. (2007). The Haua Fteah, Cyrenaica (Northeast Libya): Renewed investigations of the cave and its landscape, 2007. *Libyan Studies, 38*, 1–22.

Barker, G., Antoniadou, A., Armitage, S., Brooks, I., Candy, I., Connell, K., et al. (2010). The Cyrenaican Prehistory Project 2010: The fourth season of investigations of the Haua Fteah cave and its landscape, and further results from the 2007-2008 fieldwork. *Libyan Studies, 41*, 63–88.

Barker, G., Bennett, P., Farr, L., Hill, E., Hunt, C., Lucarini, G., et al. (2012). The Cyrenaican Prehistory Project 2012: The fifth season of investigations of the Haua Fteah cave. *Libyan Studies, 43*, 115–136.

Barker, P., & Gasse, F. (2003). New evidence for a reduced water balance in East Africa during the Last Glacial Maximum: Implication for model-data comparison. *Quaternary Science Reviews, 22*, 823–837.

Bar-Matthews, M., Marean, C. W., Jacobs, Z., Karkanas, P., Fisher, E. C., Herries, A. I. R., et al. (2010). A high resolution and continuous isotopic speleothem record of paleoclimate and paleoenvironment from 90 to 53 ka from Pinnacle Point on the south coast of South Africa. *Quaternary Science Reviews, 29*, 2131–2145.

Barton, R. N. E., Bouzouggar, J., Collcutt, S. N., Gale, R., Higham, T. F. G., Humphrey, L. T., et al. (2005). The late Upper Paleolithic occupation of the Moroccan northwest Maghreb during the Last Glacial Maximum. *African Archaeological Review, 22*, 77–100.

Barton, R. N. E., Bouzouggar, A., Bronk-Ramsey, C., Collcutt, S. N., Higham, T. F. G., Humphrey, L. T., et al. (2007). Abrupt climatic change and chronology of the Upper Paleolithic in northern and eastern Morocco. In O. Bar-Yosef, P. Mellars, C. Stringer, & K. Boyle (Eds.), *Rethinking the human revolution: New behavioural and biological perspectives on the origins and dispersal of modern humans* (pp. 177–186). Cambridge: McDonald Institute for Archaeological Research.

Barton, N., Bouzouggar, A., Humphrey, L., Berridge, P., Collcutt, S., Gale, R., et al. (2008). Human burial evidence from Hattab II Cave and the question of continuity in Late Pleistocene – Holocene mortuary practices in northwest Africa. *Cambridge Archaeological Journal, 18*, 195–214.

Barton, R. N. E., Bouzouggar, A., Collcutt, S. N., Schwenninger, J. L., & Clark-Balzan, L. (2009). OSL dating of the Aterian levels at Dar es-Soltan I (Rabat, Morocco) and implications for the dispersal of modern *Homo sapiens*. *Quaternary Science Reviews, 28*, 1914–1931.

Barton, R. N. E., Bouzouggar, A., Hogue, J. T., Lee, S., Collcutt, S. N., & Ditchfield, P. (2013). Origins of the Iberomaurusian in NW Africa: New AMS radiocarbon dating of the Middle and Later Stone Age deposits at Taforalt Cave, Morocco. *Journal of Human Evolution, 65*, 266–281.

Bassinot, F. C., Labeyrie, L. D., Vincent, E., Quidelleur, X., Shackleton, N. J., & Lancelot, Y. (1994). The astronomical theory of climate and the age of the Brunhes-Matuyama magnetic reversal. *Earth and Planetary Science Letters, 126*, 91–108.

Bayle, P., Macchiarelli, R., Trinkaus, E., Duarte, C., Mazurier, A., & Zilhão, J. (2010). Dental maturational sequence and dental tissue proportions in the early Upper Paleolithic child from Abrigo do Lagar Velho, Portugal. *Proceedings of the National Academy of Sciences of the United States of America, 107*, 1338–1342.

Beaumont, P. B. (1980). On the age of Border Cave hominids 1-5. *Paleontologia Africana, 23*, 21–33.

Beaumont, P. B., de Villiers, H., & Vogel, J. C. (1978). Modern man in sub-Saharan Africa prior to 49,000 years B.P.: A review and evaluation with particular reference to Border Cave. *South African Journal of Science, 74*, 409–419.

Behar, D. M., & The Genographic Consortium. (2008). The dawn of human matrilineal diversity. *American Journal of Human Genetics, 82*, 1130–1140.

Behrensmeyer, A. K. (2006). Climate change and human evolution. *Science, 311*, 476–478.

Ben-Ncer, A. (2004). Etude de la sépulture Ibéromaurusienne 1 d'Ifri n'Baroud (Rif oriental, Maroc). *Anthropo, 7*, 177–185.

Biittner, K. (2011). Characterization of Middle and Later Stone Age lithic artifacts from two rockshelter sites in Iringa Region, southern Tanzania. Ph.D. Dissertation, University of Alberta.

Biittner, K. M., Bushozi, P. M., & Willoughby, P. R. (2007). The Middle and Later Stone Age of the Iringa Region, southern Tanzania: An introduction. *Nyame Akuma, 68*, 62–73.

Bird, M. I., Fifield, L. K., Santos, G. M., Beaumont, P. B., Zhou, Y., di Tada, M. L., et al. (2003). Radiocarbon dating from 40 to 60 ka BP at Border Cave, South Africa. *Quaternary Science Reviews, 22*, 943–947.

Blegen, N., Tryon, C. A., Faith, J. T., Peppe, D. J., Beverly, E. J., Li, B., & Jacobs, Z. (2015). Distal tephras of the eastern Lake Victoria Basin, equatorial East Africa: Correlations, chronology, and a context for early modern humans. *Quaternary Science Reviews, 122*, 89–111.

Blome, M. W., Cohen, A. S., Tryon, C. A., Brooks, A. S., & Russell, J. (2012). The environmental context for the origins of modern human diversity: A synthesis of regional variability in African climate 150,000-30,000 years ago. *Journal of Human Evolution, 62*, 563–592.

Boaz, N. T., Pavlakis, P., & Brooks, A. S. (1990). Late Pleistocene-Holocene human remains from the Upper Semliki, Zaire. In N. T. Boaz (Ed.), *Evolution of environments and hominidae in the African Western Rift Valley* (Memoir No. 1) (pp. 273–299). Martinsville, VA: Virginia Museum of Natural History.

Bond, G., Broecker, W., Johnsen, S., McManus, J., Labeyrie, L., Jouzel, J., et al. (1993). Correlations between climate records from North Atlantic sediments and Greenland ice. *Nature, 365*, 143–147.

Bouchneb, L., & Crevecoeur, I. (2009). The inner ear of Nazlet Khater 2 (Upper Paleolithic, Egypt). *Journal of Human Evolution, 56*, 257–262.

Bousman, B. C. (2005). Coping with risk: Later Stone Age technological strategies at Blydefontein Rock Shelter, South Africa. *Journal of Anthropological Archaeology, 24*, 193–226.

Bouzouggar, A., Barton, N., Vanhaeren, M., d'Errico, F., Collcutt, S., Higham, T., et al. (2007). 82,000 year-old shell beads from North Africa and implications for the origins of modern human behavior. *Proceedings of the National Academy of Sciences of the United States of America, 104*, 9964–9969.

Bouzouggar, A., Barton, R. N. E., Blockley, S., Bronk-Ramsey, C., Collcutt, S. N., Gale, R., et al. (2008). Reevaluating the age of the Iberomaurusian in Morocco. *African Archaeological Review, 25*, 3–19.

Bowen, B. E., & Vondra, C. F. (1973). Stratigraphical relationships of the Plio-Pleistocene deposits, East Rudolf, Kenya. *Nature, 242*, 391–393.

Bräuer, G. (1984a). The Afro-European sapiens hypothesis and hominid evolution in East Asia during the late Middle Pleistocene and Upper Pleistocene. *Courier Forschungsinstitut Senckenberg, 69*, 145–165.

Bräuer, G. (1984b). A craniological approach to the origin of anatomically modern *Homo sapiens* in Africa and implications for the appearance of modern Europeans. In F. H. Smith & F. Spencer (Eds.), *The origins of modern humans: A world survey of the fossil evidence* (pp. 327–410). New York: Alan R. Liss.

Bräuer, G. (1989a). The evolution of modern humans: A comparison of the African and non-African evidence. In P. Mellars & C. Stringer (Eds.), *The human revolution* (pp. 123–154). Edinburgh: Edinburgh University Press.

Bräuer, G. (1989b). The ES-11693 hominid from West Turkana and *Homo sapiens* evolution in East Africa. In G. Giacobini (Ed.), *Hominidae* (pp. 241–245). Milan: Jaca Books.

Bräuer, G. (1992). Africa's place in the evolution of *Homo sapiens*. In G. Bräuer & F. H. Smith (Eds.), *Continuity or replacement? Controversies in Homo sapiens evolution* (pp. 83–98). Rotterdam: Balkema.

Bräuer, G. (2001). The KNM-ER 3884 hominid and the emergence of modern human anatomy in Africa. In P. V. Tobias, M. A. Rath, J. Moggi-Cecchi, & G. A. Doyle (Eds.), *Humanity from African naissance to coming millennia* (pp. 191–197). Firenze: Firenza University Press.

Bräuer, G. (2008). The origin of modern anatomy: By speciation or intraspecific evolution? *Evolutionary Anthropology, 17*, 22–37.

Bräuer, G., & Leakey, R. E. (1986a). A new archaic *Homo sapiens* cranium from Eliye Springs, West Turkana, Kenya. *Zeitschrift für Morphologie und Anthropologie, 76*, 245–252.

Bräuer, G., & Leakey, R. E. (1986b). The ES-11693 cranium from Eliye Springs, West Turkana, Kenya. *Journal of Human Evolution, 15*, 289–312.

Bräuer, G., & Mabulla, Z. P. (1996). New hominid fossil from Lake Eyasi, Tanzania. *Anthropologie, 34*, 47–53.

Bräuer, G., & Mehlman, M. J. (1988). Hominid molars from a Middle Stone Age level at the Mumba Rock Shelter, Tanzania. *American Journal of Physical Anthropology, 75*, 69–76.

Bräuer, G., & Rimbach, K. (1990). Late archaic and modern *Homo sapiens* from Europe, Africa and southwest Asia: Craniometric comparisons and phylogenetic implications. *Journal of Human Evolution, 19*, 789–807.

Bräuer, G., & Rösing, F. W. (1989). Human biological history in southern Africa. *Rassengeschichte der Menschheit, 13*, 6–137.

Bräuer, G., & Singer, R. (1996). The Klasies zygomatic bone: Archaic or modern? *Journal of Human Evolution, 30*, 161–165.

Bräuer, G., Deacon, H. J., & Zipfel, F. (1992a). Comment on the new maxillary finds from Klasies River, South Africa. *Journal of Human Evolution, 23*, 419–422.

Bräuer, G., Leakey, R. E. F., & Mbua, E. (1992b). A first report on the ER-3884 cranial remains from Ileret/East Turkana, Kenya. In G. Bräuer & F. H. Smith (Eds.), *Continuity or replacement? Controversies in Homo sapiens evolution* (pp. 111–120). Rotterdam: Balkema.

Bräuer, G., Yokoyama, Y., Falguères, C., & Mbua, E. (1997). Modern human origins backdated. *Nature, 386*, 337.

Bräuer, G., Groden, C., Delling, G., Kupczik, K., Mbua, E., & Schultz, M. (2003). Pathological alterations in the archaic *Homo sapiens* cranium from Eliye Springs, Kenya. *American Journal of Physical Anthropology, 120*, 200–204.

Bräuer, G., Groden, C., Gröning, F., Kroll, A., Kupczik, K., Mbua, E., et al. (2004). Virtual study of the endocranial morphology of the matrix-filled cranium from Eliye Springs, Kenya. *Anatomicasl Record, 276A*, 113–133.

Breuil, A. H., Teilhard de Chardin, P., & Wernert, P. (1951). Le Paléolithique du Harar. *L' Anthropologie, 55*, 219–230.

Briggs, L. C. (1955). *The Stone Age races of northwest Africa* (Bulletin of the American School of Prehistoric Research No. 18). Cambridge, MA: Peabody Museum.

Brooks, A. S. (2008). Overview of the sequence, environments and dating at Ishango and Katanda. In D. Huylebrouck (Ed.), *Ishango, 22000 and 50 years later: The cradle of mathematics?* (pp. 55–66). Brussels: Koninklijke Vlaamse Academie van Belgie voor Wetenschappen en Kunsten.

Brooks, A. S., & Robertshaw, P. (1990). The Glacial Maximum in tropical Africa: 22,000–12,000 B.P. In C. Gamble & O. Soffer (Eds.), *The World at 18 000 BP: Low latitudes*, Volume 2 (pp 121–169). London: Unwin Hyman.

Brooks, A. S., & Smith, C. C. (1987). Ishango revisited: New age determinations and cultural interpretations. *African Archaeological Review, 5*, 65–78.

Brooks, A. S., Smith, C. C., & Boaz, N. T. (1991). New human remains from Ishango, Zaire, in relation to Later Pleistocene human evolution. *American Journal of Physical Anthropology, S12*, 54–55.

Brooks, A. S., Helgren, D. M., Cramer, J. S., Franklin, A., Hornyak, W., Keating, J. M., et al. (1995). Dating and context of three Middle Stone Age sites with bone points in the Upper Semliki Valley, Zaire. *Science, 268*, 548–553.

Broom, R. (1918). The evidence afforded by the Boskop skull of a new species of primitive man (*Homo capensis*). *Anthropological Papers of the American Museum of Natural History, 23*, 63–79.

Broom, R. (1923). A contribution to the craniology of the yellow-skinned races of South Africa. *Journal of the Royal Anthropological Institute, 53*, 132–149.

Broom, R. (1929a). The Transvaal fossil human skeleton. *Nature, 123*, 415–416.

Broom, R. (1929b). Australoid element in the Koranas. *Nature, 124*, 507.

Brothwell, D. R. (1963). Evidence of early population change in Central and southern Africa: Doubts and problems. *Man, 63*, 101–104.

Brothwell, D., & Shaw, T. (1971). A Late Upper Pleistocene proto-West African Negro from Nigeria. *Man, 2*, 221–227.

Brown, F. H., & Feibel, C. S. (1986). Revision of lithostratigraphic nomenclature in the Koobi Fora region, Kenya. *Journal of the Geological Society London, 143*, 297–310.

Brown, F. H., & Fuller, C. (2008). Stratigraphy and tephra of the Kibish Formation, southwestern Ethiopia. *Journal of Human Evolution, 55*, 366–403.

Brown, F. H., McDougall, I., & Fleagle, J. G. (2012). Correlation of the KHS Tuff of the Kibish Formation to volcanic ash layers at other sites, and the age of early *Homo sapiens* (Omo I and Omo II). *Journal of Human Evolution, 63*, 577–585.

Brown, K. S., Marean, C. W., Herries, A. I. R., Jacobs, Z., Tribolo, C., Braun, D., et al. (2009). Fire as an engineering tool of early modern humans. *Science, 325*, 859–862.

Brown, K. S., Marean, C. W., Jacobs, Z., Schoville, B. J., Oestmo, S., Fisher, E. C., et al. (2012). An early and enduring advanced technology originating 71,000 years ago in South Africa. *Nature, 491*, 590–593.

Brown, P., Sutikna, T., Morwood, M. J., Soejono, R. P., Jatmiko Saptomo, E. W., et al. (2004). A new small-bodied hominin from the Late Pleistocene of Flores, Indonesia. *Nature, 431*, 1055–1061.

Bruch, A. A., Sievers, C., & Wadley, L. (2012). Quantification of climate and vegetation from southern African Middle Stone Age sites—an application using Late Pleistocene plant material from Sibudu, South Africa. *Quaternary Science Reviews, 45*, 7–17.

Bruner, E., & Manzi, G. (2002). The virtual replica of Nazlet Khater, Egypt, cranium and mandible: First results. In P. M. Vermeersch (Ed.), *Paleolithic quarrying sites in Upper and Middle Egypt* (pp. 337–346). Leuven: Leuven University Press.

Bruzek, J., Sefcakova, A., & Cerny, V. (2004). Révision du sexe des squelettes épipaléolithiques de Taforalt et d'Afalou-bou-Rhoummel par une approche probabiliste. *Antropo, 7*, 195–202.

Buckley, J. D., & Willis, E. H. (1969). Isotope radiocarbon measurements VII. *Radiocarbon, 11*, 53–105.

Butler, B. H. (1974). Skeletal remains from a Late Paleolithic site near Esna. In D. Lubell (Ed.), *The Fakhurian. A Late Paleolithic industry from Upper Egypt* (Geological Survey of Egypt, Paper No. 58) (pp. 176–183). Cairo: The Geological Survey of Egypt.

Butzer, K. W. (1969). Geological interpretation of two Pleistocene hominid sites in the Lower Omo Basin. *Nature, 222*, 1133–1135.

Butzer, K. W. (1997). Late Quaternary problems of the Egyptian Nile: Stratigraphy, environments, prehistory. *Paléorient, 23*, 151–173.

Butzer, K. W. (2004). Coastal eolian sands, paleosols, and Pleistocene geoarchaeology of the southwestern Cape, South Africa. *Journal of Archaeological Science, 31*, 1743–1781.

Butzer, K. W., & Hansen, C. L. (1968). *Desert and river in Nubia: Geomorphology and prehistoric environments at the Aswan Reservoir.* Madison: University of Wisconsin Press.

Butzer, K. W., Brown, F. H., & Thurber, D. L. (1969). Horizontal sediments of the Lower Omo Valley: The Kibish Formation. *Quaternaria, 11*, 15–29.

Butzer, K. W., Beaumont, P. B., & Vogel, J. C. (1978a). Lithostratigraphy of Border Cave, Kwazulu, South Africa: A Middle Stone Age sequence beginning c. 195,000 B.P. *Journal of Archaeological Science, 5*, 317–341.

Butzer, K., Stuckenrath, R., Bruzewicz, A., & Helgren, D. (1978b). Late Cenozoic paleoclimates of the Gaap Escarpment, Kalahari margin, South Africa. *Quaternary Research, 10*, 310–339.

Cambell, N. A. (1984). Some aspects of discrimination and allocation. In G. N. van Vark & W. W. Howells (Eds.), *Multivariate statistical methods in physical anthropology* (pp. 177–192). Dordrecht: Reidel.

Camp, A., & O'Sullivan, G. (1999). Neogene sapropels in the Mediterranean: A review. *Marine Geology, 153*, 11–28.

Camps, G. (1969). L'homme de Mechta El Arbi et sa civilisation. Contribution à l'étude des origines Guanches. *Annuario de estudios atlanticos, Madrid, 15*, 257–272.

Cann, R. L., Stoneking, M., & Wilson, A. C. (1987). Mitochondrial DNA and human evolution. *Nature, 32*, 31–36.

Carlson, D. S. (1976). Temporal variation in prehistoric Nubian crania. *American Journal of Physical Anthropology, 45*, 467–484.

Carto, S. L., Weaver, A. J., Hetherington, R., Lam, Y., & Wiebe, E. C. (2009). Out of Africa and into an ice age: On the role of global climate change in the Late Pleistocene migration of early modern humans out of Africa. *Journal of Human Evolution, 56*, 139–151.

Castañeda, I. S., Mulitza, S., Schefuß, E., dos Santos, R. A. L., Damsté, J. S. S., & Schouten, S. (2009). Wet phases in the Sahara/Sahel region and human migration patterns in North Africa. *Proceedings of the National Academy of Sciences of the United States of America, 106*, 20159–20163.

Cerling, T. E., & Quade, J. (1993). Stable carbon and oxygen isotopes in soil carbonates. In P. K. Swart, K. C. Lohmann, J. A. McKenzie, & E. S. Savin (Eds.), *Climate change in continental isotope records* (pp. 217–231). Washington, DC: American Geophysical Union.

Chamla, M. C. (1970). *Les hommes epipaleolithiques de la Columnata* (Mémoires du C.R.A.R.E. XV). Paris: Arts et Métiers Graphiques.

Chase, B. (2010). South African paleoenvironments during Marine Oxygen Isotope Stage 4: A context for the Howiesons Poort and Still Bay industries. *Journal of Archaeological Science, 37*, 1359–1366.

Chase, B. M., & Meadows, M. E. (2007). Late Quaternary dynamis of southern Africa's winter rainfall zone. *Earth Science Reviews, 84*, 103–138.

Chavaillon, J. (1982). Position chronologique des hominidés fossile d'Ethiopie. In M. A. de Lumley (Ed.), Homo erectus *et la place de l'Homme de Tautavel parmi les hominidés fossiles. Première congrès international de paléontologie humaine* (pp. 766–797). UNESCO: Colloque International du Centre National de la Récherche Scientifique.

Churcher, C. S. (1972). Late Pleistocene vertebrates from archaeological sites in the plain of Kom Ombo, Upper Egypt. *Life Sciences Contributions of the Royal Ontario Museum, 82*, 1–172.

Churchill, S. E., Pearson, O. M., Grine, F. E., Trinkaus, E., & Holliday, T. W. (1996). Morphological affinities of the proximal ulna from Klasies River Main Site: Archaic or modern? *Journal of Human Evolution, 31*, 213–237.

Clark, J. D. (1942). Further excavations (1939) at Mumbwa Caves, Northern Rhodesia. *Transactions of the Royal Society of South Africa, 29*, 133–201.

Clark, J. D. (1971). Human behavioral differences in southern Africa during the later Pleistocene. *American Anthropologist, 73*, 1211–1236.

Clark, J. D., & Brown, K. S. (2001). The Twin Rivers Kopje, Zambia: Stratigraphy, fauna, and artifact assemblages from the 1954 and 1956 excavations. *Journal of Archaeological Science, 28*, 305–330.

Clark, J. D., & Williamson, K. D. (1984). A Middle Stone Age occupation site at Porc-Epic Cave, Dire Dawa (east-central Ethiopia), part I. *African Archaeological Review, 2*, 37–64.

Clark, J. D., Beyene, Y., WoldeGabriel, G., Hart, W. K., Renne, P. R., Gilbert, H., et al. (2003). Stratigraphic, chronological and behavioural contexts of Pleistocene *Homo sapiens* from Middle Awash, Ethiopia. *Nature, 423*, 747–752.

Clark, P. U., Dyke, A. S., Shakun, J. D., Carlson, A. E., Clark, J., Wohlfarth, B., et al. (2009). The Last Glacial Maximum. *Science, 325*, 710–714.

Close, A. E. (1988). Current research and recent radiocarbon dates from northern Africa, III. *Journal of African History, 29*, 145–176.

Close, A. E., & Wendorf, F. (1990). North Africa at 18,000 BP. In C. Gamble & O. Soffer (Eds.), *The world at 18,000 BP* (Vol. 2, pp. 41–57). Low Latitudes London: Unwin Hyman.

Cohen, A. S., Stone, J. R., Beuning, K. R. M., Park, L. E., Reinthal, P. N., Dettman, D., et al. (2007). Ecological consequences of early Late Pleistocene megadroughts in tropical Africa. *Proceedings of the National Academy of Sciences of the United States of America, 104*, 16422–16427.

Cohen, P. (1996). Fitting a face to Ngaloba. *Journal of Human Evolution, 30*, 373–379.

Cole, J. M., Goldstein, S. L., deMenocal, P. B., Hemming, S. R., & Grousset, F. E. (2009). Contrasting compositions of Saharan dust in the eastern Atlantic Ocean during the last deglaciation and African Humid Period. *Earth and Planetary Science Letters, 278*, 257–266.

Collins, B. R., & Willoughby, P. R. (2010). The faunal analysis of Magubike and Mlambalasi, two MSA-LSA archaeological sites from Iringa District, Tanzania. *Journal of Taphonomy, 8*, 33–68.

Compton, J. S. (2011). Pleistocene sea-level fluctuations and human evolution on the southern coastal plain of South Africa. *Quaternary Science Reviews, 30*, 506–527.

Condemi, S., Mounier, A., Giunti, P., Lari, M., Carmelli, D., & Longo, L. (2013). Possible interbreeding in late Italian Neanderthals? New data from the Mezzena Jaw (Monti Lessini, Verona, Italy). *PLoS ONE, 8*(3), e59781. doi:10.1371/journal.pone.0059781.

Cooke, H. B. S., Malan, B. D., & Wells, L. H. (1945). Fossil man in the Lebombo Mountains, South Africa: The 'Border Cave', Ingwavuma district, Zululand. *Man, 45*, 6–13.

Copeland, L. (2000). Yabrudian and related industries: The state of research in 1996. In A. Ronen & M. Weinstein-Evron (Eds.), *Toward modern humans: Yabrudian and Micoquian, 400-50 k-years ago* (BAR International Series No. 850) (pp. 97–117). Oxford: Archaeopress.

Coplen, T. B. (1988). Normalization of oxygen and hydrogen isotope data. *Chemical Geology: Isotope Geoscience Section, 72*, 293–297.

Coplen, T. B. (1996). New guidelines for reporting stable hydrogen, carbon, and oxygen isotope-ratio data. *Geochimica et Cosmochimica Acta, 60*, 3359–3360.

Cramp, A., & O'Sullivan, G. (1999). Neogene sapropels in the Mediterranean: A review. *Marine Geology, 153*, 11–28.

Cremaschi, M., Di Lernia, S., & Garcea, E. A. A. (1998). Some insights on the Aterian in the Libyan Sahara: Chronology, environment, and archeology. *African Archaeological Review, 15*, 261–286.

Crevecoeur, I. (2008a). *Étude anthropologique du squelette du Paléolithique supérieur de Nazlet Khater 2 (Égypte). Apport à la compréhension de la variabilité passée des hommes modernes*. Leuven: Leuven University Press.

Crevecoeur, I. (2008b). Variability of Paleolithic modern humans in Africa. Future prospects of the Ishango human remain (re-)study. In D. Huylebrouck (Ed.), *Ishango, 22000 and 50 years later: The cradle of mathematics?* (pp. 87–97). Brussels: Koninklijke Vlaamse Academie van Belgie voor Wetenschapppen en Kunsten.

Crevecoeur, I. (2012). The Upper Paleolithic human remains of Nazlet Khater 2 (Egypt) and past modern human diversity. In J. J. Hublin & S. McPherron (Eds.), *Modern origins: A North African perspective* (pp. 205–221). Dordrecht: Springer.

Crevecoeur, I., & Trinkaus, E. (2004). From the Nile to the Danube: A comparison of the Nazlet Khater 2 and Oase 1 early modern human mandibles. *Anthropos (Brno), 42*, 229–239.

Crevecoeur, I., Rougier, H., Grine, F. E., & Froment, A. (2009). Modern human cranial diversity in the Late Pleistocene of Africa and Eurasia: Evidence from Nazlet Khater, Peştera cu Oase, and Hofmeyr. *American Journal of Physical Anthropology, 140*, 347–358.

Crevecoeur, I., Semal, P., Cornelissen, E., & Brooks, A. S. (2010). The Late Stone Age human remains from Ishango (Democratic Republic of Congo): Contribution to the study of the African Late Pleistocene modern human diversity. *American Journal of Physical Anthropology, S50*, 87.

Crosta, X., & Shemesh, A. (2002). Reconciling down core anticorrelation of diatom carbon and nitrogen isotopic ratios from the Southern Ocean. *Paleoceanography, 17*, 10.1-10.810.1029/2000PA000565.

Crowley, T. J. (1992). North Atlantic deep water cools the Southern Hemisphere. *Paleoceanography, 7*, 489–497.

Cruciani, F., Trombetta, B., Massaia, A., Destro-Bisol, G., Sellitto, D., & Scozzari, R. (2011). A revised root for the human Y chromosomal phylogenetic tree: The origin of patrilineal diversity in Africa. *American Journal of Human Genetics, 88*, 814–818.

Currat, M., & Excoffier, L. (2011). Strong reproductive isolation between humans and Neanderthals inferred from observed patterns of introgression. *Proceedings of the National Academy of Sciences of the United States of America, 108*, 15129–15134.

Damas, H. (1940). Observations sur des couches fossiliferes bordant la Semliki. *Revue Zoologique et Botanique Africaine, 33*, 265–272.

Dansgaard, W. (1964). Stable isotopes in precipitation. *Tellus, 16*, 466–468.

Dansgaard, W., Johnsen, S. J., Clausen, H. B., Dahl-Jensen, D., Gundestrup, N. S., Hammer, C. U., et al. (1993). Evidence for general instability of past climate from a 250-ky ice-core record. *Nature, 364*, 218–220.

Dart, R. A., & del Grande, N. (1931). The ancient iron-smelting cavern at Mumbwa. *Transactions of the Royal Society of South Africa, 19*, 379–427.

Day, M. H. (1969). Early *Homo sapiens* remains from the Omo River region of southwest Ethiopia. *Nature, 222*, 1132–1138.

Day, M. H., & Leakey, R. E. F. (1974). New evidence of the genus *Homo* from East Rudolf, Kenya (III). *American Journal of Physical Anthropology, 41*, 367–380.

Day, M. H., & Stringer, C. B. (1982). A reconsideration of the Omo Kibish remains and the *erectus-sapiens* transition. In M. A. de Lumley (Ed.). Homo erectus *et la place de l'Homme de Tautavel parmi les hominidés fossiles. Première congrès international de paléontologie humaine* (pp. 814–846). UNESCO: Colloque International du Centre National de la Récherche Scientifique.

Day, M. H., & Stringer, C. B. (1991). Les restes crânes d'Omo-Kibish et leur classification à l'intérieur du genre *Homo*. *L'Anthropologie, 95*, 573–594.

Day, M. H., Leakey, M. D., & Magori, C. C. (1980). A new fossil hominid skull (LH 18) from the Ngaloba Beds, Laetoli, Tanzania. *Nature, 284*, 55–56.

Deacon, H. J. (1992). Southern Africa and modern human origins. *Philosophical Transactions of the Royal Society London B, 337*, 177–183.

Deacon, H. J. (1995). Two Late Pleistocene-Holocene archaeological depositories from the southern Cape, South Africa. *South African Archaeology Bulletin, 50*, 121–131.

Deacon, H. J. (2008). The context of the 1967-8 sample of human remains from Cave 1 Klasies River Main Site. *South African Archaeological Society Goodwin Series, 10*, 143–149.

Deacon, H. J., & Geleijnse, V. B. (1988). The stratigraphy and sedimentology of the Main Site sequence, Klasies River, South Africa. *South African Archaeology Bulletin, 43*, 5–14.

Deacon, H. J., Talma, A. S., & Vogel, J. C. (1988). Biological and cultural development of Pleistocene people in an Old World southern continent. In J. R. Prescott (Ed.), *Early man in the southern Hemisphere* (pp. S23–S31). Adelaide: University of Adelaide, Department of Physics and Mathematical Physics.

Debénath, A. (1976). Le site de Dar-es-Soltane 2, à Rabat (Maroc). *Bulletins et Mémoires de la Société d'Anthropologie de Paris, 3*, 181–182.

Debénath, A. (1979). Découverte d'une mandibule humaine Atérienne à El Harhoura (Province de Rabat). *Bulletin d'archéologie marocaine, 12*, 1–2.

Debénath, A. (1980). Nouveaux restes humains Atériens du Maroc. *Comptes Rendus des Séances de l'Académie des Sciences (Paris) Serie D, 290*, 851–852.

Debénath, A. (2000). Le peuplement préhistorique du Maroc: Données récentes et problèmes. *L'Anthropologie, 104*, 131–145.

Debénath, A., Raynal, J. P., Roche, J., Texier, J. P., & Ferembach, D. (1986). Stratigraphie, habitat, typologie et devenir de l'Atérien marocain: Données récentes. *L'Anthropologie, 90*, 233–246.

de Heinzelin, J. (1962). Ishango. *Scientific American, 206*, 105–116.

deMenocal, P. B. (1995). Plio-Pleistocene African climate. *Science, 270*, 53–59.

deMenocal, P. B. (2004). African climate change and faunal evolution during the Pliocene-Pleistocene. *Earth and Planetary Science Letters, 220*, 3–24.

deMenocal, P., Ortiz, J., Guilderson, T., Adkins, J., Sarnthein, M., Baker, L., et al. (2000). Abrupt onset and termination of the African humid period: Rapid climate responses to gradual insolation forcing. *Quaternary Science Reviews, 19*, 347–361.

d'Errico, F., Vanhaeren, M., & Wadley, L. (2008). Possible shell beads from the Middle Stone Age layers of Sibudu Cave, South Africa. *Journal of Archaeological Science, 35*, 2675–2685.

d'Errico, F., Moreno, R. G., & Rifkin, R. F. (2012). Technological, elemental and colorimetric analysis of an engraved ochre fragment from the Middle Stone Age levels of Klasies River Cave 1, South Africa. *Journal of Archaeological Science, 39*, 942–952.

de Ruiter, D. J., Brophy, J. K., Lewis, P. J., Churchill, S. E., & Berger, L. R. (2008). Faunal assemblage composition and paleoenvironment of Plovers Lake, a Middle Stone Age locality in Gauteng Province, South Africa. *Journal of Human Evolution, 55*, 1102–1117.

de Villiers, H. (1973). Human skeletal remains from Border Cave, Ingwavuma District, KwaZulu, South Africa. *Annals of the Transvaal Museum, 28*, 229–256.

de Villiers, H. (1975). A second adult human mandible from Border Cave, Ingwavuma District, KwaZulu, South Africa. *South African Journal of Science, 72*, 212–215.

de Villiers, H., & Fatti, L. P. (1982). The antiquity of the Negro. *South African Journal of Science, 72*, 212–215.

De Vivo, B., Rolandi, G., Gans, P. B., Calvert, A., Bohrson, W. A., Spera, F. J., & Belkin, H. E. (2001). New constraints on the pyroclastic eruptive history of the Campanian volcanic Plain (Italy). *Mineralogy and Petrology, 73*, 47–65.

Domínguez-Rodrigo, M., Mabulla, A., Luque, L., Thompson, J. W., Rink, J., Bushozi, P., et al. (2008). A new archaic *Homo sapiens* fossil from Lake Eyasi, Tanzania. *Journal of Human Evolution, 54*, 899–903.

Douka, K., Grün, R., Jacobs, Z., Lane, C., Farr, L., Hunt, C., et al. (2014). The chronostratigraphy of the Haua Fteah cave (Cyrenaica, northeast Libya). *Journal of Human Evolution, 66*, 39–63.

Drake, N., & Bristow, C. (2006). Shorelines in the Sahara: Geomorphological evidence for an enhanced monsoon from paleolake Megachad. *The Holocene, 16*, 901–911.

Drennan, M. R. (1929). An Australoid skull from the Cape Flats. *Journal of the Royal Anthropological Institute, 59*, 417–427.

Duarte, C., Maurico, J., Pettitt, P. B., Souto, P., Trinkaus, E., van der Plicht, H., et al. (1999). The early Upper Paleolithic human skeleton from the Abrigo do Lagar Velho (Portugal) and modern human emergence in Iberia. *Proceedings of the National Academy of Sciences of the United States of America, 96*, 7604–7609.

Dupont, L. (2011). Orbital scale vegetation change in Africa. *Quaternary Science Reviews, 30*, 3589–3602.

Dusseldorp, G. L. (2010). Prey choice during the South African Middle Stone Age: Avoiding dangerous prey or maximizing returns? *African Archaeological Review, 27*, 107–133.

Elhaik, E., Tatarinova, T. V., Klyosov, A. A., & Graur, D. (2014). The 'extremely ancient' chromosome that isn't: A forensic bioinformatics investigation of Albert Perry's X-degenerate portion of the Y chromosome. *European Journal of Human Genetics, 2014*, 1–6. doi:10.1038/ejhg.2013.303.

Emiliani, C. (1955). Pleistocene temperatures. *Journal of Geology, 63*, 539–578.

Emiliani, C. (1966). Isotopic paleotemperatures. *Science, 154*, 851–857.

Emiliani, C., Cardini, L., Mayeda, T., McBurney, C. B. M., & Tongiorgi, E. (1964). Paleotemperature analysis of fossil shells of marine molluscs (food refuse) from the Arene Candide Cave, Italy, and the Haua Fteah Cave, Cyrenaica. In H. Craig (Ed.), *Isotopic and cosmic chemistry* (pp. 133–156). Amsterdam: North-Holland.

Ennouchi, E. (1962). Un Néanderthalien: L'homme du Jebel Irhoud (Maroc). *L'Anthropologie, 66*, 279–299.

Ennouchi, E. (1963). Les Néanderthaliens du Jebel Irhoud (Maroc). *Comptes Rendus des Séances de l'Académie des Sciences (Paris), 256*, 2459–2560.

Ennouchi, E. (1968). Le deuxième crâne de l'homme de l'Irhoud. *Annals de Paléontologie, 55*, 117–128.

Ennouchi, E. (1969). Présence d'un enfant Néanderthalien au Jebel Irhoud (Maroc). *Annals de Paléontologie, 55*, 251–265.

Epstein, S., Buchsbaum, R., Lowenstam, H. A., & Urey, H. C. (1951). Carbonate-water isotopic temperature scale. *Bulletin of the Geological Society of America, 62*, 417–426.

Epstein, S., Buchsbaum, R., Lowenstam, H. A., & Urey, H. C. (1953). Revised carbonate-water isotopic temperature scale. *Bulletin of the Geological Society of America, 64*, 1315–1326.

Eriksson, A., & Manica, A. (2011). Effect of ancient population structure on the degree of polymorphism shared between modern human populations and ancient hominins. *Proceedings of the National Academy of Sciences of the United States of America, 109*, 13956–13960.

Faith, J. T., Tryon, C., & Peppe, D. J. (2016). Environmental change, ungulate biogeography, and their implications for early human dispersals in equatorial East Africa. In S. C. Jones & B. A. Stewart (Eds.), *Africa from MIS 6-2: Population dynamics and paleoenvironments* (pp. 233–245). Dordrecht: Springer.

Farrera, I., Harrison, S. P., Prentice, I. C., Ramstein, G., Guiot, J., Bartlein, P. J., et al. (1999). Tropical climates at the Last Glacial Maximum: A new synthesis of terrestrial paleoclimate data. I. Vegetation, lake levels and geochemistry. *Climate Dynamics, 15*, 823–856.

Fatti, L. P. (1986). Discriminant analysis in prehistoric physical anthropology. In R. Singer & J. K. Lundy (Eds.), *Variation, culture and evolution in African populations* (pp. 27–34). Johannesburg: Witwatersrand University Press.

Feathers, J. K. (2002). Luminescence dating in less than ideal conditions: Case studies from Klasies River Main Site and Duinefontein, South Africa. *Journal of Archaeological Science, 29*, 177–194.

Feathers, J. K., & Bush, D. A. (2000). Luminescence dating of Middle Stone Age deposits at Die Kelders. *Journal of Human Evolution, 38*, 91–119.

Ferembach, D. (1962). La Nécropole Épipaléolithique de Taforalt (Maroc Oriental). *Étude des Squelettes Humains*. Casablanca: Edita-Casablanca.

Ferembach, D. (1965). *Diagrammes craniens sagittaux, et mensurations individuelles des squelettes Ibéromaurusiens de Taforalt (Maroc Oriental)*. Travaux du centre de recherches anthropologiques, préhistoriques, et etnographiques, Alger.

Ferembach, D. (1976a). Les restes humains Atériens de Temara (Campagne 1975). *Bulletins et Mémoires de la Société d'anthropologie de Paris, 3*, 175–180.

Ferembach, D. (1976b). Les restes humains de la grotte de Dar-es-Soltane 2 (Maroc). Campagne 1975. *Bulletins et Mémoires de la Société d'Anthropologie de Paris, 3*, 183–193.

Ferembach, D. (1998). La crâne Atérien de Témara (Maroc, Atlantique). *Bulletin d'Archeologie Marocaine, 18*, 19–66.

Ferembach, D. (1985). On the origin of the Iberomaurusians. A new hypothesis. *Journal of Human Evolution, 14*, 393–397.

Filippi, M. L., & Talbot, M. R. (2005). The paleolimnology of northern Lake Malawi over the last 25 ka based upon the elemental and stable isotopic composition of sedimentary organic matter. *Quaternary Science Reviews, 24*, 1303–1328.

Fitzsimmons, K. E., Hambach, U., Veres, D., & Iovita, R. (2013). The Campanian Ignimbrite eruption: New data on volcanic ash dispersal and its potential impact on human evolution. *PLoS ONE, 8*(6), e65839. doi:10.1371/journal.pone.0065839.

Foerster, V., Junginger, A., Langkamp, O., Gebru, T., Asrat, A., Umer, M., et al. (2012). Climatic change recorded in the sediments of the Chew Bahir basin, southern Ethiopia, during the last 45,000 years. *Quaternary International, 274*, 25–37.

Foley, R., & Lahr, M. M. (1997). Mode 3 technologies and the evolution of modern humans. *Cambridge Archaeology Journal, 7*, 3–36.

Foley, R. A., Maíllo-Fernández, J. M., & Lahr, M. M. (2013). The Middle Stone Age of the Central Sahara: Biogeographical opportunities and technological strategies in later human evolution. *Quaternary International, 300*, 153–170.

Franciscus, R. G. (2003). Internal nasal floor configuration in *Homo* with special reference to the evolution of Neandertal facial form. *Journal of Human Evolution, 44*, 701–729.

Frayer, D. W., Wolpoff, M. H., Thorne, A. G., Smith, F. H., & Pope, G. G. (1993). Theories of modern human origins: The paleontological test. *American Anthropologist, 95*, 14–50.

Frayer, D. W., Jelínek, J., Oliva, M., & Wolpoff, M. H. (2006). Aurignacian male crania, jaws, and teeth from the Mladeč caves, Moravia, Czech Republic. In M. Teschler-Nicola (Ed.), *Early modern humans at the Moravian Gate: The Mladeč caves and their remains* (pp. 185–272). Vienna: Springer.

Frazer, K. A., & The International HapMap Consortium. (2007). A second generation human haplotype map of over 3.1 million SNPs. *Nature, 449*, 851–861.

Fujita, T., Azuma, Y., Fukuyama, R., Hattori, Y., Yoshida, C., Koida, M., et al. (2004). Runx2 induces osteoblast and chondrocyte differentiation and enhances their migration by coupling with PI3K-Akt signaling. *Journal of Cell Biology, 166*, 85–95.

Gabel, C. (1963). Further human remains from the Central African Later Stone Age. *South African Archaeological Bulletin, 18*, 40–48.

Garcin, Y., Junginger, A., Melnick, D., Olago, D. O., Strecker, M. R., & Trauth, M. H. (2009). Late Pleistocene - Holocene rise and collapse of Lake Suguta, northern Kenya Rift. *Quaternary Science Reviews, 28*, 911–925.

Garrigan, D., & Hammer, M. F. (2006). Reconstructing human origins in the genomic era. *Nature Reviews Genetics, 7*, 669–680.

Garrigan, D., Kingan, S. B., Pilkington, M. M., Wilder, J. A., Cox, M. P., Soodyall, H., et al. (2007). Inferring human population sizes, divergence times and rates of gene flow from mitochondrial, X and Y chromosome resequencing data. *Genetics, 177*, 2195–2207.

Garrod, D. A. E. (1970). Pre-Aurignacian and Amudian: A comparative study of the earliest blade industries of the Near East. In K. Gripp, R. Schütrumpf, & H. Schabedissen (Eds.), *Frühe Menschheit und Umwelt* (pp. 224–229). Köln: Böhlau Verlag.

Gasse, F. (2000). Hydrological changes in the African tropics since the Last Glacial Maximum. *Quaternary Science Reviews, 19*, 189–211.

Gasse, F., Chalié, F., Vincens, A., Williams, M. A. J., & Williamson, D. (2008). Climatic patterns in equatorial and southern Africa from 30,000 to 10,000 years ago reconstructed from terrestrial and near-shore proxy data. *Quaternary Science Reviews, 27*, 2316–2340.

Gathogo, P. N., & Brown, F. H. (2006). Stratigraphy of the Koobi Fora Formation (Pliocene and Pleistocene) in the Ileret region of northern Kenya. *Journal of African Earth Sciences, 45*, 369–390.

Gibbard, P. L., & Head, M. J. (2009). IUGS ratification of the Quaternary system/period and the Pleistocene series/epoch with a base at 2.58 Ma. *Quaternaire, 20*, 411–412.

Gliganic, L. A., Jacobs, Z., Roberts, R. G., Domínquez-Rodrigo, M., & Mabulla, A. Z. P. (2012). New ages for Middle and Later Stone Age deposits at Mumba rockshelter, Tanzania: Optically stimulated luminescence dating of quartz and feldspar grains. *Journal of Human Evolution, 62*, 533–547.

Goldberg, P. (2000). Micromorphology and site formation at Die Kelders Cave 1, South Africa. *Journal of Human Evolution, 38*, 43–90.

Gonder, M. K., Mortensen, H. M., Reed, F. A., de Sousa, A., & Tishkoff, S. A. (2007). Whole-mtDNA genome sequence analysis of ancient African lineages. *Molecular Biology and Evolution, 24*, 757–768.

Gokcumen, O., Zhu, Q., Mulder, L. C. F., Iskow, R. C., Austermann, C., Scharer, C. D., et al. (2013). Balancing selection on a regulatory region exhibiting ancient variation that predates human–Neandertal divergence. *PLoS Genetics, 9*(4), e1003404. doi:10.1371/journal.pgen.1003404.

Goodwin, A. J. H. (1929). Report on the stone implements found with the Cape Flats skull. *Journal of the Royal Anthropological Institute, 59*, 429–438.

Görsdorf, J., & Eiwanger, J. (1998). Radiocarbon datings of Late Paleolithic, Epipaleolithic and Neolithic sites in northeastern Morocco. In J. Evin, C. Oberlin, J. P. Daugas & J. F. Salles (Eds.), *Actes du 3e Congrès International sur 14C et Archéologie* (Mémoires de la Societé Préhistoire Français No. 26) (pp. 365–369). Paris: Societé Préhistoire Français.

Gramly, R. M. (1976). Upper Pleistocene archaeological occurrences at site GvJm22, Lukenya Hill, Kenya. *Man, 11*, 319–344.

Gramly, R. M., & Rightmire, G. P. (1973). A fragmentary cranium and dated Later Stone Age assemblages from Lukenya Hill, Kenya. *Man, 8*, 571–579.

Greene, D. L., & Armelagos, G. J. (1972). *The Wadi Halfa Mesolithic population* (Research Report No. 11). Amherst, MA: University of Massachusetts.

Green, R. E., Malaspinas, A. S., Krause, J., Briggs, A., Johnson, P. L. F., Uhler, C., et al. (2008). A complete Neanderthal mitochondrial genome sequence determined by high throughput sequencing. *Cell, 134*, 416–426.

Green, R. E., Krause, J., Briggs, A. W., Maricic, T., Stenzel, U., Kircher, M., et al. (2010). A draft sequence of the Neanderthal genome. *Science, 328*, 710–722.

Grine, F. E. (1998). Additional human fossils from the Middle Stone Age of Die Kelders Cave 1, South Africa: 1995 excavations. *South African Journal of Science, 94*, 229–235.

Grine, F. E. (2000). Middle Stone Age human fossils from Die Kelders Cave 1, Western Cape Province, South Africa. *Journal of Human Evolution, 38*, 129–145.

Grine, F. E. (2012). Observations on Middle Stone Age human teeth from Klasies River Main Site. *South Africa Journal of Human Evolution, 63*, 750–758.

Grine, F. E., & Klein, R. G. (1985). Pleistocene and Holocene human remains from Equus Cave, South Africa. *Anthropology, 8*, 55–98.

Grine, F. E., & Klein, R. G. (1993). Late Pleistocene human remains from the Sea Harvest site, Saldanha Bay, South Africa. *South African Journal of Science, 89*, 145–152.

Grine, F. E., Klein, R. G., & Volman, T. P. (1991). Dating, archaeology and human fossils from the Middle Stone Age Layers of Die Kelders Cave 1, South Africa. *Journal of Human Evolution, 21*, 363–395.

Grine, F. E., Henshilwood, C. S., & Sealy, J. C. (2000). Human remains from Blombos Cave, South Africa: (1997–1998 excavations). *Journal of Human Evolution, 38*, 755–765.

Grine, F. E., & Henshilwood, C. S. (2002). Additional human remains from Blombos Cave, South Africa: (1999–2000 excavations). *Journal of Human Evolution, 42*, 293–302.

Grine, F. E., Pearson, O. M., Klein, R. G., & Rightmire, G. P. (1998). Additional human fossils from Klasies River Mouth, South Africa. *Journal of Human Evolution, 35*, 95–107.

Grine, F. E., Bailey, R. M., Harvati, K., Nathan, R. P., Morris, A. G., Henderson, G. M., et al. (2007). Late Pleistocene human skull from Hofmeyr, South Africa, and modern human origins. *Science, 315*, 226–229.

Grine, F. E., Gunz, P., Betti-Nash, L., Neubauer, S., & Morris, A. G. (2010). Reconstruction of the Late Pleistocene human skull from Hofmeyr, South Africa. *Journal of Human Evolution, 59*, 1–15.

Grine, F. E., Jacobs, R. L., Reed, K. E., & Plavcan, J. M. (2012). The enigmatic molar from Gondolin, South Africa: Implications for *Paranthropus* paleobiology. *Journal of Human Evolution, 63*, 597–609.

Gronau, I., Hubisz, M. J., Gulko, B., Danko, C. G., & Siepel, A. (2011). Bayesian inference of ancient human demography from individual genome sequences. *Nature Genetics, 43*, 1031–U151.

Grün, R. (2006). Direct dating of human fossils. *Yearbook of Physical Anthropology, 49*, 2–48.

Grün, R., & Beaumont, P. (2001). Border Cave revisited: A revised ESR chronology. *Journal of Human Evolution, 40*, 467–482.

Grün, R., & Stringer, C. B. (1991). Electron spin resonance dating and the evolution of modern humans. *Archaeometry, 33*, 153–199.
Grün, R., Beaumont, P. B., & Stringer, C. B. (1990a). ESR dating evidence for early modern humans at Border Cave in South Africa. *Nature, 344*, 537–539.
Grün, R., Shackleton, N. J., & Deacon, H. J. (1990b). Electron-spin-resonance dating of tooth enamel from Klasies River Mouth. *Current Anthropology, 31*, 427–432.
Grün, R., Beaumont, P., Tobias, P. V., & Eggins, S. (2003). On the age of Border Cave 5 human mandible. *Journal of Human Evolution, 45*, 155–167.
Guérin, G., Murray, A. S., Jain, M., Thomsen, K. J., & Mercier, N. (2013). How confident are we in the chronology of the transition between Howieson's Poort and Still Bay? *Journal of Human Evolution, 64*, 314–317.
Gunz, P., Bookstein, F. L., Mitteroecker, P., Stadimayr, A., Seidler, H., & Weber, G. (2009). Early modern human diversity suggests subdivided population structure and a complex out-of-Africa scenario. *Proceedings of the National Academy of Sciences of the United States of America, 106*, 6094–6098.
Gunz, P., Neubauer, S., Maureille, S., & Hublin, J. J. (2010). Brain development after birth differs between Neanderthals and modern humans. *Current Biology, 20*, R921–R922.
Hachi, S. (1996). L'Ibéromaurusien, découverte des fouilles d'Afalou (Bédjaîa, Algérie). *L'Anthropologie, 100*, 55–76.
Haile-Selassie, Y., Asfaw, B., & White, T. D. (2004). Hominid cranial remains from upper Pleistocene deposits at Aduma, Middle Awash, Ethiopia. *American Journal of Physical Anthropology, 123*, 1–10.
Halkett, D. J., Hart, T., Yates, R. J., Volman, T. P., Parkington, J. E., Orton, J., et al. (2003). First excavation of intact Middle Stone Age layers at Ysterfontein, Western Cape Province, South Africa: Implications for Middle Stone Age ecology. *Journal of Archaeological Science, 30*, 955–971.
Hammer, M. F., Karafet, T. M., Redd, A. J., Jarjanazi, H., Santachiara-Benerecetti, S., Soodyall, H., et al. (2001). Hierarchical patterns of global human Y-chromosome diversity. *Molecular Biology and Evolution, 18*, 1189–1203.
Hammer, M. F., Woerner, A. E., Mendez, F. L., Watkins, J. C., & Wall, J. D. (2011). Genetic evidence for archaic admixture in Africa. *Proceedings of the National Academy of Sciences of the United States of America, 108*, 15123–15128.
Harvati, K., & Hublin, J. J. (2012). Morphological continuity of the face in the Late Middle and Late Pleistocene hominins from northwestern Africa: A 3D geometric morphometric analysis. In J. J. Hublin & S. P. McPherron (Eds.), *Modern origins: A North African perspective* (pp. 179–188). Dordrecht: Springer.
Harvati, K., Frost, S. R., & McNulty, K. P. (2004). Neanderthal taxonomy reconsidered: Implications of 3D primate models of intra- and interspecific differences. *Proceedings of the National Academy of Sciences of the United States of America, 101*, 1147–1152.
Harvati, K., Gunz, P., & Grigorescu, D. (2007). Cioclovina (Romania): Affinities of an early modern European. *Journal of Human Evolution, 53*, 732–746.
Harvati, K., Stringer, C., Grün, R., Aubert, M., Allsworth-Jones, P., & Folorunso, C. A. (2011). The Later Stone Age calvaria from Iwo Eleru, Nigeria: Morphology and chronology. *PLoS ONE, 6*(9), e24024. doi:10.1371/journal.pone.0024024.
Harvati, K., Bauer, C. C., Grine, F. E., Benazzi, S., Ackermann, R. R., van Niekerk, K. L., & Henshilwood, C. S. (2015). A human deciduous molar from the Middle Stone Age (Howiesons Poort) of Klipdrift Shelter, South Africa. *Journal of Human Evolution, 82*, 190–196.
Haughton, S. H. (1917). Preliminary note on the ancient human skull remains from the Transvaal. *Transactions of the Royal Society of South Africa, 6*, 1–14.
Hay, R. L. (1987). Geology of the Laetoli area. In M. D. Leakey & J. M. Harris (Eds.), *Laetoli: A Pliocene site in northern Tanzania* (pp. 23–47). Oxford: Oxford University Press.
Hays, J. D., Imbrie, J., & Shackleton, N. J. (1976). Variation in the earth's orbit: Pacemaker of the ice ages. *Science, 194*, 1121–1132.
Hearty, P. J., Hollin, J. T., Neumann, A. C., O'Leary, M. J., & McCulloch, M. (2007). Global sea- level fluctuations during the Last Interglaciation (MIS 5e). *Quaternary Science Reviews, 26*, 2090–2112.
Hedges, R. E. M. (2002). Bone diagenssis: An overview of processes. *Archaeometry, 44*, 319–328.
Hendey, Q. B. (1984). Southern African Late Tertiary vertebrates. In R. G. Klein (Ed.), *Southern African prehistory and paleoenvironments* (pp. 81–106). Rotterdam: A.A. Balkema.
Henn, B. M., Gignoux, C. R., Jobin, M., Granka, J. M., Macpherson, J. M., Kidd, J. M., et al. (2011). Hunter-gatherer genomic diversity suggests a southern African origin for modern humans. *Proceedings of the National Academy of Sciences of the United States of America, 108*, 5154–5162.
Henn, B. M., Cavalli-Sforza, L. L., & Feldman, M. W. (2012). The great human expansion. *Proceedings of the National Academy of Sciences of the United States of America, 109*, 17758–17764.
Henshilwood, C. S. (2012). The Still Bay and Howiesons Poort: 'Paleolithic' techno-traditions in southern Africa. *Journal of World Prehistory, 25*, 205–237.
Henshilwood, C. S., Sealy, J. C., Yates, R. J., Cruz-Uribe, K., Goldberg, P., Grine, F. E., et al. (2001). Blombos Cave, southern Cape, South Africa: Preliminary report on the 1992–1999 excavations of the Middle Stone Age levels. *Journal of Archaeological Science, 28*, 421–448.
Henshilwood, C. S., d'Errico, F., Yates, R., Jacobs, Z., Tribolo, C., Duller, G. A. T., et al. (2002). Emergence of modern human behaviour: Middle Stone Age engravings from South Africa. *Science, 295*, 1278–1280.
Henshilwood, C. S., d'Errico, F., Vanhaeren, M., van Niekerk, K., & Jacobs, Z. (2004). Middle Stone Age shell beads from South Africa. *Science, 384*, 404.
Henshilwood, C. S., d' Errico, F., & Watts, I. (2009). Engraved ochres from the Middle Stone Age levels at Blombos Cave, South Africa. *Journal of Human Evolution, 57*, 27–47.
Henshilwood, C. S., d'Errico, F., van Niekerk, K. L., Coquinot, Y., Jacobs, Z., Lauritzen, S. E., et al. (2011). A 100,000-year-old ochre-processing workshop at Blombos Cave, South Africa. *Science, 334*, 219–222.
Henshilwood, C. S., van Niekerk, K. L., Wurz, S., Delagnes, A., Armitage, S. J., Rifkin, R. F., et al. (2014). Klipdrift Shelter, southern Cape, South Africa: Preliminary report on the Howiesons Poort layers. *Journal of Archaeological Science, 45*, 284–303.
Hessler, I., Steinke, S., Groeneveld, J., Dupont, L., & Wefer, G. (2011). Impact of abrupt climate change in the tropical southeast Atlantic during Marine Isotope Stage (MIS) 3. *Paleooceanography, 26*, PA4209. doi:10.1029/2011PA002118.
Hetherington, R., Wiebe, E., Weaver, A. J., Carto, S. L., Eby, M., & MacLoed, R. (2008). Climate, African and Beringian subaerial continental shelves, and migration of early peoples. *Quaternary International, 183*, 83–101.
Hewes, G. W., Irwin, H., Papworth, M., & Saxe, A. (1964). A new fossil human population from the Wadi Halfa area, Sudan. *Nature, 203*, 341–343.
Higham, T. F. G., Jacobi, R. M., & Bronk Ramsey, C. (2006). AMS radiocarbon dating of ancient bone using ultrafiltration. *Radiocarbon, 48*, 179–195.
Hilgren, F. J. (1991). Astronomical calibration of Gauss to Matuyama sapropels in the Mediterranean and implications for the geomagnetic polarity time scale. *Earth and Planetary Science Letters, 104*, 226–244.

Hinz, E. A., & Kohn, M. J. (2010). The effect of tissue structure and soil chemistry on trace element uptake in fossils. *Geochimica et Cosmochimica Acta, 74*, 3213–3323.

Hodgson, J. A., Bergey, C. M., & Disotell, T. R. (2010). Neanderthal genome: The ins and outs of African genetic diversity. *Current Biology, 20*, R517–R519.

Hoefs, J. (2009). *Stable isotope geochemistry* (6th ed.). Berlin: Springer.

Hoelzmann, P., Keding, B., Berke, H., Kröpelin, S., & Kruse, H. J. (2002). Environmental change and archaeology: Lake evolution and human occupation in the eastern Sahara during the Holocene. *Paleogeography, Paleoclimatology, Paleoecology, 169*, 193–217.

Holliday, T. W. (2015). Population affinities of the Jebel Sahaba skeletal sample: Limb proportion evidence. *International Journal of Osteoarchaeology, 25*, 466–476.

Houghton, K., & Thackeray, J. F. (2011). Morphometric comparisons between crania of Late Pleistocene *Homo sapiens* from Border Cave (BC 1), Tuinplaas (TP 1) and modern southern African populations. *Transactions of the Royal Society of South Africa, 66*, 159–162.

Howell, F. C. (1978). Hominidae. In V. J. Maglio & H. B. S. Cooke (Eds.), *Evolution of African mammals* (pp. 154–248). Cambridge, MA: Harvard University Press.

Hublin, J. J. (1992). Recent human evolution in northwestern Africa. *Philosophical Transactions of the Royal Society of London. Series B, Biological sciences, 337*, 185–191.

Hublin, J. J. (1993). Recent human evolution in northwestern Africa. In M. J. Aitken, C. B. Stringer, & P. Mellars (Eds.), *The origin of modern humans and the impact of chronometric dating* (pp. 118–131). Princeton: Princeton University Press.

Hublin, J. J. (2000). Modern-nonmodern hominid interactions: A Mediterranean perspective. In O. Bar-Yosef & D. R. Pilbeam (Eds.), *The geography of Neandertals and modern humans in Europe and the greater Mediterranean* (pp. 157–182). Cambridge, MA: Peabody Museum of Archaeology and Ethnology.

Hublin, J. J. (2001). Northwestern African Middle Pleistocene hominids and their bearing on the emergence of *Homo sapiens*. In L. Barham & K. Robson-Brown (Eds.), *Human roots. Africa and Asia in the Middle Pleistocene* (pp. 99–121). Bristol: Western Academic and Specialist Press.

Hublin, J. J., Verna, C., Bailey, S., Smith, T., Olejniczak, A., & Sbihi-Alaoui, F. Z. (2012). Dental evidence from the Aterian human populations of Morocco. In J. J. Hublin & S. P. McPherron (Eds.), *Modern origins: A North African perspective* (pp. 189–204). Dordrecht: Springer.

Hughes, A. R. (1990). The Tuinplaas human skeleton from the Springbok Flats, Transvaal. In G. H. Sperber (Ed.), *From apes to angels: Essays in honour of Phillip* (Vol. Tobias, pp. 197–214). New York: Wiley-Liss.

Humphrey, L. T., & Bocaege, E. (2008). Tooth evulsion in the Maghreb: Chronological and geographical patterns. *African Archaeological Review, 25*, 109–123.

Humphrey, L., Bello, S. M., Turner, E., Bouzouggar, A., & Barton, N. (2012). Iberomaurusian funerary behaviour: Evidence from Grotte des Pigeons, Taforalt, Morocco. *Journal of Human Evolution, 62*, 261–273.

Hunt, C., Davison, J., Inglis, R., Farr, L., Reynolds, T., Simpson, D., et al. (2010). Site formation processes in caves: The Holocene sediments of the Haua Fteah, Cyrenaica, Libya. *Journal of Archaeological Science, 37*, 1600–1611.

Imbrie, J. D., & McIntyre, A. (2006). SPECMAP time scale developed by Imbrie et al. 1984 based on normalized planktonic records (normalized O-18 vs time, specmap.017). doi:10.1594/PANGAEA.441706.

Imbrie, J., Hays, J. D., Martinson, D. G., McIntyre, A., Mix, A. C., Morley, J. J., et al. (1984). The orbital theory of Pleistocene climate: Support from a revised chronology of the marine $\delta^{18}O$ record. In A. Berger (Ed.), *Milankovitch and climate, Part 1* (pp. 269–305). Hingham, MA: D. Reidel.

Imbrie, J., Berger, A., Boyle, E. A., Clemens, S. C., Duffy, A., Howard, W. R., et al. (1993). On the structure and origin of major glaciations cycles. 2. The 100 000-year cycle. *Paleoceanography, 8*, 699–735.

Indriati, E., Swisher, C. C., Lepre, C., Quinn, R. L., Suriyanto, R. A., Hascaryo, A. T., et al. (2011). The age of the 20 meter Solo River terrace, Java, Indonesia and the survival of *Homo erectus* in Asia. *PLoS ONE, 6*(6), e21562. doi:10.1371/journal.pone.0021562.

Ingman, M., & Gyllensten, U. (2001). Analysis of the complete human mtDNA genome: Methodology and inferences for human evolution. *Journal of Heredity, 92*, 454–461.

Ingman, M., Kaessmann, H., Pääbo, S., & Gyllensten, U. (2000). Mitochondrial genome variation and the origin of modern humans. *Nature, 408*, 708–713.

Irish, J. (2000). The Iberomaurusian enigma: North Africa progenitor or dead end? *Journal of Human Evolution, 39*, 393–410.

Irish, J. D. (2005). Population continuity vs. discontinuity revisited: Dental affinities among Late Paleolithic through Christian-Era Nubians. *American Journal of Physical Anthropology, 128*, 520–535.

Irish, J. D., & Turner, C. G. (1990). West African dental affinity of Late Pleistocene Nubians: Peopling of the Eurafrican-South Asian triangle, II. *Homo, 41*, 42–53.

Jacobs, Z. (2010). An OSL chronology for the sedimentary deposits from Pinnacle Point Cave 13B—a punctuated presence. *Journal of Human Evolution, 59*, 289–305.

Jacobs, Z., Wintle, A. G., & Duller, G. A. T. (2003a). Optical dating of dune sand from Blombos Cave, South Africa: I—multiple grain data. *Journal of Human Evolution, 44*, 599–612.

Jacobs, Z., Duller, G. A. T., & Wintle, A. G. (2003b). Optical dating of dune sand from Blombos Cave, South Africa: II—single grain data. *Journal of Human Evolution, 44*, 613–625.

Jacobs, Z., Duller, G. A. T., Henshilwood, C. S., & Wintle, A. G. (2006). Extending the chronology of deposits at Blombos Cave, South Africa, back to 140 ka using optical dating of single and multiple grains of quartz. *Journal of Human Evolution, 51*, 255–273.

Jacobs, Z., Roberts, R. G., Galbraith, R. F., Deacon, H. J., Grün, R., Mackay, A., et al. (2008a). Ages for the Middle Stone Age of southern Africa: Implications for human behavior and dispersal. *Science, 322*, 733–735.

Jacobs, Z., Wintle, A. G., Duller, G. A. T., Roberts, R. G., & Wadley, L. (2008b). New ages for the post-Howiesons Poort, late and final Middle Stone Age at Sibudu Cave, South Africa. *Journal of Archaeological Science, 35*, 1790–1807.

Jacobs, Z., Meyer, M. C., Roberts, R. G., Aldeias, V., Dibble, H., & El Hajraoui, M. A. (2011). Single-grain OSL dating at La Grotte des Contrebandiers ('Smugglers' Cave'), Morocco: Improved age constraints for the Middle Paleolithic levels. *Journal of Archaeological Science, 38*, 3631–3643.

Jacobs, Z., Hayes, E. H., Roberts, R. G., Galbraith, R. F., & Henshilwood, C. S. (2013). An improved OSL chronology for the Still Bay layers at Blombos Cave, South Africa: Further tests of single-grain dating procedures and a re-evaluation of the timing of the Still Bay industry across southern Africa. *Journal of Archaeological Science, 40*, 579–594.

Jelinek, A. J. (1990). The Amudian in the context of the Mugharan tradition at the Tabun cave (Mt. Carmel), Israel. In P. Mellars (Ed.), *The emergence of modern humans* (pp. 81–90). Edinburgh: Edinburgh University Press.

Jerardino, A., & Marean, C. W. (2010). Shellfish gathering, marine paleoecology and modern human behavior: Perspectives from cave PP13B, Pinnacle Point, South Africa. *Journal of Human Evolution, 59*, 412–424.

Johnson, T. C., Brown, E. T., McManus, J., Barry, S., Barker, P., & Gasse, F. (2002). A high-resolution paleoclimate record spanning the past 25,000 years in southern east Africa. *Science, 296*, 113–116.

Jones, S., Antoniadou, A., Barton, H., Drake, N., Farr, L., Hunt, C., et al. (2016). Patterns of hominin occupation and cultural diversity across the Gebel Akhdar of northern Libya over the last ~ 200 kyr. In S. C. Jones & B. A. Stewart (Eds.), *Africa from MIS 6-2: Population dynamics and paleoenvironments* (pp. 77–99). Dordrecht: Springer.

Jones, T. R. (1940). Human skeletal remains from the Mumbwa Cave, Northern Rhodesia. *South African Journal of Science, 37*, 313–319.

Jouzel, J., Lorius, C., Petit, J. R., Genthon, C., Barkov, N. I., Kotlyakov, V. M., et al. (1987). Vostok Ice Core—a continuous isotope temperature record over the last climatic cycle (160,000 years). *Nature, 329*, 403–408.

Kennett, J. P., Roark, E. B., Cannariato, K. G., Ingram, B. L., & Tada, R. (2000). Latest Quaternary paleoclimatic and radiocarbon chronology, Hole 1017E, southern California margin. In M. Lyle, I. Koizumi, C. Richter & T. C. Moore (Eds.), *Proceedings of the ocean drilling program, scientific results* (Vol. 167) (pp. 249–254). College Station, TX: Ocean Drilling Program.

Kidd, J. M., Gravel, S., Byrnes, J., Moreno-Estrada, A., Musharoff, S., Bryc, K., et al. (2012). Population genetic inference from personal genome data: Impact of ancestry and admixture on human genomic variation. *American Journal of Human Genetics, 91*, 660–671.

Kivisild, T., Shen, P., Wall, D. P., Do, B., Sung, R., Davis, K., et al. (2006). The role of selection in the evolution of human mitochondrial genomes. *Genetics, 172*, 373–387.

Klein, R. G. (1970). Problems in the study of the Middle Stone Age of South Africa. *South African Archaeological Bulletin, 25*, 127–135.

Klein, R. G. (1976). The mammalian fauna of the Klasies River Mouth sites, southern Cape Province, South Africa. *South African Archaeology Bulletin, 21*, 75–98.

Klein, R. G. (1983). The Stone Age prehistory of southern Africa. *Annual Review of Anthropology, 12*, 25–48.

Klein, R. G. (1984). Later Stone Age faunal samples from Heuningneskrans Shelter (Transvaal) and Leopard's Hill Cave (Zambia). *South African Archaeological Bulletin, 39*, 109–116.

Klein, R. G. (1989). Why does skeletal part representation differ between smaller and larger bovids at Klasies River Mouth and other archeological sites? *Journal of Archaeological Science, 16*, 363–381.

Klein, R. G. (2009). *The human career: Human biological and cultural origins* (3rd ed.). Chicago: The University of Chicago Press.

Klein, R. G., & Scott, K. (1986). Re-analysis of the faunal assemblages from Haua Fteah and other Late Quaternary archaeological sites in Cyrenaican Libya. *Journal of Archaeological Science, 13*, 515–542.

Klein, R. G., & Cruz-Uribe, K. (1996). Exploitation of large bovids and seals and Middle and Later Stone Age sites in South Africa. *Journal of Human Evolution, 31*, 315–334.

Klein, R. G., Cruz-Uribe, K., & Beaumont, P. B. (1991). Environmental, ecological, and paleoanthropological implications of the Late Pleistocene mammalian fauna from Equus Cave, Northern Cape Province, South Africa. *Quaternary Research, 36*, 94–119.

Klein, R. G., Avery, G., Cruz-Uribe, K., Halkett, D. J., Parkington, J. E., Steele, T., et al. (2004). The Ysterfontein 1 Middle Stone Age site, South Africa, and early human exploitation of coastal resources. *Proceedings of the National Academy of Sciences of the United States of America, 101*, 5708–5715.

Komori, T. (2010). Regulation of osteoblast differentiation by Runx2. *Advances in Experimental Medicine and Biology, 658*, 43–49.

Krause, J., Fu, Q., Good, J. M., Viola, B., Shunkov, M. V., Derevianko, A. P., & Pääbo, S. (2010). The complete mitochondrial DNA genome of an unknown hominin from southern Siberia. *Nature, 464*, 894–897.

Kröpelin, S., Verschuren, D., Lézine, A. M., Eggermont, H., Cocquyt, C., Francus, P., et al. (2008). Climate-driven ecosystem succession in the Sahara: The past 6,000 years. *Science, 320*, 765–768.

Kuman, K., Inbar, M., & Clarke, R. J. (1999). Paleoenvironments and cultural sequence of the Florisbad Middle Stone Age hominid site, South Africa. *Journal of Archaeological Science, 26*, 1409–1425.

Kuper, R., & Kröpelin, S. (2006). Climate-controlled Holocene occupation of the Sahara: Motor of Africa's evolution. *Science, 313*, 803–807.

Kusimba, S. B. (2001). The early later Stone Age in East Africa: Excavations and lithic assemblages from Lukenya Hill. *African Archaeological Review, 18*, 77–123.

Lachance, J., Vernot, B., Elbers, C. C., Ferwerda, B., Froment, A., Bodo, J. M., et al. (2012). Evolutionary history and adaptation from high-coverage whole-genome sequences of diverse African hunter-gatherers. *Cell, 150*, 457–469.

Lahr, M. M. (1996). *The evolution of modern human diversity. A study of cranial variation*. Cambridge: Cambridge University Press.

Lalueza-Fox, C., & Gilbert, M. T. P. (2011). Paleogenomics of archaic hominins. *Current Biology, 21*, R1002–R1009. doi:10.1016/j.cub.2011.11.021.

Lam, Y. M., Pearson, O. M., & Smith, C. M. (1996). Chin morphology and sexual dimorphism in the fossil hominid mandible sample from Klasies River Mouth. *American Journal of Physical Anthropology, 100*, 545–557.

Landais, A. J., Barnola, M., Masson-Delmotte, V., Jouzel, J., Chappellaz, J., Caillon, N., et al. (2004). A continuous record of temperature evolution over a sequence of Dansgaard-Oeschger events during Marine Isotopic Stage 4 (76 to 62 kyr BP). *Geophysical Research Letters, 31*, L22211. doi:10.1029/2004GL021193.

Lane, C., Barker, G., Reynolds, T., Inglis, R., Barton, N., Albert, P., et al. (2011). Dating the Paleolithic occupation of the Haua Fteah cave, Libya, using cryptotephrochronology (Abstract 1889). XVIII INQUA Congress, Quaternary Sciences—the view from the mountains. Bern, Switzerland.

Lane, C. S., Chorn, B. T., & Johnson, T. C. (2013). Ash from the Toba supereruption in Lake Malawi shows no volcanic winter in East Africa at 75 ka. *Proceedings of the National Academy of Sciences of the United States of America, 110*, 8025–8029.

Lang, N., & Wolff, E. W. (2011). Interglacial and glacial variability from the last 800 ka in marine, ice and terrestrial archives. *Climate of the Past, 7*, 361–380.

Larrasoaña, J. C. (2012). A northeast Saharan perspective on environmental variability in North Africa and its implications for modern human origins. In J. J. Hublin & S. P. McPherron (Eds.), *Modern origins: A North African perspective* (pp. 19–34). Dordrecht: Springer.

Leakey, R. E. F. (1969). Early *Homo sapiens* remains from the Omo River region of south-west Ethiopia. *Nature, 222*, 1132–1133.

Lebamba, J., Vincens, A., & Maley, J. (2012). Pollen, vegetation change and climate at Lake Barombi Mbo (Cameroon) during the last c. 33 000 cal yr BP: A numerical approach. *Climate of the Past, 8*, 59–78.

Li, H., & Durbin, R. (2011). Inference of human population history from individual whole-genome sequences. *Nature, 475*, 403–407.

Li, J. Z., Absher, D. M., Tang, H., Southwick, A. M., Casto, A. M., Ramachandran, S., et al. (2008). Worldwide human relationships inferred from genome-wide patterns of variation. *Science, 319*, 1100–1104.

Linz, B., Balloux, F., Moodley, Y., Manica, A., Liu, H., Roumagnac, P., et al. (2007). An African origin for the intimate association between humans and *Helicobacter pylori*. *Nature, 445*, 915–918.

Lisiecki, L. E., & Raymo, M. E. (2005). A Plio-Pleistocene stack of 57 globally distributed benthic $\delta^{18}O$ records. *Paleoceanography, 20*, PA2007. doi:10.1029/2005PA001164.

Lombard, M., & Phillipson, L. (2010). Indications of bow and stone-tipped arrow use 64,000 years ago in KwaZulu-Natal, South Africa. *Antiquity, 84*, 635–648.

Louw, A. W., Brain, C. K., Vogel, J. C., Mason, R. J., & Eloff, J. F. (1969). Bushman Rock Shelter, Origstad, Eastern Transvaal: A preliminary investigation. *South African Archaeological Bulletin, 24*, 39–60.

Lowe, J., Barton, N., Blockley, S., Bronk Ramsey, C., Cullen, V. L., Davies, W., et al. (2012). Volcanic ash layers illuminate the resilience of Neanderthals and early modern humans to natural hazards. *Proceedings of the National Academy of Sciences of the United States of America, 109*, 13532–13537.

Lubell, D. (1974). *The Fakhurian. A Late Paleolithic industry from Upper Egypt* (Geological Survey of Egypt, Paper No. 58). Cairo: The Geological Survey of Egypt.

Lubell, D. (1984). Paleoenvironments and Epi-paleolithic economies in the Maghreb (20000 to 5000 B.P.). In J. D. Clark & S. A. Brandt (Eds.), *From hunters to farmers: The causes and consequences of food production in Africa.* (pp. 41–56). Berkeley: University of California Press.

Lubell, D. (2001). Late Pleistocene—Early Holocene Maghreb. In P. N. Peregrine & M. Ember (Eds.), *Encyclopedia of prehistory* (Vol. 1, pp. 129–149)., Africa New York: Kluwer.

Lyons, R. P., Scholz, C. A., Buoniconti, M. R., & Martin, M. R. (2011). Late Quaternary stratigraphic analysis of the Lake Malawi Rift, East Africa: An integration of drill-core and seismic-reflection data. *Paleogeography, Paleoclimatology, Paleoecology, 303*, 20–37.

Magori, C. C., & Day, M. H. (1983). Laetoli Hominid 18: An early *Homo sapiens* skull. *Journal of Human Evolution, 12*, 747–754.

Manega, P. C. (1993). Geochronology, geochemistry, and isotopic study of the Plio-Pleistocene hominid sites and the Ngorongoro Volcanic Highlands in northern Tanzania. Ph.D. Dissertation, University of Colorado.

Manega, P. C. (1995). New geochronological results from the Ndutu, Naisiusu, and Ngaloba Beds at Olduvai and Laetoli in northern Tanzania: Their significance for the evolution of modern humans (Abstract). Preservation and use of Olduvai Gorge, Laetoli, rock art, and other paleoanthropological resources in Tanzania, held at the Bellagio Center, Italy, June 5–9, 1995.

Manica, A., Amos, B., Balloux, F., & Hanihara, T. (2007). The effect of ancient population bottlenecks on human phenotypic variation. *Nature, 448*, 346–348.

Marean, C. W. (1992). Implications of late Quaternary mammalian fauna from Lukenya Hill (south-central Kenya) for paleoenvironmental change and faunal extinctions. *Quaternary Research, 37*, 239–255.

Marean, C. W. (2010). Pinnacle Point Cave 13B (Western Cape Province, South Africa) in context: The Cape Floral kingdom, shellfish, and modern human origins. *Journal of Human Evolution, 59*, 425–443.

Marean, C. W., Nilssen, P. J., Brown, K., Jerardino, A., & Stynder, D. (2004). Paleoanthropological investigations of Middle Stone Age sites at Pinnacle Point, Mossel Bay (South Africa): Archaeology and hominid remains from the 2000 field season. *Paleoanthropology, 2004*, 14–83.

Marean, C. W., Bar-Matthews, M., Bernatchez, J., Fisher, E., Goldberg, P., Herries, A. I., et al. (2007). Early human use of marine resources and pigment in South Africa during the Middle Pleistocene. *Nature, 449*, 905–908.

Mariotti, V., Bonfighioli, B., Facchini, F., Condemi, S., & Belcastro, M. G. (2009). Funerary practices of the Iberomaurusian population of Taforalt (Tafoughalt; Morocco, 11-12,000 BP): New hypotheses based on a grave by grave skeletal inventory and evidence of deliberate human modification of the remains. *Journal of Human Evolution, 56*, 340–354.

Mark, D. F., Petraglia, M., Smith, V. C., Morgan, L. E., Barfold, D. N., Ellis, B. S., et al. (2014). A high-precision $^{40}Ar/^{39}Ar$ age for the Young Toba Tuff and dating of ultra-distal tephra: Forcing of Quaternary climate and implications for hominin occupation of India. *Quaternary Geochronology, 21*, 90–103.

Martinson, D. G., Pisias, N. G., Hays, J. D., Imbrie, J., Moore, Theodore C., et al. (1987). Age dating and the orbital theory of the ice ages: Development of a high resolution 0 to 300,000-year chronostratigraphy. *Quaternary Research, 27*, 1–29.

Mason, R. J. (1969). Tentative interpretations of new radiocarbon dates for stone artefact assemblages from Rose Cottage Cave, O.F.S. and Bushman Rock Shelter, Tvl. *South African Archaeological Bulletin, 24*, 57–59.

McBrearty, S., & Brooks, A. S. (2000). The revolution that wasn't: A new interpretation of the origin of modern human behavior. *Journal of Human Evolution, 39*, 453–563.

McBurney, C. B. M. (1958). Evidence for the distribution in space and time of Neanderthaloids and allied strains in northern Africa. In G. H. R. von Koenigswald (Ed.), *Hundert Jahre Neanderthaler (Neanderthal Centenary 1856-1956)* (pp. 253–264). Köln: Böhlau.

McBurney, C. B. M. (1967). *The Haua Fteah (Cyrenaica) and the Stone Age of the South-East Mediterranean.* Cambridge: Cambridge University Press.

McBurney, C. B. M., Trevor, J. C., & Wells, L. H. (1953a). A fossil human mandible from a Levalloiso-Mousterian horizon in Cyrenaica. *Nature, 172*, 889–892.

McBurney, C. B. M., Trevor, J. C., & Wells, L. H. (1953b). The Haua Fteah fossil jaw. *Journal of the Royal Anthropological Institute, 83*, 71–85.

McCall, G. S. (2006). Multivariate perspectives on change and continuity in the Middle Stone Age lithics from Klasies River Mouth, South Africa. *Journal of Human Evolution, 51*, 429–439.

McCrea, J. M. (1950). On the isotopic chemistry of carbonates and a paleotemperature scale. *Journal of Chemical Physics, 18*, 849–857.

McCrossin, M. L. (1992). Human molars from later Pleistocene deposits of Witkrans Cave, Gaap escarpment, Kalahari margin. *Human Evolution, 7*, 1–10.

McDermott, F., Stringer, C., Grün, R., Williams, C. T., Din, V. K., & Hawkesworth, C. J. (1996). New late Pleistocene uranium-thorium and ESR dates for the Singa hominid (Sudan). *Journal of Human Evolution, 31*, 507–516.

McDougall, I., & Brown, F. H. (2006). Precise $^{40}Ar/^{39}Ar$ geochronology for the upper Koobi Fora Formation, Turkana Basin, northern Kenya. *Journal of the Geological Society of London, 163*, 205–220.

McDougall, I., Brown, F. H., & Fleagle, J. G. (2005). Stratigraphic placement and age of modern humans from Kibish, Ethiopia. *Nature, 433*, 733–736.

McDougall, I., Brown, F. H., & Fleagle, J. G. (2008). Sapropels and the age of hominins Omo I and II, Kibish, Ethiopia. *Journal of Human Evolution, 55*, 409–420.

Medig, M., Meier, R., Sahnouni, M., & Derradji, A. (1996). Découverte d'un crâne humain dans les niveaux Ibéromaurusiens de la grotte de Taza I, Jijel, Algérie. *Comptes Rendus de l'Academie des Sciences Paris, Série II, Sciences de la Terre et des Planetes, 323*, 825–831.

Mehlman, M. J. (1984). Archaic *Homo sapiens* at Lake Eyasi, Tanzania: Recent misrepresentations. *Journal of Human Evolution, 13*, 487–501.

Mehlman, M. J. (1987). Provenience, age and associations of archaic *Homo sapiens* crania from Lake Eyasi, Tanzania. *Journal of Archaeological Science, 14*, 133–162.

Meier, R. J., Sahnouni, M., Medig, M., & Abdelkader Derradji, A. (2003). Human skull from the Taza locality, Jijel, Algeria. *Anthropologischer Anzeiger, 61*, 129–140.

Meignen, L. (2000). Early Middle Paleolithic blade technology in southwestern Asia. *Acta Anthropologica Sinica, 19 (suppl.)*, 158–168.

Meignen, L. (2007). Middle Paleolithic blady assemblages in the Near East: A reassessment. In V. P. Lyubin (Ed.), *Caucasus and the initial human dispersals in the Old World* (pp. 133–148). Saint-Petersburg: Russian Academy of Sciences.

Mendez, F. L., Krahn, T., Schrack, B., Krahn, A. M., Veeramah, K. R., Woerner, A. E., et al. (2013). An African American paternal lineage adds an extremely ancient root to the human Y chromosome phylogenetic tree. *American Journal of Human Genetics, 92*, 454–459.

Mercader, J., & Brooks, A. S. (2001). Across forests and savannas: Later Stone Age assemblages from Ituri and Semliki, Democratic Republic of Congo. *Journal of Anthropological Research, 57*, 197–217.

Mercier, N., & Valladas, H. (2003). Reassessment of TL age estimates of burnt flints from the Paleolithic site of Tabun Cave, Israel. *Journal of Human Evolution, 45*, 401–409.

Meyer, M., Kircher, M., Gansauge, M. T., Li, H., Racimo, F., Mallick, S., et al. (2012). A high-coverage genome sequence from an archaic Denisovan individual. *Science, 338*, 222–226.

Michels, J. W., & Marean, C. W. (1984). A Middle Stone Age occupation site at Porc-Epic Cave, Dire Dawa (east-central Ethiopia), Part II. *African Archaeological Review, 2*, 64–71.

Mikdad, A., & Eiwanger, J. (2000). Recherches préhistoriques et protohistoriques dans le Rif oriental (Maroc). Rapport Préliminaire. *Beiträge zur Allgemeinen und Vergleichenden Archäologie, 20*, 109–167.

Millard, A. R. (2008). A critique of the chronometric evidence for hominid fossils: I. Africa and the Near East 500-50 ka. *Journal of Human Evolution, 54*, 848–874.

Millard, A. R., & Hedges, R. E. M. (1995). The role of the environment in uranium uptake by buried bone. *Journal of Archaeological Science, 22*, 239–250.

Miller, G. H., Beaumont, P. B., Deacon, H. J., Brooks, A. S., Hare, P. E., & Jull, A. J. T. (1999). Earliest modern humans in southern Africa dated by isoleucine epimerization in ostrich eggshell. *Quaternary Science Reviews, 18*, 1537–1548.

Miller, S. F. (1971). The age of the Nachikufan industries in Zambia. *South African Archaeological Bulletin, 26*, 143–146.

Mills, S. C., Grab, S. W., Rea, B. R., Carr, S. J., & Farrow, A. (2012). Shifting westerlies and precipitation patterns during the Late Pleistocene in southern Africa determined using glacier reconstruction and mass balance modelling. *Quaternary Science Reviews, 55*, 145–159.

Milo, R. G. (1998). Evidence for hominid predation at Klasies River Mouth, South Africa, and its implications for the behaviour of early modern humans. *Journal of Archaeological Science, 25*, 99–133.

Minugh-Purvis, N. (1993). Reexamination of the immature hominid maxilla from Tangier, Morocco. *American Journal of Physical Anthropology, 92*, 449–461.

Mitchell, P. (2008). Developing the archaeology of Marine Isotope Stage 3. *South African Archaeological Society Goodwin Series, 10*, 52–65.

Moodley, Y., Linz, B., Bond, R. P., Nieuwoudt, M., Soodyall, H., Schlebusch, C. M., et al. (2012). Age of the association between *Helicobacter pylori* and man. *PLoS Pathogens, 8*(5), e1002693.

Morgan, L. E., Renne, P. R., Taylor, R. E., & WoldeGabriel, G. (2009). Archaeological age constraints from extrusion ages of obsidian: Examples from the Middle Awash, Ethiopia. *Quaternary Geochronology, 4*, 193–203.

Morris, A. G. (1992). Biological relationships between Upper Pleistocene and Holocene populations in southern Africa. In G. Bräuer & F. H. Smith (Eds.), *Continuity or replacement: Controversies in* Homo sapiens *evolution* (pp. 131–143). Rotterdam: Balkema.

Morris, A. G. (2002). Isolation and the origin of the Khoisan: Late Pleistocene and early Holocene human evolution at the southern end of Africa. *Human Evolution, 17*, 105–114.

Morris, A. G. (2003). The myth of the East African "Bushmen". *South African Archaeological Bulletin, 58*, 85–90.

Mourre, V., Villa, P., & Henshilwood, C. (2010). Early use of pressure flaking on lithic artifacts at Blombos Cave, South Africa. *Science, 330*, 659–662.

Nami, M. (2007). Les techno-complexes Ibéromaurusiens d'Ifri El Baroud (Rif Oriental, Maroc). *Forschungen zur Archäologie Außereuropäischer Kulturen, 2*, 183–239.

Nara, M. T. (1994). Etude de la variabilité de certains caractères métriques et morphologiques des Néandertaliens. Ph.D. Dissertation, University of Bordeaux.

Noonan, J. P., Coop, G., Kudaravalli, S., Smith, D., Krause, J., Alessi, J., et al. (2006). Sequencing and analysis of Neanderthal genomic DNA. *Science, 314*, 1113–1118.

Orban, R., Semal, P., & Twiesselmann, F. (2001). Sur la biométrie des mandibules et des dents humaines d'Ishango (LSA, République démocratique du Congo). *Bulletins et mémoires de la Société d'Anthropologie de Paris, 13*(1–2), 1–13.

Parkington, J. (1990). A critique of the consensus view on the age of Howieson's Poort assemblages in South Africa. In P. Mellars (Ed.), *The emergence of modern humans* (pp. 34–55). Edinburgh: Edinburgh University Press.

Parkington, J., & Poggenpoel, C. (1987). Diepkloof Rock Shelter. In J. Parkington & M. Hall (Eds.), *Papers in prehistory of the Western Cape, South Africa* (BAR International Series No. 332) (pp. 269–293). Oxford: Archaeopress.

Partridge, T. C., deMenocal, P. B., Lorentz, S. A., Paiker, M. J., & Vogel, J. C. (1997). Orbital forcing of climate over South Africa: A 200,000-year rainfall record from the Pretoria saltpan. *Quaternary Science Reviews, 16*, 1125–1133.

Partridge, T. C., Scott, L., & Hamilton, J. E. (1999). Synthetic reconstructions of southern African environments during the Last Glacial Maximum (21-18 kyr) and the Holocene Altithermal (8-6 kyr). *Quaternary International, 57*(58), 207–214.

Pearson, O. (2000). Human remains from Twin Rivers. Appendix 8. In L. Barham (Ed.), *The Middle Stone Age of Zambia* (pp. 281–282). Bristol: Western Academic and Specialist Press.

Pearson, O. M. (2011). Integration of the genetic, anatomical and archaeological data for the African origin of modern humans: Problems and prospects. In S. C. Reynolds & A. Gallagher (Eds.), *African genesis: Perspectives on hominin evolution* (pp. 423–448). Cambridge: Cambridge University Press.

Pearson, O. M., & Grine, F. E. (1996). Morphology of the Border Cave hominid ulna and humerus. *South African Journal of Science, 92*, 231–236.

Pearson, O. M., & Grine, F. E. (1997). Re-analysis of the hominid radii from Cave of Hearths and Klasies River Mouth, South Africa. *Journal of Human Evolution, 32*, 577–592.

Pearson, O. M., Churchill, S. E., Grine, F. E., Trinkaus, E., & Holliday, T. W. (1998). Multivariate analyses of the hominid ulna from Klasies River Mouth. *Journal of Human Evolution, 34*, 653–656.

Pearson, O., Grine, F. E., Barham, L., & Stringer, C. (2000). Human remains from the Middle and Later Stone Age of Mumbwa Caves. In L. Barham (Ed.), *The Middle Stone Age of Zambia, South Central Africa* (pp. 149–164). Bristol, UK: Western Academic & Specialist Press.

Pearson, O. M., Royer, D. F., Grine, F. E., & Fleagle, J. G. (2008a). A description of the Omo I postcranial skeleton, including newly discovered fossils. *Journal of Human Evolution, 55*, 421–437.

Pearson, O. M., Fleagle, J. G., Grine, F. E., & Royer, D. F. (2008b). Further new hominin fossils from the Kibish Formation, southwestern Ethiopia. *Journal of Human Evolution, 55*, 444–447.

Peers, B., & Goodwin, A. J. H. (1953). Two caves at Kalk Bay, Cape Peninsula. Part II: Peer's Shelter B/102. *South African Archaeology Bulletin, 8*, 67–77.

Petit, J. R., Jouzel, J., Raynaud, D., Barkov, N. I., Barnola, J. M., Basile, I., et al. (1999). Climate and atmospheric history of the past 420,000 years from the Vostok ice core, Antarctica. *Nature, 399*, 429–436.

Pfeiffer, S., & Zehr, M. K. (1996). A morphological and histological study of the human humerus from Border Cave. *Journal of Human Evolution, 31*, 49–59.

Pfeiffer, S., Lazenby, R., & Thackeray, J. F. (1996). Tuinplaas: Affinity assessed from long bone cross-sectional geometry. *Kaupia, 6*, 137–143.

Pickford, M. (1986). Cainozoic paleontological sites of Western Kenya. *Münchner Geowissenschaftlich Abhandlungen, 8*, 1–151.

Pickrell, J. K., Patterson, N., Barbieri, C., Berthold, F., Gerlach, L., Güldemann, T., et al. (2012). The genetic prehistory of southern Africa. *Nature Communications, 3*, 1143. doi:10.1038/ncomms2140.

Pike, A. W. G., Eggins, S., Grün, R., & Thackeray, F. (2004). U-series dating of TP1, an almost complete human skeleton from Tuinplaas (Springbok Flats), South Africa. *South African Journal of Science, 100*, 381–383.

Pike, A. W. G., Hedges, R. E. M., & Van Calsteren, P. (2002). U-series dating of bone using the diffusion-adsorption model. *Geochimica et Cosmochimica Acta, 66*, 4268–4273.

Pinhasi, R. (2002). Biometric study of the affinities of NK—a quantitative analysis of mandible dimensions. In P. M. Vermeersch (Ed.), *Paleolithic quarrying sites in upper and middle Egypt* (pp. 283–336). Leuven: Leuven University Press.

Pinhasi, R., & Semal, P. (2000). The position of the Nazlet Khater specimen among prehistoric and modern African and Levantine populations. *Journal of Human Evolution, 39*, 269–288.

Pleurdeau, D. (2005). Human technical behavior in the African Middle Stone Age: The lithic assemblage of Porc-Épic Cave (Dire Dawa, Ethiopia). *African Archaeological Review, 22*, 177–197.

Plug, I. (2004). Resource exploitation: Animal use during the Middle Stone Age at Sibudu Cave, KwaZulu-Natal. *South African Journal of Science, 100*, 151–158.

Pokras, E. M., & Mix, A. C. (1985). Eolian evidence for spatial variability of Late Quaternary climates in tropical Africa. *Quaternary Research, 24*, 137–149.

Ponce de León, M. S., & Zollikofer, C. P. E. (2001). Neanderthal cranial ontogeny and its implications for late hominid diversity. *Nature, 412*, 534–538.

Porter, S. C. (1989). Some geological implications of average Quaternary glacial conditions. *Quaternary Research, 32*, 245–261.

Poznik, G. D., Henn, B. M., Yee, M. C., Sliwerska, E., Euskirchen, G. M., Lin, A. A., et al. (2013). Sequencing Y chromosomes resolves discrepancy in time to common ancestor of males versus females. *Science, 341*, 562–565.

Pritchard, J. K., Seielstad, M. T., Perez-Lezaun, A., & Feldman, M. W. (1999). Population growth of human Y chromosome microsatellites. *Molecular Biology and Evolution, 16*, 1791–1798.

Protsch, R. (1974). The Fish Hoek hominid: Another member of basic *Homo sapiens afer*. *Anthropologischer Anzeiger, 34*, 241–249.

Protsch, R. (1975). The absolute dating of Upper Pleistocene sub-Saharan fossil hominids and their place in human evolution. *Journal of Human Evolution, 4*, 297–322.

Protsch, R. R. R. (1977). Mumbwa: Its absolute chronology and archaeology. *Zeitschrift fur Morphologie und Anthropologie, 68*, 1–7.

Protsch, R. (1981). *The paleoanthropological finds of the Pliocene and Pleistocene. (Die Archaologischen und anthropologischen Ergebnisse der Kohl-Larsen-Expeditionen in Nord-Tanzania 1933-1939)*. Tübingen: Verlag Archaeologica Venatoria.

Protsch, R., & de Villiers, H. (1974). Bushman Rock Shelter, Origstad, Eastern Transvaal, South Africa. *Journal of Human Evolution, 3*, 387–396.

Pycraft, W. P. (1925). On the calvaria found at Boskop, Transvaal, 1913, and its relationship to Cromagnard and negroid skulls. *Journal of the Royal Anthropological Institute, 55*, 179–198.

Rak, Y. (1998). Does any Mousterian cave present evidence of two hominid species? In T. Akazawa, K. Aoki, & O. Bar-Yosef (Eds.), *Neandertals and modern humans in western Asia* (pp. 353–366). New York: Plenum.

Rampino, M. R., & Self, S. (1992). Volcanic winter and accelerated glaciations following the Toba super-eruption. *Nature, 359*, 50–52.

Rampino, M. R., & Self, S. (1993). Climate-volcanism feedback and the Toba eruption of ~74,000 years ago. *Quaternary Research, 40*, 269–280.

Reed, C. A. (1965). A human frontal bone from the Late Pleistocene of the Kom Ombo Plain, Upper Egypt. *Man, 65*, 101–104.

Reich, D., Green, R. E., Kircher, M., Krause, J., Patterson, N., Durand, E. Y., et al. (2010). Genetic history of an archaic hominin group from Denisova Cave in Siberia. *Nature, 468*, 1053–1060.

Reich, D., Patterson, M., Kircher, M., Delfin, F., Nandineni, M. R., Pugach, I., et al. (2011). Denisova admixture and the first modern human dispersals into southeast Asia and Oceania. *American Journal of Human Genetics, 89*, 516–528.

Reynolds, T. (2013). The Middle Paleolithic of Cyrenaica: Is there an Aterian at the Haua Fteah and does it matter? *Quaternary International, 300*, 171–181.

Richter, D., Moser, J., Nami, M., Eiwanger, J., & Mikdad, A. (2010). New chronometric data from Ifri n'Ammar (Morocco) and the chronostratigraphy of the Middle Paleolithic in the Western Maghreb. *Journal of Human Evolution, 59*, 672–679.

Rightmire, G. P. (1976). Relationships of Middle and Upper Pleistocene hominids from sub-Saharan Africa. *Nature, 260*, 238–240.

Rightmire, G. P. (1978). Human skeletal remains from the southern Cape Province and their bearing on the Stone Age prehistory of South Africa. *Quaternary Research, 9*, 219–230.

Rightmire, G. P. (1979). Implications of the Border Cave skeletal remains for Later Pleistocene human evolution. *Current Anthropology, 20*, 23–35.

Rightmire, G. P. (1984). *Homo sapiens* in Sub-Saharan Africa. In F. H. Smith & R. Spencer (Eds.), *The origins of modern humans: A world survey of the fossil evidence* (pp. 295–325). New York: Alan R. Liss.

Rightmire, G. P. (1998). Human evolution in the Middle Pleistocene: The role of *Homo heidelbergensis*. *Evolutionary Anthropology, 6*, 218–227.

Rightmire, G. P. (2001). Patterns of hominid evolution and dispersal in the Middle Pleistocene. *Quaternary International, 75*, 77–84.

Rightmire, G. P. (2008). *Homo* in the Middle Pleistocene: Hypodigms, variation, and species recognition. *Evolutionary Anthropology, 17*, 8–21.

Rightmire, G. P. (2009). Middle and later Pleistocene hominins in Africa and Southwest Asia. *Proceedings of the National Academy of Sciences of the United States of America, 106*, 16046–16050.

Rightmire, G. P., & Deacon, H. J. (1991). Comparative studies of Late Pleistocene human remains from Klasies River Mouth, South Africa. *Journal of Human Evolution, 20*, 131–156.

Rightmire, G. P., & Deacon, H. J. (2001). New human teeth from the Middle Stone Age deposits at Klasies River, South Africa. *Journal of Human Evolution, 41*, 535–544.

Rightmire, G. P., Deacon, H. J., Schwartz, J. H., & Tattersall, I. (2006). Human foot bones from Klasies River Main Site, South Africa. *Journal of Human Evolution, 50*, 96–103.

Roche, J. (1953). La grotte de Taforalt. *L'Anthropologie, 57*, 375–380.

Roche, J. (1963). *L'Épipaléolithique Marocaine*. Lisbon: Fondation Calouste Gulbenkian.

Roche, J. (1969). Les industries paléolithiques de la grotte de Taforalt (Maroc oriental). *Quaternaria, 11*, 89–100.

Roche, J., & Texier, J. P. (1976). Découverte de restes humains dans un niveau Atérian de la grotte des Contrebandiers à Témara, Maroc. *Comptes Rendus des Séances de l'Académie des Sciences (Paris), 282*, 45–47.

Rougier, H., Milota, Ş., Rodrigo, R., Gherase, M., Sarcină, L., Moldovan, O., et al. (2007). Peştera cu Oase 2 and the cranial morphology of early modern Europeans. *Proceedings of the National Academy of Sciences of the United States of America, 104*, 1165–1170.

Royer, D. F., Lockwood, C. A., Scott, J. E., & Grine, F. E. (2009). Size variation in early human mandibles and molars from Klasies River, South Africa: Comparison with other Middle and Late Pleistocene assemblages and with modern humans. *American Journal of Physical Anthropology, 140*, 312–323.

Saban, R. (1975). Les restes humains de Rabat (Kebibat). *Annals de Paléontologie (Vertébrés), 61*, 153–207.

Saban, R. (1977). The place of Rabat Man (Kébibat, Morocco) in human evolution. *Current Anthropology, 18*, 518–524.

Salama, N., Guillemin, K., McDaniel, T. K., Sherlock, G., Tompkins, L., & Falkow, S. (2000). A whole-genome microarray reveals genetic diversity among *Helicobacter pylori* strains. *Proceedings of the National Academy of Sciences of the United States of America, 97*, 14668–14673.

Salas, A., Richards, M., De la Fe, T., Lareu, M. V., Sobrino, B., Sanchez-Diz, P., et al. (2002). The making of the African mtDNA landscape. *American Journal of Human Genetics, 71*, 1082–1111.

Salzmann, U., Hoelzmann, P., & Morczinek, I. (2002). Late Quaternary climate and vegetation of the Sudanian Zone of northeast Nigeria. *Quaternary Research, 58*, 73–83.

Sandford, K. S. (1934). *Paleolithic man and the Nile Valley in Upper and Middle Egypt: A study of the region during Pliocene and Pleistocene times*. Chicago: University of Chicago Press.

Savage, D. K. (1983). Identifying industries in south central Africa. Ph. D. Dissertation, University of California, Berkeley.

Sawchuk, E. A., & Willoughby, P. R. (2015). Terminal Pleistocene Later Stone Age human remains from the Mlambalasi Rock Shelter, Iringa Region, southern Tanzania. *International Journal of Osteoarchaeology, 25*, 593–607.

Sawyer, S., Renauda, G., Violab, B., Hublin, J. J., Gansaugea, M. T., Shunkovd, M. V., et al. (2015). Nuclear and mitochondrial DNA sequences from two Denisovan individuals. *Proceedings of the National Academy of Sciences of the United States of America, 112*, 15696–15700.

Scerri, E. M. L. (2013). The Aterian and its place in the North African Middle Stone Age. *Quaternary International, 300*, 111–130.

Schwartz, G. T., & Tattersall, I. (2000). The human chin revisited: What is it and who has it? *Journal of Human Evolution, 38*, 367–409.

Schefuß, E., Schouten, S., Jansen, J. H. F., & Damsté, J. S. S. (2003). African vegetation controlled by tropical sea surface temperatures in the mid-Pleistocene period. *Nature, 422*, 418–421.

Schefuß, E., Schouten, S., & Schneider, R. R. (2005). Climatic controls on central African hydrology during the past 20,000 years. *Nature, 437*, 1003–1006.

Schellmann, G., & Radtke, U. (1999). Problems encountered in the determination of dose and dose rate in ESR dating of mollusk shells. *Quaternary Science Reviews, 18*, 1515–1527.

Schlebusch, C. M., Skoglund, P., Sjodin, P., Gattepaille, L. M., Hernandez, D., Jay, F., et al. (2012). Genomic variation in seven Khoe-San groups reveals adaptation and complex African history. *Science, 338*, 374–379.

Schlebusch, C. M., Lombard, M., & Soodyall, H. (2013). mtDNA control region variation affirms diversity and deep sub-structure in populations from southern Africa. *BMC Evolutionary Biology, 13*, 56. doi:10.1186/1471-2148-13-56.

Scholz, C. A., Johnson, T. C., Cohen, A. S., King, J. W., Peck, J. A., Overpeck, J. T., et al. (2007). East African megadroughts between 135 and 75 thousand years ago and bearing on early-modern human origins. *Proceedings of the National Academy of Sciences of the United States of America, 104*, 16416–16421.

Scholz, C. A., Cohen, A. S., Johnson, T. C., King, J., Talbot, M. R., & Brown, E. T. (2011). Scientific drilling in the Great Rift Valley: The 2005 Lake Malawi Scientific Drilling Project—an overview of the past 145,000 years of climate variability in southern Hemisphere East Africa. *Paleogeography, Paleoclimatology, Paleoecology, 303*, 1–4.

Schultz, H., Emeis, K. C., Erlenkeuser, H., von Rad, U., & Rolf, C. (2002). The Toba volcanic event and interstadial/stadial climates at the Marine Isotopic Stage 5 to 4 transition in the northern Indian Ocean. *Quaternary Research, 57*, 22–31.

Schuster, S. C., Miller, W., Ratan, A., Tomsho, L. P., Giardine, B., Kasson, L. R., et al. (2010). Complete Khoisan and Bantu genomes from southern Africa. *Nature, 463*, 943–947.

Schwarcz, H., & Rink, J. (2000). ESR dating of the Die Kelders Cave 1 site, South Africa. *Journal of Human Evolution, 38*, 121–128.

Schwenninger, J. L., Collcutt, S. N., Barton, N., Bouzouggar, A., Clark Balzan, L., El Hajraoui, M. A., et al. (2010). A new luminescence chronology for Aterian cave sites on the Atlantic coast of Morocco. In E. A. Garcea (Ed.), *South-eastern Mediterranean peoples between 130,000 and 10,000 years ago* (pp. 18–36). Oxford: Oxbow Books.

Scott, L. (1987). Pollen analysis of hyena coprolites and sediments from Equus Cave, Taung, Southern Kalahari (South Africa). *Quaternary Research, 28*, 144–156.

Scozzari, R., Cruciani, F., Santolamazza, P., Malaspina, P., Torroni, A., Sellitto, D., et al. (1999). Combined use of biallelic and microsatellite Y-chromosome polymorphisms to infer affinities among African populations. *American Journal of Human Genetics, 65*, 829–846.

Semino, O., Santachiara-Benerecetti, A. S., Falaschi, F., Cavalli-Sforza, L. L., & Underhill, P. A. (2002). Ethiopians and Khoisan share the deepest clades of the human Y-chromosome phylogeny. *American Journal of Human Genetics, 70*, 265–268.

Senut, B., Pickford, M., Braga, J., Marais, D., & Coppens, Y. (2000). Découverte d'un *Homo sapiens* archaïque à Oranjemund, Namibie. *Comptes Rendus des Séances de l'Académie des Sciences (Paris) Sciences de la Terre et des Planètes, 330*, 813–819.

Şenyürek, M. S. (1940). Fossil Man in Tangier. *Papers of the Peabody Museum of American Archaeology and Ethnology, Harvard University, 16*, 1–27

Shackelford, L. L. (2007). Regional variation in the postcranial robusticity of late Upper Paleolithic humans. *American Journal of Physical Anthropology, 133*, 655–668.

Shackleton, N. J. (1967). Oxygen isotope analyses and Pleistocene temperatures re-assessed. *Nature, 215*, 15–17.

Shackleton, N. J. (1969). The last interglacial in the marine and terrestrial records. *Proceedings of the Royal Society B, 174*, 135–154.

Shackleton, N. J. (1982). Stratigraphy and chronology of the Klasies River Mouth deposits: Oxygen isotope evidence. In R. Singer & J. Wymer (Eds.), *The Middle Stone Age at Klasies River Mouth in South Africa* (pp. 194–199). Chicago: University of Chicago Press.

Shackleton, N. J. (1987). Oxygen isotopes, ice volume and sea level. *Quaternary Science Reviews, 6*, 183–190.

Shackleton, N. J., & Pisias, N. G. (1985). Atmospheric carbon dioxide, orbital forcing, and climate. In E. Sundquist & W. S. Broecker (Eds.), *The carbon cycle and atmospheric CO_2: Natural variations Archean to present* (Geophysical Monograph Series No. 32) (pp. 303–317). Washington, DC: American Geophysical Union.

Shackleton, N. J., Berger, A., & Peltier, W. R. (1990). An alternative astronomical calibration of the Lower Pleistocene time scale based

on ODP Site 677. *Transactions of the Royal Society of Edinburgh, Earth Science, 81*, 251–261.
Shaw, T., & Daniels, S. G. H. (1984). Excavations at Iwo Eleru, Ondo State, Nigeria. *West African Journal of Archaeology, 14*, 1–269.
Shea, J. J., Fleagle, J. G., & Assefa, Z. (2007). Context and chronology of early *Homo sapiens* fossils from the Omo Kibish Formation, Ethiopia. In P. Mellars, K. Boyle, O. Bar-Yosef, & C. Stringer (Eds.), *Rethinking the human revolution* (pp. 153–162). Cambridge: McDonald Institute for Archaeological Research.
Sillen, A., & Morris, A. (1996). Diagenesis of bone from Border Cave: Implications for the age of the Border Cave hominids. *Journal of Human Evolution, 31*, 499–506.
Singer, R. (1993). Mythical African 'Australoids' and triangular bricks: The Cape Flats skull in retrospect. *South African Archaeological Bulletin, 48*, 105–112.
Singer, R., & Wymer, J. J. (1982). *The Middle Stone Age at Klasies River Mouth in South Africa*. Chicago: University of Chicago Press.
Smith, F. H. (1985). Continuity and change in the origin of modern *Homo sapiens*. *Zeitschrift für Morphologie und Anthropologie, 75*, 197–222.
Smith, F. H. (1992). Models and realities in modern human origins: The African fossil evidence. *Philosophical Transactions of the Royal Society B, 337*, 243–250.
Smith, F. H. (2002). Migrations, radiations and continuity: Patterns in the evolution of Middle and Late Pleistocene humans. In W. C. Hartwig (Ed.), *The primate fossil record* (pp. 437–456). Cambridge: Cambridge University Press.
Smith, F., Falsetti, A., & Donnelly, S. (1989). Modern human origins. *Yearbook of Physical Anthropology, 32*, 35–68.
Smith, P., & Shegev, M. (1988). The dentition of Nubians from Wadi Halfa, Sudan: An evolutionary perspective. *Journal of the Dental Association of South Africa, 43*, 535–541.
Smith, P., Nshimirimana, R., de Beer, F. Morris, D., Jacobson, L., Chazan, M., et al. (2012). Canteen Kopje: A new look at an old skull. *South African Journal of Science, 108*(1/2), Article 738, 9 pages. doi:10.4102/sajs.v108il/2.738.
Smith, T. M., Olejniczak, A. J., Tafforeau, P., Reid, D. J., Grine, F. E., & Hublin, J. J. (2006). Molar crown thickness, volume, and development in South African Middle Stone Age humans. *South African Journal of Science, 102*, 513–517.
Smith, T., Tafforeau, P., Reid, D., Grün, R., Eggins, S., Boutakiout, M., et al. (2007). Earliest evidence of modern human life history in North African early *Homo sapiens*. *Proceedings of the National Academy of Sciences of the United States of America, 104*, 6128–6133.
Soficaru, A., Doboş, A., & Trinkaus, E. (2006). Early modern humans from the Peştera Muierii, Baia de Fier, Romania. *Proceedings of the National Academy of Sciences of the United States of America, 103*, 17196–17201.
Soficaru, A., Petrea, C., Doboş, A., & Trinkaus, E. (2007). The human cranium from Peştera Cioclovina Uscată, Romania. Context, age, taphonomy, morphology, and paleopathology. *Current Anthropology, 48*, 611–619.
Spoor, F., & Stringer, C. (1998). Rare temporal bone pathology of the Singa calvaria from Sudan. *American Journal of Physical Anthropology, 107*, 41–50.
Spoor, F., Leakey, M. G., Antón, S. C., & Leakey, L. N. (2008). The taxonomic status of KNM-ER 42700: A reply to Baab (2008a). *Journal of Human Evolution, 55*, 747–750.
Starling, A. P., & Stock, J. T. (2007). Dental indicators of health and stress in early Egyptian and Nubian agriculturalists: A difficult transition and gradual recovery. *American Journal of Physical Anthropology, 134*, 520–528.
Stearns, C. E., & Thurber, D. L. (1965). Th^{230}/U^{234} dates of Late Pleistocene marine fossils from the Mediterranean and Moroccan littorals. *Progress in Oceanography, 4*, 293–305.
Stock, J. T., O'Neill, M. C., Ruff, C. B., Zabecki, M., Shackelford, L., & Rose, J. C. (2011). Body size, skeletal biomechanics, mobility and habitual activity from the Late Paleolithic to the Mid-Dynastic Nile Valley. In R. Pinhasi & J. T. Stock (Eds.), *Human bioarchaeology of the transition to agriculture* (pp. 347–369). New York: Wiley.
Stocker, T. F. (1998). Climate change: The seesaw effect. *Science, 282*, 61–62.
Stokes, S., & Bailey, R. (2002). Optical dating of Nazlet Safaha and Nazlet Khater 4. In P. M. Vermeersch (Ed.), *Paleolithic quarrying sites in upper and middle Egypt* (pp. 349–351). Leuven: Leuven University Press.
Stokes, S., Thomas, D. S. G., & Washington, R. (1997). Multiple episodes of aridity in southern Africa since the last interglacial period. *Nature, 388*, 154–158.
Stringer, C. B. (1974). Population relationships of later Pleistocene hominids: A multivariate study of available crania. *Journal of Archaeological Science, 1*, 317–342.
Stringer, C. B. (1978). Some problems in Middle and Upper Pleistocene hominid relationships. In D. Chivers & K. Joysey (Eds.), *Recent advances in primatology* (Vol. 3, pp. 395–418)., evolution London: Academic.
Stringer, C. B. (2002). Modern human origins: Progress and prospects. *Philosophical Transactions of the Royal Society London B, 357*, 563–579.
Stringer, C. B. (2007). The origin and dispersal of *Homo sapiens*: Our current state of knowledge. In P. Mellars, K. Boyle, O. Bar-Yosef, & C. Stringer (Eds.), *Rethinking the human revolution* (pp. 15–20). Cambridge: McDonald Institute for Archaeological Research.
Stringer, C. B., & Bräuer, G. (1994). Methods, misreading, and bias. *American Anthropologist, 96*, 416–424.
Stringer, C. B., & Andrews, P. (1988). Genetic and fossil evidence for the origin of modern humans. *Science, 239*, 1263–1268.
Stringer, C. B., Cornish, L., & Stuart-Macadam, P. (1985). Preparation and further study of the Singa skull from Sudan. *Bulletin of the British Museum Natural History (Geology), 38*, 347–358.
Stynder, D. D., Ackermann, R. R., & Sealy, J. C. (2007a). Craniofacial variation and population continuity during the South African Holocene. *American Journal of Physical Anthropology, 134*, 489–500.
Stynder, D. D., Ackermann, R. R., & Sealy, J. C. (2007b). Early to mid-Holocene South African Later Stone Age human crania exhibit a distinctly Khoesan morphological pattern. *South African Journal of Science, 103*, 349–352.
Stynder, D., Brock, F., Sealy, J., Wurz, S., Morris, A., & Volman, T. (2009). A mid-Holocene AMS ^{14}C date for the presumed Upper Pleistocene human skeleton from Peers Cave, South Africa. *Journal of Human Evolution, 56*, 431–434.
Tang, H., Siegmund, D. O., Shen, P., Oefner, P. J., & Feldman, M. W. (2002). Frequentist estimation of coalescence times from nucleotide sequence data using a tree-based partition. *Genetics, 161*, 447–459.
Tattersall, I., & Schwartz, J. H. (1999). Hominids and hybrids: The place of Neanderthals in human evolution. *Proceedings of the National Academy of Sciences of the United States of America, 96*, 7117–7119.
Teilhard de Chardin, P. (1930). Le Paléolithique en Somalie francaise et en Abyssinie. *L' Anthropologie, 40*, 331–334.
Texier, P. J., Porraz, G., Parkington, J., Rigaud, J.-P., Poggenpoel, C., Miller, C., et al. (2010). A Howiesons Poort tradition of engraving ostrich eggshell containers dated to 60,000 years ago at Diepkloof Rock Shelter, South Africa. *Proceedings of the National Academy of Science of the United States of America, 107*, 6180–6185.
Thoma, A. (1984). Morphology and affinities of the Nazlet Khater man. *Journal of Human Evolution, 13*, 287–296.
Thoma, A., & Vallois, H. V. (1977). Les dents de l'homme de Rabat. *Bulletins et Mémoires de la Societé d'Anthropologie de Paris, 4*, 31–58.

Thomas, D. S. G., & Burrough, S. L. (2012). Interpreting geoproxies of late Quaternary climate change in African drylands: Implications for understanding environmental change and early human behavior. *Quaternary International, 253*, 5–17.

Thomas, D. S. G., & Shaw, P. A. (2002). Late Quaternary environmental change in central southern Africa: New data, synthesis, issues and prospects. *Quaternary Science Reviews, 21*, 783–797.

Thomas, D. S. G., Bailey, R., Shaw, P. A., Durcan, J. A., & Singarayer, J. S. (2009). Late Quaternary highstands at Lake Chilwa, Malawi: Frequency, timing and possible forcing mechanisms in the last 44 ka. *Quaternary Science Reviews, 28*, 526–539.

Thompson, W. G., & Goldstein, S. L. (2006). A radiometric calibration of the SPECMAP timescale. *Quaternary Science Reviews, 25*, 3207–3215.

Thomson, R., Pritchard, J. K., Shen, P., Oefner, P. J., & Feldman, M. W. (2000). Recent common ancestry of human Y chromosomes: Evidence from DNA sequence data. *Proceedings of the National Academy of Sciences of the United States of America, 97*, 7360–7365.

Tierney, J. E., & deMenocal, P. B. (2013). Abrupt shifts in Horn of Africa hydroclimate since the Last Glacial Maximum. *Science, 342*, 843–846.

Tierney, J. E., Russell, J. M., Huang, Y., Damsté, J. S. S., Hopmans, E. C., & Cohen, A. S. (2008). Northern hemisphere controls on tropical southeast African climate during the past 60,000 years. *Science, 322*, 252–255.

Tierney, J. E., Lewis, S. C., Cook, B. I., LeGrande, A. N., & Schmidt, G. A. (2011). Model, proxy and isotopic perspectives on the East African Humid Period. *Earth and Planetary Science Letters, 307*, 103–112.

Tishkoff, S. A., & Verrelli, B. C. (2003). Role of evolutionary history on haplotype block structure in the human genome: Implications for disease mapping. *Current Opinion in Genetics & Development, 13*, 569–575.

Tishkoff, S. A., Gonder, M. K., Henn, B. M., Mortensen, H., Knight, A., Gignoux, C., et al. (2007). History of click-speaking populations of Africa inferred from mtDNA and Y chromosome genetic variation. *Molecular Biology and Evolution, 24*, 2180–2195.

Tishkoff, S. A., Reed, F. A., Friedlaender, F. R., Ehret, C., Ranciaro, A., Froment, A., et al. (2009). The genetic structure and history of Africans and African Americans. *Science, 324*, 1035–1044.

Tobias, P. V. (1967). The hominid skeletal remains of Haua Fteah. Appendix 1B. In C. B. M. McBurney (Ed.), *The Haua Fteah (Cyrenaica) and the Stone Age of the south-east Mediterranean* (pp. 338–352). Cambridge: Cambridge University Press.

Toerien, M. J., & Hughes, A. R. (1955). The limb bones of Springbok Flats Man. *South African Journal of Science, 52*, 125–128.

Torroni, A., Achilli, A., Macaulay, V., Richards, M., & Bandelt, H. J. (2006). Harvesting the fruit of the human mtDNA tree. *Trends in Genetics, 22*, 339–345.

Trauth, M. H., Deino, A. L., Bergner, A. G. N., & Strecker, M. R. (2003). East African climate change and orbital forcing during the last 175 kyr BP. *Earth and Planetary Science Letters, 206*, 297–313.

Trauth, M. H., Maslin, M. A., Deino, A., & Strecker, M. R. (2005). Late Cenozoic moisture history of East Africa. *Science, 309*, 2051–2053.

Trauth, M. H., Maslin, M. A., Deino, A., Bergner, A. G. N., Dühnforth, M., & Strecker, M. R. (2007). High- and low-latitude forcing of Plio-Pleistocene East African climate and human evolution. *Journal of Human Evolution, 53*, 475–486.

Tribolo, C., Mercier, N., Selo, M., Joron, J. L., Reyss, J. L., Henshilwood, C., et al. (2006). TL dating of burnt lithics from Blombos Cave (South Africa): Further evidence for the antiquity of modern human behaviour. *Archaeometry, 48*, 341–357.

Tribolo, C., Mercier, N., Valladas, H., Joron, J. L., Guibert, P., Lefrais, Y., et al. (2009). Thermoluminescence dating of a Stillbay-Howiesons Poort sequence at Diepkloof Rock Shelter (Western Cape, South Africa). *Journal of Archaeological Science, 36*, 730–739.

Tribolo, C., Mercier, N., Douville, E., Joron, J. L., Reyss, J. L., Rufer, D., et al. (2013). OSL and TL dating of the Middle Stone Age sequence at Diepkloof Rock Shelter (South Africa): A clarification. *Journal of Archaeological Science, 40*, 3401–3411.

Trinkaus, E. (1993). A note on the KNM-ER 999 hominid femur. *Journal of Human Evolution, 24*, 493–504.

Trinkaus, E. (2005). Early modern humans. *Annual Review of Anthropology, 34*, 207–230.

Trinkaus, E., Moldovan, O., Milota, Ş., Bîlgăr, A., Sarcina, L., Athreya, S., et al. (2003). An early modern human from the Peştera cu Oase, Romania. *Proceedings of the National Academy of Sciences of the United States of America, 100*, 11231–11236.

Tryon, C. A., Faith, J. T., Peppe, D. J., Fox, D. L., McNulty, K. P., Jenkins, K., et al. (2010). The Pleistocene archaeology and environments of the Wasiriya Beds, Rusinga Island, Kenya. *Journal of Human Evolution, 59*, 657–671.

Tryon, C. A., Peppe, D. J., Faith, J. T., Van Plantinga, A., Nightengale, S., Ogondo, J., et al. (2012). Late Pleistocene artefacts and fauna from Rusinga and Mfangano islands, Lake Victoria, Kenya. *Azania: Archaeological Research in Africa, 47*, 14–38.

Tryon, C. A., Crevecoeur, I., Faith, J. T., Ekshtain, R., Nivens, J., Patterson, D., et al. (2015). Late Pleistocene age and archaeological context for the hominin calvaria from GvJm-22 (Lukenya Hill, Kenya). *Proceedings of the National Academy of Sciences of the United States of America, 112*, 2682–2687.

Turner, C. G., & Markowitz, M. A. (1990). Dental discontinuity between Late Pleistocene and recent Nubians: Peopling of the Eurafrican-south Asian triangle, I. *Homo, 41*, 32–41.

Twiesselmann, F. (1958). *Les ossements humains du gîte mésolithique d'Ishango. Exploration du Parc national Albert, mission J. de Heinzelin de Braucourt, 1950* (Institut des Parcs nationaux du Congo belge, Fasc. 5). Brussels: Institut des Parcs nationaux du Congo belge.

Tzedakis, C. (2003). Timing and duration of Last Interglacial conditions in Europe: A chronicle of a changing chronology. *Quaternary Science Reviews, 22*, 763–768.

Tzedakis, P. C., Andrieu, V., de Beaulieu, J.-L., Crowhurst, S., Follieri, M., Hooghiemstra, H., et al. (1997). Comparison of terrestrial and marine records of changing climate of the last 500,000 years. *Earth and Planetary Science Letters, 150*, 171–176.

Underhill, P. A., Shen, P., Lin, A. A., Jin, L., Passarino, G., Yang, W. H., et al. (2000). Y chromosome sequence variation and the history of human populations. *Nature Genetics, 26*, 358–361.

Underhill, P. A., Passarino, G., Lin, A. A., Shen, P., Lahr, M. M., Foley, R. A., et al. (2001). The phylogeography of Y chromosome binary haplotypes and the origins of modern human populations. *Annals of Human Genetics, 65*, 43–62.

Urey, H. C. (1947). The thermodynamic properties of isotopic substances. *Journal of the Chemical Society, 1947*, 562.

Vallois, H. V. (1945). L'homme fossile de Rabat. *Comptes Rendus à l'Académie des Sciences, 221*, 669–671.

Vallois, H. V. (1951). La mandibule humaine fossile de la grotte du Porc-epic près Dire Daoua (Abyssinie). *L'Anthropologie, 55*, 231–238.

Vallois, H. V. (1969). Les hommes de Cro-Magnon et les Guanches: Les faits acquis et les hypothéses. Simposio del Cro-Magnon. *Anuario Estudias Atlanticos, Madrid, 15*, 97–119.

Vallios, H. V., & Roche, J. (1958). La mandibule Acheuléenne de Témara. *Comptes Rendus des Séances de l'Académie des Sciences (Paris), 246*, 3113–3116.

van Andel, T. H., & Tzedakis, P. C. (1996). Paleolithic landscapes of Europe and environs 150,000–25,000 years ago: An overview. *Quaternary Science Reviews, 15*, 481–500.

Van Couvering, J. (1972). Geology of Rusinga Island and correlation of the Kenya mid-Tertiary fauna. Ph.D. Dissertation, University of Cambridge.

Van Peer, P., Vermeersch, P. M., & Paulissen, E. (2010). *Chert quarrying, lithic technology, and a modern human burial at the paleolithic site of Taramsa 1, upper Egypt*. Leuven: Leuven University Press.

van Riet Lowe, C. (1929). Notes on some stone implements from Tuinplaats, Springbok Flats. *South African Journal of Science, 26*, 623–630.

van Riet Lowe, C. (1954). An artefact recovered with the Boskop calvaria. *South African Archaeological Bulletin, 9*, 135–137.

van Vark, G. N. (1984). On the determination of hominid affinities. In G. N. van Vark & W. W. Howells (Eds.), *Multivariate statistical methods in physical anthropology* (pp. 323–349). Doordrecht: D. Reidel.

van Vark, G. N., Bilsborough, A., & Dijkema, J. (1989). A further study of the morphological affinities of the Border Cave 1 cranium, with special reference to the origin of modern man. *Anthropologie et Préhistoire, 100*, 43–56.

Veeramah, K. R., Wegmann, D., Woerner, A., Mendez, F. L., Watkins, J. C., Destro-Bisol, G., et al. (2012). An early divergence of KhoeSan ancestors from those of other modern humans is supported by an ABC-based analysis of autosomal resequencing data. *Molecular Biology and Evolution, 29*, 617–630.

Vermeersch, P. M. (2002). Two Upper Paleolithic burials at Nazlet Khater. In P. M. Vermeersch (Ed.), *Paleolithic quarrying sites in upper and middle Egypt* (pp. 273–282). Leuven: Leuven University Press.

Vermeersch, P. M., Gijselings, G., & Paulissen, E. (1984). Discovery of the Nazlet Khater Man, upper Egypt. *Journal of Human Evolution, 13*, 281–286.

Vermeersch, P. M., Paulissen, E., Stokes, S., de Van Peer, P., Bie, M., Steenhoudt, F., & Missotten, S. (1997). Middle Paleolithic chert mining in Eqypt. In A. Ramos-Millán & M. A. Bustillo (Eds.), *Siliceous rocks and culture* (pp. 173–193). Granada, Spain: University of Granada.

Vermeersch, P. M., Paulissen, E., Stokes, S., Charlier, C., Van Peer, P., Stringer, C., et al. (1998). A Middle Paleolithic burial of a modern human at Taramsa Hill, Egypt. *Antiquity, 72*, 475–484.

Verna, C., Texier, P. J., Rigaud, J. P., Poggenpoel, C., & Parkington, J. (2013). The Middle Stone Age human remains from Diepkloof Rock Shelter (Western Cape, South Africa). *Journal of Archaeological Science, 40*, 3532–3541.

Vignard, E. (1928). Une nouvelle industrie lithique, le Sébilien. *Bulletin de la Société Préhistorique Française, 25*, 200–220.

Villa, P., Soriano, S., Teyssandier, N., & Wurz, S. (2010). The Howiesons Poort and MSA III at Klasies River Main Site, Cave 1A. *Journal of Archaeological Science, 37*, 630–655.

Villa, P., Soriano, S., Tsanova, T., Degano, I., Higham, T. F. G., d'Errico, F., et al. (2012). Border Cave and the beginning of the Later Stone Age in South Africa. *Proceedings of the National Academy of Sciences of the United States of America, 109*, 13208–13213.

Voelker, A. H. L., & Participants, Workshop. (2002). Global distribution of centennial-scale records for Marine Isotope Stage (MIS) 3: A database. *Quaternary Science Reviews, 21*, 1185–1212.

Vogel, J. C. (2001). Radiometric dates for the Middle Stone Age in South Africa. In P. V. Tobias, M. A. Raath, J. Moggi-Cecchi, & G. A. Doyle (Eds.), *Humanity from African naissance to coming millennia-colloquia in human biology and paleoanthropology* (pp. 261–268). Firenze: Firenza University Press.

Vogel, J. C., & Beaumont, P. (1972). Revised radiocarbon chronology for the Stone Age in South Africa. *Nature, 237*, 50–51.

Vogel, J. C., & Waterbolk, H. T. (1963). Groningen radiocarbon dates IV. *Radiocarbon, 5*, 163–202.

Vogel, J. C., Fuls, A., & Visser, E. (1986). Pretoria radiocarbon dates III. *Radiocarbon, 28*, 1133–1172.

Volman, T. P. (1984). Early prehistory of southern Africa. In R. G. Klein (Ed.), *Southern African prehistory and paleoenvironments* (pp. 169–220). Rotterdam: Balkema.

Wadley, L. (1993). The Pleistocene Later Stone Age south of the Limpopo River. *Journal of World Prehistory, 7*, 243–296.

Wadley, L. (2007). Announcing a Still Bay industry at Sibudu Cave, South Africa. *Journal of Human Evolution, 52*, 681–689.

Wadley, L., & Jacobs, Z. (2004). Sibudu Cave, KwaZulu-Natal: Background to the excavations of Middle Stone Age and Iron Age occupations. *South African Journal of Science, 100*, 145–151.

Wadley, L., & Jacobs, Z. (2006). Sibudu cave: Background to the excavations, stratigraphy and dating. *Southern African Humanities, 18*, 1–26.

Wadley, L., Hodgskiss, T., & Grant, M. (2009). Implications for complex cognition from the hafting of tools with compound adhesives in the Middle Stone Age, South Africa. *Proceedings of the National Academy of Sciences of the United States of America, 106*, 9590–9594.

Wadley, L., Sievers, C., Bamford, M., Goldberg, P., Berna, F., & Miller, C. (2011). Middle Stone Age bedding construction and settlement patterns at Sibudu, South Africa. *Science, 334*, 1388–1391.

Wall, J. D., Lohmueller, K. E., & Plagnol, V. (2009). Detecting ancient admixture and estimating demographic parameters in multiple human populations. *Molecular Biology and Evolution, 26*, 1823–1827.

Wall, J. D., Yang, M. A., Jay, F., Kim, S. K., Durand, E. Y., Stevinson, L. S., et al. (2013). Higher levels of Neanderthal ancestry in East Asians than in Europeans. *Genetics, 194*, 199–209.

Wang, Q., Tobias, P., Roberts, D., & Jacobs, Z. (2008). A re-examination of a human femur found at the Blind River site, East London, South Africa: Its age, morphology, and breakage pattern. *Anthropological Review, 71*, 43–61.

Watts, I. (2010). The pigments from Pinnacle Point Cave 13B, Western Cape, South Africa. *Journal of Human Evolution, 59*, 392–411.

Watts, W. A., Allen, J. R. M., & Huntley, B. (2000). Paleoecology of three interstadial events during oxygen-isotope Stages 3 and 4: A lacustrine record from Lago Grande di Monticchio, southern Italy. *Paleogeography, Paleoclimatology, Paleoecology, 155*, 83–93.

Weaver, T. D. (2012). Did a discrete event 200,000-100,000 years ago produce modern humans? *Journal of Human Evolution, 63*, 121–126.

Weldeab, S., Schneider, R. R., Kölling, M., & Wefer, G. (2005). Holocene African droughts relate to eastern equatorial Atlantic cooling. *Geology, 33*, 981–984.

Wells, L. H. (1935). A fossilized human femur from East London, Cape Province. *South African Journal of Science, 32*, 596–600.

Wells, L. H. (1951). The fossil human skull from Singa. *Fossil mammals of Africa* (pp. 29–42). London: British Museum of Natural History.

Wells, L H. (1952). Human crania of the Middle Stone Age in South Africa. *Proceedings of the first pan-African congress on prehistory, Nairobi, 1947* (pp. 125–133). Oxford: Blackwell.

Wells, L. H. (1959). The problem of Middle Stone Age Man in southern Africa. *Man, 244*, 158–160.

Wendorf, F. (1968). Site 117: A Nubian final Paleolithic graveyard near Jebel Sahaba, Sudan. In F. Wendorf (Ed.), *The Prehistory of Nubia* (Vol. 2, pp. 954–995). Dallas: Southern Methodist University Press.

Wendorf, F., & Schild, R. (1976). *Prehistory of the Nile Valley*. New York: Academic.

Wendorf, F., & Schild, R. (2004). Late Paleolithic warfare in Nubia: The evidence and causes. *Adumatu (Journal of the Arab World), 10*, 7–28.

Wendorf, F., Said, S., & Schild, R. (1970). Egyptian prehistory: Some new concepts. *Science, 169*, 1161–1171.

Wendorf, F., Schild, R., & Haas, H. (1979). A new radiocarbon chronology for prehistoric sites in Nubia. *Journal of Field Archaeology, 6*, 219–223.

White, T. D., Asfaw, B., DeGusta, D., Gilbert, H., Richards, G. D., Suwa, G., et al. (2003). Pleistocene *Homo sapiens* from Middle Awash, Ethiopia. *Nature, 423*, 742–747.

Williams, D. F., Thunell, R. C., Tappa, E., Rio, D., & Raffi, I. (1988). Chronology of the Pleistocene oxygen isotope record: 0–1.88 m.y. B.P. *Paleogeography, Paleoclimatology, Paleoecology, 64*, 221–240.

Willoughby, P. R. (2012). The Middle and Later Stone Age in the Iringa Region of southern Tanzania. *Quaternary International, 270*, 103–118.

Wolpoff, M. H., Frayer, D. W., & Jelínek, J. (2006). Aurignacian female crania and teeth from the Mladeč caves, Moravia, Czech Republic. In M. Teschler-Nicola (Ed.), *Early modern humans at the Moravian Gate: The Mladeč caves and their remains* (pp. 273–340). Vienna: Springer.

Woodward, A. S. (1938). A fossil skull of an ancestral Bushman from the Anglo-Egyptian Sudan. *Antiquity, 12*, 193–195.

Wright, J. D. (2000). Global climate change in marine stable isotope records. In J. S. Noller, J. M. Sowers, & W. R. Lettis (Eds.), *Quaternary geology methods and applications* (pp. 427–433). Washington, DC: American Geophysical Union.

Wrinn, P. J., & Rink, W. J. (2003). ESR dating of tooth enamel from Aterian levels at Mugharet el'Aliya (Tangier, Morocco). *Journal of Archaeological Science, 30*, 123–133.

Wurz, S. (1999). The Howiesons Poort backed artefacts from Klasies River: An argument for symbolic behavior. *South African Archaeology Bulletin, 54*, 38–50.

Wurz, S. (2002). Variability in the Middle Stone Age lithic sequence, 115,000–60,000 years ago at Klasies River, South Africa. *Journal of Archaeological Science, 29*, 1001–1015.

Wurz, S. (2008). Modern behavior at Klasies River. *South African Archaeological Society Goodwin Series, 10*, 150–156.

Wurz, S. (2012). The significance of MIS 5 shell middens on the Cape coast: A lithic perspective from Klasies River and Ysterfontein 1. *Quaternary International, 270*, 61–69.

Wymer, J. (1982). *The paleolithic age*. New York: St. Martin's Press.

Wynn, J. G. (2006). Influence of Plio-Pleistocene aridification on human evolution: Evidence from paleosols of the Turkana Basin, Kenya. *American Journal of Physical Anthropology, 123*, 106–118.

Yellen, J., Brooks, A., Helgren, D., Tappen, M., Ambrose, S., Bonnefille, R., et al. (2005). The archaeology of Aduma Middle Stone Age sites in the Awash Valley, Ethiopia. *PaleoAnthropology, 3*, 25–100.

Yokoyama, Y., Falguères, C., Sémah, F., Jacob, T., & Grün, R. (2008). Gamma-ray spectrometric dating of late *Homo erectus* skulls from Ngandong and Sambungmacan, central Java, Indonesia. *Journal of Human Evolution, 55*, 274–277.

Zhao, J. X., Hu, K., Collerson, K. D., & Xu, H. K. (2001). Thermal ionization mass spectrometry U-series dating of a hominid site near Nanjing, China. *Geology, 29*, 27–30.

Zhao, M., Dupont, L., Eglington, G., & Teece, M. (2003). *n*-Alkane and pollen reconstruction of terrestrial climate and vegetation for N. W. Africa over the last 160 kyr. *Organic Geochemistry, 34*, 131–143.

Ziegler, M., Simon, M. H., Hall, I. R., Barker, S., Stringer, S., & Zahn, R. (2013). Development of Middle Stone Age innovation linked to rapid climate change. *Nature Communications, 4*, 1905. doi:10.1038/ncomms2897.

Zielinski, G. A., Mayewski, P. A., Meeker, L. D., Whitlow, S., Twickler, M. S., & Taylor, K. (1996). Potential atmospheric impact of the Toba mega-eruption 71,000 years ago. *Geophysical Research Letters, 23*, 837–840.

[illegible] (2008). [illegible] variability in the Middle Stone Age lithic sequence [illegible] from [illegible] Klasies River, South Africa. *Journal of Archaeological Science, 35*, [illegible].

[illegible] Klasies River [illegible] *Archaeological Society* [illegible] *Series*, [illegible].

Wurz, S. (2012). The significance of MIS 5 shell middens on the Cape coast: A lithic perspective from Klasies River and Ysterfontein 1. *Quaternary International, 270*, 61–69.

Wynn, J. [illegible] New [illegible]

Wynn, J. G. (2004). Influence of Plio-Pleistocene aridification on human evolution: Evidence from paleosols of the Turkana Basin, Kenya. *American Journal of Physical Anthropology, 123*, 106–118.

Yellen, J., Brooks, A., Helgren, D., Tappen, M., Ambrose, S., Bonnefille, R., et al. (2005). The archaeology of Aduma Middle Stone Age sites in the Awash Valley, Ethiopia. *PaleoAnthropology*, [illegible]–100.

Yokoyama, Y., Falguères, C., Sémah, F., Jacob, T., & Grün, R. (2008). Gamma-ray spectrometric dating of late *Homo erectus* skulls from Ngandong and Sambungmacan, central Java, Indonesia. *Journal of Human Evolution, 55*, 274–277.

Zhao, J. X., Hu, K., Collerson, K. D., & Xu, H. K. (2001). Thermal ionization mass spectrometry U-series dating of a hominid site near Nanjing, China. *Geology, 29*, 27–30.

Zhu, M., Dupont, L., Eglinton, G., & Teece, M. (2003). [illegible] and pollen reconstruction of terrestrial climate and vegetation for N. W. Africa over the last 160 kyr. *Organic Geochemistry, 34*, [illegible]–149.

Ziegler, M., Simon, M. H., Hall, I. R., Barker, S., Stringer, C., & Zahn, R. (2013). [illegible] Middle Stone Age [illegible] *Nature Communications, 4*, 1905. [illegible]

Zollikofer, C. P. E., [illegible] (1998). [illegible] 71 000 year [illegible] *Journal of [illegible]* [illegible]–[illegible].

[illegible] & Schild, R. (2000). Late Paleolithic warfare in Nubia: The evidence and causes. *Adumatu* [illegible] [illegible]–28.

Wendorf, F., Said, R., & Schild, R. (1970). Egyptian prehistory: Some new concepts. *Science, 169*, 1161–1171.

Wendorf, F., Schild, R., & Haas, H. (1979). A new radiocarbon chronology for prehistoric sites in Nubia. *Journal of Field Archaeology, 6*, 219–223.

White, T. D., Asfaw, B., DeGusta, D., Gilbert, H., Richards, G. D., Suwa, G., et al. (2003). Pleistocene *Homo sapiens* from Middle Awash, Ethiopia. *Nature, 423*, 742–747.

Williams, D. F., Thunell, R. C., Tappa, E., Rio, D., & Raffi, I. (1988). Chronology of the Pleistocene oxygen isotope record: 0–1.88 m.y. B.P. *Palaeogeography, Palaeoclimatology, Palaeoecology, 64*, 221–240.

Willoughby, P. R. (2012). The Middle and Later Stone Age in the Iringa Region of southern Tanzania. *Quaternary International, 270*, 103–118.

Wolpoff, M. H., Frayer, D. W., & Jelinek, J. (2006). Aurignacian female crania and teeth from the Mladeč caves, Moravia, Czech Republic. In M. Teschler-Nicola (Ed.), *Early modern humans at the Moravian Gate: The Mladeč caves and their remains* (pp. 273–340). Vienna: Springer.

Woodward, A. S. (1938). A fossil skull of an ancestral Bushman from the Anglo-Egyptian Sudan. *Antiquity, 12*, 193–195.

Wright, J. L. (2005). [illegible] climatic change [illegible] networks. In [illegible] *Sciences* [illegible] [illegible] [illegible]

[illegible] [illegible] *Science* [illegible]

W[illegible] [illegible] River [illegible] [illegible]

Chapter 18
A Genetic Perspective on African Prehistory

Pedro Soares, Teresa Rito, Luísa Pereira, and Martin B. Richards

Abstract The various genetic systems (mitochondrial DNA, the Y-chromosome and the genome-wide autosomes) indicate that Africa is the most genetically diverse continent in the world and the most likely place of origin for anatomically modern humans. However, where in Africa modern humans arose and how the current genetic makeup within the continent was shaped is still open to debate. Here, we summarize the debate and focus especially on the maternally inherited mitochondrial DNA (mtDNA) and a recently revised chronology for the African mtDNA tree. We discuss the possible origin of modern humans in southern, eastern or Central Africa; the possibility of a migration from southern to eastern Africa more than 100 ka, carrying lineages within mtDNA haplogroup L0; the evidence for a climate-change-mediated population expansion in eastern Africa involving mtDNA haplogroup L3, leading to the "out-of-Africa" migration around 70–60 ka; the re-population of North Africa from the Near East around 40–30 ka suggested by mtDNA haplogroups U6 and M1; the evidence for population expansions and dispersals across the continent at the onset of the Holocene; and the impact of the Bantu dispersals in Central, eastern and southern Africa within the last few millennia.

Keywords Climate change • Migration • Molecular clock • mtDNA • Phylogeography • Population expansions

Introduction

There is a consensus across the fields of genetics, archaeology and paleoanthropology that Africa is the cradle of *Homo sapiens*. Genetically, across the genome-wide autosomal variation and the uniparental markers, the mitochondrial DNA (mtDNA) and Y-chromosome, Africa is the continent with by far the highest genetic diversity (Torroni et al. 2006; Behar et al. 2008; Li et al. 2008; Cruciani et al. 2011). However, with a few exceptions, this is where the consensus ends. Whole ranges of crucial questions remain highly controversial. These include: where in Africa modern humans first appeared; when each part of Africa was first colonized by modern humans; the impact of climate change on human populations; and how cultural and technological innovations helped shape the current genetic diversity in the African continent.

Genetics can contribute valuable insights into the origins and migrations of human populations. The rationale for this lies in the fact that mutations and recombination, the events promoting changes in the genome down the generations, are random phenomena that leave marker buoys scattered throughout the genome, each of which arose at a particular time and place. They can therefore serve as an inference tool to bracket the place and timing of evolutionary events.

The relatively small, maternally inherited mtDNA component (around 16,570 base pairs) has been heavily screened

P. Soares · T. Rito · L. Pereira
IPATIMUP (Instituto de Patologia e Imunologia Molecular da Universidade do Porto), Rua Dr. Roberto Frias s/n, 4200-465 Porto, Portugal

P. Soares
CBMA (Centre of Molecular and Environmental Biology), Department of Biology, University of Minho, Campus de Gualtar, 4710-057 Braga, Portugal

T. Rito
Life and Health Sciences Research Institute (ICVS), School of Health Sciences, University of Minho, Campus de Gualtar, 4710-057 Braga, Portugal

and

ICVS/3B's - PT Government Associate Laboratory, Braga/Guimarães, Portugal

L. Pereira
Faculdade de Medicina da Universidade do Porto, Porto, Portugal

M.B. Richards (✉)
Department of Biological Sciences, School of Applied Sciences, University of Huddersfield, Queensgate, Huddersfield, HD1 3DH, UK
e-mail: m.b.richards@hud.ac.uk

Sacha C. Jones and Brian A. Stewart (eds.), *Africa from MIS 6-2: Population Dynamics and Paleoenvironments*, Vertebrate Paleobiology and Paleoanthropology, DOI 10.1007/978-94-017-7520-5_18

worldwide, initially for short segments of the fast-evolving, non-coding control-region (>150,000 samples), and more recently for the whole-mtDNA genome. By 2013, more than 15,000 whole-mtDNA genome sequences have become publicly available, including more than 2000 from African individuals. Whole-mtDNA genomes, in particular, can resolve the details of the maternal genealogy in exquisite detail. An even greater level of genealogical resolution awaits us for the male line of descent, inscribed in the non-recombining, male-specific part of the Y-chromosome, or "MSY", as human genome sequencing proceeds apace. Although some analyses have suggested drastic reductions on the number of MSY lineages in the recent past due to the reproductive success of relatively few individuals, it may also be the case that the MSY often tracks important dispersals, leading to the spread of language families, which are much less evident from the mtDNA variation (Forster and Renfrew 2011). In the end, models erected using either the mtDNA or the MSY have to be tested against variation in the rest of the genome. The analysis of genome-wide autosomal markers, and increasingly complete human genomes, provides a more complete window onto the past that does not focus on one or other single line of descent.

Nevertheless, mtDNA led the way for archaeogenetics and remains an extremely valuable marker system. The mtDNA, like the MSY, is non-recombining, leading to the transmission down the generations of the genetic material in a block – essentially, a single DNA sequence. As it is passed down the generations mutations accumulate, leading to the formation of related clusters called haplogroups, each of which can trace its descent back to a single common ancestor. A haplogroup is effectively a named example of what evolutionary biologists call a clade – a group comprising an ancestor and all its descendants. Any ancestral node in the genealogical tree can, moreover, be dated using the "molecular clock" – a measure of how rapidly mutations accumulate over time – an approach that remains controversial, but which has seen considerable progress recently in all three systems, with the use of whole-mtDNA genomes (Soares et al. 2009), the autosomes with the development of large-scale complete human genome sequencing (Scally and Durbin 2012) and the MSY as much larger genomic tracts of the Y-chromosome start to be used (Wei et al. 2012; Francalacci et al. 2013). Thus, if a migration takes place from one region to another, new mutations unique to that region will start to accumulate there, and the age of the presence of that cluster in that region can be estimated by dating the node from which they arise. This tracking of genetic lineages (or lines of descent) in time and space by analyzing their geographic distribution and time depth is referred to as "phylogeography", and the dating of dispersals in this way in particular is called "founder analysis". The resolving power of the mtDNA genealogical tree makes mtDNA an extremely powerful tool with which to evaluate population structure and follow migrations across space and time (Torroni et al. 2006; Macaulay and Richards 2013).

Phylogeographic approaches have been criticized on a range of fronts, and have often been compared unfavourably to both more formal and supposedly robust procedures such as those based on simulation modelling or summary statistics (Nielsen and Beaumont 2009). However, hypothesis testing procedures based on evolutionary and population genetics theory have their own weaknesses; in particular, they suffer from the well-known gulf between the rejection of a null hypothesis and the inference of specific demographic scenarios (Bandelt et al. 2002). Although critics often dismiss phylogeographic analyses as "interpretative" and "story-telling", it is difficult to maintain a hard-and-fast distinction of this kind between the phylogeographic approach to population genetics and more formal procedures. In practice, all reconstructions based on inferences from the modern distribution of genetic variation are fraught with difficulties. Moreover, we are increasingly learning that these issues do not go away when ancient DNA comes into play – although this is not for the present an issue, so far as Africa is concerned.

On the other hand, the critics sometimes fail to acknowledge both the extraordinary richness of the genetic evidence – whether it be the extraordinarily fine resolving power of mtDNA or MSY phylogenies, or the incredible autosomal profiling that is now possible – and the successes that phylogeographic reconstructions have achieved in cases where the demographic history is broadly known from other lines of evidence (such as the settlement of the Americas, the Remote Pacific, and indeed southern Africa, as we discuss below). Rather than a hardline set of protocols in which demographic history is read from scratch from DNA sequence data using statistical tools of dubious reliability, we prefer an exploratory and interdisciplinary approach in which hypotheses are evaluated within the framework of models supplied by archaeology, paleoanthropology, paleoclimatology and so on.

In this chapter, we focus primarily on the patterns in the mtDNA variation and review the main respects in which the phylogeographic analysis of this particular molecule can provide information about the history of the continent, in the context of some of the autosomal and MSY work. For a broader view of the mtDNA variation, we can recommend the recent review by Rosa and Brehm (2011).

The Maternal Genealogy of Africa

The African mtDNA tree is effectively the human mtDNA tree, since the deepest two-thirds of the lineages are restricted to Africa, and the non-African lineages are only a tiny fragment of African diversity. Any phylogenetic tree comprises nodes separated by branches, in a nested array of clades and subclades – clusters of lineages that include all descendants of a given common ancestor (a subclade is simply a clade within a clade). The mtDNA tree is divided into two primary or basal clades, referred to as haplogroups L0 and L1′6, or L1′2′3′4′5′6 (Fig. 18.1) (Torroni et al. 2006; Behar et al. 2008). For historical reasons, the deep African lineages within the mtDNA tree are all prefixed with "L" – other haplogroups around the world having already claimed most other available letters (such as A through D in Native Americans, H through K and T through W in Europeans). Haplogroup labelling follows a scheme in which nested subclades within a haplogroup graced with a capital letter are then followed by a number (such as L0), and further nested subclades are given alternate letters and numbers (e.g. L0a, L0a1, L0a1b etc.) (Richards et al. 1998). Clades that include several named clades can be described by concatenating the subclades (e.g. L0a′b). Multiple concatenated clades can be abbreviated to just the first and the last, so that L1′2′3′4′5′6, for example, becomes just L1′6.

Until the emergence of haplogroup L3, roughly between 70 and 60 ka (Watson et al. 1997; Soares et al. 2012), the human mtDNA tree was bifurcating (separating into only two daughters) at every node; and nodes are usually separated from one another in time by many thousands of years. This pattern reflects small population sizes and a correspondingly high degree of genetic drift prior to this time. This in turn implies a high rate of extinction of lineages, and a corresponding loss of evidence as one extends inferences back into this period. Even so, some general conclusions can be drawn from the structure of the tree.

Phylogenetically, one-half of the human mtDNA tree – that is, L0 – seems to have a southern African distribution, and probably also origin (Figs. 18.1, 18.2a). L0 is divided between L0d (with a southern African distribution) and L0a′b′f′k where L0k is southern African and L0a′b′f has an eastern African distribution.

The other half of the human mtDNA tree has a much more complex genealogy and distribution, with subclades distributed throughout Central, eastern and West Africa, as well as (more recently) North Africa and the rest of the world. This L1′6 clade is also much more frequent overall than L0 throughout Africa – even in most of southern Africa, where L0 is found at its highest frequencies. L1′6 then splits into L1, mainly found in Central Africa (Fig. 18.2b), and/or L2′6. The latter then splits once again into L2′6 and L5. Haplogroup L5 is very rare and only found in eastern Africa. L2′6, then further divides into L2 and L3′4′6, with L2 most likely originating in Central or West Africa (Fig. 18.2c) and L3′4′6 in eastern Africa (Fig. 18.2d for L3). Haplogroup L3 also includes two major subclades that are most likely of Central or West African origin, L3b and L3d. These geographic splits in the tree represent the most ancient potential dispersals that we can detect in the mitochondrial record. L3 also includes two subclades, haplogroups M and N, which include all of the non-African (Eurasian and Australasian) diversity, excluding additional out-of-Africa dispersals in the last few millennia.

Modern Human Origins

Exactly where the so-called "anatomically modern humans" (AMH) first appeared in Africa remains a thorny problem. Perhaps the question should not even be framed in such terms; AMH may not descend from a single group of people that lived in a specific geographic location at a particular time, but may have arisen from various groups that interacted or coalesced over time (Schlebusch et al. 2012) – a kind of "multi-regionalism in one continent".

Paleoanthropology has commonly placed modern human origins in eastern Africa. The Omo 1 cranium, from the Kibish River in Ethiopia and dating to ~195 ka, is the oldest known fossil to display modern human features (McDougall et al. 2005). The remains from Herto, also in Ethiopia, date to ~160 ka (Clark et al. 2003), strengthening the case for an eastern African origin. However, the Jebel Irhoud specimens from Morocco also date to ~160 ka (Smith et al. 2007), although with some dispute over whether or not they are anatomically fully modern (Stringer 2011), and the Skhul/Qafzeh fossils from Israel date to roughly 90–135 ka (Millard 2008). In the south of the continent, the oldest known fossil is from the Klasies River Caves, with two poorly constrained pulses dating to >100 ka and 65–105 ka respectively (Deacon 1995; Millard 2008), although its status as fully modern is also contested (Rightmire et al. 2006). More archaic remains, dating to between 190 and 330 ka are found at Florisbad (Millard 2008). There is also some very early evidence in southern Africa for key elements of modern human behaviour, ~160 ka at Pinnacle Point (Marean et al. 2007). Clearly, the fossil record is extremely fragmentary, and any conclusions drawn from it regarding an eastern African origin for AMH are on shaky grounds for the time being.

Unfortunately, however, the genetic evidence does not really clarify the picture. Two genome-wide studies, one based mainly on the evidence from fast-evolving STRs (short tandem repeats) and "indels" (insertions and deletions) (Tishkoff et al. 2009), and one employing nearly 600,000 SNPs (single-nucleotide polymorphisms, or variants) (Henn et al. 2011), both pointed to a southern African origin – an early genetic legacy that left its mark particularly on the foraging and herding populations usually lumped together as "Khoisan" or "Khoe-San", belonging to the

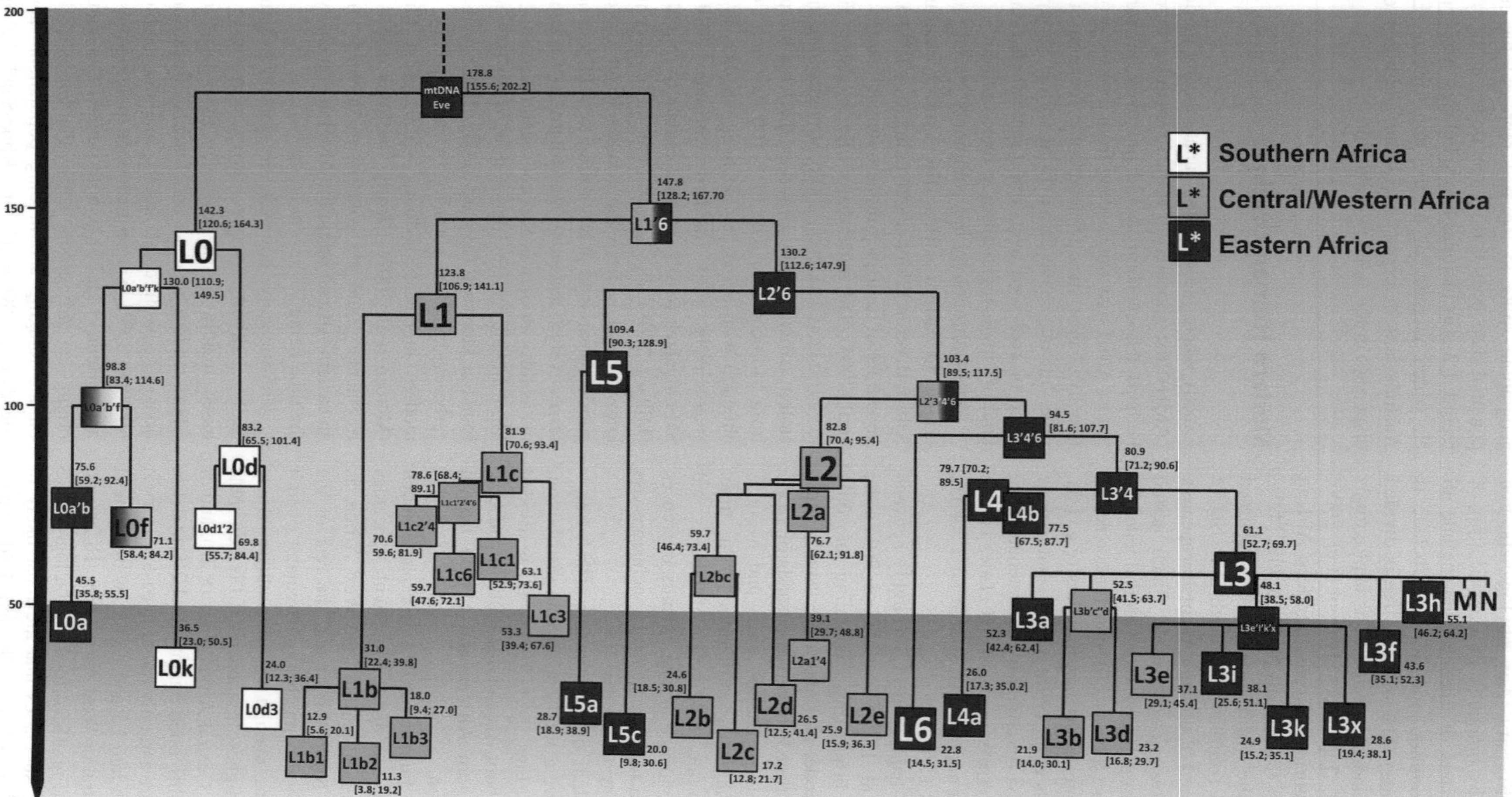

Fig. 18.1 Schematic tree of the African mtDNA diversity. Age estimates were calculated using maximum likelihood in PAML (Yang 1997) and the molecular clock corrected for purifying selection described by Soares et al. (2009)

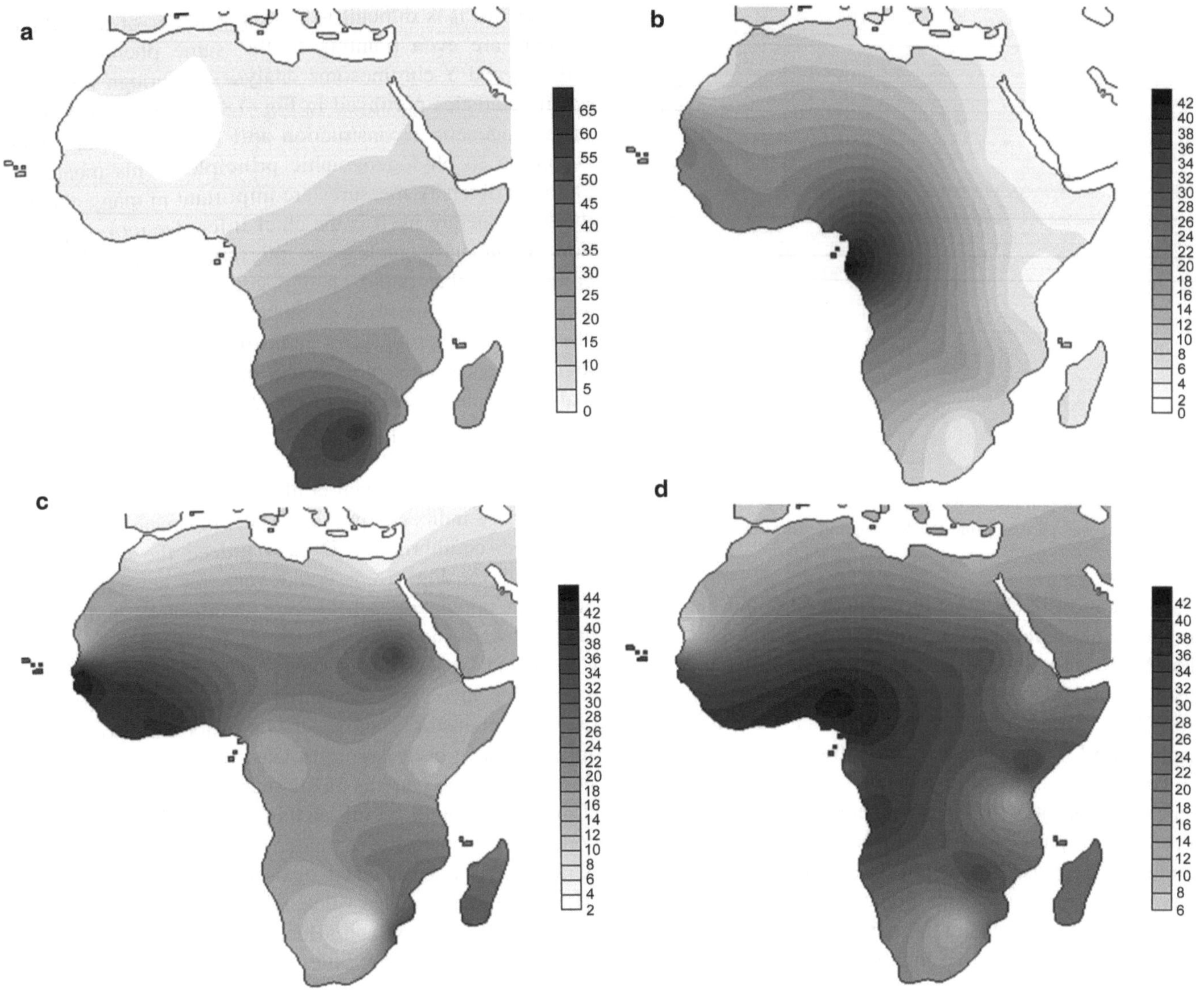

Fig. 18.2 Frequency distribution of haplogroups L0 **a**, L1 **b**, L2 **c** and L3 **d** plotted using the Kriging algorithm in Surfer software

otherwise linguistically unrelated click-consonant Khoe, Tuu and Ju families (Güldemann and Stoneking 2008).

We need to digress for a moment here. The comparative method does not confirm Greenberg's linguistic thesis of a common origin for the southern African click-consonant languages, for which he took up the biological term "Khoisan", in distinction to his three more widely accepted African linguistic phyla: Afro-Asiatic, Nilo-Saharan and Niger-Kordofanian (which includes the Bantu group). Imprecision in this regard has, unfortunately, permeated the genetic literature on Africa, generating many inaccuracies and confusions (Mitchell 2010). Strong genetic evidence has, however, accumulated for a deep common ancestry among many of these groups, in distinction to both other Africans and non-Africans. Therefore, it seems reasonable to refer to them collectively as "indigenous southern Africans" or "Khoe-San", suggested as the preferred collective name of the communities themselves (Schlebusch 2010). Nevertheless, we must emphasize that this is a distinction made on the basis of inferred geographic ancestry, akin to "West Eurasian", "East Eurasian" and "Sahulian" (Saitou 1996), and not on any linguistic basis.

So, autosomal studies have been interpreted as implying a southern source for modern humans. On the other hand, studies focused on the MSY have suggested that the root of the Y-chromosome tree may lie in Central/West Africa (Cruciani et al. 2011; Mendez et al. 2012). In terms of mtDNA, the situation is even more opaque. As noted above, the tree splits into two primary branches or clades, L0 and L1′6; the former most likely arose in southern Africa while the latter has a more northerly origin (by "northerly" we here mean simply to the north of southern Africa). It is difficult to

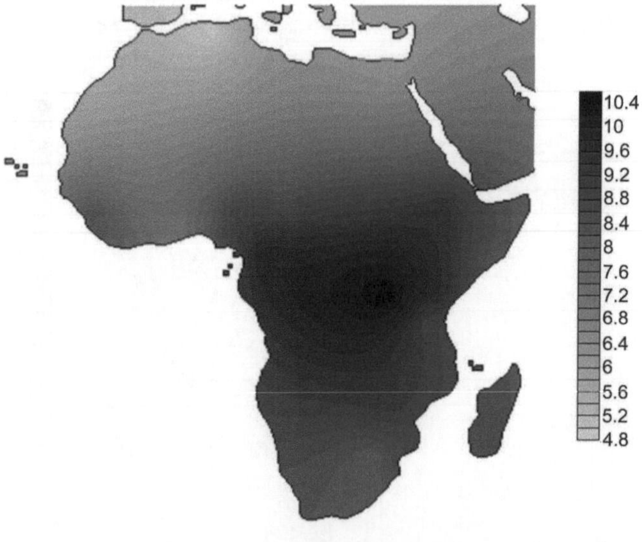

Fig. 18.3 Mean pairwise differences of the HVS-I of mtDNA across Africa. Data was plotted using the Kriging algorithm in Surfer

assess the most likely place of origin of L1′6, but the centre of gravity of its frequency distribution is Central Africa. The first bifurcation divides L1, with a Central/West African distribution (and most probably a Central African origin) and from L2′6, which includes the very rare eastern African haplogroup L5 and a West/eastern African clade (L2′3′4′6).

To try and get more of a handle on this issue from the perspective of mtDNA, we can calculate diversity measures for different parts of the continent. We map the mean number of pairwise differences on the mtDNA HVS-I (hypervariable segment I of the non-coding control region) of each available population in Africa in Fig. 18.3. These values are not likely to be highly informative, considering the time depth of the human mtDNA, and the tendency towards saturation (a levelling-off of values due to recurrent mutation) in the HVS-I beyond ~50 ka. Nevertheless, they show us which regions look especially diverse. The plotted diversity values (using the Kriging algorithm in the Surfer software) point to eastern Central Africa (roughly to the west of Lake Victoria) as the most diverse region, at the intersection of the origin of the three major clades, L0, L1 and L2′6. These three major clades have ages of ~120–140 ka when dated using maximum likelihood (Yang 1997) and the molecular clock for the whole-mtDNA genome corrected for purifying selection (Soares et al. 2009) (Fig. 18.1). This pattern might simply be generated by the meeting of very different groups of differing origins, but a Central African origin might at least appear broadly compatible with the Y-chromosome conclusions, where a recently discovered deeper root in Central/West Africa has been recently dated to ~340 ka (Batini et al. 2011a, b; Cruciani et al. 2011; Francalacci et al. 2013).

Even so, it is difficult to assess whether or not different markers are even pointing to the same phenomena. The mtDNA and Y-chromosome analyses (although not the diversity statistics portrayed in Fig. 18.3) are based primarily on phylogenetic reconstruction and conclusions are drawn according to phylogeographic principles. This means that although diversity measures are important in many contexts, the tree structure itself is the chief inference tool. A population might have undergone migrations, expansions and contractions that could change genetic diversity in many aspects, but the survival of an ancient branch not located elsewhere can provide an indication of ancestry in that region, even if other more recent clades prevail in the population. Such a case is observed in the Y-chromosome, where the deepest split is observed in Central/West Africa although the recently detected subclade that indicates this split is very rare (Cruciani et al. 2011).

Diversity indices, however, especially measures such as linkage disequilibrium (LD) and indeed the one that we employed for Fig. 18.3, could vary for several reasons, including population substructure, bottlenecks and admixture (Pritchard and Przeworski 2001). Populations in eastern, Central and West Africa probably went through major range expansions involving substantial admixture between them (as we demonstrate below), while Khoe-San populations appear to have been largely isolated until the Bantu expansion occurred, so patterns of LD such as those used by Henn et al. (2011) might be misleading. A second measure supporting their conclusion of modern human origin in southern Africa was F_{ST}, calculated against European populations (Henn et al. 2011). This value, which measures differentiation from Europeans, and which was correlated with distance, might also have little to do with modern human origins. Eastern Africa incurred a great deal of back-to-Africa gene flow from non-African populations since the out-of-Africa exodus, both in the Pleistocene and the Holocene, especially through the Arabian Peninsula (Musilova et al. 2011; Fernandes et al. 2012). Even Central and West Africa probably experienced some genetic input from non-African populations, in part due to contact with North Africa, as has also been shown for both mtDNA (Ottoni et al. 2010) and Y-chromosome variation (Cruciani et al. 2002). A more recent study focusing on southern African autosomal variation was unable to localize a geographic origin, instead pointing to a long history of "admixture and stratification" (Schlebusch et al. 2012).

A further issue is the time depth of the diversity patterns. There is little evidence for a speciation event in the emergence of *Homo sapiens*, either in the paleoanthropological record or in the genetics (Barham and Mitchell 2008). The domed cranial vault is a modern feature (Lieberman et al. 2002) that appears 150–200 ka, but it emerged gradually from more archaic forms in the previous few hundred thousand years. On the genetics side, despite much debate over the years, it seems that there is little evidence for a bottleneck at ~200 ka (Sjodin et al. 2012). There is no persuasive reason to consider that the mtDNA root dating at ~180 ka (Fig. 18.1) or the Y-chromosome root, possibly now somewhat older (Cruciani et al. 2011; Mendez et al. 2012), indicate the emergence of modern humans. What makes this speculation plausible is the proximity of these ages with the emergence of modern human features in the fossil record. However, the coalescent time of the autosomes is overall much higher than in the uniparental markers (due to a higher effective population size and therefore less drift). So, when measuring autosomal diversity, we are analyzing phenomena that most likely greatly predate modern human origins.

Although it is difficult to point to a place of origin for "mitochondrial Eve", at least by ~140 ka the extant distributions mean that it does seem likely that there were at least two groups of modern humans living in two different parts of the continent. Given the present distribution of L0d and L0k (Barbieri et al. 2013), it seems likely that L0 (dating to ~142 ka) had a southern origin while L1′6 (~148 ka) had a more northerly origin in Central or eastern Africa (Fig. 18.4a). Furthermore, the separation between these two groups is unlikely to have happened before ~180 ka, the coalescence time of the human mtDNA tree as a whole (and age of "mitochondrial Eve"), and is most likely more recent. The implication is that these two distinct groups were probably both already "anatomically modern", since there is hardly any scope for "levelling across" of anatomical features later on through admixture, given that the southern groups ancestral to modern Khoe-San populations seem to have been isolated throughout most of prehistory, at least on the maternal line of descent. The major exceptions to this isolation are two instances of gene flow, one at ~130–70 ka from southern Africa to eastern Africa, and the second only in the past 2000 years, when the Bantu expansion reached the south of the continent.

This mtDNA picture of the isolation of the two groups is also discernible in genome-wide data, for example in the neighbour-joining population tree of Tishkoff et al. (2009). Their tree separates southern African Khoe-San groups from a single group containing all of the remaining African populations with a more northerly distribution, including the tropical forest forager groups. Although, we should not read too much into a population tree, one point to note is that again Central Africans appear more basal in this tree than eastern Africans. Moreover, when they set the ADMIXTURE software (Alexander et al. 2009) to distinguish six populations, Henn et al. (2011) also obtained two distinct ancestral clusters in southern hunter-gatherers and more northerly hunter-gatherers, although the separation time of these clusters is unknown. Recent more detailed analyses have confirmed this picture (Pickrell et al. 2012; Schlebusch et al. 2012; Petersen et al. 2013). Finally, using autosomal re-sequencing data, Veeramah et al. (2012) detected an early separation between Khoe-San and other modern human populations that they dated to ~110 ka, although with large confidence intervals [52–187] ka. Other recent estimates have been similar (Schlebusch et al. 2012), although a figure of ~250–300 ka was suggested with a recent re-evaluation of the autosomal mutation rate (Scally and Durbin 2012).

Between South and East

The above-mentioned separation between southern populations and more northerly populations potentially represents the first migration registered in the human mtDNA profile, even though the direction is contentious. However, it seems likely that L0 had a southern origin. L0d is the most common clade in southern African Khoe-San populations, including both herders and hunter-gatherers, and it is also the result of the first split in L0 (Behar et al. 2008). Apart from some more recent input into eastern Africa of L0d3 sequences that we discuss below, this clade is restricted to southern African populations. This evidence for deep isolation and independent evolution across southern African indigenous populations, in genome-wide autosomes as well as the mtDNA, seems to provide some retroactive justification for the collective term "Khoe-San", despite the lack of identifiable relationships among the Khoe, Tuu and Ju language families, as discussed above (Güldemann and Stoneking 2008; Mitchell 2010).

After the branching out of L0d, there is a split within L0a′b′f′k, as L0k branches off. Again, L0k is mostly found in southern populations. One of the subclades of L0k, L0k2, has so far been seen mainly in Bantu-speaking southern African populations (Rito et al. 2013), probably the result of gene flow from the autochthonous population occupying southern Africa into Bantu speakers as they came into contact within the last ~2 ka, since L0k has never been found in more northerly African populations.

The L0a′b′f clade has a broadly eastern African origin, but the first clade to branch off, L0f, lends additional support to an origin for L0 in the south and a migration to eastern Africa, since it has a somewhat intermediate distribution (Rito et al. 2013). L0f has four subclades in the whole-mtDNA tree. One occurs only in southern Africa (albeit among Bantu speakers), but the HVS-I database allows the identification of this subclade in Zambia and Zimbabwe (i.e. in the northern part of southern Africa); one of the subclades is seen only in Tanzania (i.e. in the southern part of eastern Africa); a third is seen only in Somalia and

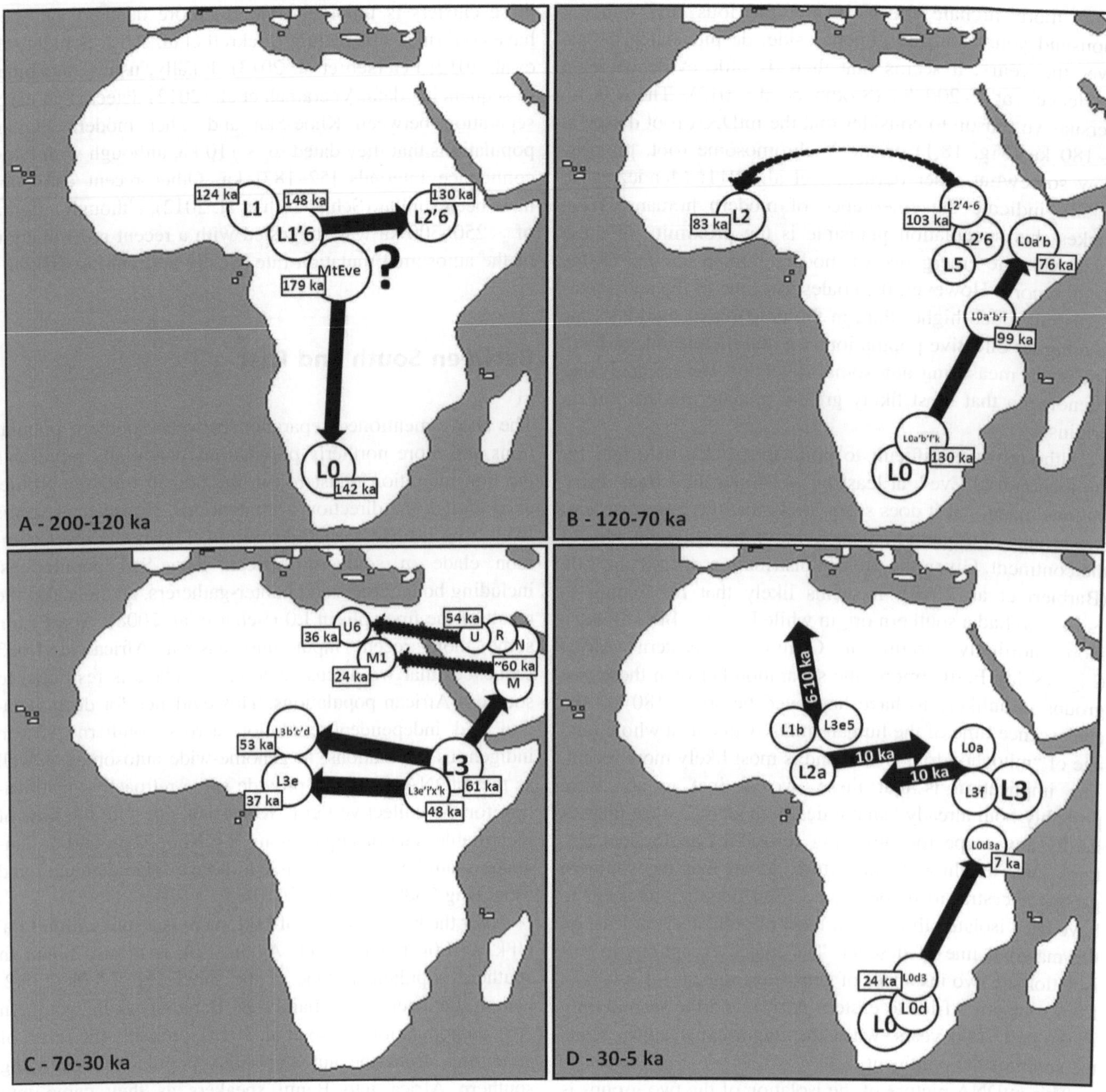

Fig. 18.4 Outline sketch of the major human dispersals within Africa suggested by mtDNA phylogeography. Arrows indicate directionality only and are not intended to represent migratory routes. Point estimates for the ages of each clade are based on the tree in Fig. 18.1. Age estimates of haplogroup U, U6 and M1 are from Soares et al. (2009), calculated using the same methodology. Estimated dispersal times displayed within the *arrows* were obtained using HVS-I data and a founder analysis approach (Richards et al. 2000). Selected time intervals were chosen solely in order to provide a clear representation. The periods are: 200–120 ka (A), 120–70 ka (B), 70–30 ka (C), 30–5 ka (D) and 5 ka to present (E)

Tanzania; and the fourth is more widely distributed through eastern Africa, and in Central Africa. The distribution suggests that L0f probably arose somewhere between southern and eastern Africa (Rito et al. 2013).

The remainder, L0a'b, has a much more northerly distribution in eastern Africa, with the rarer L0b in particular found only in Ethiopia and Kenya. Imposing a (very approximate) time depth on this reconstruction, L0 and L0a'b'f'k date to ~140 ka and 130 ka in southern Africa, and L0a'b'f dates to ~100 ka. This implies that the first steps of the south-to-east migration occurred between ~130 and 100 ka, or (less plausibly) between 100 ka and the age of

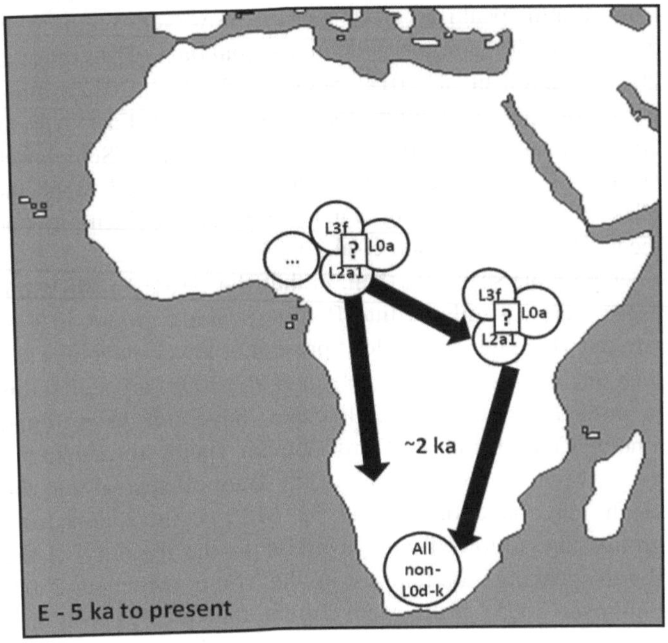

Fig. 18.4 (continued)

L0a′b if L0f arose further south (Rito et al. 2013) (Fig. 18.4b). L0a′b dates to ~75 ka, suggesting that the final leg of the dispersal into eastern Africa occurred between ~100 and 75 ka (Fig. 18.4b). Intriguingly, therefore, L0 (in the form of L0a′b or a close ancestor) may have entered eastern Africa not very long before the expansion of haplogroup L3 that would generate the out-of-Africa dispersal, ~60–70 ka (Soares et al. 2012). The timescale for the expansion from south–to-east also corresponds very roughly to the onset of renewed "megadrought" conditions in Central Africa, beginning ~115 ka, which may, paradoxically, have facilitated the expansion of human groups by creating a more open landscape in the tropical rainforest zone (Blome et al. 2012).

Considering that the earliest evidence of symbolically mediated behaviour (engraving or ornamentation) has been claimed to be in southern Africa (e.g. at Pinnacle Point, dating to ~165 ka [Marean et al. 2007]) and that this symbolically mediated behaviour was part of what was once thought of as the "human revolution" in the out-of-Africa populations, it is tempting to speculate that some aspects of modernity might have been carried by L0a′b′f migrants and eventually transmitted from them to eastern African populations carrying L3. This hypothesis could be invoked to explain the appearance of *Nassarius* bead ornamentation in North Africa where they are found by ~85 ka (Bouzouggar et al. 2007; d'Errico et al. 2009), and of elements of symbolic behaviour in the Levant, although these are present by at least 80–90 ka and possibly for longer (Millard 2008; Shea 2008). This is, of course, highly conjectural, and some dates in the north may well challenge the assumed priority of southern Africa for behavioural innovation (Barton et al. 2009).

Such a migration seems too early for the expansion of the modern Howiesons Poort industry from south to east. This industry appears to have arisen indigenously in South Africa ~65 ka from the Still Bay industry, dating to 70–74 ka, in southern Africa. Jacobs et al. (2008) and Mellars et al. (2013) have suggested that it spread into eastern Africa where similar industries are evident by ~60 ka. However, given the imprecision of genetic dating (particularly at such time depth), we should not completely dismiss the possibility, particularly given the uncertainty of the place of origin of L0f. Furthermore, several alternative scenarios might provide a channel for Howiesons Poort industries to move from south to east. The earlier dispersal might have opened up a communication channel along which cultural characteristics might have been able to flow, either via contact or by sex-biased dispersal. A signal of male gene flow from south to east might be indicated by the sharing of subclades of the deep-rooting Y-chromosome A and B haplogroups between southern African Khoe-San and Central/eastern Africa (Semino et al. 2002; Batini et al. 2011b). Alternatively, L0f might actually have arisen in southern Africa and the expansion of both this and L0a′b in eastern Africa might represent a direct, second, more recent migration from southern Africa, after ~70 ka, although the present distribution does suggest that this is rather unlikely.

It is worth noting here an apparent second, much more recent, migration involving L0 from south to east. L0d clearly has a southern African origin. L0d1′2, dating to about 70 ka, is only present in southern African, mostly Khoe-San, populations. The second subclade of L0d, L0d3, is, however, much less frequent and dates to only ~25 ka. Although it too is present in southern Africa, a single subclade, L0d3b, appears to be restricted to eastern Africa and is mostly seen in Tanzanian Sandawe (Gonder et al. 2007; Tishkoff et al. 2007), who speak a click-consonant language. The age of the clade is ~7.4 (SE 4.5) ka, based on HVS-I data (Rito et al. 2013). Although the confidence intervals are very large, and no clear linguistic evidence exists connecting southern and eastern click languages (although the possibility remains for the Sandawe [Güldemann 2008]), this genetic link to long-standing southern Khoe-San populations might suggest an expansion of individuals speaking click-consonant languages from southern Africa into eastern Africa during the early to mid-Holocene, rather than a migration in the Pleistocene followed by contraction during the Holocene, as has been previously suggested (Tishkoff et al. 2007; Güldemann and Stoneking 2008). However, this migration could have taken place at any time between the age of L0d3 at ~25 ka, and the age of the Sandawe clade (Fig. 18.4d).

A mid-Holocene south-to-east dispersal has not been identified so far in other marker systems, or in the archaeological record, although it might explain the southern African admixture detected in both East African click-consonant groups in the autosomes (Pickrell et al. 2012). However, dispersal in the reverse direction (separate from that of Bantu speakers, which we discuss below) has been proposed on the basis of MSY evidence. An east-to-south dispersal has been a long-standing hypothesis to explain the acquisition of sheep- (and subsequently cattle-) herding among southwest African foraging groups ~2 ka, slightly before the arrival of Bantu-speaking agriculturalists (Phillipson 2005). Henn et al. (2008) identified a minor MSY haplogroup that they called E3b1f-M293, now known as E1b1b1g, which was most diverse in Tanzanian Nilotic and Afro-Asiatic-speaking groups. This suggested that it arose in East Africa, but it was also present in Khoe (or Kxoe) and Ju speakers ("!Kung"), dating (very approximately) to ~2 ka. One type was even shared between Khoe speakers and the Tanzanian Sandawe, who also speak a click-consonant language. This led them to propose a direct dispersal from Tanzania to the ancestors of Khoe-speaking herders in Angola/Namibia.

Although they argued that this dispersal was independent of the slightly later Bantu dispersals, and mediated by Nilotic speakers, an alternative for the emergence of herding in southern Africa would be exchange with the leading edge of the Bantu expansion, suggested on the basis of ceramic similarities (Phillipson 2005). Few mtDNA data are available from Khoe speakers, and they have drifted to such an extent that almost every haplogroup is represented by only a single control-region sequence (Chen et al. 2000; Güldemann and Stoneking 2008), but almost every single mtDNA sequence (in a sample of Khwe; outside the southern L0d and L0k lineages) directly matches a sequence from a Bantu speaker elsewhere in Africa. With the caveat that this is merely a preliminary look at a very limited dataset, this might support acquisition from pioneer Bantu-speaking groups, rather than a separate dispersal from East Africa.

This also illustrates one significant point. Güldemann and Stoneking's (2008) suggestion that genetic drift confounds historical reconstruction applies much more strongly to the frequency-based approaches that they tend to stress, such as principal-components analysis, than to the genealogical approach emphasized here. The source of even a very heavily drifted lineage in the mtDNA genealogy can be readily identified phylogeographically, provided the source has been well sampled. This latter point is important; in this case, for example, a much better characterization of the mtDNA variation of East African Nilotic (especially Southern Nilotic) speakers would be needed to clearly distinguish the alternative hypotheses. The suggestion of a predominantly southern genetic makeup with introgression from Bantu-speaking groups does, however, seem to be consistent with genome-wide autosomal data (Tishkoff et al. 2009; Pickrell et al. 2012; Schlebusch et al. 2012); albeit with possibly a smaller contribution from East African Nilotic speakers in some Khoe-speaking groups (Schlebusch et al. 2012). An East African cluster shared with the click-consonant-speaking Sandawe is also evident in one recent data set (Petersen et al. 2013).

A possible scenario, then, is that the E1b1b1g MSY lineages were assimilated into Bantu speaking groups in East Africa (where they are indeed present at lower diversity) and then dispersed southwards. Henn et al. (2008) argue that this is unlikely, since related lineages have not been found among Central and southern African Bantu speakers; but perhaps the "pioneer phase" of the Bantu dispersal into the south may have differed in its MSY composition from groups that followed and gave rise to the majority of the Bantu-speaking populations in the south (Mitchell 2002; Phillipson 2005). If this were correct, then the dispersal may have been distinct from the main wave of Bantu expansion, but not entirely independent. Or perhaps "pastoralist" MSY lineages and "Bantu" mtDNA lineages in the Khoe speakers result from different episodes of introgression. Heterogeneity from group to group, and even individual-to-individual, is clearly very evident in the genome-wide data, emphasizing the need for larger sample sizes before drawing firm conclusions about dispersal histories. Hopefully it is clear that while this issue, like so many others, requires more work, it is likely to be clarified considerably, and in great detail, as more genetic data are brought into play.

Climate Change and the Out-of-Africa Migration

One of the most important moments in human mtDNA evolution was the emergence of haplogroup L3. This clade gave rise to all of the ancient non-African mtDNA lineages, which are entirely encompassed within haplogroups M and N (Macaulay et al. 2005). Given this, the age of L3 provides an upper bound for the out-of-Africa migration (Soares et al. 2012). L3 dates to between 60 and 70 ka with several methods (Soares et al. 2012) and is 61 ka in Fig. 18.1, but this estimate is based on only African lineages and seems to be an under-estimate due to the dramatic expansions of the L3e′i′x′k and L3b′d subclades. By chance, it seems that the most common African L3 subclades under-estimate the age of L3, whereas possibly the two non-African clades (M and N) might over-estimate it. The true age is likely to be ~70 ka, as estimated with ML when haplogroups M and N are included (Soares et al. 2009; Behar et al. 2012; Soares et al. 2012), better reflecting the four mutations between L3 and M (dating to ~55 ka) and the five between L3 and N (dating to ~60 ka).

Even so, the upper bound of the age of L3 virtually excludes an out-of-Africa dispersal (at least, for the maternal ancestry of non-Africans alive today) before ~74 ka, the time of the Mt. Toba volcanic super-eruption. Various archaeologists have proposed migrations out of Africa during MIS-5 (130–75 ka), either through the Levant (Bar-Yosef 1992) or along the southern coastal route (Petraglia et al. 2007; Armitage et al. 2011), but if these putative events left any descendants living outside Africa in the present day, there is no sign of them in the maternal line of descent. Furthermore, the Y-chromosome also suggests a post-Toba out-of-Africa dispersal (Shi et al. 2010; Mellars et al. 2013).

Scally and Durbin (2012) suggest that the autosomes indicate an earlier exit, but this is based on erroneous reasoning. The date relies on a split between Yoruba (from West Africa) and non-Africans from 90 to 130 ka. However, West Africans diverged from eastern Africans well before the latter formed the source for the non-African gene pool, and, indeed, well before the emergence of L3. A simple estimate for the split time from mtDNA data would be the divergence between L3′4′6 and L2, which is ~110 ka. In fact, though, these population divergence times are not appropriate for estimating the timing of the dispersal out of Africa. The mtDNA evidence shows that modern West Africans carry subclades of L3 (which arose since the dispersal out of Africa) living alongside more ancient lineages from L1 and L0. Even present day eastern Africans would not be representative of the source of non-African mtDNAs, since L0 and L2 lineages have arrived from the south and west, since the time of the exit, presumably accompanied by autosomal lineages that would inflate any estimate of the divergence from non-Africans. The best current autosomal estimate for the timing of the dispersal from autosomal SNPs is, rather, the divergence of Europeans and Asians, at ~40–80 ka (Scally and Durbin 2012). An estimate from autosomal microsatellites is ~56 ka, with a 95% upper bound of 67.4 ka (Prugnolle et al. 2005).

The fact that the age of the out-of-Africa mtDNA clades, M and N, is so close to the age of L3 (Macaulay et al. 2005; Soares et al. 2009; Soares et al. 2012) – probably within ~10,000 years – suggests that the diversification and expansion of L3 and the out-of-Africa expansion might be all part of a continuous demographic phenomenon (Fig. 18.4c). L3 almost certainly had an origin in eastern Africa. The large ancestral clade L2′6, dating to ~130 ka, probably originated in eastern Africa, given the extant distribution of L5, L6 and L4, although L2, dating to more than 80 ka, is very likely West African in origin. If L2′6 arose in eastern Africa, this implies a migration from eastern Africa into West Africa, crossing Central Africa between ~105 ka (the age of L2′6) and ~80 ka (the age of L2), whose signal mostly disappeared in Central Africa (Fig. 18.4b). This time frame does fit the separation between Yoruba (from West Africa) and non-Africans (originating from eastern Africans) at just over 100 ka in the recent autosomal dating referred to above (Scally and Durbin 2012). Another possibility for the route into West Africa is that this hypothetical migration occurred via the Sahel belt or North Africa. North Africa has very likely been depopulated and repopulated since the time of the Aterian industry (Bouzouggar et al. 2007; Pereira et al. 2010b; Henn et al. 2012), but shows very early evidence of symbolic behaviour in the archaeological record (Barton et al. 2009). More important for the issue of L3 origins, the clade L3′4′6 shows clear evidence of an eastern African origin. L4 and L6 are primarily present in eastern Africa and the Arabian Peninsula (Kivisild et al. 2004; Torroni et al. 2006; Behar et al. 2008) and L3 itself also has several basal clades in eastern Africa (L3a, L3e′i′k′x and L3h in Fig. 18.1). Furthermore, the Central African clades L3b′d and the two out-of-Africa clades M and N suggest an eastern African centre of gravity (Soares et al. 2012).

It has been suggested that the moister climate after ~70 ka in eastern Africa led to a dramatic increase in population size (Mellars 2006; Cohen et al. 2007; Scholz et al. 2007). This would, in turn, have given rise to the oldest clear signal of demographic expansion seen in the human mtDNA, the radiation of L3 (Behar et al. 2008; Soares et al. 2012), that led not only to the out-of-Africa expansion but also to the probable introduction of L3b′d and possibly L3e in Central Africa after 60 ka (Soares et al. 2012) (Fig. 18.4c). Therefore, the model of Scholz et al. (2007), as opposed to those that suggest an earlier successful exit in MIS 5 (Cohen et al. 2007), provides a good fit to the mtDNA chronology.

Mellars (2006) has also coupled this phase with a step-change towards behavioural modernity by analogy to the European Upper Palaeolithic, implying a single process of dispersal driving change from southern Africa into Eurasia. Although increasing evidence for modernity is most visible in southern Africa during the Middle Stone Age at about 70–80 ka (Henshilwood et al. 2009; Texier et al. 2010), neither an origin for L3 in the south (Compton 2011) nor an expansion of L3 into the south (Mellars 2006), providing a link between eastern and southern Africa in this time frame, are at all likely on the basis of the extant mtDNA distributions. L0 is the only haplogroup that could show a link between southern and eastern Africa in the period 130–70 ka. Yet, as we discussed earlier, whether a south-to-north dispersal ~70 ka is feasible is far from clear, at least for the mtDNA.

North Africa and the "Back-to-Africa" Migration

North Africa stands distinct and unique in African prehistory, as it does not carry any surviving genetic traces connecting the Middle Stone Age to the present day populations. Moreover, its mtDNA gene pool (strongly supported by Y-chromosome and autosomal data [Henn et al. 2012]) indicates that the re-population of North Africa occurred mainly from non-African populations, representing clearly a "back-to-Africa" migration (Olivieri et al. 2006; Pereira et al. 2010b; Henn et al. 2012; Bekada et al. 2013).

There is no doubt that North Africa was populated during MIS 5 and MIS 6. Some of the oldest fossils classified (at least, by some) as anatomically modern human (the Jebel Irhoud remains, dating to ~160 ka), have been found in Morocco (Smith et al. 2007). Archaeologically, there is evidence for modern symbolically mediated behaviour by at least ~80 ka (Bouzouggar et al. 2007; d'Errico et al. 2009). At the centre of the question of continuity versus discontinuity in North Africa over the last 100 ka is the identification of the bearers of the Aterian industry, recently dated to between 115 and 40 ka (Barton et al. 2009).

An analysis of the mtDNA gene pool of present day North Africans points to only two specific haplogroups with deep Pleistocene ancestry in this part of the continent: haplogroups M1 and U6 (Macaulay et al. 1999; Olivieri et al. 2006). Haplogroup M1 is a basal clade of the non-African haplogroup M, while haplogroup U6 is even more tightly embedded within the non-African haplogroup N. The second of the L3-derived non-African lineages, haplogroup N, gave rise to another large subclade that is also found worldwide, haplogroup R (Macaulay et al. 2005). Haplogroup R evolved into haplogroup U, of which one of the subclades is U6. Most probably, N originated in Arabia immediately outside Africa, soon after the exodus (Fernandes et al. 2012), and the same is likely for the R and U subclades. Although haplogroup M (aside from M1) is not found in Arabia or the Near East,[1] it may also have had an origin in the vicinity of the Arabian Gulf, alongside haplogroup N, in a glacial refuge or oasis where both M and N (and then R) diversified from L3 (Richards et al. 2006; Rose 2010; Fernandes et al. 2012).

Therefore, the oldest mtDNA lineages in North Africa came from outside Africa, most probably the Near East. M1 dates to ~26 ka and U6 to ~35 ka (Soares et al. 2009). These dates provide a lower bound for their entrance in North Africa. Since they both appear to have arisen within North Africa (Pennarun et al. 2012), the age of M (55–50 ka: Soares et al. 2009) and the age of haplogroup U (~55 ka: Soares et al. 2009), the ancestors of M1 and U6, respectively, provide upper bounds for the timing of the "back-migration". North Africa was therefore probably recolonized between 55 and 35 ka, assuming that the arrival of U6 and M1 was a single process (Fig. 18.4c) (Olivieri et al. 2006), which seems plausible but is not entirely clear (Pennarun et al. 2012). Genome-wide data also suggest that North Africa was recolonized in the Pleistocene from a Southwest Asian source, with a similar time frame for the "indigenous" North African lineages (Henn et al. 2012).

So, it is clear that the people carrying U6 and M1 lineages are not descendants of the producers of the MSA Aterian industry (Barton et al. 2009). The time of their appearance would fit with the appearance of a Eurasian-style Upper Palaeolithic blade industry, the Dabban, which appears before 40 ka in Cyrenaica (Close and Wendorf 1990; Macaulay et al. 1999; Bar-Yosef 2002; Olivieri et al. 2006; Lowe et al. 2012). Since the arrival time is not closely constrained by the presence of obvious antecedent lineages in the Near East, the ages of U6 and M1 probably depend primarily on genetic drift within North Africa, and do not make the posited association with the Dabban less plausible (Pennarun et al. 2012). Indeed, they seem to require it, if this is the only attested post-Aterian North African industry of this antiquity. Bayesian skyline plots (BSPs) of haplogroup U6 (Pereira et al. 2010b) and M1, which use genetic diversity to infer population size changes (Atkinson et al. 2008), also suggest population growth ~20–25 ka, coinciding with the beginning of the Iberomaurusian industry in the Maghreb (Blockley et al. 2006; Bouzouggar et al. 2008). The pattern is not identical for U6 and M1, possibly implying distinct trajectories (Pennarun et al. 2012); although the autosomal picture from STRUCTURE-like analyses (at least at K = 8) implies a single autochthonous North African cluster (Henn et al. 2012). This kind of analysis partitions autosomal datasets into genetic clusters, putatively representing ancestral populations, with the clusters defined by the software, but the number of clusters identified (K) defined by the user.

The Iberomaurusian in the Maghreb dates to ~22 ka and overlies a non-descript MSA flake industry, from which it is separated by a sterile layer of several thousand years, and which itself overlies the Aterian (Barton et al. 2013). The pattern with U6 and perhaps also M1 implies that the modern humans who made the MSA industries were not the ancestors of those who made the Iberomaurusian; but rather that these autochthonous North African lineages spread from further east, most likely from the makers of the Dabban industry, with an ultimately Southwest Asian ancestry – at least for U6 (the source for M1 is unresolved). The fact that the Iberomaurusian is presently dated older in the Maghreb than in Cyrenaica complicates the picture, but may either

[1]In fact Pakistan is now the most westerly place where the rest of M is found.

suggest further reverse dispersals, or exchange along the Mediterranean coastline. Alternatively, there may be older Iberomaurusian sites awaiting dating further east. In this connection, it has been suggested that the industry may date to ~19 ka at Haua Fteah in Libya (Barker et al. 2010). Further lineages were introduced from Iberia in the early Holocene, spreading subsequently from the Maghreb into the Sahel belt (Ottoni et al. 2010; Pereira et al. 2010a), and more recently (perhaps at least partly with the Arab conquests) also from the Near East (Henn et al. 2012).

The suggestion that M1 may have been introduced after the LGM, carried alongside Afro-Asiatic languages (Forster 2004; Forster and Romano 2007), seems less likely, since the major subclade M1a appears to have arisen within eastern Africa and dates to ~20 ka (Forster and Romano 2007; Pennarun et al. 2012). The overall picture suggests that the people carrying U6 were likely responsible for the production of the Dabban industry into North Africa and the subsequent spread west of the Iberomaurusian industry. The links back to a Near Eastern source and the Eurasian Upper Palaeolithic may partly explain the suggested similarities between the robust Iberomaurusian "Mechta-Afalou" burials and European Cro-Magnon remains (Irish 2000).

The Pleistocene/Holocene Transition in Africa

The late Pleistocene/early Holocene transition, ~11.5 ka, has been hypothesized from the mtDNA evidence as being a period during which major population expansions occurred across Eurasia, from Europe (Soares et al. 2010) to the Pacific (Soares et al. 2008; Soares et al. 2011). Africa is most probably not very different in this respect, as we have already suggested using both haplogroup L3 data (Soares et al. 2012) and U6 data from North Africa (Pereira et al. 2010b). For this chapter, we conducted two additional kinds of analysis in order to test this hypothesis further. One was a BSP analysis for the four major clades in Africa, L0 to L3, in order to observe which periods suggested population expansions associated with these haplogroups (Fig. 18.5). We also calculated a plot representing a random sampling of the available sequences, for comparison.

The second was a founder analysis, aiming to detect periods of migration between two regions (Richards et al. 2000). One difficulty with founder analysis is defining source and sink regions, and this is particularly acute for Africa where there has been substantial gene flow across the continent throughout prehistory. For the last ~20 ka, however, some broad haplogroup distributions within Africa can be established: L1, L2, L3b′d and L3e are found in Central/West Africa; L4, L5, L6, L0a′b′f and L3 (except L3b′d and L3e) are found in eastern Africa; and L0d and L0k are restricted to southern Africa. North Africa meanwhile harbours the "back-to-Africa" U6 and M1 lineages, with some lineages from south of the Sahara; the majority are more recent migrants from Europe and the Near East. We therefore used the founder analysis and an HVS-I dataset of nearly ten thousand African individuals to check if migrations occurred between the different regions.

We performed the following founder analyses: (a) from eastern Africa to Central Africa for the eastern African haplogroups; (b) from Central/West Africa into eastern Africa for L1, L2, L3b′d and L3e; (c) from Central, West and eastern Africa to North Africa for all the sub-Saharan lineages; and (d) from Central to West Africa for L1, L2, L3b′d and L3e. The latter is much less well-defined than the others, since the L1 and L2 tree suggests that it is difficult to identify with any confidence the source or sink for many links, but even so the founder analysis can point to common clades that expanded more extensively in the region. We then performed a final founder analysis (e), which considered all haplogroups moving into southern Africa.

The results are displayed in Fig. 18.6. We discuss them in the following pages. The more general reader may skip to the summary at the end of this section. Since, apart from the southern African founder analysis, the results correspond to only a fraction of the population profile (excluding the autochthonous haplogroups of each region over the past 20 ka) we will not refer to frequencies of the founders in the analysis, but simply indicate the major founders contributing to each migration.

The Late Glacial period in tropical Africa is now thought to have gone through a mega- drought between 16 and 17 ka comparable to that observed before 70 ka (Stager et al. 2011). From ~11.5 ka and for a few thousand years thereafter, the climate was warm and humid, except for the northern and southern extremes of the continent (Kuper and Kropelin 2006; Weldeab et al. 2007). The Holocene climatic optimum would likely have allowed populations to expand, and movements of people probably took place at this time. We detected the strongest population expansions in the time window of 15–8 ka in the random dataset and in the L3 data. The L2 data also indicated a population expansion from ~12 ka until recent times, but (probably due to poor phylogenetic resolution) it was not separated clearly from the more recent increase observable in all the data (probably due to the Bantu expansion, as discussed in the next section). Haplogroups L0 and L1 did not show any signal of expansion until the last few millennia before the present, again with the Bantu expansions. However, some subclades of these haplogroups were indeed likely to have been involved in postglacial expansions, as we discuss below.

The migration scans from eastern Africa to Central Africa (Fig. 18.6a) reveals a clear single peak at ~11 ka, closely

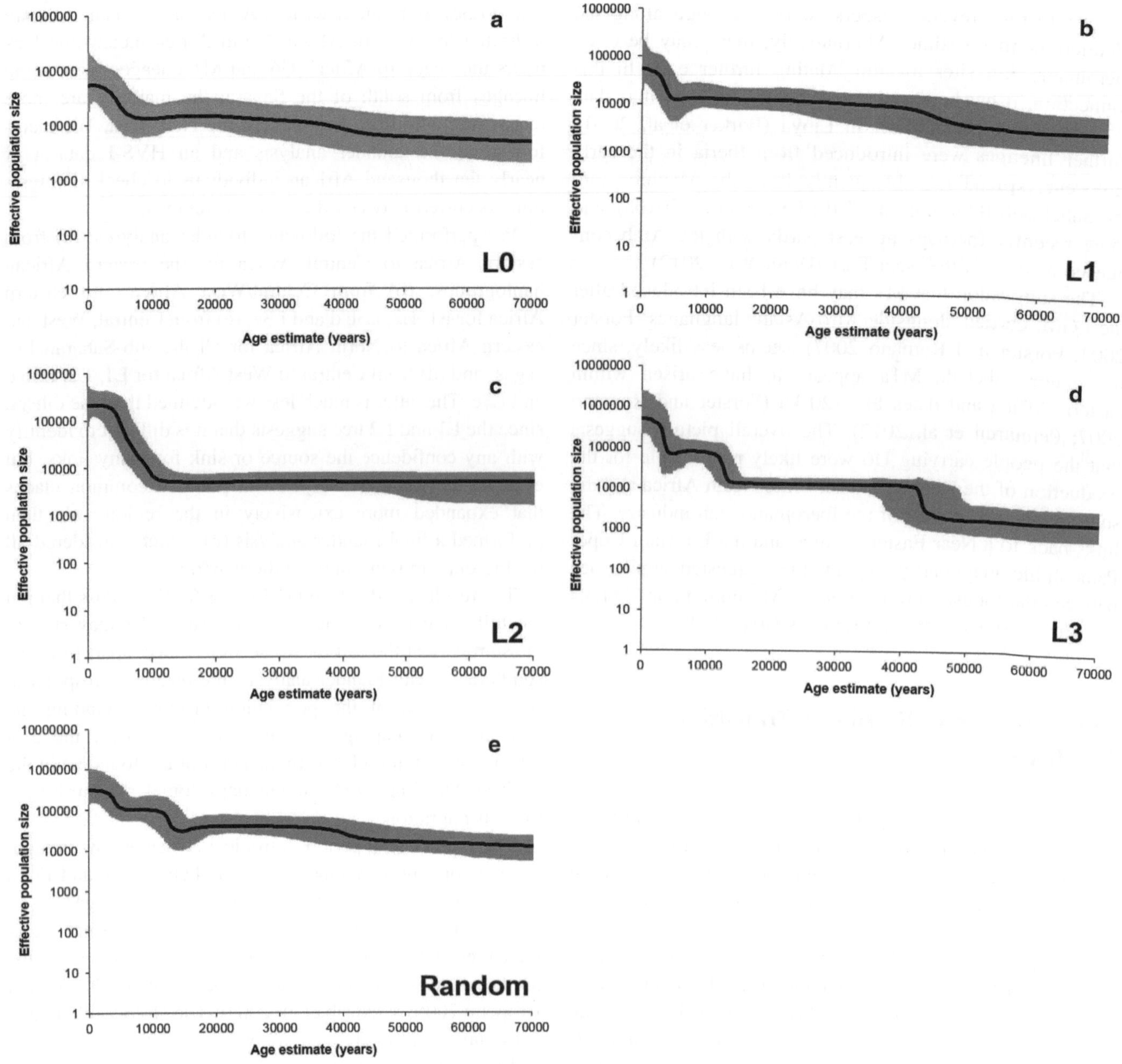

Fig. 18.5 Bayesian skyline plots of haplogroups L0 **a**, L1 **b**, L2 **c**, L3 **d** and random sample **e**

matching the onset of more humid conditions in tropical Africa. Lineages from two major clades, L3f and L0a, dominate the results for this peak (Fig. 18.4d). The distribution of L3f3 has previously been highlighted as possibly representing a pastoralist migration (Cerny et al. 2009), but the age estimates here suggest that its expansion might have been earlier. We have already pointed to L3f as a signal of postglacial expansions from eastern Africa to Central Africa (Soares et al. 2012), but the results here show that lineages from the L0a clade are equally represented in these migrations.

The analysis of the whole-mtDNA tree supports this result: the L0a1a clade, dating to about 16.5 ka, shows a mainly Central African distribution, although with some possibly basal clades in eastern Africa. Its derived subclade, L0a1a2, lives in Central Africa and dates to 13.4 [8.4–18.5] ka, similar to the time obtained in the founder analysis. The L0a1b′c′d clade has two subclades that are found in eastern Africa (L0a1c and L0a1d,) but the third (L0a1b) is Central African and dates to 14.8 [7.6–22.3] ka. In L0a2, several other lineages might also have moved in this period. L0a2b, dating to only 5.5 [1.3–9.7] ka is present in Central

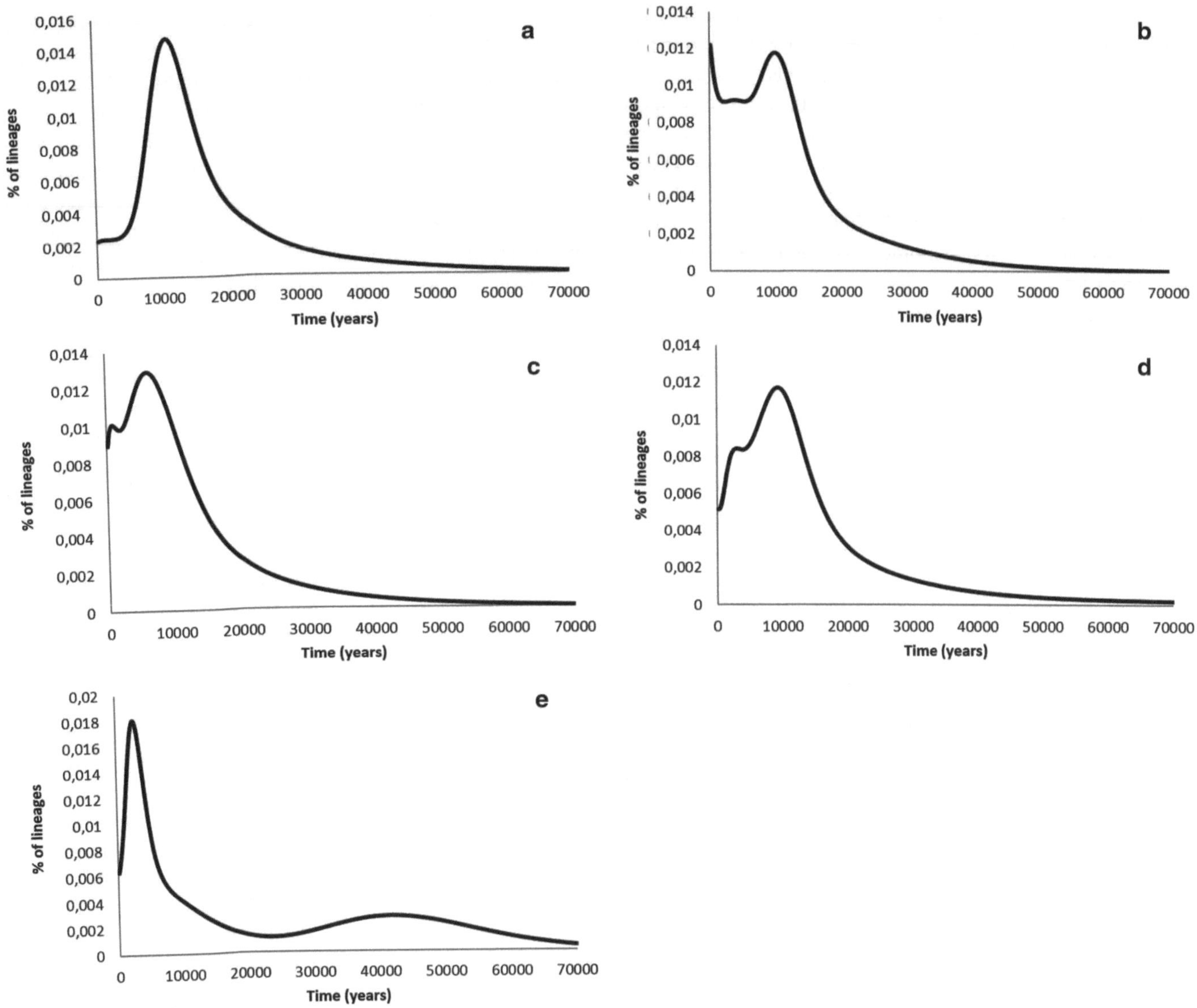

Fig. 18.6 Founder analysis migration scans for migrations from eastern Africa to Central Africa **a**; from Central/West Africa into eastern Africa **b**; from Central, West and eastern Africa to North Africa **c**; from Central Africa to West Africa **d**; and from Central/eastern Africa into southern Africa **e**

African forest forager groups, but shares a link with eastern Africa dating to ~18 ka. L0a2a1 (dating to 14.2 [7.6–21.0] ka) is mostly Central African. An important issue to note here is that, apart from L2 lineages in Central Africa, L0a lineages will be the most important mtDNA components in the migration south during the Bantu expansion.

The migration scans from Central Africa to eastern Africa (Fig. 18.6b) again indicates a major peak at ~10 ka, and a second one close to the present. A minor peak seems to be located between the two, at ~2.2 ka, visible as a small hump. We previously detected a peak containing L3b and L3d, probably related to gene flow from Bantu speakers from Central/West Africa into the east (Soares et al. 2012). These lineages still show the same signal, but the signal of haplogroup L2 is much stronger in the present dataset, and a postglacial signal at ~10 ka is much more striking in the new analysis (Fig. 18.4d). L2a1 is the lineage that provides most of the signal, more specifically the lineage or lineages carrying the variant (relative to the Cambridge Reference Sequence) at position 16189, and those carrying both the 16189 and 16192 variants.

It is difficult to check the whole-mtDNA genome tree with confidence on this issue. This is for two reasons. One is the fact that the sampling in the tree is biased (since samples are rarely selected at random for whole-mtDNA sequencing), and there are not many L2 sequences available at present from eastern Africa. The second is the fact that these HVS-I sequences match two independent subclades within haplogroup L2a1, L2a1+143+16189(+16192) and L2a1 +16189(+16192). However, the former (L2a1+143+16189, dating to 15.3 [9.9–20.8] ka) and its derived subclade with the 16192 variant (dating to 12.5 [8.3–16.8] ka) show basal eastern African (or Arabian) lineages or subclades that roughly support the founder analysis results with HVS-I data. The third most frequent founder is another subclade of L2, L2a1d, which appears to belong to eastern Africa in the whole-mtDNA genome tree and dates to 11.5 [4.7–18.6] ka.

A migration scan from tropical Africa to North Africa (Fig. 18.6c) indicates a smaller recent peak, possibly related to the recent slave trade across the Sahara (Harich et al. 2010), and a major one dating to ~6.5 ka (Fig. 18.4d). Several lineages display founder age estimates that suggest an arrival during the mid-Holocene. However, in the two founder analyses described above, only a few lineages were responsible for most of the signal. Here, in addition to two or three frequent lineages, there is also an array of lineages with intermediate frequencies which probably also entered North Africa in the Holocene, suggesting that the Sahel belt was probably home to extensive gene flow at this time (Cerny et al. 2007).

We focus on the three major founders. The most frequent is the HVS-I root type of L1b. It is difficult to know here if we are looking at several dispersing lineages or just a single major one (given the weak phylogenetic resolution of HVS-I), so the dating is uncertain. Even so, if the founder age (~9 ka) were indeed an average of several lineages, it would still be hard to reconcile with a mixture of lineages some of which arrived in the Pleistocene (which would be very unlikely due to the strength of the Sahara barrier) and some in much more recent times; so the early Holocene is still the most probable time for the arrival of L1b. L1b1a is by far the most frequent of the L1b subclades, and since it does not display any defining HVS-I motif, it matches the HVS-I founder type. L1b1a dates to 11.8 [8.5–15.1] ka in the whole- mtDNA tree, and it displays several nested star-like subclades that might suggest expansion (although not detected in the BSP analysis of L1). But very few North African sequences are yet available in the L1b whole-mtDNA tree, even though L1b can reach frequencies up to 5–10% in some North African populations.

The second most common HVS-I lineage showing an arrival in North Africa in the early Holocene is L3e5 (Soares et al. 2012; Podgorna et al. 2013). This lineage is mainly restricted to Northwest Africa, but is also found in Central Africa, where it probably arose (Podgorna et al. 2013), and it dates to ~12 [8.8–15.2] ka. The third most frequent founder is L0a1, due to its high frequency in Egypt and the vicinity. It is possible that L0a1 dispersed directly from eastern Africa into Northeast Africa, but the whole-mtDNA tree suggests that the L0a1 subclade in North Africa matches the one found in Central Africa that we mentioned above, L0a1b (in the founder analysis from eastern to Central Africa). A movement from North Africa to Central Africa is also possible. It is worth mentioning that the next two most frequent Holocene founders in North Africa also match founders detected as participating in postglacial migrations above, the L3f and the L2a1 founders that were detected in Central and eastern Africa, respectively. This implies that we are most probably detecting a single major process of expansion that spread in several directions at the same time.

A scan between Central and West Africa (Fig. 18.6d) indicates two peaks, one at ~2.5 ka and the second, more significant, at ~9.5 ka. It is difficult to judge whether or not the directionality of the migration we have imposed (Central to West Africa) is correct for many of the lineages (in particular, within L2, which seems to be of overall West African origin). Yet, it does indicate that postglacial range expansions were probably also occurring between Central and West Africa. The major contributor to this signal is the L2 HVS-I root type, which should mostly include members of L2c (again, it does not contain any defining HVS-I motif). This clade dates to 18.4 [14.4–22.6] ka and the star-like pattern, which is also observed in its major subclades, suggests an early expansion. Another major founder is the root type of L3b1, which is Central African in origin. A third is L3e4, also of Central African origin, but very poorly represented in the whole-mtDNA trees.

The final scan (Fig. 18.6e) corresponds to a full population founder analysis for southern African populations. The small hump at ~50 ka corresponds mainly to L0d. L0d is much older than this, but it is difficult to date HVS-I sequences much beyond ~40–60 ka due to saturation, where back-and-forth recurrent mutation begins to swamp the signal. In any case, the important point is that between this ancient peak and the much stronger second peak at ~2 ka there is no signal of any genetic input into southern Africa from more northerly populations whatsoever. Minor early Holocene expansions seem to have occurred in the opposite direction, from south to east, perhaps taking click-consonant languages into eastern Africa earlier in the Holocene, as discussed above.

In summary, the BSPs and the founder analyses suggest that the late Pleistocene and early Holocene were periods of major gene flow between populations in West, Central, eastern and North Africa, accompanied by population growth. Southern Africa is exceptional, suggesting by contrast a history of continuous isolation and lack of any demographic growth signal throughout this period, despite archaeological evidence for an increase in the number of sites and areas occupied in the early Holocene. A similar pattern is evident in a much larger recently published sample from across southern Africa. Minor recent growth signals in some L0d and L0k lineages may be the result of their incorporation into Bantu communities; indeed some of the rarer L0k subclades have been more commonly found in Bantu speakers today, but must have been assimilated from indigenous southern Africans within the last 2 ka or so (Barbieri et al. 2013).

Late Holocene Dispersals

The most recent major demographic phenomenon to reshape the genetic landscape of sub-Saharan Africa is thought to have been the so-called "Bantu expansion", which was attested both linguistically and archaeologically before it was investigated genetically (Heine and Nurse 2000; Pereira et al. 2001; Salas et al. 2002; Phillipson 2005). There has been a great deal of controversy about the origins and spread of the Bantu languages, which are dispersed over a huge swathe of sub-Saharan Africa from, roughly, Cameroon in the west to Kenya in the east and down to South Africa (Phillipson 2002; Eggert 2005; Phillipson 2005; Holden and Gray 2006; Marten 2006). Bantu languages have been assigned a source in the Cross-Benue region of northwest Central Africa purely on the basis of lexical comparisons, initially by Greenberg (1963), and the comparative method (which is standard in, for example, Indo-European and Austronesian studies) has never been applied. Although the Cameroon origin is widely accepted, it is not reflected in the internal diversity within the Bantu subgroup. In addition, the substructure – which has huge implications for the proposed pattern of dispersal of the speakers – has been difficult to pin down. This is probably due to rapid radiation in the Western Bantu languages (thought to be due to coastal and riverine dispersals), and extensive borrowing (likely within the context of dialect chains) in the East and eastern Central languages, followed by rapid starburst radiation once again in the south (Holden and Gray 2006).

Nevertheless, even using lexical data, and in particular by using network rather than tree models to reconstruct relationships (Holden and Gray 2006), some broad outlines are widely agreed (Nurse 1997), though circular reasoning between archaeologists and historical linguists (and now also geneticists), as has been the case with studies of Austronesian languages (Oppenheimer and Richards 2001; Eggert 2005), remains an issue in this field. The agriculturalist expansion may have started at ~5–4 ka from the region of the Cameroon/Nigeria border (Barker 2006). Yet, the early stages are attested primarily on the basis of the languages, as the forest zone has so far yielded rather little in the way of archaeological evidence. There was an initial (though archaeologically invisible) dispersal from west to east, either north or south of the rainforest, reaching the Great Lakes region of Uganda by ~3 ka. In Kenya and northern Tanzania the local communities were already herding cattle, although there is no archaeological evidence for this in Uganda. The early settlers may have been already beginning to cultivate cereals which could supplement the tuber crops (and perhaps also some cereals such as millet) presumed to have been brought from Cameroon, as well as becoming familiar in the following centuries with iron-working (Vogel 1997b; Holden 2002; Phillipson 2005; Holden and Gray 2006). With this combination of new and existing elements of a farming economy, controversy has centred primarily on the extent to which the process was fuelled by large-scale migrations or by small-scale processes involving contact and assimilation (Vansina 1995; Eggert 2005).

Populations expanding into the south have been most evident on the eastern side of the continent (Pereira et al. 2001; Phillipson 2005), within the last 2.5 ka. The signature of this in the archaeological record is widely agreed to be Phillipson's Early Iron Age "Chifumbaze complex", which arose to the west of Lake Victoria ~500–200 BC (Phillipson 2005). This expanded rapidly into Central and southern Africa over a period of a few centuries, ~2 ka, reaching Mozambique by ~1.8 ka and South Africa, where the limits of the summer rainfall belt were reached, by ~1.5 ka (Vogel 1997a; Phillipson 2005). The Eastern Bantu languages are thought to have been distributed by this "eastern stream" of dispersal. An earlier dispersal south via the river system into the rainforest from the Cameroon region by ~3.5 ka, the "western stream", may have been responsible for the spread of the Western Bantu languages as far south as Angola (Vansina 1995; Vogel 1997b; Phillipson 2005), with the two streams intermingling across Central Africa.

The Bantu expansion is clearly attested in the mtDNA record (Bandelt et al. 2001; Pereira et al. 2001; Salas et al. 2002; Plaza et al. 2004; Beleza et al. 2005), as we can see in the skyline plots discussed in the last section. There is a sharp increase in all of the major tropical African clades in the period between 1 and 4 ka (Fig. 18.4a–d). Haplogroup L2 (Fig. 5c) shows a continuum from a population expansion that began ~12 ka, but even so a population-size increment is observable in the late Holocene. Not surprisingly, a random sampling of African samples shows this same steep increment (Fig. 18.5e).

Our analyses suggest that southern Africa was mostly isolated for around the last 150 ka, since gene flow in this period probably occurred only from south to north, with the possible exception of the pastoralist dispersal from East Africa discussed above. Here, however, the effect of the Bantu expansion was massive. Bantu speakers in Mozambique carry almost entirely lineages from the north, although there is a much higher level of assimilation of L0d and L0k lineages in South Africa. L0d is common in many more southern Bantu groups and, as mentioned above, some southern lineages such as L0k2 are seen mainly in Bantu speakers, even though we can plausibly assume that they descend from much earlier southern African settlers. Khoe-San populations display the most distinctive mtDNA profiles in southern Africa, with haplogroup L0d dominating. Yet, even in these groups some minor genetic input from Bantu-speaking populations can be seen (Salas et al. 2002; Tishkoff et al. 2009).

In the founder analysis, we obtained a well-defined peak at ~2.4 ka, which, given the clear archaeological picture, provides some level of corroboration for both the method and the molecular clock we have employed, albeit slightly preceding the archaeologically dated arrival in the far south. The most common founder cluster detected was L2a1b1 (8.2%). This clade is, unfortunately, under-represented in the whole-mtDNA tree, but it does include two samples from Mozambique and one from Kenya, and dates to ~2.6 ka overall, supporting the founder analysis. A second founder cluster from L2 that also shows a significant frequency is L2a1a2 (5.1%). This clade dates to 8.6 [5.6–11.6] ka, and it also includes southern African samples, although insufficient at present to test the time of the expansion into the south using whole-mtDNA genomes. The clade has mostly West African representatives but Central and eastern African lineages are much less well-characterized at the whole-mtDNA level.

The second and fourth most common founders in the dataset are within the L0a haplogroup. One (at 4.0%) is L0a1 in the HVS-I, but most probably corresponds to L0a1b in the whole-mtDNA tree, where there is a fair number of southern African representatives. The other one, which is the more common (7.7%), is the HVS-I root of L0a. Inspection of the whole-mtDNA tree suggests that it most probably corresponds to L0a2a. These two founders were major components of a postglacial dispersal from eastern to Central Africa, so disentangling where Bantu speakers picked them up is a challenge that will require further data and analysis. It remains possible that they may be the result, at least in part, of the original Bantu arrivals in the inter-lacustrine region of eastern Africa coalescing with the indigenous Great Lakes herding populations (Salas et al. 2002). Yet, despite their ultimate eastern African ancestry, their distribution suggests that many of the lineages might have been assimilated within Central Africa during the Bantu spread southwards (Fig. 18.4e).

Broadly, the results for the expansion into the far south match the combined evidence of archaeology and linguistics quite closely. As already mentioned, this concordance (which rarely arises, except in particularly straightforward dispersal scenarios, such as the settlement of virgin territory in the Remote Pacific, or an expansion facilitated by the use of iron-working, as here) is extremely valuable for validating the genetic methodologies. This in turn comes into its own when other lines of evidence are less forthcoming. It should be noted, however, that this example could be compared to the spread of Austronesian languages in another way. The final stages of the expansion, involving profound founder effects into effectively empty space are rather clear. This is not to dismiss the role of the indigenous population, but, as we have said, the evidence suggests that southern Bantu-speaking groups largely carry genetic lineages from outside the region. However, the earlier stages of the putative linguistic dispersal (in Central Africa, as in Island Southeast Asia) are far more difficult to disentangle. Some geneticists doubt whether there was ever an "Austronesian expansion" in the sense proposed by historical linguists – that is, all the way from Taiwan to Oceania (Soares et al. 2011) – and it is not yet entirely clear whether the situation with Bantu speakers – outside southern Africa, at least – might be analogous in this regard too.

The advantage of the genetic analyses in this context is that they can also begin to illuminate the complexities of the Bantu expansion so far undisclosed by archaeology, providing a line of evidence independent of the linguistics. The founder analysis from Central/West Africa to the east establishes several possible founder lineages for Bantu speakers (the most frequent being L3e3 and L2a2), but at low levels in the sampled populations. This is not surprising, since the impact of Bantu speakers on eastern Africa, where herding populations were already taking hold, was much less profound than in the south. In West Africa, during the same period, some minor founders from Central Africa can be detected in clades L3e2, L2a, L3d and L2c, but with no suggestion of a large-scale migration, considering the small size of the peak at this time.

We can now reassess the demographic impact of the Bantu expansions, and in particular the extent of assimilation of indigenous populations in eastern and Central Africa (Newman 1995). Both earlier mtDNA analyses and genome-wide autosomal analyses have suggested a substantial eastern African input into southern African Bantu-speaking populations, implying heavy levels of assimilation of local populations after Bantu-speaking groups arrived from further west (Salas et al. 2002; Tishkoff et al. 2009), but our present analysis points to a primary source in Central Africa (although the precise location remains uncertain) and questions the scale of eastern African assimilation.

It is also very difficult to make a clear distinction between western and eastern streams, although there seems to be a more diverse array of Central African mtDNA lineages in the southwest (including a much higher level of Central African L1c lineages) and fewer instances of strong founder effect than those from the southeast (Plaza et al. 2004; Richards et al. 2004; Beleza et al. 2005). There is no trace to be seen of Khoe-San (L0d/k) introgression in this region (Plaza et al. 2004). Our analyses here emphasize the role of Central Africa, since it seems that many eastern lineages, especially within L0a, had expanded into Central Africa, as far as Cameroon, by the early Holocene.

Some Y-chromosome analyses have also suggested a more profound signal of expansion from Central/West Africa, unlike the mtDNA largely erasing the signal of earlier populations in Central Africa (Berniell-Lee et al. 2009). A correlation has been suggested between parts of the MSY tree and the linguistic tree for not only Bantu languages, but Niger–Kordofanian more generally (de Filippo et al. 2011). Yet here too, recent analyses have complicated the picture by demonstrating unexpectedly high levels of diversity in the western forest zone (Montano et al. 2011) and links between Central African forest foraging populations and southern African Bantu speakers, as with the mtDNA (Batini et al. 2011a). Without secure genetic dating for the MSY, it is difficult to test whether or not links between distinct foraging groups in Central and southern Africa, for example, are evidence for ancient common ancestry or due to assimilation and dispersal by Bantu speakers, although progress is being made (Batini et al. 2011b). The autosomal data are also throwing up intriguing complexities (Sikora et al. 2011). It appears, then, that even this very recent, and in many ways quite clearly attested, expansion presents a highly complex genetic picture that will require far more work to elucidate it clearly.

Final Remarks

As a last comment, taking a step back and looking at African mitochondrial phylogeography more generally, it is worth drawing attention to the peculiarity of the history of haplogroup L0. L0 had an origin in southern Africa ~140 ka ago, migrated into eastern Africa in the form of L0a'b'f between 130 and 75 ka, and evolved into L0a in eastern Africa ~45 ka, not long after the out-of-Africa dispersal occurred. In the early Holocene, several L0a clades migrated into Central Africa when the climate improved, and they later became an important component of the Bantu-speaking agriculturalist populations that would migrate south within the last two millennia – bringing L0 back to the south in a quite different form (L0a) to the ones that evolved *in situ* (L0d and L0k).

L0 therefore bears witness to some of the most significant events in the demographic history of sub-Saharan Africa, and is unique in this respect (Rito et al. 2013). This illustrates the challenge in reconstructing prehistoric movements in Africa from mtDNA variation, but also the extent to which, even in the face of such palimpsest-like patterns, this challenge is beginning to be met. Geneticists will never have all of the answers, and in the past they have not even always managed to address the right questions. Even so, we are convinced that with more extensive and careful sampling, along with closer interdisciplinary collaboration with archaeologists, climatologists, anthropologists and linguists, the potential for studies of genetic variation to help resolve many of the issues we have discussed here should not be under-estimated.

Acknowledgments We thank the referees, in particular Peter Mitchell, for their detailed and constructive comments, which have improved the chapter enormously, even if not quite as much as they might have liked. FCT, the Portuguese Foundation for Science and Technology, supported this work through the personal grants to P.S. (SFRH/BPD/64233/2009) and T.R. (under the project PTDC/CS-ANT/113832/2009). IPATIMUP is an Associate Laboratory of the Portuguese Ministry of Science, Technology, and Higher Education and is partially supported by FCT.

References

Alexander, D. H., Novembre, J., & Lange, K. (2009). Fast model-based estimation of ancestry in unrelated individuals. *Genome Research, 19*, 1655–1664.

Armitage, S. J., Jasim, S. A., Marks, A. E., Parker, A. G., Usik, V. I., & Uerpmann, H. P. (2011). The southern route 'Out of Africa': Evidence for an early expansion of modern humans into Arabia. *Science, 331*, 453–456.

Atkinson, Q. D., Gray, R. D., & Drummond, A. J. (2008). mtDNA variation predicts population size in humans and reveals a major southern Asian chapter in human prehistory. *Molecular Biology and Evolution, 25*, 468–474.

Bandelt, H.-J., Alves-Silva, J., Guimaraes, P. E. M., Santos, M. S., Brehm, A., Pereira, L., et al. (2001). Phylogeography of the human mitochondrial haplogroup L3e: A snapshot of African prehistory and Atlantic slave trade. *Annals of Human Genetics, 65*, 549–563.

Bandelt, H.-J., Macaulay, V. A., & Richards, M. B. (2002). What molecules can't tell us about the spread of languages and the Neolithic. In C. Renfrew & P. Bellwood (Eds.), *Examining the language-farming dispersal hypothesis* (pp. 99–107). Cambridge: McDonald Institute for Archaeological Research.

Bar-Yosef, O. (2002). The Upper Paleolithic revolution. *Annual Review of Anthropology, 31*, 363–393.

Barbieri, C., Vicente, M., Rocha, J., Mpoloka, S. W., Stoneking, M., & Pakendorf, B. (2013). Ancient substructure in early mtDNA lineages of southern Africa. *The American Journal of Human Genetics, 92*, 285–292.

Barham, L., & Mitchell, P. (2008). *The first Africans: African archaeology from the earliest toolmakers to most recent foragers*. Cambridge: Cambridge University Press.

Barker, G. (2006). *The agricultural revolution in prehistory: Why did foragers become farmers?* Oxford: Oxford University Press.

Barker, G., Antoniadou, A., Armitage, S., Brooks, I., Candy, I., Connell, K., et al. (2010). The Cyrenaican Prehistory Project 2010: The fourth season of investigations of the Haua Fteah cave and its landscape, and further results from the 2007–2009 fieldwork. *Libyan Studies, 41*, 63–88.

Barton, R. N. E., Bouzouggar, A., Collcutt, S. N., Schwenninger, J. L., & Clark-Balzan, L. (2009). OSL dating of the Aterian levels at Dar es-Soltan I (Rabat, Morocco) and implications for the dispersal of modern *Homo sapiens*. *Quaternary Science Reviews, 28*, 1914–1931.

Barton, R. N. E., Bouzouggar, A., Hogue, J. T., Lee, S., Collcutt, S. N., & Ditchfield, P. (2013). Origins of the Iberomaurusian in NW Africa: New AMS radiocarbon dating of the Middle and Later Stone Age deposits at Taforalt Cave, Morocco. *Journal of Human Evolution, 65*, 266–281.

Bar-Yosef, O. (1992). The role of Western Asia in modern human origins. *Philosophical Transactions of the Royal Society of London Series B, Biological Sciences, 337*, 193–200.

Batini, C., Lopes, J., Behar, D. M., Calafell, F., Jorde, L. B., van der Veen, L., et al. (2011a). Insights into the demographic history of African pygmies from complete mitochondrial genomes. *Molecular Biology and Evolution, 28*, 1099–1110.

Batini, C., Ferri, G., Destro-Bisol, G., Brisighelli, F., Luiselli, D., Sánchez-Diz, P., et al. (2011b). Signatures of the preagricultural peopling processes in sub-Saharan Africa as revealed by the phylogeography of early Y chromosome lineages. *Molecular Biology and Evolution, 28*, 2603–2613.

Behar, D. M., Villems, R., Soodyall, H., Blue-Smith, J., Pereira, L., Metspalu, E., et al. (2008). The dawn of human matrilineal diversity. *American Journal of Human Genetics, 82*, 1130–1140.

Behar, D. M., van Oven, M., Rosset, S., Metspalu, M., Loogvali, E. L., Silva, N. M., et al. (2012). A "Copernican" reassessment of the human mitochondrial DNA tree from its root. *American Journal of Human Genetics, 90*, 936–936.

Bekada, A., Fregel, R., Cabrera, V. M., Larruga, J. M., Pestano, J., Benhamamouch, S., et al. (2013). Introducing the Algerian mitochondrial DNA and Y-chromosome profiles into the North African landscape. *PLoS ONE, 8*, e56775.

Beleza, S., Gusmao, L., Amorim, A., Carracedo, A., & Salas, A. (2005). The genetic legacy of western Bantu migrations. *Human Genetics, 117*, 366–375.

Berniell-Lee, G., Calafell, F., Bosch, E., Heyer, E., Sica, L., Mouguiama-Daouda, P., et al. (2009). Genetic and demographic implications of the Bantu expansion: Insights from human paternal lineages. *Molecular Biology and Evolution, 26*, 1581–1589.

Blockley, S. P. E., Blockley, S. M., Donahue, R. E., Lane, C. S., Lowe, J. J., & Pollard, A. M. (2006). The chronology of abrupt climate change and Late Upper Palaeolithic human adaptation in Europe. *Journal of Quaternary Science, 21*, 575–584.

Blome, M. W., Cohen, A. S., Tryon, C. A., Brooks, A. C., & Russell, J. (2012). The environmental context for the origins of modern human diversity: A synthesis of regional variability in African climate 150,000–30,000 years ago. *Journal of Human Evolution, 62*, 563–592.

Bouzouggar, A., Barton, N., Vanhaeren, M., d'Errico, F., Collcutt, S., Higham, T., et al. (2007). 82,000-year-old shell beads from North Africa and implications for the origins of modern human behavior. *Proceedings of the National Academy of Sciences USA, 104*, 9964–9969.

Bouzouggar, A., Barton, R. N. E., Blockley, S., Bronk-Ramsey, C., Collcutt, S. N., Gale, R., et al. (2008). Reevaluating the age of the Iberomaurusian in Morocco. *African Archaeological Review, 25*, 3–19.

Černý, V., Salas, A., Hájek, M., Žaloudková, M., & Brdička, R. (2007). A bidirectional corridor in the Sahel-Sudan belt and the distinctive features of the Chad Basin populations: A history revealed by the mitochondrial DNA genome. *Annals of Human Genetics, 71*, 433–452.

Cerný, V., Fernandes, V., Costa, M. D., Hájek, M., Mulligan, C. J., et al. (2009). Migration of Chadic speaking pastoralists within Africa based on population structure of Chad Basin and phylogeography of mitochondrial L3f haplogroup. *BMC Evolutionary Biology, 9*, 63.

Chen, Y.-S., Olckers, A., Schurr, T. G., Kogelnik, A. M., Huoponen, K., & Wallace, D. C. (2000). mtDNA variation in the South African Kung and Khwe – and their genetic relationships to other African populations. *American Journal of Human Genetics, 66*, 1362–1383.

Clark, J. D., Beyene, Y., WoldeGabriel, G., Hart, W. K., Renne, P. R., Gilbert, H., et al. (2003). Stratigraphic, chronological and behavioural contexts of Pleistocene *Homo sapiens* from Middle Awash, Ethiopia. *Nature, 423*, 747–752.

Close, A. E., & Wendorf, F. (1990). North Africa at 18000 BP. In C. Gamble & O. Soffer (Eds.), *The World at 18000 BP* (Vol. 2, pp. 41–57). Low latitudes London: Unwin Hyman.

Cohen, A. S., Stone, J. R., Beuning, K. R. M., Park, L. E., Reinthal, P. N., Dettman, D., et al. (2007). Ecological consequences of early Late Pleistocene megadroughts in tropical Africa. *Proceedings of the National Academy of Sciences USA, 104*, 16422–16427.

Compton, J. S. (2011). Pleistocene sea-level fluctuations and human evolution on the southern coastal plain of South Africa. *Quaternary Science Reviews, 30*, 506–527.

Cruciani, F., Santolamazza, P., Shen, P., Macaulay, V., Moral, P., Olckers, A., et al. (2002). A back migration from Asia to sub-Saharan Africa is supported by high-resolution analysis of human Y-chromosome haplotypes. *American Journal of Human Genetics, 70*, 1197–1214.

Cruciani, F., Trombetta, B., Massaia, A., Destro-Bisol, G., Sellitto, D., & Scozzari, R. (2011). A revised root for the human Y chromosomal phylogenetic tree: The origin of patrilineal diversity in Africa. *American Journal of Human Genetics, 88*, 814–818.

d'Errico, F., Vanhaeren, M., Barton, N., Bouzouggar, A., Mienis, H., Richter, D., et al. (2009). Additional evidence on the use of personal ornaments in the Middle Paleolithic of North Africa. *Proceedings of the National Academy of Sciences USA, 106*, 16051–16056.

de Filippo, C., Barbieri, C., Whitten, M., Mpoloka, S. W., Gunnarsdóttir, E. D., Bostoen, K., et al. (2011). Y-chromosomal variation in sub-Saharan Africa: Insights into the history of Niger-Congo groups. *Molecular Biology and Evolution, 28*, 1255–1269.

Deacon, H. J. (1995). Two Late Pleistocene-Holocene archaeological depositories from the Southern Cape, South Africa. *The South African Archaeological Bulletin, 50*, 121–131.

Eggert, M. K. H. (2005). The Bantu problem and African archaeology. In A. B. Stahl (Ed.), *African archaeology* (pp. 301–326). Oxford: Blackwell.

Fernandes, V., Alshamali, F., Alves, M., Costa, M. D., Pereira, J. B., Silva, N. M., et al. (2012). The Arabian cradle: Mitochondrial relicts of the first steps along the southern route out of Africa. *American Journal of Human Genetics, 90*, 347–355.

Forster, P. (2004). Ice ages and the mitochondrial DNA chronology of human dispersals: A review. *Philosophical Transactions of the Royal Society of London. Series B, Biological Sciences, 359*, 255–264.

Forster, P., & Renfrew, C. (2011). Mother tongue and Y chromosomes. *Science, 333*, 1390–1391.

Forster, P., & Romano, V. (2007). Timing of a back-migration into Africa. *Science, 316*, 50–53.

Francalacci, P., Morelli, L., Angius, A., Berutti, R., Reinier, F., Atzeni, R., et al. (2013). Low-pass DNA sequencing of 1200 Sardinians reconstructs European Y chromosome phylogeny. *Science, 341*, 565–569.

Gonder, M. K., Mortensen, H. M., Reed, F. A., De Sousa, A., & Tishkoff, S. A. (2007). Whole-mtDNA genome sequence analysis of ancient African lineages. *Molecular Biology and Evolution, 24*, 757–768.

Greenberg, J. H. (1963). *The languages of Africa*. Bloomington: Indiana University Press.

Güldemann, T. (2008). A linguist's view: Khoe-Kwadi speakers as the earliest food-producers of southern Africa. *Southern African Humanities, 20*, 93–132.

Güldemann, T., & Stoneking, M. (2008). A historical appraisal of clicks: A linguistic and genetic population perspective. *Annual Review of Anthropology, 37*, 93–109.

Harich, N., Costa, M. D., Fernandes, V., Kandil, M., Pereira, J. B., Silva, N. M., et al. (2010). The trans-Saharan slave trade—clues from interpolation analyses and high-resolution characterization of mitochondrial DNA lineages. *BMC Evolutionary Biology, 10*, 138.

Heine, B., & Nurse, D. (Eds.). (2000). *African languages: An introduction*. Cambridge: Cambridge University Press.

Henn, B. M., Gignoux, C., Lin, A. A., Oefner, P. J., Shen, P., Scozzari, R., et al. (2008). Y-chromosomal evidence of a pastoralist migration through Tanzania to southern Africa. *Proceedings of the National Academy of Sciences USA, 105*, 10693–10698.

Henn, B. M., Gignoux, C. R., Jobin, M., Granka, J. M., MacPherson, J. M., Kidd, J. M., et al. (2011). Hunter-gatherer genomic diversity suggests a southern African origin for modern humans. *Proceedings of the National Academy of Sciences USA, 108*, 5154–5162.

Henn, B. M., Botigué, L. R., Gravel, S., Wang, W., Brisbin, A., Byrnes, J. K., et al. (2012). Genomic ancestry of North Africans supports back-to-Africa migrations. *PLoS Genetics, 8*, e1002397.

Henshilwood, C. S., d'Errico, F., & Watts, I. (2009). Engraved ochres from the Middle Stone Age levels at Blombos Cave, South Africa. *Journal of Human Evolution, 57*, 27–47.

Holden, C. J. (2002). Bantu language trees reflect the spread of farming across sub-Saharan Africa: A maximum-parsimony analysis. *Proceedings of the Royal Society of London. Series B, 269*, 793–799.

Holden, C. J., & Gray, R. D. (2006). Rapid radiation, borrowing and dialect continua in the Bantu languages. In P. Forster & C. Renfrew (Eds.), *Phylogenetic methods and the prehistory of languages* (pp. 19–31). Cambridge: McDonald Institute for Archaeological Research.

Irish, J. D. (2000). The Iberomaurusian enigma: North African progenitor or dead end? *Journal of Human Evolution, 39*, 393–410.

Jacobs, Z., Roberts, R. G., Galbraith, R. F., Deacon, H. J., Grün, R., Mackay, A., et al. (2008). Ages for the Middle Stone Age of southern Africa: Implications for human behavior and Dispersal. *Science, 322*, 733–735.

Kivisild, T., Reidla, M., Metspalu, E., Rosa, A., Brehm, A., Pennarun, E., et al. (2004). Ethiopian mitochondrial DNA heritage: Tracking gene flow across and around the Gate of Tears. *American Journal of Human Genetics, 75*, 752–770.

Kuper, R., & Kropelin, S. (2006). Climate-controlled Holocene occupation in the Sahara: Motor of Africa's evolution. *Science, 313*, 803–807.

Li, J. Z., Absher, D. M., Tang, H., Southwick, A. M., Casto, A. M., Ramachandran, S., et al. (2008). Worldwide human relationships inferred from genome-wide patterns of variation. *Science, 319*, 1100–1104.

Lieberman, D. E., McBratney, B. M., & Krovitz, G. (2002). The evolution and development of cranial form in *Homo sapiens*. *Proceedings of the National Academy of Sciences USA, 99*, 1134–1139.

Lowe, J., Barton, N., Blockley, S., Ramsey, C. B., Cullen, V. L., Davies, W., et al. (2012). Volcanic ash layers illuminate the resilience of Neanderthals and early modern humans to natural hazards. *Proceedings of the National Academy of Sciences USA, 109*, 13532–13537.

Macaulay, V., & Richards, M. B. (2013). *Mitochondrial genome sequences and phylogeographic interpretation. Encyclopedia of Life Sciences (ELS)*. Chichester: John Wiley & Sons, Ltd.

Macaulay, V., Richards, M., Hickey, E., Vega, E., Cruciani, F., Guida, V., et al. (1999). The emerging tree of west Eurasian mtDNAs: A synthesis of control-region sequences and RFLPs. *American Journal of Human Genetics, 64*, 232–249.

Macaulay, V., Hill, C., Achilli, A., Rengo, C., Clarke, D., Meehan, W., et al. (2005). Single, rapid coastal settlement of Asia revealed by analysis of complete mitochondrial genomes. *Science, 308*, 1034–1036.

Marean, C. W., Bar-Matthews, M., Bernatchez, J., Fisher, E., Goldberg, P., Herries, A. I. R., et al. (2007). Early human use of marine resources and pigment in South Africa during the Middle Pleistocene. *Nature, 449*, 905–911.

Marten, L. (2006). Bantu classification, Bantu trees and phylogenetic methods. In P. Forster & C. Renfrew (Eds.), *Phylogenetic methods and the prehistory of languages* (pp. 43–55). Cambridge: McDonald Institute for Archaeological Research.

McDougall, I., Brown, F. H., & Fleagle, J. G. (2005). Stratigraphic placement and age of modern humans from Kibish, Ethiopia. *Nature, 433*, 733–736.

Mellars, P. (2006). Why did modern human populations disperse from Africa ca. 60,000 years ago? A new model. *Proceedings of the National Academy of Sciences USA, 103*, 9381–9386.

Mellars, P., Gori, K. C., Carr, M., Soares, P. A., & Richards, M. B. (2013). Genetic and archaeological perspectives on the initial modern human colonization of southern Asia. *Proceedings of the National Academy of Sciences USA, 110*, 10699–10704.

Mendez, F. L., Krahn, T., Schrack, B., Krahn, A.-M., Veeramah, K. R., Woerner, A. E., et al. (2012). An African American paternal lineage adds an extremely ancient root to the human Y chromosome phylogenetic tree. *The American Journal of Human Genetics, 92*, 454–459.

Millard, A. R. (2008). A critique of the chronometric evidence for hominid fossils: I. Africa and the Near East 500-50 ka. *Journal of Human Evolution, 54*, 848–874.

Mitchell, P. (2002). *The archaeology of southern Africa*. Cambridge: Cambridge University Press.

Mitchell, P. (2010). Genetics and southern African prehistory: An archaeological view. *Journal of Anthropological Sciences, 88*, 73–92.

Montano, V., Ferri, G., Marcari, V., Batini, C., Anyaele, O., Destro-Bisol, G., et al. (2011). The Bantu expansion revisited: A new analysis of Y chromosome variation in central western Africa. *Molecular Ecology, 20*, 2693–2708.

Musilová, E., Fernandes, V., Silva, N. M., Soares, P., Alshamali, F., Harich, N., et al. (2011). Population history of the Red Sea – genetic exchanges between the Arabian Peninsula and East Africa signaled in the mitochondrial DNA HV1 haplogroup. *American Journal of Physical Anthropology, 145*, 592–598.

Newman, J. L. (1995). *The peopling of Africa: A geographic interpretation*. New Haven: Yale University Press.

Nielsen, R., & Beaumont, M. A. (2009). Statistical inferences in phylogeography. *Molecular Ecology, 18*, 1034–1047.

Nurse, D. (1997). Languages of southern and eastern Africa in historical perspective. In J. O. Vogel (Ed.), *Encyclopedia of precolonial Africa* (pp. 166–171). Walnut Creek: AltaMira Press.

Olivieri, A., Achilli, A., Pala, M., Battaglia, V., Fornarino, S., Al-Zahery, N., et al. (2006). The mtDNA legacy of the Levantine early Upper Palaeolithic in Africa. *Science, 314*, 1767–1770.

Oppenheimer, S., & Richards, M. (2001). Fast trains, slow boats, and the ancestry of the Polynesian islanders. *Science Progress, 84*, 157–181.

Ottoni, C., Primativo, G., Kashani, B. H., Achilli, A., Martínez-Labarga, C., Biondi, G., et al. (2010). Mitochondrial haplogroup H1 in North Africa: An early Holocene arrival from Iberia. *PLoS ONE, 5*, e13378.

Pennarun, E., Kivisild, T., Metspalu, E., Metspalu, M., Reisberg, T., Moisan, J.-P., et al. (2012). Divorcing the late Upper Palaeolithic demographic histories of mtDNA haplogroups M1 and U6 in Africa. *BMC Evolutionary Biology, 12*, 234.

Pereira, L., Macaulay, V., Torroni, A., Scozzari, R., Prata, M.-J., & Amorim, A. (2001). Prehistoric and historic traces in the mtDNA of Mozambique: Insights into the Bantu expansions and the slave trade. *Annals of Human Genetics, 65*, 439–458.

Pereira, L., Černý, V., Cerezo, M., Silva, N. M., Hájek, M., Vašíková, A., et al. (2010a). Linking the sub-Saharan and West Eurasian gene pools: Maternal and paternal heritage of the Tuareg nomads from the African Sahel. *European Journal of Human Genetics, 18*, 915–923.

Pereira, L., Silva, N. M., Franco-Duarte, R., Fernandes, V., Pereira, J. B., Costa, M. D., et al. (2010b). Population expansion in the North African Late Pleistocene signalled by mitochondrial DNA haplogroup U6. *BMC Evolutionary Biology, 10*, 390.

Petersen, D. C., Libiger, O., Tindall, E. A., Hardie, R.-A., Hannick, L. I., Glashoff, R. H., et al. (2013). Complex patterns of genomic admixture within southern Africa. *PLoS Genetics, 9*, e1003309.

Petraglia, M., Korisettar, R., Boivin, N., Clarkson, C., Ditchfield, P., Jones, S., et al. (2007). Middle Paleolithic assemblages from the Indian subcontinent before and after the Toba super-eruption. *Science, 317*, 114–116.

Phillipson, D. W. (2002). Language and farming dispersals in sub-Saharan Africa, with particular reference to the Bantu-speaking peoples. In P. Bellwood & C. Renfrew (Eds.), *Examining the farming/language dispersal hypothesis* (pp. 177–187). Cambridge: McDonald Institute for Archaeological Research.

Phillipson, D. W. (2005). *African archaeology*. Cambridge: Cambridge University Press.

Pickrell, J. K., Patterson, N., Barbieri, C., Berthold, F., Gerlach, L., Güldemann, T., et al. (2012). The genetic prehistory of southern Africa. *Nature Communications, 3*, 1143.

Plaza, S., Salas, A., Calafell, F., Corte-Real, F., Bertranpetit, J., Carracedo, Á., et al. (2004). Insights into the western Bantu dispersal: mtDNA lineage analysis in Angola. *Human Genetics, 115*, 439–447.

Podgorná, E., Soares, P., Pereira, L., & Černý, V. (2013). The genetic impact of the Lake Chad Basin population in North Africa as documented by mitochondrial diversity and internal variation of the L3e5 haplogroup. *Annals of Human Genetics, 77*, 513–523.

Pritchard, J. K., & Przeworski, M. (2001). Linkage disequilibrium in humans: Models and data. *American Journal of Human Genetics, 69*, 1–14.

Prugnolle, F., Manica, A., & Balloux, F. (2005). Geography predicts neutral genetic diversity of human populations. *Current Biology, 15*, R159–R160.

Richards, M. B., Macaulay, V. A., Bandelt, H.-J., & Sykes, B. C. (1998). Phylogeography of mitochondrial DNA in western Europe. *Annals of Human Genetics, 62*, 241–260.

Richards, M., Macaulay, V., Hickey, E., Vega, E., Sykes, B., Guida, V., et al. (2000). Tracing European founder lineages in the Near Eastern mtDNA pool. *American Journal of Human Genetics, 67*, 1251–1276.

Richards, M., Macaulay, V., Hill, C., Carracedo, A., & Salas, A. (2004). The archaeogenetics of the dispersals of the Bantu-speaking people. In M. Jones (Ed.), *Traces of ancestry: Studies in honour of Colin Renfrew* (pp. 75–87). Cambridge: McDonald Institute for Archaeological Research.

Richards, M., Bandelt, H.-J., Kivisild, T., & Oppenheimer, S. (2006). A model for the dispersal of modern humans out of Africa. In H.-J. Bandelt, V. Macaulay, & M. Richards (Eds.), *Mitochondrial DNA and the evolution of Homo sapiens* (pp. 225–265). Berlin: Springer.

Rightmire, G. P., Deacon, H. J., Schwartz, J. H., & Tattersall, I. (2006). Human foot bones from Klasies River main site, South Africa. *Journal of Human Evolution, 50*, 96–103.

Rito, T., Richards, M. B., Fernandes, V., Alshamali, F., Černý, V., Pereira, L., et al. (2013). The first modern human dispersals across Africa. *PLoS ONE, 8*(11), e80031.

Rosa, A., & Brehm, A. (2011). African human mtDNA phylogeography at-a-glance. *Journal of Anthropological Sciences, 89*, 25–58.

Rose, J. I. (2010). New light on human prehistory in the Arabo-Persian Gulf Oasis. *Current Anthropology, 51*, 849–868.

Saitou, N. (1996). Contrasting gene trees and population trees of the evolution of modern humans. In A. J. Boyce & C. G. N. Mascie-Taylor (Eds.), *Molecular biology and human diversity* (pp. 265–282). Cambridge: Cambridge University Press.

Salas, A., Richards, M., De la Fe, T., Lareu, M.-V., Sobrino, B., Sánchez-Diz, P., et al. (2002). The making of the African mtDNA landscape. *American Journal of Human Genetics, 71*, 1082–1111.

Scally, A., & Durbin, R. (2012). Revising the human mutation rate: Implications for understanding human evolution. *Nature Review Genetics, 13*, 745–753.

Schlebusch, C. M. (2010). Issues raised by use of ethnic-group names in genome study. *Nature, 464*, 487.

Schlebusch, C. M., Skoglund, P., Sjödin, P., Gattepaille, L. M., Hernandez, D., Jay, F., et al. (2012). Genomic variation in seven Khoe-San groups reveals adaptation and complex African history. *Science, 338*, 374–379.

Scholz, C. A., Johnson, T. C., Cohen, A. S., King, J. W., Peck, J. A., Overpeck, J. T., et al. (2007). East African megadroughts between 135 and 75 thousand years ago and bearing on early-modern human origins. *Proceedings of the National Academy of Sciences USA, 104*, 16416–16421.

Semino, O., Santachiara-Benerecetti, A. S., Falaschi, F., Cavalli-Sforza, L. L., & Underhill, P. A. (2002). Ethiopians and Khoisan share the deepest clades of the human Y-chromosome phylogeny. *American Journal of Human Genetics, 70*, 265–268.

Shea, J. J. (2008). Transitions or turnovers? Climatically-forced extinctions of *Homo sapiens* and Neanderthals in the east Mediterranean Levant. *Quaternary Science Reviews, 27*, 2253–2270.

Shi, W., Ayub, Q., Vermeulen, M., Shao, R.-G., Zuniga, S., Van Der Gaag, K., et al. (2010). A worldwide survey of human male demographic history based on Y-SNP and Y-STR data from the HGDP-CEPH populations. *Molecular Biology and Evolution, 27*, 385–393.

Sikora, M., Laayouni, H., Calafell, F., Comas, D., & Bertranpetit, J. (2011). A genomic analysis identifies a novel component in the genetic structure of sub-Saharan African populations. *European Journal of Human Genetics, 19*, 84–88.

Sjödin, P., Sjöstrand, A. E., Jakobsson, M., & Blum, M. G. B. (2012). Resequencing data provide no evidence for a human bottleneck in Africa during the penultimate glacial period. *Molecular Biology and Evolution, 29*, 1851–1860

Smith, T. M., Tafforeau, P., Reid, D. J., Grün, R., Eggins, S., Boutakiout, M., et al. (2007). Earliest evidence of modern human life history in North African early *Homo sapiens*. *Proceedings of the National Academy of Sciences USA, 104*, 6128–6133.

Soares, P., Trejaut, J. A., Loo, J.-H., Hill, C., Mormina, M., Lee, C.-L., et al. (2008). Climate change and postglacial human dispersals in Southeast Asia. *Molecular Biology and Evolution, 25*, 1209–1218.

Soares, P., Ermini, L., Thomson, N., Mormina, M., Rito, T., Röhl, A., et al. (2009). Correcting for purifying selection: An improved human mitochondrial molecular clock. *American Journal of Human Genetics, 84*, 740–759.

Soares, P., Achilli, A., Semino, O., Davies, W., Macaulay, V., Bandelt, H.-J., et al. (2010). The archaeogenetics of Europe. *Current Biology, 20*, R174–R183.

Soares, P., Rito, T., Trejaut, J., Mormina, M., Hill, C., Tinkler-Hundal, E., et al. (2011). Ancient voyaging and Polynesian origins. *American Journal of Human Genetics, 88*, 239–247.

Soares, P., Alshamali, F., Pereira, J. B., Fernandes, V., Silva, N. M., Afonso, C., et al. (2012). The expansion of mtDNA Haplogroup L3 within and out of Africa. *Molecular Biology and Evolution, 29*, 915–927.

Stager, J. C., Ryves, D. B., Chase, B. M., & Pausata, F. S. R. (2011). Catastrophic drought in the Afro-Asian monsoon region during Heinrich Event 1. *Science, 331*, 1299–1302.

Stringer, C. (2011). *The origin of our species*. London: Penguin.

Texier, P. J., Porraz, G., Parkington, J., Rigaud, J. P., Poggenpoel, C., Miller, C., et al. (2010). A Howiesons Poort tradition of engraving ostrich eggshell containers dated to 60,000 years ago at Diepkloof

Rock Shelter, South Africa. *Proceedings of the National Academy of Sciences USA, 107*, 6180–6185.
Tishkoff, S. A., Gonder, M. K., Henn, B. M., Mortensen, H., Knight, A., Gignoux, C., et al. (2007). History of click-speaking populations of Africa inferred from mtDNA and Y chromosome genetic variation. *Molecular Biology and Evolution, 24*, 2180–2195.
Tishkoff, S. A., Reed, F. A., Friedlaender, F. R., Ehret, C., Ranciaro, A., Froment, A., et al. (2009). The genetic structure and history of Africans and African Americans. *Science, 324*, 1035–1044.
Torroni, A., Achilli, A., Macaulay, V., Richards, M., & Bandelt, H.-J. (2006). Harvesting the fruit of the human mtDNA tree. *Trends in Genetics, 22*, 339–334
Vansina, J. (1995). New linguistic evidence and the Bantu expansion. *Journal of African History, 36*, 173–195.
Veeramah, K. R., Wegmann, D., Woerner, A., Mendez, F. L., Watkins, J. C., Destro-Bisol, G., et al. (2012). An early divergence of KhoeSan ancestors from those of other modern humans is supported by an ABC-based analysis of autosomal resequencing data. *Molecular Biology and Evolution, 29*, 617–630.
Vogel, J. O. (1997a). Bantu expansion. In J. O. Vogel (Ed.), *Encyclopedia of precolonial Africa* (pp. 435–438). Walnut Creek: AltaMira Press.
Vogel, J. O. (1997b). Eastern and South-Central African Iron Age. In J. O. Vogel (Ed.), *Encyclopedia of precolonial Africa* (pp. 439–444). Walnut Creek: AltaMira Press.
Watson, E., Forster, P., Richards, M., & Bandelt, H.-J. (1997). Mitochondrial footprints of human expansions in Africa. *American Journal of Human Genetics, 61*, 691–704.
Wei, W., Ayub, Q., Chen, Y., McCarthy, S., Hou, Y., Carbone, I., et al. (2012). A calibrated human Y-chromosomal phylogeny based on resequencing. *Genome Research, 23*, 388–395.
Weldeab, S., Lea, D. W., Schneider, R. R., & Andersen, N. (2007). 155,000 years of West African monsoon and ocean thermal evolution. *Science, 316*, 1303–1307.
Yang, Z. H. (1997). PAML: A program package for phylogenetic analysis by maximum likelihood. *Computer Applications in the Biosciences, 13*, 555–556.

Chapter 19
Africa from MIS 6-2: Where Do We Go from Here?

Peter Mitchell

Abstract This chapter seeks to identify directions along which future research into the population dynamics of African hunter-gatherer communities and the paleoenvironmental context in which they lived can be explored. These themes include: critically assessing existing datasets for their quality and chronological precision; transcending outdated stadial "block-like" thinking about the past in favor of more dynamic understandings of continuous processes of change; converting statements about broad climatic trends into more precise, quantitative statements about past temperature, and rainfall patterns in order to approximate more closely the impact of climatic change on the resources exploited by past hunter gatherers; examining the connectivity and isolation of human populations, concerting archaeological, and paleoenvironmental evidence with data from genetics and human paleontology; and addressing the challenges posed for archaeologists' use of ethnographic analogy by the recognition of "modern" forms of human cognition and behavior at much greater time depths than was once thought plausible. The chapter concludes by reflecting upon the relevance of MIS 6-2 research for African societies today.

Keywords Chronology • Climate change • Demography • Ethnographic analogy • Cognition & Behavior

P. Mitchell (✉)
School of Archaeology, University of Oxford
& St Hugh's College, Oxford, OX2 6LE, UK
e-mail: peter.mitchell@st-hughs.ox.ac.uk

P. Mitchell
GAES, University of the Witwatersrand, Johannesburg, South Africa

Introduction

When I spoke at the end of the 2010 conference, from which this volume draws its inspiration, I noted that one of its successes lay in the effectiveness with which it had drawn together such a wide diversity of speakers – colleagues who were not only working in many different parts of the African continent, but also in quite different disciplines: archaeology, for sure, but also genetics, human paleontology, and a variety of paleoenvironmental sciences. That same diversity is evident in the contributions to the present volume. In writing this concluding piece, I do not, however, address in detail the arguments already made. Rather, I identify and explore some of the pathways along which future research into the dynamics and paleoenvironmental context of the human populations who inhabited Africa during MIS 6-2 might usefully flow. Six such themes spring to mind:

1. The importance of improving the chronological precision of our existing data.
2. The importance of rigorously assessing the quality of what we think we already know.
3. The necessity of transcending stadial modes of thinking about the past in favor of more dynamic understandings of how people and climates were engaged in continuous processes of change.
4. The significance of seeking to convert statements about quite broad climatic trends or contrasts into more precise quantitative estimates of past temperature and rainfall patterns in order to approximate more closely the impacts of climate change on the resources that hunter-gatherers exploited.
5. The value of investigating – using genetic and fossil, as well as cultural, evidence – the connectivity, but also the isolation, of human populations within and between different parts of Africa.
6. Finally, the challenges posed for archaeologists' use of ethnographic analogy by the recognition of "modern"

Sacha C. Jones and Brian A. Stewart (eds.), *Africa from MIS 6-2: Population Dynamics and Paleoenvironments*, Vertebrate Paleobiology and Paleoanthropology, DOI 10.1007/978-94-017-7520-5_19

forms of human cognition and behavior at much greater time depths than we had once imagined possible.

I conclude by considering the relevance of research into Africa's past populations during MIS 6-2 for those living on the continent today.

Controlling Time and Quality

One of the founding fathers of European prehistory, Rasmus Nyerup, despondently remarked at the beginning of the nineteenth century that "everything that has come down to us from heathendom is wrapped in a thick fog; it belongs to a space of time that we cannot measure" (Daniel 1975: 38). I exaggerate only a little when noting that until a short while ago much the same could have been said of the periods discussed in the current volume. AMS technology and the development of new ultra-filtration sample preparation techniques have helped radiocarbon to dispel some of that fog, though contamination issues still bedevil its use in the later stages of MIS 3 (Higham 2011). Other chronometric methods – optically stimulated luminescence (OSL) above all, but also Thermoluminescence (TL) and Electron Spin Resonance dating – have also come of age and are increasingly deployed. Although their results continue to be debated (currently, for example, over the age of southern Africa's Howiesons Poort and Still Bay industries [Tribolo et al. 2013; cf. Jacobs et al. 2008; and see also the discussion here by Robbins et al. (2016) of OSL and TL ages at White Paintings Shelter, Botswana], the scale on which these techniques can now be applied to establish the chronology of single sites or to date comparable assemblages across many different sites represents a quantum advance for MIS 6-2 research. Nevertheless, as virtually every paper here notes, more and better dates are still very much required, not only for archaeological materials and paleoenvironmental sequences, but also for hominin fossils (Grine 2016). Securing more tightly constrained estimates of genetic divergence, with narrower confidence intervals more commensurate with those available from radiometric dating, is particularly important (Soares et al. 2016).

But refining our chronologies is not enough. We also need to have high quality evidence to date. Alongside excavating new sites, one way of achieving this is by applying state-of-the-art recovery methods and analytical techniques to sites dug with poorer stratigraphic controls decades ago: excavations by Deacon (1995) at Klasies River, or Prendergast and colleagues at Mumba Shelter, Tanzania, (Prendergast et al. 2007), along with renewed fieldwork at Taforalt and the Haua Fteah in North Africa (Barton et al. 2013; Douka et al. 2014), exemplify the importance of continuing this approach. Such sites are large and well known, but the MIS 6-2 record is also made up of a myriad of other, often smaller rock shelters, as well as open-air sites and even isolated artifacts (Cancellieri et al. 2016). A regional approach that looks at variation in site/artifact distributions across the landscape at scales geared to how foragers actually lived, rather than a narrow focus on highly productive, but contextually isolated "megasites," is sorely needed in almost all of Africa.

Assessing the quality of our existing evidence is also important, including that part of it still lying unanalyzed in museums. As Cornelissen (2016) shows in her paper on the Central African record, its careful study can still pay dividends. With regard to dating, the concept of chronometric hygiene is now well known (e.g., Sadr and Sampson 2006; Drake and Breeze 2016), but its principles can be productively extended to other sets of data. Taylor (2016) leaves literally almost no stone unturned here in his thorough reassessment of the Lupemban technocomplex, meeting a critical need if it (and no doubt some of the other entities with which we currently populate MIS 6-2) is to constitute a useful and productive concept for future research. For the fossil record, Grine's paper (2016) likewise provides an impressive state-of-the art survey, noting how readily poor quality data can end up being reused in wholly circular arguments (see his discussion of the supposed age of the Eliye Springs and Oranjemund fossils). Soares et al. (2016) recognize the same need for critical appraisal of the samples and interpretations employed by molecular geneticists, while Van Peer (2016) notes the importance of more systematic and detailed, large-scale studies of lithic technology so that we can be sure that the terminologies we use mean the same thing across sites and regions.

Putting Lego Back in Its Box

Archaeologists and others have a fondness for thinking in blocks – neat, encapsulated units that summarize a set of ideas and evidence and serve as shorthands that can be readily understood by their peers. The very concept of a "Marine Isotope Stage" is, of course, just that. And blocks are indeed useful for organizing and discussing concepts and data. But to an even greater extent than we may often think, they also control how we view the world, influencing, when not dictating, the kinds of questions that we ask and the approaches that we take to answering them. The homogenization of enormous linguistic and anthropological variability entailed in most uses of the term "Khoisan" that falsely impute uniformity across people with very complex histories is an excellent example (Güldemann and Stoneking 2008; cf. Soares et al. 2016).

Another is the transition from the Middle Stone Age (MSA) to the Later Stone Age (LSA), still not wholly disentangled from the broader topic of the emergence of "modern" behavior, with its own implication that this can be tightly defined and neatly separated off from something "non-modern," The difficulty here is the triune falsity that: (a) there is a single "MSA" entity; (b) this was followed by an equally unitary "LSA" phenomenon; and (c) only one historical pathway connected the two. In this volume, Mackay shows how inadequate such assumptions are and how much more productive it can be to take a Darwinian approach more focused on variation and the fluid, dynamic nature of technological (and, indeed, other) change within a well-dated chronological framework. Following in the tracks of Sampson (1985), Mackay also points out how the discontinuous nature of our stratigraphic sequences surreptitiously encourages us to understand change in episodic, saltatory terms: we should not compound this difficulty by retaining an overcommitment to cultural-historical pigeon-holing derived from a time before radiometric dating. Finally, he reminds us that while *we* recognize a directionality connecting past and present, this was not apparent to those who created the MIS 6-2 archaeological record and that change did not necessarily (if ever) follow a progressive, unidirectional path toward ever-growing complexity; recent discussion of what some have unduly seen as cultural "regression" in southern Africa in MIS 3 after the dizzy heights of the Still Bay and Howiesons Poort explores this further (e.g., Lombard and Parsons 2011). The conclusion to draw from all this? It is surely time to put our Lego blocks away.

Overcoming a stadial view of the past that directs us to minimizing variation *within* cultural-historical units while overemphasizing the gaps *between* them is all the more necessary since the African record increasingly seems to demonstrate the very deep historical roots and gradual nature of many of the key changes in which we are interested: pigment processing, hafting, and long-distance transport of raw materials, among others (McBrearty and Brooks 2000), as well as the biological evolution of *Homo sapiens* itself (Lahr and Foley 2016). It is also needed because it is abundantly clear that climate change was also extremely complex, far from directional in any simple, unilinear fashion, and marked by significant fluctuations on millennial, centennial, and decadal scales (Blome et al. 2012; and many of the contributors here). Understanding the links between paleoenvironmental change, human demography, and social dynamics requires us to think across all of those scales and to grasp that it is highly unlikely that ecosystems responded to them in any simple up/down, north/south, or east/west fashion. Rather, as Faith et al. (2016) suggest for East Africa, we must expect that ecologies were repeatedly resorted in novel ways, with different plant and animal species responding to climatic variation on their own terms. Their chapter and earlier work by Drake et al. (2011) make clear how this opens up considerable scope for productive conversations with botanists, zoologists, ecologists, and those studying the genetics of plants and animals.

The Weather Abroad Will Be...

Running through virtually all the papers here is a call for many more detailed terrestrial paleoenvironmental sequences. As they are captured and studied, we also need to think more carefully about the precise nature of the conditions with which people had to cope. This will not be easy: clearly, some quite major elements of climate forcing mechanisms remain imperfectly understood, for example. But we can at least set out a series of desiderata: since unpredictability may have posed the greatest test to the resilience of human communities (cf. Dugmore et al. 2007), should we look, in particular, at episodes of greatest variability in climate such as the transitions into and out of the Last Glacial (Burrough 2016; Robbins et al. 2016)? How did climate change affect the staple resources on which human communities depended? What were those resources? What were their ecological tolerance levels? What may have been the limiting factors on them and thus on people? Here, again, we have a research agenda that can be mapped out and shared across a wide disciplinary range: not just archaeology and paleoenvironmental science, but also biology and anthropology. Talking to botanists and ethnobotanists might be particularly valuable: most mentions of resources here privilege game (admittedly easier to find evidence of in the archaeological record), rather than the plants that likely underpinned forager subsistence strategies across Africa (cf. Deacon 1993), but are only rarely preserved (e.g., Mercader 2009).

Progress will, however, depend on being able to refine our understanding of paleoclimatic variation and on being able to convert broad generalizations about warmer or colder conditions, wetter or drier climates, into something more quantitative, figures that can provide at least a crude estimate of the paleotemperature and paleoprecipitation characterizing a particular place at a particular time (cf. Roberts et al. 2013; Truc et al. 2013). Where such estimates can be advanced, an extensive body of comparative ethnographic data is available to us from the very recent past on how many features of hunter-gatherer territorial organization, group size, demography, and subsistence relate to them via seasonality, rainfall, and surface water availability (e.g., Binford 2001; Kelly 2013). Though little applied thus far in an African context, they give us extensive scope for modeling the potential consequences for hunter-gatherer behavior of fluctuations across a range of possible rainfall and temperature configurations.

As we do this, and as we excavate new data out of the ground, a better understanding of the resilience of Pleistocene hunter-gatherers in Africa should emerge: how far could people be "pushed" before they had to abandon particular regions of the continent? How quickly were they able to take advantage of new opportunities as climate ameliorated? Can we find convincing evidence of social or technological innovations mapping onto episodes of environmental change in ways that allowed them to persist in places where earlier populations had been unable to do so? Conversely, which conditions were "bad" and what do we mean by that? Is, for example, the apparent absence of people from some regions of interior southern Africa over the Last Glacial Maximum (which is perplexingly at odds with their uncontested ability to survive in very arid parts of that region during the Holocene; Brooks and Robertshaw 1990) a signal of very challenging conditions? Or does it result from research bias, variation in site visibility, or different technologies, social strategies, or systems of knowledge about regional and local landscapes? Across the Equator, and earlier in time, if we see Aterians as at least partly desert-adapted (cf. Garcea 2012), then drier conditions in the Sahara suddenly become less of a barrier to human settlement and movement than if they are linked only to wetter episodes (Drake and Breeze 2016). Here, too, Lego-like understandings of past ecologies in gross spatiotemporal blocks need replacing by a more subtle appreciation of the diversity of ecological opportunities that even apparently forbidding landscapes hold for those who know them well (Cancellieri et al. 2016).

Dewar and Stewart (2016) make a similar point with respect to their work in Namaqualand: what may be globally "bad" may be locally "good," here because cooler temperatures will produce lower evapotranspiration and thus increased primary productivity, even without corresponding changes in precipitation itself (though higher rainfall also seems likely in MIS 2 and 4). Further north, Burrough (2016) makes another, not necessarily self-evident observation: the ecological possibilities of some areas (such as northern Botswana, where surface water availability largely reflects flooding of the Okavango Delta and waterways linked to it) depend more upon the climate of others far removed in space than upon local rainfall patterns. As a corollary, might we expect their population dynamics to align most closely with regions sharing the same catchment (in this case, much of south-central Africa), rather than with others often seen as geographically closer but, in fact, climatologically more distant (the rest of southern Africa)? In all these respects, building up robust local models will best allow us to develop and test ideas about the links between environmental and cultural change (Carr et al. 2016).

African *Connections*

For many of our colleagues the dominant narrative of MIS 6-2 is not really an African one at all, but the expansion beyond Africa of anatomically modern humans and their colonization *of the rest of the planet*. Yet this was but the tail end of a complex set of movements and interactions within Africa itself and it is to those *African* connections, rather than links between Africa and Eurasia, that this volume is dedicated.

Connections – and the potential for connections – between different regions of the continent go to the heart of that theme. Most graphically, perhaps, we are now gaining increasing purchase on connections across the Sahara, the corridors – sometimes very wide – along which people (and animals; Drake et al. 2011) could live and move at times of enhanced rainfall between sub-Saharan West Africa and (via the Nile) East Africa on the one hand and the North African littoral on the other (Coulthard et al. 2013; Drake and Breeze 2016). Other routes are also emerging, notably southern Africa's Orange/Senqu River, the headwaters of which may have provided an attractive refugium during particularly arid episodes (Stewart et al. 2016). Areas of increased topographic relief like the Maloti-Drakensberg Mountains also compress ecological diversity horizontally while enhancing it vertically, creating situations that people can profitably exploit by moving between different elevational zones; Lahr and Foley (2016) emphasize precisely this when discussing the Rift Valley in their survey of the East African record.

Can we identify other potential refugia, perhaps in areas thus far little researched? In South Africa, the Drakensberg Escarpment of Mpumalanga and Limpopo provinces constitutes a northerly extension of the Maloti-Drakensberg Mountains that Stewart et al. discuss, with stratigraphically problematic earlier excavations at Bushman Rock Shelter and Heuningneskrans hinting at their recurrent use over MIS 3-2 at least (Beaumont 1981; Plug 1981). Further south, did the Pondoland coast provide resources as rich and stable as those thought to have existed to its west in the Fynbos Biome of the Cape (cf. Fischer et al. 2013)? And on a much larger scale, let us recall that even today the Saharan massifs – the Tibesti, Ennedi, Ahaggar, and Aïr – receive sufficient rainfall to support pasture, fields, and relic communities of both Mediterranean and savanna species (Newby 1984). Since this was also probably repeatedly the case during the Pleistocene, did people profit from it then (Clark 1998)?

Ecological modeling can, however, only tell us where people may have preferred to move and live. How connectivity between populations was maintained is an equally important question, particularly with regard to the sharing of information and access to resources (to eat, to make tools, and to marry; Wiessner 1982). Sealy (2016) rightly emphasizes the significance of such social networks when considering how an understanding of Holocene hunter-gatherer population dynamics in southernmost Africa might inform models of such dynamics during MIS 6-2. Without falling into the trap of turning all Pleistocene Africans into mirror images of the Ju/'hoãnsi (see below), it is possible to see the long-distance movement of lithic raw materials, shell, or other items as potential evidence of similar connections much further back in time (Ambrose 2006). Obsidian is perhaps the best-studied example of this, with documented movements of >250 km in East Africa (Merrick and Nash 1994; Faith et al. 2016). Elsewhere, the movement of nonlocal silcrete into Botswana's Tsodilo Hills (Robbins et al. 2016) and silicified tuff to the mid-Saharan site of Adrar Bous (Clark et al. 2008) offers others. Finally, should we imagine people moving by water as well as overland? Within Africa, the Dufuna canoe from an early Holocene context in northeastern Nigeria is the oldest watercraft known (Breunig et al. 1996), but colonization of Sahul, the Bismarck Archipelago, and the Ryukyu Islands during MIS 3 demonstrates their much greater antiquity elsewhere, and, however flimsy, they were almost certainly involved in any crossing of the Bab el Mandeb from Africa into Arabia (Lambeck et al. 2011). Might boats have helped people institute and maintain connections along the Pleistocene waterways of a green Sahara?

While seeking connectivity we have also to consider its flip side, isolation. Are there parts of Africa where we can, at times, explore the links between people, environment, and the development of novel social and technological patterns without being concerned about inputs from outside, situations analogous to the way in which the Old World acts as a control for exploring the development of food-production, urbanism, and the state in the Americas? In the south, perhaps South Africa/Lesotho constituted one such area during arid phases in the Kalahari, though we should not assume that increased rainfall necessarily produced greater interaction: covering some 66,000 km^2, Paleolake Makgadikgadi and its associated bodies of water may (without boats) have posed a serious barrier to movement across northern Botswana (Burrough 2016). On a much bigger scale, the Sahara seems likely to have insulated the Maghreb from West Africa on multiple occasions, while within North Africa any extension of the desert to the Mediterranean coast would have likely separated Cyrenaica (discussed in this volume by Jones et al. 2016) from the Maghreb proper, and perhaps Cyrenaica from the Nile. In those circumstances, did populations endure or must we anticipate extinctions and replacements on a regional scale (Close 2002; cf. Jones et al. 2016; Soares et al. 2016)?

We also need to keep in mind how easily notions of connectivity and isolation are affected by biases in research coverage. Despite several recent studies (e.g., Mercader et al. 2009; Barham et al. 2011; Willoughby 2012), a broad swathe of south-central Africa from Angola to the Indian Ocean remains a blank on our late Pleistocene map. Given the accelerating pace of research and discovery in South Africa, the reemergence of more northerly parts of East Africa as a focus of MIS 6-2 research, and the tendency of genetic models to identify either southern or East Africa as the likely origin region for modern humans, exploring that gap becomes an important task. Other areas, of course, are also very poorly known: the Congo Basin (Cornelissen 2016; Taylor 2016) and, even more so, the whole of the forest zone of West Africa (the absence of a regional overview here is notable, but not unexpected). And in southern Africa itself, far more effort continues to be devoted to coastal/near-coastal sites (especially in the Fynbos Biome) than to the interior (Burrough 2016; Robbins et al. 2016; Stewart et al. 2016).

At a site rather than a regional level more attention should also be paid to open-air sites. Rock shelters are useful, make no mistake, for preserving sequences and organic remains, but much of what people likely did in them is sleep and eat! Most of what they had to do for a living took place instead "out-of-doors" and at some times and in many places open-air sites may be all we have. Maximizing what we can learn from patterning in artifacts found in the open may be taxing, but not all parts of Africa are predominantly erosional in nature: the sites discussed by Soriano et al. (2010) near Mali's Bandiagara Escarpment are an excellent example to the contrary, along with Florisbad in South Africa (Brink 1987), while in East Africa Katanda, with its barbed bone points, can surely not be unique (Yellen et al. 1995).

Finally, let us remember that the holistic vision we seek requires archaeological and paleoenvironmental data to be integrated with the results of paleontological and molecular genetic research. Human fossil material will, sadly, always be a rare and precious resource, the occurrence of which cannot be predicted. But both it and genetic data nevertheless suggest hypotheses of population isolation, connection, and spread that are susceptible to evaluation against archaeological and paleoenvironmental evidence. For MIS 6-2, for example, Soares et al. (2016) suggest several stimulating possibilities, including possible movements from southern into East Africa ~130–70 ka, perhaps causally linked to the megadrought identified by Scholz et al. (2007) and again less than 70 ka (to account for the dispersal there of the L0 and L0a'b mitochondrial lineages); broader expansions within Africa associated with the diversification of the L3 lineage coincident with the spread of modern humans beyond the continent ~70 ka; a movement from East into Central Africa ~60 ka introducing L3b'd (and perhaps L3e); and the spread of the U6 and M1 haplotypes into North Africa, likely from Southwest Asia, 55–33 ka. The latter can certainly find a potential archaeological linkage in the form of the arrival of the Dabban in Cyrenaica (Jones et al. 2016), though for now the others are more difficult to match against the fossil record. That fossil record, in turn, suggests that as modern humans expanded within Africa we should not be surprised if they interbred with more archaic populations, or even that some such populations persisted for some time (Grine 2016; cf. Harvati et al. 2011). It is now up to archaeologists to seek confirmation or refutation of these models in the material and, if we are lucky, paleontological records.

Anyone for *Hxaro*?

I want to turn now to the tendency in southern Africa, the region I know best, to retrodict elements of San ethnography tens of millennia into the past. This tendency has been around for some time, fuelled recently by the demonstration from Taforalt to Blombos that the supposed cognitive Rubicon near the start of Europe's Upper Paleolithic has collapsed, that people in Africa have likely had the same genetically based cognitive capacities as us for at least 100 kyr, and that the once firm cultural divide between the MSA and LSA warrants dissolution. Deacon (1995), for example, drew direct parallels between the formal design of Howiesons Poort segments and those of the mid-Holocene Wilton Industry, interpreting both as weapon armatures that, by analogy with Ju/'hoãn practice, might have been involved in reciprocal gift-exchange networks. Wadley (1987), too, once made the same suggestion, while Lewis-Williams and Pearce (2004: 5–22, 36) have argued that because the makers of Howiesons Poort tools emphasized the use of quartz (which, they argue, had shamanic associations) aspects of San cosmology might be projected back that far (but see Parkington 2005). Most recently, significant new information about the nature and antiquity of various organic artifacts from Border Cave has been hailed as "early evidence of San material culture" 43 ka (d'Errico et al. 2012).

A major difficulty here is that finding ostrich eggshell beads or what we think are arrow armatures patently does *not* prove the presence of Ju/'hoan-like *hxaro* gift-exchange networks: today such networks also include many other items; beads and arrows can be made for reasons other than exchange; and not all documented San groups practiced *hxaro* or something similar to it (Mitchell 2003). More compellingly, it seems highly probable that variation in precolonial hunter-gatherer lifeways in southern Africa far exceeded the range documented by twentieth-century ethnographers (cf. Sealy 2006) and the even narrower sample drawn upon by archaeologists seeking interpretative models. Nor should this surprise: virtually all San ethnography comes not just from a brief window in time, but from just two of southern Africa's biomes (thc Kalahari Savanna and the Karoo-Nama). Before linking Pleistocene artifacts to "San social organization, world view, and symbolic systems" (d'Errico et al. 2012: 13214), we had best therefore understand how, if at all, we can do so reliably in the present, a challenge laid down two decades ago (Barham 1992), but still not yet picked up. Otherwise, archaeologists risk creating a timeless Khoisan history in which nothing changed and from which San people today can easily be construed as living fossils, with all the damaging real-world political implications that this can entail (Reid 2005: 352).

I have written here as a southern African specialist, but the conclusion to be drawn from these paragraphs holds across the continent and returns us to a point on which I have already touched. And it is this: if what we have to say is to be heard by our colleagues working in other parts of the world, if, in other words, we are to get across our conviction that the African record is of interest for itself and not just as some swiftly dispensable prelude to the "main event" in Europe, Asia, Australia, or the Americas, then we have to find ways of contributing to hunter-gatherer research as a whole. Stunning discoveries and more effective networking between our various disciplines are not enough. For the archaeology of African MIS 6-2 hunter-gatherers to form a valuable, and valued, part of hunter-gatherer anthropology as a whole, recent work on detailing and understanding the complex cognitive steps involved in such things as heat-treating stone, manufacturing adhesives, constructing bows-and-arrows, or building traps offers one such contribution (Brown et al. 2009; Wadley et al. 2009; Wadley 2010; Lombard and Haidle 2012).

More directly germane to this volume, we can also (as I have already noted) draw upon well-established principles relating hunter-gatherer behavior to environmental variation when building models of past population dynamics. We should also not be shy of reaching beyond Africa to investigate how the patterns in colonization, connectivity, isolation, innovation, and change that we detect compare with those evident among modern humans in other parts of the world. Australian interest in comparing technological change among hunter-gatherers in arid landscapes and investigating the origins of microlithic toolkits is a good example of what I have in mind (Hiscock and O'Connor 2005; Hiscock et al. 2011). Lastly, we should also emphasize how, for the bulk of the period covered by MIS 6-2, it is only in Africa that we can see how *Homo sapiens* rose to the challenges of Pleistocene environmental change: surely a wonderful opportunity for informative comparison with the behavior of other kinds of human (Neanderthals, Denisovans, *H. floresiensis*) in Eurasia if we structure those comparisons using the kind of biome-based approach followed here.

MIS 6-2 in MIS 1 and Counting: What's the Point?

But how, I am left asking, can our research connect with today's Africa? What, if any, linkage should we seek, or strive to build, between the potentially esoteric pursuits of the international academy and the day-to-day preoccupations of contemporary Africans? Through no fault of Stewart and Jones as editors/conference organizers, those questions arise all the more because of the sadly limited African presence at Cambridge in 2010 (just four out of 53 authors) and in the papers collated here (one out of 45). Using the projects reported here as vehicles to create opportunities for African students and colleagues is a wholly necessary and right response to this situation. So too are other ventures in building increased – and increasingly sustainable – capacity for archaeological, paleoenvironmental, paleontological, and molecular genetic research within the continent, including participation in international conferences and better access to up-to-date journals and books. But on a continent where much research continues to be conducted and funded largely from the outside, a much more equal involvement in the *design and management*, rather than just the execution, of fieldwork and laboratory research is still called for: the cooperation between Moroccan and British researchers evident in ongoing work at Taforalt and other sites in northern Morocco is a good example of this (e.g., Bouzouggar et al. 2007), represented in Cambridge, though regrettably not here.

For such efforts to be sustainable they must tap into a constituency within Africa, that is, MIS 6-2 research has to be seen as valuable in and of itself, not just as a vehicle for developing a few academic careers or professions. Can it do this? One response departs from the obvious point that we all descend from people living in Africa in the early part of this period. In that sense, work on MIS 6-2 *in* Africa reaffirms humanity's common African origin and simultaneously affirms the value of the continent's cultural heritage. This requires us to look outward, beyond our own academic fellowship: contributing to school curricula that reference our African origins (Esterhuysen and Lane 2013), developing a heritage tourism that celebrates them (cf. Mabulla 2000; Bonner et al. 2007), and spreading their news through other media (is there scope for a traveling international exhibition akin to the *Ancestors* show [Delson 1985] of 30 years ago?) are all possibilities to pursue.

That pursuit, like the research discussed here, takes place under the threat of significant global climate change. Because of its aridity, fragile ecosystems, and exposure to natural disasters like flood and drought, and because of its relative poverty and poorly developed infrastructure, Africa is particularly exposed to this threat. Serious, adverse consequences loom for the economies of many African countries and the livelihoods of millions of Africans. A rise in average global temperatures of 2 °C, or more, before the end of this century is now extremely likely on all but the most optimistic scenarios (Stocker et al. 2013), scenarios that depend on an improbable abandonment of rhetoric in favor of action and of short-term political and economic gains over long-term ecological benefits. The impacts on agricultural production, water stress, human migration, public health, and biodiversity loss will be severe (World Bank 2009). And yet our understanding of what drives African climates remains uncertain, our knowledge of how global drivers produce local consequences poor, our ability to predict change accurately at subregional level weak (Conway 2009).

I have argued elsewhere that archaeologists have a professional responsibility to treat the issue of global climate change seriously (Mitchell 2008). Along with colleagues in the field of paleoenvironmental science, they are uniquely well placed to retrieve data that can identify previous episodes of climatic and ecological change and thus help assess climate models of the future by evaluating their retrodictions of what happened in the past (e.g., Willis et al. 2013). MIS 6-2 included many fluctuations in temperature, precipitation, sea-level, and other key parameters, including the two largest, most intense warmings of the last 190 kyr: the transition from MIS 6 to MIS 5 and the much more recent shifts marking the Pleistocene/Holocene boundary. By obtaining evidence of those changes, perhaps our research about human population dynamics in Africa during MIS 6-2 can make some tangible contribution to those living there today.

References

Ambrose, S. H. (2006). Howiesons Poort lithic raw material procurement patterns and the evolution of modern human behavior: A response to Minichillo (2006). *Journal of Human Evolution, 50*, 365–369.

Barham, L. S. (1992). Let's walk before we run: An appraisal of historical materialist approaches to the Later Stone Age. *South African Archaeological Bulletin, 47*, 44–51.

Barham, L. S., Phillips, W. M., Maher, B. A., Karloukovski, V., Duller, G. A. T., Jain, M., et al. (2011). The dating and interpretation of a Mode 1 site in the Luangwa Valley, Zambia. *Journal of Human Evolution, 60*, 549–570.

Barton, R. N. E., Bouzouggar, A., Hogue, J. T., Lee, S., Collcutt, S. N., & Ditchfield, P. (2013). Origins of the Iberomaurusian in NW Africa: New AMS radiocarbon dating of the Middle and Later Stone Age deposits at Taforalt Cave, Morocco. *Journal of Human Evolution, 65*, 266–281.

Beaumont, P. B. (1981). The Heuningneskrans Shelter. In E. A. Voigt (Ed.), *Guide to archaeological sites in the northern and eastern Transvaal* (pp. 133–145). Pretoria: Southern African Association of Archaeologists.

Binford, L. R. (2001). *Constructing frames of reference: An analytical method for archaeological theory building using ethnographic and environmental data sets*. Berkeley: University of California Press.

Blome, M. W., Cohen, A. S., Tryon, C. A., Brooks, A. S., & Russell, J. (2012). The environmental context for the origins of modern human diversity: A synthesis of regional variability in African climate 150,000–30,000 years ago. *Journal of Human Evolution, 62*, 563–592.

Bonner, P., Esterhuysen, A., & Jenkins, T. (Eds.). (2007). *A search for origins: Science, history and South Africa's 'Cradle of Humankind'*. Johannesburg: Wits University Press.

Bouzouggar, A., Barton, R. N. E., Vanhaeren, M., d'Errico, F., Collcutt, S., Higham, T. F. G., et al. (2007). 82,000-year-old shell beads from North Africa and implications for the origins of modern human behavior. *Proceedings of the National Academy of Sciences USA, 104*, 9964–9969.

Breunig, P., Neumann, K., & van Neer, W. (1996). New research on the Holocene settlement and environment of the Chad Basin in Nigeria. *African Archaeological Review, 13*, 111–145.

Brink, J. S. (1987). The archaeozoology of Florisbad, Orange Free State. *Memoirs van die Nasionale Museum, Bloemfontein, 24*, 1–155.

Brooks, A. S., & Robertshaw, P. T. (1990). The Glacial Maximum in tropical Africa 22-12,000 BP. In C. S Gamble & O. Soffer (Eds.), *The World at 18 000 BP: Volume II low latitudes* (pp. 121–169). London: Unwin Hyman.

Brown, J., Marean, C. W., Herries, A. I. R., Jacobs, Z., Tribolo, C., Braun, D., et al. (2009). Fire as an engineering tool of early modern humans. *Science, 325*, 859–862.

Burrough, S. L. (2016). Late Quaternary environmental change and human occupation of the southern African interior. In S. C. Jones & B. A. Stewart (Eds.), *Africa from MIS 6-2: Population dynamics and paleoenvironments* (pp. 161–174). Dordrecht: Springer.

Cancellieri, E., Cremaschi, M., Zerboni, A., & di Lernia, S. (2016). Climate, environment and population dynamics in Pleistocene Sahara. In S. C. Jones & B. A. Stewart (Eds.), *Africa from MIS 6-2: Population dynamics and paleoenvironments* (pp. 123–145). Dordrecht: Springer.

Carr, A. S., Chase, B. M., & Mackay, A. (2016). Mid to Late Quaternary landscape and environmental dynamics in the Middle Stone Age of southern South Africa. In S. C. Jones & B. A. Stewart (Eds.), *Africa from MIS 6-2: Population dynamics and paleoenvironments* (pp. 23–47). Dordrecht: Springer.

Clark, J. D. (1998). Before food production in North Africa: Overview and discussion. In S. di Lernia & G. Manzi (Eds.), *Before food production in North Africa: Questions and tools dealing with resource exploitation and population dynamics at 12000–7000 bp* (pp. 163–177). Forlì: ABACO.

Clark, J. D., Agrilla, E. J., Galloway, A., Garcea, E. A. A., Gifford-Gonzalez, D., Hall, D. N., et al. (2008). *Adrar Bous. Archaeology of a central Saharan granitic complex in Niger.* Tervuren: Royal Museum for Central Africa.

Close, A. E. (2002). Backed bladelets are a foreign country. *Archaeological Papers of the American Anthropological Association, 12*, 31–44.

Conway, G. (2009). The science of climate change in Africa: Impacts and adaptation. Grantham Institute for Climate Change, Imperial College, London, Discussion Paper No. 1. Retrieved January 14, 2014 from https://workspace.imperial.ac.uk/climatechange/public/pdfs/discussion_papers/Grantham_Institue_-_The_science_of_climate_change_in_Africa.pdf.

Cornelissen, E. (2016). The later Pleistocene in the northeastern Central African rainforest. In S. C. Jones & B. A. Stewart (Eds.), *Africa from MIS 6-2: Population dynamics and paleoenvironments* (pp. 301–319). Dordrecht: Springer.

Coulthard, T. J., Ramirez, J. A., Barton, R. N. E., Rogerson, M., & Brücher, T. (2013). Were rivers flowing across the Sahara during the Last Interglacial? Implications for human migration through Africa. *PLoS ONE, 8*(9), e74834.

Daniel, G. E. (1975). *150 years of archaeology*. London: Duckworth.

Deacon, H. J. (1993). Planting an idea: An archaeology of Stone Age gatherers in South Africa. *South African Archaeological Bulletin, 48*, 86–93.

Deacon, H. J. (1995). Two late Pleistocene-Holocene depositories from the southern Cape, South Africa. *South African Archaeological Bulletin, 50*, 121–131.

Delson, E. (Ed.). (1985). *Ancestors: The hard evidence: Proceedings of the symposium held at the American Museum of Natural History April 6–10 1984 to mark the opening of the exhibition "Ancestors: Four million years of humanity"*. New York: Alan R. Liss.

d'Errico, F., Backwell, L., Villa, P., Degano, I., Lucejko, J. J., Bamford, M. K., et al. (2012). Early evidence of San material culture represented by organic artefacts from Border Cave, South Africa. *Proceedings of the National Academy of Sciences USA, 109*, 13214–13219.

Dewar, G. & Stewart, B. A. (2016). Paleoenvironments, sea levels and land use in Namaqualand, South Africa, during MIS 6-2. In S. C. Jones & B. A. Stewart (Eds.), *Africa from MIS 6-2: Population dynamics and paleoenvironments* (pp. 195–212). Dordrecht: Springer.

Douka, K., Jacobs, Z., Lane, C., Grün, R., Farr, L., Hunt, C., et al. (2014). The chronostratigraphy of the Haua Fteah cave (Cyrenaica, northeast Libya). *Journal of Human Evolution, 66*, 39–63.

Drake, N. & Breeze, P. (2016). Climate change and modern human occupation of the Sahara from MIS 6-2. In S. C. Jones & B. A. Stewart (Eds.), *Africa from MIS 6-2: Population dynamics and paleoenvironments* (pp. 103–122). Dordrecht: Springer.

Drake, N., Blench, R. M., Armitage, S. J., Bristow, C. S., & White, K. H. (2011). Ancient watercourses and biogeography of the Sahara explain the peopling of the desert. *Proceedings of the National Academy of Sciences USA, 108*, 458–462.

Dugmore, A. J., Borthwick, D. M., Church, M. J., Dawson, A., Edwards, K. J., Keller, C., et al. (2007). The role of climate in settlement and landscape change in the North Atlantic islands: An assessment of cumulative deviations in high-resolution proxy climate records. *Human Ecology, 35*, 169–178.

Esterhuysen, A., & Lane, P. J. (2013). Archaeology and education. In P. J. Mitchell & P. J. Lane (Eds.), *The Oxford handbook of African archaeology* (pp. 239–252). Oxford: Oxford University Press.

Faith, J. T., Tryon, C., & Peppe, D. J. (2016). Environmental change, ungulate biogeography, and their implications for early human dispersals in equatorial East Africa. In S. C. Jones & B. A. Stewart (Eds.), *Africa from MIS 6-2: Population dynamics and paleoenvironments* (pp. 233–245). Dordrecht: Springer.

Fischer, E., Albert, R.-M., Botha, G., Cawthra, H. C., Esteban, I., Harris, J., et al. (2013). Archaeological reconnaissance for Middle Stone Age sites along the Pondoland coast, South Africa. *PaleoAnthropology, 2013*, 104–137.

Garcea, E. A. A. (2012). Modern human desert adaptations: A Libyan perspective on the Aterian complex. In J.-J. Hublin & S. P. McPherron (Eds.), *Modern origins: A North African perspective* (pp. 127–142). Dordrecht: Springer.

Grine, F. E. (2016). The Late Quaternary hominins of Africa: The skeletal evidence from MIS 6-2. In S. C. Jones & B. A. Stewart (Eds.), *Africa from MIS 6-2: Population dynamics and paleoenvironments* (pp. 323–381). Dordrecht: Springer.

Güldemann, T., & Stoneking, M. (2008). A historical appraisal of clicks: A linguistic and genetic population perspective. *Annual Review of Anthropology, 37*, 93–109.

Harvati, K., Stringer, C. B., Allsworth-Jones, P., & Folorunso, C. A. (2011). The Later Stone Age calvaria from Iwo Eleru, Nigeria: Morphology and chronology. *PLoS ONE, 6*(9), e24024.

Higham, T. F. G. (2011). European Middle and Upper Palaeolithic radiocarbon dates are often older than they look: Problems with previous dates and some remedies. *Antiquity, 85*, 235–249.

Hiscock, P., & O'Connor, S. (2005). Arid paradises or dangerous landscapes: A review of explanations for Paleolithic assemblage change in arid Australia and Africa. In P. Veth, M. Smith, & P. Hiscock (Eds.), *Desert peoples: Archaeological perspectives* (pp. 58–77). Oxford: Blackwell Publishing.

Hiscock, P., Clarkson, C., & Mackay, A. (2011). Big debates over little tools: Ongoing disputes over microliths on three continents. *World Archaeology, 43*, 653–664.

Jacobs, Z., Roberts, R. G., Galbraith, R. F., Barré, M., Deacon, H. J., Mackay, A., et al. (2008). Ages for Middle Stone Age innovations in southern Africa: Implications for modern human behavior and dispersal. *Science, 322*, 733–735.

Jones, S., Antoniadou, A., Barton, H., Drake, N., Farr, L., Hunt, C., et al. (2016). Patterns of hominin occupation and cultural diversity across the Gebel Akhdar of northern Libya over the last ~200 kyr. In S. C. Jones & B. A. Stewart (Eds.), *Africa from MIS 6-2: Population dynamics and paleoenvironments* (pp. 77–99). Dordrecht: Springer.

Kelly, R. L. (2013). *The lifeways of hunter-gatherers: The foraging spectrum*. Cambridge: Cambridge University Press.

Lahr, M. M. & Foley, R. (2016). Human evolution in Late Quaternary eastern Africa. In S. C. Jones & B. A. Stewart (Eds.), *Africa from MIS 6-2: Population dynamics and paleoenvironments* (pp. 215–231). Dordrecht: Springer.

Lambeck, K., Purcell, A., Flemming, N. C., Vita-Finzi, C., Alsharekh, A. M., & Bailey, G. N. (2011). Sea level and shoreline reconstructions for the Red Sea: Isostatic and tectonic considerations and implications for hominin migration out of Africa. *Quaternary Science Reviews, 30*, 3542–3574.

Lewis-Williams, J. D., & Pearce, D. G. (2004). *San spirituality: Roots, expression and social consequences*. Walnut Creek: AltaMira Press.

Lombard, M., & Haidle, N. (2012). Thinking a bow-and-arrow set: Cognitive implications of Middle Stone Age bow and stone-tipped arrow technology. *Cambridge Archaeological Journal, 22*, 237–264.

Lombard, M., & Parsons, I. (2011). What happened to the human mind after the Howiesons Poort? *Antiquity, 85*, 1433–1443.

Mackay, A. (2016). Technological change and the importance of variability: The Western Cape of South Africa from MIS 6-2. In S. C. Jones & B. A. Stewart (Eds.), *Africa from MIS 6-2: Population dynamics and paleoenvironments* (pp. 49–63). Dordrecht: Springer.

Mabulla, A. (2000). Strategy for cultural heritage management (CHM) in Africa: A case study. *African Archaeological Review, 17*, 211–233.

McBrearty, S., & Brooks, A. S. (2000). The revolution that wasn't: A new interpretation of the origin of modern human behavior. *Journal of Human Evolution, 39*, 453–563.

Mercader, J. (2009). Mozambican grass seed consumption during the Middle Stone Age. *Science, 326*, 1680–1683.

Mercader, J., Asmerom, Y., Bennett, T., Raja, M., & Skinner, A. (2009). Excavation and dating of Ngalue Cave: A Middle Stone Age site along the Niassa Rift, Mozambique. *Journal of Human Evolution, 57*, 63–74.

Merrick, H. V., & Nash, W. P. (1994). Use and movement of obsidian in the Early and Middle Stone Ages of Kenya and northern Tanzania. In S. T. Childs (Ed.), *Society, culture and technology in Africa* (pp. 29–44). Philadelphia: MASCA.

Mitchell, P. J. (2003). Anyone for *hxaro*? Thoughts on the theory and practice of exchange in southern African Later Stone Age archaeology. In P. J. Mitchell, A. Haour, & J. H. Hobart (Eds.), *Researching Africa's past: New perspectives from British archaeology* (pp. 35–43). Oxford: School of Archaeology.

Mitchell, P. J. (2008). Practising archaeology in a time of climatic catastrophe. *Antiquity, 82*, 1093–1103.

Newby, J. E. (1984). Large mammals. In L. Cloudsley-Thompson (Ed.), *The Sahara* (pp. 277–290). Oxford: Pergamon Press.

Parkington, J. E. (2005). Review of JD Lewis-Williams and DG Pearce. San spirituality: Roots, expression and social consequences. *Journal of African Archaeology, 3*, 165–166.

Plug, I. (1981). Some research results on the late Pleistocene and Holocene deposits of Bushman Rock Shelter, eastern Transvaal. *South African Archaeological Bulletin, 36*, 14–21.

Prendergast, M., Luque, L., Domínguez-Rodrigo, M., Díez-Martín, F., Mabulla, A., & Barba, R. (2007). New excavations at Mumba Rockshelter (Tanzania). *Journal of African Archaeology, 5*, 163–189.

Reid, D. A. M. (2005). Interaction, marginalization and the archaeology of the Kalahari. In A. B. Stahl (Ed.), *African archaeology: A critical introduction* (pp. 353–377). Oxford: Blackwell Publishing.

Robbins, L. H., Brook, G. A., Murphy, M. L., Ivester, A. H., & Campbell, A. C. (2016). The Kalahari during MIS 6-2 (190–12 ka): Archaeology, paleoenvironment and population dynamics. In S. C. Jones & B. A. Stewart (Eds.), *Africa from MIS 6-2: Population dynamics and paleoenvironments* (pp. 175–193). Dordrecht: Springer.

Roberts, P., Lee-Thorp, J. A., Mitchell, P. J., & Arthur, C. (2013). Stable carbon isotopic evidence for climate change across the late Pleistocene to early Holocene from Lesotho, southern Africa. *Journal of Quaternary Science, 28*, 360–369.

Sadr, K., & Sampson, C. G. (2006). Through thick and thin: Early pottery in southern Africa. *Journal of African Archaeology, 4*, 235–252.

Sampson, C. G. (1985). Review of Janette Deacon's the Later Stone Age of southernmost Africa. *South African Archaeological Bulletin, 40*, 56–58.

Scholz, C. A., Johnson, T. C., Cohen, A. S., King, J. W., Peck, J. A., Overpeck, J. T., et al. (2007). East African megadroughts between 135 and 75 thousand years ago and bearing on early-modern human origins. *Proceedings of the National Academy of Sciences USA, 104*, 16416–16421.

Sealy, J. C. (2006). Diet, mobility and settlement pattern among Holocene hunter-gatherers in southernmost Africa. *Current Anthropology, 47*, 569–595.

Sealy, J. (2016). Cultural change, demography, and the archaeology of the last 100 kyr in southern Africa. In S. C. Jones & B. A. Stewart (Eds.), *Africa from MIS 6-2: Population dynamics and paleoenvironments* (pp. 65–75). Dordrecht: Springer.

Soares, P., Rito, T., Pereira, L., & Richards, M. B. (2016). A genetic perspective on African prehistory. In S. C. Jones & B. A. Stewart (Eds.), *Africa from MIS 6-2: Population dynamics and paleoenvironments* (pp. 383–405). Dordrecht: Springer.

Soriano, S., Rasse, M., Tribolo, C., & Huysecom, E. (2010). Ounjougou: A long Middle Stone Age sequence in the Dogon country (Mali). In P. Allsworth-Jones (Ed.), *West African archaeology: New developments, new perspectives* (pp. 1–14). Oxford: Archaeopress.

Stewart, B. A., Dewar, G., Morley, M. W., Parker, A., & Allott, L. (2016). Follow the Senqu: Maloti-Drakensberg paleoenvironments and implications for modern human dispersals into mountain systems. In S. C. Jones & B. A. Stewart (Eds.), *Africa from MIS 6-2: Population dynamics and paleoenvironments* (pp. 247–271). Dordrecht: Springer.

Stocker, T. F., Qin, D., Plattner, G.-K., Tignor, M., Allen, S. K., Boschung, J., et al. (Eds.). (2013). *Climate change 2013: The physical science basis*. Cambridge: Cambridge University Press.

Taylor, N. (2016). Across rainforests and woodlands: A systematic re-appraisal of the Lupemban Middle Stone Age in Central Africa. In S. C. Jones & B. A. Stewart (Eds.), *Africa from MIS 6-2: Population dynamics and paleoenvironments* (pp. 273–299). Dordrecht: Springer.

Tribolo, C., Mercier, N., Douville, E., Loron, J.-L., Reyss, J.-L., Rufer, D., et al. (2013). OSL and TL dating of the Middle Stone Age sequence at Diepkloof Rock Shelter (South Africa): A clarification. *Journal of Archaeological Science, 40*, 3401–3411.

Truc, L., Chevalier, M., Favier, C., Cheddadi, R., Meadows, M. E., Scott, L., et al. (2013). Quantification of climate change for the last 20,000 years from Wonderkrater, South Africa: Implications for the long-term dynamics of the Intertropical Convergence Zone. *Palaeogeography, Palaeoclimatology, Palaeoecology, 386*, 575–587.

Van Peer, P. (2016). Technological systems, population dynamics and historical process in the MSA of northern Africa. In S. C. Jones & B. A. Stewart (Eds.), *Africa from MIS 6-2: Population dynamics and paleoenvironments* (pp. 147–159). Dordrecht: Springer.

Wadley, L. (1987). *Later Stone Age hunters and gatherers of the southern Transvaal: Social and ecological interpretations*. Oxford: Archaeopress.

Wadley, L. (2010). Were snares used in the Middle Stone Age and does it matter? A review and case study from Sibudu, South Africa. *Journal of Human Evolution, 58*, 179–192.

Wadley, L., Hodgskiss, T., & Grant, M. (2009). Implications for complex cognition from the hafting of tools with compound adhesives in the Middle Stone Age, South Africa. *Proceedings of the National Academy of Sciences USA, 106*, 9590–9594.

Wiessner, P. (1982). Risk, reciprocity and social influence on !Kung San economics. In E. Leacock & R. B. Lee (Eds.), *Politics and history in band societies* (pp. 61–84). Cambridge: Cambridge University Press.

Willis, K. J., Bennett, K. D., Burrough, S. L., Macias-Fauria, M., & Tovar, C. (2013). Determining the response of African biota to climate change: Using the past to model the future. *Philosophical Transactions of the Royal Society B, 368*, 20120491.

Willoughby, P. (2012). The Middle and Later Stone Age in the Iringa region of southern Tanzania. *Quaternary International, 270*, 103–118.

World Bank. (2009). *Making development climate resilient: A World Bank strategy for sub-Saharan Africa*. Washington, DC: The World Bank. Retrieved January 14, 2014 from http://siteresources.worldbank.org/INTAFRICA/Resources/ClimateChange-StrategyReport2010-Full_vNoImages.pdf.

Yellen, J. E., Brooks, A. S., Cornelissen, E., Mehlman, M., & Stewart, K. (1995). A Middle Stone Age worked bone industry from Katanda, Upper Semliki Valley, Zaïre. *Science, 268*, 553–556.

Index

Note: Page numbers followed by *f* and *t* indicate figures and tables, respectively

Sacha C. Jones and Brian A. Stewart (eds.), *Africa from MIS 6-2: Population Dynamics and Paleoenvironments*, Vertebrate Paleobiology and Paleoanthropology, DOI 10.1007/978-94-017-7520-5

K

L

M

Q

R

S

MIX
Papier aus verantwortungsvollen Quellen
Paper from responsible sources
FSC® C105338

If you have any concerns about our products,
you can contact us on
ProductSafety@springernature.com

In case Publisher is established outside the EU,
the EU authorized representative is:
Springer Nature Customer Service Center GmbH
Europaplatz 3, 69115 Heidelberg, Germany

Printed by Libri Plureos GmbH
in Hamburg, Germany